Climate Change Management

The aim of this book series is to provide an authoritative source of information on climate change management, with an emphasis on projects, case studies and practical initiatives – all of which may help to address a problem with a global scope, but the impacts of which are mostly local. As the world actively seeks ways to cope with the effects of climate change and global warming, such as floods, droughts, rising sea levels and landscape changes, there is a vital need for reliable information and data to support the efforts pursued by local governments, NGOs and other organizations to address the problems associated with climate change. This series welcomes monographs and contributed volumes written for an academic and professional audience, as well as peer-reviewed conference proceedings. Relevant topics include but are not limited to water conservation, disaster prevention and management, and agriculture, as well as regional studies and documentation of trends. Thanks to its interdisciplinary focus, the series aims to concretely contribute to a better understanding of the state-of-the-art of climate change adaptation, and of the tools with which it can be implemented on the ground.

Notes on the quality assurance and peer review of this publication

Prior to publication, the quality of the works published in this series is double blind reviewed by external referees appointed by the editor. The referees are not aware of the author's name when performing the review; the referees' names are not disclosed.

Walter Leal Filho · Marina Kovaleva ·
Fátima Alves · Ismaila Rimi Abubakar
Editors

Climate Change Strategies: Handling the Challenges of Adapting to a Changing Climate

Springer

Editors
Walter Leal Filho
Research and Transfer Centre "Sustainable Development and Climate Change Management"
Hamburg University of Applied Sciences
Hamburg, Germany

Marina Kovaleva
Research and Transfer Centre "Sustainable Development and Climate Change Management"
Hamburg University of Applied Sciences
Hamburg, Germany

Fátima Alves
Department of Life Sciences
Universidade Aberta
Lisbon, Portugal

CFE and University of Coimbra
Coimbra, Portugal

Ismaila Rimi Abubakar
College of Architecture and Planning
Imam Abdulrahman Bin Faisal University
Dammam, Saudi Arabia

ISSN 1610-2002 ISSN 1610-2010 (electronic)
Climate Change Management
ISBN 978-3-031-28727-5 ISBN 978-3-031-28728-2 (eBook)
https://doi.org/10.1007/978-3-031-28728-2

This Springer imprint is published by the registered company Springer Nature Switzerland AG
The registered company address is: Gewerbestrasse 11, 6330 Cham, Switzerland

Preface

The Intergovernmental Panel on Climate Change (IPCC) in its Sixth Assessment Report (2021) has reiterated the need for concerted actions in reducing the emissions of greenhouse gases on the one hand, and the need to intensify efforts to adapt to a changing climate on the other. The Conference of the Parties of the UN Framework Convention on Climate Change (UNFCCC) during its 27th meeting, held in Sharm el-Sheikh, Egypt in late 2022, emphasised the need for action towards achieving the world's collective climate goals as agreed under the Paris Agreement and the Convention.

There is a consensus about the urgency to systematically implement climate change strategies, and to ensure that they provide a concrete contribution towards mitigation and adaptation efforts. In this context, it is important to document and promote experiences in both areas. Against this background, this book reports on a set of works being undertaken at an international scale, where elements of climate change mitigation and adaptation are being implemented. The book is structured in two parts.

Part 1—Assessing Climate Change Impacts
Part 2—Implementing Mitigation and Adaptation Measures

By means of detailed case studies from Africa, the Caribbean, Asia, and Europe, and through a broader scale synthesis, which collates insights from across sectors and countries, the book pays special attention to evidences, which illustrate how climate change mitigation and adaptation strategies are being implemented, and indicate trends and contexts that might prove helpful in guiding future efforts.

We wish to thank all the authors who provided valuable contributions to this book, for their time and effort. We also thank many reviewers who supported the process. Their cooperation was crucial for the preparation of the book and its final publication.

We hope the experiences gathered in this book may support future initiatives on climate change mitigation and adaptation, and inspire future work in these key areas.

Hamburg, Germany — Walter Leal Filho
Hamburg, Germany — Marina Kovaleva
Lisbon, Portugal — Fátima Alves
Dammam, Saudi Arabia — Ismaila Rimi Abubakar
Summer 2023

Contents

Part I
Assessing Climate Change Impacts

Chapter 1
Impact of Climate Change and Its Adaptation in the Semi-arid Environment of Flood-Prone Dechatu Catchment, Dire Dawa, Ethiopia

Girma Berhe Adane, Asfaw Kebede Kassa, Abebe Teklu Toni, Shiwandagn Lemma Tekle, and Walter Leal Filho

Abstract This study aims to assess the future trends, and magnitude of climatic factors (temperature and rainfall) using different representative concentration pathways (RCP4.5 and RCP8.5) integrated with the identification of existing climatic adaptation practices in the Dechatu catchment, Ethiopia. Four global/derived regional climatic models (GCMs/RCMs)—CanESM2, CNRM-CM5, EC-EARTH and MIROC5—were used to understand the changes in temperature and rainfall (2025–2075); and their trend using the Trend-Free Pre-Whitening Mann–Kendall test. The highest annual daily peak rainfall was identified in all models. However, none of the rainfall data series exhibited a trend at a 5% significance level except in CanESM2 of RCP8.5. In contrast, a significant monotonic increase in temperature is forecasted in all RCMs, with higher temperature increments (1–1.50 °C in RCP8.5) in a 25-year interval. Water management, as an adaptation option to climate extremes, is considered in the semi-arid environment of Dechatu. Among others, the use of traditional hand-dug wells, flood plain excavation (locally named *Chirosh*) as drinking water sources, floodwater diversion, contour farming and dropping of shrub branches on the river banks to conserve transported soils are included. Therefore, integrating the local adaptation with conventional water management practices will reduce the impacts of climate change.

G. B. Adane (✉) · A. K. Kassa · A. T. Toni · S. L. Tekle
School of Water Resource and Environmental Engineering, Haramaya Institute of Technology, Haramaya University, Dire Dawa 138, Ethiopia
e-mail: girma.berhe@haramaya.edu.et

W. Leal Filho
Department of Natural Sciences, Manchester Metropolitan University, Chester Street, Manchester M1 5GD, UK
e-mail: walter.leal2@haw-hamburg.de

Research and Transfer Centre "Sustainable Development and Climate Change Management", Hamburg University of Applied Sciences, Ulmenliet 20, 21033 10, Hamburg, Germany

W. Leal Filho et al. (eds.), *Climate Change Strategies: Handling the Challenges of Adapting to a Changing Climate*, Climate Change Management,
https://doi.org/10.1007/978-3-031-28728-2_1

Introduction

Climate variability and change are one of the major environmental challenges of the twenty-first century, posing a threat to human well-being (Wuebbles 2018). The increase in the concentration of greenhouses gases in the atmosphere contributes to a potential change and variability in climate, particularly in precipitation and temperature (Rind et al. 1989), and global warming is indeed occurring (IPCC 2014a, b). Shewmake (2011) explained that the changes in rainfall patterns and shifts in temperature zones are expected to increase the frequency and intensity of climate-related extremes in Sub-Saharan Africa (Ringler 2008). Across the globe, an increase in frequency and intensity of extreme events like floods, droughts, heatwaves, etc. may be anticipated as a result of climate change (IPCC et al. 2021). Future climate change projection also shows that food security will be affected in the drylands of Africa and in the high mountain regions of Asia and South America (Porter et al. 2015; Mbow et al. 2019).

The rainfed agricultural practice in Africa is adversely impacted by climate change due to flooding in coastal areas, water scarcity in arid/semi-arid areas, a decline in agricultural production, worsening desertification inland with low adaptive capacity, and poor infrastructural developments which mitigate the impacts (Mendelsohn and Dinar 2003; Shewmake 2011). Furthermore, indigenous and local ecological knowledge of climate adaptations are side-lined in policymaking and implementation in Africa, particularly in indigenous communities, and climate change has therefore caused a significant strain on those communities (Leal Filho et al. 2021). Therefore, to cope with negative climatic influences, it is necessary to integrate the conventional weather forecasting systems and climate information and policies with the local and indigenous knowledge.

Teshome and Zhang (2019) have demonstrated that there is an increase in extreme drought events in Ethiopia as a result of changes and trends of climate extremes. Moreover, various studies have revealed that prolonged heat stress as a result of high-temperature extremes could impact terrestrial biota, livestock and human lives (Kilbourne 1999; Basu and Samet 2002; Argaud et al. 2007). Billi et al. (2015) have pointed out that an increase in rainfall intensity has caused an increase in the frequency of occurrence of flash floods in Dire Dawa, when compared to the land use change effects due to anthropogenic impacts.

The semi-arid climate of Dire Dawa city has experienced more than 11 major flash flood events since 1945; the 2006 flooding event caused numerous fatalities (339 human deaths) and displacements (9027) of people (Erena and Worku 2018). Consequently, to reduce and cope with the negative impacts of the climate extremes in the Dechatu catchment, it is necessary to understand the future precipitation and temperature patterns, as well as the trends and magnitude, including identification of best existing practices for water management. As such, climate adaptation mechanisms play a substantial role in planning and developing effective adaptation strategies to mitigate climate extremes.

Aims and Objectives

This chapter aims to highlight the future prediction, trends, and magnitude of climatic factors (temperature and precipitation) on a seasonal and annual basis using different emission scenarios or representative concentration pathways (RCP4.5 and RCP8.5), in line with identifying the existing climatic adaptation options—specific to water management aspects—in the flood-prone Dechatu catchment, Dire Dawa, Ethiopia.

Materials and Methods

Study Area

Dechatu Catchment (DC) is part of the Dire Dawa City Administration situated in the eastern part of Ethiopia. Geographically, the area is located between 09° 28.1″ to 09° 49.1″ N Latitude and 41° 38.1″ to 42° 19.1″ E Longitude (Fig. 1.1). The area is characterized as having a semi-arid climate with a relatively low level of rainfall, with the frequent occurrence of flash flooding from the upper catchment of Dechatu. The mean annual rainfall ranges between 812 mm in the highlands to 612 mm at the foothills of Dire Dawa city. The Dechatu catchment (149.83 km^2) includes 4 sub-watersheds (Chiri, Alla, Dure and Gogeti); it experiences river flows only when there is rainfall in a catchment, and it is characterized as a dry river bed with no flowing water on the surface.

Data Used

Four regional climate models, with a spatial resolution of 0.44° for the RCP4.5 and RCP8.5 emission scenarios, were used, as shown in Table 1.1. The ensemble mean data is available in the context of the Coordinated Regional Climate Downscaling Experiment (Giorgi et al. 2009) over Africa at 0.44° resolution for the period 1950–2100 and it has already been used throughout Africa (Nikulin et al. 2012, 2018; Gbobaniyi et al. 2014; Dosio 2017; Osima et al. 2018). The daily observed and the GCMs data from 1981 (considered as historical data) were overlapped and bias-corrected to predict the future climatic scenarios (2025–2075) of RCP4.5 and RCP8.5.

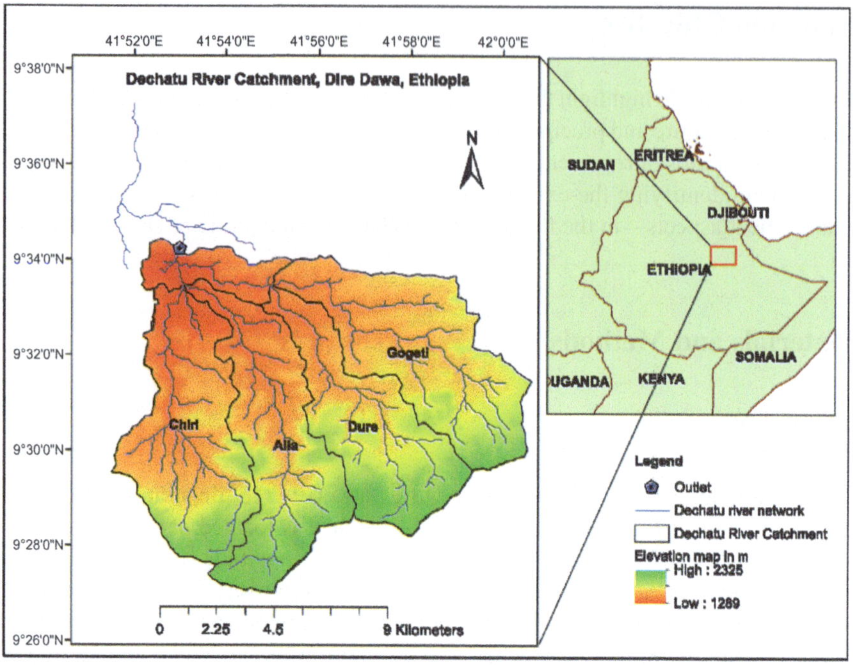

Fig. 1.1 Location map of Dechatu catchment, Dire Dawa Administration, Ethiopia

Table 1.1 Summary of the applied Regional Climate Models (RCMs) and their driving Global Climate Models (GCMs) (http://wcrpcordex.ipsl.jussieu.fr/)

GCMs	RCMs	Institution/source/	URL
CanESM2	CanRCM4_r2	Canadian Centre for Climate Modeling and Analysis (CCCma)	http://climate-modelling.canada.ca/
CNRM-CM5	CCLM4-8-17_v1	Climate Limited-area Modelling Community (CLMcom)	https://esg-dn1.nsc.liu.se/
EC-EARTH	CCLM4-8-17_v1	Climate Limited-area Modelling Community (CLMcom)	https://esg-dn1.nsc.liu.se/
MIROC5	RCA4_v1	Rossby Centre, Swedish Meteorological and Hydrological Institute (SMHI)	https://esg-dn1.nsc.liu.se/

Methods

The daily rainfall, daily peak, monthly and annual rainfall data were used where needed. In addition, the annual maximum and minimum temperature data were generated from the daily data set. The bias correction was adjusted using the past daily-observed rainfall and temperature data (1981–2014) using the power transformation function (for future rainfall data) and variance scaling methods (for future temperature data) via CMhyd software. The Tukey fence method is used to adjust the outliers of the rainfall records between 2025 and 2075. The projected temperature anomaly change (ΔT in °C) and rainfall characterization at different time-scales were analysed using various statistical methods, including the projected trend test via the Trend-Free prewhitened Mann–Kendall (TFPW-MK) test. When the time series data were not random and influenced by autocorrelation, the trend component is removed from the data and is prewhitened before the application of the trend test (Yue et al. 2002). This test analysis was performed using R-Studio Version 1.4.1103. All the graphs in these sections are plotted using Matlab R2020a and the method used for various future climatic models is summarized in Fig. 1.2.

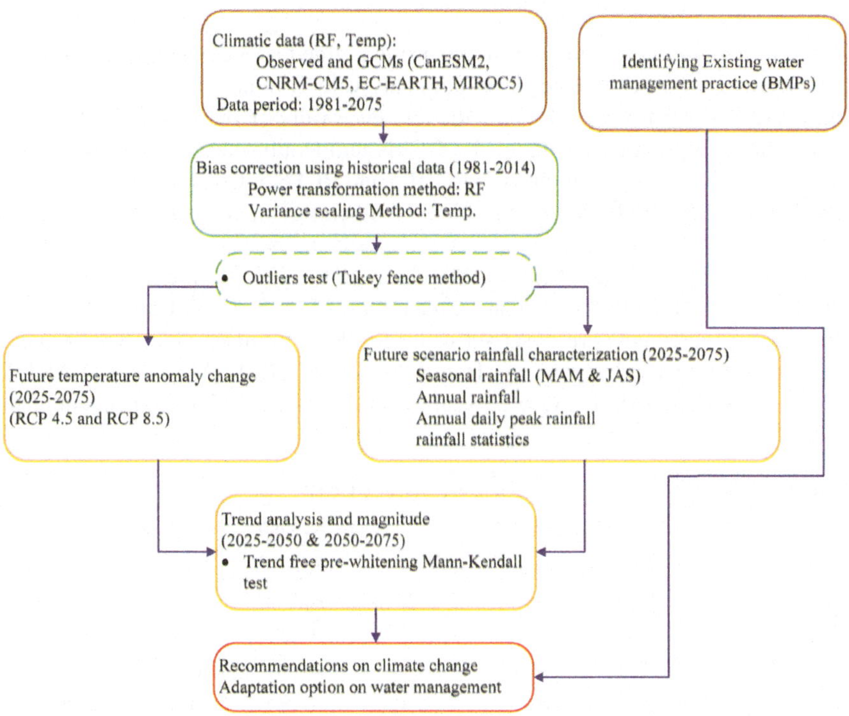

Fig. 1.2 Methodological framework used for the study

Climate Change Adaptation and Mitigation Mechanisms

In this study, various indigenous and local climate change adaptation and mitigation measures in the semi-arid catchment of Dire Dawa were identified. Major causes or challenges of climate change, including under rainfed and irrigation systems and the local mitigation practices for flash flood water management, were investigated via field surveys and the review of various documents for the catchment. In addition, aerial coverage and the types of existing best management practices (BMPs) to control soil erosion, moisture conservation and groundwater recharging mechanisms were identified using field surveys, digital elevation model (DEM) data, land use shapefile, google imageries, and high-resolution satellite imageries downloaded using SAS Planet program.

Results and Discussion

Temperature Anomaly (ΔT in °C) of DRC

The term temperature anomaly (Tanomaly) is defined as a deviation from a long-term average, with positive/negative Tanomaly values, indicating that the observed temperature was warmer/cooler than the reference value (Pyrgou et al. 2019).

The linear regression results of the GCMs temperature records show an increment in temperature anomaly in both RCP4.5 and RCP8.5. Comparing the emission scenario types, the RCP8.5 showed a higher temperature increment on a yearly basis (Table 1.2).

In every 25-year interval (2025–2050 and 2050–2075), a minimum of 0.5 °C (in MIROC5 and CNRM-CM5) and a maximum of 0.75 °C (in CanESM2) mean temperature increments are observed using RCP4.5. Similarly, the RCP8.5 mean temperature anomaly changes range from 1.00 °C (in CNRM-CM5 and EC-EARTH) to 1.50 °C in CanESM2 in every 25-year interval. Also, 1.3 °C increments in mean

Table 1.2 Temperature anomaly change in every year (°C/yr)

GCMs	RCP4.5			RCP8.5		
	T_{max}	T_{min}	Mean	T_{max}	T_{min}	Mean
CanESM2	+0.03	+0.03	+0.03	+0.07	+0.05	+0.06
CNRM-CM5	+0.03	+0.01	+0.02	+0.05	+0.03	+0.04
EC-EARTH	+0.04	+0.01	+0.03	+0.05	+0.03	+0.04
MIROC5	+0.02	+0.02	+0.02	+0.05	+0.04	+0.05

T_{max}: Maximum temperature in °C; T_{min}: Minimum temperature in °C

temperature are depicted in the MIROC5 data set for a given interval (Fig. 1.3 and Table 1.2). This indicates that the heat stress in the catchment is imminent unless appropriate biological and physical soil and water conservation measures for soil moisture retention and recharging of groundwater, including afforestation practices, are utilised in the catchment.

Future Scenario for Rainfall Analysis

The annual daily peak rainfall (ADPRF) data series of four GCMs rainfall data series showed that higher rainfall records between 2036 and 2064 will be detected in RCP4.5 of CanESM2. Similarly, in CNRM-CM5 of RCP4.5, the highest annual daily peak rainfall was predicted in 2040, 2053 and 2065. A higher forecast of rainfall records was identified in RCP8.5 of EC-EARTH (2031, 2054, 2065 and 2069) and MIROC5 (2025–2038), respectively (Fig. 1.4).

The linear regression of the ADPRF trends (RCP4.5 and RCP8.5) shows 4- and 7-mm increments in CanESM2 and 4.6- and 2.31-mm increments in MIROC5 in every 25-year interval (Table 1.3 and Fig. 1.4). EC-EARTH rainfall forecast depicted a decrease in rainfall in both RCPs. However, in the CNRM-CM5 rainfall records, it showed an alternative increment in RCP8.5 (1.03 mm) and a decrease (−2.24 mm) in every quarter of the century. In general, the annual daily peak rainfall amounts showed very high variability yearly, even though the linear regression values showed a relatively small increment or decrement in every 25-year intervals (Table 1.3).

Seasonal and Annual Rainfall Analysis

The mean annual rainfall records for various emission scenarios (RCP4.5 and RCP8.5) exhibited lower estimates in MIROC5 and maximum estimates in CanESM2 for the data analysis range of 2025–2075. The mean annual rainfall ranges between 2025 and 2050 are 421 mm/yr in MIROC5and 740 mm/yr in EC-EARTH and the lowest in MIROC5 (498 mm/yr) and the highest (754 mm/yr) in CanESM2 for RCP4.5 and RCP8.5, respectively. In addition, the lowest (in MIROC5) and the highest (in CanESM2) mean annual rainfall between 2050 and 2075 range from 489 to 714 mm/yr in RCP4.5 and 464–1009 mm/yr in RCP8.5 (Table 1.4, Figs. 1.5 and 1.6). The box plot shows the median, quantile, and range of areal-averaged rainfall using various GCM Models (2025–2075) illustrated in Figs. 1.5 and 1.6.

The bimodal rainfall characteristics of DRC are explained using the seasonal (March–April-May designated as MAM and July–August-September designated as JAS seasons) and annual rainfall conditions. The largest quantile range—between a lower and upper quantile—of rainfall was forecasted in the MAM season of CanESM2 in all four types of the model from 2025 to 2075 in both RCPs except in EC-EARTH of the RCP4.5 records of JAS season (Figs. 1.5 and 1.6). The lowest

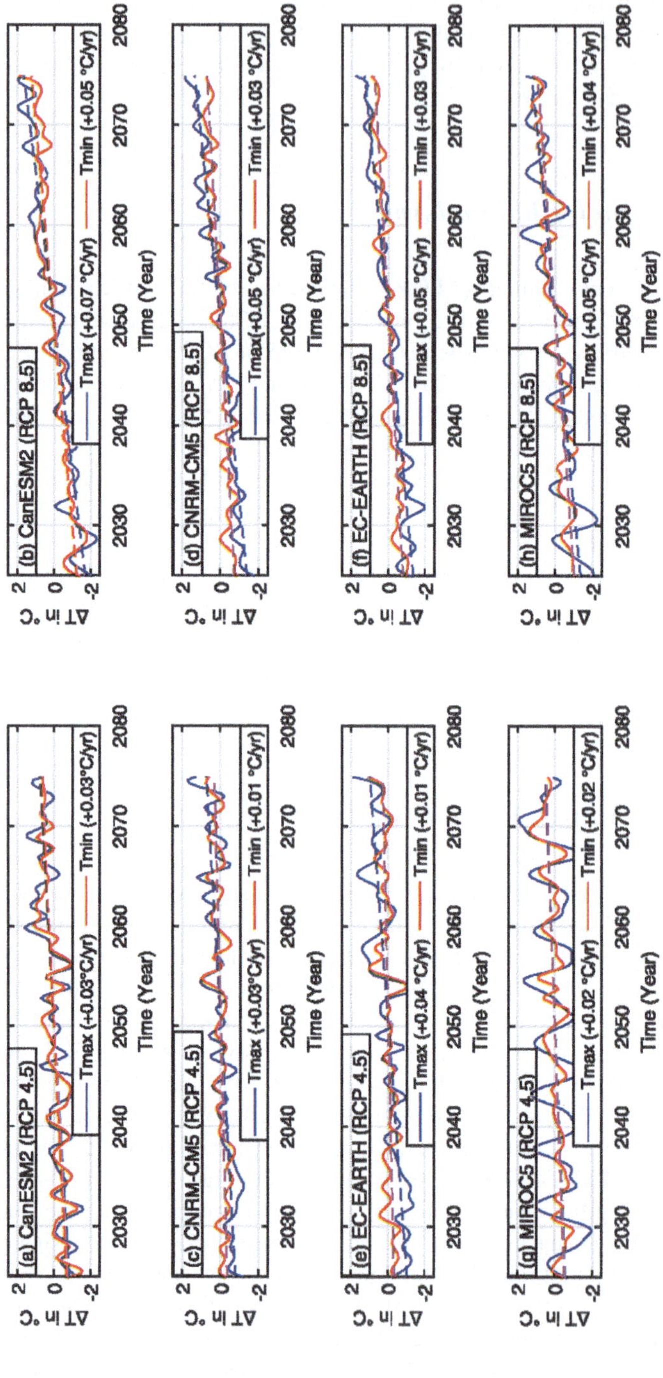

Fig. 1.3 Temperature anomaly (ΔT in °C) for future scenarios of RCP4.5 and RCP8.5

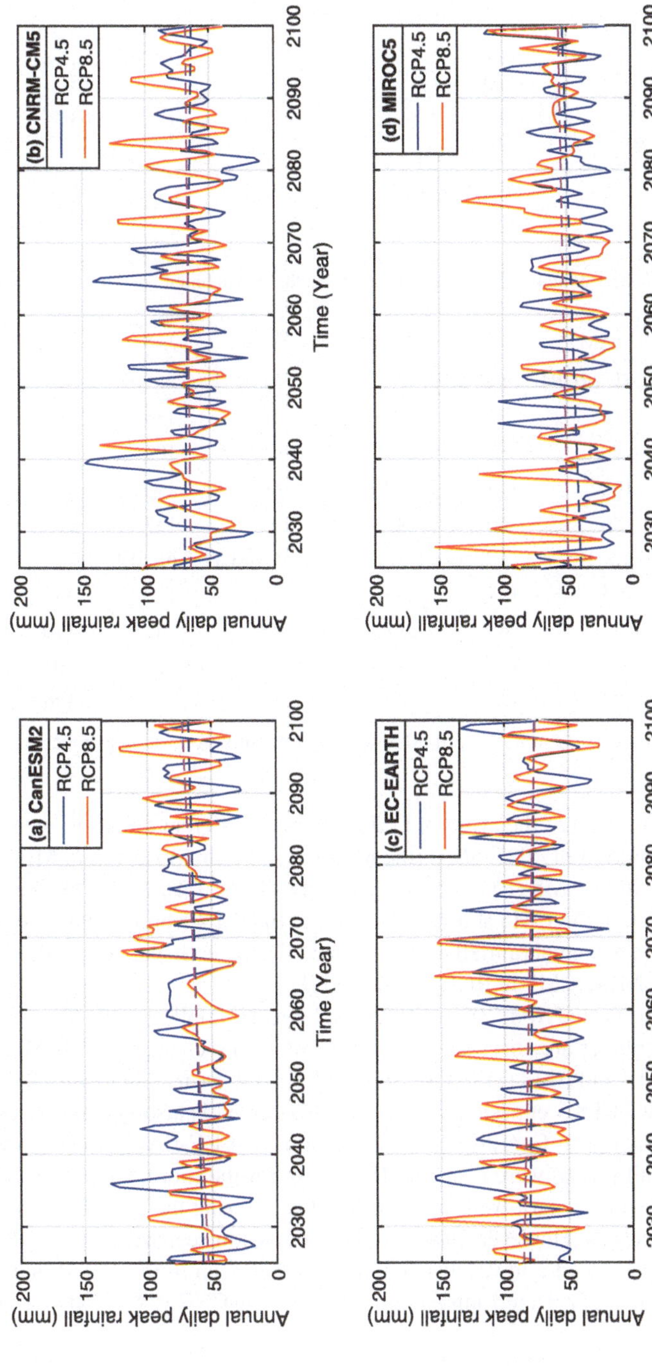

Fig. 1.4 Annual daily peak rainfall of different GCMs models (2025–2100)

Table 1.3 Linear regression of the annual daily peak rainfall trends in every 25 years (2025–2100)

GCMs	Rainfall changes in 25 years interval (mm)	
	RCP4.5	RCP8.5
CanESM2	3.59	6.55
CNRM-CM5	−2.24	1.03
EC-EARTH	−1.38	−3.17
MIROC5	4.60	2.31

Table 1.4 Seasonal and annual rainfall deviation of Dechatu catchment under various GCMs (2025–2075)

GCMs	MAM	JAS	Annual	MAM	JAS	Annual
	SD (RCP4.5 of 2025–2050)			SD (RCP8.5 of 2025–2050)		
CanESM2	219	115	326	210	136	309
CNRM-CM5	120	89	240	105	102	178
EC-EARTH	168	117	329	180	105	167
MIROC5	86	60	143	167	137	206
	SD (RCP4.5 of 2050–2075)			SD (RCP8.5 of 2050–2075)		
CanESM2	195	134	255	331	135	548
CNRM-CM5	172	88	287	158	92	170
EC-EARTH	154	107	233	152	116	201
MIROC5	105	128	198	63	72	196

SD Standard Deviation; *MAM* March–April–May seasons; *JAS* July–August–September seasons

standard deviation (SD) of the projected rainfall under different emission scenarios and GCMs for a period of 2025–2050 (the MAM season) are observed in MIROC5 (RCP4.5) and CNRM-CM5 (RCP8.5), respectively (Table 1.4). In the JAS season, the MIROC5 and CNRM-CM5 show the lowest SD in RCP4.5 and RCP8.5, respectively. Moreover, the annual rainfall records depicted a lower dispersion in MIROC5 in RCP4.5 and CNRM-CM5 in RCP8.5 (Table 1.4).

Considering the period from 2050 to 2075, the largest rainfall quantile (25–75% of the rainfall data) ranges are observed in RCP4.5 emission scenarios of CNRM-CM5 (in MAM season), CanESM2 (in JAS season) and MIROC5 (in annual rainfall data), respectively. The RCP8.5 emission scenario of this period is also explained, as the largest quantile ranges in CanESM2 (MAM season) and EC-EARTH (JAS season and annual rainfall series) were observed (Fig. 1.6). Furthermore, the lowest SD in the RCP4.5 emission scenario was noted in MIROC5 during the MAM season and annual rainfall records. However, in the RCP8.5 emission scenarios, the MIROC5 climatic data experienced a lower dispersion from the mean in the MAM and JAS seasons. Similarly, the projected annual rainfall records in CNRM-CM5 also showed a lower SD (Table 1.4).

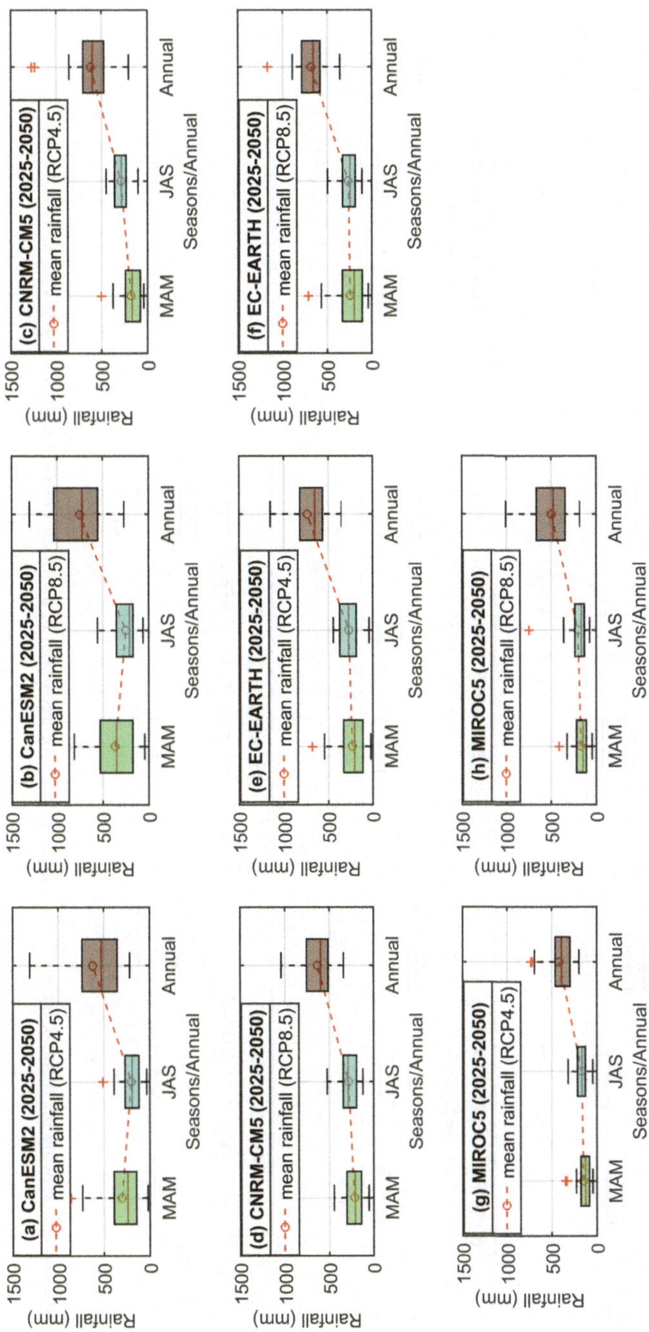

Fig. 1.5 Box plot showing the median, quantile, and range of areal-averaged rainfall using various GCM Models (2025–2050)

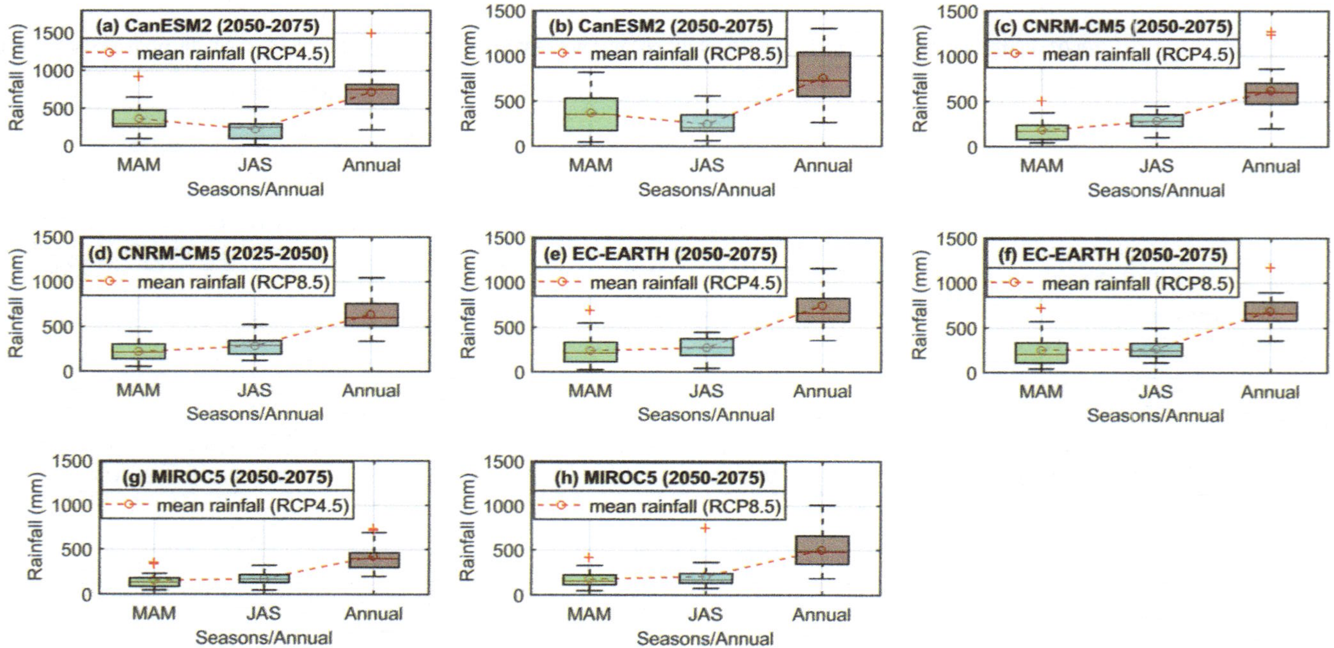

Fig. 1.6 Box plot showing the median, quantile, and range of areal-averaged rainfall using various GCM Models (2050–2075)

Table 1.5 TFPW-MK trend test for annual daily peak rainfall series for future scenarios of RCP4.5 and RCP8.5 (2025–2075)

GCM	*Z*-value	Sen's Slope	Old Sen's Slope	*P*-value	*S*	Tau
	RCP4.5					
CanESM2	0.93	0.06	0.00	0.35	205.00	0.07
CNRM-CM5	−0.16	−0.02	−0.03	0.88	−35.00	−0.01
EC-EARTH	−0.56	−0.09	−0.02	0.58	123.00	−0.04
MIROC5	1.79	0.21	0.17	0.07	393.00	0.14
	RCP8.5					
CanESM2	2.24[a]	0.26	0.23	0.02	491.00	0.18
CNRM-CM5	0.55	0.05	0.01	0.58	121.00	0.04
EC-EARTH	−0.89	−0.10	−0.11	0.37	195.00	−0.07
MIROC5	1.83	0.26	0.20	0.07	401.00	0.14

[a]Shows trend

Rainfall Trends Using the TFPW-MK Test

The trend and magnitude of the annual daily peak rainfall were tested using a modified MK (TFPW-MK) test. The lag-1 autocorrelation coefficient of annual daily peak rainfall records for the period of 2025–2075 was found to be within the recommended interval of the autocorrelation factor. Therefore, it can be concluded that the time series data of rainfall does not exhibit a significant autocorrelation. Then, testing using MK is acceptable after checking the autocorrelation of the data to avoid the biases of the MK test. At a 5% significance level, the annual daily peak rainfall records showed that none of these GCM rainfall data series exhibited a significant trend, except CanESM2 in RCP8.5, even though linear trends with increasing patterns have been detected using the linear regression model (Table 1.5). Daily peak rainfall records with increasing trend was noted in RCP4.5 (CanESM2 and MIROC5) and decreasing trends in CNRM-CM5 and EC-EARTH, respectively. Similarly, only EC-EARTH rainfall data records showed a decreasing trend in the RCP8.5 emission scenario (Table 1.5 and Fig. 1.4).

Temperature Anomaly Trends Using the TFPW-MK Test

The annual maximum and minimum temperature anomaly series for both emission scenarios were analyzed using a modified MK (TFPW-MK) test. The same procedure was followed for detecting the trends of temperature data as was used for the rainfall analysis. The time-series temperature data, except for the minimum temperature data of EC-EARTH with emission scenarios of RCP4.5, showed a significant autocorrelation at a 5% significance level. This shows that the lag-1 autocorrelation coefficient is found to be outside the interval recommended between the lower and upper bound. Therefore, pre-whitening of the data was performed and tested with MK. The minimum and maximum temperature trends of the Dechatu catchment showed that there is a significant monotonic increase in trends of temperature at a 5% significant level with the change in anomaly as explained in Fig. 1.3, Tables 1.6 and 1.7.

Table 1.6 TFPW-MK trend test for annual maximum temperature anomaly series for future scenarios of RCP4.5 and RCP8.5 (2025–2075)

GCM	*Z*-value	Sen's Slope	Old Sen's Slope	*P*-value	*S*	Tau
	RCP4.5					
CanESM2	4.63	0.03	0.03	0.00	555.00	0.45
CNRM-CM5	5.40	0.03	0.03	0.00	647.00	0.53
EC-EARTH	5.84	0.04	0.04	0.00	699.00	0.57
MIROC5	5.40	0.03	0.03	0.00	647.00	0.53
	RCP8.5					
CanESM2	7.51	0.07	0.07	0.00	899.00	0.73
CNRM-CM5	7.49	0.05	0.05	0.00	897.00	0.73
EC-EARTH	7.51	0.07	0.07	0.00	899.00	0.73
MIROC5	7.49	0.05	0.05	0.00	897.00	0.73

Table 1.7 TFPW-MK trend test for annual minimum temperature anomaly series for future scenarios of RCP4.5 and RCP8.5 (2025–2075)

GCM	*Z*-value	Sen's Slope	Old Sen's Slope	*P*-value	*S*	Tau
	RCP4.5					
CanESM2	4.85	0.03	0.03	0.00	581.00	0.47
CNRM-CM5	3.48	0.01	0.01	0.00	417.00	0.34
EC-EARTH	4.85	0.03	0.03	0.00	581.00	0.47
MIROC5	3.48	0.01	0.01	0.00	417.00	0.34
	RCP8.5					
CanESM2	7.60	0.05	0.05	0.00	909.00	0.74
CNRM-CM5	6.22	0.03	0.03	0.00	745.00	0.61
EC-EARTH	7.60	0.05	0.05	0.00	909.00	0.74
MIROC5	6.64	0.04	0.04	0.00	795.00	0.65

Climate Change Hazards, Adaptation and Mitigation Mechanisms

Climate Change Hazards

Farmers are aware of the climate change and variabilities in Eastern Ethiopia, which includes the semi-arid environment of the Dechatu catchment (Kidanu et al. 2016; Tesfaye and Seifu 2016; Mohammed 2018). This area is known for its extended dry spells in the rainy seasons, drought, extreme heat, and intensive agricultural practices in the hilly part of the catchment. It is further characterized by low productivity, a shortage of grazing land to feed cattle, and flash flooding as the direct influences of climate change in the area (Alemu 2015; Erena and Worku 2018). The projected finding reveals that there is variability in rainfall and an increase in temperature trends in the Dechatu catchment. This result concurs with the findings of Kidanu et al. (2016) using the past historical climatic trends of the Dechatu Catchment.

Water Management for Climate Adaptation and Mitigation

The heat stress and the scarcity of water during most of the productive season has forced the farmers in the region to look for the best water management options of adaptation and mitigation measures in the Dechatu catchment. However, the available soil and water management practices are limited in number and types to ensure resilience in communities that cope with the negative impact of climate change. Based on the field survey data in the catchment, the following moisture-retaining structures were identified for conserving the eroded soil and controlling runoff water, which can be used for either agricultural or recharging purposes (Table 1.8).

In comparing the entire catchment coverage (14,983 ha) from the GIS extracted data, the existing BMPs that were constructed to manage the water resources via water retention structures and mitigate the climate accounted for 17.4% (2,614 ha) of the total catchment area. The catchment is dominated by hilly topography, and the

Table 1.8 Existing best management practices (BMPs) in the Dechatu catchment

Existing BMPs	Land-use type	Area (ha)
Check dam	Bare land	31.5
	Shrubland	72.6
Terrace	Cultivated	929.5
	Cultivated *Khat*	827.2
	Shrubland	715.3
Terrace and trench	Shrubland	19.6
Trenches	Shrubland	17.8
	Total	2613.5

coverage of the existing water management practices is too small compared to the catchment size. In line with the above soil and water conservation measures (Table 1.8), the farmers use various adaptation and mitigation measures for climate change. Among these, the major practices used in the catchment include: traditional hand-dug wells to fetch the stored water in sandy soil during the rainy season and use it later in the dry period, shallow-depth excavation on the flood plain (locally named as *Chirosh*) for drinking water purpose, spate irrigation using floodwater diversion, contour farming, dropping shrub/tree branches on the river banks to conserve transported soils and cultivate the land 2–3 years later are prolifically used practices in the catchment. Some conventional water management practices, like constructing sand storage dams (currently one sand storage dam is identified), along the river course are implemented in the area with farmers' participation, regional climate and environmental offices, and other governmental and non-governmental bodies. These practices improve the resilience of the farmers to a changing climate in the region. Furthermore, the changing cropping patterns, crop types and varieties (short-duration and drought-tolerant crops), and crop diversification are also used as climate adaptation strategies as farm-level solutions in the semi-arid climate of Dire Dawa (Kidanu et al. 2016; Tesfaye and Seifu 2016).

Field et al. (2014) suggested an adaptation action that is either incremental or transformative, depending on the severity and frequency of extreme weather and climatic impacts (Loboguerrero et al. 2019). For example, with less climatic impacts, farmers in the Dechatu catchment can decide to change crop type to those requiring less water for the effective management of water resources, while more severe and frequent climatic impacts might force them to adopt transformative adaptive options. Furthermore, local solutions to climate change adaptation practices in the Dechatu catchment such as effective landscape-planning (development of small reservoirs, sediment storage dams, sand dams, irrigation development, irrigation water-saving technologies) must be implemented in a sustainable manner for all affected, including downstream users. In this catchment, groundwater is a major source of water for irrigation and other purposes; consequently, careful planning is required to avoid abstraction impacts that may cause a decrease in the groundwater reserves.

Conclusions

In this study, we assessed the future trends and magnitude of climatic parameters - mainly temperature and rainfall - and integrated them with the existing climate adaptation options for water management practices in the flood-prone Dechatu catchment, Ethiopia. The key findings of this study show that the projected rainfall exhibited high variability, but none of the rainfall data series exhibited a trend at a 5% significance level except in CanESM2 of RCP8.5. A significant monotonic increase in temperature is forecasted in all RCMs and higher in RCP8.5. This explains that heat stress in the catchment is imminent unless appropriate interventions are adopted to avert the negative impacts of climate change. Incremental or transformative adaptation action is required to ensure effective water management practices, depending on the severity and frequency of extreme weather and climatic impacts in the semi-arid catchment of Dechatu.

Recommendations

- Establishment of local hydrological, weather, and climatic information services to reduce the impact caused by flash flooding through understanding the trends and seasonal/annual variability in order to effectively plan the water resources in the catchment.
- Implementation of water-saving technologies and landscape planning to minimize the groundwater abstraction impacts.
- Traditional and local knowledge requires a deliberate framework and approach to be considered in policy development, and in order to implement climate change adaptation strategies in line with the conventional water resource management approach.
- Mainstreaming the adaptive scaling approach to reduce competition for the scarce water among sectors in the catchment can create a resilient community that adapts to the negative impacts of climate change.
- As a limitation, this paper is more focused on the technical and traditional water management practices as mitigation mechanisms to climate extremes (flood, drought, heat stress, etc.) in the Dechatu Catchment. Therefore, further investigation on other indigenous coping mechanisms—traditional weather forecasting systems, use of wild edible fruits as a nutritional supplement and traditional medicines and use of indigenous crops, among others—to climate change and how these communities use the natural resources or habitats wisely as adaptation strategies to ensure their survivals under extreme conditions.

Cross-References

- Climate change
- Rainfall variability
- Temperature anomaly
- Climate adaptation
- Climatic model.

Acknowledgements The authors gratefully acknowledge the full support and provision of funds by the Dire Dawa Administration City Manager Office and Environment, Forest and Climate Change Authority (EFCCA) of Dire Dawa City Administration with Grant Agreement No: UIID/DD/C/01/19/20 and Project ID of P163452. In addition, we are thankful to the National

Meteorological Agency of Ethiopia for providing us with data, and to Haramaya University for facilitating the administrative issues.

References

Alemu YT (2015) Flash flood hazard in Dire Dawa, Ethiopia. J Soc Sci Humanit 1:400–414

Argaud L, Ferry T, Le Q-H et al (2007) Short-and long-term outcomes of heatstroke following the 2003 heat wave in Lyon, France. Arch Intern Med 167:2177–2183. https://doi.org/10.1001/archinte.167.20.ioi70147

Basu R, Samet JM (2002) Relation between elevated ambient temperature and mortality: a review of the epidemiologic evidence. Epidemiol Rev 24:190–202. https://doi.org/10.1093/epirev/mxf007

Billi P, Alemu YT, Ciampalini R (2015) Increased frequency of flash floods in Dire Dawa, Ethiopia: change in rainfall intensity or human impact? Nat Hazards 76:1373–1394. https://doi.org/10.1007/s11069-014-1554-0

Dosio A (2017) Projection of temperature and heat waves for Africa with an ensemble of CORDEX Regional Climate Models. Clim Dyn 49:493–519. https://doi.org/10.1007/s00382-016-3355-5

Erena SH, Worku H (2018) Flood risk analysis: causes and landscape based mitigation strategies in Dire Dawa city, Ethiopia. Geoenviron Disasters 5:16. https://doi.org/10.1186/s40677-018-0110-8

Field CB, Barros VR, Dokken DJ, Mach KJ, Mastrandrea MD, Bilir TE, Chatterjee M, Ebi KL, Estrada YO, Genova, R.C.; et al. (2014) IPCC Summary for policymakers. In: Climate Change 2014: impacts, adaptation, and vulnerability, part A: global and sectoral aspects; contribution of working group II to the fifth assessment report of the intergovernmental panel on climate change. Cambridge University Press: Cambridge, UK: New York, NY, USA, pp 1–32, ISBN 978-92-9169-143-2

Gbobaniyi E, Sarr A, Sylla MB et al (2014) Climatology, annual cycle and interannual variability of precipitation and temperature in CORDEX simulations over West Africa. Int J Climatol 34:2241–2257. https://doi.org/10.1002/joc.3834

Giorgi F, Jones C, Asrar G (2009) Addressing climate information needs at the regional level: the CORDEX framework. WMO Organ Bull 58:175

IPCC (2014a) Climate change 2014–Impacts, adaptation and vulnerability: regional aspects. Cambridge University Press, Cambridge

IPCC (2014b) Climate change 2013—the physical science basis: Working Group I contribution to the fifth assessment report of the intergovernmental panel on climate change. Cambridge University Press, Cambridge

Masson-Delmotte V, Zhai P, Pirani A, Connors SL, Péan C, Berger S, Caud N, Chen Y, Goldfarb L, Gomis MI, Huang M. Climate change (2021) The physical science basis. Contribution of working group I to the sixth assessment report of the intergovernmental panel on climate change. 2021 Jun;2.

Kidanu A, Kibret K, Hajji J et al (2016) Farmers perception towards climate change and their adaptation measures in Dire Dawa Administration, Eastern Ethiopia. J Agric Ext Rural Dev 8:269–283

Kilbourne EM (1999) The spectrum of illness during heat waves. Am J Prev Med 16:359–360. https://doi.org/10.1016/S0749-3797(99)00016-1

Leal Filho W, Matandirotya NR, Lütz JM et al (2021) Impacts of climate change to African indigenous communities and examples of adaptation responses. Nat Commun 12:1–4. https://doi.org/10.1038/s41467-021-26540-0

Loboguerrero AM, Campbell BM, Cooper PJM et al (2019) Food and earth systems: priorities for climate change adaptation and mitigation for agriculture and food systems. Sustain 11:1372. https://doi.org/10.3390/su11051372

Mbow C, Rosenzweig C, Barioni LG, Benton TG, Herrero M, Krishnapillai M, Liwenga E, Pradhan P, Rivera-Ferre MG, Sapkota T, Tubiello FN, Xu Y (2019) Food security—burundi food security. In: Climate Change and Land: an IPCC special report on climate change, desertification, land degradation, sustainable land management, food security, and greenhouse gas fluxes in terrestrial ecosystems, pp 437–550

Mendelsohn R, Dinar A (2003) Climate, water, and agriculture. Land Econ 79:328–341

Mohammed Y (2018) Determinants of smallholder farmers' choices of adaptation strategies to climate variability in Rural Kebeles of Dire Dawa Administration, Ethiopia

Nikulin G, Jones C, Giorgi F et al (2012) Precipitation climatology in an ensemble of CORDEX-Africa regional climate simulations. J Clim 25:6057–6078. https://doi.org/10.1175/JCLI-D-11-00375.1

Nikulin G, Lennard C, Dosio A, et al (2018) The effects of 1.5 and 2 degrees of global warming on Africa in the CORDEX ensemble. Environ Res Lett 13:65003. https://doi.org/10.1088/1748-9326/aab1b1

Osima S, Indasi VS, Zaroug M, et al (2018) Projected climate over the Greater Horn of Africa under 1.5 °C and 2 °C global warming. Environ Res Lett 13:65004. https://doi.org/10.1088/1748-9326/aaba1b

Porter JR, Xie L, Challinor AJ, et al (2015) Food security and food production systems. Clim Chang 2014 Impacts, Adapt Vulnerability Part A Glob Sect Asp 485–534. https://doi.org/10.1017/CBO9781107415379.012

Pyrgou A, Santamouris M, Livada I, Cartalis C (2019) Retrospective analysis of summer temperature anomalies with the use of precipitation and evapotranspiration rates. Climate 7:104. https://doi.org/10.3390/cli7090104

Rind D, Goldberg R, Ruedy R (1989) Change in climate variability in the 21st century. Clim Change 14:5–37. https://doi.org/10.1007/BF00140173

Ringler C (2008) the impact of climate variability and climate change on water and food outcomes. Food Policy 1–2

Shewmake S (2011) Vulnerability and the impact of climate change in South Africa's Limpopo River Basin. SSRN Electron J 804. https://doi.org/10.2139/ssrn.1289844

Tesfaye W, Seifu L (2016) Climate change perception and choice of adaptation strategies: Empirical evidence from smallholder farmers in east Ethiopia. Int J Clim Chang Strateg Manag 8:253–270. https://doi.org/10.1108/IJCCSM-01-2014-0017

Teshome A, Zhang J (2019) Increase of extreme drought over Ethiopia under climate warming. Adv Meteorol 2019. https://doi.org/10.1155/2019/5235429

Wuebbles DJ (2018) Climate change in the 21st century: looking beyond the Paris agreement. In: Climate change management. Springer, pp 15–38

Yue S, Pilon P, Phinney B, Cavadias G (2002) The influence of autocorrelation on the ability to detect trend in hydrological series. Hydrol Process 16:1807–1829. https://doi.org/10.1002/hyp.1095

Chapter 2
Online Consumption Impact: Sustainable Practices of Young Adults Facing Climate Change

Priscila Cembranel, Jhordano Malacarne Bravim, Raissa Mariana Rita, Ana Caroline Camargo, Gabrieli Cristina Borchard, José Baltazar de Andrade Guerra, and Valeria Isabela Beattie

Abstract The purpose of this paper is to understand the differences in perspectives among young adults who practice sustainable consumption by shopping in the digital environment and to emphasize the importance of this practice as a strategy to curb climate change. The study was developed with a quantitative approach and used a survey procedure involving higher education students in the north of Santa Catarina, Brazil. The data was analyzed using frequency statistical tests, analysis of variance (ANOVA), and bivariate. Two hypotheses were tested: the first included a younger the sample of individuals with a greater emphasis on sustainable consumption and the second included women that would be more susceptive to sustainable consumption online. The study allows us to state that women are more concerned with consuming

P. Cembranel (✉) · R. M. Rita · A. C. Camargo · G. C. Borchard
University Educational Society of Santa Catarina (UNISOCIESC), Getúlio Vargas Ave, 268, Center, Jaraguá Do Sul City, Santa Catarina 89251-970, Brazil
e-mail: priscila_cembranel@yahoo.com.br

A. C. Camargo
e-mail: anacaroline.camargo@clinicorp.com

P. Cembranel
University of Contestation (UNC), Nereu Ramos Ave., 1071, Jardim do Moinho, Mafra City, Santa Catarina 89300 -000, Brazil

J. M. Bravim
Federal Institute of Education, Science and Technology Rondônia, Federal University of Parana Vilhena, BR-174 Road, Km 3 S/n Urban Zone, Vilhena City, Rondonia 76980-000, Brazil

J. B. de Andrade Guerra
Center for Sustainable Development/Research Group on Energy Efficiency and Sustainability (Greens), University of Southern Santa Catarina (UNISUL), Trajano Street, 199, Center, Florianópolis City, Santa Catarina 88130-100, Brazil
e-mail: jose.baltazarguerra@animaeducacao.com.br

V. I. Beattie
Center for Sustainable Development/Research Group on Energy Efficiency and Sustainability (Greens), Elliott School of International Affairs, George Washington University, Washington, DC 20052, USA
e-mail: vbeattie43@gwmail.gwu.edu

W. Leal Filho et al. (eds.), *Climate Change Strategies: Handling the Challenges of Adapting to a Changing Climate*, Climate Change Management,
https://doi.org/10.1007/978-3-031-28728-2_2

local products both online and locally. Additionally, there is a small group of people between the ages of 26 and 28 who are interested in knowing the company backgrounds and their managers'. This group includes, albeit timidly, young adults among who practice sustainable consumption online. Given this study, it can be concluded that individuals are advanced in terms of awareness concerning lifestyle change and can help to reduce the global impacts of climate change.

Introduction

Sustainable consumption occurs when people value a connection with the reduction of impacts on the environment. These values become visible when social changes caused by natural disasters, psychological aggressions, and medical care emergencies occur (Forbes 2017). In general, such striking events change consumption patterns and increase sustainable trends (Alexa et al. 2021).

It is known that sustainable consumption is widely discussed among young adults (Annunziata et al. 2019). This audience, affected by quarantine and social distancing, is more active on the internet and therefore, makes more online purchases (Hladikova and Hurajova 2020).

The beginning of the COVID-19 pandemic was marked by an increase in consumption based on fear and insecurity in an emergency situation. After this phase, purchases decreased (Cohen 2020). However, innovative and creative purchasing strategies emerged. Among these purchasing strategies, the increase in online purchases and an increase in awareness and a sense of responsibility in regards to sustainability issues and the impacts generated by their consumption (Degli Esposti et al. 2021).

We believe that the new generations are aware of the impacts of consumption on climate change and that this concern is expressed through the concept of sustainable consumption and attitudes that still need to evolve. This happens slowly, because with young people comes several cultural and social practices based on consumerism which hinder significant changes aimed at climate adaptation. Such as: the acquisition of goods with a lower environmental footprint, services with lower carbon emissions and appreciation of local products that are not always part of the mainstream of consumption by young people, for example.

Young people are exposed from an early age to consumption appeals. The beginning of their adult phase (from 18 years old in Brazil, which is the framework of this study) is when young people begin to make purchasing decisions. Despite bearing the influence of society, family, and friends in their behavior, young people are commonly considered drivers of the sustainability movement (Ziesemer et al. 2021). Furthermore, young adults recognize the need to buy less, maintain a healthier lifestyle, and adapt better consumption habits in the "new normal" (Korsunova et al. 2021). Thus, for this study, the following hypothesis is advanced (H1): Younger consumers are more concerned with sustainable consumption.

Studies also show that young women have been more committed to sustainability issues since the beginning of the COVID-19 pandemic, especially if they have been exposed to responsible consumption practices since their childhood (Degli Esposti et al. 2021). They stand out for consuming second-hand clothes and accessories with sustainable raw materials (Bianchi and Gonzales 2021) and are also more willing to change their eating patterns to become more sustainable (Robu et al. 2021). Thus, the second hypothesis proposed (H2) is: The female audience is more likely to consume sustainable products and services online.

This work was structured on a theoretical framework regarding sustainable, digital and young adult consumption. We then addressed the methodological aspects, the presentation of results, their discussion and then, the final considerations.

Sustainable Consumption

Consumption began in the sixteenth century, when communities exchanged consumer goods amongst themselves. At that time, consumption was restricted to meet human needs, especially food, clothing, and personal care products. However, after the industrial revolution and the development of mass production, there was a variety of products to offer. Thus, forms of consumption with other purposes were developed, such as for status, collecting, and accumulating, among others (Souza et al. 2019).

The emergence of multiple options for products and services turned consumption cheaper, trivialized, and brought consequences to society and the environment (Silva et al. 2017). One of the ways to deal with the consequences of society's excess consumerism, is to discuss sustainable consumption (Lira 2018).

Sustainable consumption advocates for reducing the use of natural resources and raw materials in the cycle of products and services (Annunziata et al. 2019). The life cycle consists of the extraction of raw material for production, product manufacturing, shipping to the final consumer, durability, and recycling (Lemos 2020). In this connection, sustainable labeling becomes a way to protect the planet and is also a market strategy (Janßen and Langen 2017). In addition, such products and services are powerful tools to inform society, since they provide consumers with an understanding of the social, ethical, and environmental impacts of their consumption (Van Loo et al. 2017).

Since the beginning of the COVID-19 pandemic, there have been changes in consumption dynamics due to social isolation. The first phase of the pandemic was marked by irrational consumption caused by the fear of a shortage of goods. After this period, some habits surveyed indicate that there was a creative and innovative consumption practices trial. People started to prepare their own food, consume more local products, give a certain preference to products with a sustainable supply chain, and buy more products online (Degli Esposti et al. 2021).

People became more aware of their relationship with nature and their consumption responsibility with regard to environmental impacts since the beginning of COVID-19. This attitude favored the adoption of ecological criteria in everyday consumption practices and accelerated the transition to sustainable consumption (Cohen 2020).

In this connection, young adults can be considered the main target of campaigns aimed at reducing consumption given that youngsters tend to reflect on the social and environmental aspects when choosing products and services for consumption (Ziesemer et al. 2021). Since the beginning of the COVID-19 pandemic, this audience has made an important observation: the economic system should be driven less by consumption (Korsunova et al. 2021).

Methodological Aspects

In order to understand the vision of the digital consumer regarding sustainable consumption, a survey was developed using a questionnaire consisting of 8 closed questions hosted in Google Form and distributed via social networks (WhatsApp, Instagram and Facebook) through the snowball technique. The target audience of the study met three inclusion criteria: be a higher education student in the north of Santa Catarina, Brazil; be between 18–28 years old, and purchase sustainable products online. Data was collected over a period of 30 days, in April 2021. A total of 218 responses were obtained.

For analytical purposes, respondents were divided into three age groups: 18–21 years old; 22–25 years old; and 26–28 years old. Then an analysis through statistical tests ANOVA and bivariate analysis with the aid of SPSS software was performed. Thus, for analytical purposes, questions were presented as variables and received the following identification Q1, Q2, Q3, etc.

Out of the total number of respondents, around 53% were male and 47% female. The age group with the most participants was composed of young people (41%) aged between 18 and 21 years, followed by 35% young adults aged between 26 and 28 years, and finally, 24% young people aged between 22 and 25 years.

Results

Out of the 218 completed questionnaires, around 53% were answered by males and 47% by females.

The age range of most participants was: between 18 and 21 years old (41%); 35% were between 26 and 28 years, and 24% between 22 and 25 years. Thus, two hypotheses were tested: the first one was that the younger the individual, the greater their concern with sustainable consumption and the second one was that women are more susceptible to sustainable consumption online.

Therefore, two variance analysis (ANOVA) were carried out using the variable Q1 ("How old are you?") as the dependent variable (factor), which can take the following values: 18–21 years; 22–25 years; and 26–28 years for the first analysis and, for the second one, the dependent variable (factor) was changed to variable Q2 ("What is your gender"), which can take the following values: Female; Male. For both analyses, the variables Q3 to Q10 were used as independent variables.

As for the verification of hypothesis 1, for the eight questions of the survey, only for the variable "Do you seek to know if the company and its managers really apply the values they claim to advocate?" (Q7). It is observed that significance was equal to less than 0.05 (Table 2.1). As for the other variables, it is possible to accept the null hypothesis so that the means of the groups that are identified by age groups are equal. That is, there is no difference in the concern with sustainable consumption in the three age groups reviewed.

In turn, in verifying hypothesis 2, for the eight analysis questions, for the variables "How often do you shop online?" (Q3) and "Do you give priority to the purchase of small outfits?" (Q10) It is observed that significance was equal to less than 0.05 (Table 2.2). For the other variables, it is possible to accept the null hypothesis so that the means of the groups that are identified by age groups are equal. That is, both the female and male audiences have the same trend towards care in the acquisition of sustainable products and services.

Sustainable Online Consumption and Gender

The first question in the questionnaire referred to the frequency with which consumers buy products online. Out of the 218 respondents, 42% stated they shop online every three months, 39% say they shop online every month, 12% said they shop online at least twice a week, and 7% said they shop online at least once a week. The male audience is the one that most often shopped online (once a month on average) and the female audience is the least often online shopper (every 3 months or more), as shown in Table 2.3.

To find out the most important factor for e-consumers when choosing their products, in the second question, the respondents were able to select their preferences based on the options presented. Among the answer choices, consumers indicated that quality (58.7%) and price (29.4%) are among the aspects with the greatest influence in their purchasing decisions, followed by the product brand (5.5%), whether the product is sustainable (3.7%) and the service provided by suppliers (2.8%).

Table 2.1 ANOVA results for testing hypothesis 1

		Sum of squares	gl	Medium square	F	Sig.
How often do you shop online?	Among groups	23.478	2	11.739	2.346	0.098
	In the groups	1075.921	215	5.004		
	Total	1099.399	217			
What is important when choosing the company that I'm going to buy a product/service from?	Among groups	7.314	2	3.657	2.764	0.065
	In the groups	284.466	215	1.323		
	Total	291.780	217			
When choosing a sustainable product. Which do you consider as the main factor?	Among groups	0.723	2	0.362	0.397	0.673
	In the groups	195.758	215	0.911		
	Total	196.482	217			
Are the company's values relevant to you at the time of purchase?	Among groups	0.454	2	0.227	0.784	0.458
	In the groups	62.207	215	0.289		
	Total	62.661	217			
Do you want to know if the company and its managers really apply the values they claim to advocate?	Among groups	1.889	2	0.944	4.350	0.014
	In the groups	46.667	215	0.217		
	Total	48.555	217			
Do you take into account the destination of the packaging when purchasing a product?	Among groups	0.380	2	0.190	0.410	0.664
	In the groups	99.625	215	0.463		
	Total	100.005	217			
Do you look for a product that has the most manual/handcrafted and personalized manufacture?	Among groups	0.804	2	0.402	0.726	0.485
	In the groups	119.068	215	0.554		
	Total	119.872	217			
Do you give priority to the purchase from small businesses?	Among groups	0.792	2	0.396	1.794	0.169
	In the groups	47.429	215	0.221		
	Total	48.220	217			

Source The authors, 2021

Table 2.2 ANOVA result for testing hypothesis 2

		Sum of squares	gl	Medium square	F	Sig.
How often do you shop online?	Among groups	52.119	1	52.119	10.750	0.001
	In the groups	1047.280	216	4.849		
	Total	1099.399	217			
What is important when choosing the company that I'm going to buy a product/service?	Among groups	1.297	1	1.297	0.964	0.327
	In the groups	290.483	216	1.345		
	Total	291.780	217			
When choosing a sustainable product, which is the main factor analyzed?	Among groups	0.233	1	0.233	0.256	0.613
	In the groups	196.249	216	0.909		
	Total	196.482	217			
Are the company's values relevant to you at the time of purchase?	Among groups	0.015	1	0.015	0.051	0.821
	In the groups	62.646	216	0.290		
	Total	62.661	217			
Do you want to know if the company and its managers really apply the values they claim to advocate?	Among groups	0.114	1	0.114	0.509	0.476
	In the groups	48.441	216	0.224		
	Total	48.555	217			
Do you take into account the destination of the packaging when purchasing a product?	Among groups	0.388	1	0.388	0.841	0.360
	In the groups	99.617	216	0.461		
	Total	100.005	217			
Are you looking for a product that has the most manual/handcrafted and personalized manufacture?	Among groups	0.625	1	0.625	1.131	0.289
	In the groups	119.247	216	0.552		
	Total	119.872	217			
Do you give priority to the purchase from small businesses?	Among groups	1.847	1	1.847	8.604	0.004
	In the groups	46.373	216	0.215		
	Total	48.220	217			

Source The authors, 2021

Table 2.3 Cross-tabulation between variables Q3 (How often do you shop online?) and Q2 (What is your gender?)

			What is your gender?		Total
			Female	Male	
How often do you shop online?	Once a month on average	Score	31	51	82
		% on How often do you shop online?	37.8%	62.2%	100.0%
	Once a month on average. Every 3 months or more	Score	3	0	3
		% on How often do you shop online?	100.0%	0.0%	100.0%
	Once a week or more	Score	3	12	15
		% on How often do you shop online?	20.0%	80.0%	100.0%
	Once every two weeks	Score	8	17	25
		% on How often do you shop online?	32.0%	68.0%	100.0%
	Once every two weeks. once a month on average	Score	0	2	2
		% on How often do you shop online?	0.0%	100.0%	100.0%
	Every 3 months or more	Score	58	33	91
		% on How often do you shop online?	63.7%	36.3%	100.0%
Total		Score	103	115	218
		% on How often do you shop online?	47.2%	52.8%	100.0%

Source The authors, 2021

Sustainable Consumption

In the ecological consumption segment, the first question refers to the factors that lead the consumer to choose a sustainable product. In the alternatives presented, the most significant criterion for the respondents was the option for durable or reusable products (46.3%). Products that have the "cruelty free" seal, that is, that are not tested on animals, follow as the second most important factor (21.1%). In third place is the option of products with renewable raw materials, which have less impact on the environment (20.2%) and finally, products that are contained in recyclable packaging (12.4%).

In the following questions, consumers are asked about their preference to buy products from small businesses, with handcrafted or personalized products, thus supporting small entrepreneurs. In addition, consumers answered whether, when purchasing, they take into account the company's values and the way in which these values are applied. About 67% of consumers claim to give priority to the purchase of sustainable products in small businesses, and 42.9% prefer to consume products made in an artisanal way, in which the female audience of the sample assessed is the one that gives priority to purchases from small businesses. Regarding the values, there is a divergence, because while 67.9% say that the company's values are indeed relevant, when it comes to really knowing if the values are applied by the company, only 33.5% of consumers bother to look for this information that may be relevant for their final selection.

In this specific case, as in the analysis of variance (ANOVA) for the test of hypothesis 1, significance was identified, which required a more detailed analysis. Since the survey respondents in the two lowest age groups are the ones who least seek information as to whether the company and its managers really apply the values they claim to advocate (Table 2.4).

In a survey on sustainable consumption, it is essential to ask consumers about the importance of the empty packaging destination. Out of the 218 respondents, 51.8% consider the packaging to be an important factor; for example, they prefer to consume products whose packaging is biodegradable or recyclable.

The test results showed that age and gender do not influence the sustainable consumption profile of young adults. It was also observed that the male public makes online purchases more frequently; females give priority to purchases from small local businesses, and people between 26–28 years of age seek to know if the company and its managers really apply the values they claim to advocate.

Table 2.4 Cross-tabulation between variables Q1 (How old are you?) and Q7 (Do you seek to know if the company and its managers really apply the values they claim to advocate?)

			Do you want to know if the company and its managers really apply the values they claim to advocate?		Total
			No	Yes	
What is your age?	18–21 years	Score	55	22	77
		% on What is your age?	71.4%	28.6%	100,0%
	22–25 years	Score	56	20	76
		% on What is your age?	73.7%	26.3%	100,0%
	26–28 years	Score	34	31	65
		% on What is your age?	52.3%	47.7%	100,0%
Total		Score	145	73	218
		% on What is your age?	66.5%	33.5%	100.0%

Source The authors, 2021

Results Analysis and Discussion

Digital consumption was better accepted by young people, especially because most of them grew up in contact with technologies. However, online shopping is affected in general by some distrust (Soares et al. 2020). One of the reasons that people do not buy online is not being able to touch the product and understand its physical characteristics before they make their choice (Nascimento et al. 2018).

By offering sustainable products and services, companies, in addition to contributing to the environment, are raising interactive awareness. For example, buyers with a sustainable consumption profile use product packaging as utensils and decorative objects. This contributes to the development of creativity, protects the environment, and strengthens the brand (Nguyen et al. 2020).

In addition, the company's values are essential for the sale of these products (Silva et al. 2017) as the organizational culture and values will impact their products. Consumers seek to purchase products from companies whose values are similar to their own (Brock et al. 2020).

Another relevant point for many consumers is the final destination after consumption. In this context, large companies started working with reverse logistics. This allows the consumer to return the product packaging and not worry about how to dispose of it (Nguyen et al. 2020). Post-consumption disposal appears as an incentive, as companies usually take some discount initiatives. These companies tend to encourage the proper disposal of the product packaging after the end of the product's life thus contributing both from the economic and environmental standpoint; when consumers return goods and materials to the production cycle, the quantities of materials discarded inappropriately are reduced (Lira 2018).

In the case of women, handcrafted products are also a form of sustainable consumption (Brock et al. 2020). In addition, the appreciation of small local businesses is also prominent among female respondents. In general, these women are concerned with the environment and maintain sustainable practices that will benefit the place where they are located besides generating income and benefits to the community. Small businesses tend to be concerned with packaging, product destination and to give priority to renewable raw materials (Nguyen et al. 2020).

Sustainable consumption is part of the SDG 12 (Responsible consumption and production). In Brazil, the goal was changed in its wording when it was proposed the articulation between federated entities (states and municipalities). For, it is believed that its implementation is guaranteed by the *Plano de Ação para Produção e Consumo Sustentáveis* (PPCS, Action Plan for Sustainable Production and Consumption) of the Ministry of Environment. Thus, the plan was launched on November 23, 2011 and incorporated the *Política Nacional sobre Mudança do Clima* (PNMC, National Policy on Climate Change), also used to comply with SDG 13 (Action against global climate change) (IPEA 2019).

The Action Plan for Sustainable Production and Consumption (PPCS) was divided into: sector pacts, government actions, voluntary initiatives, partnership actions, and task forces. In relation to sustainable consumption, the Action Plan proposed: education for sustainable consumption, retail and sustainable consumption, increased recycling, sustainable government purchases, sustainable buildings, and an environmental agent in the public administration (Ministério do Meio Ambiente 2011). Thus, we can affirm that the increased concern with consumption, production and company practices show ways to reach the goals of sustainable consumption and climate adaptation. In particular, when there are efforts aimed at valuing the local context and aligning sustainable practices by organizations and young consumers (Schäfer et al. 2021).

Final Considerations

This article was developed with the aim of understanding the differences in perception among young people who practice sustainable consumption through purchases online and establish the importance of this action as a strategy to curb climate change. To that effect, two hypotheses were established: H1: Younger consumers are more concerned with sustainable consumption and H2: The female audience is more likely to consume sustainable products and services online.

After the development of statistical tests for ANOVA it was possible to conclude that there are no significant differences between men and women in relation to sustainable consumption online; however, there are small changes that can be considered an advance in terms of awareness.

That's because age and gender do not influence the sustainable consumption profile of young adults. It was also observed that the male public purchases online more frequently, while a female public prioritizes purchases from small local businesses. In addition, the results show that people between 26 and 28 years old seek to know if the company and its managers really apply the values they claim to defend. The rest of the variables did not present significant results to establish any correlation. In this way, it is clear that although there is a movement of young people focused on sustainable consumption to promote climate adaptation, it is just beginning and comprises only young adults from 26 years old, that is, when they are closer to 30 years old.

The study offers a practical and theoretical contribution. There is a theoretical advance when stating that women are concerned with consuming local products online and that there is already a small group of people aged 26 and over that is concerned with the values of companies and their managers. This includes, albeit timidly, young adults among those who practice sustainable consumption online. Still, in a practical way, this study indicates that the adoption of sustainable practices is positive for local companies, since these can better steer their dissemination strategies and include young adults as a segment to be exploited marketwise.

With regard to sustainable consumption strategies to curb climate change, this concern emerged in Brazil in 2011 along with the Action Plan for Sustainable Production and Consumption (PPCS) incorporated by the National Climate Change Policies

(PNMC). Among many actions, the main one was the encouragement of sustainable consumption widely discussed in schools and universities. Thus, in 2021, ten years later, it is possible to see that young adults are beginning to reveal concerns about the consumption of local products and the company's values as a way of responding to climate change.

This article is limited to the studied sample and to the fact that the research was developed only through questionnaires distributed online. Therefore, as suggestions for future research, studies can be developed with audiences from other regions and countries in order to perform comparisons with this study. In addition, the development of interviews is suggested to learn more about the researched reality in order to understand the aspects that motivate sustainable consumption choices.

References

Alexa L, Apetrei A, Sapena J (2021) The COVID-19 lockdown effect on the intention to purchase sustainable brands. Sustainability 13(6):3241. https://doi.org/10.3390/su13063241

Annunziata A, Mariani A, Vecchio R (2019) Effectiveness of sustainability labels in guiding food choices: Analysis of visibility and understanding among young adults. Sustain Prod Consumption 17:108–115. https://doi.org/10.1016/j.spc.2018.09.005

Bianchi C, Gonzales M (2021) Exploring sustainable fashion consumption among eco-conscious women in Chile. Int Rev Retail Distrib Consum Res 31(4):375–392. https://doi.org/10.1080/09593969.2021.1903529

Brock A, Radtke M, Espartel L (2020) Parada obrigatória para reajuste da rota: o consumo sustentável é o destino pós COVID-19? In: Anais 13º Congresso Latino-Americano de Varejo e Consumo: "After COVID-19: building purpose through stakeholders in retailing", São Paulo, Brazil. Retrieved from: http://bibliotecadigital.fgv.br/ocs/index.php/clav/clav2020/paper/view/7460

Cohen MJ (2020) Does the COVID-19 outbreak mark the onset of a sustainable consumption transition? Sustain Sci Pract Policy 16(1):1–3. https://doi.org/10.1080/15487733.2020.1740472

Degli Esposti P, Mortara A, Roberti G (2021) Sharing and sustainable consumption in the era of COVID-19. Sustainability 13(4):1903. https://doi.org/10.3390/su13041903

Forbes SL (2017) Post-disaster consumption: analysis from the 2011: Christchurch earthquake. Int Rev Retail Distrib Consumer Res 27(1):28–42. https://doi.org/10.1080/09593969.2016.1247010

Hladikova V, Hurajova A (2020) Internet addiction in the time of the COVID-19 pandemic in young adults. In: Annual international scientific conference on marketing identity: COVID-2.0. University of St. Cyril and Methodius in Trnava, Slovakia, November, pp 129–141. Retrieved from: https://fmk.sk/download/Marketing-Identity-2020-eng.pdf

IPEA—Instituto de Pesquisa Econômica Aplicada (2019) 12. Consumo e Produção Sustentáveis. Retrieved from: https://www.ipea.gov.br/ods/ods12.html#coll_12_2

Janßen D, Langen N (2017) The bunch of sustainability labels–Do consumers differentiate? J Cleaner Prod 143:1233–1245. https://doi.org/10.1016/j.jclepro.2016.11.171

Korsunova A, Horn S, Vainio A (2021) Understanding circular economy in everyday life: Perceptions of young adults in the Finnish context. Sustain Prod Consumption 26:759–769. https://doi.org/10.1016/j.spc.2020.12.038

Lemos LV et al (2020) Compras públicas sustentáveis: Uma análise dos editais de licitação de cidades brasileiras participantes do Programa Cidades Sustentáveis. Cuadernos de Contabilidad 21:1–18. https://doi.org/10.11144/Javeriana.cc21.cpsa

Lira FT (2018) Fatores que influenciam a valorização de produtos ecológicos por consumidores ecologicamente conscientes. Revista de Gestão Social e Ambiental—RGSA, São Paulo 12(2):90–107. https://doi.org/10.24857/rgsa.v12i2.1491

Ministério do Meio Ambiente (2011) O Plano de Ação para Produção e Consumo Sustentáveis (PPCS). Retrieved from: https://antigo.mma.gov.br/responsabilidade-socioambiental/producao-e-consumo-sustentavel/plano-nacional.html

Nascimento EA, Melo EA, Lima VLAG (2018) Ice cream with functional potential added grape agro-industrial waste. J Culinary Sci Technol 16(2):128–148. https://doi.org/10.1080/15428052.2017.1363107

Nguyen AT et al (2020) A consumer definition of eco-friendly packaging. J Clean Prod 252:119792. https://doi.org/10.1016/j.jclepro.2019.119792

Robu M et al (2021) Environmental concern factors and consumers' purchase decision on the local agri-food market. Environ Eng Manage J 20(3):405–418. Retrieved from https://eemj.eu/index.php/EEMJ/article/view/4292

Schäfer M, Figueiredo MD, Iran S, Jaeger-Erben M, Silva ME, Lazaro JC, Meißner M (2021) Imitation, adaptation, or local emergency?—A cross-country comparison of social innovations for sustainable consumption in Brazil, Germany, and Iran. J Clean Prod 284:124740. https://doi.org/10.1016/j.jclepro.2020.124740

Silva MVB, dos Santos ACMZ, Petrini M, Silveira LM (2017) Promovendo o consumo sustentável: Um estudo de caso. Revista PRETEXTO 18(3):50–66. https://doi.org/10.21714/pretexto.v18i3.3989

Soares, JC et al (2020) Sustentabilidade como tema propulsor do engajamento de usuários na mídia social. Revista Eletrônica de Administração e Turismo-ReAT 14(1):40–58. https://doi.org/10.15210/REAT.V14I1.15872

Souza JS, Miyazaki VK, Enoque AG (2019) Reflexões acerca do consumo verde e sustentável na sociedade contemporânea. Cadernos EBAPE. BR 17(2):403–413. https://doi.org/10.1590/1679-395167434

Van Loo EJ, Hoefkens C, Verbeke W (2017) Healthy, sustainable and plant-based eating: Perceived (mis) match and involvement-based consumer segments as targets for future policy. Food Policy 69:46–57. https://doi.org/10.1016/j.foodpol.2017.03.001

Ziesemer F, Hüttel A, Balderjahn I (2021) Young people as drivers or inhibitors of the sustainability movement: the case of anti-consumption. J Consum Policy 44:427–453. https://doi.org/10.1007/s10603-021-09489-x

Chapter 3
Rerouting Food Waste for Climate Change Adaptation: The Paths of Research

Anne Nogueira, Fátima Alves, and Paula Vaz-Fernandes

Abstract In a context of climate change where, on the one hand, large amounts of food are lost or wasted and on the other hand climate refugees are victims of food insecurity, there has been a growing interest in food waste and in research into reducing food loss. In order to identify research trends and gaps in this area, we have conducted a systematic review in this field adopting a methodology based on the use of selected key words and temporal boundaries from 2008 to 2021. The results extracted were divided in two categories: raw material and food security. The results are discussed under the following perspectives: geographic distribution of author's institution, subject categories, and author keywords. Based on research topics found and researchers' suggestions, actual and predictable trends on food waste management were identified for food waste as a raw material and for food security. It is found that Food Loss and Waste (FLW) fields of research interest have been: (1) food waste as a source of raw materials to produce biofuels and biomaterials; and (2) food waste upcycling for human consumption as a solution to food insecurity.

Introduction

Widely known data from the Food and Agriculture Organization (FAO) of the United Nations point to the global annual food loss and waste (FLW) of 1.3 billion tonnes, which is 1/3 of the total food produced for human consumption, while 11% of the population suffers from hunger or malnutrition (FAO, n.d.-b). Also, according to the

A. Nogueira · F. Alves · P. Vaz-Fernandes
Science and Technology Department, Universidade Aberta, 1269-001 Lisbon, Portugal
e-mail: fatimaa@uab.pt

A. Nogueira (✉) · F. Alves
Department of Life Science, Faculty of Sciences and Technology, Centre for Functional Ecology-Science for People and the Planet, University of Coimbra, 3000-456 Coimbra, Portugal
e-mail: anjono@sapo.pt

P. Vaz-Fernandes
Centre for Public Administration & Public Policies, Institute of Social and Political Sciences, Lisbon University, Lisbon, Portugal 1300-663

W. Leal Filho et al. (eds.), *Climate Change Strategies: Handling the Challenges of Adapting to a Changing Climate*, Climate Change Management,
https://doi.org/10.1007/978-3-031-28728-2_3

FAO, the annual costs of the world FLW are roughly US$ 680 billion in industrialized countries and US$ 310 billion in developing countries (FAO, n.d.-b).

While researching Food Loss (FL) and Food Waste (FW) one can easily come across broadly accepted concepts (FAO 2011b) related to FL which refers to the decrease of the edible food mass throughout the part of the supply chain that leads specifically to food for human consumption. Food losses can occur during production, post-harvest and the processing stages of the food supply chain (FSC) (Parfitt, Barthel and MacNaughton 2010). When these losses occur at the end of the food chain (distribution and final consumption) they are referred to as "food waste", relating specifically to retailers and consumer behaviour (Hodges et al. 2011; Parfitt et al. 2010).

There is yet another side to the FL and FW concepts. The Food and Agriculture Organization (FAO) of the United Nations, only measures FL and FW as parts of the food chain *leading to human consumption,* excluding planned animal feed and inedible parts. Hence, all unplanned loss along the human food supply chain is accounted for as FL or FW, even when later redirected to animal feed or bioenergy production (Parfitt et al. 2010).

Food waste and food loss reduction have been one of the priorities of the international political, economic, environmental, and social agenda. Economically FLW represent an income reduction for farmers, industries, distributors, and an increase in expense for consumers. Annually, adding to capital loss, one must not forget major depletion of resources such as 250 km^3 of water, 1.4 billion hectares of land (28 percent of the world's agricultural area), energy, and labour. Moreover, associated to FLW, there is a needless production of 3.3 billion tonnes of CO_2 equivalent to Greenhouse Gas Emissions (GHGE), contributing to global warming and climate change (FAO 2011b, 2013). The International Panel on Climate Change Report of 2021 outlines the larger direct influence of global warming in climate change causing, among other negative impacts, agricultural and ecological droughts which stresses the interconnection of all FLW impacts (IPCC 2021).

As a result, the scientific interest in FLW has been increasing, as has its social, economic, and environmental sustainability. In this context, we have identified a lack of scientific literature reviews on food waste recovery for human consumption.

In response to this situation, FAO has developed the Global Initiative on Food Loss and Waste Reduction promoting a new strategic framework, adjusted to the specific needs of each region and country. Each role-player in the programme is urged to raise awareness, establish public–private partnerships, implement policies, strategies and programmes designed to bring stakeholders together to reduce food waste and loss (FAO 2015).

In September of 2015, the world leaders attended United Nations Sustainable Development Summit where the Sustainable Development Agenda until 2030 was adopted. This agenda includes 17 Sustainable Development Goals (SDG), two of which directly address FLW impacts: SDG 2 (Zero Hunger), SDG 12 (Responsible Consumption and Production), and SDG 13 (Climate Action) (UN 2015).

The SDG Zero Hunger is directly related to the concept of food security which, according to FAO, exists "when all people, at all times, have physical, social, and economic access to sufficient, safe, and nutritious food that meets their dietary needs and food preferences for an active and healthy life" (FAO 2009). Food insecurity has been on the rise. In the 2014–2019 period, 22.4% to 25.9% of the world population was affected by moderate or severe food insecurity, while millions of tonnes of food are wasted every year (UN 2020).

The European Union (EU), one of the Save Food Regions of the Global Initiative on Food Loss and Waste Reduction, estimates that the annual amount of FLW in the EU is of 88 million tonnes, with the economic value of 143 billion euros (Vittuari et al. 2016). To reduce FLW, reduce the consumption of natural resources and the associated economic costs, the EU aims to achieve the SDG proposed by the UN, amongst other actions, by implementing a European Action Plan for the Circular Economy adopted in 2015 (European Commission 2015 2016). One of the SDG targets related to FWL is target 12.3. It states: "By 2030, halve per capita global food waste at the retail and consumer levels and reduce food losses along production and supply chains, including post-harvest losses" (FAO, n.d.-a). Halving FLW benefits the environment, as it reduces GHG emissions and water consumption while also benefiting the economy avoiding the watering cost and contributing towards food security and even to eventual job creation companies that help redistribute food that would otherwise be wasted (Flanagan et al. 2019).

Upcycling food for human consumption contributes to the fulfilment of several SDG. In fact, while feeding the hungry with nutritious food, it contributes to sustainable production and consumption (preventing the need to produce more food), and avoids the environmental burdens and economic costs of food waste management. (Nogueira et al. 2021a, b).

Every year, the UN releases data on SDG progress for each country member. The 2020 SDG Report shows progress was uneven although not at the required speed and rate to meet the 2030 Agenda (UN 2020). Some of the SDG recorded gains were: (1) the fall in the number of children and youth dropping out of school; (2) the decline in the incidence of many communicable diseases; (3) improved access to safely manage drinking water; and (4) the increase in women's representation in leadership roles. However, the world has faced dramatic economic recession with rising cases of food insecurity, and natural environment depletion (UN 2020).

Reduction of FLW has raised an increasing interest by governments, non-governmental organisations, multi-national food and distribution companies and social solidarity institutions and, not surprisingly, also that of the scientific community.

During current research on FLW we found out that there is a wide amount and range of FLW data available through national and intergovernmental agencies. However, when researching for scientific literature about what is being done to tackle FLW we did not find any review articles on food surplus reuse for human consumption despite the increasing number of publications on this subject during the past years. Given the global impact of FLW on global socioeconomic and environmental sustainability, it is crucial to keep conducting studies that will further consolidate the knowledge on how to rescue and reuse FLW prioritising reuse for human consumption according to the hierarchy for waste prevention and management proposed in the Food Recovery Hierarchy Priorities (European Commission 2008; European Parliment 2019).

In this paper we aim to provide a compilation of how research to upcycle FLW is evolving to find about innovative solutions of using FLW.

Methods

To tackle this issue, in this systematic review, we have adopted a methodology based on the use of carefully chosen key words and temporal boundaries narrowed down to the past 13 years, from 2008 to 2021. The 2008 economic crisis impacted on all phases of the food chain from farm to industry, the redistribution sector, and the livelihood of families, hence the emphasis of this research on the rerouting of food waste, concentrating on solutions for food insecurity (Food and Agriculture Organization of the United Nations 2010). The body of the literature so constituted was divided into 2 categories and further analysed by types of publication, geographic distribution, and key word frequency to determine the research emphasis. We also discuss the trends on food waste management as a raw material even though industrial extraction is more material and energy demanding, and therefore contributes more to climate change.

In terms of research design, we followed guidelines that propose a review process in three stages (planning, execution and reporting) (Tranfield et al. 2003) (Fig. 3.1). In the planning phase, research keywords and terms were defined according to the aim and main topic of the study. "Food Waste" (FW) and "Food Loss" (FL) were naturally the first research keywords used.

Despite what has already been referred with regard to FL and FW, some studies opted either for their own FL and FW system boundary, methods, and definitions (Hodges et al. 2011; Parfitt et al. 2010). For this reason, other designations for FL and FW concepts such as "food surplus" and "food residue", were added to our keyword list. Because we focus on upcycling FLW we decided to make use of other research keywords such as "food waste recovery", "food recovery", "food reuse", "food rescue" and "rescued food" to ensure that food waste for human consumption research was prioritised.

To secure a broad coverage of literature about FLW reduction through upcycling, Web of Science, Scopus, Google Scholar, and b-on were used to search for papers. These platforms were selected because they provide access to a series of scientific articles and because of the search filters available. Furthermore, data extracted from at least two platforms is reported to yield a more robust and reliable bibliometric analysis (de Oliveira et al. 2019).

No geographical boundaries were set. All global, national, or local studies were included. We considered the country of the first author´s affiliation institute to be the country to publish the study.

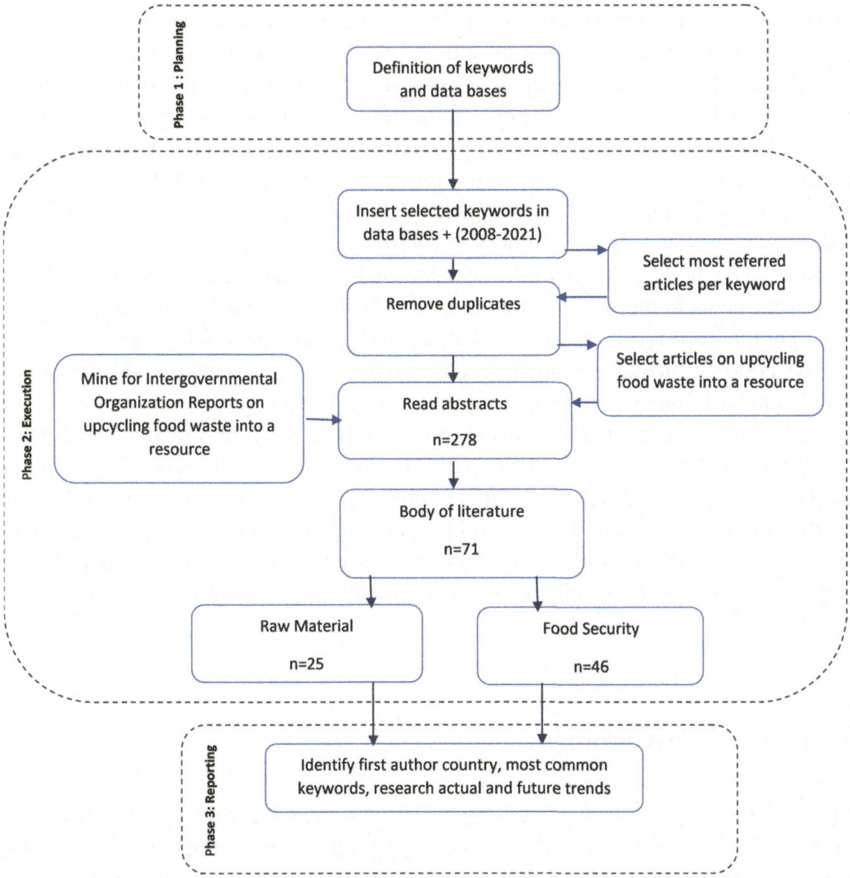

Fig. 3.1 Methodology design

We also searched for reports prepared by academic institutions and by governmental and nongovernmental organizations in the so-called grey literature. All previously mentioned keywords were used in the search for titles of publications. In addition, no specific language filter was used, and results were ordered by relevance.

All duplicates and all articles that did not focus on proposing novel systems of transforming FLW into a resource as a raw material or to tackle food insecurity, were discarded.

After reading through abstracts of roughly 280 published documents, extracted from B-On, Scopus, Web of Science and Google Academic, a body of literature of 71 publications provided an update on how to upcycle FLW. Even though our focus of interest is the reuse of food surplus and food waste for human consumption, the 71 compiled publications were divided into two main categories called Raw Material, and Food Security (Fig. 3.1).

The raw material (RM) category includes articles that focus on using FLW as a source of bioenergy or as a material source for extraction of high-value compounds or molecules to be applied in various types of industry. This is the category where, by far, most found papers fall into, confirming there is a growing interest in the use of FLW as a source a clean energy and material resource. In 2017 an extensive biometric study on food waste research (as a raw material) from 1997 to 2014 was published concluding that research on clean energy sources, treatment, valorisation, and innovative management of food waste have attracted extensive attention and will continue to be the focus point of the decision-making and policy of FW management for governments in the future (Chen et al. 2017). Thus, we believe it is important to highlight the most recent works on FLW as an energy or material resource.

The Food Security category includes articles that focus on a wide variety of issues related to the use of FLW for human consumption such as vectors and how the following are expected to evolve: recoverable food waste; practices leading to the reduction of FLW; comparative studies on the best option to reuse FLW; values and behaviours in relation to foods that contribute to the propensity to waste food in residential, institutional and commercial sectors; projects or strategies to deal with food surplus; food waste in developing countries; the environmental impact of food surplus; the economic benefits of food surplus recovery; food waste management policies.

Results and Discussion

After analysing the content of each paper, it was clear that there is an unbalanced focus on the different stages of the food supply chain. Almost all studies focus on food waste at retailing and consumer levels, while only very few address immediate postharvest losses. Considering that FAO estimates that postharvest losses can reach up to 20.7% percent in Central and South Asia, 15.7% in Northern America and Europe, 14% in Sub-Saharan Africa, 11.6% in Latin America and Caribbean, 10.8% in Northern Africa and Western Asia, with a mean value of 13.8% worldwide, this result suggests that, to tackle hunger and malnourishment, further research is needed in addressing post-harvest losses, hopefully resulting in innovative ways of tackling the problem. (FAO 2016).

Once we had gathered the literature available, we found it could be divided into four different types of documents: articles in peer review journals, books, reports, and conference proceedings. Journal articles were the most common type with 90% accounting for literature, followed by 6% for books, 3% for reports, and 1% for conference proceedings.

Journals dedicated to different, often overlapping, fields focused mainly on waste management, biotechnology, chemical engineering, environmental sciences, food sustainability research, and socio-economic sciences fields.

The few countries in which literature is published is indicative of its narrow spatial coverage. Figure 3.2 illustrates the total distribution of where the main author's institution is geographically located. It also shows that the top seven most productive countries on FLW research are the USA, Australia, Italy, the UK, and China, followed by India.

Although most existing studies on FLW are conducted in industrialized countries, (79% of main authors), there is also a clear interest in the subject in some emerging countries, (21% of main authors) indicating a worldwide interest in FLW management. The global socioeconomic crisis that began in 2008 and international policies (FAO 2015; UN, n.d., 2015; United Nations 2018) could be at the origin of the national interests of these countries. Furthermore, this distribution reflects the large values of the per capita food losses and waste, at consumption and pre-consumptions stages, in Europe, North America, Oceania, and Industrialized Asia as well as emerging countries (24). Food losses in industrialized countries are as high as in developing countries. However, in developing countries more than 40% of the food losses occur at post-harvest and processing levels, while in industrialized countries, more than 40% of the food losses occur at retail and consumer level (FAO 2011a). Moreover, national and global socioeconomic adversities were aggravated by the COVID-19 pandemic, to such an extent that it is considered the worst economic recession since the Great Depression with an increasing number of people suffering from food insecurity (UN 2020).

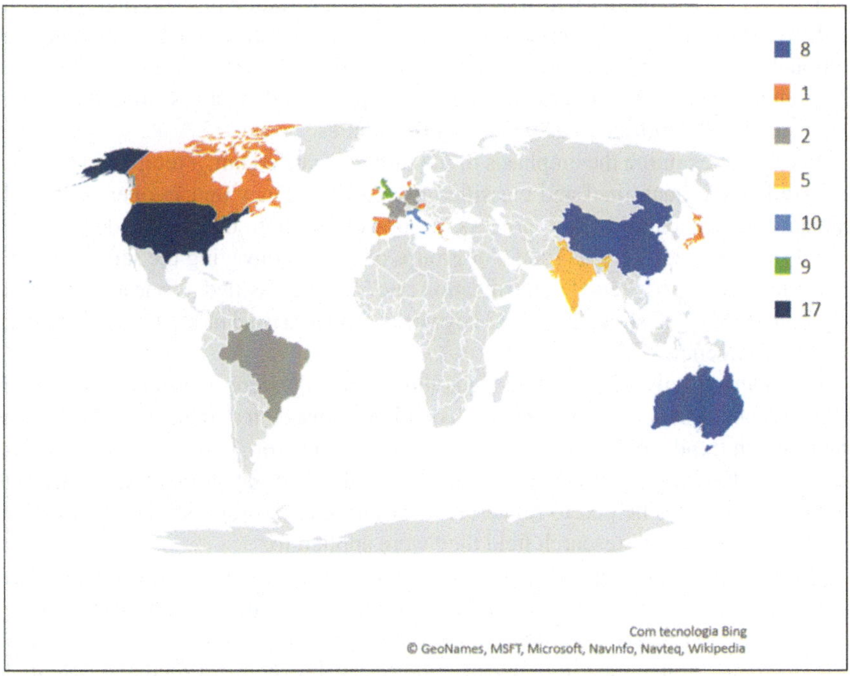

Fig. 3.2 Total geographical distribution of where the main author's institution is located

Since 2008, we have been witnessing an outburst of FLW research publications related to aspects of food waste management and bioenergy production through anaerobic digestion of FLW, as a way of both producing biofuels and manage food waste (Chen et al. 2017). From our body of literature of 71 publications, 25 fell into the Raw Material category, and 46 into the Food Security category.

The USA is one of the most productive countries, followed by Australia, Italy, and the UK. Households' food waste estimates provided by the United Nations 2021 Food Waste Index Report (United Nations Environment Programme 2021) indicate values of 59 kg/capita/year in the USA, 102 kg/capita/year in Australia, 67 kg/capita/year in Italy, and 77 kg/capita/year in the UK. FAO estimates that consumers waste 95–115 kg of food/year per capita in Europe and North America, with similar values at production to retailing level (~180 kg/capita/year) for Europe, North America and Oceania (FAO 2011a). All four countries have published in the food security category showing a large interest in upcycling food as a resource to feed food insecure populations.

The number of publications in the Raw Material category in some emerging countries such as China and India, addressing the issue of FLW on the different stages of the food supply chain as a clean source of energy and raw materials, suggests a national interest in waste management and biofuel production. Food waste estimates in India, at household level, are 20 kg/capita per year (UN n.d.). China is noteworthy as it is a very large country with a substantial internal variation in terms of population, and thus showing a wide range of food waste estimates from 64 kg/capita/year at national level but 150 kg/capita/year in urban areas (United Nations Environment Programme 2021). Due to a rapid growth, it is predictable that research in the food waste field will continue to develop in both counties.

In order to examine the emphasis of the research carried out, a total of 241 author keywords were quantified and classified by total frequency and frequency in each category. In this classification process, groups of words in different order, such as "food surplus" and "surplus food", or groups of words conveying the same concept, *e.g.* "nitrification loss" and "denitrification rate", were classified as one and the same keyword. Figure 3.3 shows all the 33 keywords that were used more than once ranked by total frequency.

Considering only 13.7% (33/241) author keywords were mentioned more than once indicates that mainstream research on FLW focused on a small area. To draw on the research trends from 2008 to 2021, we analysed the top 33 most frequent author keywords. Keeping in mind the previously established categories (Raw Material and Food Security) helped to relate broad meaning keywords like "food waste" or "sustainability" to the research field they were applied to.

Results show the top 33 keywords were related to: (1) food waste treatment and disposal (food waste (RM); anaerobic (co-) digestion, sewage-sludge, composting); (2) bioenergy production (biofuel, bioenergy, biogas, methane production, methane/biomethane, bioethanol, biorefinery, biochemical methane potential, nitrogen loss, yield); (3) food waste management (waste prevention, sustainability,

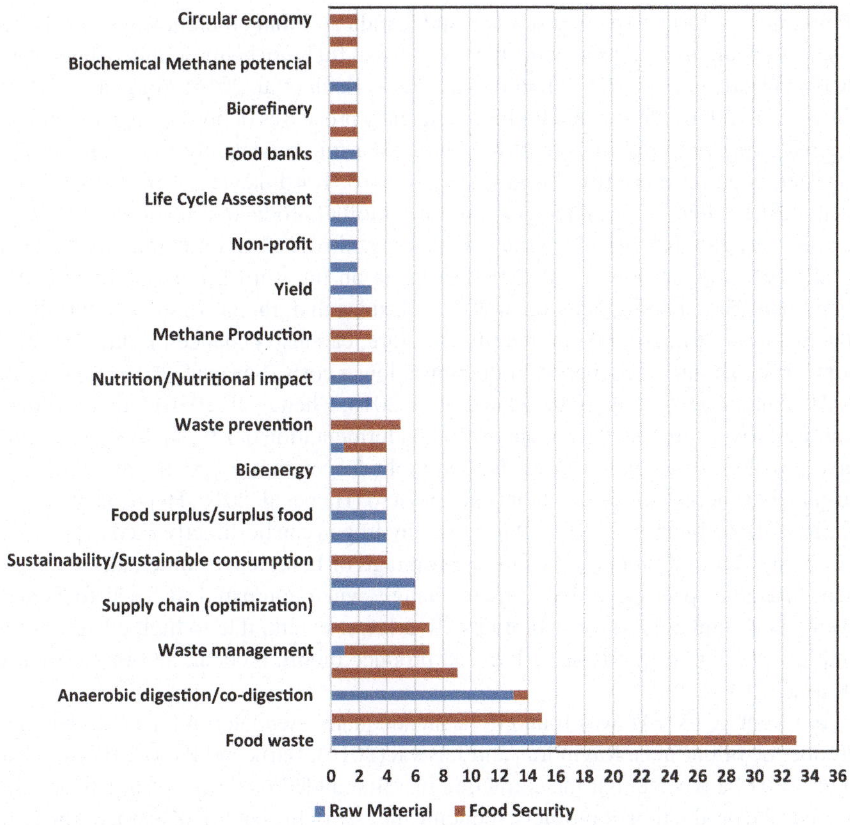

Fig. 3.3 Absolute frequency of author keywords mentioned more than once, with frequency per category (total frequency, $n = 169$)

waste management, waste minimization, food waste hierarchy, circular economy); and (4) food security (food waste (FS), food (in)security, food poverty, emergency food aid/food assistance, nutrition/dietary diversity, food banks, food donation, non-profit, food recovery, food surplus, supply chain optimization).

Food Waste as Raw Material

During the period of study for the category of Raw Material, operational conditions of bioenergy production though anaerobic digestion and co-digestion were the centre of several research papers (Chen et al. 2017; Cheng et al. 2021; Karmee 2016; Sen et al. 2016; Talan et al. 2021).

The interest in the effect of the composition of mixtures that are fed to anaerobic digesters on the yield of biomethane production for different mixtures of food waste or mixtures of food waste and sewage sludge indicates that efficient ways of solving disposal problems while replacing the use of fossil fuels are being sought (Ebner et al. 2016; Fisgativa et al. 2016; Fitamo et al. 2016; Koch et al. 2016; Yong et al. 2015; Zhang et al. 2016). The search for high efficiency processes of producing biomethane has led to the proposal of corrective chemical or micro-wave pre-treatments of the digesters feeding mixtures (Fisgativa et al. 2016; Karthikeyan et al. 2018; Zhang et al. 2016). Alternative solutions to optimize biofuel production involve controlling process variables like pH, electrical conductivity, nitrogen loss or recirculation (Chan et al. 2016; Zamanzadeh et al. 2016) or using an innovative high-solid anaerobic membrane bioreactor (Cheng et al. 2021). Besides that, monitoring and controlling the evolution of microbial community structure, can also enhance productivity and contribute to harness bioenergy production at lower costs (Sen et al. 2016; Talan et al. 2021; Zamanzadeh et al. 2016; Zhang et al. 2016; Zhen et al. 2016) Other biofuels can similarly be produced from the controlled fermentation of FW, such as bioethanol (and possibly longer chain alcohols) and biohydrogen, a most promising option of sustainable energy source due to its energy yield (Han et al. 2016; Hegde et al. 2018; Karmee 2016; Talan et al. 2021). Moreover, hydrogen can be directly used to produce electricity through fuel cells with the advantage of being environmentally friendly with limited impact on global climate change, when compared to fossil fuels and even when compared to other biofuels. To a lower extent, due to high energy costs and low efficiency, pyrolysis of FW can produce biooil, biochar, and biofertilizers (Karmee 2016).

In recent years, FW was found to be an adequate raw material for more than the production of biofuels. Ravindran and Jaiswal (2016), Barik and Paul (2017) and Xu et al. (2018) advocate that the extraction of value-added materials from FW should precede the production of biofuels, reducing the size of the landfill discarded fraction, increasing revenues and reducing co-digestion costs (Barik and Paul 2017; Ravindran and Jaiswal 2016; Xu et al. 2018). The concept of extracting biomaterial from FW adds another dimension to the keyword *biorefinery*. FW is a reservoir of carbohydrates, proteins, lipids, and other organic and inorganic substances, hence the application of the term "biorefinery" to the process of recovering material from FW using industrial processing technologies. Recycling inside the food supply chain can be an economical and environmentally sustainable way of producing renewable feedstock for industrial uses (Galanakis 2015; Girotto et al. 2015). Ongoing research already addresses several industry sectors. Recovered nutrients, molecules and metals, can either be used in the food industry, animal feed industry or pharmaceutical industry as flavourings and fragrances, antioxidants, food additives and nutraceuticals (Barik and Paul 2017; Kim et al. 2016; Ravindran and Jaiswal 2016). Other industrial sectors could benefit from the production of metabolites, enzymes, bioactive compounds, biodegradable plastics, nanoparticles or as catalysers of nitrogen removal in waste water treatment (Li et al. 2015; Ravindran and Jaiswal 2016).

All previously mentioned extractions can occur in integrated biorefineries that produce extracted chemicals with one fraction of the feedstock, and bioenergy with the remaining fraction. Even though integrated biorefineries still present many challenges they are a promising way to circular economy (Xu et al. 2018). Furthermore, integrated biorefineries have a closed loop approach to waste valorisation, contributing to promote the switch from fossil fuels to green fuels while reducing the production of greenhouse gases with corresponding positive impact on climate change.

Food Security

By analysing the top 33 keywords in the current study we found that 59.5% (frequency percentage) of them concerned food security – this shows that food security is yet another research trend.

Food insecurity is felt in developing and developed countries, and even in food surplus areas. It was found that increasing the production does not ensure food and nutritional security, which makes food insecurity a question of access to food, rather than a supply problem (Papargyropoulou et al. 2014).

Tackling both food insecurity and food waste have been the centre of several international guidelines based on the European Community Strategy on Waste Management first proposed by the European Parliament Council in 1984 (COMMISSION OF THE EUROPEAN COMMUNITIES 1989). This strategy (meanwhile adopted worldwide as the main waste management framework) is depicted as an inverted waste hierarchy pyramid (UNEP 2013). It prioritises prevention as the most favourable option and considers landfill disposal as the least favourable. In the model presented, the second-best option—reuse—is the distribution of food surplus to low income groups followed by recycling, the use of FW in animal feed and composting, with the last but one option being energy recovery of unavoidable food waste (UNEP 2013).

Even though *Prevention* through minimization of food surplus and avoidable food waste is the most preferred option of food waste hierarchy, waste prevention is the centre of very few studies. In this chapter, it is important to mention that not all FW is avoidable. Some parts of food must go to waste, such as certain peels and fruit stones (*e.g.* pineapple peel or mango stones). Considering that the percentage of avoidable food waste in some developed countries ranges from 34% (in Sweden) to 60% (in the UK), food waste prevention policies could reduce the amount of FW treated in all other hierarchy stages until disposal, with huge environmental and economic benefits (Thyberg and Tonjes 2016). Unfortunately, waste goals are usually defined in terms of recycling or diversion, rather than prevention indicators, namely SDG targets for FW reduction, defined as: "By 2030, halve global food waste per capita at the retail and consumer levels and reduce food losses along production and supply chains, including post-harvest losses" (FAO 2011b). Not only SDG provides no target specifically concerning prevention, but the only measurable target is situated at retail and consumer levels. Therefore, it is a priority to adjust policies towards food to prevent the generation of FW along the whole food supply chain (Redlingshöfer et al. 2017; Thyberg and Tonjes 2016).

In order to ensure universal access to nutritious food in adequate quantities, the change in waste prevention policies must reflect a change from a weak prevention model to a strong prevention model (Mourad 2016). Weak prevention solutions usually focus on technical or logistic innovation to improve the efficiency of consumption but are not a sustainable option from a long-term perspective (Longo-Silva et al. 2013). Moving from weak to strong prevention requires a change in governance along the food supply chain, involving all the actors—producers, retailers and consumers—in a sustainable system imbibed activity (Mourad 2016). This change of production and consumption paradigm presents implementation limitations due to constraints in individual behaviour and policy changes which are the object of several studies.

In a household environment, the required changes may be as simple as planning serving portions and changing the content of served meals or even consuming leftovers. The consumption of leftovers is regarded as a sacrifice that sustains and perpetuates family roles, reinforcing family bonds or engaging in redistribution practices (Cappellini 2009). Readjustment of serving sizes and content may also be a strategy to tackle the problem of food surplus and leftovers in school and company canteens (Evans 2012). Moreover, unavoidable surplus may be diverted from landfills in school canteens through programmes that promote the volunteer participation of students in the process of sorting uneaten food. This process facilitates food recovery while equipping the next generation with environmental stewardship habits (Prescott et al. 2020). In developing countries food waste at household level is considerably lower than at production to retail level, 40% of food loss occurring at post-harvest and processing stages (FAO 2011a). Numerous developing countries have been living in a protracted crisis with populations in need of enhanced effectiveness in terms of assistance programmes. Technology transfer and access to microcredit are needed to support and protect livelihoods, as are education programmes for farmers, better infrastructures to connect small farmers to markets, and the adoption of collective marketing. (Cicatiello et al. 2016; Hodges et al. 2011; Katz and Rivero 2015; Parfitt et al. 2010; Walia and Sanders 2019).

Another line of research that attracts researchers from countries as different as the USA (Mittal et al. 2021; Shi and Lizarondo 2021), Italy (Cicatiello et al. 2016; Cicculloet al. 2021; Muriana 2015), the Netherlands (Ciulli et al. 2020), and Australia (Nair et al. 2018; Rey et al. 2018) is the development of mathematical models to solve FW management problems at the *Reuse* stage of the FW hierarchy inverted pyramid.

Food waste or food surplus is often redistributed to the food insecure populations through food banks and other non-profit organizations. One of the first identified problems of these type of organizations is the cost of food transportation from the food source (*e.g.* farmers, retail distribution centres) to the food redistribution centres. A 2013 report by the World Resources Institute suggests that the transportation obstacle can be overcome by establishing additional food redistribution centres and optimizing FW picking up and delivery routes (Lipinski et al. 2013). Humanitarian logistics optimization has evolved through the use of different mathematical algorithms (very often based on linear regression), presenting novel approaches for the food rescue and delivery problems. It is important to highlight that the optimization of pickup and delivery routes, while lowering operating costs, also reduces the environmental impact of the pickup process and alleviates climate burdens due to food surplus (Hiç et al. 2016).

Rey et al. (2018) found that using fewer vehicles is likely to help improve the fairness of FW allocation, thus proposing a solution of using a single large capacity vehicle with longer routes to ensure enough food supply (Rey et al. 2018). Recruiting additional donors and volunteers was also found to be an efficient strategy to decrease economical costs of food relief logistics (Mousa and Freeland-Graves 2017; Phillips et al. 2013; Shi and Lizarondo 2021).

Food rescue and redistribution, of perishable foods may be more economically and environmentally costly and complex than landfill and composting or even than acquiring, storing and redistributing dried staple foods (Krishnan et al. 2020; Philip et al. 2017; Reynolds et al. 2015). It also requires infrastructures to safeguard food hygiene and safety standards and particularly when redistributing frozen and chilled food (Alexander and Smaje 2008; Garrone et al. 2014). Nevertheless, procuring food for the food insecure is a lower cost method than purchasing meals directly, with improved nutritious quality when compared with the dried and staple foods usually distributed in foodbanks (Philip et al. 2017). A review article published in 2019 documented the need for additional evaluation of food rescue interventions, with consistent metrics, but still suggesting promising effects, including large quantities of food rescued and served clients, positive return on investment, and decreased environmental burden (Hecht and Neff 2019). Life Cycle Analysis results suggest that, apart from reducing environmental pressure and capital loss, food waste rescue may also generate work opportunities, while conditioning stakeholders' behaviour (Santagata et al. 2021).

To ensure high stakeholder satisfaction and the sustainability of charitable networks it is vital to understand the relationship between supply and demand (Hecht and Neff 2019; Krishnan et al. 2020; Nair et al. 2018). In FW management and redistribution this means being able to forecast food donations while correctly estimating food demand, minimizing the fraction of recovered food that is not used for human consumption (Ciulli et al. 2020; Garrone et al. 2014). A software based on structural equation modelling and neural networks was found to provide improved demand estimation when compared with multiple linear regression (Nair et al. 2017).

Researchers also focus on modelling the suppliers' side of FW redistribution to determine the optimal shelf life and quantity of products to be removed from distribution shelves and shipped to non-profit organizations (or livestock feed). Sustainability of the donation process means that the optimal point found must correspond to maximum profit for the retailer (Giuseppe et al. 2014; Muriana 2015). Implementation of circular economy at retail level can also be achieved through the adoption of repositioning, reallocating, reacting, re-engineering, and relating practices, with economic, social and environmental benefits (Huang et al. 2021).

Digital platforms have recently emerged as a response to the need for circularity brokers. These innovative tools facilitate the brokerage of food surplus supply and demand all along the food supply chain, as well as transportation and human resources logistics (Cicculo et al. 2021; Mittal et al. 2021; Nair et al. 2018, 2017; Shi and Lizarondo 2021).

Although it may seem that recovery and redistribution of food waste and food surplus is a contemporary form of socioenvironmental activism supported by the whole scientific community, this issue is not that linear (Edwards and Mercer 2012). If some researchers consider that alternative food networks are a way to ensure food justice and environmental sustainability, through social innovation, in need of a special set of tools, resources and laws, others have another set of concerns (Baglioni et al. 2017; Lindberg et al. 2014; Planchenstainer 2013). A first concern regards the pertinence of the charitable food sector, in those policies targeting which determinants of food insecurity should be implemented to reduce demand for emergency food, together with governmental regulations on FW minimization, rather than relying on the power and initiative of civil society organizations. Empowering community governance only, depoliticizes food insecurity issues and transfers the responsibility of rescuing and redistributing food waste and food surplus to local community members (Lindberg et al. 2014; Warshawsky 2015; Wingrove et al. 2017).

A second concern is related to the rescue and redistribution system where each party at each transfer point passes on property waste disposal obligations (Alexander and Smaje 2008). This may lead to situations where parties might be more interested in avoiding disposal costs than providing the food insecure with nutritious food, transferring food in poor hygiene and safety conditions.

Food Waste Future Research Trends

Due to the ever-increasing population, resource, and commodity limitations, to climate change, and unpredictable scenarios such as the COVID-19 pandemic, we will see an economic, social, and environmental increase in the price of food, leading to the need for more efficient food production and consumption systems (Aldaco et al. 2020; Pollard et al. 2018). Thus, in addition to the ongoing research paths, as another result of this literature review, we have identified possible future FLW research trends.

Affluent countries are likely to develop responses to FSC waste trends such as: (1) circular economy implemented in the legislation of all countries; (2) new strategies to reuse FW as a bioenergy and biomaterial source; (3) clear labelling and consumer education campaigns; (4) better policies on FCS waste and FW taxation; (5) public and private partnerships; (6) better demand forecast; (7) optimized (shared) logistics; (8) domestic kitchen technologies; (9) better packaging; and (10) technology transfer to BRIC countries (Brazil, Russia, India and China) (Cicatiello et al. 2016; Facchini et al. 2018; Hodges et al. 2011; Katz and Rivero 2015; Parfitt et al. 2010; Walia and Sanders 2019). In order to reduce the impacts of FW significantly and generate social and environmental benefits, less developed countries will have to implement: (1) widespread education programmes directed at farmers; (2) better infrastructures to connect small farmers to markets; (3) adoption of collective marketing; (4) improve access to technologies through microcredit; and (5) financial incentive opportunities that improve efficiency along the FSC (Hodges et al. 2011).

Conclusions

This compilation provides an overview on the evolution of FLW throughout the investigation period of 2008/2021. Based on the selected publications, the following conclusions were drawn from the current study: (1) FLW as raw material research, mainly concentrated in China, India and the USA, focuses on the fields of chemical engineering, energy fuels, environmental engineering, biotechnology, and microbiology applied to engineering; (2) A second FLW research field, taking place mainly is Australia, European countries and the USA, focuses on food security, FW upcycling for human consumption improvements in terms of humanitarian logistics, concerning supply, transportation, conservation and volunteerism; (3) For lower and medium income countries FLW resides mainly at production to retail level where farmer support programmes could promote environmental and social improvements; and (4) There is a lack of research focusing on measuring the potential of rescued and redistributed perishable FW as a way of tackling food insecurity.

Valorisation of food waste turning it into a material or bioenergy source requires know-how and more time, energy and other material consumption than the process of upcycling food as is. To reduce climate change resulting from the environmental burdens of food waste, we advocate that FW upcycling through technological innovation should be sponsored towards developing solutions for inevitable and non-edible parts of FW.

To address food insecurity, the future decision-making and policy of FW management should prioritise targeting determinants of food insecurity which in turn would reduce the food aid demand. Furthermore, climate change causing food insecurity and malnourishment should also be addressed with policies aiming at changing the food production and consumption paradigm throughout the food chain.

The main limitation of this review is that it cannot comprise an in-depth study of all subjects. Issues such as policies, laws, and technical particularities of all the processes to transform food waste into a material or energy source were approached in a descriptive way since the aim was to identify the different paths of food waste research.

Funding This work was carried out at the R&D Unit Centre for Functional Ecology–Science for People and the Planet (CFE), with reference UIDB/04004/2020, financed by FCT/MCTES through national funds (PIDDAC).

References

Aldaco R, Hoehn D, Laso J, Margallo M, Ruiz-Salmón J, Cristobal J, … Vazquez-Rowe I (2020) Food waste management during the COVID-19 outbreak: a holistic climate, economic and nutritional approach. Science of the Total Environ 742:140524. https://doi.org/10.1016/j.scitotenv.2020.140524

Alexander C, Smaje C (2008) Surplus retail food redistribution: an analysis of a third sector model. Resour Conserv Recycl. https://doi.org/10.1016/j.resconrec.2008.07.009

Baglioni S, De Pieri B, Tallarico T (2017) Surplus food recovery and food aid: the pivotal role of non-profit organisations. Insights From Italy and Germany. Voluntas. https://doi.org/10.1007/s11266-016-9746-8

Barik S, Paul KK (2017) Potential reuse of kitchen food waste. J Environ Chem Eng. https://doi.org/10.1016/j.jece.2016.11.026

Cappellini B (2009) The sacrifice of re-use: The travels of leftovers and family relations. J Consum Behav. https://doi.org/10.1002/cb.299

Chan MT, Selvam A, Wong JWC (2016) Reducing nitrogen loss and salinity during 'struvite' food waste composting by zeolite amendment. Biores Technol 200:838–844. https://doi.org/10.1016/J.BIORTECH.2015.10.093

Chen H, Jiang W, Yang Y, Yang Y, Man X (2017) State of the art on food waste research: a bibliometrics study from 1997 to 2014. J Clean Prod. https://doi.org/10.1016/j.jclepro.2015.11.085

Cheng H, Li Y, Hu Y, Guo G, Cong M, Xiao B, Li YY (2021) Bioenergy recovery from methanogenic co-digestion of food waste and sewage sludge by a high-solid anaerobic membrane bioreactor (AnMBR): mass balance and energy potential. Biores Technol 326(January):124754. https://doi.org/10.1016/j.biortech.2021.124754

Cicatiello C, Franco S, Pancino B, Blasi E (2016) The value of food waste: An exploratory study on retailing. J Retail Consum Serv 30:96–104. https://doi.org/10.1016/j.jretconser.2016.01.004

Ciccullo F, Cagliano R, Bartezzaghi G, Perego A (2021) Implementing the circular economy paradigm in the agri-food supply chain: the role of food waste prevention technologies. Resour, Conserv and Recycl 164(January 2020):105114. https://doi.org/10.1016/j.resconrec.2020.105114

Ciulli F, Kolk A, Boe-Lillegraven S (2020) Circularity Brokers: Digital Platform Organizations and Waste Recovery in Food Supply Chains. J Bus Ethics 167(2):299–331. https://doi.org/10.1007/s10551-019-04160-5

Commission of the European Communities (1989) A community strategy for waste management, SEC/89/934 (final). http://aei.pitt.edu/5679/1/5679.pdf

de Oliveira JO, da Silva FF, Juliani F, Barbosa LCFM, Nunhes TV (2019) Bibliometric method for mapping the state-of-the-art and identifying research gaps and trends in literature: an essential instrument to support the development of scientific projects. Scientometrics Recent Adv:1–20. https://doi.org/10.5772/intechopen.85856

Ebner JH, Labatut RA, Lodge JS, Williamson AA, Trabold TA (2016) Anaerobic co-digestion of commercial food waste and dairy manure: characterizing biochemical parameters and synergistic effects. Waste Manage 52:286–294. https://doi.org/10.1016/J.WASMAN.2016.03.046

Edwards F, Mercer D (2012) Food waste in Australia: the freegan response. The Sociological Review 60(S2):174–191. https://doi.org/10.1111/1467-954X.12044

European Commission (2008) Waste framework directive. Retrieved August 2, 2021, from https://ec.europa.eu/environment/topics/waste-and-recycling/waste-framework-directive_en

European Commission (2015) EUR-Lex - 52015DC0614 - EN - EUR-Lex. Retrieved March 5, 2019, from https://eur-lex.europa.eu/legal-content/EN/TXT/?uri=CELEX:52015DC0614

European Commission (2016) EU actions against food waste | Food Safety. Retrieved March 5, 2019, from https://ec.europa.eu/food/safety/food_waste/eu_actions_en

European Parliment (2019) *Decision (EU) 2019/1004 of 7 June 2019*

Evans D (2012) Binning, gifting and recovery: The conduits of disposal in household food consumption. Environ and Planning D: Society and Space. https://doi.org/10.1068/d22210

Facchini E, Iacovidou E, Gronow J, Voulvoulis N (2018) Food flows in the United Kingdom: the potential of surplus food redistribution to reduce waste. J Air Waste Manag Assoc 68(9):887–899. https://doi.org/10.1080/10962247.2017.1405854

FAO. (n.d.-a). Goal 12 .:. Sustainable Development Knowledge Platform. Retrieved March 5, 2019, from https://sustainabledevelopment.un.org/sdg12

FAO (n.d.-b) Key facts on food loss and waste you should know! | SAVE FOOD: global initiative on food loss and waste reduction | Food and agriculture organization of the United Nations. Retrieved March 5, 2019, from http://www.fao.org/save-food/resources/keyfindings/en/

FAO (2009) Declaration of the world summit on food security World Summit on Food Security. Rome. http://www.fao.org/fileadmin/templates/wsfs/Summit/Docs/Declaration/WSFS09_Draft_Declaration.pdf

FAO (2011a) FAO report on extent of food losses and waste. In Global food losses and food waste. http://www.fao.org/docrep/014/mb060e/mb060e02.pdf

FAO (2011b) Global food losses and food waste—extent, causes and prevention

FAO (2013) Food wastage footprint. In Fao. www.fao.org/publications

FAO (2015) Global inItIatIve on food loss and waste reductIon 2015. http://cait.wri.org

FAO (2016) Sustainable development goals—Indicator 12.3.1 - Global food loss and waste. Retrieved August 1, 2021, from http://www.fao.org/sustainable-development-goals/indicators/1231/en/

Fisgativa H, Tremier A, Dabert P (2016) Characterizing the variability of food waste quality: a need for efficient valorisation through anaerobic digestion. Waste Manage 50:264–274. https://doi.org/10.1016/J.WASMAN.2016.01.041

Fitamo T, Boldrin A, Boe K, Angelidaki I, Scheutz C (2016) Co-digestion of food and garden waste with mixed sludge from wastewater treatment in continuously stirred tank reactors. Biores Technol 206:245–254. https://doi.org/10.1016/j.biortech.2016.01.085

Flanagan K, Robertson K, Hanson C (2019) Reducing food loss & waste. https://files.wri.org/d8/s3fs-public/reducing-food-loss-waste-global-action-agenda-executive-summary.pdf

Food and Agriculture Organization of the United Nations (2010) The state of food insecurity in the world addressing food insecurity in protracted crises 2010 Key messages. In: Notes. http://www.fao.org/docrep/013/i1683e/i1683e.pdf

Galanakis CM (2015) Food waste recovery: processing techologies and industrial techniques. Academic Press. https://doi.org/10.13140/RG.2.1.2145.0005

Garrone P, Melacini M, Perego A (2014) Surplus food recovery and donation in Italy: The upstream process. British Food Journal. https://doi.org/10.1108/BFJ-02-2014-0076

Girotto F, Alibardi L, Cossu R (2015) Food waste generation and industrial uses: a review. Waste Manage. https://doi.org/10.1016/j.wasman.2015.06.008

Giuseppe A, Mario E, Cinzia M (2014) Economic benefits from food recovery at the retail stage: An application to Italian food chains. Waste Manage. https://doi.org/10.1016/j.wasman.2014.02.018

Han W, Fang J, Liu Z, Tang J (2016) Techno-economic evaluation of a combined bioprocess for fermentative hydrogen production from food waste. Biores Technol 202:107–112. https://doi.org/10.1016/J.BIORTECH.2015.11.072

Hecht AA, Neff RA (2019) Food rescue intervention evaluations: a systematic review. Sustainability (Switzerland) 11(23). https://doi.org/10.3390/su11236718

Hegde S, Lodge JS, Trabold TA (2018) Characteristics of food processing wastes and their use in sustainable alcohol production. Renew Sustain Energy Rev. https://doi.org/10.1016/j.rser.2017.07.012

Hiç C, Pradhan P, Rybski D, Kropp JP (2016) Food Surplus and Its Climate Burdens. Environ Sci Technol. https://doi.org/10.1021/acs.est.5b05088

Hodges RJ, Buzby JC, Bennett B (2011) Postharvest losses and waste in developed and less developed countries: opportunities to improve resource use. J Agric Sci. https://doi.org/10.1017/S0021859610000936

Huang IY, Manning L, James KL, Grigoriadis V, Millington A, Wood V, Ward S (2021) Food waste management: a review of retailers' business practices and their implications for sustainable value. J Clean Prod 285:125484. https://doi.org/10.1016/j.jclepro.2020.125484

IPCC (2021) IPCC, 2021: summary for policymakers. In: Climate change 2021: the physical science basis. Contribution of working group I to the sixth assessment report of the intergovernmental panel on climate change. https://www.ipcc.ch/report/ar6/wg1/downloads/report/IPCC_AR6_WGI_SPM_final.pdf

Karmee SK (2016) Liquid biofuels from food waste: Current trends, prospect and limitation. Renew Sustain Energy Rev. https://doi.org/10.1016/j.rser.2015.09.041

Karthikeyan OP, Trably E, Mehariya S, Bernet N, Wong JWC, Carrere H (2018) Pretreatment of food waste for methane and hydrogen recovery: a review. Biores Technol 249:1025–1039. https://doi.org/10.1016/j.biortech.2017.09.105

Katz L, Rivero S (2015) Food rescue services, barriers, and recommendations in Santa Clara County. www.foodshift.net

Kim MH, zu Ermgassen EKHJ, Salemdeeb R, Al-Tabbaa A, Balmford A (2016) Environmental and health impacts of using food waste as animal feed: a comparative analysis of food waste management options. J of Clean Prod. https://doi.org/10.1016/j.jclepro.2016.05.049

Koch K, Plabst M, Schmidt A, Helmreich B, Drewes JE (2016) Co-digestion of food waste in a municipal wastewater treatment plant: comparison of batch tests and full-scale experiences. Waste Manage 47:28–33. https://doi.org/10.1016/J.WASMAN.2015.04.022

Krishnan R, Agarwal R, Bajada C, Arshinder K (2020) Redesigning a food supply chain for environmental sustainability—An analysis of resource use and recovery. J Clean Prod 242:118374. https://doi.org/10.1016/j.jclepro.2019.118374

Li Y, Tang J, Zhang Y, Cheng Z, Wang XC (2015) Effect of fermentation liquid from food waste as a carbon source for enhancing denitrification in wastewater treatment. Chemosphere. https://doi.org/10.1016/j.chemosphere.2015.09.036

Lindberg R, Lawrence M, Gold L, Friel S (2014) Food rescue—an Australian example. British Food Journal. https://doi.org/10.1108/BFJ-01-2014-0053

Lipinski B, Hanson C, Lomax J, Kitinoja L, Waite R, Searchinger T (2013) Installment 2 of "Creating a sustainable food future" reducing food loss and waste. http://www.worldresourcesreport.org.

Longo-Silva G, Toloni M, Rodrigues S, Rocha A, de Taddei JA, A. C. (2013) Qualitative evaluation of the menu and plate waste in public day care centers in são paulo city, brazil. Revista De Nutricao 26(2):135–144. https://doi.org/10.1590/S1415-52732013000200002

Mittal A, Oran Gibson N, Krejci CC, Marusak AA (2021) Crowd-shipping for urban food rescue logistics. Int J Phys Distrib Logist Manag 51(5):486–507. https://doi.org/10.1108/IJPDLM-01-2020-0001

Mourad M (2016) Recycling, recovering and preventing "food waste": competing solutions for food systems sustainability in the United States and France. J Clean Prod. https://doi.org/10.1016/j.jclepro.2016.03.084

Mousa TY, Freeland-Graves JH (2017) Organizations of food redistribution and rescue. Public Health. https://doi.org/10.1016/j.puhe.2017.07.031

Muriana C (2015) Effectiveness of the food recovery at the retailing stage under shelf life uncertainty: an application to Italian food chains. Waste Manage. https://doi.org/10.1016/j.wasman.2015.03.028

Nair DJ, Grzybowska H, Fu Y, Dixit VV (2018) Scheduling and routing models for food rescue and delivery operations. Socioecon Plann Sci 63:18–32. https://doi.org/10.1016/j.seps.2017.06.003

Nair DJ, Rashidi TH, Dixit VV (2017) Estimating surplus food supply for food rescue and delivery operations. Socioecon Plann Sci 57:73–83. https://doi.org/10.1016/J.SEPS.2016.09.004

Nogueira A, Alves F, Vaz-Fernandes P (2021a) The Contribution of Up-Cycled Food Waste to a Balanced Diet of Low-Income Households. Sustainability 13(9):4779. https://doi.org/10.3390/su13094779

Nogueira A, Alves F, Vaz-Fernandes P (2021b) The nutritional content of rescued food conveyed by a food aid organization. Int. J. Environ. Health Res Public Health 18(22). https://doi.org/10.3390/ijerph182212212

Papargyropoulou E, Lozano R, Steinberg J, Wright N, bin Ujang Z (2014) The food waste hierarchy as a framework for the management of food surplus and food waste. J Clean Prod 76:106–115. https://doi.org/10.1016/j.jclepro.2014.04.020

Parfitt J, Barthel M, MacNaughton S (2010) Food waste within food supply chains: quantification and potential for change to 2050. Philosophical Transactions of the Royal Society B: Biological Sciences. https://doi.org/10.1098/rstb.2010.0126

Philip D, Hod-Ovadia S, Troen AM (2017) A technical and policy case study of large-scale rescue and redistribution of perishable foods by the "leket Israel" food bank. Food Nutr Bull. https://doi.org/10.1177/0379572117692440

Phillips C, Hoenigman R, Higbee B, Reed T (2013) Understanding the sustainability of retail food recovery. PLoS ONE. https://doi.org/10.1371/journal.pone.0075530

Planchenstainer F (2013) "They collected what was left of the scraps": food surplus as an opportunity and its legal incentives. In: SSRN. https://doi.org/10.2139/ssrn.2221709

Pollard CM, Mackintosh B, Campbell C, Kerr D, Begley A, Jancey J, … Booth S (2018) Charitable food systems' capacity to address food insecurity: an Australian capital city audit. Int J of Environ Res and Public Health 15(6). https://doi.org/10.3390/ijerph15061249

Prescott MP, Grove A, Bunning M, Cunningham-Sabo L (2020) A systems examination of school food recovery in Northern Colorado. Resour Conserv Recycl 154(September):104529. https://doi.org/10.1016/j.resconrec.2019.104529

Ravindran R, Jaiswal AK (2016) Exploitation of food industry waste for high-value products. Trends Biotechnol. https://doi.org/10.1016/j.tibtech.2015.10.008

Redlingshöfer B, Coudurier B, Georget M (2017) Quantifying food loss during primary production and processing in France. J Clean Prod. https://doi.org/10.1016/j.jclepro.2017.06.173

Rey D, Almi'ani K, Nair DJ (2018) Exact and heuristic algorithms for finding envy-free allocations in food rescue pickup and delivery logistics. Transp Res Part E: Logist and Transp Rev. https://doi.org/10.1016/j.tre.2018.02.001

Reynolds CJ, Piantadosi J, Boland J (2015) Rescuing food from the organics waste stream to feed the food insecure: an economic and environmental assessment of australian food rescue operations using environmentally extended waste input-output analysis. Sustainability (switzerland). https://doi.org/10.3390/su7044707

Santagata R, Ripa M, Genovese A, Ulgiati S (2021) Food waste recovery pathways: challenges and opportunities for an emerging bio-based circular economy. A systematic review and an assessment. J Clean Prod 286:125490. https://doi.org/10.1016/j.jclepro.2020.125490

Sen B, Aravind J, Kanmani P, Lay C-H (2016) State of the art and future concept of food waste fermentation to bioenergy. Renew Sustain Energy Rev 53:547–557. https://doi.org/10.1016/j.rser.2015.08.065

Shi ZR, Lizarondo L (2021) A recommender system for crowdsourcing food rescue platforms. In: IWWWC Committee (ed) Ljubljana, Slovenia: Creative Commons CC-BY 4.0 License. https://doi.org/10.1145/3442381.3449787

Talan A, Tiwari B, Yadav B, Tyagi RD, Wong JWC, Drogui P (2021) Food waste valorization: energy production using novel integrated systems. Bioresour Technol 322(October 2020):124538. https://doi.org/10.1016/j.biortech.2020.124538

Thyberg KL, Tonjes DJ (2016) Drivers of food waste and their implications for sustainable policy development. Resour Conserv Recycl. https://doi.org/10.1016/j.resconrec.2015.11.016

Tranfield D, Denyer D, Smart P (2003) Towards a methodology for developing evidence-informed management knowledge by means of systematic review. Br J Manag 14(3):207–222. https://doi.org/10.1111/1467-8551.00375

UN (n.d.) States members of the United Nations and states members of specialized agencies. Retrieved September 3, 2020, from https://sustainabledevelopment.un.org/memberstates
UN (2015) About the sustainable development goals—United Nations sustainable development. Retrieved March 5, 2019, from https://www.un.org/sustainabledevelopment/sustainable-development-goals/
UN (2020) The sustainable development goals report 2020.https://sdgs.un.org/sites/default/files/2020-09/The-Sustainable-Development-Goals-Report-2020.pdf
UNEP (2013) Home subscribe archive contact municipal solid waste: is it garbage or gold? https://na.unep.net/geas/archive/pdfs/GEAS_Oct2013_Waste.pdf
United Nations (2018) World economic situation and prospects 2018. http://www.escwa.un.org/main/contact.asp
United Nations Environment Programme (2021) Food waste index report 2021. In: Unep
Vittuari M, Azzurro P, Gaiani S, Gheoldus M, Burgos S, Aramyan L, … Timmermans T (2016) Recommendations and guidelines for a common European food waste policy framework. In: Fusions. https://doi.org/10.1016/j.jclinane.2015.08.012
Walia B, Sanders S (2019) Curbing food waste: a review of recent policy and action in the USA. Renewable Agric Food Syst 34(2):169–177. https://doi.org/10.1017/S1742170517000400
Warshawsky DN (2015) The devolution of urban food waste governance: case study of food rescue in Los Angeles. Cities. https://doi.org/10.1016/j.cities.2015.06.006
Wingrove K, Barbour L, Palermo C (2017) Exploring nutrition capacity in Australia's charitable food sector. Nutr Diet. https://doi.org/10.1111/1747-0080.12284
Xu F, Yang L, Ge X, Li Y (2018) Anaerobic digestion of food waste—Challenges and opportunities. Biores Technol 247:1047–1058. https://doi.org/10.1016/j.biortech.2017.09.020
Yong Z, Dong Y, Zhang X, Tan T (2015) Anaerobic co-digestion of food waste and straw for biogas production. Renewable Energy 78:527–530. https://doi.org/10.1016/J.RENENE.2015.01.033
Zamanzadeh M, Hagen LH, Svensson K, Linjordet R, Horn SJ (2016) Anaerobic digestion of food waste – Effect of recirculation and temperature on performance and microbiology. Water Res 96:246–254. https://doi.org/10.1016/J.WATRES.2016.03.058
Zhang J, Lv C, Tong J, Liu J, Liu J, Yu D, … Wei Y (2016) Optimization and microbial community analysis of anaerobic co-digestion of food waste and sewage sludge based on microwave pretreatment. Bioresour Technol. https://doi.org/10.1016/j.biortech.2015.10.037
Zhen G, Lu X, Kobayashi T, Kumar G, Xu K (2016) Anaerobic co-digestion on improving methane production from mixed microalgae (Scenedesmus sp., Chlorella sp.) and food waste: kinetic modelling and synergistic impact evaluation. Chem Eng J 299:332–341. https://doi.org/10.1016/j.cej.2016.04.118

Chapter 4
Modeling Climate Resilient Economic Development

Anett Großmann, Markus Flaute, Christian Lutz, Frank Hohmann, and Maximilian Banning

Abstract Climate change is challenging governments around the world to incorporate mitigation and adaptation measures into long-term development strategies. As one response, national E3 (economy-energy-emission) models have been developed for Georgia and Kazakhstan to allow determining the macroeconomic impacts of climate change and adaptation measures by model users in the respective country. This can significantly improve long-term planning by including climate change and adaption in this process. The chapter first describes the basic model structure, then explains how policies can be implemented as well as what the respective information and data needs are. In a second step, the country models e3.ge and e3.kz are applied to assess the effects of investment in additional irrigation systems and windbreaks that protect agriculture against either droughts or winds, which are expected to occur more often and more pronounced due to climate change. The quantifications of the economy-wide impacts in both countries are built on sector- and country-specific cost–benefit analyses. Scenario analysis is applied to compare a scenario with the adaptation measure against a climate change scenario without adaptation. Model results for the period up to 2050 show positive effects on GDP and employment for the adaptation measures. Besides agriculture, other industries as construction and services will benefit from adaptation in terms of production and employment.

A. Großmann (✉) · M. Flaute · C. Lutz · F. Hohmann · M. Banning
Institute of Economic Structures Research (GWS mbH), Heinrichstraße 30, 49080 Osnabrück, Germany
e-mail: grossmann@gws-os.com

C. Lutz
e-mail: lutz@gws-os.com

F. Hohmann
e-mail: hohmann@gws-os.com

M. Banning
e-mail: banning@gws-os.com

W. Leal Filho et al. (eds.), *Climate Change Strategies: Handling the Challenges of Adapting to a Changing Climate*, Climate Change Management,
https://doi.org/10.1007/978-3-031-28728-2_4

Introduction

Climate change is a major and urgent challenge for the environmental and socioeconomic development. At the recent UN climate summit in Glasgow (COP26), the importance of both emission reductions to limit global temperature rise and adaptation to climate change was emphasized.

Although participating countries revised and strengthened climate plans to curb their CO_2 emissions, current global efforts are not enough to reach the Paris Agreement's targets. Without additional measures continuous global warming is likely in the twenty-first century.

According to climate scientists, climate change cannot be fully prevented as it already has and will have impacts on the people, the environment, and the economy. Countries are facing gradual climate change impacts such as increasing temperatures and changes in precipitation leading for example to melting glaciers and sea level rise. The risk of disruptive change is also rising.

The "Global Climate Risk Index" (Eckstein et al. 2021) shows that many countries are also threatened by recurring extreme whether events (EWEs) like droughts, floods, heat waves and storms which are likely to occur more frequently and more intensely.

Kazakhstan and Georgia, the countries under consideration, are vulnerable to various climate change impacts. Droughts are for example a severe risk for both countries resulting in yield losses. Other EWEs such as floods and heavy winds destroy energy, road and building infrastructure and thus cause economic costs to repair the damage. Increasing temperatures adversely impact human health as well as energy demand and supply. Apart from these direct impacts, further losses result from e.g., impaired production due to power outages.

Adaptation to climate change is of major importance to reduce the adverse impacts of climate change while exploiting the benefits of climate change. Various adaptation options and evaluations exist for key economic sectors and climate hazards. Macroeconomic impacts and intersectoral effects of climate change and adaptation which go beyond single economic sectors analyses are often still not considered.

Policymakers need powerful tools to evaluate the macroeconomic impacts of sector-specific economic risks and benefits (awareness raising) as well as different sectoral adaptation strategies (preparedness) to be able to initiate the transition to a climate-resilient economy and integrate findings in long-term strategies. Environmentally extended economic models in combination with scenario analysis support policymakers with these issues.

An extensive exchange with partners and experts in Kazakhstan and Georgia, as well as the cooperation between the respective ministries in the countries, local research institutes, GWS and GIZ resulted in the development of the country specific E3 models within the project "Policy Advice for Climate Resilient Economic Development" (CRED) financed by the German Federal Ministry for the Environment, Nature Conservation, Nuclear Safety and Consumer Protection (BMUV) and implemented by Deutsche Gesellschaft für Internationale Zusammenarbeit (GIZ).

Jointly with the country partners, the model approach has been selected and national economic and climate data were compiled. This was followed by the development and application of the E3 models to climate change and adaptation related issues including intensive capacity building for the partners.

Climate change and adaptation scenarios were designed comprising information and data on the most relevant climate hazards, their sector-specific impacts as well as suitable adaptation options. These scenarios were analyzed with the E3 models regarding their long-term macroeconomic impacts. A comparison of scenario results allows policymakers to identify those adaptation measures that are highly effective and beneficial for the economy, employment, and the environment (win–win options).

Finally, the macroeconomic results are fed into stakeholder discussions and policy processes (Dekens and Hammill 2021) to support them with evidence-based adaptation planning.

Macroeconomic Modeling of Climate Change Adaptation

Various approaches to estimate the macroeconomic effects of climate change are described in the literature. A starting point have been calculations by William Nordhaus. They led to the development of one of the first integrated assessment models (IAMs) DICE (Dynamic Integrated Climate-Economy, Nordhaus 1992), which attempted to represent the interrelationships between climate change and the global economy in a dynamic model.

Nikas et al. (2019) differentiate six classes of climate-economy models building on various classifications in the literature. The first IAMs have been optimal growth models, which represent the economy in a single sector. Damage costs enter these models as simple equations, e.g., the economic damage is parameterized as a square function of the temperature increase due to climate change. Other total models are computable general equilibrium models (CGE) and macro-econometric models, which represent the economic development according to different economic sectors including the linkages between these sectors and with climate. As partial models, the other model types are not suitable for the integrated socio-economic analysis of climate change impacts.

According to the Network for Greening the Financial System (NGFS 2020), economic models assessing climate risks can be similarly divided into integrated climate-economy models and adapted macroeconomic models. Economic models can be distinguished according to their underlying economic theory explaining the functioning of and interaction within an economy. Basically, these are CGE, static Input–Output (IO) and macro-econometric (or dynamic) IO models (Lehr et al. 2020; Máñez et al. 2016; NGFS 2020; Pollitt and Mercure 2018).

CGE models follow neoclassical theory and are supply-side driven. They are optimization models and characterized by the assumption of market clearing, fully flexible prices, and immediate substitution. In this regard, CGE models are suitable for long-term analyses assuming functioning markets, but climate change and adaptation costs tend to be underestimated (Botzen et al. 2019; OECD 2015). In the EU PESETA project, the impacts of climate change are modeled in the CGE model GEM-E3 either through damage to the capital stock, sectoral productivity losses or as welfare losses of private households (Feyen et al. 2020). The latter may result from additional energy demand for cooling or involuntary additional expenditures for the repair of flood damage.

Static IO models are based on IO tables which provide a detailed view on inter-industry linkages and the demand supply relationship (Miller and Blair 2009; United Nations 2018). IO models go back to Leontief (1986) who mathematically illustrated the effects of additional demand in a single industry and its economy-wide direct and indirect impacts. The static IO approach is appropriate for short-term analysis due to its constant economic structure. It is used among others for (natural) disaster analyses. In contrast to CGE models, immediate substitution does not exist. Long-term adaptation processes cannot be represented in a static IO model due to lacking consideration of adjustment processes over time. Adaptation costs tend to be over-estimated as this model type does not allow for substitution processes when confronted with higher costs (Botzen et al. 2019; Lehr et al. 2020; Máñez Costa et al. 2016).

Macro-econometric (or dynamic) IO models (Almon 1991, 2014; West 1995) build upon the advantages of static IO models but largely resolve their limitations and inherent assumptions, amongst others the absence of time and of capacity constraints. Prices indicate shortages due to capacity constraints. Due to the explicit consideration of time in dynamic models, they can reflect the economic development year by year and can therefore show the temporal adjustment path of recovery periods from climate change effects and adjustment processes of adaptation. Like static IO models, dynamic IO models are typically demand-side driven. However, demand is determined endogenously and not given exogenously. Income, which is influenced by the current labor market situation and consumer prices, is an important determinant for consumer demand.

Macro-econometric IO models rely on a comprehensive data set that allows to model volume and price reactions based on empirical estimations as opposed to CGE models that are using parameters calibrated to a base year. Trends which were econometrically detected for the past are assumed to be valid in the future and relax the assumption of a constant economic structure and/or import dependency—this approach is much more realistic for a mid- to long-term projection (Meyer and Ahlert 2019). Future technological changes and innovations may be considered and make the model more useful to analyze structural changes (Mercure et al. 2019). Nevertheless, the assumption of constant parameters (which are derived from past observations) is continuously less valid with increasing distance in time.

The macro-econometric IO modelling approach is e.g., applied to evaluate climate change impacts and adaptation measures in Germany using the model PANTA RHEI (Lehr et al. 2016), for the EU islands (Leon et al. 2021) and at EU level using the models GINFORS (Lehr et al. 2018; Aaheim et al. 2015; European Commission 2021) and E3ME (Cambridge Econometrics 2019). Damages and losses from

EWEs were collected by screening literature and damage databases e.g., from reinsurance companies, translated into model variables and then implemented as economic impulses into the models. Damages were modeled, for example, as a reduction in the capital stock in the machinery and real estate sectors. The non-usability of transport infrastructure due to flooding leads to higher costs (economic losses) due to the use of other means of transport and routes, which were captured in the model as well. Productivity losses were modeled by higher imports in the respective sectors, so that the lower production level is at least partially compensated (Lehr et al. 2016, 2020).

Methodologies

Excel-Based E3 Models for Georgia and Kazakhstan

Based on the presented modeling approaches, it is obvious that there are different methods for modeling economic impacts of climate change and adaptation measures. Each approach has its strengths and weaknesses (Keen 2020; Keppo et al. 2021). As of now, there is no common standard approach. Thus, multiple models that complement each other are sometimes used simultaneously (e.g., Feyen et al. 2020; Leon et al. 2021).

Basically, the most important requirements for an economic model to map climate change can be defined as follows (Großmann et al. 2021): It must consider the main economic impacts (e.g., productivity and income losses), the economic sectors directly affected by climate change (e.g., agriculture, energy), and intersectoral dependencies. In addition, such an economic model must capture long-term macroeconomic developments regarding future climate change impacts and adjustment reactions in the years following a climate event.

For Georgia and Kazakhstan, the macro-econometric (dynamic) IO modeling approach is a suitable solution which fulfills the necessary requirements. This economic model is extended to an E3 (economy, energy, emission) model, so that it is also possible to identify synergies and trade-offs of adaptation and mitigation strategies as well as Nationally Determined Contribution goals.

For both countries, such a projection and simulation model has been developed jointly with Georgian and Kazakh partners to evaluate the macroeconomic impacts of climate change and adaptation measures until 2050 in a holistic and consistent model framework. The E3 models contain three interlinked model parts: the economic model (1) and the energy (2) and emissions (3) module (Fig. 4.1).

Each model part is based on a national, official, and up-to-date dataset given as time series which allows to derive country-specific parameters empirically. The climate module describes the impact of projected climate scenarios informed by empirics of past and current climate related damages in the countries.

The **economic model** (1) is based on the INFORUM approach (Almon 1991, 2014; West 1995) and features post-Keynesian properties. The core of the economic

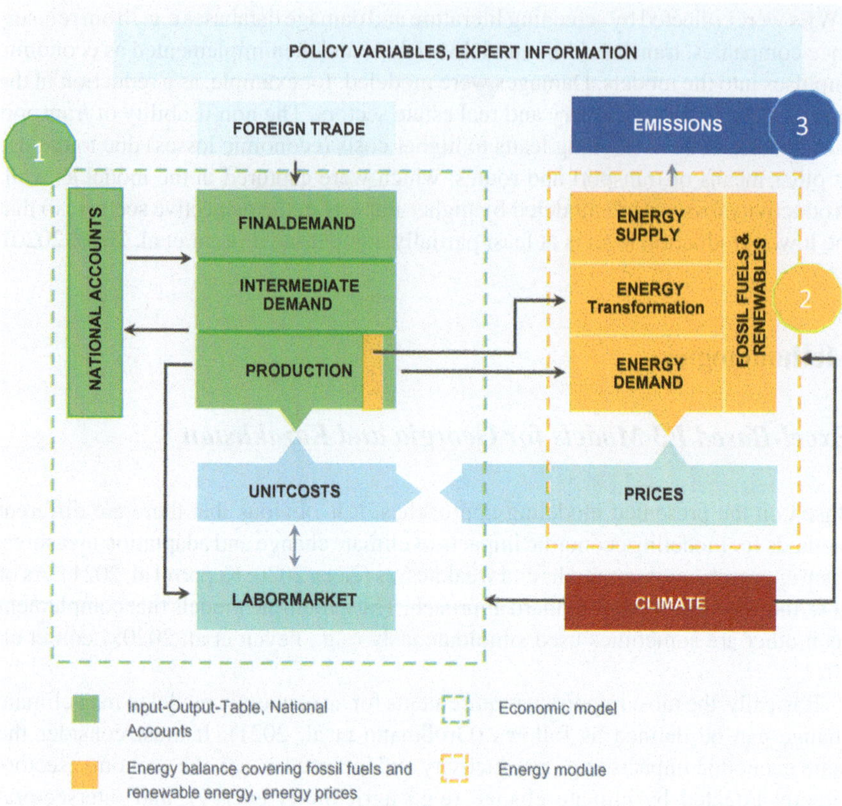

Fig. 4.1 E3 model overview (*Source* Großmann et al. 2021)

model is the IO framework representing the economic sectors, their interlinkages as well as the domestic and foreign factors contributing to economic growth. The relationship between demand and supply is represented by the Leontief production function.

Supply and price elements are considered as well to account for supply constraints possibly caused by EWEs. Production costs for each economic sector are derived from costs to be paid for intermediate goods and primary inputs (such as labor costs) as given in the IO table. Production prices are determined based on these costs.

Labor market indicators such as number of jobs, wages and income are part of the model to monitor sector-specific labor productivity, employment, and wealth. Labor demand follows the economy activity in the sectors considering labor productivity. The macroeconomic wage rate is derived from the Phillips curve approach taking the overall labor productivity and labor scarcity indicator—measured as the ratio of population at working age and total labor demand—into account.

The modelling approach which covers not only quantity effects but also income and price effects provides multipliers that determine the dynamics of the system:

- Leontief multiplier: Shows the direct and indirect effects of demand changes (e.g., consumption, investments) on production;
- Employment and income multiplier: Increased production leads to more jobs and thus higher incomes resulting in higher demand (induced effect);
- Investment accelerator: Indicates the necessary investments to maintain the capital stock needed for production based on the demand for goods.

The **energy module** is based on national energy balances depicting the energy supply, transformation, and demand for different fossil fuels and renewables in physical terms. It describes the relations within the energy sector more comprehensively than the economic model. The energy demand is modeled in detail for the largest consumers such as industry, private households, and transport. Key drivers of sector-specific energy demand are the sector-specific economic activity, the respective energy intensity of the production processes and energy price developments. The energy demand of private households is estimated with population. The energy supply is determined by the energy demand of all sectors. Energy is either produced domestically or imported. Primary energy use is covered for power generation as well as heat generation.

The **emissions module** comprises the energy-related CO_2 emissions which are linked to the combustion-related energy consumption of fossil fuels. Reductions in the use of fossil fuels due to expansion of renewable energy or increased energy efficiency are visible as CO_2 savings.

The requirements for data and the model approach are kept moderate for a sustainable solution which is also important for the model ownership in the countries. However, the model approach is flexible, can be expanded in many ways and allows for integrating expert inputs.

Additionally, the E3 models are developed in Microsoft (MS) Excel using the model building framework DIOM-X. The framework is built upon the MS Office programming language Visual Basic for Applications (VBA) and was developed for creating **D**ynamic **I**nput–**O**utput **M**odels in Excel (DIOM-X) (Großmann and Hohmann 2019). Besides MS Excel, "R" or EViews can be used to perform regression analysis which is beyond the options provided by MS Excel. The DIOM-X modelling framework provides the tools to exchange the dataset, regression equations with their parameters automatically between both programs.

The full model database, model equations, scenario settings and scenario results are stored in a single Excel workbook to ensure that all aspects of the model can be examined, adjusted, and extended. Model users conduct scenario analysis by adjusting the values of model variables in one Excel worksheet. Thus, there is no need to learn programming.

Climate Change Adaptation Scenarios as "What-If"—Impact Analysis Under Uncertainty

Scenario (or "what-if") analysis is a method for dealing with the uncertainties of the future. Scenarios are consistent sets of assumptions to test how the future might evolve under certain conditions ("if"). "What" comprises the economy-wide impacts and consequences resulting from the assumptions made. Thus, a scenario helps to better understand what could happen and who or what is affected and how? However, scenarios should not be considered as precise forecasts.

Various policy options and measures can be analyzed by conducting scenario analysis depending on the main purpose of the model application and key research question. The E3 models were developed to answer questions such as: What are the macroeconomic effects of sector-specific climate change impacts and adaptation options?

With the help of scenarios, sector-specific climate change impacts and adaptation measures are implemented in the model by adjusting appropriate model variables such as productivity or investments. Then the model quantifies the direct, indirect, and induced impacts on other economic sectors, the macroeconomy and emissions.

The comparison of model results from different scenarios to a reference scenario not including a certain policy or measure helps to identify the option which is appropriate for a particular issue. Policymakers need to identify and prioritize those criteria (e.g., GDP or employment effects) that are most important for them to finally select the better or best policy option(s).

Implementing Climate Change Adaptation Impacts in E3 Models

Climate change and socio-economic impacts, changes, and responses in the form of mitigation and adaptation are mutually dependent. While it is already difficult to estimate the frequency and intensity of climate events, it is even more challenging to quantify their economic impacts (Brasseur et al. 2017). Therefore, representing these interactions and relationships in simulation models is a challenging task that involves a high degree of uncertainty. To get an idea of the possible future economic impacts of climate change and in particular of EWEs, the E3 models in combination with scenario analysis are applied.

Usually, economists derive future developments from past observations but that is hardly possible for climate change and adaptation issues due to the following reasons:

- Climate change and its economic impacts can often not be observed in (official) statistics.

- Climate change impacts may not be detected as such because repairing climate change damages potentially results in positive GDP effects (so called defensive spending).
- The damage may have been avoided or reduced by existing adaptation measures.
- Economic and climate models are operating on different temporal and spatial scales. While climate models have a high spatial resolution and a long-term horizon, the current E3 models considering economic development at national level and have a mid-term to long-term perspective until 2050.
- Climate models are complex and very computation-intensive while the E3 models for Georgia and Kazakhstan use a simplified, less computation-intensive approach and run simulations in less than a minute on an average desktop computer or laptop.

For implementing the economic impacts of climate change and adaptation measures, a four-step approach is applied (Großmann et al. 2021; Flaute et al. 2021).

The main prerequisites for this approach are

- Climate scenarios for Georgia and Kazakhstan with a regional breakdown and the future evolution of climate hazards,
- Identification of relevant interfaces and affect chains of climate hazards (Fritzsche et al. 2014),
- Sector-specific and quantified damage data (as well as benefits, if appropriate).
- Cost–Benefit Analyses (CBA) of sector-specific adaptation measures.

Following this approach helps to understand the economic impacts of climate change and how potential adaptation measures support to minimize or even avoid these effects.

1. **Identification of EWEs and their effects**

   Modeling future impacts of climate change on the national economy requires linkages between climate projections and country- and sector-specific economic damages. Climate experts from the University of the Balearic Islands provided the occurrence and intensity of country-specific climate hazards (e.g., drought, heatwave, flooding) for the RCP2.6 and RCP8.5 scenarios for selected vulnerable areas and locations (Navarro and Jordà 2021). The frequency of a climate hazard is either derived from past observations or expert knowledge.

   Since no official climate damage register exists in Georgia and Kazakhstan, the economic damages from past and current EWEs are collected by screening scientific (national and international) literature, media, and expert surveys. These damage data serve as benchmarks for estimating future climate change impacts. Adjustments will be made to the benchmarks in scenarios to reflect the expected intensity changes of climate hazards.
2. **Translation of EWEs into model variables**

   The identified climate change effects need to be translated into E3 model variables. Basically, the initial impacts of climate hazards are implemented as effects on human behavior (e.g., consumption expenditures for health care), production factors (e.g., labor productivity) and / or infrastructure (e.g., investments to repair damages).

3. **Possible adaptation options** preventing/minimizing the damages or taking advantage of opportunities that may arise are identified by screening national and international literature as well as through discussions with experts. CBAs are important cornerstones to identify the costs and benefits of sectoral adaptation measures. The benefits are implemented into the E3 models as the reverse impacts of climate change damages, costs are usually integrated as investments and depending on how the investments are financed, prices or government expenditures are affected as well.
4. **For evaluating the impacts of adaptation measures**, a climate change scenario without adaptation measures and the respective adaptation scenario are compared. Usually, there is more than one adaptation option (e.g., irrigation systems or drought-resistant crops). Scenario results help to detect the option(s) with high effectiveness and positive effects on the economy and the environment. Selection criteria might be the biggest avoided damages, employment effects or synergies with other strategies such as mitigation which needs to be prioritized by policymakers.

Results from Climate Change and Adaptation Scenarios in Agriculture

Investing in Irrigation Systems—A Case Study from Kazakhstan

Droughts impact agriculture, especially the cultivation of rain-fed wheat in North-Kazakhstan (World Bank 2015, 2016; MNE et al. 2017). In southern and eastern regions of Kazakhstan, the risk of flooding due to glacial melt is more likely in the medium term and by mid-century water supply will be threatened. Soil degradation, desertification, pest, and disease outbreaks are the main issues associated with these climate change impacts resulting in lower agricultural productivity.

Agriculture is one of the most important sectors in Kazakhstan contributing 5% of GDP and employing 1.2 million people, respectively about 13% of the workforce in 2019 (Bureau of National Statistics of the Republic of Kazakhstan 2021). Grain production of agro-holdings and large-scale farms is the most important segment of agriculture with wheat being a major export good (UNDP 2019). In the rest of the country, livestock and vegetable are cultivated mainly by small-scale farmers in rural areas (OECD 2020).

Climate change is likely to exacerbate the already volatile production and income risks. Due to shifts of climate zones, some regions may profit whereas others suffer from negative impacts due to changed precipitation patterns.

UNDP (2020) estimates the decline of wheat yield losses to amount 33% of the current potential by 2030 and 12% by 2050. The grazing capacity is likely to decline resulting in a livestock productivity reduction of 10% by 2030 to maximum 20%

by 2050 of the current potential. In contrast, sunflower seed yields are expected to benefit from climate warming leading to an increase of 8% by 2030 and around 4% by 2050 compared to current gross output. Overall, crop production is more risk-prone than livestock (World Bank 2016).

Crop losses lower Kazakhstan's export opportunities while dependence on agricultural imports increases to ensure food security. With more severe droughts and without adaptation, jobs and incomes are at risk not only in agriculture, but also in industries dependent on agriculture.

Climate change adaptation in agriculture is a priority area in Kazakhstan according to the New Environmental Code (Article 313) adopted in 2021. Farmers have different options to adapt to climate change such as usage of water-efficient technologies, cultivation of drought-resistant crops and restoring of water infrastructure. Soil conservation can be improved by moisture saving technologies such as conservation agriculture and no-till farming (UNDP 2020; World Bank 2016). Other options include improved crop protection through fertilization and selective breeding to control pests and diseases. Livestock productivity can also be increased through rotational grazing which aims to avoid overgrazing.

While each of these specific practices helps to at least partially offset yield losses caused by climate change, insurance against crop failures can partly compensate farmers, but does not prevent losses.

An option to reduce drought impacts in agriculture is the (re-)construction of water canals and the installation of water-saving technologies such as drip irrigation. The associated costs and benefits are shown in Table 4.1. This data is used as input for the Kazakh e3.kz model to analyze the macroeconomic impacts of this measure. The government is expected to subsidize these investments at the expense of other expenditures.

The macroeconomic effects of the investments in water canals, reservoirs and drop irrigation systems in agriculture are positive according to Fig. 4.2. The increased construction activity and higher crop yields due to an expanded irrigated area have a positive impact on GDP which is up to 1.2% per year above a scenario with droughts and without adaptation.

There are only a few Kazakh producers of drip irrigation systems. Therefore, most of these must be imported, mainly from Europe, Israel, or China (EBRD et al. 2018). Higher imports of maximum 1.1% per year have a negative economic impact,

Table 4.1 Key assumptions for investments in irrigation systems

Adaptation measures	Cumulated investment (2022–2050)	Annual adaptation benefits (in terms of higher agricultural output)
(Re-)construction of water canals and reservoirs	2,894 Billion Tenge	537 Billion Tenge
Installation of drip irrigation	105 Billion Tenge	47 Billion Tenge

Source Astana Times (2019), EBRD et al. (2018), Kazakh Government (2020)

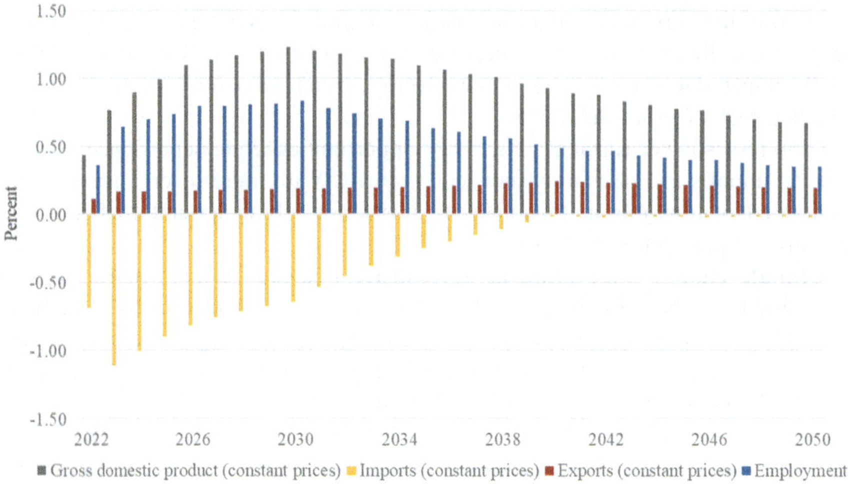

Fig. 4.2 "Irrigation" scenario: key impacts, 2022–2050, deviations from a "drought" scenario in percent (*Source* Own figure based on e3.kz results)

but this is compensated by the benefits of the adaptation measure. The agricultural imports are lower. Export opportunities are no longer fully impaired by droughts and are up to 0.24% per year higher. The agriculture and construction sectors profit the most but also several other industries providing necessary inputs such as concrete or further processing agricultural goods.

Additional jobs in the construction sector are created during the construction period. Afterwards, regular maintenance and replacement investments are necessary, which preserve jobs. Permanent jobs and income are created in agriculture by restored and additional irrigated land. In total, a maximum of 78,000 additional jobs respectively 0.8% per year will be created compared to a situation where no adaptation is done, and droughts occur.

Overall, the measures reduce the water scarcity and increase agricultural productivity not only during droughts. Apart from additional jobs, food security can be improved.

Investing in Wind Breaks—A Case Study from Georgia

Wind impacts in agriculture: Agriculture is a key sector for the Georgian economy, employing about 40% of the workforce, primarily as self-employed farmers. At the same time, agriculture is vulnerable to climate change. Thus, climate change does not only increase the risk for agriculture, but, due to its importance for the Georgian economy, impacts economic and social welfare (MEPA 2017).

Several climatic trends are already observable today and will continue to be even more noticeable in the future. Statistically increasing trends in mean annual temperature, significant increases in the duration of heatwaves, droughts, and heavy winds, and changing patterns of precipitation throughout the country are only some of the climate challenges. Besides others, these may lead to a displacement of agro-climatic zones, desertification tendencies, reduction of arable land and reduction of crop productivity.

In recent years, wind erosion was a major environmental problem in dry land areas, causing land degradation, reducing crop yields, and affecting sector development (MEPA 2017). In 2000, low wheat productivity was caused by high speed of winds in spring, "sweeping" the seeds and humus from the fields. In 2015–2016 in Samegrelo, hot winds destroyed the harvest of hazelnuts by 80–90%. Harmful diseases were spread by the wind, which in the following years also negatively influenced the yields. Strong wind can cause lodging down of maize. Thus, wind is not only a risk in the period of blooming, but also in other periods of further growth, as the growing plants are more and more exposed to the wind. This makes it necessary to arrange windbreak zones (MEPA 2017).

Windbreaks as an adaptation option in agriculture: Several adaptation measures can be implemented to reduce the negative effects of climate change on the agricultural production and to even increase the crop yields over time. Adapted varieties and new crop types can better resist to the new climatic threats. In combination with adapted cultivation methods and other structural measures (irrigation systems, hail protection nets, hail guns, windbreaks) soil conservation and water saving can be achieved.

Windbreaks reduce wind speeds over fields, resulting in soil conservation and thus providing additional protection for the plants. Furthermore, they can be implemented not only as bushes, but with useful crops and trees to provide additional food security. Being implemented in the right location and in a functional way, windbreaks not only provide agricultural benefits but also economic, environmental, and social benefits (Smith et al. 2021).

The windbreaks are usually placed at the edge of the fields (IBiS 2019). Windbreaks were already implemented during Soviet time, but they have been cut down and used as firewood during the energy crises in the 1990s. Nowadays, fire and grazing cattle are the biggest threats for windbreaks (IBiS 2019).

The reactivation and re-construction of the windbreak system requires not only seedlings for trees and bushes, but also additional machinery, plastic, and agricultural services. Since average income in agriculture is low, the role of the government to finance the reactivation of windbreaks becomes crucial. The maintenance of windbreaks also requires additional financial resources (IBiS 2019). It is assumed that half of the investment is financed by the government and half by private households, partly crowding out other expenditures.

Table 4.2 summarizes the costs and benefits of implementing windbreaks in Georgia. Plants, plastic coverage, agricultural services and machinery are needed to set up the windbreaks. Depending on the crops, the increase in yields varies. In total, an increase of 17.8% is assumed.

Table 4.2 Key assumptions for investments in windbreaks Georgia

Adaptation measures	Investment per year (2022–2050)	Annual adaptation benefits per (in terms of higher crop yields)
Installation of Windbreaks	• Plants: 6 million GEL • Plastics: 4 million GEL • Agricultural services: 5.2 million GEL • Machinery: 2 million GEL	• Total (weighted by share in agriculture): 17.8%

Source Geostat (2020), Moore (n.d.)

Assuming heavy winds every five years starting in 2025, implementing windbreaks results in the following economy-wide effects.

In total, the GDP increases by up to 1.4% in one year in the period from 2022 to 2050 (Fig. 4.3). The agricultural sector benefits on the one hand from the additional demand for seedlings and agricultural services, and on the other hand the greatest impact stems from the increase in crop yields in all years due to the windbreaks (Table 4.2). The additional demand for plastics to cover the seedlings increase the production in that respective sector. Total imports are affected on the one hand from additional imports due to higher consumption and investment and on the other hand from reduced imports of agricultural products. In the years with a heavy wind event, the effects are also positive, but smaller than in the other years, since the windbreaks cannot avoid all the damages caused by the wind.

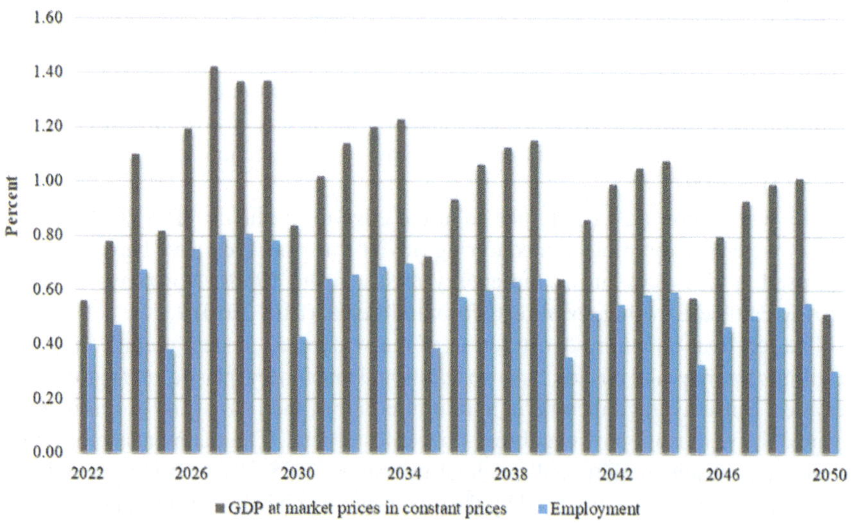

Fig. 4.3 "Windbreaks" scenario in Georgia: key impacts, 2022–2050, deviations from a "wind" scenario in percent (*Source* Own figure based on e3.ge results)

Additional employment is created especially in those sectors that benefit from the implementation of windbreaks. The additional production and services in the agricultural sector require a higher workforce. Rubber and plastic producing sector and the transportation sector also benefit in terms of employment. With an increasing productivity expected in the respective economic sectors, additional employment is needed, which decreases over time, but the overall effect remains clearly positive. Since more people have a regular income, the money is spent for consumption purposes (increasing by up to 1.1%), thus, creating additional production and employment effects.

Conclusions and Outlook

The Excel-based E3 models are powerful tools which support in understanding and quantifying the economy-wide impacts of climate change and of adaptation measures by conducting scenario analysis. This approach goes beyond a classic CBA which usually analyzes the economic impacts for a single sector.

The scenario results show that climate change has an impact on agriculture as well as on other sectors along the value chain. The adverse impacts of climate change can be limited by appropriate adaptation options. Evaluation and comparison of various adaptation options help to find favorable solutions. However, other criteria such as health aspects and ecosystem services must be considered to get a more comprehensive evaluation.

Other key economic sectors such as energy and tourism are at risk as well and should be evaluated intensively regarding climate damages and suitable adaptation measures (for examples see Großmann et al. 2021; Flaute et al. 2021). An extensive, systemic data collection of damages caused by climate change and CBAs for adaptation options would support an evidence-based, macroeconomic analysis. Other scenarios based on these exemplary scenarios can be conducted to show for example the impacts of financing adaptation measures through international funds. Prospects for funding of such measures in developing countries are good given that industrialized countries renewed their promise in Glasgow to support measures which tackle causes and consequences of climate change in the future. With this support, the macroeconomic effects of these measures for the recipient countries would be even better.

Model results trigger the necessary process of coordination and implementation with national institutions and stakeholders to identify those adaptation measures that are suitable to prevent from climate change impacts and have positive effects on the economy, employment, and the environment. In this regard, the E3 models support evidence-based policy making and thus help to design "better" adaptation policies.

A success factor for the implementation of such a modeling tool is the simplified, transparent, and easy to use Excel-based model framework in conjunction with intensive capacity building. Transferring full ownership of the model to the respective partners in the country allows for continuous application, update, and expansion.

The simplified E3 modeling approach has its limitations. The top-down approach does not allow structural changes to be modeled endogenously, as it is possible in integrated bottom-up models such as PANTA RHEI (Lehr et al. 2020). Therefore, it is particularly important to integrate expert knowledge and information from sector-specific bottom-up models.

Moreover, the projections of the E3 models rely on comprehensive data sets, empirical estimations, and expert knowledge. The assumptions of estimated but constant parameters are less and less valid with longer projection periods. The same is true for calibrated parameters in CGE models. Scenario analysis helps to deal with the inherent uncertainty of the future. Model parameters can be varied in scenarios which then lead to other development paths. However, a projection should not be interpreted as a precise forecast of the future.

Furthermore, the analysis mainly focuses on economy-wide impacts of climate change in one respective country. However, climate change leads to damages and impacts worldwide, which also may affect international supply chains. The consideration of such effects requires detailed information on trade flows and global climate impacts, which the model cannot represent in its current form.

Nonetheless, the experiences from the CRED project shows that the Excel-based models and the coaching concept serve as a blueprint to evaluate climate change related issues which can be applied to other countries.

Acknowledgements This work has been produced in the scope of the project "Policy Advice for Climate Resilient Economic Development" (CRED) implemented by Deutsche Gesellschaft für Internationale Zusammenarbeit (GIZ) GmbH on behalf of the German Federal Ministry for the Environment, Nature Conservation, Nuclear Safety and Consumer Protection (BMUV) as part of the International Climate Initiative (IKI). Responsibility for the contents of this chapter lies entirely with the authors and can in no way be taken to reflect the views of GIZ.

References

Aaheim A, Ahlert G, Meyer M, Meyer B, Orlov A, Heyndrickx C (2015) Integration of top-down and bottom-up analyses of adaptation to climate change in Europe—the cases of energy, transport and health. European Community's 7th Framework Programme under Grant Agreement No. 308620 (Project ToPDAd), Deliverable D3.4, Oslo

Almon C (1991) The INFORUM approach to interindustry modeling. Econ Syst Res 3(1):1–7

Almon C (2014) The craft of economic modelling—Part 1, 5th edn. Department of Economics, University of Maryland

Astana Times (2019) Kazakh government to increase irrigated land area to 3.5 million hectares. Retrieved on 10 February 2022 from https://astanatimes.com/2019/01/kazakh-government-to-increase-irrigated-land-area-to-3-5-million-hectares/

Botzen W, Deschenes O, Sanders M (2019) The economic impacts of natural disasters: a review of models and empirical studies. Rev Environ Econ Policy 13(2):167–188

Brasseur G, Jacob D, Schuck-Zöller S (2017) Klimawandel in Deutschland. Entwicklung, Folgen, Risiken und Perspektiven. Springer Spektrum

Bureau of National Statistics of the Republic of Kazakhstan (2021) Statistics of labor and employment. Retrieved on 10 February 2022 from https://stat.gov.kz/official/industry/25/statistic/7

Cambridge Econometrics (2019) Committee on climate change. A consistent set of socioeconomic dimensions for the CCRA3 Evidence Report research projects. Final report, June 2019

Dekens J, Hammill A (2021) Climate economic modelling. Using climate economic modelling for sustainable economic development. A practitioner's guide. Published by GIZ on behalf of the Federal Ministry for the Environment, Nature Conservation and Nuclear Safety. Retrieved on 10 February 2022 from https://www.giz.de/de/downloads/giz2021-en-climate-economic-modelling-practitioners-guide.pdf

Eckstein D, Künzel V, Schäfer L (2021) Global climate risk index 2021. Who suffers most from extreme weather events? Weather-related loss events in 2019 and 2000–2019. Briefing Paper, Bonn, Berlin. Retrieved on 10 February 2022 from https://germanwatch.org/sites/default/files/Global%20Climate%20Risk%20Index%202021_2.pdf

European Bank for Reconstruction and Development (EBRD), Finance and Technology Transfer Centre for Climate Change (FINTECC), Food and Agriculture Organization of the United Nations (FAO) (2018) Adoption of climate technologies in Kazakhstan's agrifood sector. Retrieved on 10 February 2022 from http://www.eastagri.org/docs/group/512/KAZ_ACT_25.10.18.pdf

European Commission (2021) Impact Assessment Report. Forging a climate-resilient Europe—the new EU Strategy on adaptation to climate change, Brussels, 24.02.2021. Retrieved on 10 February 2022 from https://eur-lex.europa.eu/legal-content/EN/TXT/PDF/?uri=CELEX:52021SC0025&from=EN

Feyen L, Ciscar JC, Gosling S, Ibarreta D, Soria A (ed) (2020) Climate change impacts and adaptation in Europe. JRC PESETA IV final report. EUR 30180EN, Publications Office of the European Union, Luxembourg, ISBN 978-92-76-18123-1. https://doi.org/10.2760/171121, JRC119178. Retrieved on 10 February 2022 from https://ec.europa.eu/jrc/sites/jrcsh/files/pesetaiv_summary_final_report.pdf

Flaute M, Reuschel S, Lutz, C, Banning M, Hohmann F (2021) Supporting climate resilient economic development in Georgia—application of the e3.ge model to analyze the economy-wide impacts of climate change and adaptation. On behalf of the Federal Ministry for the Environment, Nature Conservation and Nuclear Safety (BMU). Deutsche Gesellschaft für Internationale Zusammenarbeit (GIZ) GmbH, Bonn and Eschborn, Germany

Fritzsche K, Schneiderbauer S, Bubeck P, Kienberger S, Buth M, Zebisch M, Kahlenborn W (2014) The vulnerability sourcebook concept and guidelines for standardised vulnerability assessments. Published by GIZ. Retrieved on 10 February 2022 from https://www.adaptationcommunity.net/download/va/vulnerability-guides-manuals-reports/vuln_source_2017_EN.pdf

Geostat (2020) Agriculture of Georgia 2019. Preliminary data on plant growing. National Statistics Office of Georgia, Tbilisi

Großmann A, Hohmann F, Lutz, C, Reuschel S (2021) Supporting climate resilient economic development in Kazakhstan—application of the e3.kz model to analyze the economy-wide impacts of climate change adaptation. On behalf of the Federal Ministry for the Environment, Nature Conservation and Nuclear Safety (BMU). Deutsche Gesellschaft für Internationale Zusammenarbeit (GIZ) GmbH, Bonn and Eschborn, Germany

Großmann A, Hohmann F (2019) Static and dynamic input-output modelling with Microsoft Excel. SHAIO Conference Paper 2019, Osnabrück

IBiS (2019) Approach for "Rehabilitation of Windbreaks in East Georgia". Integrated Biodiversity Management, South Caucasus

Kazakh Government (2020) Eighty-eight thousand jobs to be created during modernization of irrigation infrastructure in Kazakhstan. Retrieved on 10 February 2022 from https://www.primeminister.kz/en/news/v-hode-modernizacii-irrigacionnoy-infrastruktury-kazahstana-budet-sozdano-88-tys-rabochih-mest-73818

Keen S (2020) The appallingly bad neoclassical economics of climate change. Globalizations. https://doi.org/10.1080/14747731.2020.1807856

Keppo I, Butnar I, Bauer N, Caspani M, Edelenbosch O, Emmerling J, Fragkos P, Guivarch C, Harmsen M, Lefèvre J, Le Gallic T, Leimbach M, McDowall W, Mercure JF, Schaeffer R, Trutnevyte E, Wagner F (2021) Exploring the possibility space: taking stock of the diverse capabilities and gaps in integrated assessment models. Environ Res Lett 16:053006

Lehr U, Flaute M, Ahmann L, Nieters A, Wolff C, Hirschfeld J, Welling M, Gall A, Kersting J, Mahlbacher M, von Möllendorff C (2020) Vertiefte ökonomische Analyse einzelner Politikinstrumente und Maßnahmen zur Anpassung an den Klimawandel – Abschlussbericht (In-depth economic analysis of individual policy instruments and adaptation measures to climate change—final report). Climate Change 43/2020, Dessau-Roßlau

Lehr U, Meyer M, Figini P, Ahrends B (2018) Work package 3: Climate change vulnerability assessment framework and complex impact chains. Deliverable 3.4. Specifications for climate, climate impacts and economic modelling—report on modelling chains

Lehr U, Nieters A, Drosdowski T (2016) Extreme weather events and the German economy: the potential for climate change adaptation. In Leal Filho W, Musa H, Cavan G, O'Hare P, Seixas J (ed) Climate change, adaptation, resilience and hazards, Springer, pp 125–141

Leon et al (2021) Downscaling climate change impacts, socio-economic implications and alternative adaptation pathways for islands and outermost regions. McGraw-Hill, Madrid. https://doi.org/10.5281/zenodo.5141549

Leontief W (1986) Input-output economics, 2nd edn. Oxford University Press, New York

Máñez Costa M, Rechid D, Bieritz L, Lutz C, Nieters A, Stöver B, Jahn M, Rische MC, Schulze S, Yadegar E, Hirschfeld J, Schröder A, Hirte G, Langer S, Tscharaktschiew S (2016) Synthesis of existing regional and sectoral economic modelling and its possible integration with regional earth system models in the context of climate modelling. Report 27. Climate Service Center, Hamburg. Retrieved on 10 February 2022 from https://epub.sub.uni-hamburg.de/epub/volltexte/2017/69272/pdf/report27.pdf

MEPA (2017) Climate change national adaptation plan for Georgia's agriculture sector. Ministry of Environment and Natural Resources Protection of Georgia, Tbilisi

Mercure JF, Knobloch F, Pollitt H, Paroussos L, Scrieciu S, Lewney R (2019) Modelling innovation and the macroeconomics of low carbon transitions: theory, perspectives and practical use. Climate Policy 19(8):1019–1037. https://doi.org/10.1080/14693062.2019.1617665

Meyer B, Ahlert G (2019) Imperfect markets and the properties of macro-economic-environmental models as tools for policy evaluation. Ecol Econ 155:809–887. https://doi.org/10.1016/j.ecolecon.2017.06.017

Miller RE, Blair PD (2009) Input-output analysis—foundations and extensions, 2nd edn. Cambridge University Press

Ministry of Energy of the Republic of Kazakhstan, United Nations Development Programme in Kazakhstan, Global Environment Facility (MNE et al. 2017) (2017) Seventh National Communication and Third Biennal report of the Republic of Kazakhstan to the UNFCCC

Moore L (n.d.) Economics of windbreaks. USDA-NRCS

Navarro JS, Jordà G (2021) Climatic hazards evolution in the 21stcentury in the frame of the CRED project. Mediterranean Institute for Advanced Studies (IME-DEA, UIB-CSIC), University of the Balearic Islands (UIB)

Network for Greening the Financial System (NGFS) (2020) Technical document. Guide to climate scenario analysis for central banks and supervisors

Nikas A, Doukas H, Papandreou A (2019) A detailed overview and consistent classification of climate-economy models. In: Doukas H, Flamos A, Lieu J (eds) Understanding risks and uncertainties in energy and climate policy. Springer, Cham. https://doi.org/10.1007/978-3-030-03152-7_1

Nordhaus WD (1992) The "DICE" Model: background and structure of a dynamic integrated climate-economy model of the economics of global warming. Cowles Foundation for Research in Economics at Yale University, Discussion Paper Nr. 1009, 1992

Organisation for Economic Co-operation and Development (OECD) (2015) The economic consequences of climate change. OECD Publishing, Paris. http://dx.doi.org/https://doi.org/10.1787/9789264235410-en

Organisation for Economic Co-operation and Development (OECD) (2020) Kazakhstan, in agricultural policy monitoring and evaluation 2020. OECD Publishing, Paris. https://doi.org/10.1787/d3c7bdcf-en

Pollitt H, Mercure JF (2018) The role of money and the financial sector in energy-economy models used for assessing climate and energy policy. Climate Policy 18(2):184–197. https://doi.org/10.1080/14693062.2016.1277685

Smith MM, Bentrup G, Kellerman T, MacFarland K, Straight R, Ameyaw L (2021) Windbreaks in the United States: a systematic review of producer-reported benefits, challenges, management activities and drivers of adoption. Agric Syst 187:103032

United Nations (2018) Handbook on supply, use and input-output tables with extensions and applications. Department of Economic and Social Affairs, Statistics Division, New York

United Nations Development Programme (UNDP) (2019) Building the climate resilience of grain farming in the Northern Kazakhstan

United Nations Development Programme (UNDP) (2020) Summary analytical report on the assessment of economic losses in the agricultural sectors. UNDP project "Development of the Eighth National Communication of the Republic of Kazakhstan under the UNFCCC and preparation of two (fourth and fifth) biennial reports"

West GR (1995) Comparison of input-output, IO-Econometric and CGE impact models at the regional level. Econ Syst Res 7(2):209–227

World Bank (2015) Kazakhstan: nationwide assessment of climate-change related risks and formulation of mitigation strategy. Policy and Institutional Directions for Bolstering Climate Resilience in the Agriculture, Forestry and Energy Sectors. Joint Economic Research Program (JERP). Retrieved on 10 February 2022 from https://openknowledge.worldbank.org/bitstream/handle/10986/22488/Kazakhstan000N0mitigation0strategy.pdf?sequence=1

World Bank (2016) Kazakhstan. Agricultural sector risk assessment. Retrieved on 10 February 2022 from https://documents1.worldbank.org/curated/en/422491467991944802/pdf/103076-WP-KZ-P154004-Box394863B-PUBLIC-ASRA.pdf

Chapter 5
Local Context of Climate Change Adaptation in the South-Western Coastal Regions of Bangladesh

M. Ashrafuzzaman, Carla Gomes, and João Guerra

Abstract This study was conducted in 12 unions of the Shyamnagar upazila in Shatkira District, located in the south-western coastal region of Bangladesh (SWCRB). The inhabitants of the SWCRB are affected by different climate-influenced events such as high-intensity cyclones, saltwater intrusion, sea-level rise, and weather pattern-affected agriculture. This study focused on how the local inhabitants are coping with climate change using multilevel adaptation. A mixed approach of data collection, including quantitative and qualitative data, was followed for both primary and secondary sources. Individual level data collection, key informant interviews, close-ended questions, focus groups, life history of SWCRB residents and workshops were used to understand vulnerability and social perceptions at the local level. The findings indicated that multiple adaptation practices are employed by people in SWCRB, such as rainwater harvesting, plantation of different rice varieties, gardening of indigenous vegetables, and pond sand filtering. However, the construction of multipurpose cyclone shelters along with coastal afforestation contribute towards building resilience in the SWRCB from the socio-economic and environmental perspectives. Therefore, this study would help to find the most adequate strategy towards climate change adaptation and sustainability.

M. Ashrafuzzaman (✉)
Candidate Climate Change and Sustainable Development Policies, Nova University of Lisbon, Lisbon, Portugal
e-mail: md.a@edu.ulisboa.pt

M. Ashrafuzzaman · C. Gomes · J. Guerra
Institute of Social Sciences, University of Lisbon, Av. Prof. Aníbal Bettencourt 9, 1600-189 Lisbon, Portugal
e-mail: carla.gomes@ics.ulisboa.pt

J. Guerra
e-mail: joao.guerra@ics.ulisboa.pt

M. Ashrafuzzaman
Department of Geography, University of Valencia, Valencia, Spain

University of East Anglia, Norwich, UK

Department of Anthropology, University of Chittagong, Chattogram, Bangladesh

W. Leal Filho et al. (eds.), *Climate Change Strategies: Handling the Challenges of Adapting to a Changing Climate*, Climate Change Management,
https://doi.org/10.1007/978-3-031-28728-2_5

Introduction

Adaptation is the process by which local communities, such as those in the south-western coastal region of Bangladesh (SWCRB), deal with new challenges such as climate change. Specifically, it involves a holistic approach to increase resilience towards shocks and risks, which may not consequently cause greater or equally negative impacts on the community (Mistry and Berardi 2016). During the twentieth century, major climate change events took place, including rising sea level, temperature increase, changing rainfall patterns, and reduction of ice and snow (IPCC 2022), causing direct drastic impacts on the global ecosystem, even in the twenty-first centuries.

The concept of adaptation is considerably new in the field of climate change as the mitigation of Greenhouse gases (GHGs) has received significant attention from policymakers and scientists, in the past two decades. Since the Marrakech Accords of 2001, adaptation methods have gained momentum in climate change negotiations (Eguavoen et al. 2015). However, the report from the Intergovernmental Panel on Climate Change (IPCC) on the significance of adaptation stated that adaptation keeps a footmark in diminishing acquaintance and vulnerability to climate change.

Environmental adaptation requires evolutionary progressions, and, for humans, this comprises pre-emptive or sensitive natural transformations. With climate change, for instance, there are features that influence adaptational changes, especially when it comes to socio-ecological arrangements of these changes (IPCC 2022). Adaptation is defined as the process to adjust to real or anticipated climate changes and its impacts, with preparation of damage or gain of useful prospects. In a natural context, adaptation is the procedure of adjustment to real climate and its consequences through human interference (IPCC 2022). Thus, adaptation is an option for the people who will be affected by climate change, such as the inhabitants of the South Asian region (Aryal et al. 2020).

The IPCC working group II's latest report on Impacts, Adaptation, and Vulnerability (IPCC 2022) clarifies different types of adaptation: Incremental or gradual, slow rate adaptation; adaptation that sustains or progresses the crux and veracity of a structure at a specified gauge; transformational or sweeping adaptation; adaptation that changes traits of socio-ecological settings due to climate change; options/choices in adaptation; Accessible tactics and procedures for adaptation focusing on physical, environmental, official, or interactive activities; capacity/scopes of adaptation; for living beings, establishments, and other entities to conform with advantages/disadvantages and impacts of climate change to reduce climatic events, though protection from hazards through adaptive arrangements is tricky to accomplish.

SWCRB, due to its socio-economic and geomorphic location, remains the most vulnerable to sea-level rise (SLR) (Brammer 2014; Hossain and Szabo 2017). By the year 2100, 0.53–0.97 m of SLR is projected across 37 coastal stations, with a global SLR of 0.09–0.88 m (Asirul et al. 2019). Approximately 1.54 million people would face the damaging effects of SLR of 1 m by the year 2070, and ~13 million people would be facing permanent relocation (World Bank 2013). Multiple impacts of climate change would severely affect the poorest nations, especially their food and water sources. It would affect millions of people and consequently harm the economic growth of countries. From an international perspective, Bangladesh is often viewed as one of the most vulnerable nations to climate change (Chen and Mueller 2018); SWCRB would be one of the most affected due to its sensitive hydro-geophysical location and socioeconomic conditions (Akter and Ahmed 2020; Mehzabin and Mondal 2021; Kabir et al. 2021). The region already has a vast number of victims of climate change. At the same time, these regions have become a testing ground for small-scale adaptation measures, which are particularly suitable for small communities.

This study reviews local adaptation strategies along with potential impacts of SLR in the SWCRB (Alam et al. 2017; Asirul et al. 2019). Furthermore, this paper showcases possible sustainable adaptable measures and local adaptation in SWCRB. Its findings are based on a literature review, primary field data collection, focus group discussions (FGDs), case studies, interviews and population surveys. The aim of the study was to observe the local adaptation practices in the SWCRB. The specific objectives were to lineup current adaptation practices and identify sustainable adaptation practices in SWCRB.

Materials and Methods

Study Area

The study area was Shyamnagar upazila situated in Kulna Division in the Satkhira District in Bangladesh. There are 46,592 households in the area, spread across 1968.24 km^2 (Fig. 5.1). The main rivers that surround these areas are Kobadak, Rayamangal, Kholepetua, Kobadak, Arapangachhia, Hariabhanga, Malancha, and Chuna (BPC 2001). Landless farmers accounted for 19%, small farmers accounted for 30%, marginal farmers accounted for 28%, middle-class farmers accounted for 16.5%, and wealthy people accounted for 6.5% among the local farming community; the arable land per capita was 0.13 ha (Cultural Survey Report of Satkhira District 2007).

The SWCRB is a part of the Ganges Delta and is composed of alluvial soil carried by the upstream water. This part mostly consists of coastal wetlands (70% of the landmass) and is connected by a network of rivers in the Bay of Bengal. This coastal region forms the lowest landmass (0–30 cm mean sea level), is part of the delta of the extended Himalayan drainage ecosystem, and is highly vulnerable to multiple threats from climate change, such as hurricanes, storm surges, floods, and tsunamis (Mallick et al. 2017). The Sundarbans mangrove forest, which are adjacent to the world's largest mangrove forest and a UNESCO world heritage site, protect this region from tidal surges (Brammer 2014). Natural calamities such as tidal surges,

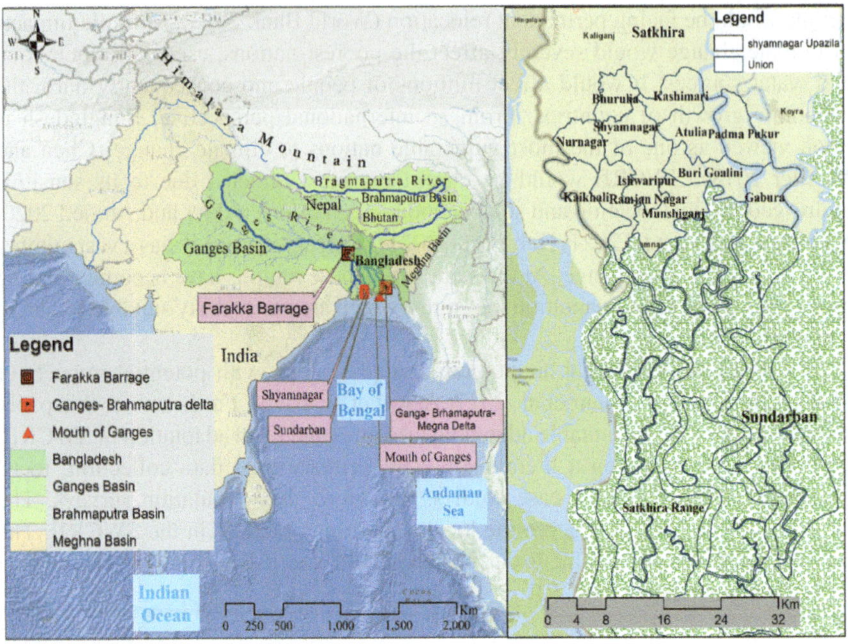

Fig. 5.1 Geographical location of south-western coastal region of Bangladesh

cyclones, land subsidence, and water logging are a common occurrence in this part of Bangladesh, and they play a major role in the lives and livelihood of the people (Roy et al. 2017).

Data Collection

Both primary and secondary data were collected for the study. At Shaymnagar upazila, key informant interviews (Ngo et al. 2019), case studies (Yin 1994), workshops, and 26 FGDs were organized, including members from the local neighborhood and innovators, especially during the field visit; additionally, an open discussion was conducted. A detailed, close-ended questionnaire all-inclusive of economic, social, and environmental aspects of the current adaptation protocol was used for data collection.

Following a participatory approach (Burdon et al. 2019), the community-based climate adaptation practices in the exposed coastal area were identified. We ensured the collection of comprehensive ideas and qualitative information from all types of participants, including under-privileged and vulnerable communities, and government and non-government actors.

Following the literature review and initial desk study, a broad range of participatory rural appraisal tools (Chambers 1994) were used, including detailed interviews, interviews with key informants and FGDs (Morgan 1997; Wibeck et al. 2007). Numerous community meetings with multi-stakeholders were conducted at all the unions, where the stakeholders were asked to present innovative adaptation processes. The FGDs and meetings were on a voluntary basis and held at common and convenient places where the concerned parties were able to discuss issues and express their ideas and concerns independently. The first author used a recorder to record the meeting, as well as before the discussion explained the purpose of the session. All members were encouraged to talk freely and ask questions at any time during the session, and subsequently, individual consent was obtained. In the seasonal livelihood calendar, the age, occupations, income and education of local people and the available livelihood assets were noted. With the consent of the participants, a climate change adaptation checklist was used, demonstrating other benefits. This prompted a guided joint discussion to develop a common understanding of the pressing issues. The members of the community meeting included Union Parishad representatives, local women and school teachers, farmers of both genders, members of the civil society and non-governmental organizations (NGOs), and journalists.

Quantitative Sample Size Determination

A reliable examining approach was employed for the study (Table 5.1). A 95% certainty level was considered to test the example, delegated similarly for every union and appropriated proportionately as per populace. Respondents were selected based on their gender, age, physical challenges (if any), ethnicity, minority status, and economic status.

The testing approach and measurable formula used has been presented below:

Table 5.1 Detail of quantitative sample size distribution

Type of respondents	Union coverage	*N*	*n*	Female numbers	Male numbers	Percentage of youth
Risk and hazard analysis	12	318,254	387	189	198	35.5
Social vulnerability and adaptation analysis	09	242,392	320	98	222	25.5

where *N* is the target population size; *n* is sample size; *e* or the admissible error in the estimate is 5%

$$n = \frac{z^2.p.q.N}{z^2.p.q + (N - 1)e^2} \tag{5.1}$$

whereas,

n = Sample size
N = Target population size
e = Admissible error in the estimate
p = Proportion of defectiveness or success for the indicator
$q = 1 - p$
z = Standard normal variable at the given level of significance.

For ensuring representative sample size from each union, equal sample size was used. Sample size distribution through stratified random sampling is provided in Table 5.1.

Closed-ended questions were used to collect quantitative data from respondents in two levels including 707 random households in a multi-stage sampling design: (1) 387 households surveyed for risk and vulnerability assessment—performed in 12 unions of Shyamnagar upazila; (2) 320 households surveyed to study local adaptation to climate change—performed in nine unions of Shyamnagar upazila.

Limitation of the Study

Our investigation did not reflect the organizational adaptations required by societies in response to climate change. Susceptibilities determined by organizations only replicate on a timescale, reflecting social weakness, creating outlines of adaptive arrangement trade-offs, and prevents flexibility in adaptation. Thus, barriers and restrictions to climate change are socially built, but cannot be rebuilt without effort. However, there will be restrictions to adaptation if the social fluctuations are sluggish in response to climate change, which, with more alterations, creates additional implications. A number of hypotheses reflect barriers and boundaries in adaptation. For example, the financial limits to adaptation (Eisenack et al. 2014) can be understood when the financial adaptation surpasses the expenses from effects. This was not included in our investigation, however. We did include optimistic selections of adaptation for people's reactions and susceptibilities to threats in our systemic investigation.

Concept of Adaptation

Adaptation to climate change is defined as the process of adjusting behavior or financial structure, inclusive of its short- and long-term effects. Adaptation to climate change is the process through which the vulnerable group can lower the impact of climate change on their livelihood and attain maximum opportunities for employment

(Carman and Zint 2020). In 2007, the IPCC defined adaptation as the environment's adjustment, both natural and built, in reaction to the predicted and observed climatic conditions and their negative effects, mitigating the harm or improving the exploitation of useful opportunities. Broadly, adaptation can be categorized as anticipatory, planned, and self-directed (O'Brien et al. 2014). The most recent report from (IPCC 2022) outlines various types of adaptation, as for example: incremental adaptation or gradual and slow rate adaptation: adaptation that sustains the root and veracity of a structure or progression at a specified scale. Transformational adaptation or comprehensive adaptation: adaptation that alters the essential traits of a socio-ecological setting in expectation of climate change and its consequences (IPCC 2022). Adaptation options/choices: the collection of tactics and procedures that are accessible and suitable for focusing on adaptation. They comprise an extensive series of activities that can be characterized as physical, official, environmental or interactive. Adaptive capacity/scopes: the capability of locations, establishments, humans and other entities to fine-tune with likely impairment, to yield benefit of prospects, or to respond to impacts. The adaptation process for the impact of climate change which will be performing the purposes of reducing climatic events but cannot be protected from unbearable hazards through adaptive arrangements (IPCC 2022).

Framework of Adaptation Practices in the South Western Coastal Region of Bangladesh

Anthropological investigation proposes that the most robust and adaptive social units over extended periods may be the domestic ones, such as household circles and extended families, instead of the public or state (Thornton and Manasfi 2010). Households adjust to patterned yet varied ways to the strains of contemporary weather, food, water, and financial emergencies (West 2009). The exclusive ability of families to be involved in procedures of extension and disintegration as defined by ecological constraints is a critical factor in their adaptability.

Smit et al. (2000) offers an applied method for categorizing features of traditional adjustment centered on who/what has to adjust (the scheme of concentration), what they have to acclimate to (the incentive), and how they familiarize (the practices and methods). The scheme of adaptation can differ from an entire system or nation to individuals or species. It may be adjusting to longstanding mean climate inconsistency, climate extravagances, forthcoming climate change, or the risks and opportunities of climate incentives, among other things (Thornton and Manasfi 2010). The adaptation procedure itself can differ in resolution, timing, temporal and spatial range, and arrangement (technological, collective, and influential).

Considering the active and intricate traits of adjustment in the population of SWCRB, a multifold strategy that revolves around eight crucial processes of broad-scale human adaptation was presented. These processes were mobility, exchange, allocating, sharing, variation, strengthening, novelty, and revival (Workshops, 2017–2019). Each procedure has a distinguishing unique motivational base. The above adjustable procedures include local acquaintances, acquiring lessons, and organizations at numerous measures, from the most rudimentary families in SWCRB (Table 5.2). Therefore, adjustable procedures must be assumed as both factors of modification and responses to adjustment in the native setting.

Table 5.2 Local adaptation and behavior in reaction to climate change in societies of SWCRB

Adaptation practice	Narrative	Illustration
Mobility	Periodic movement or perpetual relocation to evade peril or in quest of better conditions	Seasonal, temporary and permanent migration
Interchange	The movement of material and symbolic goods and services between people	Handover of customary information across to generations, e.g., house construction. Providing information for land uses, e.g., farmstead gardening and mangrove plantation (UDMC) (WDMC)
Allocating	Managing the spread or ingesting of limited or acute assets among the people	Recovery relief from Government and NGO
Sharing	Spreading or involving of new livelihood options across social clusters	Allocation of new scientific information and expertise, Local knowledge concerning climate change adaptation
Variation	Diversifying food, earnings production tactics, and specialty to augment livings	Modifying agricultural excellence and profession, such as shrimp agriculture and women entrepreneurship saline endurable rice
Strengthening	Growing the obtainability of assets by increasing their harvest	Individual yields in winter and saline endurable vegetables in
Novelty	New, scheme that arises to deal with a definite necessity	Biofuel for culinary need, and renewable energy practice,
Revival	Systematized reconfiguration of dogmas and performances to decrease strain and generate a more pleasing culture	Utilizing conventional information, association, and skills. Embankment project 1960–1980, 'Sorjan method agriculture; National Adaptation Plan of Action (NAPA) and the Bangladesh Climate Change Strategies and Action Plan (BCCSAP)'

Workshops (2017–2019)

Results

Climate Risk and Vulnerability in South West Coastal Region of Bangladesh's

Disaster risk is defined by the probability of catastrophic consequences, or predicted loss of lives, property, livelihoods, and disruption of the economy (or damage to the environment) as a result of collisions between man-made or natural-induced hazards and vulnerable conditions, such as Cyclones Bhola in 1970, Sidr in 2007 and Aila in 2009 in coastal Bangladesh (Hadi et al. 2021). Hazards materialize into disaster risk when they collide with social, physical, economic, and environmental vulnerabilities. A hazard is an occurrence, a phenomenon or an event with potential to cause death, destruction of environment or loss of property (Haase 2013; Smith 2013).

Vulnerability is defined as the susceptibility to hazard of a community and prevailing factors—socio-economic, political, and physical—that negatively affect the capability to respond to such events (Nur and Shrestha 2017). The combined susceptibility of a community to possible damage by a specific hazard within a particular time-frame is referred to as community-based disaster risk (Davies et al. 2015). Assessment and management of the risk rooted in physical, environmental, and socioeconomic vulnerabilities should be a continuous process (Fekete et al. 2010; Wisner et al. 2014). Disaster risk management measures include mapping and estimating hazards, determining risk elements and their susceptibility to specific hazards, and developing actions for risk reduction (Menoni et al. 2012).

The calculation for risk is presented as in Tate 2012:

$$\text{Risk} = \text{Hazard} \times \text{Vulnerability} \tag{5.2}$$

Additionally, risk is calculated as a part of capacity, vulnerability, and hazard (Van Riet 2009). People dependent on the Sundarban forest for their livelihoods have switched their professions because of the climatic disaster. Numerous negative impacts, especially for coastal communities have been predicted in the aftermath of climate change. Most respondents also face challenges for their livelihoods, such as saline intrusion, tidal surge, heavy and poor rainfall, rapid flooding, resulting in local-based adaptation practices. They have to repair or build a raised, strong pond wall to block saline water, as well as use netting to protect crabs from heavy rainfall and inundation. Flooding by saline water during a storm surge or breach of a coastal embankment affects parts of tidal and estuarine coastal plains. Generally, salt intrusion from flood water penetrates deep into the soil in the pre-monsoon season as the soil is relatively dry (Ahammed et al. 2018).

Figure 5.3 represents the scenario of loss or damage which has been conceived by the respondents during hazards in the last 50 years. Majority of the respondents mentioned that agriculture and fish farming have decreased because of hazards such as salinity intrusion and flash floods in the last 50 years.

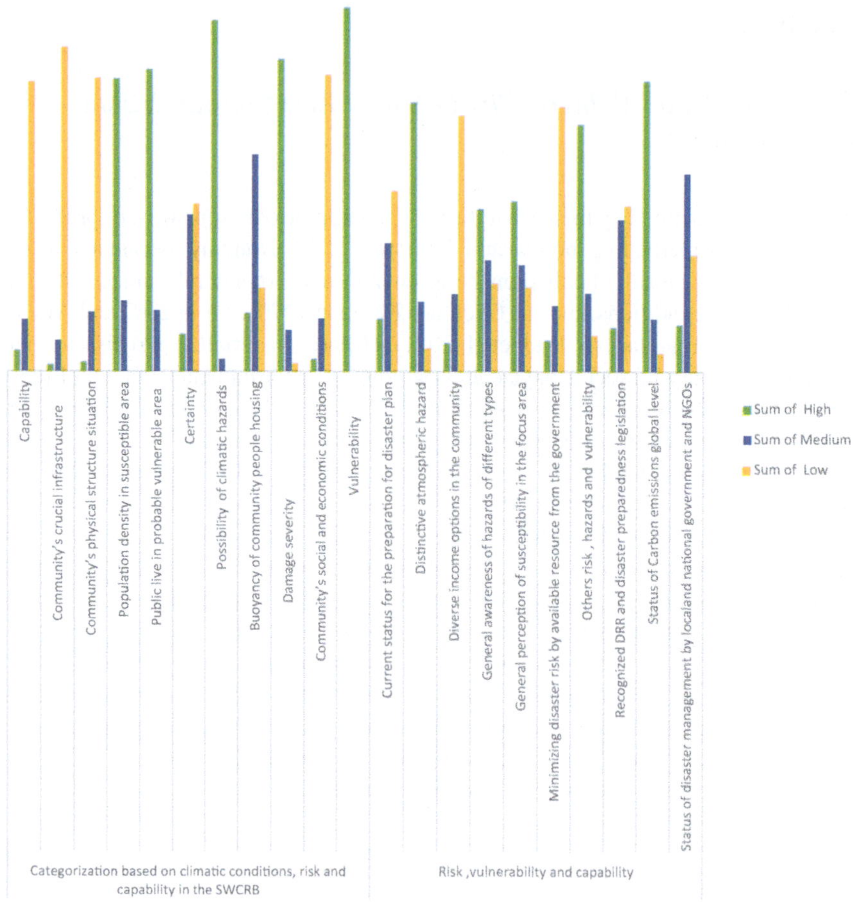

Fig. 5.2 Sorting based on climatic occurrence, vulnerability, risk and ability in the SWCRB ($n =$ 387). From Fig. 5.2, 387 respondents voted each component of the figure to represent sum of high, medium, and low

Adaptation

Confrontation and Preparedness for Disasters

Bangladesh has seen a tremendous improvement in confronting disasters and developing preparedness. However, many are unaware of the interpretation of the warnings. Though the majority of people received early warnings of the storms, 4% were unable to understand the differences in the signals.

Figure 5.4 represents the respondent's perception of the early warning system used in the study area. The early warning system is an essential initiative to alert people about upcoming natural disasters. In this regard, 95% of the respondents were

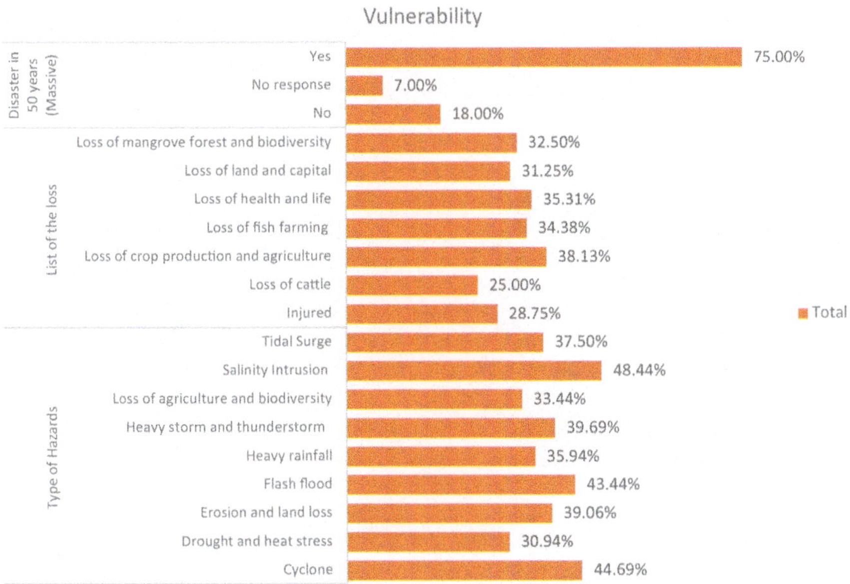

Fig. 5.3 Strong or massive disasters in recent 50 years and types of hazards with loss ($n = 320$)

aware of the early warning system. A total of 61% of the respondents confirmed the reception of early warnings through announcements about the upcoming disaster, while 22% of the respondents mentioned that Cyclone Preparedness Programme (CPP) volunteers play a great role in conveying early warnings about disasters and the mode of action to be undertaken. Besides all these, people get early warnings through radio and television before a disaster occurs. When the respondents have been asked if they understand the signals of climate change, 79% of the respondents responded affirmatively. Most of the respondents indicated that signal seven of the early warning system indicates leaving the house and going to the cyclone center, while 3.8% of the respondents indicated that pregnant mothers, children and elderly people should go to the cyclone center as soon as possible. Approximately 57.8% of the respondents have stated that they went to a shelter center during the disaster period, while 10% of the respondents mentioned staying in their own house. Besides this, 25% and 7.2% of the respondents have indicated that they went to Bandh and to relatives' houses during the last disaster.

Adaptation by Agriculture, Fish Farming and Structural Level

In the SWCRB out of total population, 43% of the population being under the poverty line and is prone to multi-hazards (Islam 2015). Technology like homestead and plinth raising, high pond boundary, harvesting of rainwater, earthen embankment, cultivation of saline tolerant fish, crab fattening, and cultivating salt systems were

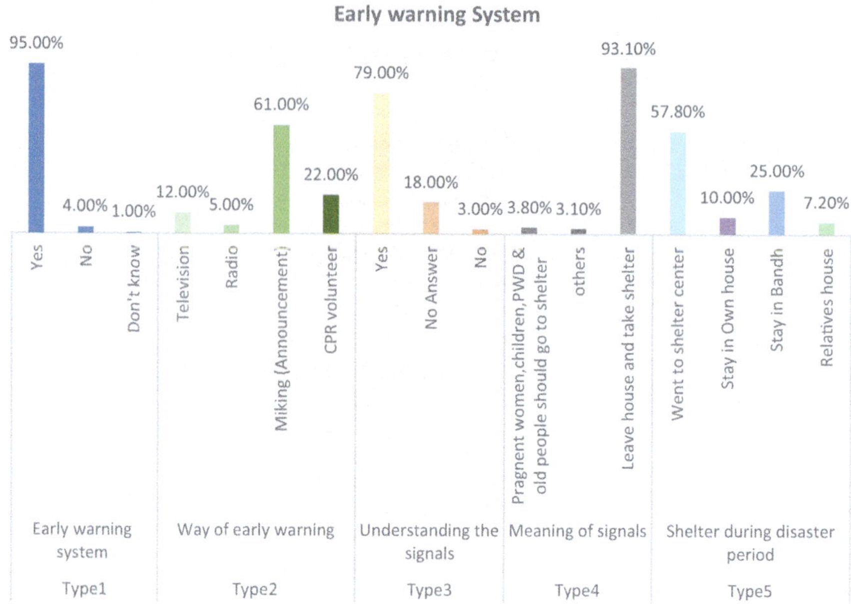

Fig. 5.4 Adaptation through the early warning system ($n = 320$)

also useful and effective ways under the study to reduce climatic risks as well as to ensure climate resilient adaptation practices. As mitigation of the effects is not possible, adaptation to the impacts of climate change is a natural choice for the residents of SWRCB. The existing adaptation practices (Table 5.3) are categorized based on sectors and seasonality.

Interviewees mentioned planting Napier grass, Keora plants from Sunderabans on fallow land, lowlands, and homesteads around farm dikes and also cultivate small-scale fish, especially Baghdad, at the household or community level. They collect immature and soft crabs from the streams in the Sundarbans and raise them in ponds until they reach a marketable size (150 g) and strong enough for livelihood (fish farming). They also grow mile in salt-rich areas. Another adaptation strategy is planting vegetables and spices on the embankments of farmland, shrimp ponds, and homestead fish ponds. At the household level, locals grow gourds, pumpkins, seaweed, sweet pumpkins, and other vegetables in suspended earthen pots and baskets, in addition to growing vegetables on floating beds in flooded areas. People have also restarted raising poultry and ducks to increase household income and meet the protein needs of the family.

People in SWCRB prefer shrimp-fish and prawn-fish polyculture in freshwater and saltwater ponds, respectively, at family and community levels. They also grow Goalpata on the river floodplains or in homesteads around ditches and lowlands. In addition, one of the adaptation strategies is to bypass the flood season and plant rice, especially during the overlap period of kharif-1 and kharif-2 (May to mid-August).

Table 5.3 Local adaptation events that are reducing climatic risk as well as exploring climate-resilient livelihoods in vulnerable communities

Adaptation	Adaptation segment	Participant code/informer under 12 unions
Adapt to salinity	Grass Cultivation, Cage aquaculture, Crab fattening	Participant 33, Participant 34, Participant 35, Participant 36
Adapt to cyclones, storm surges, tides, floods, waterlogging and high salinity, severe erosion in the rainy season	Keora Nursery, Mele (Reed) Cultivation, Cattle Raising, Shrimp-fish mixed cultivation, Growing local rice variety (BR 28), Goalpata Cultivation, Keora Cultivation, Vegetables cultivation on raised mound with concrete wall, Fish vegetables combined cultivation, Eter Paja (Household Level Brick-kiln), Dyke cropping, Hanging vegetables, Hydroponics, Poultry farming, Cropping on raised mound-vegetables, Homestead gardening, Combined cultivation of fish and hanging vegetables, Aman rice and fish (Bagda, Golda, Rui, and katla) combined cultivation	Participant 37, Participant 38, Participant 39, Participant 40, Participant 41, Participant 42, Participant 43, Participant 44, Participant 45, Participant 46, Participant47, Participant48, Participant 49, Participant 50, Participant 51
Alternative income generation, Innovative income generation	Apiculture—Bee keeping and honey, Vegetables cultivation on the banks of homestead fish pond, Vegetables cultivation on homestead yard, Shrimp cultivation at homestead	Participant 52, Participant 53, Participant 54, Participant 55, Participant 56, Participant 57, Participant 58
Adapt to drought	Local rice variety (T-Aus.)	Participant 59, Participant 60, Participant 61, Participant 62, Participant 63, Participant 64
Adapt to the salinity of groundwater and even most surface water bodies	Purification of pond water using govt. supported filter, Purification of pond water using traditional knowledge, Rainwater harvesting in rectangular concrete tank, Rainwater harvesting through hanging canvas while raining, Pond Filter	Participant 65, Participant 66, Participant 67, Participant 68, Participant 69
Structural Adaptation to pests and insects	Gola (a granary, a storehouse for grain)	Participant 70, Participant 71, Participant 72, Participant 73

(continued)

Table 5.3 (continued)

Adaptation	Adaptation segment	Participant code/informer under 12 unions
Structural Adaptation to strong force of cyclone wind and tidal surge, riverbank erosion, coastal flooding and salinity	School cum cyclone shelter, Bamboo made piling, Bamboo cage—to protect riverbanks, Raising plinth	Participant 74, Participant 75, Participant 76, Participant 77, Participant 78, Participant 79, Participant 80, Participant 81, Participant 82
Structural—Adaptation	Lower house roofs for houses with stronger disaster resistance (Low house)-Lower house roofs as much as possible to avoid the destructive power of strong winds in hurricanes	Participant 83, Participant 84 Participant 85, Participant 86, Participant 87, Participant 88, Participant 89, Participant 90, Participant 91, Participant 92
Structural—Adaptation	Earth embankment for protection of salt water and embankment is built by the local community	Participant 93, Participant 94, Participant 95, Participant 96, Participant 97, Participant 98, Participant 99

FGD, Interview, Workshops (2017–2019)

Different vegetable types (such as vegetables, bitter gourd, lady fingers, peppers, broccoli, cabbage, radishes) are grown in the elevated yard with fish in the same swamp. One corner of the plot is raised high enough to avoid being submerged by normal coastal flooding. Raising the plinth of a house is an important measure to respond to flood and tidal surge hazards as it reduces flooding, and allows for an opportunity to prepare compost and start vegetable gardening and poultry farming.

In response to salinity-vulnerable hazards, fattening the crab is another original livelihood adaptation procedure for the inhabitants of exposed SWCRB. The owner of the crab fattening facility can meet the needs of their own consumption, ensuring the nutritional security through own resources. It also provides an opportunity for extra earnings in the local market due to the high demand for these products. The lime-treated pond walls prevent saline intrusion and store rainwater, which improves the environment around the area and creates an opportunity for tree plantation. An example of alternative livelihood adaptation is found in the mono-sex Tilapia fish, as it responds well to salinity-prone climatic hazards of the SWCRB with its higher tolerance to salinity compared with other fish. Thus, it is a beneficial fish to farm, and the raw materials required for this are locally available throughout the year. Furthermore, one production of fish cultivation only takes three to four months and is continued all year round, only being restricted by water availability (Fig. 5.5).

Ninety-eight respondents use harvested rainwater for drinking and household activities. The deep and shallow water tube-well was not a successful approach due to excessive salinity. A considerable segment of households in this region purchases drinking water. Considering this, harvesting rainwater is an important ingenious livelihood option to ensure safe drinking water availability in coastal areas. Water

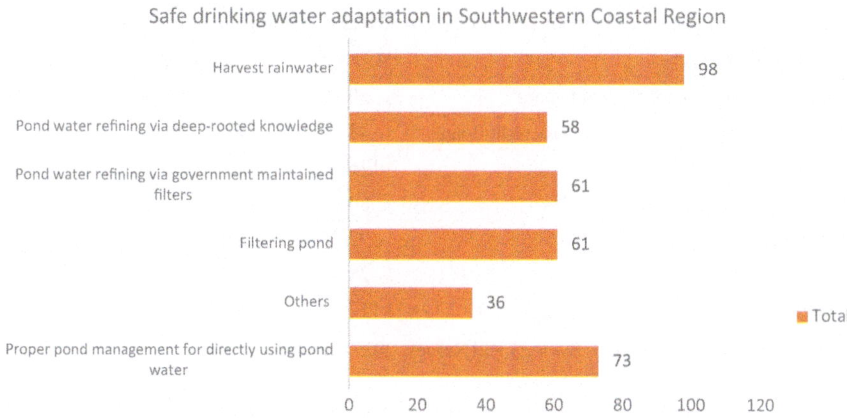

	Proper pond management for directly using pond water	Others	Filtering pond	Pond water refining via government maintained filters	Pond water refining via deep-rooted knowledge	Harvest rainwater
Total	73	36	61	61	58	98

Fig. 5.5 Existing practices for safe drinking water adaptation ($n = 387$)

collected during the monsoon is used only for drinking and is preserved for the next five months to use through the drought season. Seventy-one respondents used pond water through proper pond management, including heightening the pond bank, removing unnecessary trees and plants from the pond.

Sixty-one respondents use pond water refining via government-maintained filters. Through these filters, pond water can be easily purified for drinking and other household activities. Additionally, 61 respondents use filtering ponds for their household drinking water needs. Fifty-eight respondents use pond water for their daily household activities via deep-rooted knowledge of multi-layer water purification.

Due to different climatic disasters, complicated hydro-geological conditions, and adverse water quality, the exposed areas and islands and the coastal belt is identified as a crisis area with water shortage. The saline intrusion from the tidal surge causes severe problems for drinking and domestic use of water. Water from nearby ponds, canals, ghers and rivers is mainly used for domestic purposes while pond is the source of potable water. To prevent salt water ingress and coastal flooding, the height of the pond boundary has been raised by around one meter (Table 5.4).

Adaptive Alternatives in Relation to Adverse Situations

Occupational changes: People dependent on the Sundarban forest for their livelihood had to change their professions due to catastrophic destruction within the Sundarban forest (Mozumder et al. 2018). Simultaneously, people working in the

Table 5.4 Local adaptive alternatives in relation to adverse situations in the societies of SWCRB

Adaptive alternatives	Count[a]	Percentage of responses
Financial management for health care	50	15.6
Alternative adaption through microcredit /loan	88	27.5
Occupational changes	176	55
Migration as an alternative process	80	25
Social networking	92	28.8
Savings	28	8.75
Others	112	35

[a]Multiple Response: $N = 320$, *Source* Field Survey, 2017–2019

profession of farming shrimp needed diversification of their income due to destruction of farms and with no possibility to recover their capital. As a consequence, workers were relocated to cities nearby as they were displaced (Ahmed et al. 2019).

Financial management for health care: Respondents resorted to adaptive measures for alternative means to fulfil their healthcare through diverse sources of finance, including loans, selling ornaments and cattle and alternatives such as traditional medicinal practices, self-medication, and discussion with friends and families. According to Chowdhury et al. (2020), the reduction of capacity for decreasing vulnerabilities is correlated to informal and formal forms of economic, technical, social, cultural, and physical obstacles. Therefore, it can be stated that these factors are hindering the ability to adapt to climate change, and this is also consistent with the findings of the research.

Migration as an alternative process: Migration is common after cyclone disasters, especially in the middle-income groups, with there being at least one migration per family. The low-income groups receive donations from both government and non-government organizations, removing the need for migration. Seasonal migrations have increased, as respondents are more frequently moving to other towns and villages. There, the work consists of physical labor, such as working in the rice fields or making bricks. This migration is mainly temporary, though it is important to consider the probability of permanent migration. For example, it is common for at least one family member to migrate to earn money for their survival, though the family, generally, relies on them staying in the village. Permanent migrations, however, have increased and are reflecting an influx of people from coastal villages to towns for survival. These migrations are results from both economic reasons and scarcity of land. Furthermore, migrations do not only reflect people moving from one village or town, but also within their own village. Natural disasters also create lifestyle changes for villagers when they are forced to move from agricultural professions to fishing. For instance, villagers mainly catch fish from the Sundarban, but the over-use of this natural source has forced the government to implement policies for a safer ecosystem. Thus, the inland fishing is mainly controlled by the rich, as the villagers are forced to move to coastal farming. Furthermore, villagers face a decline in food, work, and life necessities, as their land has been lost to saline water or erosion, resulting in migration.

Alternative adaption through microcredit: Through the FGD, Interview and workshops we came to know that in the poor socioeconomic status area like SWCRB is where the microfinance (MFI) comes in, as it is a financial service that provides support (i.e. small loans, insurance, savings accounts) to customers that do not have access to traditional financial services, mainly due to poverty. Microfinance, thus, play an important role in climate change, as it has a broad delivery infrastructure in the country with a good reputation for reliable service delivery, and can support climate change activities. For instance, certain microfinance promotes innovative and sustainable livelihood solutions, and can run vocational training, and awareness building programs. This helps build human and social capitals, and increases resilience of local communities. Moreover, households and communities can make decentralized decision, as there is high volume and limited values of the services by microfinance. microfinance also promote social inclusion and equity for women, making microfinance important catalysts for community-based adaptation in Bangladesh. Thus, providing microfinance to people who are most vulnerable to the impacts of climate change will enhance the adaptations for local communities. Microfinance can help these vulnerable communities in Bangladesh by creating climate sensitive schemes and mainstream climate change and disaster risk reduction in their operations. For instance, providing different insurances, such as agricultural insurance, can help provide financial options to combat climate-induced loss and damage.

Besides, few respondents mentioned that 70% of all households, especially those afflicted by damage, took loans, including microcredits offered by NGOs to allow them to go back to their earlier professions. The repayment system was observed to be too strict. Only 10% of people were capable of surviving on their savings and assets prior to the disaster.

Structural Adaptation

Sixty-two percent of the respondents answered affirmatively for preparedness against hazards at a household level (Fig. 5.6). Dry food collection (26%), moveable woven, savings, collecting firewood, tiding house with bamboo (15%) and rope (16.9%), first aid box collection, tree plantation (13.1%) were adopted as preparedness at the household level. The mitigation steps at the household and community level involve mangrove plantation (29.19), renovating embankments (27.5%), improving sluice gates and river canals of coastal areas (18.1%), and using renewable energy (9.7%).

Other adaptation measures include schools with cyclone shelters, domestic brick kilns, granary (warehouse), goalghor (cow shed), bamboo piles (with bamboo fences), raising the base/plunch, lowering the house roofs, and fencing with bamboo, bran, and Gab tree fences (Fig. 5.7).

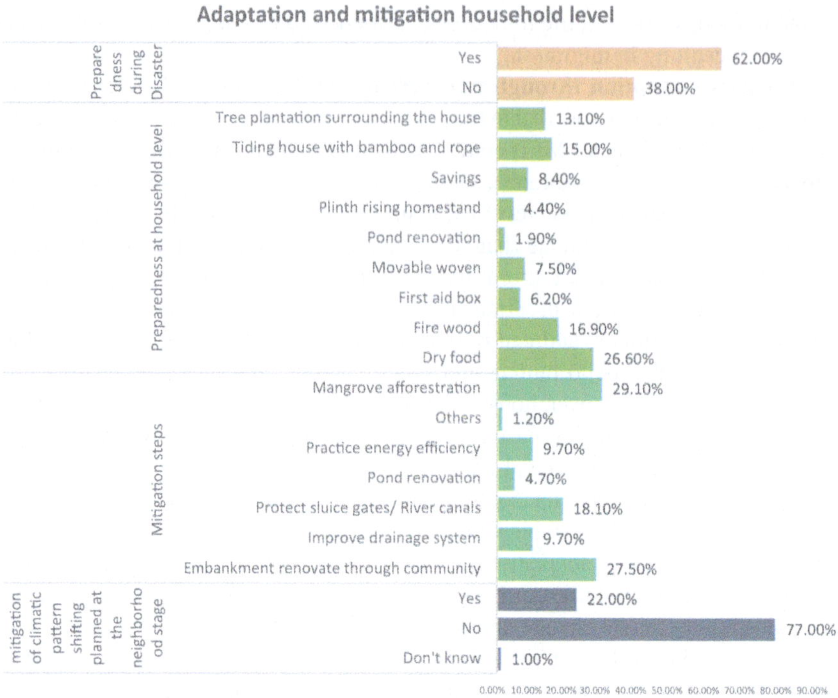

Fig. 5.6 Adaptation and mitigation household level ($n = 320$)

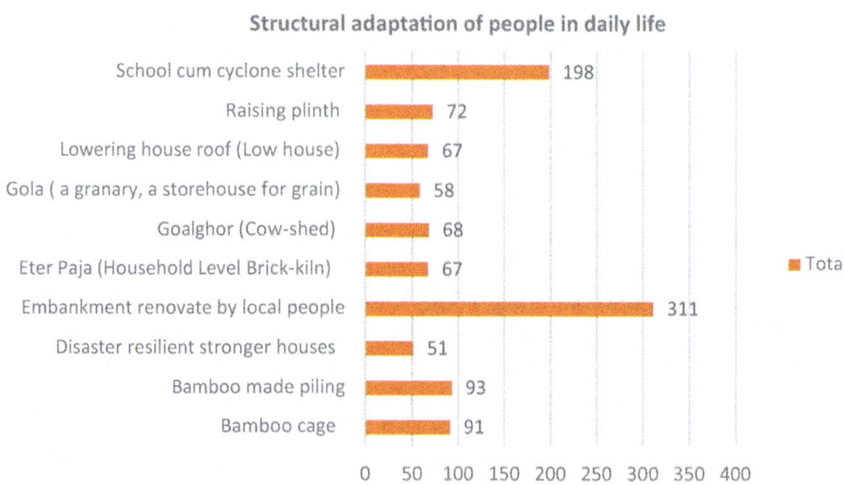

Fig. 5.7 Structural adaptation of people daily life ($n = 387$), Multiple response

Fig. 5.8 **a** A recently built embankment after cyclone Amphan through community people for coastal flood protection; **b** a multipurpose cyclone shelter center; and **c** coastal mangrove arboriculture

Coastal Embankment Rehabilitation

Prime adaptation approach against flooding and cyclonic storm surg is the development of earthen embankments along the rivers and also community people are improving the parallel to the coastline for benefit agricultural production and flood protection (MoEF 2015). The community people are increasing the height of the embankments after the cyclone Amphan and which was important for reduce salinity, safe drinking water and health. Community people are developing plantations to strengthen the sloping face of the embankment under the social forestry regime (Fig. 5.8).

Construction of a Multipurpose Cyclone Shelter

In Bangladesh, multipurpose cyclone shelters are planned to accommodate 2,000 people with various facilities, including heightened plinths, toilet installation on higher grounds, tube well installation, food storage, separation of washrooms on the basis of gender, and segregation of livestock (Asirul et al. 2019).

Bank Protection

Bank protection includes seawall and revetment to protect against coastal erosion (hard procedures; (Zaman and Mondal 2020; Crawford et al. 2020), which requires long-term maintenance and huge investment.

Adaptation Through Coastal Afforestation

Recently, flood mitigation measures in the coast zone have been enhanced by increasing plantation (190,000 ha mangroves till 2010 to 192,395 ha in 2013; 8690 ha of non-mangrove, 2873 ha of goalpata, and 12,127 km land strip) in the coastal area and also along the embankments by the Bangladesh Forest Department (MoEF 2015; Islam and Rahman 2015). The actual area under coastal plantation now stands at 61,574 hectares due to destruction (BFD and UNDP 2018).

One of respondent from Gabura Union said that Plantation prevents erosion on the coast, traps sediment, lowers the speed of the wind and water velocity, in addition to improving forest cover, alleviating degradation to the environment, serving as an efficient carbon sink, offering shelter and breeding ground for wildlife and fisheries, serving as source of livelihood, and adding to the aesthetic value of the region.

Mainstreaming Community-Based Adaptation

The process of mainstream adaptation revolves around a coordinated and systematic effort across stakeholders and institutions, as well as the integration of development plans on a local level for the purpose of reducing vulnerabilities caused by climate change and promoting adequate adaptation strategies (Roy 2018).

The Bangladeshi government has taken all the steps necessary to facilitate the implementation of mainstreaming throughout all the sectors (DoE 2015), including capacity-building, sprained awareness, advocacy, and orientation at various levels of operation (Seddik0079 et al. 2020), and community-based adaptations (Roy 2018).

Community-Based Adaptation

CBA encompasses the needs, knowledge, capacities, and priorities to strengthen people's resolve to prepare and endure the effects of climate change (Reid et al. 2009). CBA encompasses many factors and is involved in participatory programs to improve, strengthen, and reduce vulnerabilities of livelihoods to improve resilience to climate-related disasters. It also involves disciplines such as ecosystems, livelihoods, food security, and infrastructure. This approach acknowledges cultures and societies through climatic impacts and environmental knowledge (Seddiky et al. 2020) to empower inhabitants to be able to take their own decisions and implement them. Therefore, it is a community-centric partnership that exists between communities and institutions or stakeholders and is an approach approved by many organizational stakeholders (Ensor et al. 2014). Certain initiatives taken at the local level include local radio, youth empowerment, awareness through inclusion in local curriculum, early warnings, alternative livelihoods, afforestation and reforestation schemes, coastal resilience through ecosystem functions, promotion of salt-tolerant varieties, renovating boats and houses, potable water at low prices, and clean energy production. Satkhira Unnayan Sangstha, a regional NGO undertakes activities such as reed farming, promoting indigenous rice varieties, implementing watercourse cropping, commercialize salt-tolerant grass, harvest rainwater, providing artificial aquifers, establish social networks, empowering women, and increasing public awareness (FGD, Workshops, interview-2017–2021).

Institutional Adaptation

Priority areas for investment in coastal areas of Bangladesh include cyclone shelters, coastal level projects, coastal green belt projects, Sundarbans biodiversity conservation projects, different disaster management projects, construction of dykes, development of polders, and cyclone shelters (MoEF 2009). Bangladesh developed a National Coastal Zone Policy in 2005 to coordinate coastal zone policies across ministries, departments, and agencies to coordinate their activities. The CZP proposed an institutional framework, recommended the protection of the coastline from soil erosion, flooding, and storm surges through a seaside embankment, which was later adopted by Bangladesh's National Adaptation Action Plan (MoWR 2005). In addition, the Water Resources Planning Organization implemented a project to develop an Integrated Coastal Zone Management Plan for 2002–2006 to develop capacity and prioritize actions for coastal development in Bangladesh (WARPO 2005).

To address the issue of climate change vulnerability, the skills and practices of the local community are of prime importance in Bangladesh. The Bangladesh National Adaptation Program of Action and Bangladesh Climate Change Strategies and Action Plans (BCCSAP) have recognized areas affected by the climate and recommended urgent and immediate actions, but there has been a lack of initiative at the primary level to address this issue (MoEF 2009). Greater awareness in the poor and vulnerable communities, access to information and knowledge, engagement motivation, ability and skill to build local action for adaptation, use of proper technology and institutional support to address the issue at local and regional level can be used to improve awareness. Climate smart policies and investments toward making the country more resilient to the effects of changing climate conditions, including loss of property, habitat, and infrastructure, are required for adaptation (Ensor et al. 2014).

Figures 5.9 and 5.10 provide the during and after adaptation measures provided by the government and the types of initiatives taken. 75% and 40% of the total respondents have received assistance and first aid, respectively, from the government, while 80% received relief after the period.

A total of 87% of the respondents were also aware of the adaptation and mitigation initiatives taken by the government and the NGOs. These initiatives include renovating embankments, tree plantation, providing safe drinking water, and use of renewable energy.

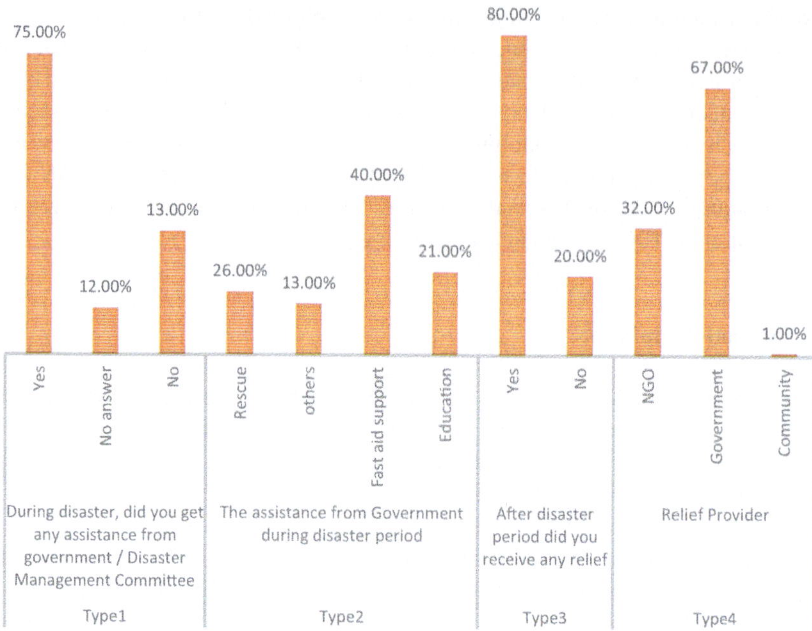

Fig. 5.9 During and after disaster adaptation through government and NGO ($N = 320$)

Public Anticipation of Their Upcoming Adjustment Mechanism

Figure 5.11 depicts the inhabitant's anticipation of the SWCRB's upcoming adjustment mechanism. These include raising the elevation of the ridge (embankment), availability of pure drinking water, rainwater collection system, provision of pond sand filters to ensure uncontaminated drinking water, availability of government hospital, obtainability of agricultural apparatus, soil fecundity by exploitation of organic fertilizer, planting salt-tolerant trees were some of the anticipated adjustment measures.

Discussions Through FGD, Workshop, Interview, and Case Study

The coastal region of Bangladesh has been vulnerable to cyclones, floods, tidal waves, tornados, water logging, salinity, downpour, drought, river erosion, hail, and earthquakes. A depression in the Centre of the Bay of Bengal turns into a cyclone and assaults the coastal belt of the country, leading to storm and tidal wave inundation.

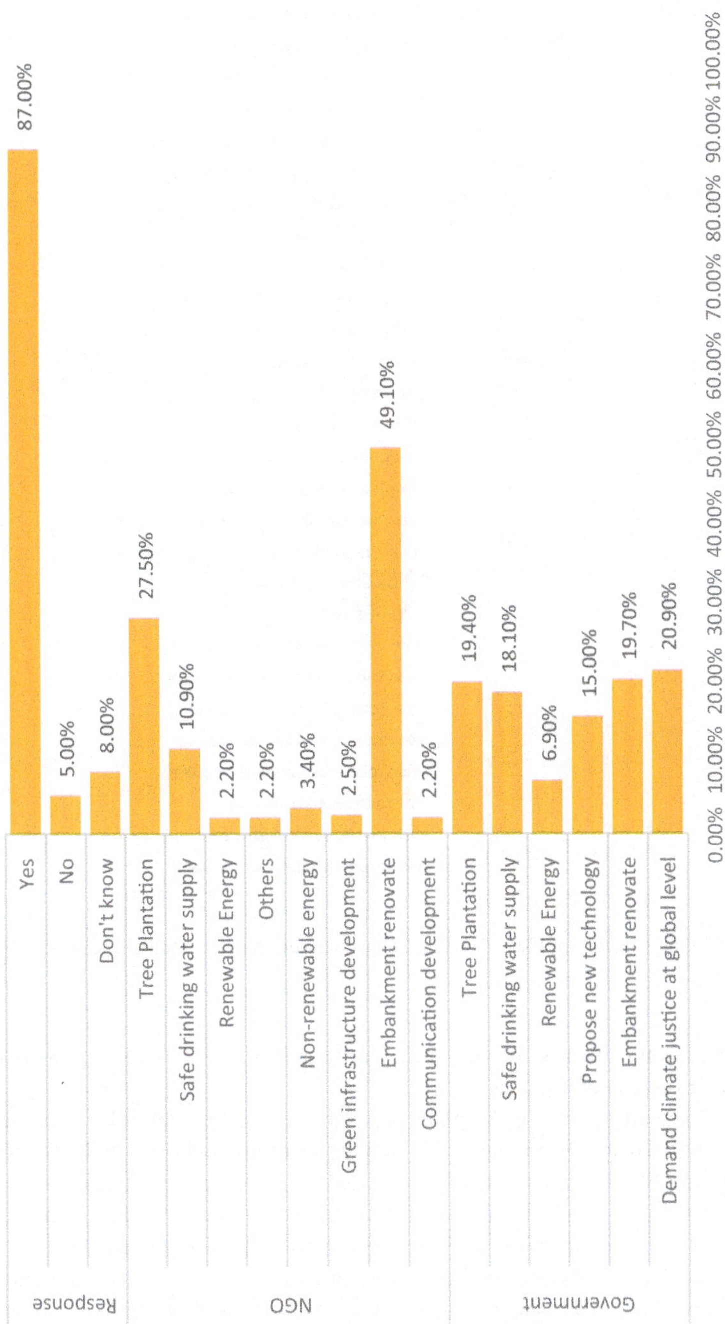

Fig. 5.10 Types of adaptation and mitigation initiatives taken by NGOs and government ($N = 320$)

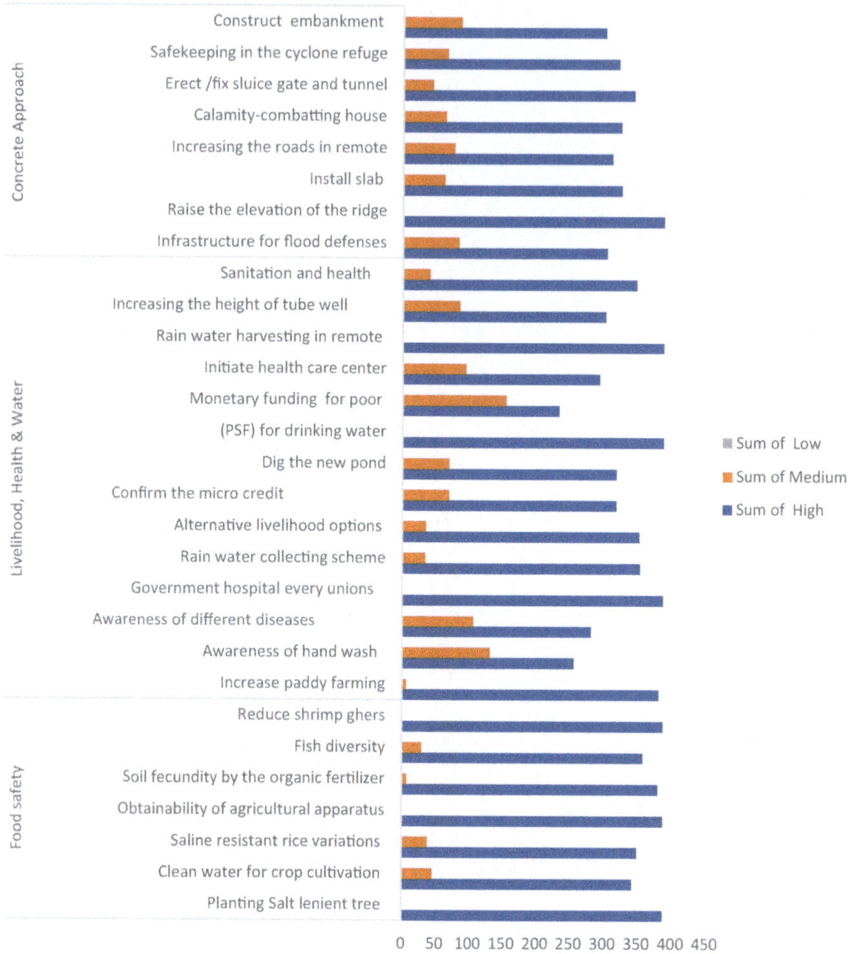

Fig. 5.11 Public anticipation of their upcoming adjustment mechanism, multiple response ($N = 387$)

When people are not able to cope with the hazards and calamities, they migrate from the coastal regions to other places.

Due to the assault of cyclone Amphan, the dam made by the Water Development Board got destroyed and the village of Golakhali, near the Sundarban under Shyamnagar Upazila of Shatkhira district suffered extensive flooding. People rescued from the flood were taken to the shelter although this has become a periodical phenomenon. The income has also reduced due to the fish and crab farms being inundated.

The coastal area of Satkhira had been experiencing strong winds and rain due to the impact of cyclone Amphan. The rivers alongside the Sundarbans coast become turbulent, and the tidal water also increased. Warnings were issued throughout the country and the inhabitants were brought to the shelter. More than 200,000 people took shelter in 1645 shelter homes in Satkhira, including the 145 shelters in Shyamnagar. The rest of the people took refuge in various safe places, such as schools and mosques. The four worst cyclone-affected upazilas of Satkhira were Shyamnagar, Asashuni, Kaliganj, and Satkhira Sadar. A constructed embankment and tube well could not prevent the flooding, and destruction of houses and power lines.

Garbage, filth, and stench surrounded the inflicted areas. The biggest problem in coastal areas was the lack of drinking water due to submergence of freshwater ponds and tube wells in salt water. Water from tube wells was saline and arsenic-contaminated. Flooding has also resulted in unemployment. Potable water was provided by NGOs at the corners of every village, which was still insufficient to meet needs. Adaptation measures to face these challenges include renovating the drinking water ponds, building dam, and assistance for the embankment from the local people.

Funding was provided for construction of houses, additional grain production, and animal husbandry. The Water Development Board was also working to protect the banks of the river by building dams to prevent floods, install culverts and Swiss gate control systems for irrigation, and flood preparedness, providing awareness and repeated warnings. During the catastrophes, the inhabitants of the coastal region took refuge at the shelter centers, even though the participants did not initially pay heed to the message about the potential danger and the need to go to the shelter.

During the catastrophe, the government, Union disaster management committee (UDMC), Ward disaster management committee (WDMC), volunteer organizations, the defense force, and local people helped in the evacuation of people to shelters. The ministry also kept 1,500 medical teams ready with abundant water purification tablets and oral saline. As a precaution, the local government engineering department and the public health engineering department cancelled leaves of absence for all staff in the 13 upazilas of the coastal region. The government employees were instructed to not leave their workstations and were also directed to maintain communication with the administration. The local government engineering department opened up a monitoring cell to take effective action after the storm to repair roads and culverts and to keep them moving.

The Army was deployed to remove the trees that fell into roads, obstructing traffic and smoothening out the process after the cyclone Bulbul hit Shatkhira. Medical and rescue work was also carried out, while providing emergency relief and medical services. The Atulia Union Disaster Management Committee in 2018–2019 supplied tank for rainwater, sand filter renewal and tube well installation, supplying micronutrient powder to children between 6 and 23 months, and training programs for skill enhancement. Disaster risk reduction includes providing training on disaster reduction, organizing pot music sessions, repairing the muddy roads, digging pinds and canals, improving social accountability, and improving education, health, family planning, agriculture, fisheries and livestock, disaster management, social welfare, sewerage development, settling family disputes, and women and child development. UNICEF also undertook activities including providing nutrition knowledge, training on water and sewerage, child protection including awareness about the dangers of early marriage for girls under the age of 18.

Analysis

The concept of adaptation methods is still relatively new in the field of climate change. An important role in adaptation is played by the local adaptation approach to reducing vulnerability to climate change by using early warning systems as an immediate coping strategy. Several studies that have looked into local-level adaptation to climate change in Bangladesh (Rahman and Alam 2016; Rabbani et al. 2018; Hoque et al. 2019) identified that the practice of this knowledge helps the community decide about ways to respond to changes in the environment and ways to adapt and improve their life conditions. Local knowledge has also proved important and complements warning systems. Paul and Routray (2013) mentioned that people could predict storm surges from a drop in temperature without waiting for the weather signal.

Few studies have identified practices and countermeasures to overcome the sea level rise variability and change-related challenges. Examples of such are: floating vegetation in water-logged conditions (Dedekorkut-Howes et al. 2020), constructing a platform that floats during a flood for cattle refuge, constructing earthen embankments with the support of bamboo fence and grass plantation to reduce water erosion of the soil (Afroz et al. 2016), introducing better drainage to prevent saline encroachment into the soil, and rescheduling cultivation time in an effort to avoid the flood (Hussain et al. 2019). It was observed that the native rice varieties have the potential to withstand salinity, drought, tidal surge and water logging, which is very much expected under the climate change scenarios regimes (Nahar 2019).

Another local-based adaptation involves elevating a large area of land to solve the serious drainage congestion and waterlogging caused by the sedimentation of major rivers. Integrated farming of crops, fish and livestock, cultivating salt-tolerant varieties plays an important role in increasing food production and income. Raising cultural awareness, creating earthen embankments to prevent saline water intrusion, preparation of vulnerability map, swimming lessons for children, raising awareness of climate change through Bangla folk songs are other measures. Mitigation measures including sluice gates, cyclone shelters, and embankments are vital to lowering disaster risk at the community level. Providing financial resources also improves hazard preparedness in addition to means like self-employment, vocational training, and micro-business opportunities. In addition, several initiatives including architectural and other measures are in place to lower the vulnerability to disaster in this area. For example, a research study by Alam and Chowdhury (2010) Enhanced earthen stoves in SWCRB which are significant for the novelty of socio-economic change in Shyamnagar Upazila. The innovative earthen stove is an upgraded and altered version of the traditional stove that is primarily used by the villagers. Innovative stoves are made with good mud soils in the kitchen or outside the kitchen. Tiny-price ingredients and simple technology are useful to make it safe, easy, low-priced and operative. SWCRB has been one of the pioneers in successfully implementing local-based climate change adaptation.

Conclusion

This study addressed two major topics related to adaptation to climate change: vulnerability to climate risks and adaptation strategies, both state-led and locally-based. In the perspective of risk and vulnerability people residing in the south-western coastal region of Bangladesh are anguished by extreme climatic perils, which gave rise to enormous hindrances for their livelihoods, life choices and socio-economic development. Community engagement in adaptation is recognized as an effective way to reduce climate change vulnerability, especially SLR vulnerability, and increase community resilience worldwide. The impacts of climate change occur primarily at the local level, so local adaptation practices are important for proper resilience; thus, adaptation plans and strategies should be managed and implemented locally. It is essential that the vulnerability of areas be considered while devising these plans.

In a nutshell, we identified a number of local adaptation practices that have been employed by vulnerable populations in response to climate change, which are influenced by the societal culture of SWCRB, such as agility, interchange, allocation, sharing, variation, magnification, novelty, and revival are the sturdy connections to the effective attainment of SDGs (Table 5.2). In this study we evaluated the perils through the qualitative and quantitative method, and we observed that people of the SWCRB through numerous plans, activities, approaches, strategies to reduce the consequences of risks, the costs of perils and vulnerability, have taken countless effort. The adaptation approaches include early warning systems, saline tolerant agriculture, rainwater harvesting, secured drinking water adaptation, raising homestead and plinth, salt-tolerant fish farming, heightening pond boundaries, structural adaptation and increasing community resilience. Adaptive substitutes are implemented in connection with opposing circumstances, structural adaptation in domestic setting and public domain. Adaptation initiatives comprise modernizing the drinking water filter, construction of dams, and community people engagement during the catastrophe for the construction of embankments.

A range of adaptation initiatives are undertaken by diverse governmental organizations and NGOs with the aid of state and global organizations like respondents have acknowledged, e.g., distribution of drinking water facilitated by NGOs. Throughout the calamity, the government, disaster management committee, volunteer organizations, the defense force, and local people aid in the evacuation of individuals to shelters which is also a commendable adaptation tactic in the local setting. In Fig. 5.11, 387 respondents selected imminent adaptation choices as a adaptation limit such as increasing the elevation of the ridge (embankment), obtainability of unadulterated consumable water, and accessibility of public hospital. Further, for houses with lack of capital, both governmental and non-governmental agencies could provide micro credit and training to farmers on adaptation strategies and skill development.

References

Afroz S, Cramb R, Grunbuhel C (2016) Collective management of water resources in Coastal Bangladesh: formal and substantive approaches. Hum Ecol 44(1):17–31
Ahmed I, Ayeb-Karlsson S, van der Geest K, Huq S, Jordan JC (2019) Climate change, environmental stress and loss of livelihoods can push people towards illegal activities: a case study from coastal Bangladesh. Climate Dev 11(10):907–917
Ahammed SJ, Chung ES, Shahid S (2018) Parametric assessment of pre-monsoon agricultural water scarcity in Bangladesh. Sustainability 10(3):819
Alam GM, Alam K, Mushtaq S (2017) Climate change perceptions and local adaptation strategies of hazard-prone rural households in Bangladesh. Clim Risk Manag 17:52–63
Alam SN, Chowdhury SJ (2010) Improved earthen stoves in coastal areas in Bangladesh: economic, ecological and socio-cultural evaluation. Biomass Bioenerg 34(12):1954–1960
Akter S, Ahmed KR (2020) Insight and explore farming adaptation measures to support sustainable development goal 2 in the southwest coastal region of Bangladesh. Environ Dev Sustain. https://doi.org/10.1007/s10668-020-00778-y
Asirul HM, Habibur R, Dilara R (2019) An evaluation of sea level rise vulnerability and resilience strategy to climate change in the coastline of Bangladesh. Int J Environ Sci Nat Res 18(2):555983
Aryal JP, Sapkota TB, Khurana R, Khatri-Chhetri A, Rahut DB, Jat ML (2020) Climate change and agriculture in South Asia: adaptation options in smallholder production systems. Environ Dev Sustain 22(6):5045–5075
Burdon D, Potts T, McKinley E, Lew S, Shilland R, Gormley K, Thomson S, Forster R (2019) Expanding the role of participatory mapping to assess ecosystem service provision in local coastal environments. Ecosyst Serv 39:101009
BFD and UNDP (2018) Co-management and benefit-sharing from coastal afforestation. Integrating Community-Based Adaptation into Afforestation and Reforestation (ICBA-AR) Programme in Bangladesh. Bangladesh Forest Department, Ministry of Environment, Forest and Climate Change and UNDP Bangladesh, Dhaka
(BPC) Bangladesh Population Census (2001) Bangladesh bureau of statistics. https://in.top10place.com/shyamnagar-upazila-1066643260.html
Brammer H (2014) Bangladesh's dynamic coastal regions and sea-level rise. Climate Risk Manage 1:51–62
Carman JP, Zint MT (2020) Defining and classifying personal and household climate change adaptation behaviors. Glob Environ Chang 61:102062
Chambers R (1994) Participatory rural appraisal (PRA): analysis of Experience. World Dev 22(9):1253–1268
Crawford TW, Islam MS, Rahman MK, Paul BK, Curtis S, Miah M, Islam M (2020) Coastal erosion and human perceptions of revetment protection in the Lower Meghna Estuary of Bangladesh. Remote Sensing 12(18):3108
Chowdhury MA, Hasan MK, Hasan MR, Younos TB (2020) Climate change impacts and adaptations on health of Internally Displaced People (IDP): an exploratory study on coastal areas of Bangladesh. Heliyon 6(9):e05018
Cultural survey report of Satkhira District (2007) Cultural survey report of upazilas of Satkhira District 2007. http://wikimapia.org/14577704/Shyamnagar-Upazila-HQ
Chen J, Mueller V (2018) Coastal climate change, soil salinity and human migration in Bangladesh. Nat Clim Chang 8(11):981–985
Davies T, Beaven S, Conradson D, Densmore A, Gaillard JC, Johnston D, Milledge D, Oven K, Petley D, Rigg J, Robinson T (2015) Towards disaster resilience: a scenario-based approach to co-producing and integrating hazard and risk knowledge. Int J Disaster Risk Reduction 13:242–247
Dedekorkut-Howes A, Torabi E, Howes M (2020) When the tide gets high: a review of adaptive responses to sea level rise and coastal flooding. J Environ Planning Manage 63(12):2102–2143

DoE (2015) Fifth national report of Bangladesh to the convention on biological diversity. Department of Environment (DoE), Ministry of Environment and Forests, Government of the People's Republic of Bangladesh, Dhaka

Eguavoen I, Schulz K, De Wit S, Weisser F, Müller-Mahn D (2015) Political dimensions of climate change adaptation: conceptual reflections and African examples. In: Leal Filho W (ed) Handbook of climate change adaptation. Springer-Verlag, Berlin, pp 1183–1199

Ensor J, Berger R, Huq S (2014) Community-based adaptation to climate change: emerging lessons. Practical Action Publishing, Rugby, UK

Eisenack K, Moser SC, Hoffmann E, Klein RJ, Oberlack C, Pechan A, Rotter M, Termeer CJ (2014) Explaining and overcoming barriers to climate change adaptation. Nat Clim Chang 4(10):867–872

Fekete A, Damm M, Birkmann J (2010) Scales as a challenge for vulnerability assessment. Nat Hazards 55(3):729–747

Haase D (2013) Participatory modelling of vulnerability and adaptive capacity in flood risk management. Nat Hazards 67(1):77–97

Hadi T, Islam MS, Richter D, Fakhruddin BS (2021) Seeking shelter: the factors that influence refuge since Cyclone Gorky in the Coastal Area of Bangladesh. Prog Disaster Sci 11:100179

Hossain MS, Szabo S (2017) Understanding the socio-ecological system of wetlands. In: Prusty BAK, Chandra R, Azeez PA (eds) Wetland science: perspectives from South Asia. Springer, New Delhi, pp 285–300

Hussain MS, Abd-Elhamid HF, Javadi AA, Sherif MM (2019) Management of seawater intrusion in coastal aquifers: a review. Water 11(12):2467

Hoque MZ, Cui S, Lilai X, Islam I, Ali G, Tang J (2019) Resilience of coastal communities to climate change in Bangladesh: research gaps and future directions. Watershed Ecol Environ 1:42–56

IPCC (2022) Climate change 2021: the physical science basis. Contribution of Working Group I to the Sixth Assessment Report of the Intergovernmental Panel on Climate Change [Masson-Delmotte V, Zhai P, Pirani A, Connors SL, Péan C, Berger S, Caud N, Chen Y, Goldfarb L, Gomis MI, Huang M, Leitzell K, Lonnoy E, Matthews JBR, Maycock TK, Waterfield T, Yelekçi O, Yu R, Zhou B (eds)]. Cambridge University Press, in press

IPCC (2007) Climate change 2007: synthesis report. Cambridge University Press, Cambridge, UK, Intergovernmental Panel on Climate Change

Islam MN (2015) Community based adaptation to climate change in the exposed coastal areas of Bangladesh. In: Proceedings of the 5th international conference on water and flood management (ICWFM-2015) (pp. 591–598)

Islam SA, Rahman MM (2015) Coastal afforestation in Bangladesh to combat climate change induced hazards. J Sci Technol Environ Inform 2(1):13–25

Kabir AM, Amin N, Roy KM, Hossain S (2021) Determinants of climate change adaptation strategies in the coastal zone of Bangladesh: implications for adaptation to climate change in developing countries. Mitig Adapt Strat Glob Change 26(7):1–25

Mallick B, Ahmed B, Vogt J (2017) Living with the risks of cyclone disasters in the south-western coastal region of Bangladesh. Environments 4(1):13

Menoni S, Molinari D, Parker D, Ballio F, Tapsell S (2012) Assessing multifaceted vulnerability and resilience in order to design risk-mitigation strategies. Nat Hazards 64(3):2057–2082

Mistry J, Berardi A (2016) Bridging indigenous and scientific knowledge: local ecological knowledge must be placed at the centre of environmental governance. Science 352(6291):1274–1275

MoEF (2009) Bangladesh climate change strategy and action plan. Ministry of Environment and Forest, Government of Republic Bangladesh, Dhaka

MoWR (2005) Coastal zone policy. Ministry of Water Resources, Government of People's Republic of Bangladesh

MoEF (2015) Intended Nationally Determined Contributions (INDC). Ministry of Environment and Forest (MoEF), Government of the People' Republic of Bangladesh, Dhaka

Mozumder MMH, Shamsuzzaman MM, Rashed-Un-Nabi M, Karim E (2018) Social-ecological dynamics of the small scale fisheries in Sundarban Mangrove Forest, Bangladesh. Aquac Fish 3(1):38–49

Morgan DL (1997) Planning and research design for focus groups. Focus Groups Qual Res 16(10.4135):9781412984287

Mehzabin S, Mondal MS (2021) Assessing impact of climate variability in Southwest Coastal Bangladesh using livelihood vulnerability index. Climate 9(7):107. https://doi.org/10.3390/cli9070107

Ngo DP, Le MD, Tran NLD, Wassmann R, Sander BO (2019) Key informant analysis for climate smart agriculture practices in Tra Hat Village, Vietnam. CCAFS Working Paper No. 280. CGIAR Research Program on Climate Change, Agriculture and Food Security (CCAFS), Wageningen, the Netherlands. www.ccafs.cgiar.org

Nahar N (2019) Assessment of storm surge induced vulnerability and changing adaptation strategies in the south-west coast of Bangladesh (Doctoral dissertation, University of Dhaka)

Nur I, Shrestha KK (2017) An integrative perspective on community vulnerability to flooding in cities of developing countries. Procedia Eng 198:958–967

O'Brien K, Eriksen S, Inderberg TH, Sygna L (2014) Climate change and development. In Climate change adaptation and development: transforming paradigms and practices, p 273

Paul SK, Routray JK (2013) An analysis of the causes of non-responses to cyclone warnings and the use of indigenous knowledge for cyclone forecasting in Bangladesh. Climate change and disaster risk management. Springer, Berlin, Heidelberg, pp 15–39

Rahman MH, Alam K (2016) Forest dependent indigenous communities' perception and adaptation to climate change through local knowledge in the protected area—a Bangladesh case study. Climate 4(1):12

Rabbani G, Munira S, Saif S (2018) Coastal community adaptation to climate change-Induced salinity intrusion in Bangladesh. In: Agricultural economics—current issues. IntechOpen

Roy R (2018) Evaluating the suitability of community-based adaptation: a case study of Bangladesh. In: Handbook of climate change communication, vol 1. Springer, Cham, pp. 39–59

Roy K, Gain AK, Mallick B, Vogt J (2017) Social, hydro-ecological and climatic change in the southwest coastal region of Bangladesh. Reg Environ Change 17(7):1895–1906

Reid H, Alam M, Berger R, Cannon T, Huq S, Milligan A (2009) Community-based adaptation to climate change: An overview. Participatory Learn action 60(1): 11–33

Smit B, Burton B, Klein RJT, Wandel J (2000) An anatomy of adaptation to climate change and variability. Clim Change 45:223–251

Smith K (2013) Environmental hazards: assessing risk and reducing disaster. Routledge, UK

Seddiky MA, Giggins H, Gajendran T (2020) International principles of disaster risk reduction informing NGOs strategies for community based DRR mainstreaming: the Bangladesh context. Int J Disaster Risk Reduction 48:101580

Tate E (2012) Social vulnerability indices: a comparative assessment using uncertainty and sensitivity analysis. Nat Hazards 63(2):325–347

Thornton TF, Manasfi N (2010) Adaptation–genuine and spurious: demystifying adaptation processes in relation to climate change. Environ Soc 1(1):132–155

Van Riet G (2009) Disaster risk assessment in South Africa: some current challenges. South African Rev Sociol 40(2):194–208

Wibeck V, Dahlgren MA, Öberg G (2007) Learning in focus groups: an analytical dimension for enhancing focus group research. Qual Res 7(2):249–267

World Bank (2013) 40 turn down the heat: climate extremes, regional impacts, and the case for resilience. The World Bank, New York

WARPO [Water Resources Planning Organization] (2005) Project completion report of Integrated Coastal Zone Management Plan (ICZMP), Dhaka

West CT (2009) Domestic transitions, desiccation, agricultural intensification, and livelihood diversification among rural households on the Central Plateau, Burkina Faso. Am Anthropologist 111(3):275–288

Wisner B, Blaikie P, Cannon T, Davis I (2014) At risk: natural hazards, peoples vulnerability and disasters. Routledge, UK

Yin R (1994) Case study research: design and methods, 2nd edn. Sage Publishing, Thousand Oaks, CA

Zaman S, Mondal MS (2020) Risk-based determination of polder height against storm surge Hazard in the south-west coastal area of Bangladesh. Prog Disaster Sci 8:100131

Chapter 6
Vulnerability Assessment on Agriculture in East Nusa Tenggara

Mariana Silaen, Yudiandra Yuwono, Cynthia Ismail, Amanda Ramadhani, and Takeshi Takama

Abstract Despite the region's vulnerability, climate change adaptation strategies in poor contexts such as Asia are limited. A vulnerability assessment was conducted in one of the poor areas of Indonesia, East Nusa Tenggara (NTT) Province. The study demonstrates the future of the NTT agricultural sector, focusing on estate crops, particularly coffee. The assessment applied a three-domain function of exposure, sensitivity, and adaptive capacity based on (1) climate data and land suitability; (2) the commodity production data; and (3) socioeconomic, policy, and institutional context, respectively. This study utilises climate data from as the exposure and the land suitability projection (the sensitivity) was obtained from modelling by Matlab and ArcGIS. As a semi-arid region, the viability of NTT's agricultural lands is projected to decline due to increasingly prominent climate-change impacts. The result showed divergent temperature rises and precipitation patterns across the islands. Consequently, the suitable land would gradually decline for all estate crops in the region; hence, bringing these farmers out of vulnerability requires multi-prolonged climate-resilient agricultural practices. In addressing climate change, this study gives a holistic recommendation to increase the adaptive capacity by strengthening extension support, crops diversification, institutional change, added-value-chain, and the co-finance should be considered in the poor context like NTT province.

Introduction

Climate change impacts have become more severe across Indonesia, particularly in rural areas' economic activities and environmental changes. Because the area is an archipelago, the Indonesian population mostly lives along the coastlines, including Sumatra, Java, and Nusa Tenggara Islands. the coastal communities are at high

M. Silaen · Y. Yuwono · C. Ismail (✉) · A. Ramadhani · T. Takama
Sustainability & Resilience (Su-Re.Co), Bali, Indonesia 80351
e-mail: CynthiaJuwita.Ismail@autonoma.cat

C. Ismail
Department of Geography, Universidad Autònoma de Barcelona, 08193 Barcelona, Spain

W. Leal Filho et al. (eds.), *Climate Change Strategies: Handling the Challenges of Adapting to a Changing Climate*, Climate Change Management,
https://doi.org/10.1007/978-3-031-28728-2_6

risk of suffering from negative effects and natural disasters, where flood, landslide, tornado, and drought were the top four disasters (Kuswanto et al. 2019). The shift in weather patterns is the main cause of worsening droughts and wildfires in the last two decades, negatively impacting the agricultural sectors and food security (Mayer et al. 2011). Climate change has exacerbated the vulnerability of East Nusa Tenggara at an alarming rate because it is island province and one of the major crop producers in Indonesia.

East Nusa Tenggara, or *Nusa Tenggara Timur* (NTT), is located within the lesser Sunda Island in the south-eastern part of Indonesia (see Fig. 6.1). This province includes more than 1,100 islands, with the largest islands being Flores, Sumba, Timor, and Alor. With over 5.3 million people, the province is divided into 21 districts *(kab.)* and one autonomous city (BPS NTT 2021). Climatically, NTT is categorised as a semi-arid region that experiences four months of wet season (January–March and December) and eight months of dry season with moderate rainfall (BPS 2018). This climate shift significantly impacts the livelihood of the NTT population, particularly farmers. In 2020, the labour force reached more than 2.8 million people, around 51% of whom was absorbed by the agriculture, forestry, and fishery sector and contributed 28.51% to the province's gross regional domestic product (GRDP) (BPS NTT 2021). However, recent data places NTT as the third poorest province nationally after Papua and West Papua, with 21% living under the poverty threshold (BPS NTT 2021). Although agriculture is the highest contributor to industrial, trade and service sectors, agriculture's GRDP has decreased slightly since 2016. With rising concern about NTT's agriculture, this paper entails vulnerability assessment focusing on coffee crop combined with a sustainable livelihood framework to assess the projected impact and adaptive capacity required for NTT.

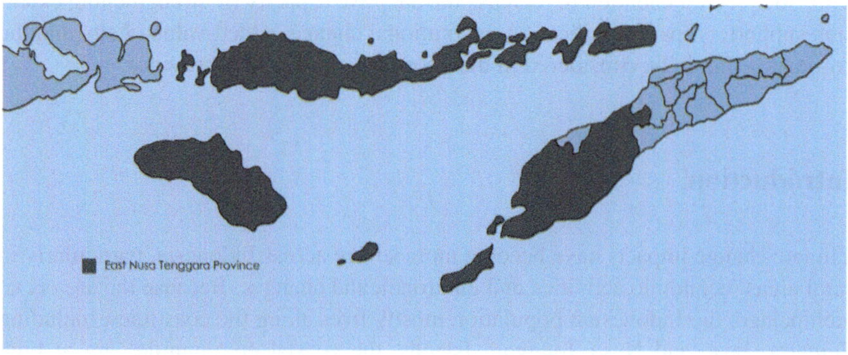

Fig. 6.1 Map of East Nusa Tenggara

Materials and Method

Vulnerability Concepts

Climate change adaptation has two concepts of vulnerability, encompassing outcome vulnerability and contextual vulnerability (IPCC 2007; O'Brien 2011; Takama et al. 2016). The perspective of outcome vulnerability is the overall effect of climate change, its impacts, and adaptive capacity. Meanwhile, contextual vulnerability assumes the main influence of potential characteristics of issues, context, purpose, and system. This study is based on the outcome vulnerability framework defined by the IPCC's Fourth Assessment Report (AR4), which has been re-interpreted in the Fifth Assessment Report of the IPCC (AR5) with an alternative approach and terminology. However, since the updated framework tends to approach the disaster risk concept, this study opted for the AR4 definition.

Vulnerability, according to AR4, is "…the degree to which a system is susceptible to and unable to cope with, adverse effects of climate change, including climate variability and extremes" (IPCC 2007: 6). Based on this definition, vulnerability is further defined as a function of three domains, comprising exposure, sensitivity, and adaptive capacity.

$$\text{Vulnerability} = \text{f(exposure, sensitivity, adaptive capacity)}$$

As the natures and the measures of these domains differ, each shall be assessed separately. For instance, exposure must be explored by considering the climatic analysis, including the current, historical, and projected climate change and variability and locations of the potential climate variations. Meanwhile, sensitivity can be investigated through the susceptibility level of the condition in an interference, such as agricultural productivity, cropping pattern input use, income inequality, or even employment migration (IPCC 2007). In terms of adaptive capacity, this domain is defined as the ability of humans or communities to adjust or cope with climate change variability.

Framework of Adaptive Capacity

There are different frameworks to assess adaptive capacity: the sustainable livelihood framework based on assets (DfID 1999), paired with institutions and political contexts that influence their availability and mobility (Smit and Wandel 2006). This framework is chosen as it encompasses both a bottom-up (livelihood of the farmers) and top-down (support from the government is considered from policy and institutional context) perspective (Nelson et al. 2009) to assess adaptive capacity. The livelihood assets comprise:

- Human capital encompasses skills, knowledge, and support labour activities to achieve livelihood objectives, such as health and education, as these two factors contribute to poverty.
- Natural capital comprises the natural resources that are beneficial and essential to achieving the livelihood objective, such as the atmospheric environment and biodiversity.
- Physical capital is the basic infrastructure and producer goods that can help achieve the livelihood objective. Infrastructure is the modification of the existing environment to help support the welfare; meanwhile, the producer goods include the equipment or tool that people use to support their functionality.
- Financial capital is the financial resources to reach the aim of livelihood and contribute to consumption and production, with two main sources: (a) available stocks, such as bank deposits, livestock, and jewellery and (b) regular inflows of money, such as salary and pension.

Vulnerability Assessment

Exposure is defined as changes in temperature and precipitation between current conditions and the future projection. Climate data and facts as part of exposure were collected from literature and modelling activities. The modelling was conducted with a downscaled WRF model and post-processed Matlab and ArcGIS. At its core, climate data includes the average value of the maximum and minimum temperature and precipitation. The climate data utilises the gridded data from Meteorological, Climatological, and Geophysical Agency of Indonesia (BMKG). The results for each sample point are then processed using categorised symbols into unique values and grouped into five ranges (very low, low, moderate, high, and very high). The reference period used in ArcGIS modelling is 1979–2004.

The model has three scenarios: the baseline scenario, RCP 4.5 and RCP 8.5. The baseline scenario ranges between 1970 and 2004, where the activities flow in line with the business as usual. The RCP 4.5 and RCP 8.5 scenarios present the projection of three different time ranges, encompassing near future (2010–2039), mid-future (2040–2069), and far-end (2070–2099), as explained in (Table 6.1).

The second component, sensitivity, is translated through literature review and land suitability maps of the crop to climate variables derived from the FAO (1981) framework, presented in baseline condition and at the projected time frame. The suitability will be divided into four categories: S1 (highly suitable), S2 (moderately suitable), S3 (marginally suitable), and N (not suitable). In addition to four livelihood assets, the adaptive capacity of the farmers will also explore policy context and existing institutional support from local, national, and international aspects. These dimensions of adaptive capacity will be scoured through literature reviews and national data statistics.

Table 6.1 Characteristics of RCP 4.5 and RCP 8.5 scenarios (Riahi et al. 2011; Thomson et al. 2011)

	RCP 4.5	RCP 8.5
Scenario type	Stabilisation scenario	Failed mitigation scenario
Simulation	Stabilisation to 4.5 W/m^2 at stabilisation after 2100	Rising radiative forcing pathway up to 8.5 W/m^2 in 2100
Assumptions	Existence of effective climate policies	Rapid population growth, low-income rate, inadequate technological shift and energy development, long-term energy demand, increasing GHG emissions

Table 6.2 Comparison of productivity of key crops in NTT to national level

Commodity	Production area (ha)	Productivity in NTT (ton/ha)	National average productivity (ton/ha)	Average production growth per year in NTT (%)
Coffee	73,251[a]	0.325[c]	0.608[d]	6.62[b]

Source [a] (BPS NTT 2019a), [b] (BPS NTT 2020), [c] (BPS NTT 2019b) [d] (BPS 2020)

Context: Current Conditions of Agriculture Sector in NTT

Agriculture has a big role in driving economic activity of NTT. The top five economic contributor of 2020 are staples crops production (7.75% to the province GRDP), livestock services and production (9.89%), fishery (5.75%), horticulture crops (2.2%), and estate crops production (2.34%) (BPS NTT 2021). Ninety-nine percent of staple crops and 82.5% of horticulture crops produced were consumed locally. On the contrary, estate crops such as coffee, cacao, cashew, and coconuts account for almost 70% of exported commodities of the combined agriculture, forestry, and fishery industries. Estate crops are one of leading subsectors that absorbs the majority of the smallholder farmers in NTT. However, East Nusa Tenggara lags far behind the estate crops' average productivity compared to other productive provinces in Indonesia, as shown in (Table 6.2). Therefore, coffee was selected to be investigated in the region in correlation with the climate change impacts.

Coffee

Coffee is one of Indonesia's major plantation commodities. It grows on almost all islands in Indonesia and involves smallholder farmers. As one of the main coffee crop regions, NTT holds 57% of the national coffee production area with more than fifty-one thousand coffee farmers (DFAT 2016). From 2016 to 2019, the coffee production in NTT increased up to 6.62% per year and continued to grow every year (BPS NTT 2021). There are two types of coffee grown, (1) Robusta that is mainly grown at low altitudes and (2) Arabica that is grown at high altitudes and mountainous areas. Among other regions, Flores Island, particularly East Manggarai, and Ngada regency are the biggest Arabica coffee producing areas, with more than 84% of the province coffee production. Thus, it is internationally known as *kopi flores*, traditionally managed without artificial agrochemicals (BPS NTT 2021). However, due to lack of knowledge and limited access to technology for Good Agricultural Practices (GAP) and post-harvest handling, the coffee productivity in NTT is lower than the national level, as shown in Table 6.1 (DFAT 2016). Therefore, coffee farmers there are still dependent on traditional cultivation methods resulting in low quality and yields are highly vulnerable. The climate also potentially jeopardises the production quality and quantity, causing many coffee-growing regions to become unsuitable for coffee plantations.

Result and Discussions

Exposure

Observed Climate in NTT

Indonesia has a balanced number of months for both wet and dry seasons, but the southern regions like NTT have a longer dry season. Based on 1981 and 2010 historical data from Global Precipitation Climatology Centre (Schneider et al. 2011), NTT annual average precipitation is approximately 1,696 mm, 36% lower than the national average rainfall (Fig. 6.2). Meanwhile, Vijendra K. Boken et al. (2005) previously estimated that Sumba and Eastern Flores have relatively longer dry seasons, lasting 25 dekads (one dekad equals 10 days).

While the area experiences a longer dry season, the precipitation varies across the region during the wet and dry season (see Fig. 6.3). Higher precipitation during the wet season is mostly identified at the western part of Flores Island, the northern part of West Timor, the northern area and some eastern parts of Sumba Island.

The temperature also varies across the islands. The maximum temperature of NTT areas, such as Sumba Island, Ende, and some parts of West Timor, has reached more than 35.3 °C (see Fig. 6.4. Climate in NTT 1970–2004 (3) average maximum temperature (left); (4) average minimum temperature (right).). The maximum and minimum temperature maps also show that the southern coasts of each island, the smaller islands, and narrower lands experience relatively lower temperature variability, which is from 26.3 to 31.5 °C. The opposite occurs on the midlands and northern coasts, where the temperature range is larger, around 15.7° difference. Sikka and East Flores district have the lowest precipitation and temperature variability during all seasons (see Table 6.3. Climate in NTT 1970–2004 Precipitation and Average Temperature level according to District). Due to longer dry season and

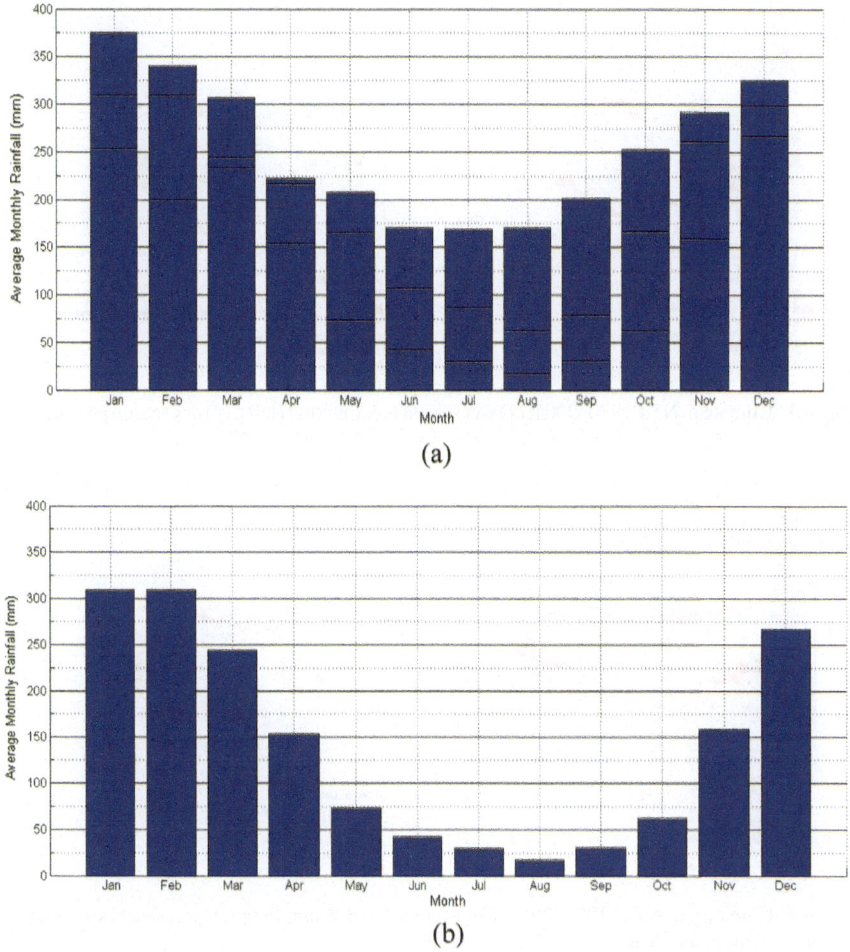

Fig. 6.2 Indonesia **a** and NTT **b** monthly average rainfall in 30 years between 1981 and 2010

low precipitation, NTT could lead to severe drought; difficulties in access to clean water and agriculture leading to crop failure, reduced income, food shortage, and potentially malnutrition and other diseases.

Projected Climate in NTT Province

Echoing the ArcGIS findings above, most areas in NTT will reach even higher maximum temperatures if RCP 4.5 and 8.5 scenarios occur, as demonstrated in Fig. 6.5. In the observed climate, the moderate temperature is around 32–34.2 °C; whereas through the projected climate, most areas may experience 35 °C in both

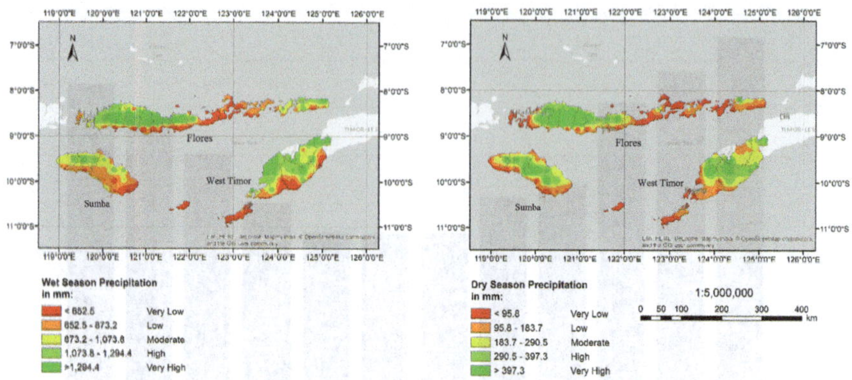

Fig. 6.3 Climate in NTT 1970–2004: (1) wet season precipitation (left); (2) dry season precipitation (right)

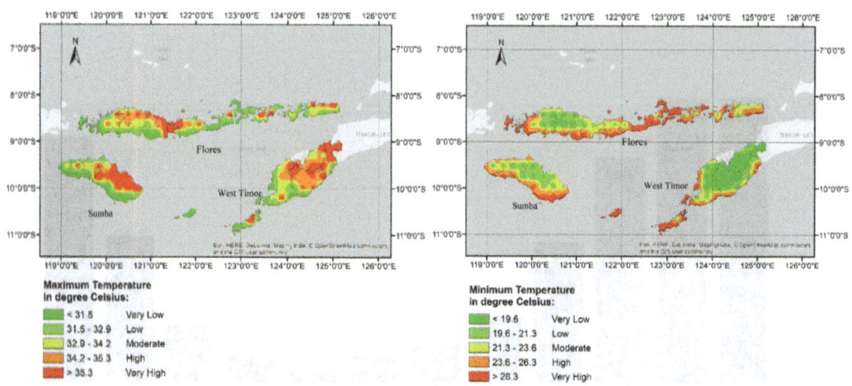

Fig. 6.4 Climate in NTT 1970–2004 (3) average maximum temperature (left); (4) average minimum temperature (right)

scenarios for 2040–2069, with Belu overshooting the 40 °C mark. As shown in Fig. 6.5. Projection of maximum temperature in NTT (2040–2069) the maximum temperature is even higher if RCP 8.5 occurs, except Southwest Sumba.

The same ArcGIS modelling study also claims that a divergent rainfall pattern may occur in NTT. It is shown in Fig. 6.6. Projection of precipitation trends in NTT that rainfall projections vary depending on districts. Overall, the result implies that the rainfall is projected to be much lower in scenario RCP 8.5 than RCP 4.5. Therefore, based on the highest maximum temperature and lower precipitation projection, some districts, such as South Timor and East Sumba, are highly susceptible to drought. An increase in rainfall is likely to happen in Manggarai, East Manggarai, and Sikka, hence, floods and landslides are likely to occur in those areas.

The projection shows the declining precipitation in NTT in both wet (December–January–February, Fig. 6.7) and dry (June–July–August, Fig. 6.8) seasons. Note that

Table 6.3 Climate in NTT 1970–2004 precipitation and average temperature level according to district

Category	Wet season precipitation	Dry season precipitation
Very low	Sikka, Flores Timur, Rote Ndao, TTS	Sikka, Flores Timur, Rote Ndao, TTS
Low	Lembata, Sumba Timur	Lembata, Alor, Kupang
Moderate	Alor, Sumba Barat, TTU	Sumba Barat, Timur Tengah Utara (TTU)
High	Manggarai Barat	Manggarai Barat, Sumba Timur
Very high	Manggarai, Ngada, Ende, Kupang	Manggarai, Ngada, Ende
Category	**Max temperature**	**Min temperature**
Very low	Flores Timur	Kupang, TTU, TTS
Low	Ngada, Sikka, Alor, Rote Ndao	Manggarai Barat, Manggarai, Ngada, Sumba Timur
Moderate	Manggarai Barat, Sumba Barat, Kupang	Ende, Alor, Sumba Barat
High	Manggarai, TTS	Lembata
Very high	Ende, Lembata, Sumba Timur, TTU	Sikka, Flores Timur, Rote Ndao

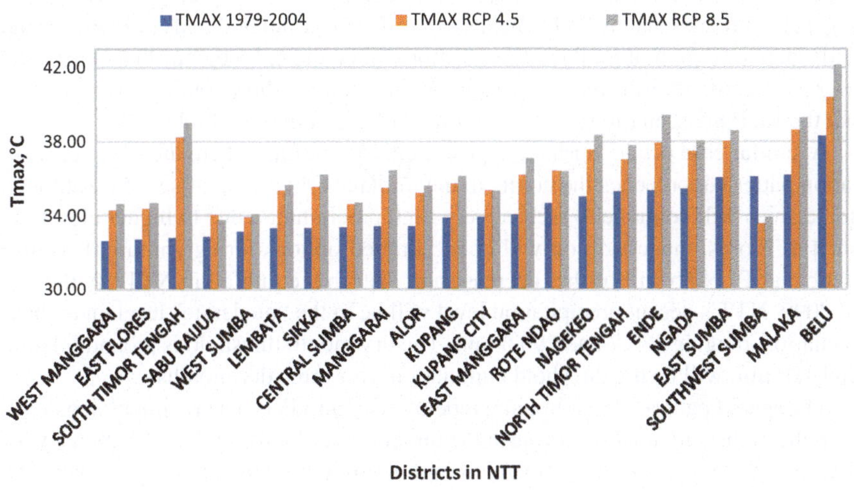

Fig. 6.5 Projection of maximum temperature in NTT (2040–2069)

the predominant pattern of wet season precipitation decrease begins from the southern coast of Sumba, Flores, and West Timor Island before approaching the midlands. The island region of Kupang, Rote Ndao, East Flores, and Lembata is already considered drier than other parts of NTT due to the geographical locations. If the decline in precipitation continues, the amount of arable land left will also diminish. In Fig. 6.7a, by 2039, only the northern part of Flores and West Timor will be left in the midland of Sumba as the regions with higher precipitation. Fig. 6.7c shows that by 2069,

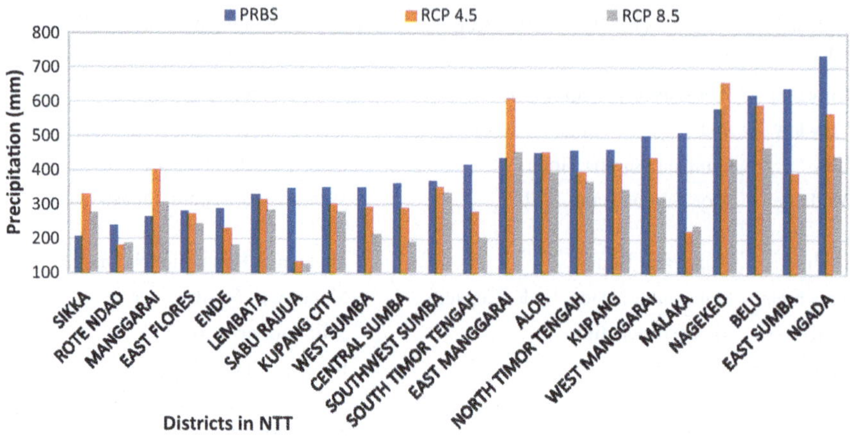

Fig. 6.6 Projection of precipitation trends in NTT

only the northern part of Flores Island and lesser West Timor northern coasts and a small area in Sumba Island will experience moderate to high precipitation. Finally, Fig. 6.7e indicates that by 2099, the lesser north coasts and midlands of Flores Island and a narrow region of midland West Timor will be the only regions that experience average precipitation during wet seasons. Both scenarios show similar coverage areas but the RCP 8.5 scenario shows lower range of precipitation (Table 6.4).

A similar case occurs during the projected dry seasons, where the lower precipitation pattern approaches the coasts to the midlands (Fig. 6.8; Table 6.5). Although some areas will continue to receive very high precipitation due to higher elevation, such as Flores, this condition will create limited options for agriculture. It is also critical as El Niño will further decrease and disturb the rainfall in NTT. By the end of 2099, NTT's decent precipitation level will be in the "moderate" level under both scenarios, ranging 183.7 and 290.5 mm. The dry month threshold is around 60 mm, and 180 mm will be the threshold cumulation over three dry months.

The spatial pattern of maximum temperature contradicts the precipitation pattern, with the highest number occurring at the midlands and/or the higher elevation levels. Two patterns of maximum temperature and minimum temperature are shown. The maximum temperature tends to be the lowest at the outer parts of the provinces, reaching 31.5 °C or less, particularly the southern coasts and narrower lands. Interestingly, the maximum temperature in the midlands can be very high, up to 35.3 °C or more. The increasing propagation will start from the midlands and radiate to the coastal zones (Table 6.6). However, heat transfers between sea and land and the coasts will remain the areas with lower maximum temperatures than other NTT areas (Fig. 6.9).

While the minimum temperature spatial figure shows the opposite pattern, it is coherent with the precipitation pattern (Fig. 6.10. Minimum temperature projection in various time ranges under RCP 4.5 [left] and RCP 8.5 [right] scenarios). The temperature range between the coasts and midlands and/or the mountainous regions

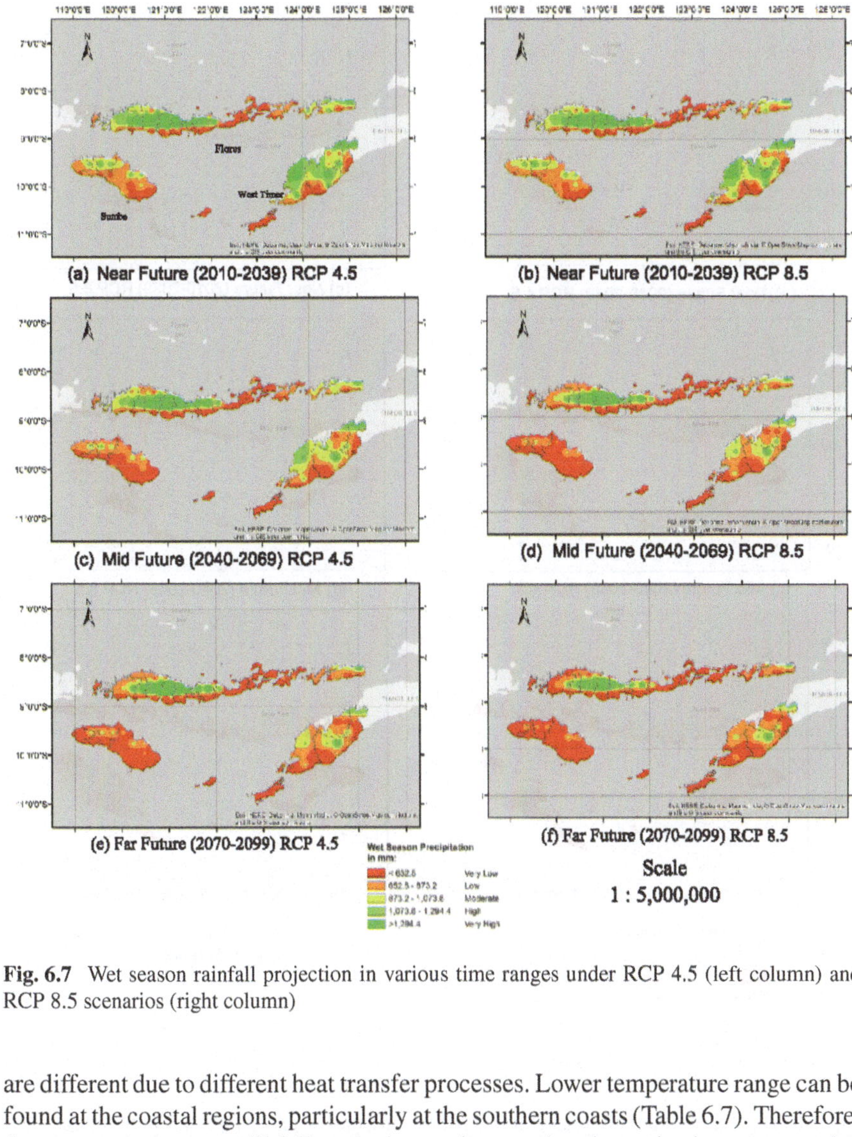

Fig. 6.7 Wet season rainfall projection in various time ranges under RCP 4.5 (left column) and RCP 8.5 scenarios (right column)

are different due to different heat transfer processes. Lower temperature range can be found at the coastal regions, particularly at the southern coasts (Table 6.7). Therefore, the mean temperature will follow the increasing trend as the projection suggests due to increasing maximum and minimum temperatures. This information can help the agriculture sector to determine suitable crops.

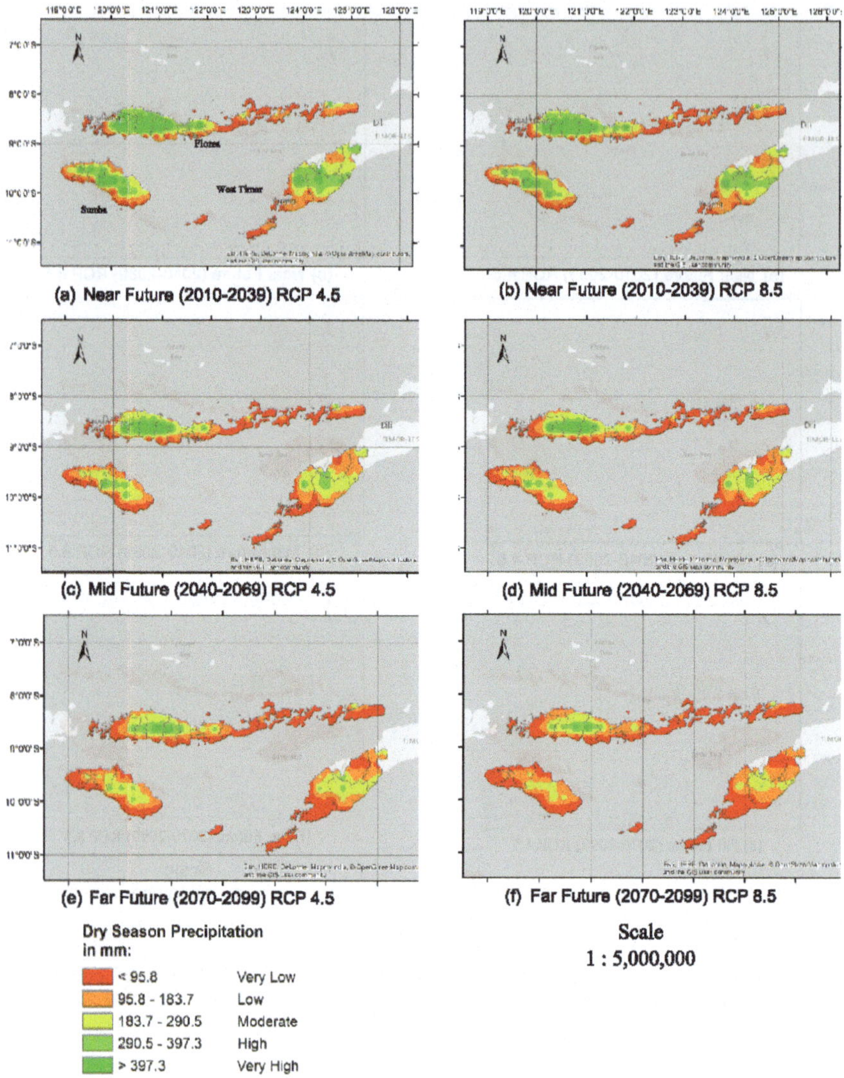

Fig. 6.8 Dry season rainfall projection in various time ranges under RCP 4.5 (left column) and RCP 8.5 (right column) scenarios

Table 6.4 Descriptive wet season rainfall projection in various time ranges under RCP 4.5 and RCP 8.5 scenarios

Precipitation (wet season)	RCP 4.5	RCP 8.5
Near Future (2010–2039)	*Low precipitation*: southern Sumba, south eastern and northern Timor, eastern and southern part of Flores, small islands in the south *High precipitation*: northern Sumba, Northern Flores	The wet and dry regions are relatively similar to those shown under the RCP 4.5 scenario, with expanded areas: *Low precipitation*: north-western part of Flores and the western part of Sumba *Moderate precipitation*: middle part of West Timor following the topography relief of the lowlands
Mid Future (2040–2069)	*Low precipitation*: a few areas of the northern part of Flores, larger Sumba, lowlands Timor, eastern part of Flores *High precipitation*: midland Flores and Highland Timor	Western coasts of Flores start experiencing drier conditions than the RCP 4.5 scenario projects Almost the whole Sumba will receive very low rainfall, meanwhile the lowland areas that receive low rainfall will expand
Far Future (2070–2099)	The projection shows similar condition to Mid Future projection under the RCP 8.5 scenario	The north-western coast of Flores will receive what is categorised as very low rainfall and the northern coast of West Timor will become drier. The only wet areas remaining are the highlands

Table 6.5 Descriptive dry season rainfall projection in various time ranges under RCP 4.5 and RCP 8.5 scenarios

Precipitation (dry season)	RCP 4.5	RCP 8.5
Near Future (2010–2039)	*Very low precipitation*: the west and east coast of Sumba, as well as the southern coast of Flores and West Timor Narrow lands and small islands will remain the areas with very low precipitation	Similar to RCP 4.5 scenario, with more emphasised and extended dry areas. Some areas with very high precipitation under the RCP 4.5 scenario will receive moderate rainfall
Mid Future (2040–2069)	Low precipitation areas will expand. The contrast of precipitation level in each island will be higher lower than the Near Future time frame	Similar to RCP 4.5 scenario, but with larger dry areas. Topographic influence is more prominent in this time frame
Far Future (2070–2099)	Only small areas are receiving a very high amount of rainfall and the midlands will mainly remain the wetter areas	Mountainous areas at Flores Island will become the only areas receiving the highest amount of rainfall during the dry season; meanwhile the midlands of Sumba and West Timor will receive moderate rainfall

Table 6.6 Descriptive maximum temperature projection in various time ranges under RCP 4.5 and RCP 8.5 scenarios

Maximum temperature	RCP 4.5	RCP 8.5
Near Future (2010–2039)	The lowest maximum temperature will occur at the coasts (in Flores at the southern coasts), small islands, and narrow lands	Similar to RCP 4.5 scenario, with a small expansion of very high maximum temperature
Mid Future (2040–2069)	Very high maximum temperature will start propagating toward the lowlands. Southern and coasts will experience the lowest and moderate maximum temperature	Similar to RCP 4.5 scenario, with a slight difference at the eastern part of Flores. The maximum temperature will increase to the moderate level, while remaining very low under the RCP 4.5 scenario. Overall spatial change will occur towards the coasts
Far Future (2070–2099)	The areas near the coasts will start experiencing very high maximum temperature	This projection shows the most notable change of maximum temperature. The region will experience very high maximum temperature, except the small islands, narrow islands, some parts of western Sumba, and a small area at the southern Flores

Sensitivity

Coffee

In NTT, the Robusta coffee is more suitable in the Sumba Island and West Timor than in Flores Island (Fig. 6.10). While Arabica coffee has higher market value, the crop is more sensitive to temperature rise than Robusta coffee. Hence, this study selected Robusta for further suitability analysis, as it is more widely practised in NTT and its climate change impact indicates the severity for Arabica coffee. Based on the projected climate data, divergent rainfall and temperature variability promote damage conditions to coffee crops in NTT. Robusta coffee requires 3–4 months of dry months to grow at the worst case (marginally suitable; S3) and 1–2 months of dry months to grow at the optimum level. Too little rainfall interferes with plant growth and fruiting, while too heavy rainfall causes coffee defects such as mould growth, disease, and excessive fermentation (Kath et al. 2021). Continuous temperature rise will also accelerate the ripening of fruit development, which eventually will degrade the coffee bean quality (Davis et al. 2012). In 2099, these islands will no longer be suitable due to decreased water availability in NTT and the number of dry months (Fig. 6.11).

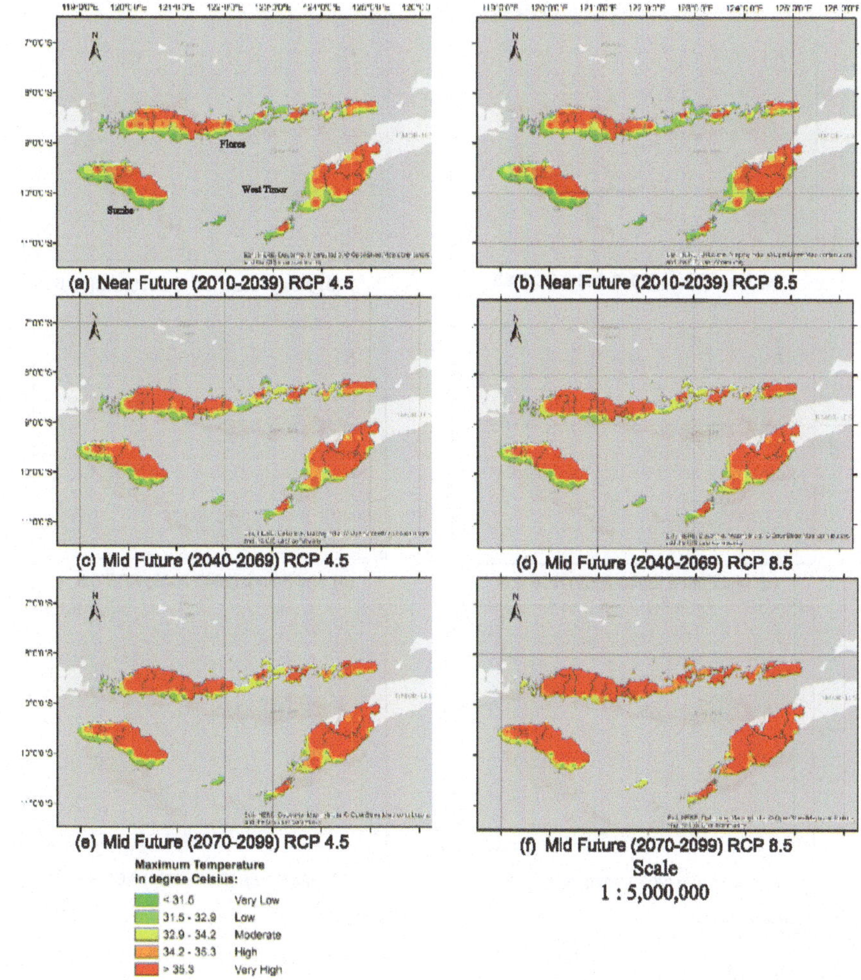

Fig. 6.9 Maximum temperature projection in various time ranges under RCP 4.5 (left) and RCP 8.5 (right) scenarios

Adaptive Capacity

Human Capital

Farmers in Indonesia tend to manage the crop planting by themselves based on their traditional knowledge and practices (Sarjana 2010). However, as climate change has become harder to predict, agricultural workers meet many constraints. The farmers' lack of understanding about climate information from the national meteorological

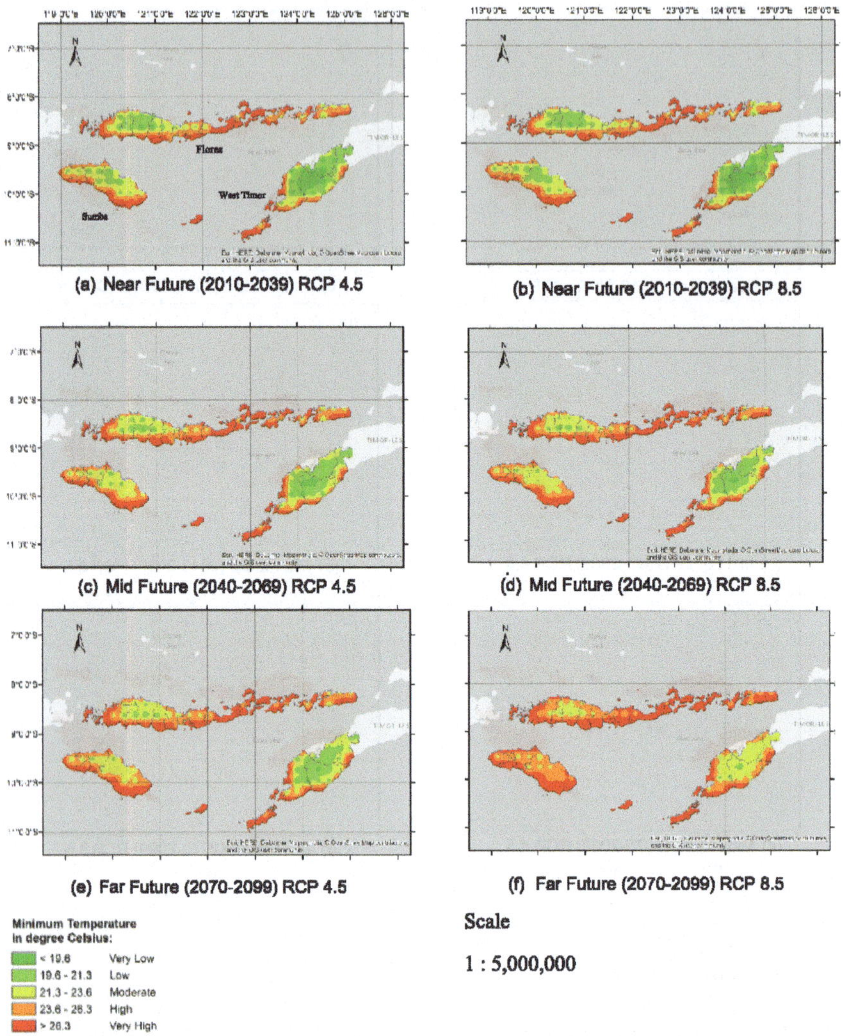

Fig. 6.10 Minimum temperature projection in various time ranges under RCP 4.5 (left) and RCP 8.5 (right) scenarios

agency, climate-smart farming practices, and financial aspects have become underlying challenges of building adaptive capacity in NTT (Sarjana 2010, DFAT 2016). These challenges relate to the fact that most households' heads are primary school-level graduates. Although 70% of households in NTT have access to weather update through public television or radio, the public did not utilise the forecast due to low accuracy and lack of adaptive capacity building to change agricultural practices from the government (Kuswanto et al. 2019).

Table 6.7 Descriptive minimum temperature projection in various time ranges under RCP 4.5 and RCP 8.5 scenarios

Minimum temperature	RCP 4.5	RCP 8.5
Near Future (2010–2039)	The lowest minimum temperature occurs in the midlands and mountainous regions	A slight expansion area with very low temperature at the Sumba midlands
Mid Future (2040–2069)	Most areas will shift to the moderate level. Large areas at Flores and West Timor's midlands will still experience very low minimum temperature	Wider expansion of increasing minimum temperature toward the midlands
Far Future (2070–2099)	Similar spatial pattern to the Mid Future RCP 8.5 projection; however, only a few regions on the Flores midlands will experience very low minimum temperature. The very low minimum temperature will remain occurring at West Timor	The minimum temperature in West Timor will increase significantly (will become at the moderate level) and in the midlands of Flores

In terms of health, malnutrition poses serious concerns to numbers children under the age of five aged who suffered malnutrition in El Niño in 2002 and 2005, which are three times higher than in a non-El Niño year (UNDP, UN ESCAP, RIMES, APCC, OCHA 2017) to become: In places affected by El Nio / La Nia-related phenomena—most notably in South-Eastern Asia and South Asia—there have been recent decreases in food availability and increases in food costs. For example, in Village Bene, the food shortages occurred during the recent El Niño in 2016 because the water shortage halted the growth of nutritious plants, and it might continue until before the next harvest season (Vaessen 2016). There are potentially around 15,000 people living in 5 villages in South Timor Tengah experiencing severe crop failure. Other regencies, such as East Flores, East Sumba, Central Sumba, and West Sumba also had similar conditions with South Timor Tengah (Bonasir 2015; Saptohutomo 2015).

Social Capital

As the main source of NTT livelihood, various social resources have been formed to pursue agricultural practices. The information was gathered through field visits and interviews in 2016 at resources across Flores Island.

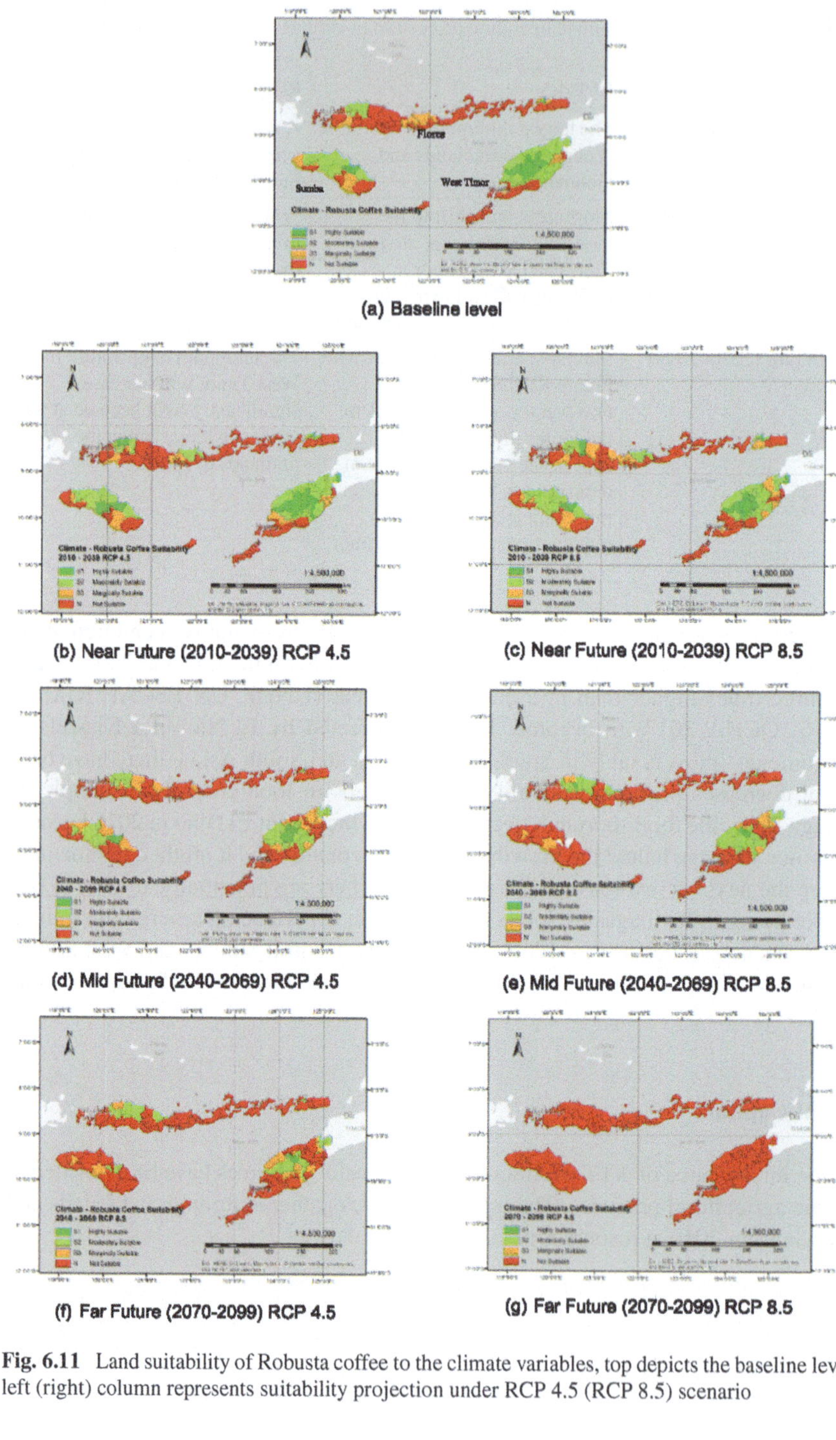

Fig. 6.11 Land suitability of Robusta coffee to the climate variables, top depicts the baseline level, left (right) column represents suitability projection under RCP 4.5 (RCP 8.5) scenario

Ethnic and Religious Groups

Basic values in each ethnic group include family relations, tolerance, and togetherness inherited by their ancestors. The group takes authority in their community's behavioural pattern to prevent conflict, enhance partnership, and maintain peace. The social capital is used as the communities' tool to address their problems if the government does not intervene too deeply. In some regions, a Religion Leader Communication Forum has major roles in reducing the conflicts among religious groups, enhancing relationships, and facilitating communication.

Farmers Groups

Historically, the farmer group is based on top-down government initiative through the Department of Estate Crops and Horticulture under the Ministry of Agriculture. In Manggarai regency alone, government has established 81 farmers groups. However, farmers groups do not have a significant role in the chain due to the nature of liberalised market system that enabled farmers to access the marketing channel independently. The association mainly provides training and technical assistance in coffee cultivation and contributes to the government's machinery or financial grant. In June 2014, there were 20,112 farmers divided into 2,321 farmers' groups (gapoktan) and dominated by beginner class (83.74%).

Savings and Loan Cooperatives

In NTT, there are many saving and loan cooperatives in which farmers depend on this cooperative to support their livelihood. The saving and loan cooperative started from the small one with around 20 members until the big one such as the primary cooperative of saving and loan Pintu Air in Sikka. Pintu Air Cooperative is a primary cooperative for savings and loans. This cooperative was established in 2004 with 80,637 members, consisting of 39,479 male members and 41,158 female members. Other saving and loan cooperatives need to be further checked during the field visit.

Commodity Cooperatives

Only a few cooperatives work for commodity, while others usually start from saving and loan cooperatives or multipurpose cooperative. The list below is of cooperatives that work on commodity or has a business unit that deals with commodity. There are potential cooperatives in NTT that play a significant role in the agriculture sector in NTT.

a. ASNIKOM (cooperative)

Asnikom merges the concept of association and cooperatives with 1,062 members. It connects marketing agreements with coffee exporters, in which each member pays fees and receives a profit share. Asnikom carries out three marketing channels after receiving coffee beans from the members: (1) Direct selling to exporter, (2) Selling through auction, (3) Selling to the individual buyer (café e.g., Anomali coffee, individual exporter). Non-member farmers can sell the coffee beans with a selling price similar to the open market price and not receive profit shares at the end of the year. The cooperative gains a slight margin from purchasing and selling price to maintain the operational cost.

b. Kopsi

Formed in 2015, Kopsi is a cooperative focused on coffee production and sales with around 150 members. Previously nurtured by VECO under PRISMA fund and with the help of Tananua, a local NGO, Kopsi connects with 30 coffee producing villages in Ende. Both Arabica and Robusta farmers under Kopsi produced almost 4 tonnes of coffee in 2016. Most of the arabica is processed in a full-washed method while the robusta is treated traditionally. The farmers' plantation is mostly under an intercropping system with candlenuts and cloves. Kopsi had almost 10 farmers who are well trained in post-harvesting and sorting, with their buying the coffee cherries from other farmers. With no funds available, Kopsi's activities are halted and only continued with consignment with farmers to sell the coffee to a reputable market. With only household scale fermentation drums for each farmer, the lack of proper storage and drying facility caused production delay, leaving Kopsi unable to fulfil buyers' quality level and quantity demand. Kopsi had its own incentivised system for each value-adding activity done by the members or farmers' group that could potentially be improved for all coffee-producing villages in Ende. However, its unreliable management requires improved facilities and business models to advance the value-adding process by producing roasted coffee.

Natural Capital

Water availability, hilly topography, and low soil organic matter content are the main challenges for agriculture development in NTT (Suriadi et al. 2021). On top of being highly vulnerable to erosion, the slash-and-burn method as a common agricultural practice is accused of causing land damage; particularly the forest loss and declining soil productivity (Hidayatullah 2008). There was no evidence of any smallholder agroforestry resulting in forest loss unless supported by any governmental programmes (Fisher 2010). As a semi-arid region, NTT is also often exposed to drought and water shortage. During the dry season in 2003, NTT's water availability balance was a deficit. The increasing drought severity due to El Niño, declining trend, and erratic rainfall worsened access to water availability, clean water, and sanitation, leading to harvest failure, food shortage, and income reduction. This situation will exacerbate poverty due to the reduced productivity of the livelihood, environment, and nature, which even caused death to many villagers (Vaessen 2016). To anticipate the seasonal El-Niño issue, the communities in NTT have access to a reservoir called "embung," which provides water for household consumption, irrigation, and cattle needs (Widiyono 2010). However, due to the limited technical capacity of the "embungs," many of them are predicted to become dry (Bere 2017).

Physical Capital

Improvement in cultivation and processing technology was also on the line of adapting to climate change (Takama et al. 2016). Although farmers have been encouraged to use new technologies, they hardly accept innovations due to their high dependence on their knowledge. Among many reasons, findings identify scepticism and pride in their method (Thrupp 1989). This means no matter how much aid is provided to them, the reduced vulnerability will not be achieved due to capital discrepancy (agricultural aid as physical capital, scientists as social capital, and knowledge as human capital) (Jacobs et al. 2015). On the other hand, the provision of knowledge by the scientists did not guarantee to reduce the vulnerability since the farmers did not own financial capital to apply and operate the technology. Thus, farmers' engagement is crucial in physical capital development as their traditional practices follow modest physical capital.

The needs and capability of the farmers in operating the solution are critical. For instance, a government-built initiative to address water scarcity, "embung" fulfils the communities' water intake and irrigational needs during the dry season (Pusat Analisis Sosial Ekonomi dan Kebijakan Pertanian 2010). However, due to the poor construction and operation, the reservoirs built by the government become dry during the dry terms (Bere 2017). The identified issues of "embung" in Oemasi, Kupang, such as water delivery and efficiency issue from the storage to the utilisation regions may occur in the whole province (Widiyono 2010).

Financial Capital

NTT receives funding allocation from the Ministry of Agriculture to finance district agricultural services implementation (MOF 2015). However, the farming practice did not continue after procuring agricultural equipment. The farmers group believes that the local agency lacked in the fund disbursement monitoring and farmer selection, particularly in choosing committed and hard-working farmers instead of new farmers who are only in for the money (Sura 2020). Especially since NTT is considered too risky for profit-oriented private sector entities, private commodity buyers are likely to continue to support farmers only in locations with profit-generation prospects. The government and donors are likely to continue to offer piecemeal support without a multiple-barrier-removal approach in mind. The increasing vulnerability of the farmers also includes the secretive money lenders, with the loan-sharking system, which is one of the most common microfinance practices in Indonesia. The loan sharks mostly operate at the markets and have face-to-face interactions with the farmers in the villa. This loan is an accessible way to obtain an immediate loan without collateral. However, the issue lies in the very high interest offered by the loan sharks, resulting in financial insecurity among the farmers, which is usually occurs in poor and developing economies (Hull and James 2012).

Farmers' income depends highly on the seasonal pattern (Khandker and Mahmud 2012). The farmers in NTT, who are mostly subsistence farmers, gain their income by selling their cash crops annually, constrained by steep slopes, unpredictable rainfall, and drought (Smits and Mthembu 2012). Most villagers' major cash income source is from the agricultural production system, mostly maise and livestock production that family labour can only handle with traditional practices (UNDP 2015). However, this situation is not sufficient to fulfil the need of the households across the season due to frequent post-harvest losses (van de Fliert et al. 2008). When farmers receive reduced income due to crop failure, farmers respond by reducing their food expenditure (Kuswanto et al. 2019).

Institutional and Policy Context

In the updated Nationally Determined Contribution (NDC) in 2020, sustainable agriculture and plantations were included as station programmes for economic resilience (UNFCCC 2021). In the RAN API or National Action Plan on Climate Change Adaptation, issues were categorised into five domains: (1) economic resilience, which also encompassed energy and food security; (2) life system resilience, which consisted of some sub-sectors, like health and infrastructure; (3) ecosystem resilience; (4) special regional resilience, including the coastal areas and small islands; and (5) supporting systems that could directly or indirectly assist other mentioned domains. Based on Minister Decree SK Menteri PPN/Kepala Bappenas No Kep.38/M.PPN/HK/03/2012, RAN API works under the Climate Change Action (CCA) Coordination Team. Related to the proposal, the climate change adaptation program considers economic resilience very important, especially on food security. NTT, which has serious issues on food security, has become the priority for the national adaptation strategy. Under Ministry of Agriculture Decree No. 64.1/KPTS/RC.110/J/12/2017, Indonesia has adopted technical instructions to diversify staple food beyond rice and wheat. However, the production side has not changed, with rice fields expanding (CIFOR 2021). With the updated NDC finalised in 2021, more revisions and enactment of adaptation-related national and sub-national level provisions will follow.

Ministry of Environment and Forestry, Ministry of Agriculture, and Ministry of National Planning and Development Agency (BAPPENAS) have actively, with local government and stakeholders, resulted in Presidential Regulation 59/2017 that initiates sustainable food production system to ensure food access for all (Arif 2020). As a commitment to end hunger, address malnutrition, and increase agricultural activity in sustainable ways (SDG 2), combined with SDG no. 13.1 on strengthening resilience and capacity, it has also translated climate smart agriculture training into

NTT local mid-term development 2018–2023 (RPJMD). As a result, NTT integrated an adaptation curriculum for local universities and trained 587 agriculture workers and 17,092 farmers through farmers field school programs conducted in collaboration with BMKG (BAPPENAS 2019).

With regard to SDG and national targets, international development agencies have been actively involved in climate change adaptation efforts. A UNDP-led project called (Strategic Planning and Action to strengthen Climate Resilience in NTT (SPARC) has developed 120 community action plans and 21 pilot projects based on RAN API, in which Bank NTT played a crucial role in channelling and leveraging financing to vulnerable communities (UNDP 2016). As a result, SPARC becomes a major reference for the NTT Medium-term plan (RPJMP) 2013–2018 (MOF 2015). During the same years, the Australian-Indonesia Partnership for Rural Economic Development, in collaboration with the Ministry of National Development and Planning, also initiates to boost farmers income by addressing market constraints and policy advocacy to increase competitiveness in Eastern Indonesia (DFAT 2016). Those organisations are key actors in promoting climate change adaptation in top-down approach.

Discussion and Conclusions

Based on the vulnerability assessment and the adaptive capacity, NTT farmers are highly vulnerable to climate change. The current farming system and overall capital are sensitive to climate change because of the limited adoption of climate-resilient production for the highly potential estate crops. Moreover, a substantial proportion in NTT derives their revenue from rain-fed, extremely small-scale farming, putting their livelihoods at risk from climate change. Through precipitation and temperature analysis, coffee crop suitability maps indicate that most arable land will be reduced drastically in both scenarios, RCP 4.5 (stabilisation) and RCP 8.5 (failed mitigation). Without climate-smart extension support, farmers are highly risk-averse to adopting new approaches due to the low baseline development of the province, high prevalence of poverty, and limited social safety nets. Given the cross-sectoral barriers, this study encouraged adaptation strategies should be a combination of incremental, system, and transformative adaptation are an effective way to achieve the objective as proposed by Rickards and Howden (2012). However, addressing adverse climate change impacts requires a transformative adaptation.

Given the complex barriers, a dominant social capital that embodies a wide existing network of agricultural cooperatives and farmers groups could be an asset for an extensive transformative adaptation by utilising the farmers network in strengthening their knowledge and learning capacity with support from other actors such as government, development agencies, and non-governmental agencies. Currently, the climate information still emphasised rice production and neglected high-value estate crop production like coffee. Ultimately, funding allocation to local NTT development plan is crucial to achieve sustainable and resilient coffee production. The limited institutional, policy, physical, and financial capital that are not intact to adapt to the projected condition may exacerbate farmers' livelihood insecurity during one external shock like a prolonged dry season. Thus, further research is necessary to explore the management and funding support for NTT financial literacy and marketing knowledge capacity building for coffee farmers. The capacity of private and non-government entity networks will be strengthened as a critical agent that works with both farmers and the government, especially in farmer climate-smart agriculture training in production and post-production. The project's changes in climate risk transfer are expected to transform how climate risks are shared between vulnerable farmers and a third-party entity. Since the forecast is perceived as inaccurate and requires high literacy, integrating this vulnerability assessment into the climate smart agriculture training will be one strategy to engage and prioritise NTT farmers in pursuit of local adaptation measures and updated policies. Although this study emphasized that climate information should be extensively communicated among farmers, additional empirical research is still required to determine whether this intervention could trigger transformative adaptation.

References

Arif S (2020) Strategic review of food security and nutrition in Indonesia: 2019–2020 update. SMERU Research Institute, Jakarta.

Badan Pusat Statistik (2018) Provinsi nusa tenggara timur dalam angka 2018. Jakarta.

Bappenas (2019) Voluntary national reviews (VNR) empowering people and ensuring inclusiveness and equality. Jakarta, Indonesia. https://sustainabledevelopment.un.org/content/documents/23803INDONESIA_Final_Cetak_VNR_2019_Indonesia_Rev2.pdf

Bere SM (2017, April 29) Banyak Embung di NTT Terancam Mengalami Kekeringan. Kompas. http://properti.kompas.com/read/2017/04/29/123122121/banyak.embung.di.ntt.terancam.mengalami.kekeringan

Boken VK, Cracknell AP, Heathcote RL (2005) Monitoring and predicting agricultural drought: a global study. Oxford University Press. https://books.google.co.id/books?hl=id&lr=&id=y6r9jUTzJiYC&oi=fnd&pg=PA330&dq=vulnerability+of+cashew+to+climate+change+in+indonesia&ots=slGtE4cIYc&sig=iyPk-u8MI0pYPmbTNrNIyJsreTc&redir_esc=y#v=onepage&q&f=false

Bonasir R (2015) Kemarau panjang perparah gizi buruk di NTT. BBC Indonesia. East Nusa Tenggara. http://www.bbc.com/indonesia/berita_indonesia/2015/07/150720_indonesia_ntt_pangan

BPS (2020) Jenis Tanaman Perkebunan Rakyat. BPS - Statistics Indonesia, Jakarta. https://www.bps.go.id/site/resultTab

BPS NTT (2019a) Luas Areal Tanaman Perkebunan Menurut Kabupaten/Kota (Hektar), 2019. BPS - Statistic of Nusa Tenggara Timur Province, Kupang. https://ntt.bps.go.id/indicator/54/58/1/luas-areal-tanaman-perkebunan-menurut-kabupaten-kota.html

BPS NTT (2019b) Produksi Tanaman Perkebunan Menurut Kabupaten/Kota (Ton), 2019. Kupang. https://ntt.bps.go.id/indicator/54/57/1/produksi-tanaman-perkebunan-menurut-kabupaten-kota.html

BPS NTT (2020) Statistik Pertanian Provinsi Nusa Tenggara Timur 2020. BPS - Statistic of Nusa Tenggara Timur Province, Kupang. https://ntt.bps.go.id/publication/2021/09/28/5f0995beae43547eb4594e75/statistik-pertanian-nusa-tenggara-timur-2020.html

BPS NTT (2021) Statistik Daerah Provinsi Nusa Tenggara Timur 2021. BPS - Statistic of Nusa Tenggara Timur Province, Kupang. https://ntt.bps.go.id/publication/2021/09/27/468e8ddf8b7303fee541f816/statistik-daerah-provinsi-nusa-tenggara-timur-2021.html

CIFOR (2021) Linking food, nutrition and the environment in Indonesia a perspective on sustainable food systems. Bogor, Indonesia. https://doi.org/10.17528/cifor/008070. https://www.cifor.org/publications/pdf_files/brief/8070-Brief_Linking-Nutrition.pdf

Davis AP, Gole TW, Baena S, Moat J (2012) The impact of climate change on indigenous Arabica coffee (Coffea arabica): predicting Future trends and identifying priorities. PLoS ONE 7(11):e47981. https://doi.org/10.1371/journal.pone.0047981

DFAT (Australian) Department of Foreign Affairs and Trade, & BAPPENAS (2016) Progress report and implementation plan: promoting rural income through support for markets in agriculture. Indonesia. https://aip-prisma.or.id/data/public/uploaded_file/18.04.17_04042017-PRISMA%20PRIP%202016%20Semester%201-Public%20Version.pdf

DfID (1999) Sustainable livelihoods guidance sheets. Department for International Development. https://www.ennonline.net/dfidsustainableliving

FAO (Ed.) (1981) A framework for land evaluation (2. print). ILRI Publication.

Fisher RP (2010) Socialising the pixel, a mixed method approach to assessing the state of forests in West Timor. Charles Dawin University, Darwin. https://researchers.cdu.edu.au/en/studentTheses/socialising-the-pixel-a-mixed-methods-approach-to-assessing-the-s

Hidayatullah M (2008) Rehabilitasi Lahan dan Hutan di NTT (Translated: Land and Forest Rehabilitation in East Nusa Tenggara). Info Hutan 5(1):17–24

Hull E, James D (2012) Introduction: popular economies in South Africa. Africa 82(1):1–19. https://doi.org/10.1017/S0001972011000696

IPCC (Ed.) (2007) Climate change 2007: impacts, adaptation and vulnerability: contribution of Working Group II to the fourth assessment report of the Intergovernmental Panel on Climate Change. Cambridge University Press.

Jacobs B, Nelson R, Kuruppu N, Leith P (2015) An adaptive capacity guide book: assessing, building and evaluating the capacity of communities to adapt in a changing climate. Southern Slopes Climate Change Adaptation Research Partnership (SCARP), University of Technology Sydney and University of Tasmania.

Kath J, Mittahalli Byrareddy V, Mushtaq S, Craparo A, Porcel M (2021) Temperature and rainfall impacts on robusta coffee bean characteristics. Clim Risk Manag 32:100281. https://doi.org/10.1016/j.crm.2021.100281

Khandker SR, Mahmud W (2012) Seasonal hunger and public policies: evidence from Northwest Bangladesh. The World Bank, Washington DC.

Kuswanto H, Hibatullah F, Soedjono ES (2019) Perception of weather and seasonal drought forecasts and its impact on livelihood in East Nusa Tenggara Indonesia. Heliyon 5(8):e02360. https://doi.org/10.1016/j.heliyon.2019.e02360

Mayer J, Ryan A, Aspinall E (2011) Climate change and Indonesia. Charles Dawin University, Inside Indonesia

Ministry of Finance, Republic of Indonesia, Center for Climate Finance and Multilateral Policy, and Fiscal Policy Agency (2015) Provincial climate public expenditure and institutional review (CPEIR). Indonesia, Jakarta. https://www.climatefinance-developmenteffectiveness.org/sites/default/files/documents/03_31_17/Climate-Public-Expenditure-and-Institutional-Review-East-Nusa-Tenggara-of-Indonesia.pdf

Nelson GC, Rosegrant MW, Koo J, Robertson R, Sulser T, Zhu T, Ringler C, Msangi S, Palazzo A, Batka M, Magalhaes M, Valmonte-Santos R, Ewing M, Lee D (2009) Climate change: Impact on agriculture and costs of adaptation. International Food Policy Research Institute, Washington, DC. https://doi.org/10.2499/0896295354

O'Brien K (2011) Responding to environmental change: a new age for human geography? Prog Hum Geogr 35(4). https://doi.org/10.1177/0309132510377573

Pusat Analisis Sosial Ekonomi dan Kebijakan Pertanian (2010) Strategi Petani Kupang Menanggulangi Kekurangan Air. 32(3).

Riahi K, Rao S, Krey V, Cho C, Chirkov V, Fischer G, Kindermann G, Nakicenovic N, Rafaj P (2011) RCP 8.5—a scenario of comparatively high greenhouse gas emissions. Clim Change 109:33–57. https://doi.org/10.1007/s10584-011-0149-y

Rickards L, Howden SM (2012) Transformational adaptation: agriculture and climate change. Crop Pasture Sci 63:240–250. https://doi.org/10.1071/CP11172

Saptohutomo AP (2015) Lima desa di NTT terancam kelaparan karena gagal panen. Merdeka.Com. https://www.merdeka.com/peristiwa/lima-desa-di-ntt-terancam-kelaparan-karena-gagal-panen.html

Sarjana IMB (2010) Governance for food security: the case of Indonesia in decentralization era. Maastricht Graduate School of Governance, Maastricht.

Schneider U, Becker A, Finger P, Meyer-Christoffer A, Rudolf B, Ziese M (2011) GPCC full data reanalysis version 6.0 at 0.5°: monthly land-surface precipitation from rain-gauges built on GTS-based and historic data. https://doi.org/10.5676/DWD_GPCC/FD_M_V7_050

Smit B, Wandel J (2006) Adaptation, adaptive capacity and vulnerability. Glob Environ Chang 16(3):282–292. https://doi.org/10.1016/j.gloenvcha.2006.03.008

Smits J, Mthembu T (2012) Impact of microcredit on poverty in Eastern Indonesia: dealing with selection bias'. Norwegian University of Life Sciences.

Sura HO (2020) Young NTT farmers pin hopes on microcredit program to expand their agriculture business. The Jakarta Post. Retrieved December 17, 2021, from https://www.thejakartapost.com/news/2020/03/07/young-ntt-farmers-pin-hopes-on-microcredit-program-to-expand-their-agriculture-business.html

Suriadi A, Mulyani A, Hadiawati L, Suratman. (2021) Biophysical characteristics of dry-climate upland and agriculture development challenges in West Nusa Tenggara and East Nusa Tenggara Provinces. IOP Conference Series: Earth and Environmental Science 648(1):012014. https://doi.org/10.1088/1755-1315/648/1/012014

Takama T, Aldrian E, Kusumaningtyas SDA, Sulistya W (2016) Identified vulnerability contexts for a paddy production assessment with climate change in Bali, Indonesia. Clim Dev 9(2):110–123. http://www.tandfonline.com/doi/abs/10.1080/17565529.2016.1167658

Thomson AM, Calvin KV, Smith SJ, Kyle GP, Volke A, Patel P, Delgado-Arias S, Bond-Lamberty B, Wise MA, Clarke LE, Edmonds JA (2011) RCP4.5: a pathway for stabilization of radiative forcing by 2100. Clim Change 109(1–2):77–94. https://doi.org/10.1007/s10584-011-0151-4

Thrupp LA (1989) Legitimizing local knowledge: from displacement to empowerment for third world people. Agric Hum Values 13–24.

UNDP (2015) The state of Indonesia's renewable energy. http://www.id.undp.org/content/indonesia/en/home/presscenter/articles/2015/07/22/the-state-of-indonesia-s-renewable-energy.html

UNDP (2016) UNDP fact sheet: strategic planning and action to strengthen climate resilience of Rural Communities in Nusa Tenggara Timor Province (SPARC). https://www.id.undp.org/content/dam/indonesia/docs/envi/project/SPARC%20-%20Results%20Sheet.pdf

UNDP, UN ESCAP, RIMES, APCC, OCHA (2017) Enhancing resilience to extreme climate events: lessons from the 2015-2016 El Niño Event in Asia and the Pacific. UNDP. https://www.unescap.org/sites/default/files/El%20Nino%20report-%20finalized%20ESCAP07082017.pdf

UNFCCC (2021) Nationally determined contributions under the Paris agreement: synthesis report by the secretariat. Glasgow. https://unfccc.int/sites/default/files/resource/cma2021_08_adv_1.pdf

Vaessen S (2016) El Niño causing food shortage in Indonesia's West Timor. http://www.aljazeera.com/news/2016/02/el-nino-causing-food-shortage-indonesia-west-timor-160229063051094.html

van de Fliert E, Christiana B, Hendayana R, Murray-Prior R (2008) Pilot roll-out: adaptive research in farmers' worlds. Ext Farm Syst J 6(1):63–71. https://www.csu.edu.au/__data/assets/pdf_file/0009/109638/EFS_Journal_v06_n01_07_deFliert_et_al.pdf

Widiyono W (2010) Upaya Peningkatan Efisiensi Pemanfaatan Air Embung di NTT: Studi Kasus Embung Oemasi—Kupang. Jurnal Hidrosfir Indonesia 5(3):1–11

Chapter 7
National Integrated Coastal Zone Management Frameworks Need to Adapt

Fernanda Terra Stori, Anne Marie O'Hagan, and Cathal O'Mahony

Abstract Many coastal areas in the world are experiencing coastal erosion and flooding, which is likely to escalate due to sea-level rise and an increase in the frequency and intensity of storms, however, some countries still lack a dedicated policy to guide the application of adaptation solutions specific to the coastal zone. This remains the case in Ireland, a country with more than 7,000 km of shoreline, much of which is threatened by climate change. There was observed a gap in studies that critically analyse the integration of coastal and climate legal frameworks and proposed adaptation solutions to support decision-making in the face of climate hazards in coastal areas. This study analysed Ireland's coastal-climate adaptation governance system by examining 26 documents at national level (laws, policies and plans) related to coastal management (7); planning and development (6); and climate change (13). The investigation revealed that whilst the State has an extensive legislative framework, there is no dedicated policy or legal approach to guide the implementation of coastal-climate adaptation solutions. Generic guidance on the implementation of certain types of adaptation solutions and differences in the adopted terminology were observed. The lack of integration between the analysed frameworks hampers the effective implementation of climate adaptation solutions in the Irish coastal zone. Integrated Coastal Zone Management policy at the national level is recommended in order to provide clarity on how to implement ecosystem-based and technical solutions; appropriate funding support; and coordination mechanisms to ensure consistency in the governance system necessary to the implementation of coastal climate adaptation solutions. This would provide a solid policy basis for Local Authorities to implement adaptation solutions which strengthen coastal resilience, enable public participation, learning and democratic vision building in

F. T. Stori (✉) · A. M. O'Hagan · C. O'Mahony
MaREI the SFI Research Centre for Energy, Climate and Marine, Beaufort Building, Environmental Research Institute, University College Cork, Ringaskiddy Co. Cork, P43 C573, Ireland
e-mail: f.terra.stori@gmail.com

A. M. O'Hagan
e-mail: A.OHagan@ucc.ie

C. O'Mahony
e-mail: c.omahony@ucc.ie

W. Leal Filho et al. (eds.), *Climate Change Strategies: Handling the Challenges of Adapting to a Changing Climate*, Climate Change Management,
https://doi.org/10.1007/978-3-031-28728-2_7

the process, and ultimately deliver more effective planning and management for sustainable social-ecological systems.

Introduction

Many coastal areas in the world are experiencing coastal erosion and flooding, which is likely to increase due to expected impacts from climate change, including sea-level rise and increase in frequency and intensity of storms (IPCC 2021). However, some countries still lack a legal framework to guide the employment of adaptation solutions in the coastal zone — this is the case in Ireland.

Ireland is an island state with over 7,000 km of shoreline, which is home to more than 50% of its population (Devoy 2008; Flood et al. 2020). Currently, the governance system comprises a number of laws, policies and plans that address coastal and climate issues individually or as a set of norms, but which are not wholly specific to the coast. Despite an evident reliance on coastal areas to maintain economically important sectors, such as fisheries, tourism, port activities and many others (DHLGH 2021), Ireland still lacks a policy framework to support the implementation of an Integrated Coastal Zone Management (ICZM) approach as encouraged by the scientific literature and advocated by European Union and international policies. The ICZM approach is widely recognized as a participatory and continuous planning process that should address economic, social, environmental and governance topics in coastal areas (GESAMP 1996; Cicin-Sain 1993). A policy framework based on ICZM principles can contribute to addressing coastal erosion and flooding, and preparing coastal communities to adapt to expected impacts of climate change.

This chapter explores the existing laws, policies, and plans relevant to progress coastal-climate adaptation in Ireland and discusses the opportunities and weaknesses of the current governance system. The chapter starts with this context-setting introduction and a brief contextualization of current evidence about sea-level rise globally and in Ireland, then presents the methodological approach, a discussion of the current coastal climate adaptation governance system, followed by a discussion of the limitations of the work, and concludes with a number of considerations necessary to advance a national ICZM policy that, amongst other subjects on coastal sustainability, should provide for coastal climate adaptation solutions in Ireland.

Current Global Evidence About Sea-Level Rise

Sea Level Rise (SLR) globally is a result of combined thermal expansion from ocean warming and land-based ice melting (IPCC 2019, 2021; Lindsey 2020). Studies observed an increase of around 21–24 cm in global mean sea level (GMSL) since 1880 (Lindsey 2020), reaching an average of 3.7 mm per year from 2006–2018 (IPCC 2021). SLR leads to coastal inundations, storm floods, coastal erosion, loss of nesting beaches, encroachment of tidal waters into estuaries and river systems, displacement of coastal lowlands and wetlands, as well as contamination of freshwater reserves and food crops (UN 2017).

According to the IPCC AR6 report, it is certain that GMSL will continue to rise over the twenty-first century (99–100% probability) (IPCC 2021). It is likely that GMSL will rise to around 0.5 m above the reference level by 2100 and reach up to 3 m by 2300 if the most optimistic scenario is considered (IPCC 2021). However, if the current high emissions scenario persists, global mean temperature may exceed 4.0 °C (±1 °C) from the pre-industrial baseline, and GMSL may rise more than 1 m by 2100 and reach more than 7 m by 2300 (IPCC 2021).

The current principal causes of coastal flooding and erosion are an increased frequency of severe storms and gradual SLR (Devoy 2008; IPCC 2019; Flood et al. 2020). These threats will bring critical future social and economic consequences for vulnerable coastal communities. It is estimated that currently around 10% of the world's population (more than 600 million people) live in areas that are less than 10 m above sea level (UN 2017; Bassetti 2020). It is estimated that areas where around 300 million people live will fall below the elevation of an average annual coastal flood by 2050, and that by 2100, areas where around 200 million people live could be constantly below the mean sea level (Kulp and Strauss 2019).

Greater storm events and SLR will undoubtedly impact economies and affect hundreds of millions of people living in coastal cities. The Global Commission on Adaptation (GCA) estimates that by 2050 recovery measures may cost more than $1 trillion each year globally (GCA 2019). Reducing emissions to zero, avoiding rises in global temperature and, consequently hampering SLR, will augment possibilities for adaptation in social-ecological systems, such as in low-lying coastal areas and small island developing states (IPCC 2018a). However, considering the IPCC most recent report, GMSL will continue to rise even if the targets of the Paris Agreement are reached (IPCC 2021). Therefore, mitigation and adaptation actions are increasingly urgent. According to the IPCC, adaptation is "*the process of adjustment to actual or expected climate and its effects, in order to moderate harm or exploit beneficial opportunities*" (IPCC 2018b).

The Irish Case

The impacts of climate change are expected around the coast of Ireland, where the most immediate risks are the increase in intensity and frequency of coastal storms, subsequent flooding and SLR (Flood et al. 2020). SLR combined with potential increases in storm surges, will lead to coastal flooding, erosion of beaches and cliffs, and consequent damage to coastal ecological systems (Dwyer 2012; Flood and Sweeney 2012; Flood et al. 2020).

Studies estimate that circa 350 km^2 of the Irish coast is vulnerable under a SLR of 1 m, and circa 600 km^2 is vulnerable under a SLR of 3 m (Flood et al. 2020). Estimates predict that up to 2 million people may be impacted by coastal erosion by

2050 when considering regions within 5 km from the sea, including Ireland's major cities, industries and infrastructure, such as electricity network, transport systems and water supplies (Government of Ireland 2020). Costs related to property insurance claims will potentially reach circa €1.1 billion under the 1 m scenario, increasing to circa €2.1 billion under the 3 m scenario (Flood 2012; Flood and Sweeney 2012).

Many coastal areas in Ireland face erosion and flooding as a consequence of climate hazards, including rural areas (e.g., Maharees, County Kerry), peri-urban areas (e.g., Youghal, County Cork), and urbanised areas (e.g., Portrane, Fingal/North County Dublin) (Cronin et al. 2017; Farrell 2021; Smith et al. 2021; Stori and O'Mahony 2021). The lack of a dedicated ICZM policy or supporting legal framework to guide the application of adaptation solutions focused on the coastal zone is hampering climate action and, in many cases, is further exacerbating the problems (Stori and O'Mahony 2021).

Methodological Approach

In order to provide a comprehensive understanding of challenges and opportunities of coastal climate adaptation governance system in Ireland, an examination of documents at the national level including laws, policies, and plans was undertaken. Ireland was selected as a case study since it is an island-state country with an extensive coastline facing widespread coastal flooding and erosion issues (Cronin et al. 2017; Devoy 2008; Flood et al. 2020). A recognised analytical gap on a thorough compilation and critical analysis of the current national coastal climate governance system was evident. Such analysis would provide clearer guidance on the employment of adaptation solutions and support the implementation of adaptation projects at the local level.

Information from legislative instruments, policies and plans addressing aspects of coastal zone management and climate change adaptation was garnered from government sources, research reports, journal papers, and coastal flooding and erosion assessment reports. A total of 26 documents were selected for further analysis. These documents were organised into the following domains: Coastal Management (7 documents); Planning and Development (6 documents); and Climate Change (13 documents). Laws, policies and plans related to Biodiversity were presented within the climate change section, since they provide the basis for the Biodiversity Climate Change Sectoral Adaptation Plan (DCHG 2019).

The documents analysed ranged in time from 1925 to 2021.[1] All the relevant Irish legislation was examined from their original version to their latest amendments (Acts and statutory instruments). Considering policies and plans, this study examined the most recent ones, from 2000 to December 2021.

[1] A visual communication of this examination is available online on the platforms *ArcGIS StoryMaps* and *TimeGraphics*—Available on: http://www.ccatproject.eu/coastal-climate-adaptation-Ireland/.

The main governance aspects of the laws, policies and plans were examined using a qualitative approach incorporating their titles, year of publication, main objectives and identifying the departments/agencies responsible for their implementation. In addition, a thorough search for provisions relating to coastal climate adaptation solutions was carried out and the typology of adaptation solution adopted was identified (such as engineering (hard) solutions/grey adaptation; nature/ecosystem-based solutions/green adaptation; soft solutions; and displacement solutions). Each of the following sections presents the outcomes of this qualitative analysis according to the given domains Coastal Management (Sect. "National Coastal Management Legislation, Policies and Plans"); Planning and Development (Sect. "National Planning and Development Legislation, Policies and Plans"); and Climate Change (Sect. "National Climate Change Legislation, Policies and Plans"). Section "Limitations of the Work" presents the limitations of this research.

Coastal Climate Adaptation in Ireland

The examination revealed that Ireland has a relatively limited legal framework that is directly applicable to coastal issues and that national ICZM policy is absent. Instead, the country relies on a range of plans and policies, supported by specific legal instruments that deal with various aspects of coastal zone planning and management. The categories reflected were organized thematically into Coastal Management; Planning and Development; and Climate Change (including Biodiversity); and will be discussed hereafter.

National Coastal Management Legislation, Policies and Plans

Ireland does not have any dedicated Integrated Coastal Zone Management legal instrument, plan or policy, however, a number of existing instruments impact upon how the coast is managed. The Foreshore Acts, 1933–2011 require that leases and licences be obtained from the Minister for Housing, Local Government and Heritage for the carrying out of works and placing structures on the State-owned foreshore[2] (including sea defence works such as walls, piers, groynes, etc.). Projects likely to have significant effects on the environment are subject to an Environmental Impact Assessment (EIA) before receiving the licence and the Acts provide for public consultation as part of the decision-making process. The Foreshore Acts also contain provisions relating to the removal/deposit of beach material, public use of the foreshore, and erection/removal of structures on the foreshore. Whilst useful in

[2] The foreshore is defined as the shore and the seabed, below the line of high water of ordinary or medium tides to the 12 nautical miles of the sea, including tidal river/estuaries, channels, creeks, and bays.

specific situations, the instrument was not designed to facilitate integrated coastal management.

The Coast Protection Act, 1963 specifies the procedures for the implementation of coastal protection schemes in cases of encroachment by the sea and clarifies competencies between the responsible department at the national level (Office of Public Works/OPW) and Local Authorities (LAs). In 2009, the OPW published the Minor Flood Mitigation Works and Coastal Protection Scheme, (OPW 2009) a document that guides the provision of funding to LAs for the installation of minor[3] flood mitigation and coastal protection works and/or to studies addressing localised problems. The Scheme indicates that LAs should deliver an appropriate coastal erosion risk management plan indicating appropriate options and measures such as 'Do Nothing', 'Do Minimum', 'Hold the Line', 'Advance the Line', and 'Managed Realignment' options. This is the only policy that provides guidance for the implementation of coastal climate adaptation solutions, even though with an incomplete and non-statutory approach.

In order to provide information to support decision making for a better management of risks related to coastal erosion and flooding in Ireland, the OPW commissioned the Irish Coastal Protection Strategy Study (ICPSS) (2003–2013). The study provided maps regarding current and future flood and erosion scenarios (up to 2100) and assessed the existing and potential risks for each sector of the Irish coast (OPW 2010). However, until the publication of this study, the government has made no progress in developing coastal frameworks to prepare for continuing changes in the climate. To fill this gap, the Government launched the National Coastal Change Management Strategy Steering Group in 2020, with the aim to develop a framework to guide decision-making for improved coastline management and to mitigate the risks from increasing storm surge events and coastal erosion (Government of Ireland 2020). This group is made up of senior officials from various departments of the national government as well as the County and City Management Association. It is assumed that the Covid-19 global pandemic has impacted upon the original timelines for the work of this group.

Following the EU's adoption of the Maritime Spatial Planning Directive (2014/89/EU), Ireland launched its National Marine Planning Framework (NMPF) in 2021. The NMPF outlines the government's vision, objectives and policies for marine-based human uses activities, detailing how these should interact with each other to safeguard the sustainability of marine resources up to 2040 (DHLGH 2021). This framework is operated via the Maritime Area Planning Act, 2021 (MAPA), which provides a statutory basis for the planning and management of the Irish marine area, sets out a process for the designation of maritime plan areas, and modernises the consent system for marine developments. MAPA creates a new nearshore area — an area that extends from the high water mark up to three nautical miles seaward where 'Coastal Planning Authorities' (CPA) will assume certain responsibilities, such as the evaluation of applications for developments that do not require EIAs and Appropriate

[3] Projects that cost no more than €750 k.

Assessments, the granting of development consents, and enforcement and compliance. However, the new Act does not provide for Integrated Coastal Zone Management or the management of coastal erosion and flooding, other than to update the provisions relating to emergency works in the maritime area, which includes works relating to sea defences. Integrated Coastal Zone Management requires actions that go beyond the nearshore area, including solutions further inland. There is also limited consideration of how the marine planning and land planning systems will operate in parallel or integrated.

National Planning and Development Legislation, Policies and Plans

Planning and development legislation, policies and plans set out the planning and investment strategies to provide national and local governments with detailed guidance on building a sustainable future in Ireland, including coastal and climate actions. The Local Government Acts, 1925–2019, define the structures, functions, powers and duties of local government with respect to land area planning; economic development; housing; community development; and environment (such as noise, waste and air pollution). The Acts provide Local Authorities (LAs) with the power to enact bye-laws, which can be used, amongst other things, to encourage protection of localized coastal environments and may assist climate mitigation and adaptation. By stipulating enforcement and associated fines, bye-laws, for instance, prevent trampling on dune systems, control litter pollution and animal grazing on coastal environments (O'Mahony et al. 2012).

The Planning and Development Acts, 2000–2021 form the foundation of the terrestrial planning system and set out the details to be contained in development plans at national level (such as the National Planning Framework), regional level (such as the Regional Spatial and Economic Strategy), and local levels (County Development Plans and Local Area Plans). The Acts emphasise the need to carry out flood risk assessment to regulate, control and restrict developments in flooding risk areas; and to regulate, control and restrict developments in areas at risk of coastal erosion and threatened by other natural hazards.

A number of strategic planning frameworks have been published to strengthen and future proof the planning system. The National Planning Framework (NPF)—Project Ireland 2040 (Government of Ireland 2018a) and its predecessor, the National Spatial Strategy 2002–2020 (Government of Ireland 2002), are the most recent frameworks, published since the beginning of this century. These set out the objectives for Ireland's future spatial development (under the National Development Plans) and highlight environmental challenges, including the management of coastal and marine resources, policy response to coastal erosion and flood risks and the conservation of biodiversity and natural heritage sites. Particularly, the current framework, the

NPF, uses the United Nations Sustainable Development Goals (SDGs) to guide long-term planning, comprising waste management, sustainable management of water and additional environmental resources. It encourages a sequence of climate actions in order to tackle climate change in line with the National Mitigation Plan (DCCAE 2017) and the National Adaptation Framework (DCCAE 2018a) (presented in Sect. "National Climate Change Legislation, Policies and Plans"). The NPF suggests the use of grey adaptation and green adaptation solutions to deal with the impacts of climate change nationwide, and define these respectively as:

> *Grey adaptation which typically involves technical or engineering-oriented responses to climatic impacts, such as the construction of sea walls in response to a sea level rise.*
>
> *Green adaptation which seeks to use ecological properties to enhance the resilience of human and natural systems in the face of climate change, such as creation of green spaces and parks to enable better management of urban micro-climates.* (Government of Ireland 2018a, p. 120)

The National Development Plan (NDP) is a national level strategic roadmap aimed at delivering long-term economic, social and environmental development goals during a specified time period. The NDPs provide for investment in projects for sustainable development such as climate change strategies and flood risk management. The 2007–2013 NDP ring-fenced over €23 million to conduct the Irish Coastal Protection Strategy Study (ICPSS) and fund schemes to protect the coast from flooding and erosion (Government of Ireland 2007). These investments include assistance with planning and development, guidance and procedures for scheme selection, risk evaluation, and hard and soft engineering projects. The current NDP (2018–2027) identified as a priority to invest in the transition to a low-carbon and climate-resilient society and approximately €21.8 billion should be invested to achieve this strategic outcome (Government of Ireland 2018b). It also directs investment towards actions focused on flood risk management and recommends flood relief schemes to minimise the impacts flooding on coastal communities. However, the NDP does not explicitly provide for investments dealing with coastal erosion alone or more integrated approaches to coastal management (Stori and O'Mahony 2021).

National Climate Change Legislation, Policies and Plans

Initial steps to address climate change were made in 2014 with the publication of the National Policy Position on Climate Action and Low Carbon Development, which aimed to deliver a high-level policy guidance for the government adoption and implementation of strategies to support the State to reach a low carbon economy by 2050 (Government of Ireland 2014). This policy offers a long-term vision with a goal to reduce carbon dioxide (CO_2) emissions by at least 80% by 2050 (with reference to 1990 concentrations) considering sectors such as built environment, power generation and transport; as well as carbon neutrality in the land-use and agriculture sectors.

This policy recommends the adoption of a succession of national plans over the period, including the development of climate adaptation and mitigation frameworks.

Therefore, the Climate Action and Low Carbon Development Acts, 2015–2021 was published and it is the core legislation to institute the national climate framework. The Act's objective is "*to reduce the extent of further global warming, pursue and achieve, by no later than the end of the year 2050, the transition to a climate resilient, biodiversity rich, environmentally sustainable and climate neutral economy*" (Government of Ireland 2021a, p. 8). The 2021 Act brought advances to the original Act by including biodiversity and Net Zero approaches within its scope. The Acts set out definitions and the legal basis for the National Adaptation Framework, the National Mitigation Plan, Sectoral Adaptation Plans, the National Long Term Climate Action Strategy, and required LAs to formulate their Local Climate Action Plans including adaptation and mitigation actions. The Acts establish the Climate Change Advisory Council and encourage citizen participation in the formulation of the aforementioned plans.

In 2012, Ireland published its first National Climate Change Adaptation Framework (NCCAF), based on the recommendations of the EU White Paper on Adaptation, and indicated a two-phased adaptation approach to Ireland. The first stage focused on the identification of climate change vulnerabilities at the national level, based on potential impacts when considering the adaptive capacity at that moment. The second stage focused on the elaboration and implementation of sectoral and local adaptation plans. The NCCAF acknowledged the Local Development Plans as the principal strategy to implement climate adaptation solutions at the local level (DECLG 2012). Subsequently, after the enactment of the Climate Action and Low Carbon Development Act in 2015, the current National Adaptation Framework (NAF) published in 2018, sets out the international policy context for climate and brought forward the up-to-date objectives for low carbon, climate resilience and sustainable development in Ireland (DCCAE 2018a). This framework provides guidance for the implementation of adaptation actions in different sectors and by LAs in their jurisdiction. However, it does not propose projects or adaptation actions for the sectors and local jurisdiction, considering that these should be prepared under the sectoral and local plans. The NAF is currently under review (May–July 2022) in order to upgrade the framework in line with the requirements of the Climate Action and Low Carbon Development Act, 2021. This review aims to guide national adaptation priorities over the coming years, so there is scope to progress on a range of adaptation solutions in future iterations.

In 2019, the first national Climate Action Plan (CAP) formally declared a "Climate and Biodiversity Emergency" in Ireland. The Plan consisted of a government guidance proposing a series of policy actions to accomplish by 2030 a "*net zero carbon energy systems objective for Irish society and in the process, create a resilient, vibrant and sustainable country*" (Government of Ireland 2019, p. 8). It outlined climate threats across diverse sectors by recognising the scale and nature of the challenges. It set out the governance structure responsible to deliver transformations, including the establishment of a Climate Action Council and a Climate Action Delivery Board. The Plan recognised the upcoming impacts of climate change on the Irish coastal zone and highlighted the demand for adaptation solutions to assist in addressing those impacts. A new CAP was published in November 2021 setting a goal of a 51% reduction in emissions by 2030 and achieving net zero by 2050. The 2021 Plan details a series of actions required to achieve those climate targets and indicates emissions reductions targets for each economic sector, such as transport, electricity, agriculture and built environment. A section on climate adaptation covers 9 pages of the 211 page Plan. Main actions related to coastal erosion and flooding are to "*develop and publish coastal vulnerability mapping and coastal erosion databases for the east and south coasts of Ireland; delivery of a national implementation strategy for nature-based solutions to the management of rainwater and surface water runoff in urban areas; and develop flood forecasting capability, including the development of coastal and marine monitoring and predictive capability*" (Government of Ireland 2021b, pp. 204–205). However, the CAP does not provide clear deadlines for carrying out these actions.

First published in 2016, then updated in 2018, the Local Authority Adaptation Strategy Development Guidelines (DCCAE 2018b) provides a rational step-by-step methodology to adaptation planning in order to support the LAs in preparing their adaptation schemes and ensure that Local Plans agree with the Sectoral Adaptation Plans. It recognizes that LAs maintain close relations with local communities and are the first to respond to crises, so these are considered the best level of government to implement adaptation actions on the ground. Four Climate Action Regional Offices (CAROs) were established in 2018 to assist LAs to progress on their adaptation schemes and support their implementation.

The Sectoral Planning Guidelines for Climate Change Adaptation (DCCAE 2018c) were prepared to support different Government Departments to develop their statutory Sectoral Adaptation Plans related to their core competency, ensuring the adoption of a rational approach to adaptation planning by the key sectors. Sectors were required to develop their plans according to a proposed six-step planning cycle while also following the general requirements stipulated by the Acts and the NAF. According to the Sectoral Planning Guidelines, adaptation encompasses hands-on actions to reduce vulnerability to the negative impacts of climate change and improve opportunities or benefits. These comprise a variety of actions that can be categorised as grey, green, and soft and may range from simple solutions to large-scale projects:

> *Grey adaptation: It involves technical or engineering solutions to climate impacts, examples include raising roads where flooding is expected to occur.*
>
> *Green adaptation: It seeks to utilise ecological properties to enhance the resilience of human and natural systems to climate change impacts. For example, increasing green space in urban areas could provide areas for retention of floodwaters and significantly ameliorate the impacts of rising surface temperatures resulting from climate change;*
>
> *Soft adaptation: It involves alteration in behaviour, regulation or system of management, examples include: Extending timeframes of plans further into the future; zoning development away from sensitive areas; and instituting or strengthening building codes in hazard prone areas.* (DCCAE 2018c, pp. 39–40)

In total, 12 Sectoral Adaptation Plans were prepared, including one on "Flood Risk Management" and another on "Biodiversity" which provide particularly relevant adaptation approach for coastal zones.

The first Flood Risk Management, Climate Change Sectoral Adaptation Plan was published in 2015 under the guidance of the NCCAF (2012) and the second, which is currently in operation, was published in 2019 in line with the NAF (2018). The 2019 Flood Risk Management Plan updates the former Plan by incorporating more recent information on climate change, particularly the impacts on flood risk management (OPW 2019). This includes progress made on the actions proposed by the former adaptation plan, especially in the areas of research and assessment (such as the information generated through the Catchment Flood Risk Assessment and Management "CFRAM" Programme[4] (OPW 2018)), and on planning and implementation of flood relief schemes. The government focus on such plans and programmes evidence that solving flooding issues is a priority over solving problems posed by sea level rise (such as coastal erosion).

The Biodiversity Climate Change Sectoral Adaptation Plan (BCCSAP) was published in 2019 taking account of the Wildlife Acts, 1976–2012 and the National Biodiversity Action Plan (NBAP). The Wildlife Acts, 1976–2012, are the main national legislation for wildlife conservation and control of activities that may adversely affect wildlife in light of Ireland's commitment to the UN Convention on Biological Diversity (CBD). The NBAPs (published in 2002, 2011 and 2017) outline the objectives, targets and actions that should be accomplished to realise Ireland's vision for biodiversity conservation. The plan in operation (NBAP 2017–2021) postulates the vision that "*biodiversity and ecosystems in Ireland are conserved and restored, delivering benefits essential for all sectors of society and that Ireland contributes to efforts to halt the loss of biodiversity and the degradation of ecosystems in the EU and globally*" (DCHG 2017, p. 8). By considering both seminal frameworks for biodiversity, the main goal of the BCCSAP is "*to protect biodiversity from the impacts of climate change and to conserve and manage ecosystems so that they deliver services that increase the adaptive capacity of people and biodiversity while also contributing to climate change mitigation*" (DCHG 2019, p. 9). The BCCSAP highlights that biodiversity is highly vulnerable to the impacts of climate change whilst playing a key role in climate mitigation and adaptation. It acknowledges coastal ecosystems also highly vulnerable to the impacts of climate change owing to the effects of changes in temperature combined with further impacts of sea level rise. The Plan highlights the need to "*develop an integrated coastal management strategy which includes ecosystem-based adaptation actions to manage climate risk and build resilience to climate change*" (DCHG 2019, p. 49). The Plan recommends employing appropriate adaptation solutions to reduce vulnerability to the impacts of

[4] The CFRAM Programme is an extensive flood risk assessment performed by the OPW in cooperation with LAs, comprising a comprehensive assessment of 300 areas considered to be at high risk from flooding, including 90 areas located in coastal regions (OPW 2018). The study encompassed six CFRAM projects across 29 river basins and other site-specific projects.

climate change, such as the implementation of nature-based solutions jointly with screening for maladaptation in order to enhance the likelihood of mutually beneficial climate actions (DCHG 2019).

Limitations of the Work

Coastal zones in Ireland are subject to a wide variety of legislation, policies and plans, operated by different actors at different governance scales and this has been the case for many decades. With increasing awareness of climate change impacts and international legal commitments, adaptation actions have had to be embedded into this already fragmented coastal management system. The analysis conducted as part of this work indicates certain shortcomings resulting from this fragmentation, for example, contradictions between adaptation solutions proposed by different sectors, different priorities because of wider non-climate related policy objectives, and the rapidly changing policy environment that currently characterises adaptation and climate change policy more generally.

One other potential limitation of this work is that it focuses entirely on stated law and policy objectives and less on the actual implementation of adaptation actions. This is important and relevant for two reasons: firstly, to implement adaptation options anywhere there needs to be a solid policy basis and sometimes this can be difficult to determine given the range of applicable legislation and policies. Secondly, it is possible that where a specific adaptation option is being considered, the responsible actors may meet and discuss with other regulatory stakeholders to ensure there are no issues. This type of case specific scenario is not considered in this documentary review. Our analysis seeks to highlight the complexity of implementing adaptation in coastal zones where there is no policy basis for integrated management and uncovers a diverse range of actors which ultimately makes actions more difficult.

Final Considerations

Whilst Ireland has an extensive legal and policy framework, the current governance system has proven to be unresponsive to past and current coastal changes in many localities along the Irish coast, evidenced by the lack of an assertive approach to the coastal zone which is hindering fast decision-making to urgent coastal problems. Government is increasingly focusing and strengthening their approach to management of the impacts of climate change, including strengthened policy for adaptation across multiple sectors. In the majority of cases, this requires policy changes at sectoral level, as evidenced by the sectoral adaptation plans published. However, a problem arises in relation to coasts, as there is no national coastal management plan or policy and consequently nowhere into which adaptation solutions can be incorporated. This restricts the ability of local authorities, who usually assume the day-to-day management of coastal areas, to undertake adaptation solutions and might go some way towards explaining why different approaches to coastal issues occur in different parts of the country.

Ireland has made progress in the management of climate hazards in recent years; however, priority is clearly given to countrywide flooding issues to the detriment of coastal erosion management and perhaps wider coastal management. Coastal erosion remains a problem for many coastal communities and Local Authorities have been managing the problem on a reactive basis instead of a preventive basis due to the lack of a coherent approach at national scale.

Generic guidance on the implementation of certain types of adaptation solutions and differences in the terminology adopted were observed, revealing inconsistencies even among contemporaneous documents. A streamlined approach that facilitates a strategic and structured process for implementing coastal climate adaptation solutions on the ground, such as engineering (hard) solutions/grey adaptation (hold the line; advance the line), nature or ecosystem-based solutions/green adaptation, displacement solutions (such as managed retreat), and soft solutions such as preparedness and response systems is not reflected in any of the law or policy documents analysed. An additional issue could be contradictions between the various sectoral adaptation plans when viewed through a climate adaptation lens, which emphasises the need for an institutional body with the role of overseeing the implementation of adaptation solutions. This role could be allocated to the CAROs, but would require further resourcing.

An Integrated Coastal Zone Management policy at national level is recommended in order to provide clarity on how to implement ecosystem-based and technical solutions, appropriate funding support, and coordination mechanisms to ensure consistency in the governance system necessary to support the implementation of coastal climate adaptation solutions. It should be based on integrated, participatory and ecosystem-based approaches, which include coastal climate adaptation strategies and solutions as cross cutting themes interlinked to other coastal matters (biodiversity conservation, management of hydrographic basins, urban development and housing, tourism and leisure, and others). This would then provide a solid policy basis for local authorities to implement adaptation solutions, which strengthen coastal resilience, enable public participation, learning and democratic vision building in the process, and ultimately deliver more effective planning and management for sustainable social-ecological systems.

In terms of future research related to this topic, there is significant scope to conduct an in-depth analysis of the application of the current governance system at a local scale focusing on specific coastal adaptation challenges. This could take the form of developing and piloting a step-by-step methodology on how to incorporate adaptation into ICZM at various governance scales to demonstrate consistency and coherency across policy ambitions, also involving actors at all levels.

References

Bassetti F (2020) Preparing for rising sea levels. Foresight CMCC. https://www.climateforesight.eu/articles/preparing-for-rising-sea-levels/. Accessed 9 Nov 2020

Cicin-Sain B (1993) Sustainable development and integrated coastal management. Ocean Coast Manag 21:11–43. https://doi.org/10.1016/0964-5691(93)90019-U

Cronin A, Kandrot S, Gault J, Devoy R, O'Hagan, AM, O'Mahony C, Nuyts S, (2017) Local authority coastal erosion policy and practice audit. MaREI Centre, Environmental Research Institute, University College Cork. Commissioned by Fingal County Council. 126p. https://www.fingal.ie/sites/default/files/2020-11/local-authority-coastal-erosion-policy-and-practice-audit-final_0.pdf. Accessed 25 Sep 2020

DCCAE/Department of Communications, Climate Action & Environment (2017) National mitigation plan. 200p. www.dccae.ie/documents/National%20Mitigation%20Plan%202017.pdf. Accessed 22 Jan 2021

DCCAE/Department of Communications, Climate Action & Environment (2018a) National adaptation framework: planning for a climate resilient Ireland. 110p. https://www.gov.ie/en/publication/fbe331-national-adaptation-framework/. Accessed 22 Jan 2021

DCCAE/Department of Communications, Climate Action & Environment (2018b) Local authority adaptation strategy development guidelines. 65p. https://www.gov.ie/en/publication/41066-local-authority-adaptation-strategy-development-guidelines/. Accessed 28 Jul 2021

DCCAE/Department of Communications, Climate Action & Environment (2018c) Sectoral planning guidelines for climate change adaptation. 65p. https://www.gov.ie/en/collection/51df3-sectoral-adaptation-planning/. Accessed 20 Feb 2020

DCHG/Department of Culture Heritage and the Gaeltacht (2017) National biodiversity action plan. 88p. https://www.npws.ie/sites/default/files/publications/pdf/National%20Biodiversity%20Action%20Plan%20English.pdf. Accessed 22 Jan 2021

DCHG/Department of Culture, Heritage and the Gaeltacht (2019) Biodiversity climate change sectoral adaptation plan, prepared under the national adaptation framework. 87p. https://www.npws.ie/sites/default/files/publications/pdf/Biodiversity-Climate-Change-Sectoral-Adaptation-Plan.pdf. Accessed 8 Apr 2021

DECLG/Department of the Environment, Community and Local Government (2012) National climate change adaptation framework. 73p. https://www.gov.ie/en/publication/df8e2-national-climate-change-adaptation-framework/. Accessed 20 Feb 2020

Devoy RJN (2008) Coastal vulnerability and the implications of sea-level rise for Ireland. J of Coastal Res 24(2):325–341. West Palm Beach (Florida), ISSN 0749–0208. https://doi.org/10.2112/07A-0007.1

DHLGH/Department of Housing, Local Government and Heritage (2021) National marine planning framework. 210p. https://www.gov.ie/en/publication/60e57-national-marine-planning-framework/. Accessed 28 Jul 2021

Dwyer N (2012) The status of Ireland's climate 2012. Environmental Protection Agency, Johnstown Castle, Ireland. http://www.epa.ie/pubs/reports/research/climate/CCRP26%20-%20Status%20of%20Ireland%27s%20Climate%202012.pdf. Accessed 25 Nov 2021

Farrell EJ (2021) Case study: maharees conservation association. In: Management of the coasts and the marine environment. In: Devoy R, Cummins V, Brunt B, Bartlett D, Kandrot S (eds) The coastal atlas of Ireland. Cork University Press. pp.755–759. ISBN 10: 1782054510

Flood S (2012) Climate change and potential economic impacts in Ireland: the case for adaptation. PhD thesis. Maynooth University, Ireland. http://mural.maynoothuniversity.ie/ethesis/view/ . Accessed 25 November 2021

Flood S, Sweeney J (2012) Quantifying impacts of potential sea-level rise scenarios on Irish coastal cities. In Otto-Zimmermann K (ed) Resilient cities 2. Local sustainability, Vol. 2. Springer, Dordrecht, the Netherlands. https://doi.org/10.1007/978-94-007-0785-6

Flood S, Paterson S, O'Connor E, O'Dwyer B, Whyte H, Le Tissier M, Gault J (2020) National risk assessment of impacts of climate change: bridging the gap to adaptation action. EPA Research Report (2016-CCRP-MS.39). 78p. ISBN: 978–1–84095–948–2

GCA/Global Commission on Adaptation (2019) Adapt now: a global call for leadership on climate resilience. Global Center on Adaptation and World Resources Institute. 90p. https://gca.org/global-commission-on-adaptation/report. Accessed 29 Oct 2019

GESAMP (1996) The contributions of science to integrated coastal management. GESAMP Reports and Studies, n.61. FAO, Rome 65p. ISBN 92–5–103856–2

Government of Ireland (2002) The national spatial strategy 2002–2020. 160p. http://www.housing.old.gov.ie/planning/national-spatial-strategy/national-spatial-strategy. Accessed 22 Jan 2021

Government of Ireland (2007) National development plan (NDP) "Transforming Ireland—a better quality of life for all" (2007–2013). 265p. http://www.socialinclusion.ie/documents/NationalDevelopmentPlan2007-2013.pdf. Accessed 09 Feb 2021

Government of Ireland (2014) National policy position on climate action and low carbon development. https://www.gov.ie/en/publication/6f393-national-climate-policy-position/. Accessed 23 Feb 2021

Government of Ireland (2018a) National planning framework—project Ireland 2040. 182p. https://npf.ie/wp-content/uploads/Project-Ireland-2040-NPF.pdf. Accessed 13 Dec 2019

Government of Ireland (2018b) National development plan (2018–2027). 109p. https://assets.gov.ie/19240/62af938dce404ed68380e268d7e9a5bb.pdf. Accessed 12 Aug 2020

Government of Ireland (2019) Climate action plan. 150p. https://www.gov.ie/en/publication/ccb2e0-the-climate-action-plan-2019/. Accessed 22 Jan 2020

Government of Ireland (2020) National coastal change management strategy steering group meets for first time. https://www.gov.ie/en/press-release/7b418-national-coastal-change-management-strategy-steering-group-meets-for-first-time/. Accessed 3 Sep 2020

Government of Ireland (2021a) Climate action and low carbon development (Amendment) Act 2021. http://www.irishstatutebook.ie/eli/2021a/act/32/enacted/en/print.html?printonload=true. Accessed 27 Jul 2021

Government of Ireland (2021b) Climate action plan 2021: securing our future. 208p. https://www.gov.ie/en/publication/6223e-climate-action-plan-2021b/. Accessed 05 Nov 2021

IPCC (2018a) Summary for policymakers. In: Masson-Delmotte V, Zhai P, Pörtner H-O, Roberts D, Skea J, Shukla PR, Pirani A, Moufouma-Okia W, Péan C, Pidcock R, Connors S, Matthews JBR, Chen Y, Zhou X, Gomis MI, Lonnoy E, Maycock T, Tignor M, Waterfield T (eds) Global warming of 1.5°C. An IPCC special report on the impacts of global warming of 1.5°C above pre-industrial levels and related global greenhouse gas emission pathways, in the context of strengthening the global response to the threat of climate change, sustainable development, and efforts to eradicate poverty, World Meteorological Organization, Geneva, Switzerland, 32 p. https://www.ipcc.ch/sr15/chapter/spm/. Accessed 5 Jan 2021

IPCC (2018b) Annex I: Glossary In: Matthews JBR (ed) In: Masson-Delmotte V, Zhai P, Pörtner H-O, Roberts D, Skea J, Shukla PR, Pirani A, Moufouma-Okia W, Péan C, Pidcock R, Connors S, Matthews JBR, Chen Y, Zhou X, Gomis MI, Lonnoy E, Maycock T, Tignor M, Waterfield T (eds) Global Warming of 1.5°C. An IPCC Special Report on the impacts of global warming of 1.5°C above pre-industrial levels and related global greenhouse gas emission pathways, in the context of strengthening the global response to the threat of climate change, sustainable development, and efforts to eradicate poverty 24p. https://www.ipcc.ch/site/assets/uploads/sites/2/2019/06/SR15_AnnexI_Glossary.pdf. Accessed 5 Jan 2021

IPCC (2019) Summary for policymakers. In: Pörtner H-O, Roberts DC, Masson-Delmotte V, Zhai P,Tignor M, Poloczanska E, Mintenbeck K, Nicolai M, Okem A, Petzold J, Rama B, Weyer N (eds) IPCC special report on the ocean and cryosphere in a changing climate, Geneva: IPCC. 36p. https://report.ipcc.ch/srocc/pdf/SROCC_SPM_Approved.pdf. Accessed 15 Apr 2020

IPCC (2021) Summary for policymakers. In Masson-Delmotte VP, Zhai A, Pirani SL, Connors, CPan S, Berger N, Caud Y, Chen L, Goldfarb MI, Gomis M, Huang K, Leitzell E, Lonnoy JBR, Matthews TK, Maycock T, Waterfield O, Yelek iRYu, Zhou B (eds) Climate change 2021:

the physical science basis. Contribution of working group I to the sixth assessment report of the intergovernmental panel on climate change. Cambridge University Press. 42p. https://www.ipcc.ch/report/ar6/wg1/. Accessed 25 Aug 2021

Kulp SA, Strauss BH (2019) New elevation data triple estimates of global vulnerability to sea-level rise and coastal flooding. Nat Commun 10:4844. https://doi.org/10.1038/s41467-019-12808-z

Lindsey R (2020) Climate change: global sea level. NOAA Climate.gov. https://www.climate.gov/news-features/understanding-climate/climate-change-global-sea-level. Accessed 17 Dec 2020

O'Mahony C, O'Hagan AM, Meaney E (2012) A review of beach bye-law usage in supporting coastal management in Ireland. Coast Manag 40(5):461–483. https://doi.org/10.1080/08920753.2012.709462

OPW/Office of Public Works (2009) Minor flood mitigation works and coastal protection scheme. 11p. https://www.gov.ie/en/publication/0e3b3d-minor-flood-mitigation-works-and-coastal-protection-scheme/. Accessed 28 Jul 2021

OPW/Office of Public Works (2010) Irish coastal protection strategy study—phase 3. North East Coast. Technical Report. 115p. https://www.gov.ie/en/collection/d52a5e-irish-coastal-protection-strategy-study-phase-3-north-east-coast/#main-technical-report. Accessed 02 Jun 2021

OPW/Office of Public Works (2018) Catchment flood risk assessment and management (CFRAM) programme. https://www.gov.ie/en/collection/396090-catchment-flood-risk-assessment-and-management-plans/. Accessed 19 Mar 2021

OPW/Office of Public Works (2019) Flood risk management—climate change sectoral adaptation plan. 146p. https://www.gov.ie/pdf/?file=assets.gov.ie/46534/3575554721374f7ab6840ee11b8b066a.pdf#page=1. Accessed 15 Jan2021

Smith G, LeTissier M, O'Hagan AM, Farrell EJ (2021) Policy coherence for climate change adaptation at the land-sea interface in Ireland. Plan Pract Res 37:173–188. https://doi.org/10.1080/02697459.2021.1991657

Stori FT, O' Mahony C (2021) Coastal climate adaptation in Ireland: the effects of climate change in Portrane (Fingal, Co. Dublin) and future perspectives. MaREI the SFI Research Centre for Energy, Climate and Marine; Environmental Research Institute; University College Cork. Coastal Communities Adapting Together—CCAT Project. 202p. https://cora.ucc.ie/handle/10468/12264. Accessed 23 Nov 2021

UN/United Nations (2017) The ocean conference fact sheet. https://www.un.org/sustainabledevelopment/wp-content/uploads/2017/05/Ocean-fact-sheet-package.pdf. Accessed 10 Dec 2020

Chapter 8
The Influence of Climate Factors in the Distribution of Birds

Walter Leal Filho, Newton R. Matandirotya, and M. Mahendiran

Abstract In atmospheric sciences, teleconnections refer to the persistent cyclical features/oscillations that produce abnormal weather, temperature, and precipitation patterns co-occurring over vast, seemingly unrelated areas, affecting continent-wide processes ranging from bird migration to plant reproduction. In this work, we explore the effects of climate change, particularly the climatic indices on the population dynamics of birds and the relationships of the teleconnection patterns on the bird migrations and distribution by critically examining the published literature on this interdisciplinary field. We also suggest a broader roadmap to advance the interdisciplinary research on climate change connecting bird migration studies in the following years ahead.

W. Leal Filho (✉)
Research and Transfer Centre "Sustainable Development and Climate Change Management", Hamburg University of Applied Sciences, Ulmenliet 20, 21033 Hamburg, Germany
e-mail: walter.leal2@haw-hamburg.de

Department of Natural Sciences, Manchester Metropolitan University, Chester Street, Manchester M1 5GD, UK

N. R. Matandirotya
Department of Geosciences, Faculty of Science, Nelson Mandela University, Port Elizabeth 6000, South Africa

Centre for Climate Change Adaptation and Resilience, Kgotso Development Trust, P.O. Box 5, Beitbridge, Zimbabwe

M. Mahendiran
Division of Wetland Ecology, Sálim Ali Centre for Ornithology and Natural History (SACON), Anaikatti (PO), Coimbatore, Tamil Nadu 641108, India

W. Leal Filho et al. (eds.), *Climate Change Strategies: Handling the Challenges of Adapting to a Changing Climate*, Climate Change Management,
https://doi.org/10.1007/978-3-031-28728-2_8

Introduction

The Sixth Assessment Report (AR6) by the Intergovernmental Panel on Climate Change (IPCC) predicts that the global mean temperatures will increase by 1.5–2 °C from pre-industrial levels, warming the climate of Earth at a rate unprecedented in 2,000 years, resulting in more frequent weather extremes, threatening the survival of flora and fauna (IPCC 2022). Climate change is an important driver of biodiversity loss across regions worldwide. The migratory systems of birds are among the ones that are highly affected by the rapid changes in climate across the world, and therefore, climate-driven changes pose crucial challenges for conservation and policy decisions on migratory species (Gill et al. 2019). Current data on the decline of migratory bird populations worldwide on all major flyways, has driven international calls for combined action, which is a much-needed effort at this time (Runge et al. 2015). In this paper, we explore the effects of climate change, chiefly the climatic indices, on the population dynamics of birds and the relationships of the teleconnection patterns on bird migrations and distribution, by critically examining the published studies in this interdisciplinary field. We also suggest a broader roadmap to advance this interdisciplinary research on climate change and bird migration in the years ahead.

As shown in Fig. 8.1 there are various elements that are directly or indirectly climate change-related which influence birds' migration and behaviours including climate hazards, habitat depletion, lack of protective policies and urbanization.

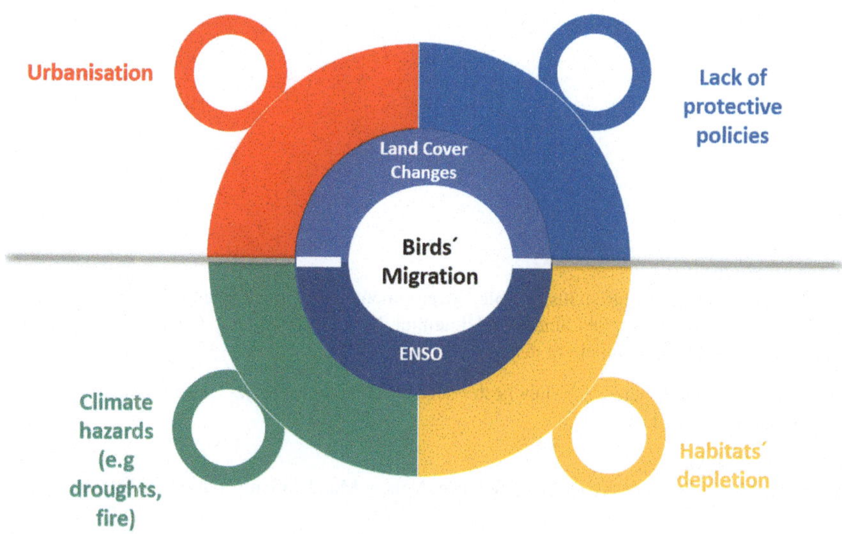

Fig. 8.1 Some of the climate-related elements that influence birds' migration (*Source* Authors)

Methods

We carried out an exhaustive literature review, and the pieces of literature we cited comprise the bulk of work carried out concerning climate change hypotheses connecting bird migration and teleconnections. The literature selection was primarily based on the study of birds with the following attributes (a) long-term study (b) data covering large geographic areas (c) studies connecting climate change patterns and teleconnections. Whenever we encountered any study at local or highly confounded by non-climatic global change factors, we excluded them.

We reviewed and synthesized information on the following topics: phenology and population dynamics, distribution of migratory birds, connections with teleconnections, influence of atmospheric factors and influences of climatic indices on bird migration and distribution. We did not perform any meta-analysis during this study, but signals of such requirements are highlighted in this work. We searched using academic databases, namely Scopus, Web of Science, PubMed, ScienceDirect, and JSTOR, with keywords such as "bird migration," "phenology," "teleconnections," and "Climate Change." Thousands of references are available on selected keywords from databases; however, only less than 500 citations were available in PubMed on the keyword "teleconnections," which indicates that this topic of research is the latest one and has constantly been increasing since 2011, which is the key of this work.

Results and Discussion

(A) Effects of climate change on population dynamics of birds

Climate change plays a vital role in the population dynamics of vertebrates, particularly birds, through the influence on their demographic parameters, such as survival, breeding success, and migration of individuals (Feldstein 2000; Rivalan et al. 2010). Hence, the studies that use birds as experimental models could easily quantify the effects of climate change and their inter-relationships. The early onset of spring on the breeding grounds corresponds to the early return of migrating birds from the wintering grounds; however, the climate indices of wintering grounds, particularly in African regions, highly influence the timing of spring migration (Remisiewicz and Underhill 2022). Climate change has produced numerous shifts in the abundance and distribution of species in the past (Gill et al. 2019). The advancement in phenological traits, including range shifts averaging 6.1 km per decade towards the pole or meters per decade upward, and the mean advancement of spring events by 2.3 days per decade was confirmed by meta-analyses (Parmesan and Yohe 2003; Knudsen et al. 2011). Based on mid-range climate-warming scenarios for 2050, it is estimated that 15–37% of sampled species committed to extinction (Thomas et al. 2004).

(b) Effects of climate change on the distribution of birds

Climate change affects biodiversity at an unprecedented rate; the human exploitation of Earth for agriculture, forestry, fisheries, and urbanization is equally catastrophic (Møller et al. 2013). Therefore, it is equally crucial to establish the linkages between climate and land-use changes. There is ample observational evidence of migratory birds becoming sedentary, wintering closer to their breeding grounds, and abandoning distant parts of the former wintering range (Knudsen et al. 2011). Whenever the migratory species fail to adjust to climate change, it often faces stiff competition from resident birds, which produces fatal interactions that lead to the mortality of migratory species (Samplonius and Both 2019). Nevertheless, we know little about departure dates from wintering areas wherein the bio-logging could be a potential solution for studying the whole migration phenomenon.

Interestingly, the long-distance migrants, particularly those crossing ecological barriers (Fig. 8.1), often show a smaller phenological response to climate change than short-distance migrants because of strong stabilizing selection on timing of arrival that often resulted in rigid endogenous control of migration synchronized by photoperiod in long-distance migrants. In contrast, the short-distance migrants allow adjustments of departure and arrival schedules according to weather conditions at their destination (Rivalan et al. 2010; Knudsen et al. 2011; Berthold 2001). Even the effects of local changes have a cascading influence on the abundance and distribution of species across ranges, where some species tend to relocate territories by shifting elevations and latitudinal distributions (Nunez et al. 2019; Alves et al. 2019).

Climate change poses a big problem to migratory birds (Figs. 8.2 and 8.3). It is negatively influencing their dynamics, and in some cases, their populations are declining, which is becoming a major conservation challenge (Møller 2008; Vickery et al. 2014; Gilroy et al. 2016; Studds et al. 2017). Nevertheless, the birds with greater within-population variability in migratory movements and destinations, termed as migratory diversity, help and facilitate species responses to climate change and equally might be more resilient to environmental change (Alves et al. 2019; Gilroy et al. 2016).

(c) Influences of teleconnection patterns on birds migration and distribution

In atmospheric science, climatic anomalies relating at a considerable distance, typically thousands of kilometers, are referred to as teleconnections. The climate signals propagate through the atmosphere and act as synchronizing agents leading to teleconnection patterns. Thus the teleconnections produce persistent cyclical features/oscillations producing abnormal weather, temperature, and precipitation patterns co-occurring over vast, seemingly unrelated areas (Feldstein 2000). The dominant teleconnection/oscillations are the El Niño–Southern Oscillation (ENSO, including warm El Niño, neutral phase, and cold La Niña phase [Fig. 8.3]), the Indian Ocean Dipole (IOD), North Atlantic Oscillation (NAO), Pacific–North American (PNA), Pacific Decadal Oscillation (PDO), Northern Pacific Oscillation (NPO), Arctic Oscillations (AO), Scandinavian Pattern (SCP), Southern Annular Mode (SAM), and Tropical Intra-seasonal Oscillation or Madden–Julian Oscillation (MJO)

Fig. 8.2 Some of the long-distance migration of bird species and their migratory routes showed in different colors (Berthold 2001), and the landcover using the ESA WorldCover project 2020. See their details in the figure and text

which create events of extreme natural disasters in both hemispheres (Feldstein 2000; Steptoe et al. 2018).

The IOD is a measure of change in Sea Surface Temperature (SST), often considered the counterpart of Pacific El Niño and La Niña (Steptoe et al. 2018). Among the oscillations, the MJO is the shortest with two opposing phases, namely the enhanced rainfall phase and the other suppressed rainfall phase producing a dipole that move eastwards with the disturbance of clouds, rainfall, winds, and pressure and traverse the planet in the tropics in 30–60 days on average. The ENSO events modulate the MJO activity on a regional scale, and therefore the seasonal MJO conditions could be anticipated from a predicted ESNO state (Dasgupta et al. 2021).

Global climate variability, otherwise termed atmospheric hazards, often makes climatic anomalies in terrestrial and marine systems that result in contrasting patterns in temperature or precipitation anomalies concurrent at two/more diverse geographic locations affecting continent-wide processes from bird migration to plant reproduction (Zuckerberg et al. 2020). Across the globe, areas for each teleconnection in the pie charts represent the relative contribution to each atmospheric hazard associated

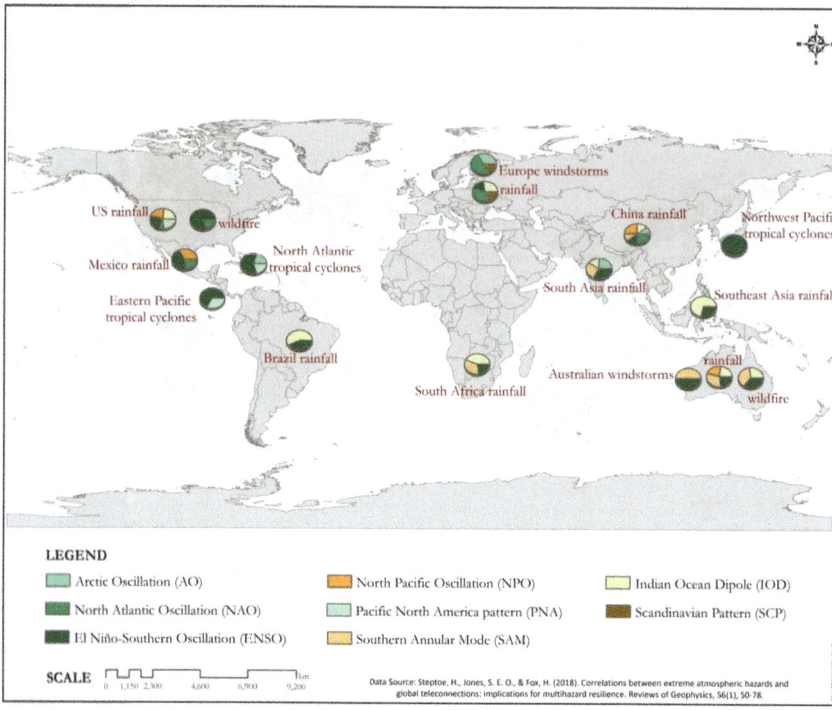

Fig. 8.3 Global distribution of some important (climate variability) atmospheric hazards and corresponding climatic drivers (teleconnections). The diagram is modified from Steptoe et al. (2018) with the permission. The areas in the pie chart for each teleconnection represent the relative contribution to each atmospheric hazard

with each region (Fig. 8.3). For example, five teleconnections drive the south Asia rainfall, and two teleconnections decide Brazil's rainfall. Given the number of teleconnection patterns operating across the globe, it is vital to further explore the links that connect the migratory patterns and population dynamics of birds with atmospheric hazards. When connecting the information in Figs. 8.2 and 8.3, it is possible to appreciate the number of teleconnections operating worldwide and their effects on long-distance migratory birds, say for example, the distributional range of the Short-tailed Shearwater (Fig. 8.3). Therefore, the role of teleconnections across the migrating range of Short-tailed shearwater in shaping its biology is inevitable and predictable (Price et al. 2021).

The teleconnections make the physical-biological linkage mechanisms between a climate signal and populations, and it directly & indirectly affects fish & birds through the primary production and even cascades through the food chain (Rivalan et al. 2010; Steptoe et al. 2018). SST is linked to the vertical mixing of the water masses, for it controls the quality of the physicochemical environment, ultimately pelagic primary production. The vertical mixing layer and high SST reduction are generally poor growing conditions for zooplankton in which the marine pelagic systems mostly have bottom-up regulation. The effect of climate forcing on zooplankton would cascade upwards through the food web. These environmental conditions, combined with climatic variables, affect the whole food chain by changing food quality or availability at all trophic levels (Rivalan et al. 2010).

The teleconnection has both the nature of dipole and quadrapole in their spatial structure. The NAO on one side over Greenland and the other in midlatitudes is the best example of a dipole, while the PNA has a quadrapole with a center over the subtropical Pacific near Hawaii, south of the Aleutian Islands, north-western Canada, and the south-eastern United States (Feldstein 2000). Through the teleconnection/climatic indices, one could measure changes in spatial patterns in the atmosphere linking remote locations across the globe. Therefore, understanding the teleconnection/climatic indices is essential for advancing migration and phenology theories, particularly the underlying determining factors for long-distance bird migrations (Steptoe et al. 2018; Zuckerberg et al. 2020; Wynn et al. 2022).

(d) **The influence of atmospheric factors on directing migration**

Atmospheric winds (e.g., 300-hPa meridional wind) and their directions and velocity equally play an important role in bird migration (Feldstein 2000, 2003; Hameed et al. 2009). The bio-logging studies helped understand how atmospheric conditions create freeways, detours, and tailbacks for migration birds across the globe (Shamoun-Baranes et al. 2017). A recent review suggested that more research is needed to separate the effects of various drivers (tailwinds) and impacts (enhanced food availability) on the bird migration patterns (Knudsen et al. 2011). Details of bird migration strategies, namely how individual birds may adjust/stop, when, where, and how to fly in response to atmospheric and geographic conditions, assume importance (Wynn et al. 2022; Feldstein 2003; Shamoun-Baranes et al. 2017). Changes in wind conditions become detrimental, especially for species that depend on the wind to cross ecological barriers, which often result in the reduction of suitable habitats, as observed in the Oriental honey-buzzard in its autumn migration (Nourani et al. 1854). Inter-Tropical Convergence Zone over the Indian sub-continent produces south-west and retreating (northeast) monsoons that aids the migratory movements of species such as the Great Cormorants (*Phalacrocorax carbo*) and Indian Cormorant (*Phalacrocorax fuscicollis*) (Mahendiran 2010, 2016).

The effects of local weather en route on the migration, the migrating approaches used by birds (Knudsen et al. 2011) and the sense of the magnetoreception of long-distance migratory birds (Wynn et al. 2022) are crucial in understanding the phenomenon of migration. Many long-distance migrants such as Arctic tern (*Sterna paradisaea*), Pacific Golden Plover (*Pluvialis fulva*), Swainson's Hawk (*Buteo swainsoni*), and Short-tailed Shearwater (*Ardenna tenuirostris*) are some ideal model species to understand the role of magnetoreception, i.e. a sense allows birds to detect the Earth's magnetic field (Fig. 8.2).

The next-generation radar (NEXRAD) is a network of 160 high-resolution S-band Doppler weather radars jointly operated by National Weather Services (NWS), the federal Aviation Administration (FAA) and the U.S. Air Force. Using the NEXRAD archive, the bird migration forecast model explores the associations between atmospheric conditions and bird migration intensity at a continental scale, explaining up to 81% variation in migration intensity across the United States (Doren and Horton 2018). A real-time analysis, based on actual nocturnal bird migration intensities detected by the United States weather surveillance radar network, shows its close association with local sunset and sunrise. The e-bird databases, citizen's science projects, and ecological networks like National Ecological Observatory Network have become more spatially distributed and recognized, revealing the broad patterns that link the ecological and atmospheric drivers and help us make several conservation actions (Bauer et al. 2019; Liechti et al. 2019; La Sorte and Somveille 2021).

(e) **Influences of the climatic indices on the bird migration and distribution**

Studies exploring the relationship between climatic indices (NAO, ESNO, IOD) and avian population dynamics are at the forefront (Rivalan et al. 2010; Møller et al. 2013; Berthold 2001; Zuckerberg et al. 2020; Feldstein 2003). The El Niño-Southern Oscillation (ENSO) has been linked to the survival of northern birds wintering in Africa and the short-distance migrants showing advanced arrival dates in years with a positive NAO index. There was a negative relationship between birds' total migratory population size and winter ENSO and MJO (Adams 2014). At the same time, the IOD determines climate change impacts on the East African ecosystem, particularly on avian life cycles (Tryjanowski et al. 2013).

The short-tailed shearwaters (*Ardenna tenuirostris*) are one of the long-distance migratory birds (Fig. 8.1), breeding in Tasmania and Australia, and their breeding success was influenced by large-scale climate oscillations (NPO, PDO, & ENSO) and weather conditions of the preceding year (Price et al. 2021). Another long-term breeding study on the great tit (*Parus major*) showed the influence of teleconnections (ENSO, NAO, AO), often more capable of predicting and explaining ecological changes than local-scale climatic indices (Laczi et al. 2019). A study using imputation models indicated that the preceding year's weather conditions significantly impacted the current year's nesting abundance of long waders (Dwevedi et al. 2021). However, imputation models are affected by the averaging effects, often failing to capture the subtle teleconnections related to nesting failures. In heronries of the Indian subcontinent, often nesting failures are simplistically exclusively attributed to the proximal overt factors, viz., monsoon failure, and subsequent water scarcity (Mahendiran 2010), instead of exploring from the teleconnection angle. Even though the vulture die-off due to diclofenac is well known, the potential role of climate in the recent crash of the Indian Vulture population across a broad region of western Rajasthan, India, apparently by ENSO events, is also equally compelling (Hall et al. 2012).

The IOD is considered an independent oscillation pattern viewed earlier as an artifact of the ENSO. The conditions of SST during the positive and negative phases of IOD have always been quite in contrast across the globe (Fig. 8.4a, b). The abundance of the red-footed booby (*Sula sula*) and wedge-tailed shearwater (*Ardenna pacifica*) strengthened with positive IOD (Perez-Correa et al. 2020). The large albatrosses generally use the wind to reduce flight costs; the weakening of prevailing winds during a negative IOD induced higher energetic costs of flight, with a possible negative impact on their survival rate and breeding success (Rivalan et al. 2010). A potential species to establish the mechanism of the ecological dipole is the short-tailed shearwaters (*Ardenna tenuirostris*), for it is one of the long-distance migratory birds, breeding in Tasmania and Australia and wintering in Artic and Alaskan regions (Fig. 8.2). Their breeding success has been highly influenced by large-scale climate oscillations (NPO, PDO, & ENSO) and weather conditions of the preceding year (Price et al. 2021).

Local (temperature and precipitation) and large scale climatic phenomena (El Niño Southern Oscillation (ENSO), North Atlantic Oscillation (NAO) affect ecological patterns (e.g., geographical ranges, abundance, diversity) and processes (e.g., primary productivity, birth and death rates) in both marine and terrestrial ecosystems (Rivalan et al. 2010; Zuckerberg et al. 2020; Feldstein 2003; Hall et al. 2012). Egg-laying dates were closely related to the NAO-index variation, and at the same time, the NAO affected birds' breeding (Price et al. 2021; Perez-Correa et al. 2020). However, meta-analysis shows that the biological meaning of the NAO as a general predictor for climate change effects on the migration phenology of birds seems highly questionable (Haest et al. 2018).

Equally, the global climate variability often makes climate dipoles in terrestrial and marine systems as contrasting patterns in temperature or precipitation anomalies concurrent at two diverse geographic locations affecting continent-wide processes from bird migration to plant reproduction producing ecological dipoles (Zuckerberg et al. 2020). One could identify ecological dipoles by applying space–time analytical approaches to biological entities, particularly bird migration patterns and climatological observations at continental scales (Fig. 8.2). The irruption patterns of Pine Siskins and movements of boreal bird populations in North America across the continent at biennial to decadal periodicities reportedly coincide with dipoles of temperature and precipitation anomalies (Strong et al. 2015).

Conclusions

This paper has presented evidences of the influences of climate change on birds' migration patterns. Despite numerous studies cited in the above sections that connect birds with climatic indices (ENSO, IOD, NAO), more meta-analysis focusing on hypothesis testing on the various relationship between climatic indices and the avian population dynamics across different world regions is required. The concept of magnetoreception is getting established well in birds (Wynn et al. 2022), and research on these using bio-logging tools would add more information to questions, such as when and where birds stop during migration.

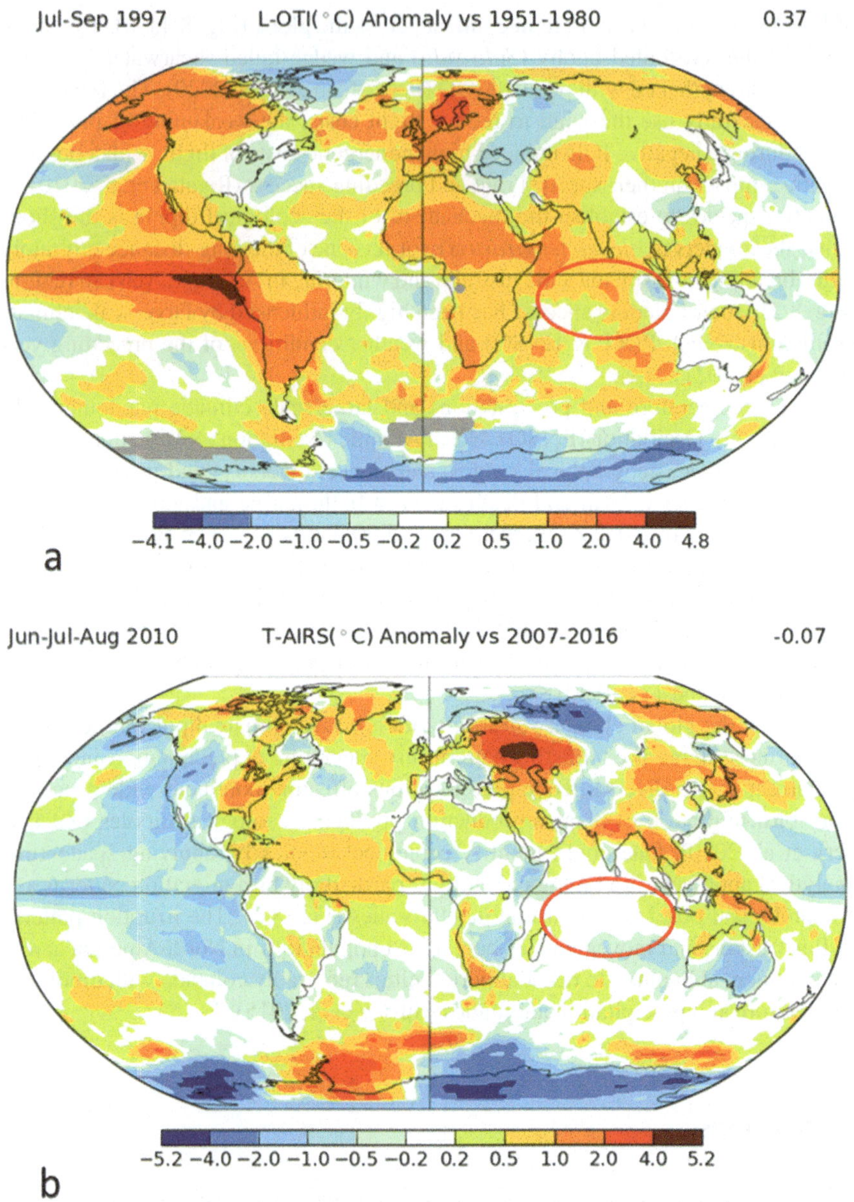

Fig. 8.4 **a** World map of the 1997 summer months plotted with Surface Sea Temperature (SST) anomalies indicating the strong El Niño events and a strong positive Indian Ocean Dipole (IOD) represented as the red circle, which resulted from an enhanced effect of MJO in the previous year. **b** The World map plotted with the SST anomalies of 2010 indicates strong La Niña where the negative IOD is marked as a red circle. The maps were prepared based on NASA GISS Surface Temperature Analysis version 4 data (GISTEMP Team [2021]: GISS Surface Temperature Analysis (GISTEMP), version 4. NASA Goddard Institute for Space Studies. Dataset accessed 2021-11-30 at data.giss.nasa.gov/gistemp/)

Many opportunities lie in utilizing existing infrastructure and technologies, such as satellites and radars, to enrich bird migration and monitoring capabilities. The data such as radar or satellite bird monitoring data or citizen size data, if these were utilized, can have significant scientific and positive societal contributions (Bauer et al. 2019; Liechti et al. 2019; La Sorte and Somveille 2021). Weather radars already can detect high-flying migrants, including insects and birds (Hu et al. 2020; Dokter et al. 2018). The citizen size data also contributes equally to making practical conservation goals and policy decisions. Drawing on citizen size data, a widespread population decline in birds over the past half-century was reported (Rosenberg et al. 2019). Such results have significant implications for ecosystem integrity that faces threats to the wildlife and native ecosystem on which they depend.

Conservationists and policymakers should take more advantage of the available technologies (e.g., radar and satellite data) to monitor migration patterns and integrate their use with climatic models to better understand birds' migration and distribution. Future research should also encourage studies that identify existing ecological dipoles and measure how climate change alters population abundance and their distribution (Steptoe et al. 2018; Zuckerberg et al. 2020; Strong et al. 2015).

Acknowledgements This study was supported by the International Climate Change Information and Research Programme (ICCIRP) at Hamburg University of Applied Science, Germany, Centre for Climate Change Adaptation and Resilience-Kgotso Development Trust of Zimbabwe, and the Sálim Ali Centre for Ornithology and Natural History (SACON) of India. It is part of the "100 papers to accelerate climate change mitigation and adaptation" initiative.

References

Adams EM (2014) Using migration monitoring data to assess bird population status and behavior in a changing environment. The University of Maine

Alves JA, Gunnarsson TG, Sutherland WJ, Potts PM, Gill JA (2019) Linking warming effects on phenology, demography, and range expansion in a migratory bird population. Ecol Evol 9(5):2365–2375. https://doi.org/10.1002/ece3.4746

Bauer S, Shamoun-Baranes J, Nilsson C, Farnsworth A, Kelly JF, Reynolds DR, Dokter AM, Krauel JF, Petterson LB, Horton KG, Chapman JW (2019) The grand challenges of migration ecology that radar aeroecology can help answer. Ecography 42(5):861–875. https://doi.org/10.1111/ecog.04083

Berthold P (2001) Bird migration: a general survey. Oxford University Press on Demand

Dasgupta P, Roxy M, Chattopadhyay R, Naidu C, Metya A (2021) Interannual variability of the frequency of MJO phases and its association with two types of ENSO. Sci Rep 11(1):1–16

Dokter AM, Farnsworth A, Fink D, Ruiz-Gutierrez V, Hochachka WM, La Sorte FA, Robinson OJ, Rosenberg KV, Kelling S (2018) Seasonal abundance and survival of North America's migratory Avifauna determined by Weather Radar. Nat Ecol Evol 2(10):1603–1609

Dwevedi R, Deo V, Sethy J, Gupta R (2021) Performance of imputation-based models in predicting breeding population trend of a near-threatened bird in changing water regime: a 36-year long-term case study of Painted Stork, Mycteria Leucocephala. J Appl Nat Sci 13: 1072–1082. https://doi.org/10.31018/jans.v13i3.2876

El Feldstein SB (2003) Ni~no/Southern oscillation and North American land birds. Birding 35:606–614

Feldstein SB (2000) The timescale, power spectra, and climate noise properties of teleconnection patterns. J Clim 13(24):4430–4440

Gill JA, Alves JA, Gunnarsson TG (2019) Mechanisms driving phenological and range change in migratory species. Philos Trans R Soc Lond B Biol Sci 374(1781):20180047. https://doi.org/10.1098/rstb.2018.0047

Gilroy JJ, Gill JA, Butchart SHM, Jones VR, Franco AMA (2016) Migratory diversity predicts population declines in birds. Ecol Lett 19(3):308–317. https://doi.org/10.1111/ele.12569

Haest B, Hüppop O, Bairlein F (2018) The influence of weather on Avian Spring migration phenology: what, where and when? Glob Change Biol 24(12):5769–5788. https://doi.org/10.1111/gcb.14450

Hall JC, Chhangani AK, Waite TA, Hamilton IM (2012) The impacts of La Niña-induced drought on Indian vulture gyps Indicus populations in Western Rajasthan. Bird Conserv Int 22(3):247–259. https://doi.org/10.1017/S0959270911000232

Hameed S, Norwood HH, Flanagan M, Feldstein S, Yang C (2009) EARTH interactions. EARTH Interact 13

Hu R, Gu Y, Luo M, Lu Z, Wei M, Zhong J (2020) Shifts in Bird Ranges and Conservation Priorities in China under Climate Change. PLoS One 15(10):e0240225. https://doi.org/10.1371/journal.pone.0240225

IPCC (2022) Climate change 2022: Mitigation of climate change. Contribution of working group III to the sixth assessment report of the intergovernmental panel on climate change. In: Shukla PR, Skea J, Slade R, Al Khourdajie A, van Diemen R, McCollum D, Pathak M, Some S, Vyas P, Fradera R, Belkacemi M, Hasija A, Lisboa G, Luz S, Malley J (ed). Cambridge University Press, Cambridge, UK and New York, NY, USA

Knudsen E, Lindén A, Both C, Jonzén N, Pulido F, Saino N, Sutherland WJ, Bach LA, Coppack T, Ergon T, Gienapp P, Gill JA, Gordo O, Hedenström A, Lehikoinen E, Marra PP, Møller AP, Nilsson ALK, Péron G, Ranta E, Rubolini D, Sparks TH, Spina F, Studds CE, Saether SA, Tryjanowski P, Stenseth NC (2011) Challenging claims in the study of migratory birds and climate change. Biol Rev 86(4):928–946. https://doi.org/10.1111/j.1469-185X.2011.00179.x

La Sorte FA, Somveille M (2021) The Island biogeography of the EBird citizen-science programme. J Biogeogr 48(3):628–638. https://doi.org/10.1111/jbi.14026

Laczi M, Garamszegi LZ, Hegyi G, Herényi M, Ilyés G, Könczey R, Nagy G, Pongrácz R, Rosivall B, Szöllősi E (2019) Teleconnections and local weather orchestrate the reproduction of tit species in the Carpathian Basin. J Avian Biol 50(12)

Liechti F, Aschwanden J, Blew J, Boos M, Brabant R, Dokter AM, Kosarev V, Lukach M, Maruri M, Reyniers M, Schekler I, Schmaljohann H, Schmid B, Weisshaupt N, Sapir N (2019) Cross-calibration of different radar systems for monitoring Nocturnal Bird migration across Europe and the Near East. Ecography 42(5):887–898. https://doi.org/10.1111/ecog.04041

Mahendiran MA (2010) Comparative field study on the ecology of Cormorants in the Delhi Region. Ph.D. Thesis, Delhi University, New Delhi

Mahendiran M (2016) Coexistence of three Sympatric Cormorants (Phalacrocorax Spp.); Partitioning of time as an ecological resource. R Soc Open Sci 3(5):160175. https://doi.org/10.1098/rsos.160175

Møller AP (2008) Climate change and micro-geographic variation in laying date. Oecologia 155(4):845–857. https://doi.org/10.1007/s00442-007-0944-3

Møller AP, Merino S, Soler JJ, Antonov A, Badás EP, Calero-Torralbo MA, de Lope F, Eeva T, Figuerola J, Flensted-Jensen E, Garamszegi LZ, González-Braojos S, Gwinner H, Hanssen SA, Heylen D, Ilmonen P, Klarborg K, Korpimäki E, Martínez J, Martínez-de la Puente J, Marzal A, Matthysen E, Matyjasiak P, Molina-Morales M, Moreno J, Mousseau TA, Nielsen JT, Pap PL, Rivero-de Aguilar J, Shurulinkov P, Slagsvold T, Szép T, Szöllősi E, Török J, Vaclav R, Valera F, Ziane N (2013) Assessing the Effects of Climate on Host-Parasite interactions: a comparative study of European Birds and their parasites. PloS One 8(12):e82886. https://doi.org/10.1371/journal.pone.0082886

Nourani E, Yamaguchi NM, Higuchi H (1854) Climate change alters the optimal wind-dependent flight routes of an Avian Migrant. Proc R Soc B Biol Sci 2017(284):20170149. https://doi.org/10.1098/rspb.2017.0149

Nunez S, Arets E, Alkemade R, Verwer C, Leemans R (2019) Assessing the impacts of climate change on biodiversity: is below 2 °C enough? Clim Change 154(3–4):351–365. https://doi.org/10.1007/s10584-019-02420-x

Parmesan C, Yohe G (2003) A globally coherent fingerprint of climate change impacts across natural systems. Nature 421(6918):37–42. https://doi.org/10.1038/nature01286

Perez-Correa J, Carr P, Meeuwig JJ, Koldewey HJ, Letessier TB (2020) Climate oscillation and the invasion of Alien species influence the oceanic distribution of seabirds. Ecol Evol 10(17):9339–9357. https://doi.org/10.1002/ece3.6621

Price CA, Emery TJ, Hartmann K, Woehler EJ, Monash R, Hindell MA (2021) Inter-annual and inter-colony variability in breeding performance of four colonies of short-tailed shearwaters. J Exp Mar Biol Ecol 537:151498

Remisiewicz M, Underhill LG (2022) Climate in Africa sequentially shapes spring passage of Willow Warbler Phylloscopus Trochilus across the Baltic Coast. PeerJ 10:e12964. https://doi.org/10.7717/peerj.12964

Rivalan P, Barbraud C, Inchausti P, Weimerskirch H (2010) Combined impacts of longline fisheries and climate on the persistence of the Amsterdam Albatross diomedia Amsterdamensis: Impact of fishery and climate on Amsterdam Albatross. Ibis 152(1):6–18. https://doi.org/10.1111/j.1474-919X.2009.00977.x

Rosenberg KV, Dokter AM, Blancher PJ, Sauer JR, Smith AC, Smith PA, Stanton JC, Panjabi A, Helft L, Parr M, Marra PP (2019) Decline of the North American Avifauna. Science 366(6461):120–124. https://doi.org/10.1126/science.aaw1313

Runge CA, Watson JEM, Butchart SHM, Hanson JO, Possingham HP, Fuller RA (2015) protected areas and global conservation of migratory birds. Science 350(6265):1255–1258. https://doi.org/10.1126/science.aac9180

Samplonius JM, Both C (2019) Climate change may affect fatal competition between two bird species. Curr Biol 29(2):327-331.e2. https://doi.org/10.1016/j.cub.2018.11.063

Shamoun-Baranes J, Liechti F, Vansteelant WMG (2017) Atmospheric conditions create freeways, detours and tailbacks for migrating birds. J Comp Physiol A 203(6–7):509–529. https://doi.org/10.1007/s00359-017-1181-9

Steptoe H, Jones S, Fox H (2018) Correlations between extreme atmospheric hazards and global teleconnections: implications for multihazard resilience. Rev Geophys 56(1):50–78

Strong C, Zuckerberg B, Betancourt JL, Koenig WD (2015) Climatic dipoles drive two principal modes of North American Boreal Bird irruption. Proc Natl Acad Sci 112(21):E2795–E2802

Studds CE, Kendall BE, Murray NJ, Wilson HB, Rogers DI, Clemens RS, Gosbell K, Hassell CJ, Jessop R, Melville DS, Milton DA, Minton CDT, Possingham HP, Riegen AC, Straw P, Woehler EJ, Fuller RA (2017) Rapid population decline in migratory Shorebirds Relying on Yellow Sea Tidal Mudflats as Stopover Sites. Nat Commun 8(1):14895. https://doi.org/10.1038/ncomms14895

Thomas CD, Cameron A, Green RE, Bakkenes M, Beaumont LJ, Collingham YC, Erasmus BFN, De Siqueira MF, Grainger A, Hannah L, Hughes L, Huntley B, Van Jaarsveld AS, Midgley GF, Miles L, Ortega-Huerta MA, Peterson AT, Phillips OL, Williams SE (2004) Extinction risk from climate change. Nature 427(6970):145–148. https://doi.org/10.1038/nature02121

Tryjanowski P, Stenseth N, Matysioková B (2013) The Indian Ocean dipole as an indicator of climatic conditions affecting European Birds. Clim Res 57(1):45–49. https://doi.org/10.3354/cr01162

Van Doren BM, Horton KG (2018) A continental system for forecasting bird migration. Science 361(6407):1115–1118

Vickery JA, Ewing SR, Smith KW, Pain DJ, Bairlein F, Škorpilová J, Gregory RD (2014) The decline of Afro-Palaearctic Migrants and an Assessment Of Potential Causes. Ibis 156(1):1–22. https://doi.org/10.1111/ibi.12118
Wynn J, Padget O, Mouritsen H, Morford J, Jaggers P, Guilford T (2022) Magnetic stop signs signal a European Songbird's arrival at the breeding site after migration. Science 375(6579):446–449
Zuckerberg B, Strong C, LaMontagne JM, St. George S, Betancourt JL, Koenig WD (2020) Climate dipoles as continental drivers of plant and animal populations. Trends Ecol Evol 35(5):440–453. https://doi.org/10.1016/j.tree.2020.01.010

Chapter 9
Effect of Extreme Climatic Events on Plant-Pollinator Interactions in Blueberry

Helena Castro, Hugo Gaspar, João Loureiro, and Sílvia Castro

Abstract Pollination is a key ecosystem service that is, however, under threat due to multiple environmental pressures, such as climate change, compromising crop production. The main goal of this study was to investigate how extreme events due to climate change affect flower traits and plant-pollinator interactions, and how this impacts fruit production, using the insect-dependent blueberry crop as study system. For this, we set up a controlled pot experiment using two blueberry cultivars (Bluecrop and Duke). At the time of bud swelling, half of the plants (12 per cultivar) were placed for two weeks in a glasshouse under stress conditions (no water and increased temperature), while the other half remained outdoors and watered. At flowering, flower traits were measured, and plants were exposed to pollinators; the identity of pollinators visiting blueberry flowers was registered as well as their behavior and the number of flowers visited. Later, mature fruits were randomly collected and weighed individually. Results showed that in our study site the most frequent visitor of blueberry flowers was *Anthophora plumipes* (Fabricius, 1781). Results also showed that stress conditions did not affect flower traits and insect pollinator visitation rates, regardless of blueberry cultivar, but affected insect preferences for Bluecrop cultivar, with *A. plumipes* preferring control over stressed plants. However, for Duke cultivar, control plants produced heavier fruits than plants under stress conditions. Our study provides some insights into the effects of climate changes on plant-pollinator interactions, but further research is necessary to better understand the impacts of climate change on plant-pollinator interactions and how this may impact food production.

H. Castro (✉) · H. Gaspar · J. Loureiro · S. Castro
Centre for Functional Ecology, TERRA Associate Laboratory, Department of Life Sciences, University of Coimbra, Calçada Martim de Freitas, 3000-456 Coimbra, Portugal
e-mail: hecastro@ci.uc.pt

H. Gaspar
e-mail: hgaspar@uc.pt

J. Loureiro
e-mail: jloureiro@bot.uc.pt

S. Castro
e-mail: scastro@bot.uc.pt

W. Leal Filho et al. (eds.), *Climate Change Strategies: Handling the Challenges of Adapting to a Changing Climate*, Climate Change Management,
https://doi.org/10.1007/978-3-031-28728-2_9

Introduction

Pollination is a key ecosystem service maintaining the stability of agricultural food production, with animal pollination affecting the yield and quality of over 75% of crops worldwide (Klein et al. 2007). However, pollination services are under threat due to multiple environmental pressures. Land use changes, such as fragmentation and agriculture intensification, pesticide use, biological invasions and eutrophication have been shown to negatively impact plant-pollinator interactions (Potts et al. 2010). Climate changes may be a further threat to pollination services by altering plant and pollinator traits, phenologies and behavior, causing phenological mismatches in plant-pollinator interactions (Gérard et al. 2020; Keeler et al. 2021). Climate prediction trends for the Mediterranean region point to higher drought, increased inter-annual variability of precipitation and increases in extreme climatic events such as heatwaves and severe droughts (IPCC 2021). According to the IPCC (2021) report, Portugal is already experiencing climate change, and climate projections point to an increase in drought stress. This will bring challenges to the agricultural sector with expected increases in water demand and decreases in crop productivity. Further challenges arising from climate change with negative impacts on crop production are related to disrupting plant mutualistic relations with their pollinators (Keeler et al. 2021). For example, water stress has been shown to decrease the quantity and quality of nectar and the quality of pollen with effects on the survival and productivity of developing honeybees and bumblebees (Wilson Rankin et al. 2020). However, most of this information comes from natural systems. The effect of climate change on the pollination of crops has been poorly addressed so far, despite its relevance for food production and security. This chapter addresses the impacts of climate change on flower traits and plant-pollinator interactions, using the insect-dependent blueberry crop as a study system.

Climate change leading to water deficit and increased temperature cause an increase in plant physiological stress, affecting crop production both directly and indirectly. Direct effects include the decrease in resources available for investment in reproduction, including flower and fruit production (Eziz et al. 2017). Information from published studies, mostly performed in natural ecosystems, indicates that climate change, in particular increased drought and temperature, influences flower visual traits and olfactory cues (*e.g.*, Burkle and Runyon 2017; Descamps et al. 2020a; Gallagher and Campbell 2017). Producing and maintaining flowers requires carbon, nutrients and water, representing a considerable cost to plants and implying trade-offs with vegetative growth (Obeso 2002). Therefore, under water stress, plants tend to produce fewer and smaller flowers (Kuppler and Kotowska 2021). While for flower visual traits drought stress seems to consistently have negative effects, for nectar, pollen and olfactory clues, which are not directly linked to transpiration and water losses, the effects of drought stress are less clear and its effects are species dependent, and may reflect species adaptations to drought (Burkle and Runyon 2016; Phillips et al. 2018). For example, reported effects of drought stress on nectar quantity and sugar content range from decrease to no change (*e.g.*, Descamps et al. 2018, 2020a; Phillips et al. 2018; Rering et al. 2020), depending on the species studied.

The number of flowers, as well as their size and colour, are important cues for pollinator attraction, while nectar, pollen and flower morphology are important determinants of flower handling and pollinator's efficiency (Parachnowitsch et al. 2019). Factors, such as water stress, that affect these visual and olfactory cues used by pollinators when searching for food, as well as reward availability, have the potential to change the behavior, preferences and fidelity of insect pollinators (Klatt et al. 2013) and indirectly impact fruit production. However, few studies have addressed pollinator responses to changes in flower traits and floral rewards resulting from drought stress, with results pointing to variable effects on pollinator visitation rates regardless of the negative effects on such traits (Burkle and Runyon 2016; Descamps et al. 2018; Glenny et al. 2018; Rering et al. 2020; Kuppler et al. 2021). Additionally, the response of pollinators to drought mediated effects on flower traits and rewards may also be dependent on the pollinator species (Burkle and Runyon 2016; Kuppler et al. 2021).

While it is clear that water stress will directly and negatively affect crop growth and yield, the impact on plant-pollinator interactions and pollination services to crops remains uncertain. Additionally, and despite its relevance, floral traits and pollinator attraction in the context of climate change are still poorly studied (Byers 2017) and empirical studies explicitly focusing on the effects of climate change on pollination services to crops are almost inexistent (Vaissière et al. 2011). However, evaluating the effects of drought on flower traits and plant fitness and how changes in flower traits affects plant-pollinator interactions and fruit production and quality is key to understand how crops could cope with predicted future climate changes.

Blueberries require insect pollination to produce marketable fruits (Klein et al. 2007). Although this crop presents varying levels of self-fertility (depending on the species, cultivar and genotype), they are primarily outcrossing, showing larger fruit size and earlier fruit ripening when cross-pollinated (Dogterom et al. 2000; Song and Hancock 2011; Taber and Olmstead 2016). Blueberries are nectar rewarding plants, bearing bell-shaped flowers, with poricidal anthers, and with nectar producing structures located at the basis of the corolla. Its cultivars present particular floral attributes (Rodriguez-Saona et al. 2011; Huber 2016), which can also be affected by different water availabilities. Flower size and morphology limit pollinators' access to rewards, limiting potential pollinator species, and may encourage nectar-robbing behavior (Courcelles et al. 2013).

Considering all this, in this work, our main objective was to investigate how extreme events due to climate change affect flower traits and plant-pollinator interactions, and how this impacts fruit production, using the insect-dependent blueberry crop as study system. To achieve this, we set up a water-controlled pot experiment and quantified flower traits, plant-pollinator interactions, and fruit traits. We hypothesize that stress conditions: (1) will lead to changes in flower traits relevant for insect attraction; (2) that these changes will affect interactions with pollinators, and (3) that this will have an impact on fruit production and fruit weight.

Methods

Experimental Design

Two blueberry cultivars commonly grown in Portugal with similar flowering times, Duke and Bluecrop, were selected for this study. A total of 48 plants (24 per cultivar), were purchased at a nursery shop specialized in blueberry plants. The plants were two-years old, as they is the stage at which they are available at nurseries, and were transplanted into 6L pots filled with professional blueberry suitable soil and placed in the Botanical Garden of the University of Coimbra. At the beginning of bud swelling, 12 plants of each cultivar were placed in a glasshouse until the beginning of flowering, which took two weeks. During this period, plants were not watered, and they experienced an increase by 5.1 °C in mean air temperature when compared to 12 control plants of each cultivar, which remained outdoors. At the end of these two weeks, we observed the opening of the first flowers in plants inside the greenhouse. At this stage all the plants were moved to an open patch dominated by small sized grasses, in the Botanical Garden (lat. 40.206403°, long. −8.425170°, 65 m a.s.l.) and were exposed to pollinators.

Flower Morphology and Rewards

For all the 48 plants in the experiment, one, young but fully open, flower per plant was selected for morphological measurements. Blueberry flowers are urn-shaped, and we recorded measurements that capture the variation of this shape among cultivars. Following the methodology described by Courcelles et al. (2013), we measured corolla length, diameter at widest area, and diameter of corolla opening (flower throat). Measurements were made using a digital caliper. Floral volume (corolla volume) was calculated considering the shape of a cylinder and using the length of the corolla and the diameter at the widest portion.

Nectar amount was quantified in the morning on flowers bagged the day before, using micro-capillaries and following Dafni et al. (2005). The percentage of sugars (°Brix) was determined with a hand-held refractometer.

Pollinator's Observation

Pollinator observations followed standard methodologies (Dafni et al. 2005). Pollinators were observed at the peak of flowering, between 22nd of March and 12th of April 2020, on sunny days with temperatures above 13 °C, i.e., weather conditions favorable for pollinators activity. Visits were recorded for 10 min periods, at different times of the day from 9 A.M. to 5 P.M. (CET + 1 h), totaling 13 h 18 min of net observation. We recorded the number of individual insects that visited the flowers, the insect species, and the treatment, cultivar and number of flowers visited. Additionally, the total number of open flowers per treatment and cultivar were also recorded.

Indexes of floral preference and constancy were calculated for the main pollinator species [*Anthophora plumipes* (Fabricius, 1781)], excluding visits with less than three visited flowers (Dafni et al. 2005; Castro et al. 2020). Floral preference was calculated for each cultivar, as the ratio between the number of visits to plants under control conditions and the total number of visits. Values of 0.5 indicate no preference by the pollinator, values of 0 indicate preference for plants under stress conditions, and values of 1 indicate preference for plants under control conditions. Floral constancy was calculated for each cultivar as the ratio between the number of movements within the same treatment and the total number of movements during the visit. A value of 0 indicates alternating foraging behavior, a value of 0.5 indicates random foraging behavior and a value of 1 indicates constancy in foraging behavior within treatment.

Fruit Sampling and Processing

The percentage of flowers that set fruit (fruit set) was calculated as the ratio between the mean number of fruits produced per inflorescence and the mean number of flowers per inflorescence, by counting the number of flowers and fruits in five inflorescences per plant.

All fruits produced by the plants were collected when ripe, counted and weighed in an analytical scale (accuracy 0.1 mg). To determine mean fruit weight a subset of 15 fruits per plant was taken and each fruit was weighed individually.

Data Analysis

A *t*-test was used to explore differences between control and stress treatments within each blueberry cultivar in flower morphology, floral rewards and fruit parameters (fruit set, number of fruits per plant and fruit weight). The *t*-test was also used to explore differences in flower traits between blueberry varieties. For this analysis we pooled control and stressed plants of each cultivar, as no significant differences were found between treatments (see 'Results').

Floral preference and constancy indices were used as a measure of pollinator behavior and were analyzed as the deviation from 0.5 which represents the randomly expected value (Dafni et al. 2005) using a one-sample *t*-test.

All analyses were done using *R* version 3.3.2 (Core Development Team 2016) and differences were considered statistically significant at $P < 0.05$.

Results

Flower morphology differed between blueberry cultivars, but it was not affected by stress conditions (Figs. 9.1 and 9.2), with exception of the corolla volume in Duke cultivar, where stress led to marginally significantly larger volume than control conditions (Table 9.1). Duke flowers had significantly larger corolla throat and wider corollas than Bluecrop flowers, which is reflected in a larger corolla volume in Duke flowers when compared to Bluecrop (Fig. 9.2; Table 9.1).

The mean number of flowers per inflorescence was not significantly affected by stress conditions, although Bluecrop plants under stress conditions produced a higher number of flowers than control plants (Table 9.2). The number of inflorescences per plant was not significantly affected by stress conditions (Table 9.2), although, for both cultivars, there was a tendency for a higher number of inflorescences in stressed plants. We did not find significant differences in the mean number of flowers per inflorescence ($t = 1.089$, $P = 0.285$) or total number of inflorescences ($t = -0.405$, $P = 0.689$) between cultivars.

Stress conditions triggered earlier flowering in Bluecrop plants, but not in Duke plants (Fig. 9.3).

Nectar volume and sugar content were not affected by stress conditions, as shown, for both cultivars, by the lack of significant differences between control and stressed plants (Fig. 9.4, Table 9.1). Despite not significant, stressed plants presented a slight decrease in nectar volume and a slight increase in sugar content when compared to control plants.

Overall, there were no significant differences in nectar volume and sugar content (Table 9.1) between blueberry cultivars, but there was a trend for higher nectar volume in Bluecrop (5.04 ± 0.95 µl) than in Duke (4.09 ± 0.53 µl), and for higher sugar content in Duke ($19.79 \pm 0.95\%$) than in Bluecrop ($16.23 \pm 0.53\%$).

Fig. 9.1 Blueberry flowers from Bluecrop (a) and Duke (b) cultivars

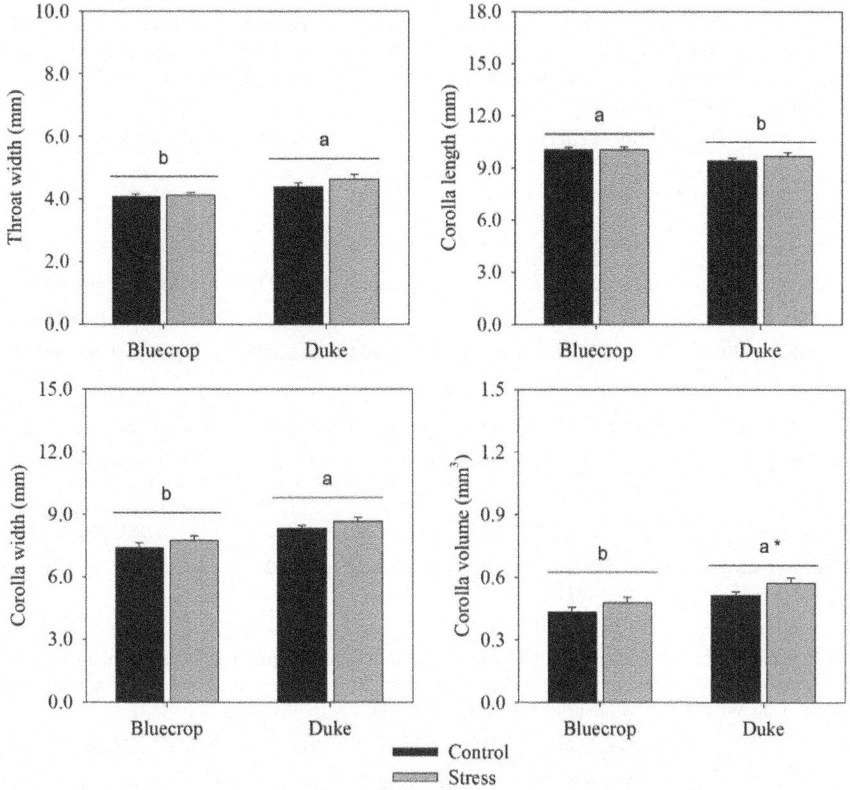

Fig. 9.2 Mean (±SE) throat width (upper-left), corolla length (upper-right), corolla width (down-left) and corolla volume (down-right) of blueberry cultivars, Bluecrop and Duke, under control and stress conditions. Different letters indicate significant differences between blueberry varieties ($P < 0.05$) after a t-test, while * indicates marginal significant differences between control and stress plants within cultivar Duke ($P = 0.09$)

A total of seven pollinator species were observed visiting blueberry flowers, but the main pollinator visiting blueberry flowers was *Anthophora plumipes*, accounting for 93.7% of all visits (Table 9.3).

The percentage of visited flowers was not affected by the imposed conditions, although for Bluecrop, pollinators tended to visit a higher percentage of flowers from plants under control conditions (Fig. 9.5).

Pollinator behavior as indicated by floral preference and constancy indices showed that *A. plumipes* for Bluecrop preferred control plants while, for Duke, it showed no preference (Table 9.4). Data also shows that *A. plumipes* preferred Bluecrop plants rather than those from Duke cultivar (preference index = 0.40, P-value = 0.009). Regarding constancy, *A. plumipes* significantly visited flowers within the same treatment rather than among treatments (Table 9.3).

Table 9.1 Statistical results (*t*-test and *P*-value) from *t*-tests used to test differences in flower traits, nectar volume and sugar, and fruit weight between stressed and control Bluecrop and Duke plants and between cultivars. Significant differences ($P < 0.05$) and marginally significant differences ($P < 0.05$) are highlighted in bold

	Treatment effect		Cultivar effect
	Bluecrop	Duke	
Flower traits			
Throat width	$t = -0.322$, $P = 0.751$	$t = -1.214$, $P = 0.238$	**$t = -3.509$, $P = 0.001$**
Corolla length	$t = 0.031$, $P = 0.976$	$t = -1.115$, $P = 0.278$	**$t = 2.974$, $P = 0.005$**
Corolla width	$t = -1.116$, $P = 0.277$	$t = -1.354$, $P = 0.192$	**$t = -4.509$, $P < 0.001$**
Corolla volume	$t = -1.238$, $P = 0.230$	**$t = -1.792$, $P = 0.088$**	**$t = 3.555$, $P = 0.001$**
Nectar			
Nectar volume	$t = -0.810$, $P = 0.441$	$t = 0.412$, $P = 0.688$	$t = 0.952$, $P = 0.352$
Nectar sugar	$t = -1.513$, $P = 0.156$	$t = -0.625$, $P = 0.545$	$t = -1.863$, $P = 0.076$
Fruit			
Fruit set	$t = 0.2936$, $P = 0.772$	$t = -1.580$, $P = 0.129$	$t = -0.963$, $P = 0.341$
Fruit weight	$t = -1.585$, $P = 0.128$	**$t = 2.736$, $P = 0.012$**	$t = 0.642$, $P = 0.524$

Table 9.2 Number of flowers per inflorescence and number of inflorescences per plant for both blueberry cultivars under control and stress conditions. Results are given as mean ± standard error of the mean (SE). The *P* value after a *t*-test comparison is also provided

Cultivar		No. of flowers/inflorescence		No. of inflorescences/plant	
		Mean ± SE	*t*-test, *P* value	Mean ± SE	*t*-test, *P* value
Bluecrop	Control	7.00 ± 0.50	$t = -1.800$, $P = 0.085$	7.83 ± 1.40	$t = -1.344$, $P = 0.196$
	Stress	8.23 ± 0.47		11.42 ± 2.27	
Duke	Control	7.04 ± 0.18	$t = -0.060$, $P = 0.952$	9.17 ± 0.69	$t = -1.988$, $P = 0.063$
	Stress	7.06 ± 0.21		11.92 ± 1.20	

Also, for Duke cultivar the mean number of fruits harvested in stressed plants (60.42 ± 7.15) was significantly higher ($t = -2.369$, $P = 0.030$) than that of control plants (40.83 ± 4.15). For Bluecrop cultivar the mean number of fruits harvested from stressed (50.50 ± 8.44) and control (40.08 ± 6.98) plants was not significantly different ($t = -0.952$, $P = 0.352$).

Fruit set was not significantly affected by the imposed conditions for either blueberry cultivar (Fig. 9.6, Table 9.1). Regarding fruit weight, control plants of Duke cultivar, yielded significantly heavier fruits than those exposed to stress (Fig. 9.7, Table 9.1).

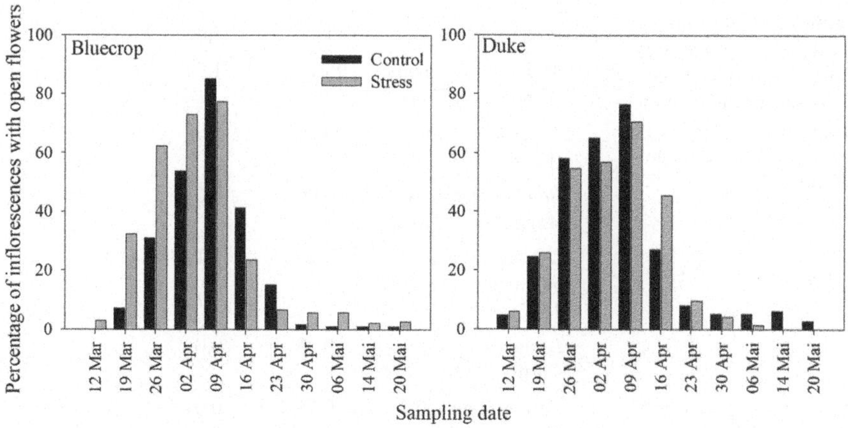

Fig. 9.3 Flowering phenology of blueberry cultivars, Bluecrop and Duke, under control and stress conditions

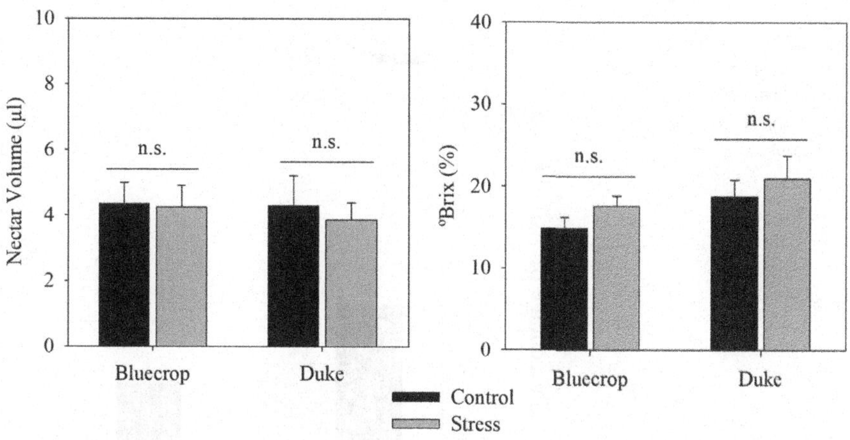

Fig. 9.4 Mean (±SE) of nectar volume (left) and sugar content (ºBrix; right) of the flowers of blueberry cultivars, Bluecrop and Duke, under control and stress conditions

Discussion

Flower number and density are considered important visual cues for pollinator attraction, and it has been shown in the literature that a decrease in such traits may bear consequences for pollinator visitation rates and the species of pollinators visiting the flowers (Kuppler and Kotowska 2021). Flower production requires plant resources and is a source of water loss under water stress conditions. Therefore, plants under water deficit are expected to produce fewer and smaller flowers, as often reported

Table 9.3 List of insect species visiting blueberry flowers, overall percentage of visits carried out by each species and abundance per unit of observation time (10 min)

Insect species	Visits (%)	Abundance per 10 min
Anthophora plumipes (Fabricius, 1781)	93.71	1.74
Andrena nigroaenea (Kirby, 1802)	0.75	0.03
Bombus terrestris (Linnaeus, 1758)	0.90	0.01
Lasioglossum malachurum (Kirby, 1802)	3.29	0.19
Lasioglossum sp.1 Curtis, 1833	0.30	0.03
Lasioglossum sp.2 Curtis, 1833	0.75	0.04
Vespula sp. Thomson, 1869	0.30	0.01

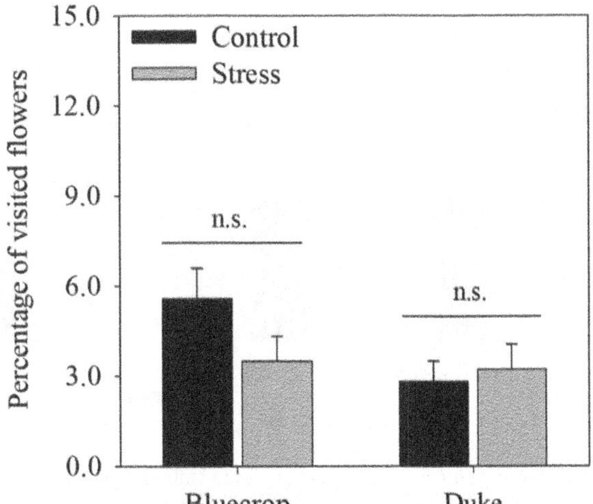

Fig. 9.5 Percentage of flowers visited for blueberry cultivars Bluecrop and Duke under control and stress conditions

in studies evaluating the effect of water stress on floral traits (*e.g.*, Descamps et al. 2020b; Gallagher and Campbell 2017; Kuppler et al. 2021). In our study, water deficit did not affect flower size significantly, contrary to our hypothesis. A few factors may have contributed for this results: (1) the water stress treatment was imposed before flowering; a recent meta-analysis by Kuppler and Kotowska (2021) on the effects of water stress on flower traits showed that a reduction in water availability before flowering has a lower impact than if applied at the beginning of flowering; (2) Blueberry flower production follows a dormancy period and is mainly dependent on stored resources, as vegetative growth starts simultaneously or even after flower bud

Table 9.4 Pollinator behavior as indicated by floral preference and constancy indices of the main pollinator of blueberry cultivars, Bluecrop and Duke, under control and stress conditions. Results are given as mean ± standard error of the mean (SE). Floral preference: values of 0.5 indicate no preference by the pollinator, values of 0 indicate preference for plants under stress conditions, and values of 1 indicate preference for plants under control conditions. Floral constancy: values of 0 indicate alternating foraging behavior, values of 0.5 indicate random foraging behavior and values of 1 indicate constancy in foraging behavior within treatment. The *P* value after a *t*-test comparison is also provided. Significant differences at $P < 0.05$ are highlighted in bold

Species	Preference index				Constancy index			
	Bluecrop		Duke		Bluecrop		Duke	
	Mean (±SE)	*t*-test, *P*-value	Mean (±SE)	*t*-test, *P*-value	Mean (±SE)	*t*-test, *P*-value	Mean (±SE)	*t*-test, *P*-value
A. plumipes	**0.61 ± 0.05**	***t* = 2.099, *P* = 0.042**	0.49 ± 0.05	*t* = −0–104, *P* = 0.918	**0.84 ± 0.04**	***t* = 8.801, *P* < 0.001**	**0.88 ± 0.03**	***t* = 11.432, *P* < 0.001**

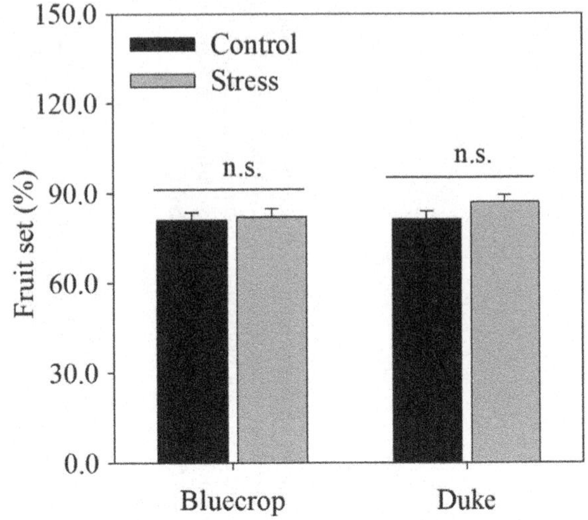

Fig. 9.6 Mean (±SE) fruit set for blueberry cultivars, Bluecrop and Duke, under control and stress conditions

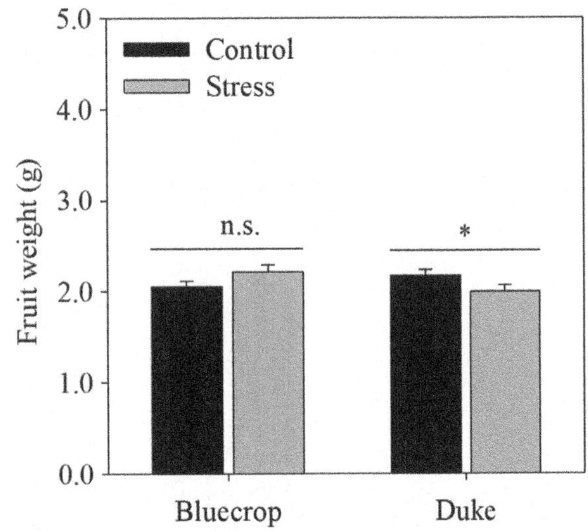

Fig. 9.7 Mean (±SE) of fruit weight for blueberry varieties, Bluecrop and Duke, under control and stress conditions. *Indicates significant differences ($P <$ 0.05) between Duke plants under control and under stress conditions after a *t*-test

growth; (3) Being late winter, the humidity in the air was high and soil water evaporation low. The combination of points 2 and 3 may have resulted in a very mild water stress (difference between control and stress was around 8%; data not shown), which combined with the application of water stress before flowering have likely resulted in no or low effect of water stress on the floral traits of blueberry cultivars Bluecrop and Duke.

Floral morphology differed significantly between both cultivars. Duke flowers presented a significantly larger corolla throat and wider corollas than Bluecrop flowers. This was expected from intrinsic traits of the cultivars and similar results were obtained by other authors (*e.g.*, Courcelles et al. 2013; Huber 2016) for the same cultivars. The differences in corolla size between blueberry cultivars, however, did not affect pollinator visitation rate. This result is in contrast with those of the study by Courcelles et al. (2013), which showed higher visitation rates by honeybees and bumblebees to Duke than to Bluecrop, and with the authors pointing to a relationship with the larger corolla of Duke. However, the response to flower traits may be related with differences in the pollinator community present during the flowering period in the surrounding landscape and interacting with the blueberry flowers, which in our case was scarcely represented by bumblebees and lacked honeybees. Indeed, at our study site, the main pollinator was *A. plumipes*, which is an insect with long proboscis and relatively large body, being mostly active early in the flowering season (Ornai and Keasar 2020), when blueberries flower in this geographic region. Species of the same genus have been reported as pollinators of blueberries under field conditions and were even used as managed pollinators in lowbush blueberry [e.g., *A. pilipes villosula* Smith, 1854 (Bushmann and Drummond 2020)]. The second most common pollinators in our study belong to the genus *Lasioglossum* Curtis, 1833, here represented by small bodied species that can fit inside blueberry flowers, regardless of the cultivar (author's observation). Thus, the behavior of the small *Lasioglossum* spp. observed here were not impacted by throat width of blueberry flowers.

The observation of plant-pollinator interactions shows two main behavioral approaches by floral visitors to blueberry flowers. First, big body insects with long proboscis (such as *A. plumipes*) would be able to reach the nectar legitimately and contact with the pollen inside the flower. Second, relatively small body insects with small proboscis (such as *Lasioglossum* spp.) have no restriction by the corolla and easily access nectar and pollen rewards, being potentially involved in blueberry pollination. The different behaviors of the flower visitors have an impact in pollination efficiency and reproductive fitness (*e.g.*, Castro et al. 2008, 2013).

Other important note was the high prevalence of male individuals of *A. plumipes* (author's observation), the main pollinator at the study site. These individuals usually start flying earlier in the season than females (Michener 2007) in an active search for matting partners (and collecting mainly nectar), contrarily to females that actively collect pollen (Michener 2007). Even though this data was not actively collected, male and female *A. plumipes* may thus have different pollination efficiencies impacting differently blueberry fruit production.

The main pollinator in our study, *A. plumipes*, preferred control plants for Bluecrop, while it showed no preference in Duke. *Anthophora plumipes*, also preferred Bluecrop plants rather than Duke. Although non-significant, the slight differences in nectar volume and sugar content found between cultivars may have contributed for the differences in *A. plumipes* behavior. Additionally, water deficits have been shown to affect floral volatile emissions and composition in natural communities, as well as other aspects of nectar composition such as secondary metabolites (Glenny et al. 2018; Descamps et al. 2021), and although we cannot confirm that these occurred in our study, it may be factor influencing *A. plumipes* behavior in our study. Variations in both nectar and floral volatile amount and composition have been linked to changes in pollinator behavior and preferences (Burkle and Runyon 2017; Parachnowitsch et al. 2019), and should be considered in future studies.

Nevertheless, the differences found in pollinator visitation were not reflected in differences in fruit production, as no significant differences in fruit set and in fruit weight were found between control and stressed plants of Bluecrop cultivar. Similarly, we did not find significant differences in fruit set between stressed and control Duke plants. This suggests that visitation by the pollinator's community to blueberry plants was not affected to the point of causing pollination deficits in stressed plants. However, we found that fruit weight was affected in Duke cultivar, with control plants yielding heavier fruits. Also, despite non-significant, fruit set and the number of inflorescences were higher in stressed than in control plants, which resulted in a significantly higher number of fruits. Considering that the plant has limited resources, producing more fruits implies investing less resources in each of them, which resulted in lighter fruits. However, considering our data, and the fact that such differences in fruit weight and number were observed in Duke cultivar where there was no preference of pollinators for a particular set of plants, it is not possible to associate fitness differences to pollinator's behavior.

Increased temperatures or heat pulses out of season resulting from climate changes may lead to early flowering of plants (Gérard et al. 2020), including crop species, with potentially negative consequences for crop production. Early flowering may result in plants missing part of the activity window of their main pollinators, which may lead to consequences for fruit production (Gérard et al. 2020). Such plant behavior may also lead to two potentially co-occurring consequences with strong impacts on crop production: (1) flower loss due to late frosts; (2) heterogeneity in flowering that may affect pollination success (Inouye 2008; Gérard et al. 2020), negatively affecting fruit production and quality, and causing heterogeneity in fruit ripening, and subsequently, increasing harvesting costs. In our study, increased temperature triggered earlier flowering in stressed Bluecrop plants, but not in Duke. Duke is considered an early flowering cultivar, even in relation to Bluecrop (Huber 2016) even though both cultivars overlap in their flowering period. The earlier flowering trait of Duke cultivar may have buffered the effect the stress imposed in this study. The response of flowering phenology to temperature increases has been shown to affect flowering phenology (Gérard et al. 2020), but it also seems to vary with plant species even within the same genus. For instance, flowering phenology was mostly unaffected in *Echium vulgare* L., while the opposite was observed in *E. plantagineum* L. (Descamps et al. 2020b).

Conclusions

We provide some insights into the effects of climate changes on plant-pollinator interactions. However, one should bear in mind that the blueberry individuals we used in our study, were small two-year-old plants entering the first productive year and therefore, a conservative approach was used when applying stress to these plants to reduce the risk of losing them due to the imposed stress. As referred above, further studies should involve drought periods during flowering, which are thought to have stronger effects on floral traits (Kuppler and Kotowska 2021). Also, there are inherent limitations related with the experimental design and its effects on pollination, as it is difficult to induce a similar stress on wild pollinators, which are usually studied in natural populations. Nevertheless, our study draws attention to key issues in pollination services to crops under climate change.

References

Burkle LA, Runyon JB (2016) Drought and leaf herbivory influence floral volatiles and pollinator attraction. Glob Chang Biol 22:1644–1654. https://doi.org/10.1111/gcb.13149

Burkle LA, Runyon JB (2017) The smell of environmental change: Using floral scent to explain shifts in pollinator attraction. Appl Plant Sci 5:1600123. https://doi.org/10.3732/apps.1600123

Bushmann SL, Drummond FA (2020) Analysis of pollination services provided by Wild and Managed bees (Apoidea) in Wild Blueberry (Vaccinium angustifolium Aiton) production in Maine, USA, with a literature review. Agronomy 10:1413. https://doi.org/10.3390/agronomy10091413

Byers DL (2017) Studying plant-pollinator interactions in a changing climate: a review of approaches. Appl Plant Sci 5:1700012. https://doi.org/10.3732/apps.1700012

Castro M, Loureiro J, Husband BC, Castro S (2020) The role of multiple reproductive barriers: strong post-pollination interactions govern cytotype isolation in a tetraploid-octoploid contact zone. Ann Bot 126:991–1003. https://doi.org/10.1093/aob/mcaa084

Castro S, Silveira P, Navarro L (2008) Consequences of nectar robbing for the fitness of a threatened plant species. Plant Ecol 199:201–208. https://doi.org/10.1007/sll258-008-9424-z

Castro S, Loureiro J, Ferrero V, Navarro L (2013) So many visitors and so few pollinators: variation in insect frequency and effectiveness governs the reproductive success of an endemic milkwort. Plant Ecol 214:1233–1245

Core Development Team R (2016) A language and environment for statistical computing. R Foundation for Statistical Computing, Vienna

Courcelles DMM, Button L, Elle E (2013) Bee visit rates vary with floral morphology among highbush blueberry cultivars (Vaccinium corymbosum L.). J Appl Entomol 137:693–701. https://doi.org/10.1111/jen.12059

Dafni A, Kevan PG, Husband BC (2005) Pratical pollination ecology. Enviroquest, LTA

Descamps C, Quinet M, Baijot A, Jacquemart AL (2018) Temperature and water stress affect plant–pollinator interactions in Borago officinalis (Boraginaceae). Ecol Evol 8:3443–3456. https://doi.org/10.1002/ece3.3914

Descamps C, Marée S, Hugon S et al (2020a) Species-specific responses to combined water stress and increasing temperatures in two bee-pollinated congeners (Echium, Boraginaceae). Ecol Evol 10:6549–6561. https://doi.org/10.1002/ece3.6389

Descamps C, Quinet M, Jacquemart A (2020b) The effects of drought on plantpollinator interactions: what to expect? Environ Exp Bot 104297. https://doi.org/10.1016/j.envexpbot.2020.104297

Descamps C, Quinet M, Jacquemart AL (2021) Climate change-induced stress reduce quantity and alter composition of nectar and pollen from a bee-pollinated species (Borago officinalis, Boraginaceae). Front Plant Sci 12:1–12. https://doi.org/10.3389/fpls.2021.755843

Dogterom MH, Winston ML, Mukai A (2000) Effect of pollen load size and source (self, outcross) on seed and fruit production in highbush blueberry cv. 'Bluecrop' (VACCINIUM CORYMBOSUM; Ericaceae). Am J Bot 87:1584–1591. https://doi.org/10.2307/2656734

Eziz A, Yan Z, Tian D et al (2017) Drought effect on plant biomass allocation: a meta-analysis. Ecol Evol 7:11002–11010. https://doi.org/10.1002/ece3.3630

Gallagher MK, Campbell DR (2017) Shifts in water availability mediate plant-pollinator interactions. New Phytol 215:792–802. https://doi.org/10.1111/nph.14602

Gérard M, Vanderplanck M, Wood T, Michez D (2020) Global warming and plant-pollinator mismatches. Emerg Top Life Sci 4:77–86. https://doi.org/10.1042/ETLS20190139

Glenny WR, Runyon JB, Burkle LA (2018) Drought and increased CO2 alter floral visual and olfactory traits with context-dependent effects on pollinator visitation. New Phytol 220:785–798. https://doi.org/10.1111/nph.15081

Huber G (2016) An investigation of highbush blueberry floral biology and reproductive success in British Columbia. Doctoral dissertation, University of British Columbia

Inouye DW (2008) Effects of climate change on phenology, frost damage, and floral abundance of montane wildflowers. Ecology 89:353–362

IPCC Climate Change (2021) The physical science basis. Contribution of working group I to the sixth assessment report of the intergovernmental panel on climate change.

Keeler AM, Rose-Person A, Rafferty NE (2021) From the ground up: building predictions for how climate change will affect belowground mutualisms, floral traits, and bee behavior. Clim Chang Ecol 1:100013. https://doi.org/10.1016/j.ecochg.2021.100013

Klatt BK, Burmeister C, Westphal C et al (2013) Flower volatiles, crop varieties and bee responses. PLoS One 8:e72724. https://doi.org/10.1371/journal.pone.0072724

Klein A-M, Vaissière BE, Cane JH et al (2007) Importance of pollinators in changing landscapes for world crops. Proc R Soc B Biol Sci 274:303–313. https://doi.org/10.1098/rspb.2006.3721

Kuppler J, Kotowska MM (2021) A meta-analysis of responses in floral traits and flower–visitor interactions to water deficit. Glob Chang Biol 27:3095–3108. https://doi.org/10.1111/gcb.15621

Kuppler J, Wieland J, Junker RR, Ayasse M (2021) Drought-induced reduction in flower size and abundance correlates with reduced flower visits by bumble bees. AoB Plants 13:1–8. https://doi.org/10.1093/aobpla/plab001

Michener CD (2007) The bees of the world, 2nd edn. The Johns Hopkins University Press

Obeso JR (2002) The costs of reproduction in plants. New Phytol 155:321–348

Ornai A, Keasar T (2020) Floral complexity traits as predictors of plant-bee interactions in a Mediterranean pollination web. Plants 9:1432. https://doi.org/10.3390/plants9111432

Parachnowitsch AL, Manson JS, Sletvold N (2019) Evolutionary ecology of nectar. Ann Bot 123:247–261. https://doi.org/10.1093/aob/mcy132

Phillips BB, Shaw RF, Holland MJ et al (2018) Drought reduces floral resources for pollinators. Glob Chang Biol 24:3226–3235. https://doi.org/10.1111/gcb.14130

Potts SG, Biesmeijer JC, Kremen C et al (2010) Global pollinator declines: trends, impacts and drivers. Trends Ecol Evol 25:345–353. https://doi.org/10.1016/j.tree.2010.01.007

Rering CC, Franco JG, Yeater KM, Mallinger RE (2020) Drought stress alters floral volatiles and reduces floral rewards, pollinator activity, and seed set in a global plant. Ecosphere 11:e03254. https://doi.org/10.1002/ecs2.3254

Rodriguez-Saona C, Parra L, Quiroz A, Isaacs R (2011) Variation in highbush blueberry floral volatile profiles as a function of pollination status, cultivar, time of day and flower part: implications for flower visitation by bees. Ann Bot 107:1377–1390. https://doi.org/10.1093/aob/mcr077

Song G-Q, Hancock JF (2011) Vaccinium. In: Kole C (ed) Wild crop relatives: genomic and breeding resources. Springer, Berlin Heidelberg, Berlin, Heidelberg, pp 197–221

Taber SK, Olmstead JW (2016) Impact of cross- and self-pollination on fruit set, fruit size, seed number, and harvest timing among 13 Southern Highbush Blueberry Cultivars. Horttechnology 26:213–219. https://doi.org/10.21273/HORTTECH.26.2.213

Vaissière B, Freitas B, Gemmill-Herren B (2011) Protocol to detect and assess pollination deficits in crops: a handbook for its use

Wilson Rankin EE, Barney SK, Lozano GE (2020) Reduced water negatively impacts social bee survival and productivity via shifts in floral nutrition. J Insect Sci 20:1–8. https://doi.org/10.1093/jisesa/ieaa114

Chapter 10
Understanding the Politics of Climate Change in Zimbabwe

Munyaradzi A. Dzvimbo, Abraham R. Matamanda, Samuel Adelabu, Adriaan Van der Walt, and Albert Mawonde

Abstract This article analyses how politics and livelihoods are challenged and affected by global climate change while paying particular attention to how traditionally formed, non-climatic factors are manipulated within Zimbabwe's climate change policy framework. In this context, the article will proffer legislation and laws as tools of governance that enable climate change policies and legislature to be implemented within the confines of the existing constitution. The methods will be drawn from an extensive literature review between 2011 and 2022 and document analysis to report the interconnectedness of climate change and politics. After critiquing the vulnerability theoretical framework and applying qualitative methods, the article will discuss and trace the influence of politics on how climate policies and laws impact livelihoods. The findings of this article are two-fold. Firstly, local communities and ordinary people are not involved in climate change decisions and policy-formulation processes whilst they are the most vulnerable. Secondly, the government, as the critical decision-maker, must adhere to international climate change conventions and develop homegrown policies with local consultations.

Introduction

Zimbabwe is a developing country experiencing calamities induced by climate change, and the consequence of this phenomenon is adversely affecting the majority of its population. Several Zimbabweans engage in farming as a backbone of the economy, but due to climate change, the country has been grappling with prolonged droughts, heatwaves and intermittent floods (Matsa 2020). These climate-induced disasters have resulted in food insecurity, forced migration, poverty, malnutrition,

M. A. Dzvimbo (✉) · A. R. Matamanda · S. Adelabu · A. Van der Walt
Department of Geography, University of the Free State, 205 Nelson Mandela Drive, Park West, Bloemfontein 9301, South Africa
e-mail: munyadzvimbo@gmail.com

A. Mawonde
College of Agriculture and Environmental Science, University of South Africa, 28 Pioneer Ave, Florida Park Roodepoort, Johannesburg 1709, South Africa

W. Leal Filho et al. (eds.), *Climate Change Strategies: Handling the Challenges of Adapting to a Changing Climate*, Climate Change Management,
https://doi.org/10.1007/978-3-031-28728-2_10

school dropouts, and poor living standards in the country (Nhapi 2021). The Food and Agricultural Organisation (FAO) (2016) opines that climate change is a danger to food access worldwide as it disturbs all proportions of food security, such as accessibility, availability, stability and utilisation. The majority of the affected people by climate change lack an adequate understanding of the root problem causing climate change, and they are often left out in climate change policy-making and decisions as grassroots people (Chirisa and Matamanda 2022). The government of Zimbabwe relies more on a top-down approach to governance where citizens do not have a say in policy formulation issues (Nyama and Mukwada 2022) and climate politics. This scenario exacerbates climate change challenges as vulnerable people fail to adapt and appropriately respond to them due to an absence of sound public participation in climate issues (Thomas et al. 2019).

The climate crisis is a political question; hence, the government is part of the climate question and the obstacle to a solution or the enabler to the solution (Dodman and Mitlin 2015). Therefore, the solution to the climate question requires the political power of labourers and the oppressed sections of the society to participate in the policy-making processes. The policy-making process must take into cognisance local resources that are not homogenous but heterogeneous and, along the process dismiss a "one size fit all" climate change policy that negates salient localised climate change challenges faced by a variety of societies in Zimbabwe. Climate change politics is centralised and intensified under capitalism (Dodman and Mitlin 2015; Tsabora 2019; Zhanda et al. 2021). Although Zimbabwe is a democratic country, its climate change dynamics adhere to Western Capitalist Climate Change Policies adopted at various conferences and fora. These capitalistic climate change ideologies sometimes do not address climate change challenges at local levels as they often target supranational scale or regional, or catchment area scales that ignores poor rural people (Broto 2017), or the "bottom billion" (Beddington 2009). Capitalist countries take the capital's side in the labour-capital and nature-capital contradictions. So, the State is on one side of both contradictions. As the central political apparatus, the government is the partner of capital in ecological destruction as it gears for policies that favour economic growth over environmental concerns, especially for the poor and vulnerable sections of society (Allen et al. 2018). Mainly, the capitalist State is organised to dominate the capitalist class on labouring classes and nature. All policies, climate, energy, transportation, mining, industry, construction, land use, agriculture, animal husbandry, and other country policies, are decided by the government's legislative and executive organs without public consultation (Chatiza 2019; Coban 2021).

These policies are made to look out for the interests of the capitalists, not the benefit of labourers and nature (Coban 2021). In almost every country, the government sets and implements climate change policies. However, as various countries draw up climate change policies, the capitalists in the form of investors and developers have a good influence on how to protect their interests in the form of wealth gathering, product making, residues, cost-effectiveness, private property and labour manipulation (Michaelowa 2013; Mashizha et al. 2017; Hudson 2021). Furthermore, the intertwined elements also include severe product consumption, individualisation of environmental accountability and high energy consumption shaped by these progressions.

The results of rapid wealth accumulation at the expense of the environment cause resource over-exploitation, ignorance of local communities' plights and climate change (Singh et al. 2021). In light of the above, energy from fossil sources (coal-fired power stations) is causing climate change, while there is no clear road map for Zimbabwe's renewable energy plan (Howells et al. 2021). Although Zimbabwe has a National Renewable Energy Policy of 2019, which advocates for geothermal, wind, hydro and biomass energy, the country is challenged by a lack of capital and relies on private capital funded by Capitalist countries (National Renewable Energy Policy 2019). According to the NREP (2019), Zimbabwe's renewable energy policy is limited by inadequate technology, a lack of renewable studies, a preliminary renewable tariff plan and a regulatory framework.

Even though Zimbabwe alluded to developing a Nationally Determined Contribution (NDC) to diminish GHG emissions, the country is focussing on ambitious hydroelectricity power generation under the vulnerability of climate change due to reduced precipitation (Howells et al. 2021). This scenario may cause the country to continuously rely on coal-fired power generation, hence reversing the ultimate goal to reduce carbon emissions by 2030 as stated in the country's climate change masterplan called Zimbabwe's National Climate Change Response Strategy (ZNCCRS) of 2017.

Under the ZNCCRS (2014) strategic objectives, objective (b) states that the country promotes the use and exploitation of natural resources sustainably and limits carbon emissions in all spheres of the economy by steering the adoption of green energy infrastructure that is not carbon-intensive. However, it appears as if the country is struggling to secure green and renewable energy funding as it is celebrating the extension of Hwange Coal Power Station Units by 600 MW via a Chinese company called Sinohydro (nsenergybusiness.com). In contrast to the Hwange Coal Power Extension deal, the Chinese are reported to have halted the project halfway before its completion as the superpower adopted a policy of green energy financing (Banya et al. 2022). Using coal as an energy source is causing environmental degradation as it is altering the natural carbon cycle and the collapse of ecosystems. Similarly, the climate crisis is also political because countries have not taken the necessary measures to prevent labourers and the oppressed sections of society from the adverse effects of climate change. The government has not established a solid and financially backed organisational structure to tackle climate change and allocate necessary climate change budgets that transform the economy towards a circular and green economy (UNFCC 2015; Mawonde and Togo 2021).

Methods

This study is qualitative as it explores legislation and laws as tools of climate change governance in Zimbabwe. As such it adopted a case study research design in which document analysis was applied to examine reports and policies on the interconnectedness of climate change and politics. The methods used were drawn from an extensive literature review between 2011 and 2022.

Results

Are Policies Beneficial or a Political Coy in Climate Change Consensus?

Government is the principal actor in the international climate regime. The government adopts the decisions taken on an international scale or remains outside climate agreements and implements or does not implement some of the agreements adopted through laws and policies (Coban 2021). Government bureaucracies under capitalism give all permits and licences that lead to the plunder of nature by capital. In the audits, companies' activities against regulations are sometimes ignored to maintain good relations with investors, especially in less economically developed countries. Besides, as the executive organ, the president or the government in the parliamentary system controls the army, police, and gendarmerie, the "repressive state apparatuses "used to control natural resources and its people (Aseh 2011). The government needs a paradigm shift from simply signing and ratifying international conventions on climate change to implementing such policies meaningfully in a sustainable manner for the benefit of all.

Government politics focusing on GHGs emissions alone cannot solve all facets of climate change. Moreover, GHGs emissions focussed by the government cannot be solved by a "one size fit all" policy. Capitalism as a framework for economic development adopted by several governments damages the biophysical environment through mining, construction, timber, tourism and similar product production activities (White 2018). All these activities encourage deforestation, reducing trees as carbon sinks to control carbon dioxide levels in the atmosphere. In fact, the government must encourage planting more trees along with economic growth to solve deforestation and climate change (Singh et al. 2021). In total, the deliberate rate of planting trees by various people between 2010 and 2020 was half the rate at which people cut down trees. In simpler terms, the rate at which afforestation occurs is far below the rate at which natural forests are being destroyed by people (FAO 2020). The time taken by various ecosystems' lifecycles differs from one element to another. The carbon life cycle is an example of an ecosystem life cycle that differs in circulation rate to the rate at which capitalist industries and investments inject it into the atmosphere. Another instance is the long-time taken by fossil fuel to form as opposed to a short period of combustion to produce energy in industries, factories and coal power stations. Unfortunately, climate change impacts as a result of fossil fuel combustion last for a very long time in the range of five to seven hundred years (Poole et al. 2019). This situation creates long-lasting effects of global warming and climate change, while the energy used from the fuels has since been forgotten (Brinkmann 2021).

The climate impact of methane gas is twenty times greater than carbon dioxide. However, the atmospheric life of methane is twenty years. At the same time, carbon dioxide remains in the atmosphere for hundreds of years (Ramanathan 2020). Therefore, drastically limiting methane emissions in fossil and livestock farms will have a positive atmospheric effect in a short time. In a capitalist economy, fossil fuel reserves utilisation has been shortened while the accumulation of GHGs in the atmosphere increased in comparison to the past. Coban (2021) argued that "a problematic situation exists between the time of nature of time to accumulate wealth because of the cyclical processes of the global world evolving over millions of years and the need for rapid production, distribution and profit by capital". The challenges faced by the world in terms of climate change are historical. However, hegemonic climate politics have no yesterday and tomorrow (Coban 2021).

Historical Climate Change Political Dynamics

Often, in historical terms, events that transpired during the past are overlooked to underestimate the historical climate liability that the well-known establishments of rich countries have created for a hundred and fifty years (Coban 2021). Climate change uncertainties threaten years to come, and the world is fast approaching climate disaster and destruction with the likelihood of a six-degree temperature rise (Wallace-Wells 2019). There is an international call for all countries to put in immense effort to reduce global warming and climate change (Chappell et al. 2018; Williamson et al. 2018; Peter 2018). The emphasis on doomsday no-tomorrows is applied to the people to embrace solutions of the ruling climate politics, which asserts to overturn that destiny.

As a result, it is advised that it is possible to avert catastrophe, for example, the ratification of the Paris Agreement and its implementation in various countries (Bodansky 2016; Dzvimbo et al. 2017a, b; Coban 2021). The catastrophic impacts of climate change come from the linkages between the risk of climate destruction and the disposition to be affected by the destruction. The extent of the effects of climate change is also linked to the societal segments that will be negatively impacted and subjected to, for example, extreme weather events or rainfall variability patterns and droughts. To support the previous argument, one can argue that the impacts of climate change are hinged on the past; one cannot separate climate change calamities from negative impacts that occurred in the past and climate change is produced from past events (Compton 2020). Lately, several ecologically related organisations have lost hope in the United Nations climate deliberations and negotiations. Climate Activists regard the United Nations climate negotiations as weak and non-binding as they prefer continual change (Glasgow Agreement 2020; Chirisa and Matamanda 2022).

The global corporations of the socialist and communist parties operating under various names are not effectively leading solid operations and campaigns on the climate change catastrophe question. Climate change has seen the mushrooming of social movements at the grassroots level, such as Greta Thunberg's idea for Friday school strikes. These activist movements started in 2018 and 2019 due to vibrant climate change-oriented students (Han and Ahn 2020; Zhanda et al. 2021). These students pursued and championed cleaner production methods and sustainable development broadly. As such, Greta Thunberg's students encouraged people to treat climate action as everyone's business, as the effects are threatening human lives globally. Unfortunately, ecologists and climate change organisations lack the autonomy to contest the United Nations climate change diplomacy.

Zimbabwe's Climate Change Policy Framework

Zimbabwe is drafting a climate change policy, and there is hope and optimism that such a policy will translate to meaningful climate programmes, plans and projects. Table 10.1 shows climate change policies in Zimbabwe to date.

Zimbabwe ratified the United Nations Framework Convention on Climate Change (UNFCCC) in 1992 at the Rio Earth Summit, showing commitment to climate change action in all facets of its economy. Zimbabwe produced its first communication to the UNFCCC in 1998 detailing its commitment to lower greenhouse gas emissions and possibly serious challenges that global climate change might have on the country through its various economic activities such as mining, agriculture, transport, energy emissions, rural impacts and forestry (Zimbabwe's Initial Communication on Climate Change 1998). As a follow-up to the Initial Communication on Climate Change in 1998, Zimbabwe produced its Second National Communication in 2013, emphasising reduced climate change pollutants and enhanced sustainable living conditions of citizens from all spheres of life and economic activities (Dodman and Mitlin 2015).

Table 10.1 Zimbabwe climate change policies

Name of climate change policy	Year
Zimbabwe Initial National Communication to the UNFCCC	1998
Zimbabwe Second National Communication of Climate Change to the UNFCCC	2013
Zimbabwe's National Climate Change Response Strategy	2014
Zimbabwe Third National Communication to UNFCCC	2016
Zimbabwe Fourth National Communication and First Biennial Update to the Report UNFCCC	2020

Source Authors

In 2014, Zimbabwe produced its National Climate Change Response Strategy. The response strategy has seven central pillars in response to climate change issues. These pillars include (1) Adaptation and Disaster Risk Management; (2) Mitigation and low carbon development strategies (LCDS); (3) Capacity to effect: adaptation and mitigation, climate change communication, education and raising awareness, research and development, appropriate institutions to address climate change issues; (4) Governance framework; (5) Finance and Investment; (6) Technology development and transfer, including infrastructure and (7) Communication and advocacy; information management and dissemination (Zimbabwe's National Climate Change Response Strategy 2014). The country's national climate change policy is a comprehensive document that considers the inclusive nature of an intervention strategy that includes all major stakeholders and supporting frameworks emanating from various international protocols.

In its commitment to a brighter climate future, Zimbabwe produced a Third National Communication to the United Nations Framework Convention on Climate Change in 2016. The production included a multi-actor approach ranging from various ministries in the country, NGOs, academics, civil society, Zimbabwe Meteorological Services Department and Zimbabwe National Statistics Agency, to mention a few. The overarching aim of this document is to notify the Convention Parties, relevant climate change stakeholders, government policy planners and decision-makers, various field specialists, and the public at large of the impacts of climate change in Zimbabwe. The Communication also delivers Zimbabwe's GHG emissions whilst outlining mitigatory measures to be taken to avert the impacts caused by climate change (ZTNCUNFCC 2016).

In 2020, Zimbabwe produced its First Biennial Update Report (BUR1) to the UNFCCC. Through this report, the country provides updated climate change information on national climate change conditions, GHG profile, and mitigatory plans and outlines finance, technology and capacity-building needs and support for a better climate-resilient society (ZFBURUNFCCC 2020).

Apart from these national climate change policies, according to Dodman and Mitlin (2015), Zimbabwe has a National Climate Change Office housed in the Ministry of Environment and funded by the United Nations Development Programme (UNDP). Its mandate is to liaise with the UNFCCC to produce various national climate change communication documents. The country has a National Climate Change Task Team housed in the Office of the President, with a permanent secretary handling its affairs at the highest level. The task team has a climate change budget from the national government and is responsible for coordinating all relevant government structures to design, plan and execute a solid national climate change strategy that focuses on the country's climate change challenges (Dodman and Mitlin 2015). In addition, Zimbabwe has a National Climate Change Steering Committee. This committee is responsible for facilitating a more comprehensive range of climate change participants from government and civil society in response to formulating national climate change response strategies.

While Zimbabwe is good at crafting sound climate change policies among others, it suffers from implementation paralysis; hence it is yet to be seen if the climate change policy will be different in any way. Significant climate policy needs broader collective changes due to its long-duration nature. This may be challenged by those in the engineering field who propose and believe in scientific and quantitative methods as solutions to GHG emissions reduction instead of policies compartmentalised in the government's storage files without proper implementation (Dzvimbo et al. 2017a, b).

Methods such as the scientific geological storage and sequestration of carbon-captured fossil fuel combustion or geo-engineering are becoming progressively more prevalent (Huisingh et al. 2015); hence, Zimbabwe's government needs to spearhead such projects with the help of the private sector and emissions markets. This approach is technologically vibrant as carbon emissions mitigation markets work well if the government ensures scarcity on the market. This scenario leads to a price and an effective fine or punishment for non-complying companies. Climate change policies must be treated as essential set standards to ensure better living standards for people as stipulated by sustainable development goal 13 (climate action). Climate change policies must not be seen or regarded as policies that conflict with humanity or divisive policies (Thomas 2021; Ojha et al. 2016). This paradigm drifting in climate change policies means that the government should act as an "enabling state" by ensuring an adequate, inclusive and integrated climate change policy framework that enables a smooth operationalisation of climate change adaptation and mitigation (Clarvis et al. 2014).

Climate change activists and individuals concerned about the climate are constantly warning the world about an impending severe, catastrophic climate challenge that requires swift and drastic adaptation and responsive measures to be implemented (Han and Ahn 2020; Lomborg 2020). On the other side of the coin, climate doubters suspect and doubt the reality or the seriousness of global warming. They usually contend contrary to far-reaching climate policy or find climate policy not relevant at all (Renn 2011; Lo 2014). For example, in the United States, climate-doubting people have relatively huge media attention as the media believe in hearing both sides of the story (Boykoff and Boykoff 2004). As a result, a balanced bias is created as the small climate change cohort attains much media coverage as opposed to their non-sceptical proponents.

Discussion

The Dynamics of Global Climate Change Policies and Frameworks

Studies show the mismatch between the policies and the necessary emission reduction target. For example, following the Paris Climate Agreement signed in 2015, the state parties to the agreement should update and upgrade greenhouse gas emission reduction targets for 2030 (FAO 2016; Zhanda et al. 2021). It is critical to note that if the temperature rise is to hold at 1.5 degrees by 2100, it is necessary to reduce the parties' emissions by 55%, according to their first pledges. However, according to a report that examines the declaration of the 75 states who reported a renovated target, there is only 2.8% of additional reduction pledges from these countries compared to their targets five years ago (UNFCCC 2021: 5; Fransen et al. 2021). This is a mere drop in the ocean. Besides, the countries' total emissions have increased since 1992, when the United Nations Climate Convention was signed. Annual emissions have increased by 60%, while the global cumulative emission doubled from 1990 to 2015 (Matamanda et al. 2017). Let alone the reduction of emissions, the hegemonic climate policy measures have doubled cumulative output and thus clearly indicate their ineffectuality (Çoban 2021).

Questioning the Politics of Climate Change

Climate politics is evolving over time and may be entering a new era (Zhang and Li 2018). Several inquiries and questions have been asked on how to respond to climate change and who are the majority of victims of the change (Davis 2020). However, it is known now that climate change affects human health directly and indirectly and increases human mental health challenges like solastalgia (Bourque and Cunsolo Willox 2014). Youthful climate change activists and movements are playing a significant role in imparting climate change awareness and education; the landscape of climate action seems to be overwhelmed by unintentional impacts as pragmatic evidence emerges of greening projects' adverse impacts, for example, causing people's displacement (UN-Habitat 2020). However, Davies (2021) noted that despite international concern about climate change issues, the world seems to be losing the battle to keep GHG emissions low to a reasonable amount. Developing countries suffer the most from the failure to adapt to climate change (Tadesse 2010; Burke et al. 2018).

The Covid-19 occurrence in 2020 has negatively affected global economies, but its impact on the environment is unclear (International Energy Agency 2020; Khan et al. 2020). According to Cheval et al. (2020), efforts to reduce poverty and inequality have been compromised by the arrival of the Covid-19 pandemic. Covid-19 has triggered the need to avail stimulus packages to assist peoples' standard of living. The bulk of the stimulus package was also channelled to infrastructural developments such as information technology, hospital expansion, purchasing personal protective equipment, and climate emergency (Shah et al. 2020). Part of the Covid-19 stimulus package was meant to support climate emergency under the auspices of the "Green New Deal" and just transitions. Dzvimbo et al. (2017a, b) and Samper et al. (2021) argued that terms like the "Green New Deal" may reflect a new paradigm shift towards climate change or are easily a continuation of old climate change politics wrapped in a different piece of glittering paper. There has been a comprehensive, widely held response to the clarion call to confront anthropogenic climate change on a global scale (Scott 2021). The ecological footprint presents a gloomy picture that threatens all dimensions of sustainable development.

Beyond Climate Change Impacts and Challenges

The impacts of climate change are complicated to quantify and predict; however, it is estimated that anthropogenic climate change causes over 300,000 deaths per year. The death is related to hunger, poor nutrition, famine and food insecurity (Paslakis et al. 2021). The challenges caused by the negative impacts of climate change have a ripple effect (Levner and Ptuskin 2018). Human migration in the form of forced or voluntary migration is one of the most significant results of environmental degradation and resource overexploitation (Parrish et al. 2020). Environmental degradation and over-exploitation of natural resources, especially plants and trees, directly result from drought, low rainfall and a shift in weather patterns. The result is human-animal conflict and human-to-human conflict over boundaries and the availability of fresh water and arable land (Everard et al. 2020).

Compared to other catastrophic events that have confronted the world, climate change can be regarded as one of the greatest calamities humanities ever faced (Gurría and Leape 2009; Dzvimbo et al. 2017a, b; Conversi 2020). Considering the diabolical implications of keeping up with the current GHG emission levels, let alone increasing them (Khan and Munira 2021), it is scary that less effort has been put into altering global production-consumption levels except for endless talks yielding meagre results. In an attempt to limit GHG emissions by the UNFCCC, the Kyoto Protocol educated nations about the calamities of climate change and the need to reduce GHG emissions but failed to govern and regulate various countries' GHG emissions as it favoured continual change and phasing out of chemically toxic ozone depleting substance (Tabassum 2021).

Climate change is an urgent crisis with a devastating effect on the development of intact radical forms of political commitment and engagement within the Camp for Climate Action (Howes et al. 2015). Whilst it is premature to challenge the climate change problem as a matter of urgency and concern to us all, the appeal of urgency encourages the liberal carbon consensus (Hornsey and Fielding 2020). The IPCC Fourth Assessment Report and the New Economics Foundation's "100 months" report noted a rise in GHG emissions as time progressed (IPCC 2007). However, in these reports, the IPCC emphasised the need to deal with climate change first while other contented issues were set aside for tomorrow's deliberations.

Trade-Offs and Rethinking Climate Change Integration

The United Nations Framework Convention on Climate Change (UNFCCC) met in Copenhagen for the Fifteenth Conference of Parties (COP15) (Dzvimbo et al. 2017a, b; Kinley et al. 2021). The COP 15 meeting emanated from the 1992 Rio Summit on the Environment and Development, also known as the Rio Earth Summit. The highest meeting and renowned of them is Kyoto in 1997, where the disreputable 'Kyoto Protocol' was adopted to curb increased greenhouse gas emissions. Under the Kyoto Protocol, numerous carbon reduction strategies were proposed and adopted. These include the Cap and Trade and Clean Development Mechanism, emissions and trading and joint implementation (Michaelowa et al. 2019). The COP15 was elevated as a champion in solving climate change impacts by reducing GHGs emissions. The media and other communication led the positive Kyoto Protocol campaign, but a few years later, it proved to be weak ratification of climate change policy (Ruddock 2009; Costa 2016).

There are concerns in the literature about the Copenhagen Climate Change Conference of 2009 on whether the conference tried to solve climate change and embrace a sustainable way of living or promoted the starting of a new cycle of wealth accumulation. This Copenhagen crisis is a tug-of-war between Copenhagen from above and Copenhagen from below. The Copenhagen above and below is a clash over 'justice', a battle of values. The Capitalists view maintaining and extending their modes of production and resource exploitation in a drive to accumulate wealth as fast as possible, whereas the 'below' wants to embrace sustainable development principles in wealth accumulation growth. Capitalists regard the Copenhagen Climate Change Conference of 2009 as an opportunity to restore faith in the capitalist system and representative democracy amid political and economic crises (Klein 2015).

Study Limitations

The study was limited by time constraints on the researcher's side. The researchers felt that more time was needed to gather more grey literature to support the study with more facts and examples. However, due to work commitments, the researchers struggled to coordinate effectively. Nonetheless, the researchers gathered valuable secondary data from a few available resources to develop a comprehensive line of argument.

Conclusion

The chapter outlined climate change politics in Zimbabwe and detailed hegemonic climate change policies adopted from various United Nations conferences. Zimbabwe drafted various localised climate change policies without many implementations of those policies on the ground. Ordinary Zimbabweans are not consulted on climate change policies, and the country relies on a top–down approach making it difficult for ordinary citizens to participate in climate change adaptation and coping mechanisms. While climate change policies in Zimbabwe appear very good on paper, the country is challenged by a lack of finance, international funding, technology and real timeframes to embrace projects which support the green economy transition. Climate change is affecting rainfall patterns in Zimbabwe, reducing the amount of water required at green power stations such as Kariba Hydro Power Station. Climate change takes the country back to relying more on producing energy through non-renewable resources in the form of coal at Hwange Power Station. The expansion of Hwange Power Station is a testimony to Zimbabwe's government's challenges to adopting clean sources of energy. Despite being a signatory to international conventions on climate change, Zimbabwe still has a long way to go in domesticating and crafting policies and plans. Much is yet to be realised from the climate change policy in Zimbabwe.

References

Allen JS, Longo SB, Shriver TE (2018) Politics, the state, and sea level rise: the treadmill of production and structural selectivity in North Carolina's coastal resource commission. Sociol Q 59(2):320–337

Aseh N (2011) Ideologies, governance and the public sphere in Cameroon. Africa Development 36(1):169–220

Banya and Reid (Reuters, 2022). In Zimbabwe, coal power project seeks other backing after China's U-turn. https://www.reuters.com/business/energy/zimbabwe-coal-power-project-seeks-other-backing-after-chinas-u-turn-2022-03-30/. Accessed 17 Mar 2022

Beddington J (2009) Food, energy, water and the climate: a perfect storm of global events. World Dev, 1–9

Bodansky D (2016) The legal character of the Paris Agreement. Rev European Comparative Int Environ Law 25(2):142–150

Bourque F, Cunsolo Willox A (2014) Climate change: the next challenge for public mental health? Int Rev Psychiatry 26(4):415–422

Boykoff M, Boykoff J (2004) Balance as bias: global warming and the US prestige press. Glob Environ Change 14:125–136

Brinkmann R (2021) Our climate change challenge. Practical sustainability. Palgrave Macmillan, Cham, pp 13–36

Broto VC (2017) Urban governance and the politics of climate change. World Dev 93:1–15

Burke SE, Sanson AV, Van Hoorn J (2018) The psychological effects of climate change on children. Curr Psychiatry Rep 20(5):1–8

Chappell A, Lee JA, Baddock M, Gill TE, Herrick JE, Leys J, Marticorena B, Petherick L, Schepanski K, Tatarko J, Telfer M (2018) A clarion call for aeolian research to engage with global land degradation and climate change. Aeol Res 32:A1–A3

Chatiza K (2019) Cyclone Idai in Zimbabwe: an analysis of policy implications for post-disaster institutional development to strengthen disaster risk management. https://oxfamilibrary.openrepository.com/bitstream/handle/10546/620892/bp-impact-response-cyclone-idai-zimbabwe-071119-en.pdf. Accessed 7 Sept 2022

Cheval S, Mihai Adamescu C, Georgiadis T, Herrnegger M, Piticar A, Legates DR (2020) Observed and potential impacts of the COVID-19 pandemic on the environment. Int J Environ Res Public Health 17(11):4140

Chirisa I, Matamanda AR (2022) Science communication for climate change disaster risk management and environmental education in Africa. In *Research anthology on environmental and societal impacts of climate change*. IGI Global, pp 636–652

Clarvis MH, Fatichi S, Allan A, Fuhrer J, Stoffel M, Romerio F, Gaudard L, Burlando P, Beniston M, Xoplaki E, Toreti A (2014) Governing and managing water resources under changing hydro-climatic contexts: the case of the upper Rhone basin. Environ Sci Policy 43:56–67

Coban,A (2021) People's climate politics against hegemonic climate politics published in Abstrakt Dergi (Dossier 9), November 26, 2021 Part I - http://www.abstraktdergi.net/peoples-climate-politics-against-hegemonic-climate-politics-1/. Accessed 12 Jan 2022

Compton C (2020) The temporality of disaster: data, the emergency, and climate change. Anthropocenes Human Inhuman Posthuman, 1(1). https://www.anthropocenes.net/article/id/667/

Conversi D (2020) The ultimate challenge: nationalism and climate change. Nationalities Papers 48(4):625–636

Costa O (2016) Beijing after Kyoto? The EU and the new climate in climate negotiations. EU Policy Responses to a Shifting Multilateral System. Palgrave Macmillan, London, pp 115–133

Davis DS (2020) Studying human responses to environmental change: trends and trajectories of archaeological research. Environ Archaeol 25(4):367–380

Dodman D, Mitlin D (2015) The national and local politics of climate change adaptation in Zimbabwe. Climate Dev 7(3):223–234

Dzvimbo MA, Monga M, Mashiza TM (2017a) Perpetual 'outcasts': the impact of climate change on rural children in Zimbabwe. J Pub Admin Dev Altern JPADA 2(2):60–69

Dzvimbo MA, Mashizha TM, Monga M, Ncube C (2017b) Conservation agriculture and climate change: implications for sustainable rural development in Sanyati, Zimbabwe. J Soc Dev Sci 8(2):38–46

Everard M, Johnston P, Santillo D, Staddon C (2020) The role of ecosystems in mitigation and management of Covid-19 and other zoonoses. Environ Sci Policy 111:7–17

FAO (2016) The state of food and agriculture, climate change food and agriculture. https://www.fao.org/3/i6030e/i6030e.pdf. Accessed 8 Oct 2022

FAO (2020) Food and agriculture organisation, global forest resources assessment 2020: key findings, Rome, 2020. https://www.fao.org/forest-resources-assessment/2020/en/. Accessed 12 July 2022

Fransen T, Waskow D, Thwaites J, Seymour F, Dagnet Y (2021) Leaders Summit on climate offers jolt of momentum for global action. https://www.wri.org/insights/leaders-summit-climate-offers-jolt-momentum-global-action. Accessed 8 Oct 2022

Glasgow Agreement (2020) Glasgow climate pact. https://unfccc.int/sites/default/files/resource/cop26_auv_2f_cover_decision.pdf. Accessed 7 Oct 2022

Gurría A, Leape J (2009) Climate change: the biggest threat to economic review. OECD Observer. http://www.oecdobserver.org/news/fullstory.php/aid/3074/Climate_change:_the_biggest_threat_to_economic_recovery.html

Han H, Ahn SW (2020) Youth mobilization to stop global climate change: narratives and impact. Sustainability 12(10):4127

Hornsey MJ, Fielding KS (2020) Understanding (and reducing) inaction on climate change. Soc Issues Policy Rev 14(1):3–35
Howes M, Tangney P, Reis K, Grant-Smith D, Heazle M, Bosomworth K, Burton P (2015) Towards networked governance: improving interagency communication and collaboration for disaster risk management and climate change adaptation in Australia. J Environ Planning Manage 58(5):757–776
Howells M, Boehlert B, Benitez PC (2021) Potential climate change risks to meeting Zimbabwe's NDC goals and how to become resilient. Energies 14(18):5827
Hudson R (2021) Capitalism, contradictions, crises: pushing back the limits to capital or breaching the capacity of the planetary ecosystem? Area Development and Policy 6(2):123–142
Huisingh D, Zhang Z, Moore JC, Qiao Q, Li Q (2015) Recent advances in carbon emissions reduction: policies, technologies, monitoring, assessment and modeling. J Clean Prod 103:1–12
International Energy Agency (2020) World energy outlook 2020. https://iea.blob.core.windows.net/assets/a72d8abf-de08-4385-8711-b8a062d6124a/WEO2020.pdf. Accessed 11 Sept 2022
IPCC, Intergovernmental Panel on Climate Change, Climate Change 2007: Climate change 2007. Impacts, adaptation and vulnerability. https://www.ipcc.ch/site/assets/uploads/2018/03/ar4_wg2_full_report.pdf. Accessed 12 June 2022
Khan MR, Munira S (2021) Climate change adaptation as a global public good: implications for financing. Clim Change 167(3):1–18
Khan K, Zhao H, Zhang H, Yang H, Shah MH, Jahanger A (2020) The impact of COVID-19 pandemic on stock markets: an empirical analysis of world major stock indices. J Asian Finance Econ Bus 7(7):463–474
Kinley R, Cutajar MZ, de Boer Y, Figueres C (2021) Beyond good intentions, to urgent action: former UNFCCC leaders take stock of thirty years of international climate change negotiations. Climate Policy 21(5):593–603
Klein N (2015) This changes everything: capitalism vs. the climate. Simon and Schuster, New York, United States of America
Levner E, Ptuskin A (2018) Entropy-based model for the ripple effect: managing environmental risks in supply chains. Int J Prod Res 56(7):2539–2551
Lo AY (2014) The right to doubt: climate-change scepticism and asserted rights to private property. Environmental Politics 23(4):549–569
Lomborg B (2020) False alarm: how climate change panic costs us trillions, hurts the poor, and fails to fix the planet. Hachette UK
Mashizha TM, Monga M, Dzvimbo MA (2017) Improving livelihoods of resettled farmers through development of a knowledge base on climate change in Mhondoro-Ngezi District, Zimbabwe. Int J Sustain Dev Res 3(2):18–26
Matamanda AR, Dzvimbo MA, Kadebu RT (2017) Climate change and infrastructure planning: implications for sustainable urban management in Harare, Zimbabwe. J Pub Admin Dev Altern 2(1):95–111
Matsa M (2020) Climate change and agriculture in Zimbabwe. Springer Nature Switzerland AG
Mawonde A, Togo M (2021) Challenges of involving students in campus SDGs-related practices in an ODeL context: the case of the University of South Africa (Unisa). Int J Sustain Higher Edu
Michaelowa A (2013) The politics of climate change in Germany: ambition versus lobby power. Wiley Interdisciplinary Rev Climate Change 4(4):315–320
Michaelowa A, Shishlov I, Brescia D (2019) Evolution of international carbon markets: lessons for the Paris Agreement. Wiley Interdisciplinary Rev Climate Change 10(6):e613
National Renewable Energy Policy (2019) Zimbabwe ministry of energy and power development. https://www.zera.co.zw/National_Renewable_Energy_Policy_Final.pdf. Accessed 11 July
Nhapi TG (2021) Perspectives on safeguarding children in sustainable disaster mitigation in Zimbabwe. Cyclones in Southern Africa. Springer, Cham, pp 81–93
NS Energy (2022) Hwange power station expansion. https://www.nsenergybusiness.com/projects/hwange-power-station-expansion/. Accessed 12 Feb 2022

Nyama V, Mukwada G (2022) Factors affecting citizen participation in local development planning in Murewa District, Zimbabwe. J Asian Afr Stud, p 00219096211069643

Ojha HR, Ghimire S, Pain A, Nightingale A, Khatri DB, Dhungana H (2016) Policy without politics: technocratic control of climate change adaptation policy making in Nepal. Climate Policy 16(4):415–433

Parrish R, Colbourn T, Lauriola P, Leonardi G, Hajat S, Zeka A (2020) A critical analysis of the drivers of human migration patterns in the presence of climate change: a new conceptual model. Int J Environ Res Public Health 17(17):6036

Paslakis G, Dimitropoulos G, Katzman DK (2021) A call to action to address COVID-19–induced global food insecurity to prevent hunger, malnutrition, and eating pathology. Nutr Rev 79(1):114–116

Peter SC (2018) Reduction of CO2 to chemicals and fuels: a solution to global warming and energy crisis. ACS Energy Lett 3(7):1557–1561

Poole JA, Barnes CS, Demain JG, Bernstein JA, Padukudru MA, Sheehan WJ, Fogelbach GG et al (2019) Impact of weather and climate change with indoor and outdoor air quality in asthma: a Work Group Report of the AAAAI Environmental Exposure and Respiratory Health Committee. J Allergy Clin Immunol 143(5):1702–1710

Ramanathan V (2020) Climate change, air pollution, and health: common sources, similar impacts, and common solutions. In Health of People, Health of Planet and Our Responsibility. Springer, Cham, pp 49–59

Renn O (2011) The social amplification/attenuation of risk framework: application to climate change. Wiley Interdiscip Rev Climate Change 2(2):154–169

Ruddock J (2009) Transcript of Parliamentary debate on COP15. http://www.actoncopenhagen.decc.gov.uk/en/ambition/achievements/november/transcriptparliamentary-debate. Accessed 22 July 2022

Samper JA, Schockling A, Islar M (2021) Climate politics in green deals: exposing the political frontiers of the European Green Deal. Polit Gov 9(2):8–16

Scott D (2021) Sustainable tourism and the grand challenge of climate change. Sustainability 13(4):1966

Shah AUM, Safri SNA, Thevadas R, Noordin NK, Abd Rahman A, Sekawi Z, Ideris A, Sultan MTH (2020) COVID-19 outbreak in Malaysia: actions taken by the Malaysian government. Int J Infect Dis 97:108–116

Singh C, Chauhan N, Upadhyay SK, Singh R, Rani A (2021) The Himalayan natural resources: challenges and conservation for sustainable development. J Pharmacogn Phytochem 10(1):1643–1648

Tabassum N (2021) Climate capitalism or carbon colonialism? The critical features of climate change adaptation and mitigation policies. In Handbook of development policy. Edward Elgar Publishing, pp 313–323

Tadesse D (2010) The impact of climate change in Africa. Institute for Security Studies Papers, 220, p 20

Thomas A (2021) Framing the just transition: how international trade unions engage with UN climate negotiations. Glob Environ Chang 70:102347

Thomas K, Hardy RD, Lazrus H, Mendez M, Orlove B, Rivera-Collazo I, Roberts JT, Rockman M, Warner BP, Winthrop R (2019) Explaining differential vulnerability to climate change: a social science review. Wiley Interdiscip Rev Climate Change 10(2):e565

Tsabora J (2019) The constitutional state and traditionalism under the 2013 Zimbabwean Constitution: a critique. Raoul Wallenberg Institute of Human Rights and Humanitarian Law. https://ir.uz.ac.zw/handle/10646/3979. Accessed 11 Sept 2022

United Nations Framework Convention on Climate Change (2015). Paris Agreement. https://www.un.org/en/development/desa/population/migration/generalassembly/docs/globalcompact/FCCC_CP_2015_10_Add.1.pdf. Accessed 8 Oct 2022

United Nations Climate Change Annual Report (2021) https://unfccc.int/sites/default/files/resource/UNFCCC_Annual_Report_2021.pdf. Accessed 11 Sept 2022

UN-Habitat (2020) World cities report 2020: the value of sustainable urbanization. UN-Habitat, Nairobi. Retrieved from https://unhabitat.org/World. Accessed 12 June 2022

Wallace-Wells D (2019) The uninhabitable earth: a story of the future. Penguin UK

White R (2018) Climate change criminology. Bristol University Press, United Kingdom

Williamson K, Satre-Meloy A, Velasco K, Green K (2018) Climate change needs behavior change: making the case for behavioral solutions to reduce global warming. Arlington, VA: Rare

Zhanda K, Dzvimbo MA, Chitongo L (2021) Children climate change activism and protests in Africa: reflections and lessons from Greta Thunberg. Bull Sci Technol Soc 41(4):87–98

Zhang L, Li X (2018) Changing institutions for environmental policy and politics in New Era China. Chin J Pop Resour Environ 16(3):242–251

Zimbabwe's First Biennial Update Report to the United Nations Framework Convention on Climate Change (2020) Ministry of environment, climate, tourism and hospitality industry. https://www4.unfccc.int/sites/SubmissionsStaging/NationalReports/Documents/60235741_Zimbabwe-BUR1-1-Zimbabwe%20Biennial%20Update%20Report%201.pdf. Accessed 14 Apr 2022

Zimbabwe's Initial National Communication on Climate Change (1998) Prepared for the United Nations framework convention on climate change. https://unfccc.int/sites/default/files/resource/zimnc1.pdf. Accessed 17 Apr 2022

Zimbabwe's National Climate Change Response Strategy (2014) http://www.climatechange.org.zw/sites/default/files/National%20Climate%20Change%20Response%20Strategy.pdf. Accessed 12 Feb 2022

Zimbabwe's National Climate Change Response Strategy (2017) Government of Zimbabwe Ministry of environment, water and climate. http://ncuwash.org/newfour/wp-content/uploads/2017/08/Climate-Change-Response-Strategy.pdf. Accessed 8 Oct 2022

Zimbabwe Third National Communication to the United Nations Framework Convention on Climate Change (2016) United Nations framework convention on climate change. https://unfccc.int/sites/default/files/resource/zwenc3.pdf. Accessed 17 Apr 2022

Chapter 11
Geospatial Surveillance of Vicissitudes Observed in the Land Surface Temperature of the Federal Capital of Pakistan Over Three Decades

Amna Butt

Abstract The provision of geospatial surveillance for the assessment of the built environment's bearing on urban microclimate is of paramount significance, yet it is very complicated. The amalgamation of remote sensing techniques and tools has made it possible to appraise the impacts of the built environment on micro-climatic conditions in the past. The present study, therefore, deployed various tools to determine the extent of risk the built environment posed to Islamabad in the past three decades and predict the future in the prevalence of current practices. The calibration of the past landscape was done by per-pixel signature-based hybrid classification performed in the ArcGIS environment. The next step was a calculation of the selected micro-climate variable i.e. Land Surface Temperature (LST) via a mono-window NDVI-based algorithm. The outcomes of both analyses were correlated in the SPSS environment to determine which land use category correlated strongly with the LST. The research outcomes highlighted that the urban expansion in the capital city was mostly at the expense of vegetation. The urban areas of Islamabad expanded from covering 10% of the study area in 1988 to about 23% area in 2018 respectively while the vegetation cover shrunk from 54 to 36% in the same period. This landscape change contributed significantly to the rise of land surface temperature (maximum) from 24 to 31 °C in Islamabad as evident by the strong correlation coefficient obtained (0.999*) between LST and Normalized Difference Built-up Index (NDBI) calculated for the capital city. These findings further indicate a future where the effects of landscape alterations on the land surface temperature will be devastating especially if the current practices continue to persist. In the absence of strong policy interventions, these outcomes indicate a highly unsustainable future. However, this research could be used as a reference point to develop, test, and implement developmental plans and policies that establish appropriate ratios for different land use types to ensure the sustainability of the environment and the availability of its associated amenities.

A. Butt (✉)
Department of Environmental Sciences, Fatima Jinnah Women University, The Mall, Rawalpindi, Pakistan
e-mail: ambutt91@yahoo.com

W. Leal Filho et al. (eds.), *Climate Change Strategies: Handling the Challenges of Adapting to a Changing Climate*, Climate Change Management,
https://doi.org/10.1007/978-3-031-28728-2_11

Introduction

The built environment is a term that encompasses not only the assortment of buildings (for instance the places where we live or work); it rather includes the physical outcome of various socio-economic as well as environmental processes, which pertain strongly to societal needs and standards. Thereby, the built environment can be coined "urban influence" whether that influence is on public health, physical activity, transportation network, or any other physical feature or process of life. The unique biophysical features of the built environment differ from that of the surrounding rural areas and contribute more significantly to alterations in urban microclimate (Guha et al. 2020).

Everything apparent in a built environment is designed by someone to look natural but is rarely so. Even the natural reserves and parks are devised with the aesthetic beauty and comfort of an urbanite in mind. These changes in the natural environment are primarily the outcomes of the altered land cover of an area. Therefore, the first step in determining the extent of the risk posed by the built environment to the land surface temperature is understanding the local built environment itself. A wholesome account of the local built environment of an area, as well as the alterations observed in that area over the areas, is of paramount significance as many aspects of the built environment impact us in a variety of ways and the impact of each aspect on people's life is subtle, but the combined effect poses significant causes for concern.

Climatic Implications of Built Environment

Most of the components of the built environment, whether they be biophysical or socioeconomic contribute significantly to the changes in climatic conditions. Such an environment comprises unique biophysical features such as biotic and energy exchange patterns, as well as hydrological and atmospheric cycles that are relative to surrounding areas. Fluctuation in these exchange patterns is primarily attributed to the land cover dynamics of an area (such as the replacement of vegetated surfaces with paved ones leads to access runoff of rainwater and thus changes the local hydrology) and contributes towards the changes in local climatic conditions (Bendib et al. 2017).

Additionally, the built environment is the leading cause of raised levels of land surface temperature in metropolitan cities all over the world, making it responsible for creating urban heat islands. For instance, exhaust gasses from vehicles emit greenhouse gasses that are responsible for raising the earth's temperature. Paved surfaces, new infrastructure, and construction operations on the other hand lead to increased runoff rate, altered evapotranspiration rate as well as land surface temperature elevation (as mentioned above). Therefore, changes in the land use/land cover pattern in a built environment cause deforestation, agriculture intensification, and rapid urbanization which both, directly and indirectly, are responsible for land surface temperature elevation (by increasing CO_2 level in the atmosphere). Furthermore, the shaping of infrastructure for both buildings as well as transportation purposes is affecting the access to green spaces and is responsible for habitat disruption of environmentally imperative species.

This is a serious cause of concern for a country like Pakistan because the built environment of Pakistan has changed significantly over the decades due to rapid urbanization. Major vicissitudes have been observed in the built environment of both the urban and rural areas of Pakistan in the last few decades, most especially in the metropolitan areas. These vicissitudes are associated with immense socio-economic, demographic, and technological transformation (Khan et al. 2020; Belokrenitsky 2017). The driving force behind this rapid urbanization in Pakistan was different in different eras. Before the 1980s, urbanization was driven by industrialization-based modernization. Whereas after the 1980s, the agglomeration of the population within cities' boundaries, urban periphery, and townships were responsible for the present state of urbanization. Furthermore, because of the colossal migration from rural to urban areas, more than millions of housing units are required to be built in the urban areas of Pakistan on yearly basis (Qayyum 2019).

The rampant shift in land use and the associated climatic implications are more prominent in the Federal capital of Pakistan because it has been subjected to Urban agglomeration more than many other cities because of its vicinity, aesthetic beauty, and economic potential. This fact was also highlighted by the study conducted by Naeem et al. (2018) that comparatively analyzed the association of urban greenspaces of Beijing and Islamabad with LST. The outcomes suggested that both the configuration and composition of the urban greenspaces influence the distribution of LST in urban centers. The study indicated that cool surfaces can be increased or the LST of urban centers can be decreased if the number of green spaces available in these centers is increased.

Role of Geospatial Tools in Built-Environment Impact Assessment

Assessment of the impacts of climate change on the built environment or the impacts of various aspects of the built environment on the regional climate has been the focus of numerous studies over the past (Mukherjee et al. 2016; Feizizadeh et al. 2013). Recent research has placed intensive emphasis on how to design the future built environment to cope better with the changing climate and lately, how to reduce the impacts of rapidly increasing Land Surface Temperature (LST), which is an outcome of rapid urbanization (Millward et al. 2014; Li et al. 2012). To that effect, Remote sensing and GIS tools have played a key role in determining the level of land surface temperatures and subsequent development of the urban heat island effect in different parts of the world as evident by the substantial amount of literature available on the topic (Gohain et al. 2020; Logan et al. 2020; Ayanlade 2017; Bendib et al. 2017, etc.).

Gohain et al. (2020) associated the intensification of LST in Pune city with land cover changes (from 1990 to 2019) by using satellite data of different Spatio-temporal resolutions. Logan et al. (2020) appraised the import and influence of urban land surface characteristics on land surface temperature (remotely sensed) for four cities in the U.S. The study used night-time thermal satellite imagery of moderate resolution along with certain statistical models to accomplish this task. The outcomes of the research highlighted the important role played by both impervious and vegetative surfaces in association with LST.

Khan et al. (2020) endeavored to determine the impact of land use dynamics on Relative Land Surface Temperature (RLST). The study classified land use dynamics via a random forest classifier and retrieved RLST via a standardized radiative transfer equation. The study found the increase in Urban Heat Island (UHI) effect due to continuous anthropogenic activities and the conversion of natural surfaces to impervious ones, which is a trend that is increasing day by day.

Ayanlade (2017), assessed various land classification methods and their implications on variations in land surface temperature by using remotely sensed data in a case study based in Nigeria. It was established that many Spatio-temporal, spectral and environmental factors can determine which of the remote sensing methods will work best to effectively determine the land surface temperature. Thereby, the choice of the method should be made by keeping all these factors into consideration.

Bendib et al. (2017) in another study employed the Mono-Window (MW) algorithm on Landsat 8 TIRS for the retrieval of land surface temperature. The algorithm required surface emissivity and spectral radiance as inputs which were obtained with the help of the Normalized Difference Vegetation Index (NDVI) and band 10 of the imagery respectively.

Sun et al. (2015) performed an intercomparison of LST retrieved from two separate sources namely: MODerate Resolution Imaging Spectroradiometer (MODIS) and Geostationary Operational Environmental Satellites (GOES). The differences in the LST calculated via both satellites, which operate on different algorithms, were more pronounced during daytime than nighttime. This difference was attributed to the land surface properties (such as surface emissivity and vegetation cover) as well as the anisotropy in satellite viewing geometry.

Mukherjee et al. (2016) in their research debated that both the built environment and urban heat islands generated as a consequence of the Spatio-temporal changes in the built environment are of great importance for urban policy and planning. The study claimed that the variation in land surface temperature within the built environs reported in the respective research was due to different land use practices with the highest temperature in urban areas followed closely by suburbs and the lowest (relative) observed in rural areas.

Bearing the insights gained from the above literature, the fundamental objective of the present research was to employ geospatial modeling tools to appraise the changes in the built environment of the Federal Capital over the past decades and determine the extent of risk such change poses to the urban microclimate (more specifically, the land surface temperature). To that effect, the present research employed an amalgamation of Remote sensing and GIS tools to determine the extent of risk the built environment posed to local climatic conditions (the specific objectives that were drawn for this research are outlined in the subsequent section).

Specific Objectives

Employ geospatial modeling tools to appraise the changes in the built environment of Islamabad over the past decades—This objective was specifically designed to determine the past and present trends of land cover and land use in the city with the help of remotely sensed data (Landsat 4–5 TM and Landsat 8 OLI imagery) and GIS tools (per-pixel signature-based hybrid classification). The maps generated by the geospatial appraisal of the changes observed in the built environment were thereby used as a baseline or foundation in the present study on which the subsequent objectives relied.

Determine which land use classes are associated most with the vicissitudes in land surface temperature—After the comprehensive appraisal of vicissitudes observed in the landscape of Islamabad over the decades (1988–2018), it was imperative to evaluate each land use category against the Land Surface Temperature calculated for each year by the means of the Mono-window NDVI-based algorithm. This endeavor gave a clear account of which land use class or category correlated even slightly with the calculated LST.

Determine the extent of the risk land use change posed to current climatic conditions in terms of elevated land surface temperature—To accomplish the final and ultimately significant objective of the present research, two indices were further developed, namely Normalized Difference Built-up Index (NDBI) and Normalized Difference Water Index (NDWI). These indices were then correlated with the results of the LST to support the previous objectives that indicated that built-up areas and vegetation play a crucial role in determining the urban microclimate of a region, especially in terms of the elevated Land Surface Temperature.

Materials and Methods

The general methodology adopted to fulfill the study objectives consisted of the following steps: Satellite data collection and processing, Hybrid classification, Accuracy assessment, Land Surface Temperature (LST) calculation, and Correlating LST outcomes with different Indices (see Fig. 11.1 for the details).

The satellite data for the present research was acquired from Landsat 4–5 TM (years 1988, 1998, and 2008) and Landsat 8 OLI (the year 2018) and was processed for radiometric and atmospheric corrections in the ArcGIS environment. The next step of the research involved the quantification of built urban areas, water-covered areas, agricultural areas, and barren, and vegetation-covered areas of Islamabad for

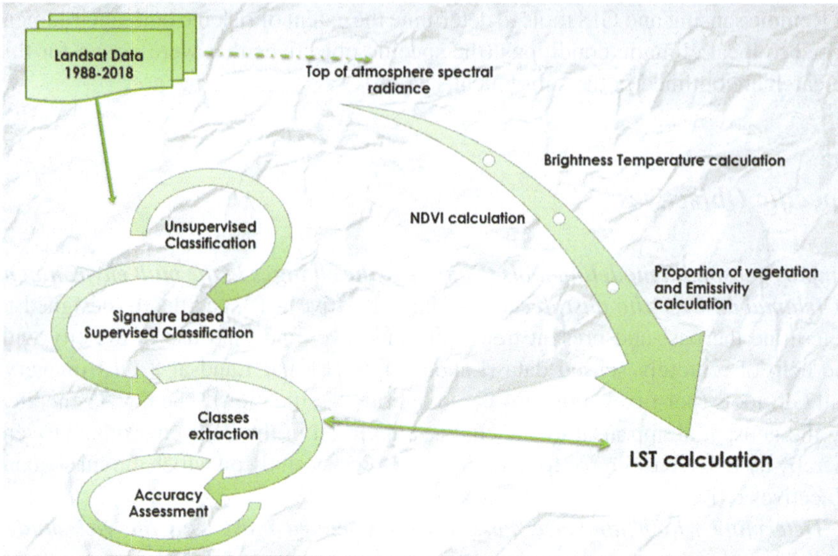

Fig. 11.1 The methodology adopted to assess the relationship between land use change and Land Surface Temperature elevation

the years 1988, 1998, 2008, and 2018. The outcomes obtained for each land use class were then compared to observe the shift in the land use pattern over the years. This was done by the signature-based hybrid classification method. This method combined the outcomes of software-defined unsupervised classification with user-defined maximum likelihood supervised classification to generate an output of 5 land use classes, namely Agriculture, Water, Vegetation, Soil/Rocks, and Settlements/Built area. Out of these, the Vegetation and Settlements/Built area class was further extracted to be correlated with the LST outcomes as the previous research suggested that they correlated significantly with the LST and thus played a crucial role in determining the LST of an area as opposed to water bodies and barren lands.

For LST estimation, the mono-window NDVI-based algorithm (Cartalis et al. 2016) was employed for which the thermal infrared bands were extracted from the imagery. Then, the data obtained from the bands was converted into radiance by utilizing thermal rescaling radiance factors. This radiance was used to calculate brightness temperature by using thermal constants available in the metadata file. Additionally, land surface emissivity was derived from the emitted radiance and LST maps were generated.

The formulae for these calculations are given below:

1. Landsat data were rescaled to calculate the top of atmosphere spectral radiance by using the following formula:

$$L_\lambda = M_L Q_{cal} + A_L \tag{11.1}$$

where L_λ represented TOA spectral radiance calculated in Watts/(m_2 * srad * μm), M_L was a band-specific multiplicative rescaling factor obtained from the metadata and A_L represented a band-specific additive rescaling factor and was also obtained from the metadata file. Q_{cal} on the other hand was a symbol for the band's DN value also known as quantized and calibrated standard product pixel values.

2. In the second step, Top of Atmosphere (TOA) brightness was calculated by using the following equation:

$$\mathrm{BT} = (\mathrm{K2}/(\ln(\mathrm{K1}/\mathrm{L}) + 1)) - 273.15 \tag{11.2}$$

where BT is brightness temperature, K1 and K2 are band-specific thermal conversion constants obtained from the metadata file, ln is the log and L is TOA radiance calculated in the previous step.

3. The next step included calculation of the Normalized Difference Vegetation Index (NDVI) using Red and Infra-Red bands.
4. Using the output of step 3, the proportion of vegetation (Pv) was calculated as follows:

$$\mathrm{Pv} = \mathrm{Square}\ ((\mathrm{NDVI} - \mathrm{NDVImin})/(\mathrm{NDVImax} - \mathrm{NDVImin})) \tag{11.3}$$

5. Following that, the emissivity was calculated by employing the formula:

$$\varepsilon = 0.004 * Pv + 0.986 \tag{11.4}$$

6. The last step of the methodology was calculation of LST by adding the values obtained for all the above factors in the formula given below:

$$\mathrm{LST} = (\mathrm{BT}/(1 + \mathrm{w} * (\mathrm{BT}/1.4388) * \mathrm{Ln}(\varepsilon))) \tag{11.5}$$

where w represented the wavelength of emitted radiance (11.5 μm).

The assessment of the impact of the built environment on LST was carried out via calculating the Normalized Difference Built-up Index (NDBI) (methodology adopted: Guha et al. 2020) and correlating its outcomes with those obtained for LST. The Index was calculated by using the following formula (vi):

NDBI could be calculated by reflectance SWIR band (5 or 7 of Landsat 4–5/6 or 7 of Landsat 8) and NIR band (4 of Landsat 4–5 and 5 of Landsat 8). In the raster calculator write:

$$\mathrm{NDBI} = \mathrm{SWIR} - \mathrm{NIR}/\mathrm{SWIR} + \mathrm{NIR} \tag{11.6}$$

where reflectance SWIR band (band 5 or 7 of Landsat 4–5 TM and band 7 of Landsat 8 Oli) and NIR band for Landsat 4–5 was band 4 and in the case of Landsat 8, it was band 5.

The justification of the study outcomes was provided by yet another index i.e., the Normalized Difference Water Index (NDWI) which was calculated using the following formula:

$$\mathrm{NDWI} = \mathrm{NIR} - \mathrm{SWIR}/\mathrm{NIR} + \mathrm{SWIR} \tag{11.7}$$

The outcomes of both these indices were calculated in the ArcGIS environment and correlated with the LST outcomes in SPSS. The resultant maps that were generated based on the calculations in ArcGIS were compared to provide a better understanding of the Spatio-temporal variations observed for the city over the decades.

Results and Discussion

Policymakers actively tend to seek to direct urbanization to uphold the quality of life. Urban ecosystems and microclimates play a crucial role in this process as they impact the well-being of local populations and the long-term sustainability of the cities in one way or another (Sander 2016). However, it is challenging to factor in the impact of land use policies or changes on the microclimate using traditional methods. The outcomes of the present research provided the means to assess how different anthropogenic and biogeophysical variables might be responsible for shaping the landscape of today and the resultant impact on the urban microclimate and consequently well-being.

Land Use Dynamics

To determine the direct and indirect impact of the built environment on Land Surface Temperature elevation, it was imperative to accurately calibrate the record of land use in the study area. Therefore, hybrid classification was performed on Landsat imagery from 1988, 1998, 2008, and 2018. The accuracy of Islamabad's classification was above 80% for overall images and above 70% for each land use class in any given year (the details are given in Table 11.1).

As shown in Figs. 11.2 and 11.3, five land use classes were delineated for each city, namely: Agriculture (pink color), Built-up area (red color), Soil/Rocks (yellow), Vegetation (green), and Water (blue). The area covered by each class in Islamabad over the years is given in Table 11.2. However, for future analyses and correlations, only the area covered by the Settlements and Vegetation class was taken under consideration.

Table 11.1 Class-wise classification accuracy and Kappa statistics for the study area

Sr	Year	Overall accuracy %	Kappa statistics	Class type wise accuracy %				
				Agriculture	Built-up	Soil	Vegetation	Water
1	1988	82.47	0.7647	78.57	87.10	84.85	86.54	100
2	1998	82.55	0.7669	72.73	85.29	92.86	82.35	86.67
3	2008	84.00	0.7857	76.67	91.67	87.50	83.87	100
4	2018	87	0.8297	82.14	80	100	88	75

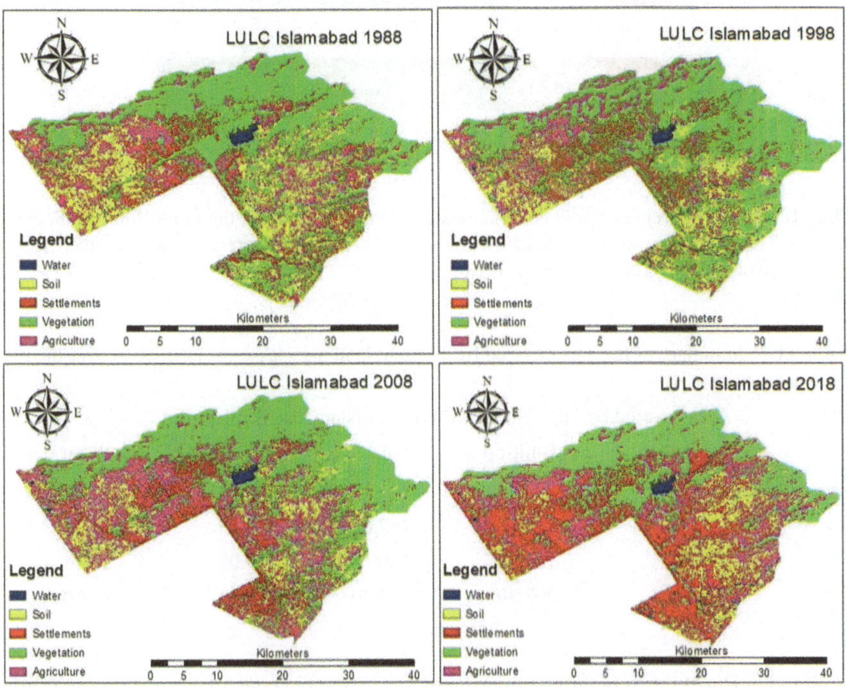

Fig. 11.2 Land use map of Islamabad for different years showing all five land use classes i.e. Agriculture (pink), Settlements/Built-up (red), Soil (yellow), Vegetation (green), and Water (blue) class

The percentage area covered by each land use class in Islamabad over the years is given in Fig. 11.3. The figures given in Table 11.2 gave a clear indication of the urban area expansion in the capital city of Pakistan over the years, whereas the in-depth analysis of the map showing Spatio-temporal dynamics of land use indicated that this expansion was not only at the expense of surrounding vegetative areas, but also put a considerable strain on the water reserves of the city which depleted steadily over the decades.

However, for future analyses and correlations, only the area covered by the Settlements and Vegetation class was taken under consideration. Observing the trends

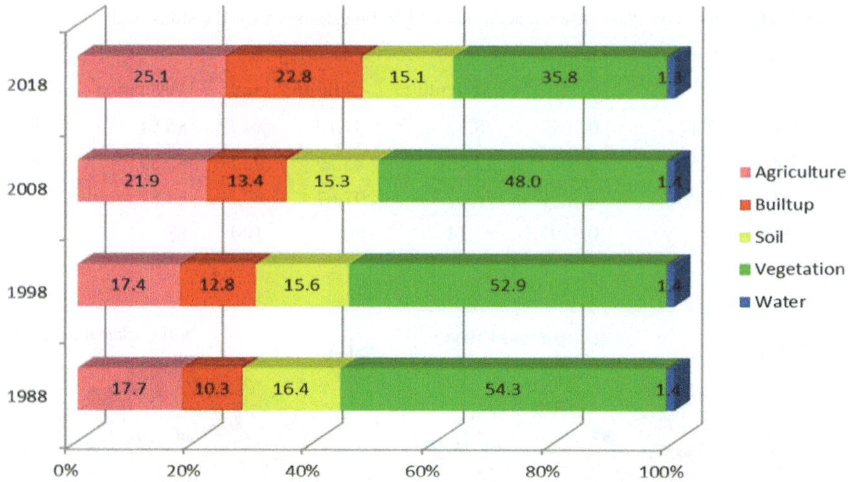

Fig. 11.3 Graphical representation of Islamabad's land use classes for the years 1988, 1998, 2008, and 2018 showing five different land use types of the city in different decades as well as the percentage of area covered by them

Table 11.2 Summary of the area covered by different land use classes of Islamabad over the past three decades

Year	Area covered by different LULC types of Islamabad Sq Km				
	Agriculture	Built-up	Soil	Vegetation	Water
1988	162.13	93.74	149.54	495.85	12.67
1998	158.78	116.78	142.67	483.23	12.47
2008	200.12	122.14	140.18	439.05	12.44
2018	229.10	208.39	137.80	326.53	12.20

depicted in Fig. 11.1, it became apparent that over the past decades, settlements/built-up areas expanded in the southwest and southeast areas of Islamabad as well as in the vicinity of Rawal lake and on the outskirts of Margalla hills. This was reflected in the percentage increase given in Fig. 11.2 as well. Contrarily, the area covered by higher vegetation decreased over the years predominantly in the northwest, northeast, and southeast regions.

Furthermore, the shrinkage of vegetation in these areas was predominantly because of the increasing demands for the development of infrastructure to accommodate the ever-increasing population in the city, especially near the urban centers and the vicinity of the major water reservoirs. Also, the added strain was placed on the vegetation by the construction of motorways, highways, and a general broadening of the road networks to increase the accessibility of the population to the available ecological and economic amenities. These outcomes were generally in line with the previous studies conducted in the area, such as the one conducted by Hassan et al. (2016) which deliberated the land use dynamics of Islamabad from 1992 through 2012. The study highlighted the economic growth of the federal capital as the main driver behind the general trend of the environmental degradation for the sake of urban expansion observed in the city, followed closely by the population growth, as well as changing climatic conditions, and mass migration from surrounding rural areas.

Another research conducted by Butt (2015) focused on assessing the environmental implications of rapid urbanization and deforestation in the Rawal watershed (which is mainly located in the capital city). It was found that Spatio-temporal modifications of the land parcels in the area were greatly influenced by a shifting pattern of land use in the area which consequently put great pressure on the available resources. Similarly, Butt et al. (2015) also stressed that the accelerated rate of land transformations in the vicinity of Islamabad needs to be catered to properly developed management plans and conservation strategies to ensure the protection and continued availability of the water, forest reserves, and soil resources. However, not much research conducted in the area has put any considerable focus on assessing the impact of changing landscape on the land's surface temperature and what it might mean for the population.

The outcomes presented in the subsequent sections provide great and valuable insight regarding the land surface temperature of the study area. Furthermore, the following research outcomes highlight how the dynamics of land use in the study area correlate with land surface temperature vicissitudes recorded by the satellite data over the decades, thus providing comprehensive proof of the impact land use change has on the microclimatic conditions of the city.

Temporal Land Surface Temperature (LST) Fluctuations

The outcomes of LST analyses (given in Figs. 11.3 and 11.4) for the last two decades (1998–2018) revealed an overall increasing trend over the entire period however the increase was exponential over the last decade. The presence of vegetation, whether it was in the form of agriculture, shrublands, or the thick forest of Margalla, was responsible for the cooling of Earth's surface because of evapotranspiration. This was reflected in the strong negative correlation it showed with the land surface temperature in the city (1.000** at 0.01 interval).

As evident from the graph given below (Fig. 11.4), the minimum value of temperature for Islamabad was 6.8 °C in 1998 which increased up to 10.2 °C in 2008 and 17.1 °C in 2018. The maximum temperature recorded by the satellite however was 23.9 °C in 1998 which rose slightly in the following decade (24.8 °C) but showed an exponential increase in the last decade where the highest recorded value was 31.4 °C in October 2018.

The previous study conducted for the area by Waseem and Kayyam (2019) highlighted the loss of vegetation as the sole responsible factor for the increase in Land Surface Temperature where the study recorded the loss of vegetation (including all

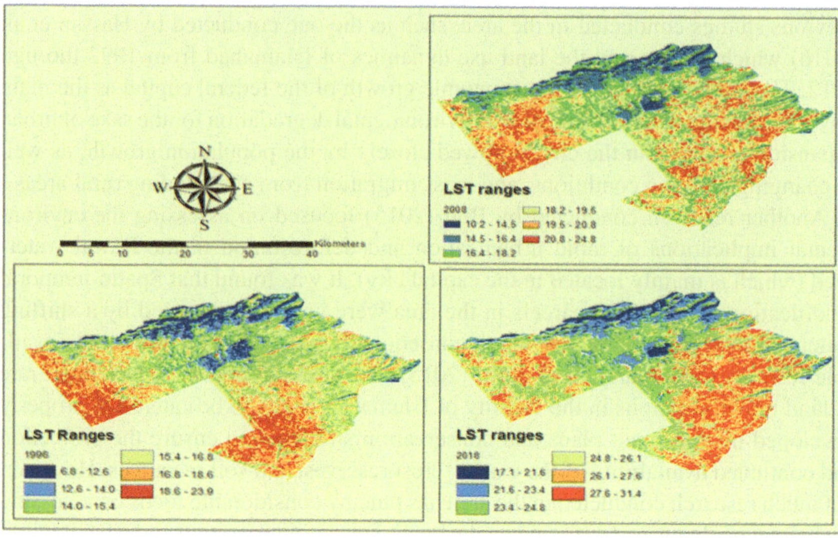

Fig. 11.4 Land surface temperature maps of Islamabad showing spatiotemporal variations over two decades

forms of vegetation) for the past 25 years (1992–2017) and correlated its impact on the LST. However, the study did not take into account the role the Built environment played in determining the recorded vicissitudes in LST over the decades. The subsequent section of the present research, however, highlights how the vicinity of urban or rural centers contributed to the rise or fall of LST.

Built Environment's Contribution to the Observed LST Vicissitudes

The effect of land use changes on land surface temperature varied with the land use type. However, these effects were more prominent in the urbanized areas. The presence of vegetation was responsible for the cooling of the Earth's surface because of evapotranspiration. This was reflected in the strong negative correlation it showed with the land surface temperature (1.000** at 0.01 interval). Contrarily, Fig. 11.5 deliberated that fallow land/soil, as well as impervious surfaces, contributed significantly to the increase in land surface temperature with a very strong positive correlation between built-up areas and land surface temperature (Fig. 11.5 and Table 11.3). Similar results were attained by Aslam et al. (2021) for Islamabad. The study considered recent years i.e., 2013 and 2019, and explored the Spatio-temporal dynamics of LST. The study also assessed the Local Climatic Zone (LCZ) and correlated the outcomes with NDBI and NDVI. Bokaie et al. (2016) also obtained comparable outcomes for Tehran. It could thereby be debated that not only the LST be decreased by increasing the number of green spaces in or nearby urban centers as recommended by Naeem et al. (2018) but also by decreasing the number of impervious surfaces and increasing waterbodies in the vicinity of urban centers (discussed in detail below).

Both the figure and the table reveal that the overall expansion of built-up areas in the city and subsequent loss of vegetation land until 2018 was responsible for raising the land surface temperature quite significantly (more so in the last decade) and if the same trend is continued, it is predicted that the mean surface temperature could rise to 30 °C by 2028 (this prediction is made for October). The primary reason for this rise in the temperature is the increase in surface albedo which reportedly rises when vegetation cover is replaced by or is converted into another land cover especially impervious surfaces with great reflecting potential (Kirchner 1984).

Nevertheless, reviewing Fig. 11.5 (which compared LST with NDBI outcomes) it was revealed that although the overall temperature of the city centers or urban hubs was higher than that of other areas, it was lower compared to the surrounding rural zones of the cities during the daytime. This phenomenon is known as Urban Cool

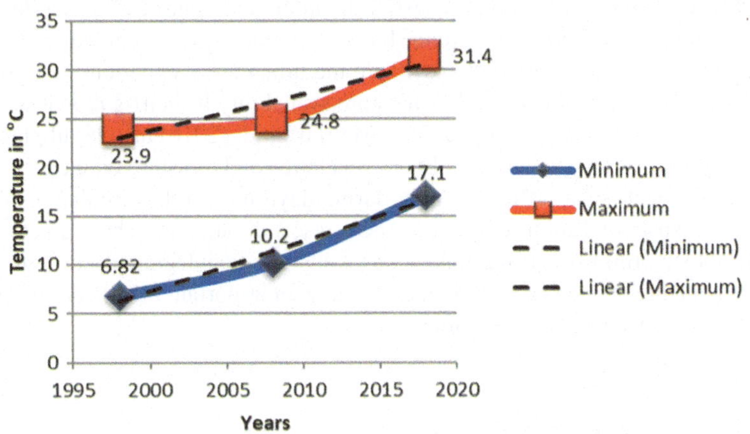

Fig. 11.5 Land surface temperature graph showing the minimum and maximum values of satellite calculated temperature and the general trend of increase observed for Islamabad

Table 11.3 Tabular representation of the correlation between Normalized Difference Built-up Index (NDBI) and LST

NDBI and LST correlations			
LST maximum	NDBI maximum		
LST Maximum Sig. (2-tailed) N	Correlation coefficient	1	0.999[a]
	0.029		
	3	3	
NDBI Maximum Sig. (2-tailed) N	Correlation coefficient	0.999[a]	1
	0.029		
	3	3	

[a]Correlation is significant at the 0.05 level (2-tailed)

Island (UCI) effect and is observed in many other cities around the world, typically in semi-arid and dry sub-humid climatic regions such as Pune, India (Gohain et al. 2020), Hong Kong, China (Yang et al. 2017); Erbil, Iraq (Rasul et al. 2015), etc. The causes for this include lack of moisture and vegetation in the areas surrounding urban centers or in rudimentary developments. The lack of moisture availability to evaporate in the areas surrounding urban centers, whether they be rudimentary developments or agricultural lands, made them behave as hot barren fields (during the daytime) and evidently, more water evaporates from the urban centers (during the day) thus making two phenomena to play a role in defining the city's climate at a time i.e., Urban Cool Island effect within an Urban Heat Island (Fig. 11.7).

This assumption could be verified from Fig. 11.6 which shows that the areas with low moisture content were the areas with the highest temperatures. It was also evident that over the years, generally these areas where the land surface temperature was recorded by the satellite to be the highest, were away from the urban centers in the daytime and near rudimentary developments instead. Outcomes of the research conducted by Kumar et al. (2017) and Martins et al. (2016) corroborated these findings, where Kumar et al. (2017) assessed the dominant control of agriculture and irrigation on the UHI effect in India. The study found a great contrast in the rural and urban land surface temperatures where the urban areas experienced a positive UHI effect. The study attributed it to moisture availability from irrigation near urban centers due to better irrigation practices. Martins et al. (2016) also highlighted the major influence of the availability of both vegetation and moisture on the mitigation of the urban heat island effect, notably during daytimes, in the simulations drawn for a new district of Toulouse, France. These research outcomes show great opportunities for supporting decision-makers on taking specific integrated actions towards truly sustainable neighborhoods besides creating an important urban cool island for the urban pedestrians during daytimes.

Limitations of the Study

The present study was focused largely on assessing the impacts of the built environment on the land surface temperature. It highlighted the significance of urban green spaces and parks for land surface temperature mitigation. However, the present research only considered anthropogenic and biogeophysical variables and not considered economy-related (landscape valuation) and policy-related variables (i.e. land use zoning, infrastructure policy) due to quantification and data availability issues. Therefore, it is recommended that future studies may include these factors to get more realistic outcomes as these factors can significantly alter the rate of urban expansion.

Besides, the present research considered only the effect of future landscape changes on the land surface temperature. It is recommended that future studies should consider the impacts of such changes on other microclimatic variables and ecological amenities.

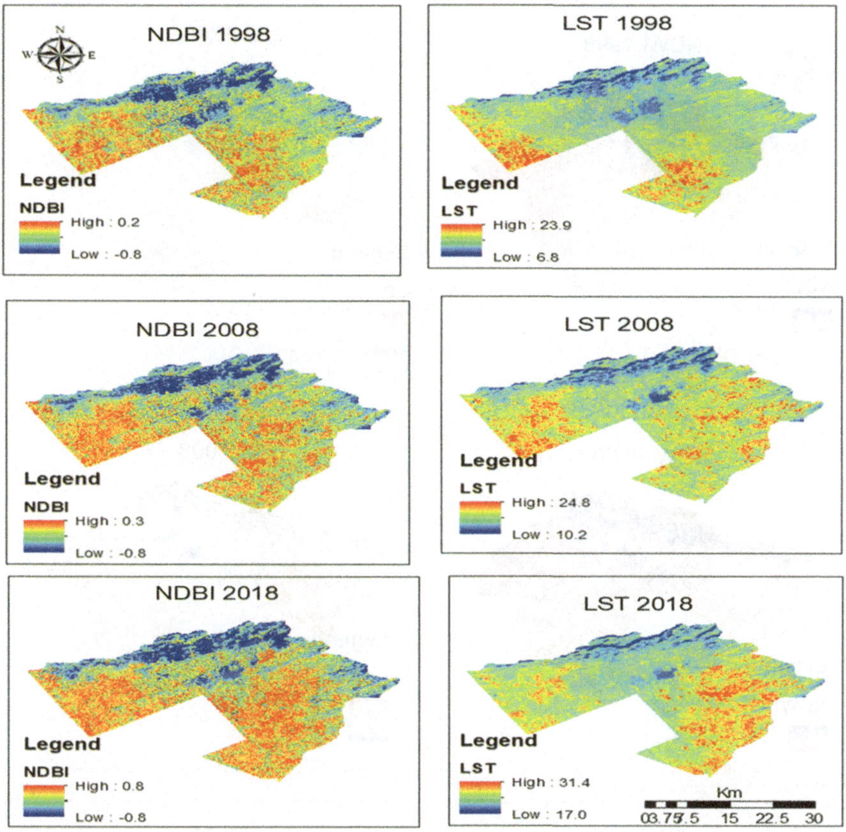

Fig. 11.6 Correlation between spatiotemporal variation in NDBI and LST outcomes in Islamabad

Conclusion

Given the study outcomes, it could be concluded that the haphazard expansion of urban areas and subsequent loss of vegetation not only poses a serious cause for concern in the present but also predicts a future where these negative trends lead to environmental unsustainability. If the LST continues to increase at this alarming rate, then very shortly we might see a reclassification of the climatic region of the federal capital. This is a serious cause of concern as it is a clear indicator of global warming. Furthermore, these trends of haphazard urban expansion and water and vegetation degradation are unlikely to change unless the government intervenes strongly. Without intervention, overuse of major land resources (urban vegetation) and subsequent land degradation may lead to an uncertain future. It is the need of the hour that while planning the future urban layout of Islamabad city, the construction of broader green belts, designated green spaces and artificial reservoirs within the urban center is ensured. In addition, it would also be quite beneficial if the natural reserves are properly conserved and all incursions in them be strictly regulated.

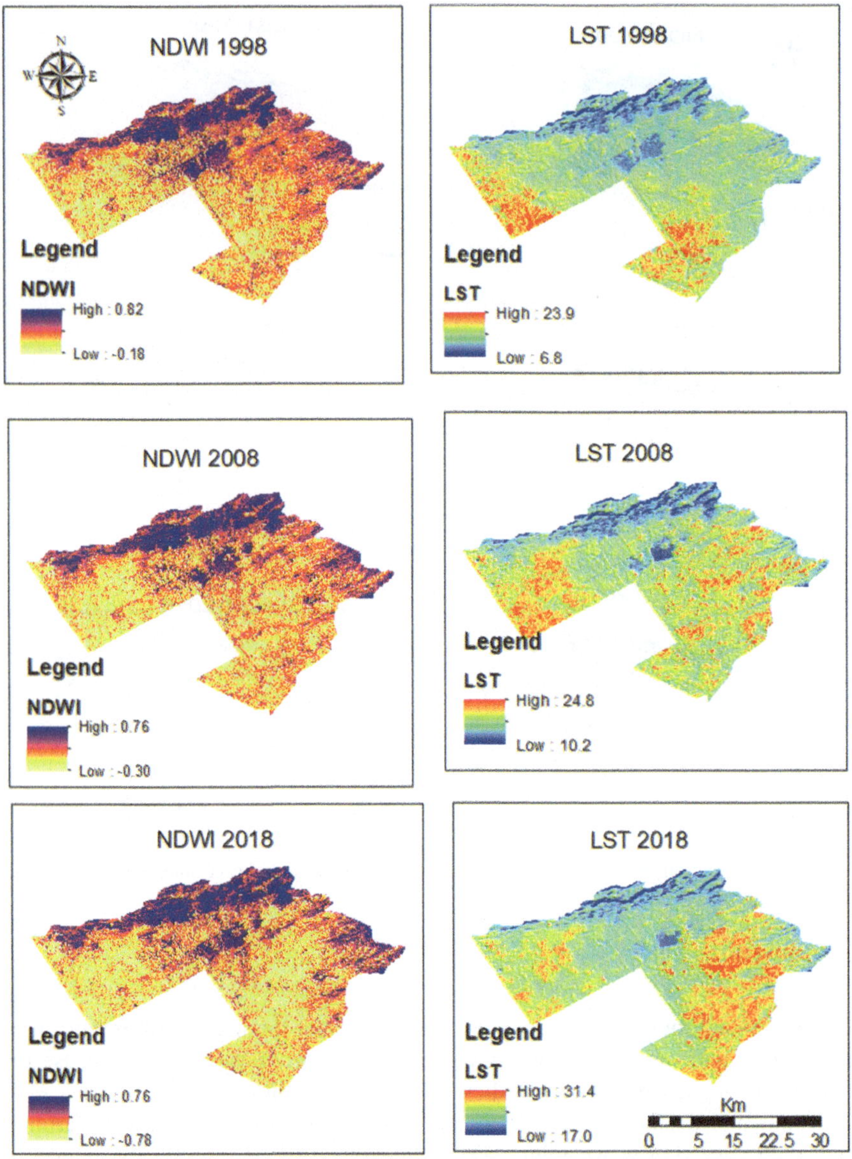

Fig. 11.7 A comparison of Spatio-temporal variations in NDWI and LST for Islamabad city

Recommendations for Future Adaptations

To address the issues highlighted above, the study recommends that future city development plans should establish an appropriate ratio for different land use types such that they dedicate at least 20% of the total city area for tree canopies and open green spaces including green yards/roadside vegetation/shrubs that should be considered as a separate planning requirement. Reservation of potential green surfaces to that percentage of land will allow satisfactory use of local vegetation to encourage evapotranspiration and ensure shading as effective cooling strategies for not only public but also private spaces.

Correspondingly, vegetation in all its forms is crucial to the sustainability of the city environment, not to mention the lack of it reduces the cooling effects provided by evapotranspiration and shading, and affects carbon dioxide removal. Steps should be taken to protect vegetation biomass, diversity, and distribution. The cutting of forests should be strictly regulated.

Building roofs should be made cool by placing vegetation on them to make green roofs. Additionally, construction activities should be based on green technology to minimize damage to soil and consequently the vegetation depending on it.

Efforts should be made to reduce current impervious surfaces made of asphalt and concrete since they are likely to introduce new overheating zones. At present, using cool paving materials is not envisaged primarily for the reason of high investment cost. Another recommendation in this regard is the adaptation of built form strategies, including altering built height, street width, and building density. However, since the existing roads and streets could not be modified or widened (without costing a lot of money), the proposed strategies could be adopted successfully for future development projects.

Additionally, improvement in the awareness campaigns regarding the importance of maintaining urban microclimate both among the decision-makers and public by the researchers holds great future potential and thus to be encouraged.

Introduction of new water conservation strategies is also recommended to maximize the urban cooling effect in city centers with water evaporation from ponds, fountains, and lake surfaces.

References

Aslam A, Rana IA, Bhatti SS (2021) The spatiotemporal dynamics of urbanisation and local climate: a case study of Islamabad Pakistan. Environ Imp Assess Rev 91:106666

Ayanlade A (2017) Remote sensing approaches for land use and land surface temperature assessment: a review of methods. Int J Imag Dat Fus: 1–23

Belokrenitsky VY (2017) Present stage of urbanization in Pakistan: some features and problems. Mus Persp 11(3):2–18

Bendib A, Dridi H, Kalla MI (2017) Contribution of Landsat 8 data for the estimation of land surface temperature in Batna city Eastern Algeria. Geocarto Int 32(5):503–513

Bokaie M, Zarkesh MK, Arasteh PD, Hosseini A (2016) Assessment of urban heat island based on the relationship between land surface temperature and land use/land cover in Tehran. Sustain Cities Soc 23:94–104

Butt A (2015) Hazards to environmental health of Rawal watershed due to rapid urbanization and deforestation. Mid Eas J Bus 10 (1)

Butt A, Shabbir R, Ahmad SS, Aziz N, Nawaz M, Shah MTA (2015) Land cover classification and change detection analysis of Rawal watershed using remote sensing data. J Biodivers Environ Sci 6(1):236–248

Cartalis C, Santamouris M, Nyktarakis G, Polydoros A, Th M (2016) Assessing the interlinks between urbanization, the built environment and the thermal environment in view of smart and sustainable urban development: a demonstration application for Athens. Int J Ear Environ Sci. https://doi.org/10.15344/2456-351X/2016/107

Feizizadeh B, Blaschke T, Nazmfar H, Akbari E, Kohbanani HR (2013) Monitoring land surface temperature relationship to land use/land cover from satellite imagery in Maraqeh County Iran. J Environ Plan and Manag 56(9):1290–1315

Gohain KJ, Mohammad P, Goswami A (2020) Assessing the impact of land use land cover changes on land surface temperature over Pune city, India. Quat Int. https://doi.org/10.1016/j.quaint.2020.04.052

Guha S, Govil H, Gill N, Dey A (2020) A long-term seasonal analysis on the relationship between LST and NDBI using Landsat data. Quat Int 575–576(2):249–258. https://doi.org/10.1016/j.quaint.2020.06.041

Hassan Z, Shabbir R, Ahmad SS, Malik AH, Aziz N, Butt A, Erum S (2016) Dynamics of land use and land cover change (LULCC) using geospatial techniques: a case study of Islamabad Pakistan. Springerplus 5(1):1–11

Khan MS, Ullah S, Sun T, Rehman AU, Chen L (2020) Land-use/land-cover changes and its contribution to urban heat Island: a case study of Islamabad Pakistan. Sustain 12(9):3861

Kirchner M (1984) Influence of different land use on some parameters of the energy and water balance. Prog in Biomet 3:65–74

Kumar R, Mishra V, Buzan J, Kumar R, Shindell D, Huber M (2017) Dominant control of agriculture and irrigation on urban heat island in India. Sci Rep 7(1):1–10

Li K, Lin B, Jiang D (2012) A new urban planning approach for heat Island study at the community scale. J Heat Isl Inst Int 7:50–54

Logan TM, Zaitchik B, Guikema S, Nisbet A (2020) Night and day: the influence and relative importance of urban characteristics on remotely sensed land surface temperature. Rem Sen Environ 247:111861

Martins TA, Adolphe L, Bonhomme M, Bonneaud F, Faraut S, Ginestet S, Guyard W (2016) Impact of urban cool Island measures on outdoor climate and pedestrian comfort: simulations for a new district of Toulouse, France. Sustain Cit Soc 26:9–26

Millward AA, Torchia M, Laursen AE, Rothman LD (2014) Vegetation placement for summer built surface temperature moderation in an urban microclimate. Environ Manag 53(6):1043–1057

Mukherjee S, Joshi PK, Garg RD (2016) Analysis of urban built-up areas and surface urban heat island using downscaled MODIS derived land surface temperature data. Geocarto Int. https://doi.org/10.1080/10106049.2016.1222634

Naeem S, Cao C, Qazi WA, Zamani M, Wei C, Acharya BK, Rehman AU (2018) Studying the association between green space characteristics and land surface temperature for sustainable urban environments: an analysis of Beijing and Islamabad. ISPRS Intl J Geo-Inf 7(2):38

Qayyum F (2019, June 11) A car-unfriendly Pakistan? Yes, that's how you make livable cities. Retrieved on October 12, 2020, from DAWN: https://www.dawn.com/news/1448181

Rasul A, Balzter H, Smith C (2015) Spatial variation of the daytime Surface Urban Cool Island during the dry season in Erbil, Iraqi Kurdistan, from Landsat 8. Urb Clim 14:176–186

Sander HA (2016) Assessing impacts on urban greenspace, waterways, and vegetation in urban planning. J Environ Plan Manag 59(3):461–479

Sun D, Yu Y, Yang H, Fang L, Liu Q, Shi J (2015) A case study for intercomparison of land surface temperature retrieved from GOES and MODIS. Int J Dig Ear 8(6):476–494. https://doi.org/10.1080/17538947.2014.906509

Waseem S, Khayyam U (2019) Loss of vegetative cover and increased land surface temperature: a case study of Islamabad, Pakistan. J Clean pro 234:972–983

Yang X, Li Y, Luo Z, Chan PW (2017) The urban cool island phenomenon in a high-rise high-density city and its mechanisms. Int J Climat 37(2):890–904

Chapter 12
Mapping Multi-decadal Mangrove Forest Change in the Philippines: Vegetation Extent and Impacts of Anthropogenic and Climate-Related Factors

Alvin B. Baloloy, Kayziel P. Martinez, Ariel C. Blanco, Margaux Elijah P. Neri, Kristina Di V. Ticman, Diana Faith Burgos, Jeark A. Principe, Rosalie B. Reyes, Severino G. Salmo III, and Kazuo Nadaoka

Abstract There is a continuous decline of mangrove forests in the Philippines due to anthropogenic activities and natural disturbances. Through the years, monitoring of mangrove extent was done as part of the many local efforts to manage mangrove forests. However, existing mangrove cover estimates were generated by varied sources with different methodologies and classification techniques. This study aims to utilize a standardized method to detect multi-decadal spatio-temporal mangrove extent with the use of Landsat-derived Mangrove Vegetation Index (MVI) calculated in Google Earth Engine (GEE). Mangroves were mapped per region and areal changes were calculated from the period 2000 to 2020. A decline of 29,000 hectares was recorded from 2000 to 2020, observed in 12 out of the 17 Philippine administrative regions. The variations in rainfall, typhoons, sea surface height, temperature, and occurrence of land use/cover conversion were seen to have impacted the structure and extent of mangrove communities. Results show that higher multi-decadal mangrove losses were observed in regions with higher precipitation change ($r = 0.68$ to 0.99), higher maximum sea surface temperature ($r = 0.39$), more frequent typhoons ($r = 0.43$), and those exposed to extreme heating ($r = 0.47$) and precipitation events ($r = 0.51$). Areas that were least devastated by typhoons (fewer than 10 typhoons) have recorded increased vegetation cover such as in Northern Mindanao (Region 10), Autonomous Region in Muslim Mindanao (ARMM), and Zamboanga

A. B. Baloloy (✉) · K. P. Martinez · A. C. Blanco · M. E. P. Neri · K. Di V. Ticman · D. F. Burgos · J. A. Principe · R. B. Reyes
Department of Geodetic Engineering (DGE), University of the Philippines, 1101 Diliman, Quezon City, Philippines
e-mail: abbaloloy@up.edu.ph

K. P. Martinez
e-mail: kpmartinez1@up.edu.ph

A. C. Blanco
e-mail: acblanco@up.edu.ph

M. E. P. Neri
e-mail: mpneri1@up.edu.ph

W. Leal Filho et al. (eds.), *Climate Change Strategies: Handling the Challenges of Adapting to a Changing Climate*, Climate Change Management,
https://doi.org/10.1007/978-3-031-28728-2_12

Peninsula (Region 9). Reduction in precipitation can increase salinity which lowers seedling survival, growth rates, and productivity; while the impact of typhoons is mainly attributed to physical damages brought by strong waves and wind. Some of the previous mangrove areas were found to be converted to fishponds and built-up areas brought by the increasing demands for the economical use and development of land. Understanding the impacts of these natural and human-induced drivers will aid in formulating effective conservation and resource management measures. Further, the proposed standardized mapping workflow allowed detailed analysis of these impacts as reflected by the spatio-temporal changes in mangrove extent.

Introduction

There is a continuous decline of mangrove forests in the Philippines due to climate-related factors and anthropogenic activities. The human-induced factors include mainly conversion to aquaculture, urban development, large scale deforestation, and lack of coastal protection measures. The decline of Philippine mangroves from 1950 to 1970s was largely attributed to the conversion to fishponds, and these were well documented (Primavera 1995, 2000). It was only on the later part of the 1970s that the national government fully understood the economic and environmental importance of mangroves. Since then, various programs and policy initiatives were undertaken to recover the damages brought by land use conversion.

D. F. Burgos
e-mail: dmburgos@up.edu.ph

J. A. Principe
e-mail: japrincipe@up.edu.ph

R. B. Reyes
e-mail: rbreyes3@up.edu.ph

A. C. Blanco · J. A. Principe · R. B. Reyes
DGE and Training Center for Applied Geodesy and Photogrammetry, University of the Philippines Diliman, 1101 Quezon City, Philippines

A. C. Blanco
Space Information Infrastructure Bureau, Philippine Space Agency, 1101 Quezon City, Philippines

S. G. Salmo III
Institute of Biology, University of the Philippines, Diliman, 1101 Quezon City, Philippines
e-mail: sgsalmo@up.edu.ph

K. Nadaoka
School of Environment and Society, Tokyo Institute of Technology, Meguro-Ku, Tokyo 152-8550, Japan
e-mail: nadaoka.k.aa@m.titech.ac.jp

Meanwhile, climate-induced factors such as climate variability were reported to greatly impact the species composition, adaptation to salinity, survival rate, and productivity of mangroves around the world (Ghosh et al. 2017). Among the factors that can influence mangrove forests, multiple studies highlighted the effects of climatic variables such as changes in temperature, rainfall variation, sea-level rise, frequency of typhoons, high water events, and concentration of atmospheric gases (Alongi 2015; Duke et al. 1998; Ghosh et al. 2017; Willard and Bernhardt 2011). Changes in the level of these variables are associated with climate change as continued global warming intensifies global water cycle, monsoon precipitation, and the severity of extreme events (IPCC 2021). The effect of typhoons to mangroves is an area of interest because the country is a known gateway for tropical cyclones coming from the Pacific Ocean. Annually, there are around 20 tropical cyclones that enter the Philippine Area of Responsibility (PAR) including typhoons like Haiyan, one of the strongest typhoons that devastated the country in 2013 which has caused damage to the mangrove forests within and near its track (Buitre et al. 2019).

The impacts of climate-related and human-induced factors to mangrove forest can be known by analyzing the trend or changes in the area or extent. Mangrove extent statistics can be compared through the use of historical and recent mangrove extent data. In the country, varying trends of mangrove area were observed throughout the decades. Larger mangrove areas were reported in earlier years between 1918 (450,000 Ha) to 1968 (448,310) with field estimates reported by then Philippine Council for Agriculture, Forestry and Natural Resources Research and Development and L.M. Lawas, respectively. Relatively smaller yearly estimates were reported by the Bureau of Forest Development (now the Forest Management Bureau) from 1969 to 1984 (295,190–233,514 Ha). From year 1990 onwards, a significant shift from field-based mapping to Remote Sensing-based (RS) approaches was observed in the Philippines, providing more rapid and less expensive approach. These includes the estimates of Long and Giri (2011), Long et al. (2014), NAMRIA (every 5 years), and Global Mangrove Watch (Bunting et al. 2018). These mangrove cover estimates have different levels of accuracy based on factors such as the spatial and temporal resolutions of the satellite images, image classification technique, and the quality and quantity of ground validation data.

Despite the existing methodologies in mangrove extent mapping, a standardized mangrove mapping workflow is still needed to accurately detect the impacts of climate and human-induced drivers to mangrove forest's extent. For example, the estimates accounted for Philippine mangroves were obtained from different institutions and resources which utilized different methods, satellite data, and validation protocols. Comparing statistics among these different sources may not actually capture the actual trend of local areal changes. To solve this, new methodologies were being developed including novel methodologies for rapid extent estimation using mangrove indices. The Mangrove Vegetation Index (MVI) was proposed by Baloloy et al. (2020) to rapidly classify mangrove cover and separate the same from non-mangrove classes such as bare soil, built-up, terrestrial forests, grassland, clouds, and water. MVI measures the probability of a pixel to be a 'mangrove' by extracting the greenness and moisture information from the green, NIR, and SWIR satellite image bands. The range of MVI values may vary depending on factors such as land cover classes, climatic conditions, or tidal conditions. MVI was previously utilized to map the 2019 mangrove extent of the Philippines using Sentinel-2 data (Baloloy et al. 2020). The application of MVI for historical mapping using Landsat imagery offers opportunity for an accurate, long-tern analysis of mangrove trends and in detecting the potential contribution or effect of climate and human-induced factors to the decadal changes in extent.

This study aims to utilize a satellite remote sensing method to detect multi-decadal spatio-temporal mangrove extent (2000–2020) in the Philippines with the use of Landsat-derived MVI. Decadal values of factors that may affect mangrove extent will be obtained, including the following variables: rainfall, sea surface temperature (SST), sea surface height (SSH), number of typhoons, air-sea climate exposures, and LULC changes. These environmental and climate-related factors will be correlated with the decadal mangrove statistics to determine their respective impacts. Understanding the effect of these drivers will aid in formulating effective conservation and resource management measures.

The mangrove extent estimates calculated in this study were derived from Landsat data with no ground validation conducted. The mapping method adapted, however, was previously applied in Sentinel-2-based mangrove mapping in the Philippines with high accuracy results. The study period is limited between years 2000 and 2020 only, thus discussions on pre-2000 mangrove estimates and land cover conversion activities were based on published references. Moreover, only major climate-related factors were considered in this study, selected based on their reported impacts to vegetation health and the availability of local data with long temporal records. The variables presented here are also the major climatic variables highlighted in the IPCC 2021 report with a global concern: sea surface temperature, precipitation, sea level rise, and the occurrence of tropical cyclones and other extreme weather events.

Materials and Methods

The information needed to assess the impacts of human-induced and climate-related factors to mangrove extent are divided mainly of two categories. Firstly, historical mangrove extents were generated (2000–2020) as the main indicators of the impacts from the environmental variables. This mangrove extents were obtained using Remote Sensing data and techniques, with Landsat as the main satellite data source. Novel methodologies (e.g., MVI-based mapping) and platforms such as Google Earth Engine were utilized to hasten the generation of a mangrove extent data. The second main data needed is the collated climate-related and human induced factors. Among the many variables that may affect the extent and health of mangrove forests, significant variables were selected based on previous related studies (Alongi 2015; Buitre et al. 2019; Duke et al. 1998; Friess et al. 2012; Ghosh et al. 2017; Krauss et al. 2014; Willard and Bernhardt 2011; Ximenes et al. 2016).

Mangrove Extent Mapping

Multi-decadal mangrove maps of the Philippines (2000, 2010, and 2020) were generated using the MVI formula implemented in GEE (Fig. 12.1). Atmospherically corrected Sentinel-2 images were downloaded and used to generate the MVI raster layers based on the identified optimal minimum threshold and a fixed maximum threshold of 20. MVI is a new simplified index for fast and accurate mapping of mangrove extent from remotely-sensed images. The Mangrove Vegetation Index (MVI) equation is in the form:

$$\text{MVI} = (\text{NIR} - \text{Green})/(\text{SWIR1} - \text{Green}) \tag{12.1}$$

where NIR, Green, and SWIR1 are the near-infrared, green, and shortwave infrared-1 reflectance values. The | NIR – Green | enhances the differences of vegetation greenness between mangrove pixels and other vegetation, while | SWIR1 – Green | captures the distinct moisture of mangrove pixels compared to non-mangrove pixels. In this study, the equivalent bands in Landsat 8, 7, and 5 were utilized. The example formula for Landsat-8 MVI (Eq. 12.2) is written as:

$$\text{Landsat-8 MVI} = (\text{B8} - \text{B3})/(\text{B6}-\text{B3}) \tag{12.2}$$

where Bands 3, 5, and 6 are the equivalent green, NIR and SWIR1 bands in Landsat-8, respectively.

Generation of MVI layer, threshold selection, and quality-checking were all implemented in GEE. GEE is a powerful web-platform for cloud-based processing of remote sensing data on large scales, providing a variety of constantly updated data sets, thus no download of raw imagery is required. The study specifically utilized Code Editor, a web-based integrated development environment (IDE) which can be accessed at: https://code.earthengine.google.com. After generating the MVI layers, outputs were exported as a raster file for data cleaning and area calculation in ArcGIS™. Cleaning of noise pixels was done by overlaying the MVI raster on the false color composite display of Sentinel-2 image (RGB: B11-B8-B4).

The annual mangrove map statistics was calculated per Philippine administrative region as shown in Fig. 12.2. The country has a total of 17 regions, but only 16 were used in this study excluding the Cordillera Administrative Region (CAR)

Fig. 12.1 Summary of the workflow for generating the decadal mangrove maps and statistics. More detailed information on MVI-based mapping was described in Baloloy et al. (2020). Here, the MVI was first applied to Sentinel-2 imageries to map the 2019 mangrove extent in the Philippines

since the region is in higher elevation areas with no mangrove forest in its administrative boundary. Re-calculation of mangrove statistics per region was implemented in ArcGIS™ with the zonal statistics tool. The region shapefile was obtained from the National Mapping and Resource Information Authority (NAMRIA). In addition, decadal statistics on the nationwide constant mangrove cover, cover gain, and cover lost were mapped and calculated.

Climate-Related Variables

There were five selected climate-related or climate-induced variables considered in this study namely annual rainfall, sea surface temperature (SST), sea surface height (SSH), typhoon data, and air-sea climate exposure.

The annual rainfall data was obtained from Climate Hazards Group InfraRed Precipitation with Station data (CHIRPS), a 35+ year quasi-global rainfall data set. CHIRPS is produced at 0.05×0.05 degree spatial resolution. The CHIRPS dataset was developed on previous approaches to smart interpolation techniques and high-resolution precipitation estimates based on infrared Cold Cloud Duration datasets (Funk et al. 2015). Satellite information were incorporated in the CHIRPS data in three ways: by producing high resolution precipitation climatologies from satellite; by using CCD fields to estimate monthly and pentadal precipitation anomalies; and by using satellite precipitation fields to estimate local distance decay functions (Funk et al. 2015). The CHIRPS data was utilized by significant number of studies on climate and rainfall trend modelling (Beck et al. 2017; Mu et al. 2021) while some studies were conducted to validate this dataset (Shrestha et al. 2017; Rivera et al. 2018). The rainfall data for years 2000–2020 was downloaded and accessed from CHIRPS website: https://www.chc.ucsb.edu/data/chirps.

The sea surface height dataset used are actual measurement from tide gauges in the country, acquired from NAMRIA and processed in text file format. One tide gauge station per region was selected. Correction was applied to reference the height data with the WGS84 ellipsoid. Meanwhile, Group for High Resolution Sea Surface Temperature (GHRSST) Level 4 sea surface temperature dataset was used in determining the effect of SST to the mangrove extent decadal trend. GHRSST is an analysis based upon nighttime GHRSST L2P skin and subskin SST observations from several instruments (e.g., AMSR-E, MODIS, AVHRR). This dataset was widely used as a validation data to other satellite-derived SSTs such as those derived from Landsat (Donlon et al. 2007; Jang and Park 2019). High-resolution SST analysis based on GHRSST SSTs have been adopted internationally by operational agencies. For this study, the dataset was obtained from the Physical Oceanography Distributed Archive Center (PODAAC) of NASA's Jet Propulsion Laboratory: https://podaac.jpl.nasa.gov/. The downloaded grids are those within or nearest the tide gauges stations in the respective Philippine administrative regions. The available dataset is from year 2003 to 2020.

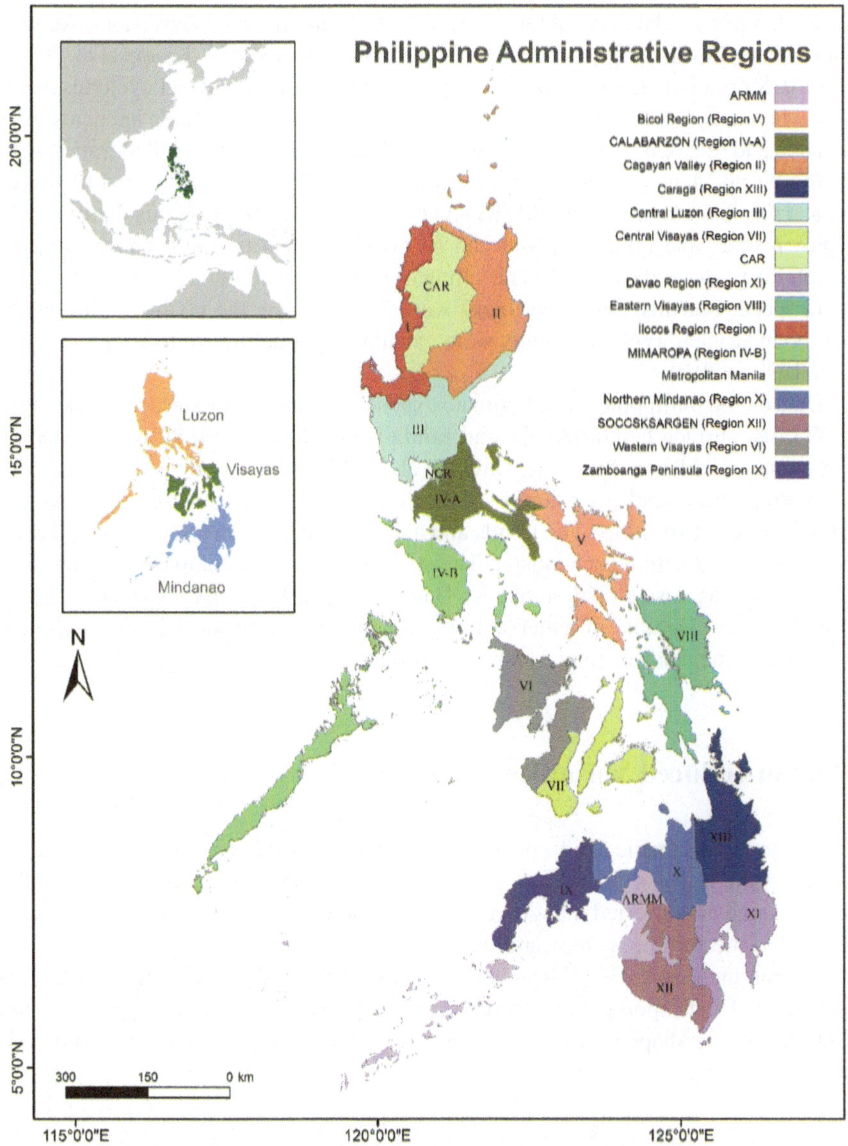

Fig. 12.2 The Philippine administrative regions used as the smallest unit for mangrove extent comparison between regions and between decades. Administrative regions are composed of provinces and/or independent cities. Philippines has three main island groups: Luzon, Visayas, and Mindanao (lower left inset), consists of varying number of administrative regions. The climate-related factor variables were also collated by region to facilitate comparison with the mangrove statistics. Analysis and presentation of results were discussed using this unit

Typhoon track data was obtained from NOAA International Best Track Archive for Climate Stewardship (IBTrACS), filtered from 1999 to 2019 (Knapp et al. 2018). IBTrACS project is the most complete global collection of tropical cyclones available wherein recent and historical tropical cyclone data from multiple agencies were collated and released publicly. IBTrACS was developed through a collaboration with all the World Meteorological Organization (WMO) Regional Specialized Meteorological Centers around the world (Knapp et al. 2010, 2018). From this dataset, only typhoon track and typhoon count can be obtained; data on the typhoon strength or intensity are not included. The climatologies and counts are based solely on wind speed. For this study, the tracks were filtered using the national administrative boundary and typhoon counts were calculated per administrative regions from year 1999 to 2019.

Data on the Philippine air-sea climate exposures, namely, sea surface temperature (SST), sea surface height (SSH), and rainfall (David et al. 2015), were added as important variables in this study. David et al. (2015) utilized remote sensing data to categorize the archipelagic waters of the Philippines into distinct clusters of historical air-sea climate exposures. The trends and anomalies of SST, precipitation, and SSH were calculated within each cluster. The statistics were then compared amongst the clusters and against global statistics (David et al. 2015). This secondary dataset is a significant variable in understanding the impact of this natural air-sea climate exposure to the changes in our mangrove forests.

Human-Induced Variables

The change in specific land cover type per region is the main indicator used in detecting the impacts of anthropogenic activities to mangrove extent. Specifically, the total area of converted mangrove cover to fishponds and built-up structures were utilized as the anthropogenic variable metrics.

For this analysis, the NAMRIA land cover data for years 2006 and 2015 (Fig. 12.3) were used. The fishponds and built-up shapefile for year 2015 were clipped with the 2006 mangrove shapefile to identify the areas that were previously detected with mangrove cover.

Variable Metrics and Regression

The five selected variables (rainfall, typhoon count, SST, SSH, and air-sea climate exposure) can signify different impacts to mangrove extent based on the temporal and spatial aggregation of their statistics. To simplify and focus the analysis, only significant variable metrics were utilized to later regress with the decadal mangrove extent data. These metrics are shown in Table 12.1.

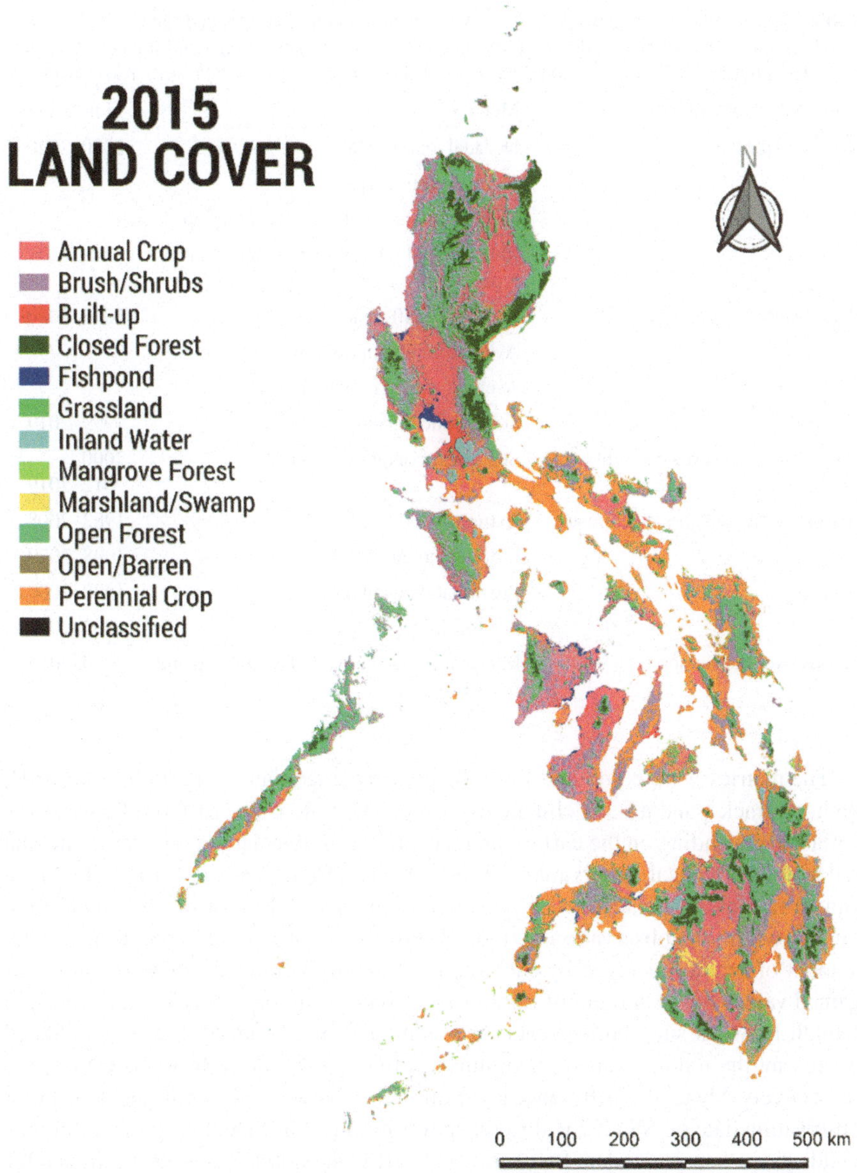

Fig. 12.3 The 2015 Land Cover Map of the Philippines from NAMRIA, the central mapping agency of the Philippine government. The 12 land cover classes were derived from Landsat-8 images processed in e-Cognition software for image segmentation techniques

Table 12.1 Specific metrics used for the environmental and human-induced factors. Selection of variables and metrics were guided by published reports on climate change and the corresponding impacts on mangrove forests. Availability of local and remotely-sensed data were also considered

Environmental factors	Metrics	Time frame
Rainfall (mm)	Decadal total rainfall	2000–2020
	Mean annual rainfall	
	Mean annual rainfall increment/region	
	Mean annual rainfall increment/climate type	
Sea surface temperature (°C)	Mean SST difference	2003–2020
	Average annual minimum SST	
	Average annual max SST	
Typhoon	Total typhoon count	1999–2019
Tide gauge sea surface height (m)	Sea surface height difference	2000, 2012–2019
Historical air-sea climate exposure	Extreme heating events	1982–2008
	Extreme precipitation	1998–2009
	Precipitation changes	1998–2009
	Sea level rise	1992–2008
Land cover conversion	Mangroves area converted to built-up and fishponds (Ha)	2000–2015

The metrics were computed from the pre-processed data on rainfall, SST, SSH, typhoon tracks, and air-sea climate exposure. The time frame differed between the variables depending on the earliest and the latest available dataset. However, all data fall within the decal observation period which is from year 2000 to 2020. Mean annual rainfall was calculated by computing the decadal mean of the annual total rainfall within the observation period. Meanwhile, total rainfall is the decadal total amount of rainfall received by each region. Rainfall increment refers to the lost or gained values in the amount of rainfall every other year, or simply the difference in rainfall between succeeding years. Minimum and maximum SST were calculated by getting the minimum and maximum recorded daily values throughout the year, respectively. Mean SST difference is the difference between the earliest and the latest observation data on SST. SSH difference was obtained by calculating the difference in tide gauge-measured data for years 2019, 2012, and 2000. The metrics utilized for the historical air-sea climate exposures were the parameters regressed with remote sensing data.

Data correlation was done through linear regression as applied to the mangrove extent and environmental variables. Variables such as rainfall and SST include all available annual data between the observation period and not only for the earliest (2000) and latest year (2020). For the air-sea climate exposure, the data used was already processed by the primary source (David et al. 2015) and is ready for correlation between the mangrove extent data. This dataset only covered the years 1992–2009, or half of the study period considered in this study. Thus, one limitation in adopting.

This dataset for correlation with mangrove trend is the assumed linear behavior of the climate-related hazards after year 2009. All correlations were made between datasets aggregated to regional levels, while additional correlation was done between mangrove extent and rainfall increment per climate type. This was carried by identifying first the dominant climate type category of each region and grouping all regions belonging to the same category.

Decadal Mangrove Trend in the Philippines

Maps and Statistics

Mangrove extent maps and statistics were generated for years 2000, 2010, and 2020 (Fig. 12.4). Varying trends were observed wherein the mangrove extent is highest in 2000 (294,026 Ha), followed by 2020 (264,818 Ha), and lowest in 2010 (230,597 Ha).

By comparing maps on a national scale, only the general trend could be observed. There are variations on the level of extent gain or loss through data aggregation by administrative regions, which will be correlated later with the respective human-induced and climate-related variables. Regional statistics showed that in most regions, the highest mangrove areas were recorded in 2000, most declined in 2010, and either increased or decreased in 2020 (Fig. 12.5). Consistently higher

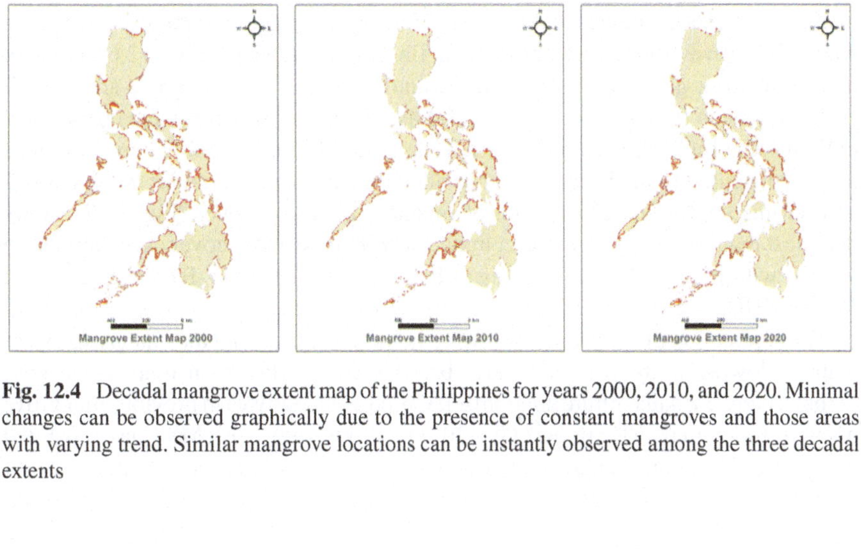

Fig. 12.4 Decadal mangrove extent map of the Philippines for years 2000, 2010, and 2020. Minimal changes can be observed graphically due to the presence of constant mangroves and those areas with varying trend. Similar mangrove locations can be instantly observed among the three decadal extents

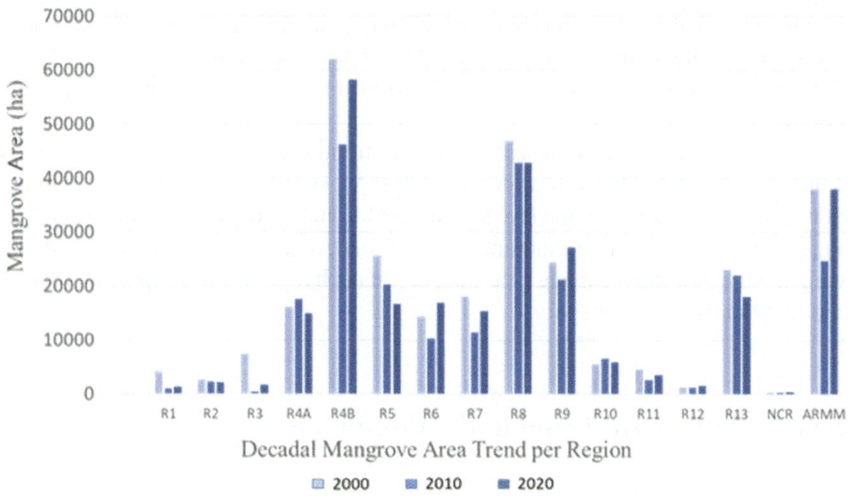

Fig. 12.5 Decadal mangrove extent regional statistics of the Philippines for year 2000, 2010, and 2020

estimates were recorded in Region 4B (MIMAROPA), Region 8 (Eastern Visayas), Region 9 (Zamboanga Peninsula), and the Autonomous Region in Muslim Mindanao (ARMM). These regions are located in the Visayas and Mindanao islands. Few mangroves are located in regions within Luzon such as Region 1 (Ilocos region), Region 2 (Cagayan Valley), and Region 3 (Central Luzon). A decline of 29,000 hectares was recorded from 2000 to 2020, and the declining trend was observed in 12 out of the 17 Philippine administrative regions. From these observations, we can say that the general trend of the mangroves in the country is still decreasing.

In addition to the decadal map, cover maps of the gained, loss, or retained mangroves were generated (Figs. 12.6 and 12.7). These maps show if the changes in mangrove forests took place within the same area or coverage, or if there are new areas where mangroves have grown or extended. Based on these maps (Figs. 12.6 and 12.7), mangrove cover gains were observed within or near existing mangrove areas signifying expansion from its previous extent. Meanwhile, cover losses were pronounced only in some regions, such as Eastern Visayas and Bicol Region, where variations maybe attributed to the human-induced and climate-related variables being considered in this study. The distribution of retained mangrove cover (Fig. 12.7) shows that most mangrove forests in the county can still thrive and recover especially those located within dense forests such as in Palawan (in Region IV-B) and Siargao (in Region XIII).

The application of MVI was considered helpful in generating the mangrove extent maps following a standardized index-based method. Based on quality-checking outputs, there are aquaculture areas misclassified as mangroves in previous mangrove maps in the country (Long and Giri 2011). In this study, the MVI has separated mangroves from non-mangrove areas, although further validation activities are still

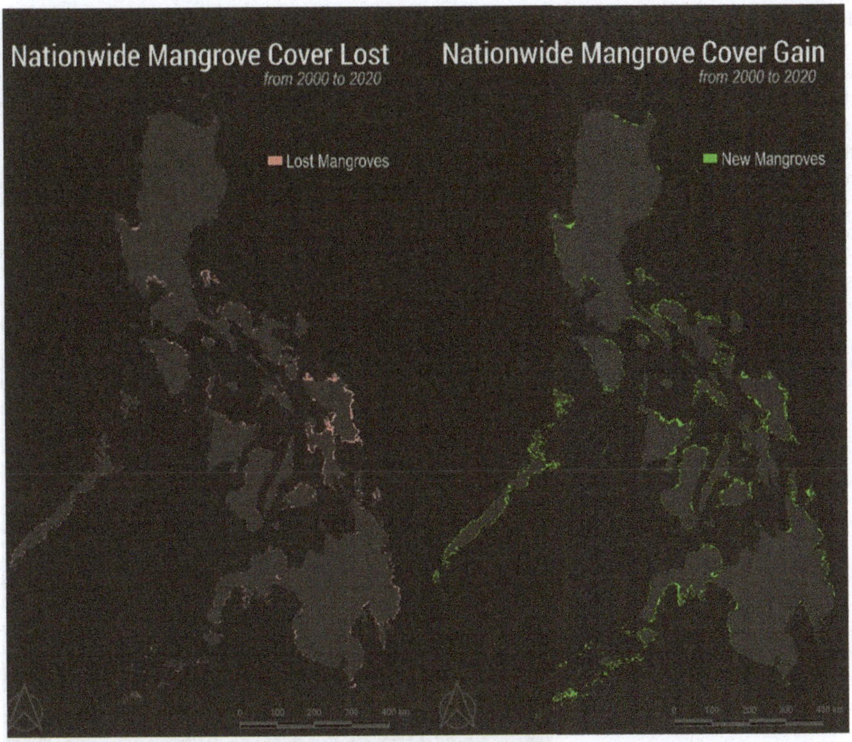

Fig. 12.6 Nationwide mangrove cover lost (left) and cover gained from year 2000 to 2020

needed. In terms of decadal trend of mangrove extent in the Philippines from other references, the result of the current study is comparable with the mangrove trend from Global Mangrove Watch (Bunting et al. 2018) from 1996 to 2016. GMW also reported a general decreasing trend, although the reported mangrove loss is relatively lower (9,000 Ha) than the lost calculated in this study for 2000 to 2020 (29,000 Ha). Meanwhile, in the Long and Giri study, a decreasing trend was also observed between the earliest (1990) and latest (2010) estimates, with a loss of 28,172 Ha (Long and Giri 2011). The differences in the observation periods between among these studies may have affected the variations in the lost mangrove estimates, in addition to the differences in the satellite data and applied methodologies.

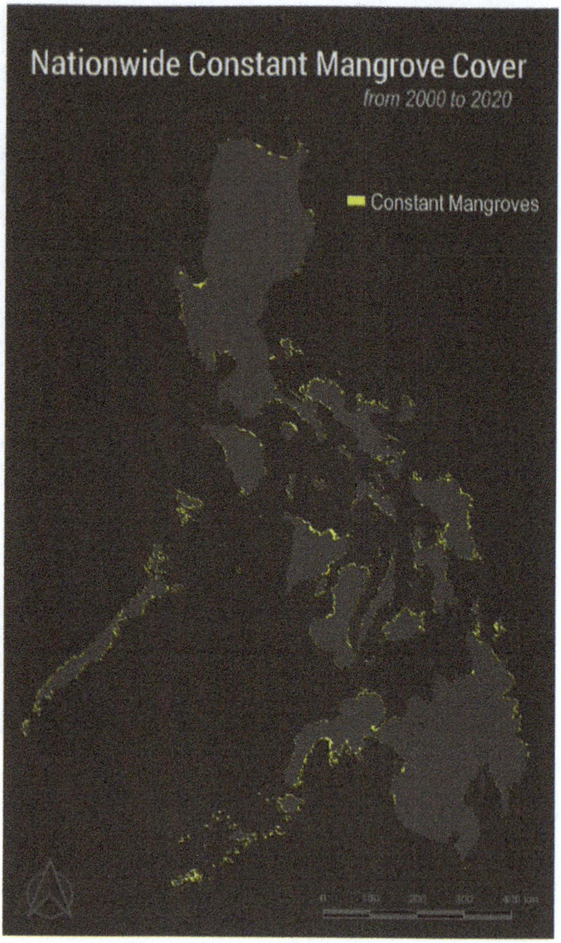

Fig. 12.7 Nationwide retained mangrove cover from year 2000 to 2020. The distribution of retained mangroves was observed throughout the country, even to areas frequently visited by typhoons. This may signify natural resiliency and regrowth in mangrove forests, or by human-assisted forest recovery

Impacts of Climate Related Variables

Effect of Rainfall on Mangrove Extent

Rainfall in the Philippines is influenced by the northeast (NE) monsoon and southwest (SW) monsoon and the climate regions, which are based on monsoonal rainy seasons (Coronas 1920). Matsumoto et al. (2020) used the TRMM 3B42 rainfall 1998–2013 to analyze climatological seasonal changes of rainfall in the Philippines. Results from this study show that rainfall in the Philippines is influenced both by the NE and SW monsoon. During the SW monsoon season, the west coast of the Philippines receives relatively higher rainfall compared with the east coast of the country. This is

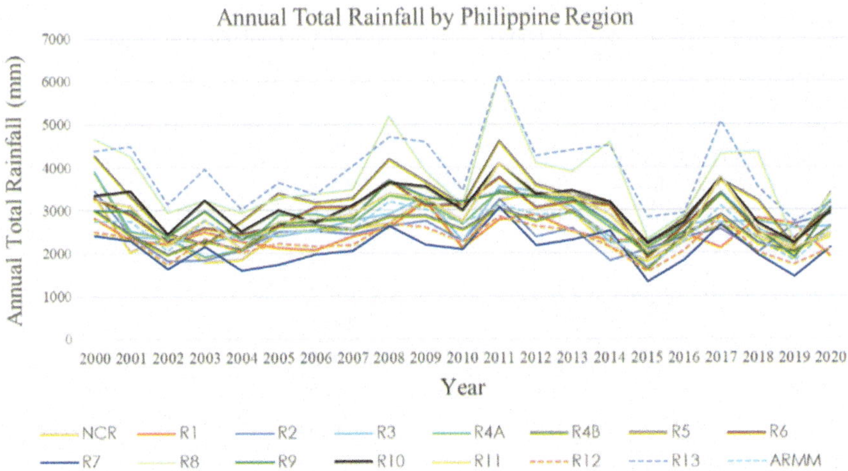

Fig. 12.8 Annual total rainfall (mm) in each Philippine region from 2000 to 2020. Most of the regions have similar annual trends as influenced by climatic patterns such as El Niño and La Niña

attributable to tropical cyclone (TC) activities that can also enhance SWM rainfall. In addition to monsoon, rainfall distribution in the Philippines is also influenced by the location of the mountain systems. The mean annual rainfall of the Philippines varies from 965 to 4,064 mm annually (PAGASA).

Based on CHIRPS data, the annual total rainfall was calculated as shown in Fig. 12.8.

Relatively higher amounts of annual total rainfall were recorded in years 2000, 2008, 2011, and 2017. High amounts of rainfall were received by regions 13, 8, 4B, 10, and 11. The rainfall statistics among regions follows similar trend, especially during the La Niña years such as 2007–2008 and 2017–2018 wherein a significant increase in rainfall was observed across all regions.

The decadal rainfall difference shows varying results among regions (Table 12.2). Almost all regions have lower precipitation in the later year (2020) than in year 2000. Highest decadal rainfall differences were calculated in regions 8, 13, 5, and 3, ranging from −883 mm to −1,320 mm. Only one region (region 9) recorded an increase in annual rainfall amount (+63 mm) after two decades.

In this study, mangrove extent data was regressed with rainfall metrics including decadal total rainfall, mean annual rainfall, and mean annual rainfall increment (Fig. 12.9).

Decadal total rainfall, mean annual increment, and mean annual rainfall show positive correlation with mangrove loss. Between the rainfall metrics, the mean annual rainfall increment gave the highest correlation ($r = 0.68$). This implies that the higher the decline in mean annual rainfall (higher variability), the higher the loss in mangrove area (Fig. 12.9). Changes in rainfall patterns have significant impacts

Table 12.2 Decadal rainfall difference from 2000 to 2020 calculated using CHIRPS data. Negative difference implies decreased amount of rainfall in 2020 compared to value two decades earlier

Region	Decadal rainfall difference (mm)
R1	−873.34
R2	−557.03
R3	−1292.61
R4A	−676.96
R4B	−379.64
R5	−883.32
R6	−250.43
R7	−284.7
R8	−1320.57
R9	63.83
R10	−390.64
R11	−753.46
R12	−388.22
R13	−1232.83
NCR	−951.33
ARMM	−158.11

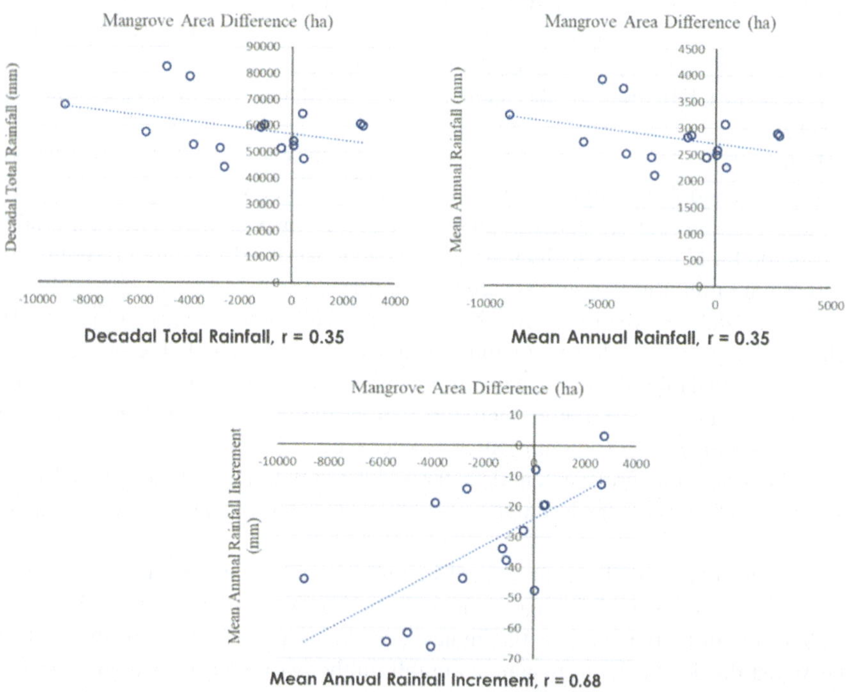

Fig. 12.9 Correlation results between mangrove extent and rainfall variable metrics. The mean annual rainfall increment generated higher correlation with the mangrove extent loss data

on mangrove growth, extent, and spatial distribution and have been suggested as a mechanism for the landward growth of mangroves (Gilman et al. 2008; Eslami-Andargoli et al. 2009). Decrease in rainfall also resulted to mangrove loss in other countries as reported by previous studies (Ghosh et al. 2017; Eslami-Andargoli et al. 2009).

Reduced precipitation over an area decreases water input to groundwater and lesser freshwater surface water input to mangroves, thereby increasing salinity. The increase in soil salinity induces pore water salinity, resulting in increased tissue salt levels in mangrove trees, which then decreases water availability and consequently reduces productivity and low seedling survival (Gilman et al. 2008). Decreased precipitation and rise in sea levels could result in salinity and inundation stresses (Salmo and Juanico 2015) and changing competition between species of mangrove (Ellison 2000). Mangrove population located in highly saline and frequently inundated sites may eventually collapse despite indicators of being healthy during the early stages of its development (Salmo and Juanico 2015). Meanwhile, the increase in precipitation is known to have a significant positive relationship with mangrove areas due to the landward expansion of mangrove (Eslami-Andargoli et al. 2009).

Mangrove productivity is increased as a result of decreased pore water salinity and sulfate concentration due to increased precipitation (Gilman et al. 2008). However, this increase in growth rate as a response to increase in precipitation could be species-specific (Krauss et al. 2014). Therefore, species that can adapt or tolerate such condition have competitive advantage over non-adaptive species.

Further analysis was done by identifying first the major climate type in each region, and clustering together the regions with similar climate type (Fig. 12.10). This analysis greatly improved the correlation results as the climate types were mainly characterized based on the occurrence and frequency of precipitation. The mangrove extent data was also re-grouped based on the regions belonging to the same climate type.

By clustering first the regions by climate type, correlation results for all rainfall metrics greatly increased (Table 12.3). The mean annual rainfall increment has an R-value of 0.99 compared to the 0.68 R-value when rainfall metrics were aggregated directly by region. The correlation coefficients for mean annual and decadal total rainfall also increased. Among the Philippine climate types, higher mean annual increments and higher mangrove losses were observed in areas classified under Types I and II. These areas have longer and more pronounced wet season.

In the aggregated precipitation metrics by climate type, it was observed that the mean annual rainfall increment is significantly lower in areas under Type III, possibly due to the occurrence of less pronounced wet season.

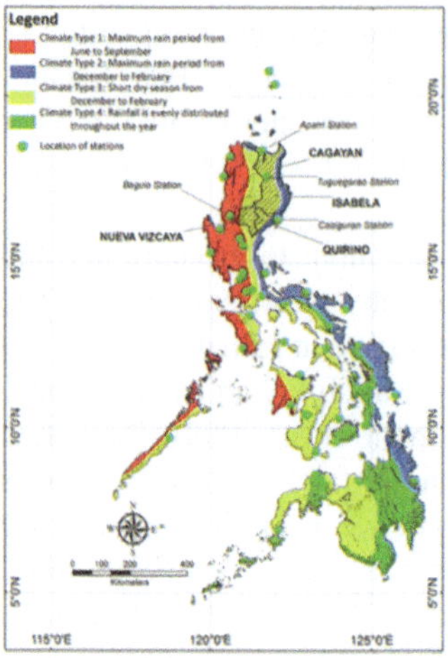

Region	Major Climate Type	Other Climate Type
R1	Type I	N/A
R2	Type II	Type III
R3	Type I	Type II, III
R4A	Type II	Type I,II
R4B	Type I	Type III
R5	Type II	Type III, IV
R6	Type III	Type I
R7	Type III	Type IV
R8	Type II	Type IV
R9	Type III	Type IV
R10	Type III	Type IV
R11	Type IV	N/A
R12	Type IV	Type III
R13	Type II	Type IV
NCR	Type I	N/A
ARMM	Type IV	Type III

Fig. 12.10 The Philippine climate type map (left, Basconcillo et al. 2016) and the dominant climate type for each of the administrative regions. Most of the regions have other climate types aside from the major climate, but identification of the dominant type was necessary to group regions per climate category

Effect of Typhoon Frequency on Mangrove Extent

The typhoon tracks data obtained from NOAA International Best Track Archive are shown in Fig. 12.11 and Table 12.4. These tracks were recorded from year 1999–2019.

There are relatively fewer typhoons that devastated the regions in Mindanao island or those regions in the southern part of the country (Fig. 12.11). Most of the typhoon tracks are recorded in Luzon and in the eastern portion of Philippines such as Eastern Visayas and Bicol Region.

Based on the mangrove loss and typhoon data statistics, areas that were least devastated by typhoons (fewer than 10 typhoons, gray-shaded rows in Table 12.4) have recorded increased vegetation cover such as in Northern Mindanao (Region 10), and Zamboanga Peninsula (Region 9). Regions with greater frequency of typhoons have declined mangrove areas, including regions 3, 4B, 5, and 8.

Results of linear regression (Fig. 12.12) show moderate positive correlation between typhoon count and mangrove loss ($r = 0.43$). Typhoon frequency was positively correlated with mangrove loss mainly due to the destructive impacts of typhoons such as lower mangrove productivity, canopy defoliation, uprooting of trees, and rapid decrease in soil elevation (Cahoon et al. 2003; Paling et al. 2008; Ward et al. 2016). Similar correlation results between typhoon frequency and mangrove structure were reported by other studies (Adame et al. 2013; Simard et al. 2019; Lin et al. 2020; Rovai et al. 2016).

Table 12.3 Correlation between mangrove extent and rainfall metrics aggregated by major climate type. Since climate types were defined based on the amount, timing, and distribution of rainfall, correlating mangrove loss and rainfall with this unit generally improved the correlation results

Climate Type	Climate Type Description	Mangrove Area Difference (ha)	Mean Annual Rainfall (mm/year)	Mean Annual Rainfall Increment (mm)	Decadal Total Precipitation (mm)
Type I	Has two pronounced seasons: dry from November to April and wet throughout the rest of the year.	−3082.75	2550.52	−43.71	53560.94
Type II	Characterized by the absence of a dry season but with a very pronounced maximum rain period from November to January	−3898.60	3236.81	−46.71	67972.98
Type III	Seasons are not very pronounced but are relatively dry from November to April and wet during the rest of the year.	793.50	2737.73	−10.77	57492.28
Type IV	Characterized by a more or less even distribution of rainfall	−186.67	2574.39	−21.66	54062.16
Mangrove Extent Difference vs Rainfall Variable		Multiple r	**0.51**	**0.99**	**0.51**

Typhoon-induced damages in mangrove forests are well documented in the Philippines, such as studies assessing the impact of Typhoon Haiyan in 2013 (Long et al. 2016; Buitre et al. 2019) and Typhoon Chan-hom in 2009 (Salmo et al. 2013). Multiple studies were conducted to monitor the changes in typhoon frequency and intensity in the country. DOST-PAGASA (2018) reported a slightly decreasing number of typhoons while a slightly increasing number of intense typhoons. Global reports also cited low confidence on the multi-decadal to centennial trends in the frequency of typhoons, while increase in the occurrence of stronger typhoons was certain in the past four decades (IPCC 2021).

Effect of Sea Surface Temperature on Mangrove Extent

The SST data for years 2003 and 2020 for selected Philippine regions are shown in Fig. 12.13. Higher SST values was recorded in 2020 for all regions, especially during the drier months of May to August. Closer SST values between the two dataset were recorded within the months of January and February.

Fig. 12.11 The Philippine typhoon track from year 2009 to 2019 (*Source* NOAA IBTrACS). The occurrence of typhoons (blue lines) was heavily concentrated in Luzon and Visayas islands

Among the SST metrics, the average annual maximum SST generated the highest correlation with mangrove loss (r = 0.39) (Fig. 12.15). Lower correlation values were obtained with the mean SST difference (r = 0.22) and lowest with the average annual minimum SST (r = 0.11) (Fig. 12.14). The higher the annual maximum SST, the higher the loss in mangrove area. Each mangrove species grows within a specific SST threshold and exceeding this limit may result to inhibition of physiological

Table 12.4 Number of Typhoons and the corresponding mangrove area difference per region

REGION CODE	REGION	Mangrove Area Difference (ha)	Number of Typhoons (1999-2019)
R1	Ilocos Region	–2784	36
R2	Cagayan Valley	–409	51
R3	Central Luzon	–5738	26
R4A	CALABARZON	–1211	19
R4B	MIMAROPA	–3857	36
R5	Bicol Region	–8940	28
R6	Western Visayas	2657	25
R7	Central Visayas	–2628	20
R8	Eastern Visayas	–4007	33
R9	Zamboanga Peninsula	2748	**4**
R10	Northern Mindanao	397	**9**
R11	Davao Region	–1061	8
R12	SOCCSKSARGEN	439	**3**
R13	Caraga	–4926	19
NCR	Metropolitan Manila	46	**7**
ARMM	ARMM	62	**2**

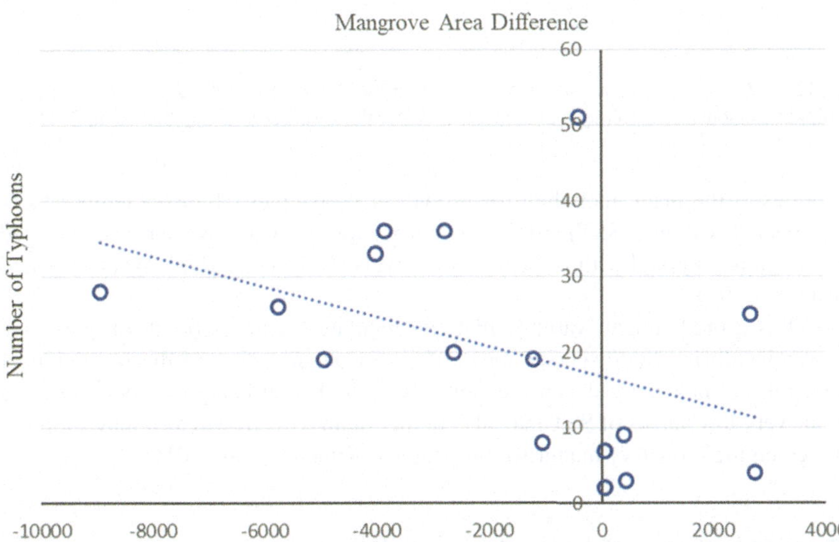

Fig. 12.12 Correlation between mangrove area difference and typhoon track count per region from 1999–2019. A positive correlation was observed between these variables due to the direct and non-direct impacts of extreme weather events to any vegetated land cover

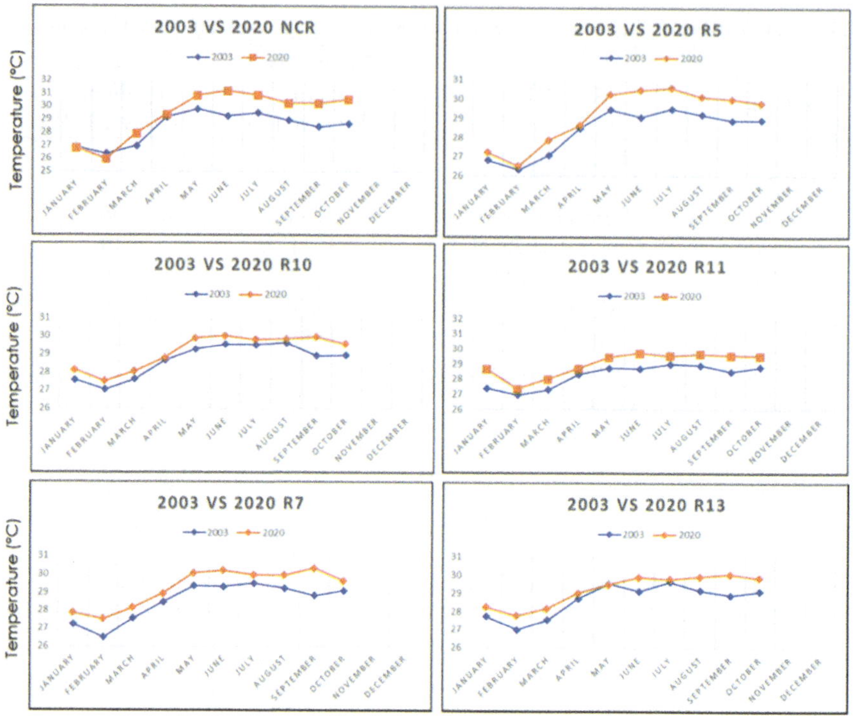

Fig. 12.13 Comparison between the Average Monthly SST (°C) in 2003 and 2020 in selected Philippine regions. Consistently higher monthly SST values were recorded in 2020 than 2003

functions of the plant, including the growth of propagules. High sea temperatures (as reflected with max SST) induced mangrove loss as mangrove survival and seed germination is limited within specific SST levels (Duke et al. 1998; Woodroffe and Grindrod 1991).

SST is a predominant variable that can regulate seedling growth of mangrove propagules during the dispersal stage. The success or failure of the formation of propagules in new lands depends essentially on SST after being transported by sea water. Very low values of SST may also hinder mangrove growth as it may suppress and lessen the growth of mangrove propagules (Ximenes et al. 2016).

Effect of Sea Surface Height on Mangrove Extent

The differences in SSH between the observation period were calculated and shown in Table 12.5. The greatest increase is plus 0.18 recorded in region 11 (Davao Region), while the greatest decline (0.17) was recorded in region 9 (Zamboanga Peninsula) with minus 0.17 in the SSH data. Most of the regions within the Luzon island have increased SSH values.

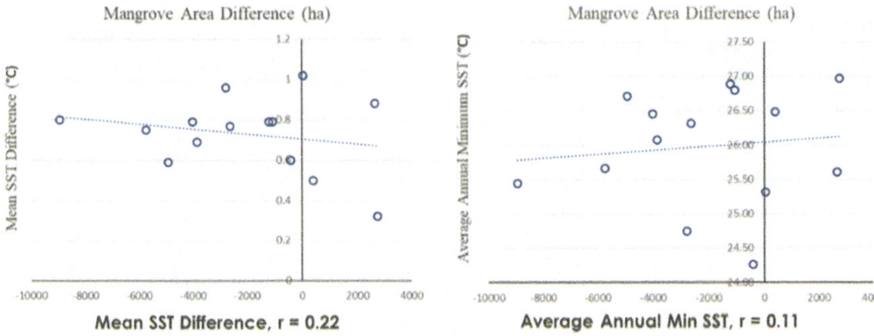

Fig. 12.14 Correlation results between mangrove extent and Mean SST and Average Annual Minimum SST. The SST metrics gave the lowest correlation to mangrove loss among the climate-related factors considered in this study

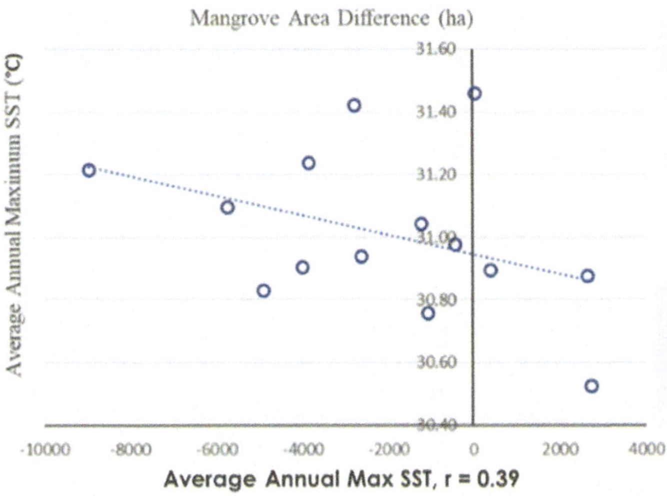

Fig. 12.15 Correlation results between mangrove extent and Average Annual Minimum SST

Sea level rise, which is regionally variable, is likely to have varied yet minimal impacts on mangroves locally (Ward et al. 2016). In this study, mangrove extent data was regressed with the difference in SSH within the administrative regions. Results showed a very low positive correlation (r = 0.16) between mangrove area difference and sea surface height difference (Fig. 12.16). Previous studies also reported very minimal impacts of SSH (Ward et al. 2016; Friess et al. 2012).

It was reported by previous studies that mangroves can adapt to sea-level rise if it occurs slowly enough. With minimal changes in sea surface level relative to mangrove surface, and other factors such as salinity, period, and depth of inundation will also remain constant, the mangrove margins and the inhabiting mangrove community will remain the same (Gilman 2006). Change in elevation within mangrove forests

Table 12.5 SSH difference and the corresponding mangrove area difference per region

Region	Mangrove extent difference (ha)	Sea surface height difference (m)
R1	−2784	−0.17
R2	−409	0.02
R3	−5738	0.00
R4A	−1211	0.10
R4B	−3857	−0.05
R5	−8940	−0.07
R6	2657	−0.02
R7	−2628	0.03
R8	−4007	−0.10
R9	2748	−0.17
R10	397	−0.10
R11	−4926	0.18
R13	−4926	−0.04
NCR	48	0.01

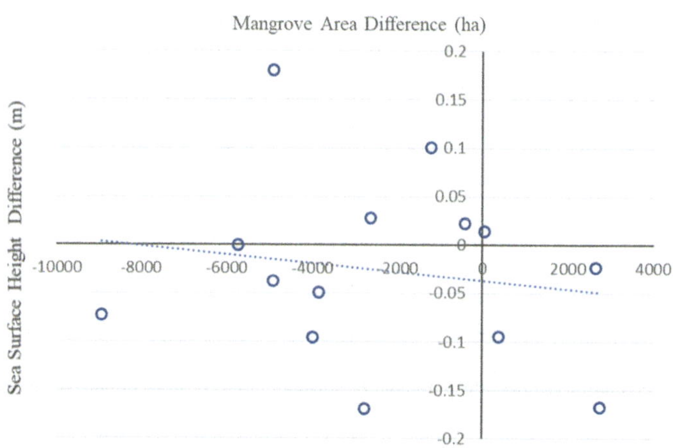

Fig. 12.16 Correlation results between mangrove extent and SSH decadal difference (m)

normally occurs at very slow rates which leads to directional changes over long time periods (Krauss et al. 2014). The two-decade observation in this study may not be long enough to capture these changes in sea surface heights. In comparison, the global mean sea level increase between 1901 and 2018 is 20 m, with an average increase rate of 3.7 mm yr^{-1} between 2006 and 2018 (IPCC 2021).

To persist, mangrove ecosystems must adapt to rising sea level by growing vertically through vertical accretion or deposition on the soil surface. Previous studies show that during periods of slower historical rise, mangrove soils were able to adapt to sea-level rise by building vertically through sediment deposition and peat formation, and mangrove forests expanded in many locations (Krauss et al. 2014; McKee et al. 2007; Willard and Bernhardt 2011).

Effect of Historical Air-Sea Climate Exposure on Mangrove Extent

The Philippines is vulnerable to the ill-effects of climate change due to its extreme exposure to increasing ocean temperature, extreme rainfall events and sea level rise. Such extreme conditions can also have a negative effect on the coastal habitats such as mangroves, seagrasses and corals (David et al. 2015; Lovelock et al. 2015). The study of David et al. (2015) provided the different levels of exposure to these variables as a significant input to national vulnerability assessments.

The mangrove extent data was regressed with the level of relative exposure per climatic variable which include increasing ocean temperature, extreme heating events, extreme precipitation, rainfall changes, and sea level rise. The level of exposure varies among the regions and among the climatic variables (Table 12.6).

Table 12.6 Relative exposure level and the corresponding mangrove area difference per region. A value of 1 implies lower level of exposure while 5 implies higher level of exposure to each of the climatic variables below

		Level of relative exposure per climatic variable (Highest: 5)				
Region	Mangrove area difference	Increasing ocean temperature	Extreme heating events	Extreme precipitation	Rainfall changes	Sea level rise
R1	−2784	3	5	5	5	5
R2	−409	3	5	2	3	4
R3	−5738	3	5	2	3	4
R4A	−1211	2	4	1	5	5
R4B	−3857	2	4	1	5	5
R5	−8940	3	4	4	4	4
R6	2657	3	3	1	3	5
R7	−2628	3	3	1	3	5
R8	−4007	5	4	4	3	4
R9	2748	3	3	1	3	5
R10	397	5	4	4	3	4
R11	−1061	5	4	3	1	5
R12	439	3	3	1	3	5
R13	−4926	5	4	4	3	4
NCR	48	2	4	1	5	5
ARMM	62	3	3	1	3	5

Table 12.7 Selected climatic variables and the correlation results with mangrove loss. Regions with higher relative exposures recorded higher mangrove area loss, especially those exposed to extreme heating events, extreme rainfall, and occurrence of sea level rise

Historical climatic variable	Multiple R
Increasing ocean temperature	0.1
Extreme heating events	0.47
Extreme rainfall	0.51
Rainfall changes	0.19
Sea level rise	0.54

	Main Climate-Related Variables									Exposure Ranking with Climatic Variables				
	Decadal Total Rainfall	Mean Annual Rainfall	Mean Annual Rainfall Increment/ Region	Mean Annual Rainfall Increment/ Climate Type	Total Typhoon	Mean SST Difference	Average Annual Minimum SST	Average Annual Max SST	Sea Level Height Difference	Increasing Ocean Temperature	Extreme Heating Events	Extreme Precipit ation	Precipitation Changes	Sea Level Rise
Multiple R	0.35	0.35	0.68	0.99	0.43	0.22	0.11	0.39	0.16	0.1	0.47	0.51	0.19	0.54

Fig. 12.17 Correlation of climate-related variables with mangrove extent loss (2020–2000)

Losses in mangrove covers were observed in regions with high exposure to extreme precipitation (r = 0.51), extreme heating (r = 0.47), and sea level rise (r = 0.54) (Table 12.7).

The summary of correlation between the climate-related variable and mangrove loss is shown in Fig. 12.17. Numbers highlighted in blue are variables with relatively higher correlation results.

Impacts of Human-Induced Variables

Mangrove Forest Conversion to Built-Up

A total of 200 Ha of mangroves were converted to built-up from years 2000 to 2020 (Table 12.8). This is only 0.1% of the total decadal mangrove loss (29,208 Ha). The highest area converted to built-up was recorded in region 7 (62 Ha).

Mangrove Forest Conversion to Fishponds

A total of 1,409 Ha of mangroves were converted to fishponds from years 2000 to 2020 (Table 12.9). This is less than 1% of the total decadal mangrove loss (29,208 Ha). Highest percentage of converted area to fishponds was recorded in regions 10 and 12 (8.5%). It must be noted that in the Philippines, mangrove conversion to fishponds largely took place in 1951–1988 (Friess et al. 2019; Primavera 1995) in which the rate of fishpond expansion after this period is lower based on NAMRIA data. Less

Table 12.8 Mangrove area converted to built-up cover in each Philippine region

REGION CODE	REGION	NAMRIA LAND COVER DATA	
		Mangrove Area (ha) Converted to Built-up	% Converted Area
R1	Ilocos Region	*No mangroves detected in 2006 NAMRIA LC*	
R2	Cagayan Valley	6.26	0.59
R3	Central Luzon	*No intersection*	
R4A	CALABARZON	0.49	0.01
R4B	MIMAROPA	36.69	0.14
R5	Bicol Region	6.69	0.04
R6	Western Visayas	8.58	0.10
R7	Central Visayas	62.58	0.67
R8	Eastern Visayas	23.24	0.09
R9	Zamboanga Peninsula	4.39	0.04
R10	Northern Mindanao	6.49	0.26
R11	Davao Region	0.15	0.01
R12	SOCCSKSARGEN	0.53	0.13
R13	Caraga	41.87	0.40
NCR	Metropolitan Manila	*No mangroves detected in 2006 NAMRIA LC*	
ARMM	ARMM	2.00	0.03
TOTAL		**199.96**	**0.14**

mangrove cover lost from fishpond conversion were also reported by NAMRIA between 2006 and 2015, with an area that is only 7% of the total converted areas prior to 2006. Declining rates of loss may be attributed to already reduced cover, improved data quality, and to national and international conservation policies (Primavera 1995, 2000). Further improvement and longer analysis of aquaculture maps can provide more coherent results.

The highest built-up area conversion from mangroves was recorded in Central Visayas (63 Ha) and lowest in Davao region (0.15 Ha). Meanwhile, the highest fishpond area conversion from mangroves was recorded in Zamboanga Peninsula (330 Ha) and lowest in Davao region (2.8 Ha). Limitations in the resolution of satellite images used have resulted to some regions with no reported mangroves like Metro Manila and Ilocos Region, specifically on the 2006 land cover data.

Conclusions

To understand the impacts of human-induced and climatic variables on mangroves, relationship between these factors must be identified both in the temporal and spatial scales. In this study, decadal mangrove extent was the metric used in describing the potential impact of selected human and climate-related variables on the mangrove

Table 12.9 Mangrove area converted to fishponds cover in each Philippine region

REGION CODE	REGION	NAMRIA LAND COVER DATA	
		Mangrove Area (ha) Converted to Fishponds	% Converted Area
R1	Ilocos Region	*No mangroves detected in 2006 NAMRIA LC*	
R2	Cagayan Valley	4.05	0.38
R3	Central Luzon	48.46	0.73
R4A	CALABARZON	73.23	0.84
R4B	MIMAROPA	158.33	0.59
R5	Bicol Region	72.51	0.45
R6	Western Visayas	86.64	1.02
R7	Central Visayas	151.14	1.61
R8	Eastern Visayas	151.42	0.62
R9	Zamboanga Peninsula	330.47	3.21
R10	Northern Mindanao	212.83	8.52
R11	Davao Region	2.83	0.10
R12	SOCCSKSARGEN	34.75	8.47
R13	Caraga	82.52	0.78
NCR	Metropolitan Manila	*No mangroves detected in 2006 NAMRIA LC*	
ARMM	ARMM	*No intersection*	
TOTAL		**1409.19**	**1.01**

forests in the Philippines. Multi-decadal mangrove extent maps were generated with the use of Landsat-derived Mangrove Vegetation Index layers. Although there are already existing historical estimates of mangrove area, a standardized index-based mapping workflow is needed in reliably detecting actual changes on the ground. An overall trend of decreasing mangrove extent was observed in the Philippines from 2000 to 2020 with the annual losses surpassing the annual gains. The extent of vegetation gain and loss varies per region, and this was associated with the variables considered in this study.

Variations in the climate-related variables have all impacted the extent of mangrove communities. Linear regression results highlight significant relationships between major climate variables and the decadal extent of mangrove forests. Among these variables, the mean annual rainfall increment, annual max SST, and number of typhoons have the greatest correlation and thus considered as the main drivers of decadal mangrove loss in the country. This observation agrees with previous studies on the correlation of mangrove forest statistics with rainfall variability, maximum SST, and typhoon frequency. In addition, historical exposures to sea level rise and extreme heating and precipitation events were positively correlated with the decline in mangrove areas. The level of correlation of climatic variables with mangrove loss also varies per region and per climate type.

There is a minimal percentage of mangroves that were converted to built-up and fishpond areas during the period considered. Around 1% of the total mangroves were converted to fishponds. Less mangroves loss from fishpond conversion were also reported by NAMRIA between 2006 and 2015. Most of the mangrove-to-fishpond conversion in the country took place a long time before the period used in this study (1950s to 1970s). After the said period, land use conversion still took place but is now controlled and limited due to the strict regulations and continuous monitoring of concerned Philippine agencies.

This study highlighted the vulnerability of Philippine mangrove forests to natural and climate-related factors. Results of this study provide valuable information to better manage and protect mangrove forests especially those located in areas that are frequently visited by typhoons, with declining annual rainfall, and areas exposed to maximum temperature during the dry months. Identifying the main threat to mangrove per region will also help in selecting effective region-specific interventions.

Recommendations

- The results of this study can be utilized in drawing site-specific intervention to the local threats on mangroves. For instance, stronger naturally growing frontline
- mangrove species must be selected whenever there are replanting activities within regions frequently visited by typhoons. Species that are more tolerant to increasing sea temperatures must be selected within regions having high SSTs. Further, the methodologies used in this study can be applied to provincial level for a more spatially detailed analysis.
- Improvement and continuation of effective mangrove management and monitoring strategies and programs, including regulation of mangrove to fishpond conversion.
- Capacity-building of local government officials on the use of MVI and other remote sensing data for rapid and accurate monitoring of mangrove forest.
- Dissemination of information to the target community on the importance and vulnerability of mangroves, and the steps to protect and conserve these resources.
- This study assessed the impacts of climatic variable using selected metrics only. There are other potential metrics that can be included in further studies such as land subsidence and typhoon intensity. For human-induced metrics, historical data on aquaculture conversion is needed to capture more extended observations.

References

Adame MF, Zaldívar-Jimenez A, Teutli C, Caamal JP, Andueza MT, López-Adame H, Cano R, Hernández-Arana HA, Torres-Lara R, Herrera-Silveira JA (2013) Drivers of mangrove litterfall within a karstic region affected by frequent hurricanes. Biotropica 45:147–154. https://doi.org/10.1111/btp.12000

Alongi DM (2015) The impact of climate change on mangrove forests. Curr Clim Change Rep 1(1):30–39. https://doi.org/10.1007/s40641-015-0002-x

Basconcillo J, Lucero A, Solis A (2016) Statistically downscaled projected changes in seasonal mean temperature and rainfall in Cagayan Valley, Philippines. J Meteorol Soc Jpn 94A:151–164. https://doi.org/10.2151/jmsj.2015-058

Beck HE, Van Dijk AIJM, Levizzani V, Schellekens J, Miralles D, Martens B, De Roo A (2017) MSWEP: 3-hourly 0.25° global gridded precipitation (1979–2015) by merging gauge, satellite, and reanalysis data. Hydrol Earth Syst Sci 21:589–615

Buitre MJC, Zhang H, Lin H (2019) The Mangrove forests change and impacts from tropical cyclones in the Philippines using time series satellite imagery. Remote Sens 11(6):688. https://doi.org/10.3390/rs11060688

Bunting P, Rosenqvist A, Lucas R, Rebelo L-M, Hilarides L, Thomas N, Hardy A, Itoh T, Shimada M, Finlayson CM (2018) The global mangrove watch—a new 2010 global baseline of mangrove extent. Remote Sens 10:1669. https://doi.org/10.3390/rs10101669

Cahoon DR, Hensel P, Rybczyk J, McKee KL, Proffitt E, Perez BC (2003) Mass tree mortality leads to mangrove peat collapse at Bay Islands, Honduras after Hurricane Mitch. J Ecol 91:1093–1105. https://doi.org/10.1046/j.1365-2745.2003.00841.x

Coronas J (1920) The climate and weather of the Philippines, 1903 to 1918. Bureau of Printing, Manila, Philippines

David LT, Borja, Del Rosario R, Peñaflor EL, Cordero-Bailey K, Villanoy CL, Aliño PM et al (2015) Developing a Philippine climate ocean typology as input to national vulnerability assessments. ACRS

Donlon C, Robinson I, Casey KS, Vazquez-Cuervo J, Armstrong E, Arino O, Gentemann C, May D, LeBorgne P, Piollé J et al (2007) The global ocean data assimilation experiment high-resolution sea surface temperature pilot project. Bull Am Meteorol Soc 88:1197–1213. https://doi.org/10.1175/BAMS-88-8-1197

DOST-PAGASA (2018) Observed and projected climate change in the Philippines. Philippines Atmospheric, Geophysical and Astronomical Services Administration, Quezon City, Philippines, 36 pp

Duke NC, Ball MC, Ellison JC (1998) Factors influencing biodiversity and distributional gradients in mangroves. Glob Ecol Biogeogr 7:27–47. https://doi.org/10.1111/j.1466-8238.1998.00269.x

Ellison J (2000) How South Pacific mangroves may respond to predicted climate change and sea level rise. In: Gillespie A, Burns W (eds) Climate change in the South Pacific: impacts and responses in Australia, New Zealand, and Small Islands States. Kluwer Academic Publishers, Dordrecht, Netherlands (Chapter 15), pp 289–301

Eslami-Andargoli L, Dale P, Sipe N, Chaseling J (2009) Mangrove expansion and rainfall patterns in Moreton Bay, Southeast Queensland, Australia. Estuar Coast Shelf Sci 85:292–298. https://doi.org/10.1016/j.ecss.2009.08.011

Friess DA, Krauss KW, Horstman EM, Balke T, Bouma TJ, Galli D, Webb EL (2012) Are all intertidal wetlands naturally created equal? Bottlenecks, thresholds and knowledge gaps to mangrove and saltmarsh ecosystems. Biol Rev 87:346–366. https://doi.org/10.1111/j.1469-185X.2011.00198.x

Friess DA, Rogers K, Lovelock CE et al (2019) The state of the world's mangrove forests: past, present, and future. Annu Rev Environ Resour 44:89–115. https://doi.org/10.1146/annurev-environ-101718-033302

Funk C, Peterson P, Landsfeld M et al (2015) The climate hazards infrared precipitation with stations—a new environmental record for monitoring extremes. Sci Data 2:150066. https://doi.org/10.1038/sdata.2015.66

Ghosh M, Kumar L, Roy C (2017) Climate variability and mangrove cover dynamics at species level in the Sundarbans, Bangladesh. Sustainability 9(5):805. https://doi.org/10.3390/su9050805

Gilman E (ed) (2006) Proceedings of the symposium on mangrove responses to relative sea-level rise and other climate change effects, 13 July 2006. Catchments to Coast. The Society of Wetland Scientists 27th international conference, 9–14 July 2006, Cairns Convention Centre, Cairns, Australia. Published by the Western Pacific Regional Fishery Management Council, Honolulu, USA

Gilman E, Ellison J, Duke N, Fiel C (2008) Threats to mangroves from climate change and adaptation options: a review. Aquat Bot 89:237–250. https://doi.org/10.1016/j.aquabot.2007.12.009

IPCC (2021) Climate change 2021: The physical science basis. Contribution of working group I to the sixth assessment report of the intergovernmental panel on climate change [Masson-Delmotte, V., P. Zhai, A. Pirani, S.L. Connors, C. Péan, S. Berger, N. Caud, Y. Chen, L. Goldfarb, M.I. Gomis, M. Huang, K. Leitzell, E. Lonnoy, J.B.R. Matthews, T.K. Maycock, T. Waterfield, O. Yelekçi, R. Yu, and B. Zhou (eds.)]. Cambridge University Press, Cambridge, United Kingdom and New York, NY, USA, In press. https://doi.org/10.1017/9781009157896, https://www.ipcc.ch/report/ar6/wg1/about/how-to-cite-this-report/

Jang JC, Park KA (2019) High-resolution sea surface temperature retrieval from Landsat 8 OLI/TIRS data at coastal regions. Remote Sens 11(22):2687. https://doi.org/10.3390/rs11222687

Knapp KR, Kruk MC, Levinson DH, Diamond HJ, Neumann CJ (2010) The international best track archive for climate stewardship (IBTrACS): unifying tropical cyclone best track data. Bull Am Meteor Soc 91:363–376. https://doi.org/10.1175/2009BAMS2755.1

Knapp KR, Diamond HJ, Kossin JP, Kruk MC, Schreck CJ (2018) International best track archive for climate stewardship (IBTrACS) Project, Version 4 [indicate subset used]. NOAA National Centers for Environmental Information. https://doi.org/10.25921/82ty-9e16

Krauss K, McKee K, Lovelock C, Cahoon D, Saintilan N, Reef R, Chen L (2014) How mangrove forests adjust to rising sea level. New Phytol 202:19–34. https://doi.org/10.1111/nph.12605

Lawas LM (1974). Economic study on alternative uses of mangrove swamps: Bakawan production or fish ponds. In: Proceedings of Indo-Pacific Fishery Council, pp 65–69. 15th Session, 18–27 October 1972, Wellington, New Zealand, Section 2 Bangkok, FAO

Lin TC, Hogan JA, Chang CT (2020) Tropical cyclone ecology: a scale-link perspective. Trends Ecol Evol 7:594–604. https://doi.org/10.1016/j.tree.2020.02.012

Long JB, Giri C (2011) Mapping the Philippines' mangrove forests using Landsat imagery. Sensors 11(3):2972–2981. https://doi.org/10.3390/s110302972

Long J, Napton D, Giri C, Graesser J (2014) A mapping and monitoring assessment of the Philippines' mangrove forests from 1990 to 2010. J Coastal Res 30(2):260–271. https://doi.org/10.2112/JCOASTRES-D-13-00057.1

Long J, Giri C, Primavera JH, Trivedi M (2016) Damage and recovery assessment of the Philippines' mangroves following super typhoon Haiyan. Mar Pollut Bull 109:734–743. https://doi.org/10.1016/j.marpolbul.2016.06.080

Lovelock C, Cahoon D, Friess D et al (2015) The vulnerability of Indo-Pacific mangrove forests to sea-level rise. Nature 526:559–563. https://doi.org/10.1038/nature15538

Matsumoto J, Olaguera LMP, Nguyen-Le D, Kubota H, Villafuerte II MQ (2020) Climatological seasonal changes of wind and rainfall in the Philippines. Int J Climatol:1–15.https://doi.org/10.1002/joc.6492

McKee KL, Cahoon DR, Feller IC (2007) Caribbean mangroves adjust to rising sealevel through biotic controls on change in soil elevation. Glob Ecol Biogeogr 16:545–556. https://doi.org/10.1111/j.1466-8238.2007.00317.x

Mu Y, Biggs T, Shen S (2021) Satellite-based precipitation estimates using a dense rain gauge network over the Southwestern Brazilian Amazon: Implication for identifying trends in dry season rainfall. Atmos Res 261:105741. https://doi.org/10.1016/j.atmosres.2021.105741

NAMRIA (2015) National Mapping and Resource Information Authority. https://www.namria.gov.ph/

PAGASA (2018) Philippine Atmospheric, Geophysical and Astronomical Services Administration. Official website: https://www.pagasa.dost.gov.ph/

Paling EI, Kobryn HT, Humphreys G (2008) Assessing the extent of mangrove change caused by Cyclone Vance in the eastern Exmouth Gulf, northwestern Australia. Estuar Coast Shelf Sci 77:603–613. https://doi.org/10.1016/j.ecss.2007.10.019

Philippine Council for Agriculture, Forestry and Natural Resources Research and Development (1991) The Philippines recommends for mangrove production and harvesting. Philippines Recommends series No. 74. PCARRD/DENR, 96 pp

Primavera JH (1995) Mangroves and Brackishwater pond culture in the Philippines. Hydrobiologia 295:303–309. https://doi.org/10.1007/BF00029137

Primavera JH (2000) Development and conservation of Philippine mangroves: institutional issues. Ecol Econ 35(1):91–106. https://doi.org/10.1016/S0921-8009(00)00170-1

Rivera JA, Marianetti G, Hinrichs S (2018) Validation of CHIRPS precipitation dataset along the Central Andes of Argentina. Atmos Res 213:437–449. https://doi.org/10.1016/j.atmosres.2018.06.023

Rovai AS, Riul P, Twilley RR, Castañeda-Moya E, Rivera-Monroy VH, Williams AA, Simard M, Cifuentes-Jara M, Lewis RR, Crooks S, Horta PA, Schaeffer-Novelli Y, Cintrón M, Pozo-Cajas P, Pagliosa PR (2016) Scaling mangrove aboveground biomass from site-level to continental-scale. Glob Ecol Biogeogr 25:286–298. https://doi.org/10.1111/geb.12409

Salmo S, Juanico DE (2015) An individual-based model of long-term forest growth and carbon sequestration in planted mangroves under salinity and inundation stresses. Int J Philippine Sci Technol 8:31–35. https://doi.org/10.18191/2015-08-2-019

Salmo S, Lovelock CE, Duke N (2013) Assessment of vegetation and soil conditions in restored mangroves interrupted by severe tropical typhoon 'Chan-hom' in the Philippines. Hydrobiologia 733(1):85–102. https://doi.org/10.1007/S10750-013-1766-4

Shrestha NK, Qamer FM, Pedreros D, Murthy MSRR, Wahid SM, Shrestha M (2017) Evaluating the accuracy of Climate Hazard Group (CHG) satellite rainfall estimates for precipitation based drought monitoring in Koshi basin, Nepal. J Hydrol Reg Stud 13:138–151. https://doi.org/10.1016/j.ejrh.2017.08.004

Simard M, Fatoyinbo T, Smetanka C, Rivera-Monroy VH, Castaneda-Mova E, Thomas N, Van Der Stocken T (2019) Global mangrove distribution, aboveground biomass, and canopy height. https://doi.org/10.3334/ORNLDAAC/1665

Ward RD, Friess DA, Day RH, MacKenzie RA (2016) Impacts of climate change on mangrove ecosystems: a region by region overview. Ecosyst Health and Sust 2(4):e01211. https://doi.org/10.1002/ehs2.1211

Willard DA, Bernhardt CE (2011) Impacts of past climate and sea level change on Everglades wetlands: placing a century of anthropogenic change into a late-Holocene context. Clim Change 107:59–80. https://doi.org/10.1007/s10584-011-0078-9

Woodroffe CD, Grindrod J (1991) Mangrove biogeography: the role of quaternary environmental and sea-level change. J Biogeogr 18:479–492. https://doi.org/10.2307/2845685

Ximenes A, Maeda E, Arcoverde G, Dahdouh-Guebas F (2016) Spatial assessment of the bioclimatic and environmental factors driving mangrove tree species' distribution along the Brazilian coastline. Remote Sens 8(6):451. https://doi.org/10.3390/rs8060451

Chapter 13
Anthropogenic Climate Change in the Mangrove-Dominated Indian Sundarbans: Spatio-temporal Analyses, Future Trends, and Recommendations for Mitigation and Adaptation

Sangita Agarwal, Pritam Mukherjee, Mourani Sinha, Johannes M. Luetz, and Abhijit Mitra

Abstract Anthropogenic climate change is among the most defining challenges confronting the current era, with impacts already affecting both the health of the environment and the functioning of virtually every sector of society. In the present study, four environmental indicators were used as proxies to analyse the footprints of climate change in the mangrove-dominated deltaic region of the Indian Sundarbans, namely surface air temperature, near-surface atmospheric carbon dioxide (CO_2), surface water salinity, and surface water pH. The study grouped the mangrove ecosystem into three distinct zones (western, central, and eastern sectors) and analysed 37 years of data (1984–2020). The results indicate that although both air temperature and atmospheric CO_2 concentrations have risen uniformly in all three sectors of the study area, increases in aquatic salinity have not been uniform. While the western and eastern parts of the Indian Sundarbans exhibited a lowering of salinity, the

S. Agarwal
Department of Applied Science, RCC Institute of Information Technology, Canal South Road, Beliaghata, Kolkata 700015, India

P. Mukherjee (✉)
Department of Oceanography, Techno India University, West Bengal, EM-4, Sector V, Salt Lake, Kolkata 700091, India
e-mail: mukherjee.pritam14@gmail.com

M. Sinha
Department of Mathematics, Techno India University, West Bengal, EM-4, Sector V, Salt Lake, Kolkata 700091, India

J. M. Luetz
School of Social Sciences, University of New South Wales, Sydney, NSW 2052, Australia

School of Law and Society, University of the Sunshine Coast, Maroochydore, QLD 4556, Australia

Graduate Research School, Alphacrucis University College, Brisbane, QLD 4102, Australia

A. Mitra
Department of Marine Science, University of Calcutta, 35 B. C. Road, Kolkata 700019, India

W. Leal Filho et al. (eds.), *Climate Change Strategies: Handling the Challenges of Adapting to a Changing Climate*, Climate Change Management,
https://doi.org/10.1007/978-3-031-28728-2_13

central region displayed a gradual increase in aquatic salinity. These differences may be explained by divergent dilution factors and siltation-linked freshwater obstructions. Aquatic pH has decreased uniformly in all three sectors of the present study area. Two-way analysis of variance (ANOVA) reflects a distinctive climate change signal influencing all four primary environmental parameters. Additionally, artificial intelligence (AI) technology was employed to evaluate future aspects of these environmental variables using nonlinear autoregressive (NAR) networks. In synthesis, the climate change signal in the Indian Sundarbans region is mixed with "noise" that can be linked to other human inputs, including urbanisation, industrialisation, and changes in land-use patterns (shrimp culture, tourism), among others. Notably, an alarming situation is predicted to occur around the middle of the century. The study recommends the implementation of both mitigative and adaptive measures to counteract the adverse impacts of anthropogenic climate change. Recommendations encompass low-carbon alternative livelihoods, mangrove plantations, and rainwater harvesting, among others.

Introduction

Climate change can be broadly categorised into natural and anthropogenic causes. The natural factors may include volcanic activity, forest fires, continental drift over long timescales, astronomical events, and variations in solar energy output due to magnetic storms in the sun. The anthropogenic factors are linked to human activities that began with the advent of the industrial revolution and introduced sweeping changes to the environment (Mitra 2013). Sustained human-induced changes to nature have dramatically increased concentrations of carbon dioxide (CO_2) and other greenhouse gases (GHGs) in the atmosphere, thus amplifying the Earth's natural greenhouse effect (Hansen et al. 2006; Leal Filho et al. 2021a). Based on several years of Intergovernmental Panel on Climate Change (IPCC) assessment reports, one of the most important effects of climate change is the significant rise in global mean surface temperatures over the last 50 years (IPCC 2001, 2007, 2022). Significantly, climate change has been accelerating over recent decades, with global average temperatures rising by about 0.2 °C per decade since 1975 (Hansen et al. 2006; IPCC 2022). Globally, temperatures have already increased by more than 1 °C since 1850, and climatologists have warned of significant further warming this century by 1.1–6.4 °C (Majra and Gur 2009; IPCC 2022).

Coastal and estuarine regions are vulnerable to climate change because of increases in surface water temperature, unnatural depressions and erratic weather phenomena, cyclonic storms, sea-level rises, and concomitant seawater intrusions. The progressive degradation of "blue carbon" ecosystems over recent decades has increased CO_2 concentrations at the local level (Mitra 2013; Mitra and Zaman 2014, 2015, 2016; Agarwal et al. 2016; Mitra and Sundaresan 2016; Pal et al. 2016; Zaman et al. 2016; Mitra et al. 2017).

"Blue carbon" refers to the stored carbon reservoir in the marine and estuarine systems comprising sediment, water, and coastal vegetations, including mangroves, salt marshes, seaweeds, seagrass meadows, and several other halophytes, which sequester and store more carbon per unit area than terrestrial forests. The ability of these blue carbon reserves to remove CO_2 from the atmosphere makes them important net carbon sinks and effective natural environmental safeguards. In addition, these ecosystems provide essential benefits for climate change adaptation for coastal communities in the Bay of Bengal and beyond, including in areas of coastal protection and food security, among others (Luetz 2008, 2018; Mitra 2020). Degradation or destruction of these vital carbon sinks through human activities (directly or indirectly) and natural events (cyclones, storm surges, wave actions, etc.) causes the release of the stored carbon from the soil and water and results in the emissions of CO_2 back to the atmosphere, thus contributing to climate change. Dedicated conservation, restoration, and remediation efforts can make these ecosystems long-lasting carbon sinks by ensuring that no new CO_2 emissions arise from their degradation and loss while at the same time enabling new carbon sequestration to occur through restoring previous carbon-rich brackish water, coastal, and marine habitats.[1]

The Sundarbans mangrove forest is known for its rich gene pool, which sustains some 34 species of true mangroves and other mangrove associate species. The blue carbon community of the Indian Sundarban ecosystem includes both salt marshes (dominated by *Porteresia coarctata*) and mangroves (Mitra and Zaman 2015). Mangroves are halophytic (salt-tolerant) floral plants that experience inundation and exposure twice a day during high and low tides; therefore, they are unique in comparison to the normal terrestrial flora. They provide a variety of essential ecosystem services. Besides acting as a powerful carbon sink (Fig. 13.1), mangroves prevent soil erosion, serve as a buffer against cyclonic storms and wave actions, provide nursery beds for fishes, crabs, oysters, etc., and offer habitat for a plethora of species and biotic communities, including microbes (Chakraborty and Choudhury 1985; Mitra and Choudhury 1993). In addition, mangroves are known to bioaccumulate toxic heavy metals in their body tissues and thus act as a critical player in phytoremediation (Mitra 2020; Mitra et al. 1987, 1992; Saha et al. 1999). Finally, mangroves are noted for their disaster mitigating propensities (Luetz 2008; Menéndez et al. 2020), and there is a case to promote their conservation towards better disaster preparedness, resilience, and risk reduction through mainstreaming locally relevant disaster risk education (Luetz 2008, 2020; Luetz and Sultana 2019).

Notwithstanding the manifold benefits of mangroves, rapid urbanisation, and industrialisation in and around the city of Kolkata, Howrah, and Haldia, along with increased salinity, are progressively posing threats to the overall biodiversity in the Indian Sundarban region. The name "Sundarbans" originates from the once-dominant mangrove plant "Sundari" (*Heritiera fomes*), which is presently on the verge of extinction in the hypersaline pockets of the Indian Sundarbans owing chiefly to rises in sea level and concomitant increases in soil salinity. Inundation of land during high tides with saline seawater and seawater ingress through drains, creeks, and rivers

[1] https://www.iucn.org/resources/issues-briefs/blue-carbon.

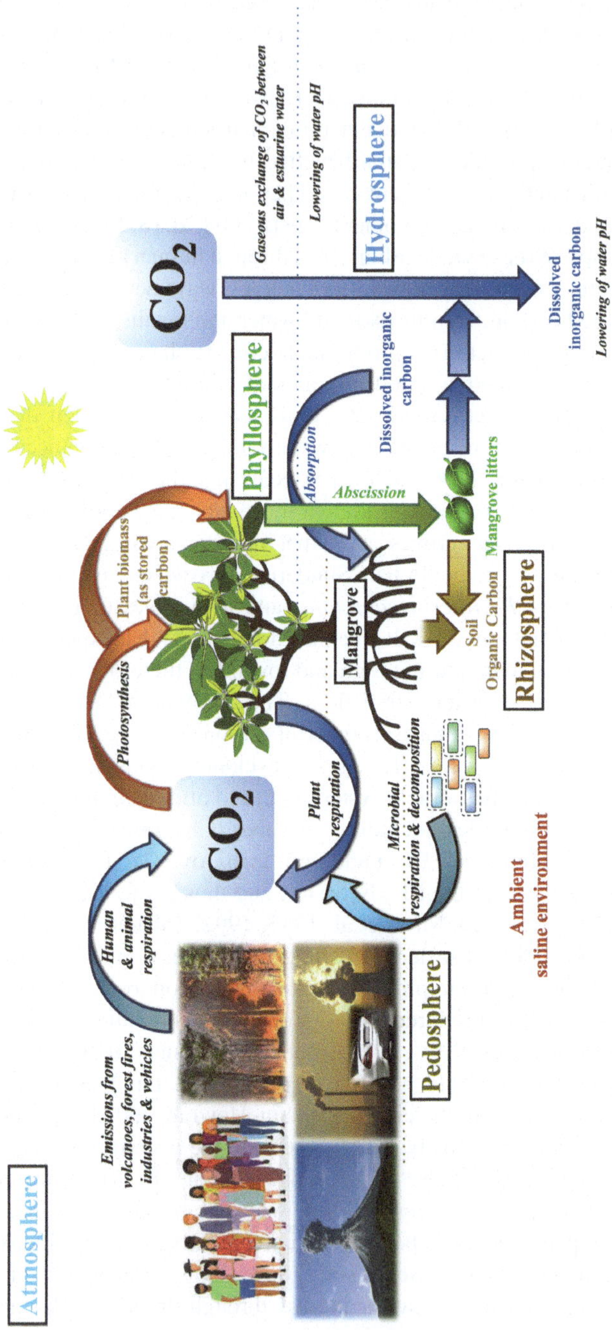

Fig. 13.1 Illustration of mangrove vegetation as a significant carbon sink

cause salinisation of the estuarine soils. Throughout the deltaic Sundarban region, shrimp farms and mangrove-based tourism are meaningful livelihoods on a regional scale. In spite of these benefits, the forests have come under significant pressure from human activities and can no longer serve as a substantial carbon sink. Furthermore, the multiplication and modernisation of ports and harbours in coastal zones and river mouths (estuaries) have put additional pressure on blue carbon ecosystems, mainly mangroves (Banerjee et al. 2002, 2003, 2013).

Against this background, this chapter explores the trend of surface air temperatures, near-surface atmospheric CO_2 concentrations, surface water salinity, and surface water pH in the mangrove-dominated deltaic region of the Indian Sundarbans. To this end, the present research analysed 37 years of data (1984–2020) across three distinct zones, namely the western, central, and eastern sectors. In addition, it used a nonlinear autoregressive (NAR) model and artificial intelligence (AI) technology to forecast dominant trends for these four environmental indicators until the year 2050. Recommendations for climate change mitigation and adaptation are also proposed.

Research Rationale

The Indian Sundarban region has been experiencing considerable population pressure, with currently 4.4 million people inhabiting an area of 9,630 sq. km. Moreover, rapid urbanisation, increased fishing activities, and the proliferation of shrimp farms, brick kilns, fish landing stations, and tourism units have combined with increases in salinity in the region to put additional pressure on mangroves (Fig. 13.2). This situation makes it beneficial to better understand how the interplay of human activities and climate change impacts this unique environment.

Aims and Objectives of This Study

To generate and synthesise new knowledge, this study monitored the footprints of climate change using the following environmental indicators:

a. Surface air temperature
b. Near-surface atmospheric CO_2 levels
c. Surface water salinity
d. Surface water pH.

To do so, it queried and analysed 37 years of data from a database (1984–2020). This data is used as baseline information in this chapter.

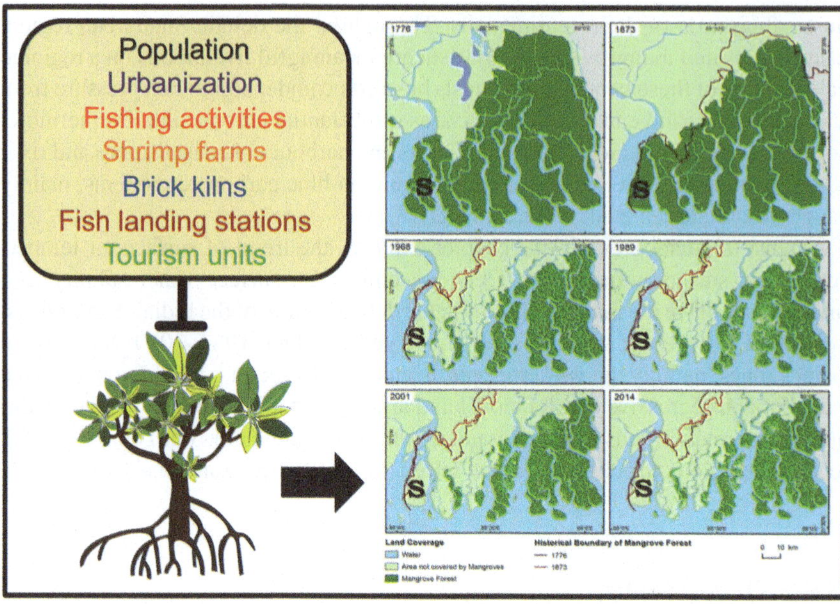

Fig. 13.2 Responsible anthropogenic factors for the temporal changes in the mangrove vegetation of Indian Sundarbans. "S" represents Sagar Island (336 sq. km), the largest island in the Indian Sundarban region. In this article, it serves as a symbolic representation that highlights the significant temporal variations affecting this blue carbon study site. Modified after Ghosh et al. (2015) (*Draft and cartography* S., Schmidt, T., Fickert, A., Ghosh & M., Nüsser. *Sources* 1. Maps Rennell (1780), Statistical Account of Bengal (1875). 2. Satellite Images Corona 1968/02/06, Landsat 5 1989/01/19, Landsat 7 2001/01/04, Landsat 8 2014/03/05 [USGS])

Materials and Methods

Selection of the Study Sites

The Indian Sundarban deltaic region covers 9,630 sq. km of territory and comprises 104 islands. Three sampling stations were identified for data collection, one in each of the three sectors, namely the western, central, and eastern sectors (Table 13.1 and Fig. 13.3). These three sectors of the Indian Sundarbans were demarcated based on a secondary data bank on surface water salinity (Mitra et al. 2009a, b, 2015; Mitra 2013; Sengupta et al. 2013; Mitra and Zaman 2021). The selected stations [Kakdwip (Stn. 1) (21° 52′ 22.68″ N; 88° 11′ 58.61″ E), Canning (Stn. 2) (22° 19′ 03.20″ N; 88° 41′ 04.43″ E), and Chandkhali (Stn. 3) (21° 51′ 13.59″ N; 89° 00′ 44.68″ E)] are considered in this study as the representatives of the three sectors of the Indian Sundarbans (Table 13.1; Fig. 13.3). The map showing the location of the selected sites is presented in Fig. 13.3. The main reasons for the selection of these sampling stations are based on observations that reflect significant variations in respect of (1) anthropogenic influences, (2) aquatic salinity, and (3) density and distribution of mangroves (Table 13.1).

Table 13.1 List of selected sampling sites in the study area along with their geographical coordinates and major drivers

Sector	Degree of aquatic salinity	Station name and number	Geographical coordinates	Major drivers				
				Fish landing	Tourism	Brick kilns	Shrimp farming	Mangrove vegetations
Western	Hyposaline	Kakdwip (Stn. 1)	21° 52′ 22.68″ N; 88° 11′ 58.61″ E	+++	+++	+++	+++	+
Central	Hypersaline	Canning (Stn. 2)	22° 19′ 03.20″ N; 88° 41′ 04.43″ E	++	++	+	++	+
Eastern	Hyposaline	Chandkhali (Stn. 3)	21° 51′ 13.59″ N; 89° 00′ 44.68″ E	–	+	–	–	+++

Activities with the highest magnitude (+++); activities with moderate magnitude (++); activities with the lowest magnitude (+); absence of any activities (–); "Stn." denotes a station

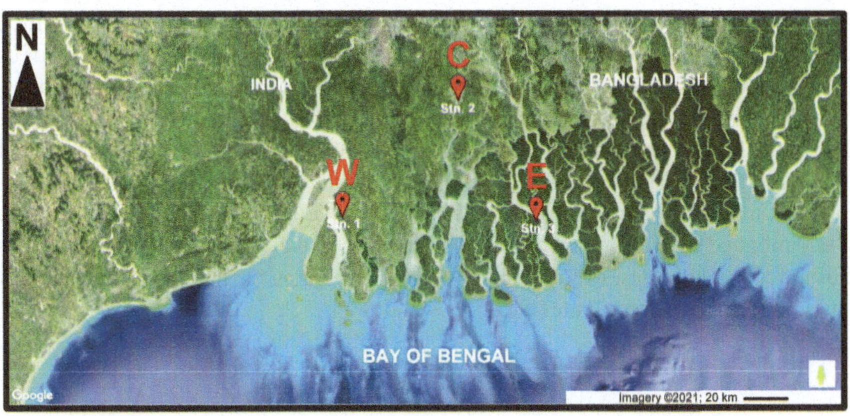

Fig. 13.3 Satellite map of the study area. Location of sector-wise sampling stations in Indian Sundarbans; the dark green-coloured clustered patch indicates the mangrove vegetation. Stn. 1 reflects Kakdwip [in the hyposaline western (W) sector], Stn. 2 reflects Canning [in the hypersaline central (C) sector], and Stn. 3 reflects Chandkhali [in the hyposaline eastern (E) sector]. "N" denotes north (direction). The map was created using the online Map Maker software (https://maps.co/gis/) (© 2021 Google)

Environmental Parameter Analyses in Selected Stations

Analyses of the environmental parameters were carried out at the selected Indian Sundarban stations, as detailed below. The data collections were performed every year (1984–2020) in three fixed months, namely during May (pre-monsoon; summer season), September (monsoon; rainy season), and December (post-monsoon; winter season). The data presented in this study reflect the mean of three seasons for every year for every sampling station. It should also be noted that both data collection and analytical methods did not change from 1984 to 2020.

Analysis of Surface Air Temperature

The surface air temperature (in °C) was analysed using a sensor-based thermometer. The readings were taken from ten different sampling spots (10 m apart) for each sampling station, and the average values of ten readings were considered for further interpretation.

Analysis of Near-Surface Atmospheric Carbon Dioxide

CO_2 concentrations (in ppm) were measured at 18 m above ground at Kakdwip (western sector), Canning (central sector), and Chandkhali (eastern sector) with a non-dispersive infra-red (NDIR) detector or gas analyser (LI6262, LI-COR Inc., USA). An instrument with an air intake system was installed at 1 m above the roof of the experimental building. A diaphragm pump was used for drawing the ambient air through a ¼ inch Teflon tube at a rate of 10 l min^{-1}, with maximum air being vented. The water vapor was removed from the drawn air sample. Then, the dry sampled air was introduced into the NDIR analyser's sample cell, and the output voltage of the gas analyser was integrated at an interval of five minutes. The NDIR analyser was calibrated at every four-hour interval by successively inserting (injecting) four calibrated working gases (340, 380, 410, and 450 ppm CO_2 in dry air) into the gas analyser's cell for five minutes each.

Analysis of Surface Water Salinity

The surface water salinity (in psu) was measured using an optical refractometer (Atago, Japan) and cross-checked in the laboratory following Mohr-Knudsen's method (Strickland and Parsons 1972). The correction factor was determined by titrating a silver nitrate ($AgNO_3$) solution against International Standard Sea Water (I.A.P.S.O. Standard Sea Water Service, Charlottenlund Slot, Denmark; chlorinity = 19.376‰). Real-time data through field sampling were simultaneously collected from 1984 to 2020 from the three selected sites in the lower Gangetic region during high tide conditions. For each site, at least 30 samplings were done at a distance of 500 m from each other, and then the average of 30 readings was considered for further statistical analyses.

Analysis of Surface Water pH

The systematic sampling to determine the surface water pH in the estuarine waters of the three Indian Sundarban sectors was done during the high tide period in the three selected stations for 37 consecutive years (1984–2020). A portable pH meter (Model BST-BT65; sensitivity = ± 0.01) was used to measure the pH of the surface water. Twenty-five readings were noted from each station, and their average was considered for further statistical interpretations.

Predicting Environmental Parameter Trends

The NAR model is a basic form of AI technology that can predict the overall trend of an environmental parameter using time-series data (Chow and Leung 1996). In this study, the NAR model was used for the time-series prediction or forecasting of the four environmental indicators mentioned earlier as described in Bhattacharya and Sinha 2021. It consists of an input layer, hidden layers (where the algorithm is trained), and an output layer. The general architecture of a representative neural network training is highlighted in Fig. 13.4. For the present study, a data bank of 35 years (1984–2018) was used to train the model. In contrast, the data sets from 2019 and 2020 were used to validate the trained model. The NAR model was used to forecast the data until the year 2050; this futuristic approach was adopted to envisage appropriate adaptive and mitigative measures for climate change.

Statistical Analyses

Two-way analysis of variance (ANOVA) was employed to assess variations in air temperature, atmospheric CO_2, aquatic salinity, and aquatic pH across the three

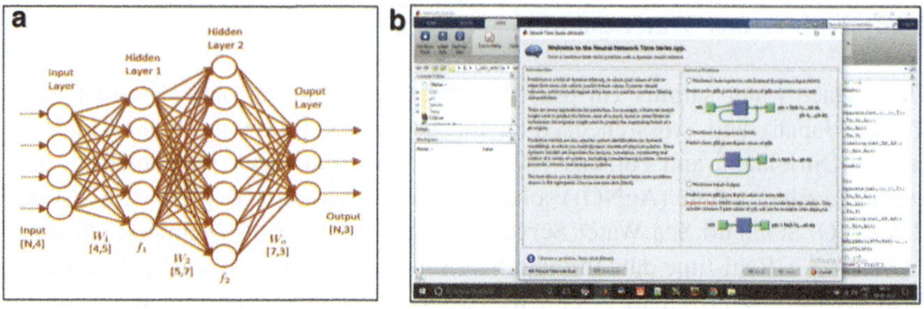

Fig. 13.4 Time-series prediction of the environmental indicators using a discrete, nonlinear autoregressive (NAR) model. **a** Illustration of a basic neural network, which mimics the neural network of a human brain and comprises an input layer, one or more hidden layers (where the algorithm is trained), and an output layer. **b** Representative screenshot of the "Neural Network Time Series" app for evaluating the trend of air temperature, atmospheric carbon dioxide (CO_2), aquatic salinity, and aquatic pH in the three Indian Sundarban sectors. The NAR model was used to predict trends until 2050 to support effective mitigation and adaptation measures

sectors and years; a *p*-value of <0.001 was considered significant. Correlation coefficients (r values) were determined to elucidate the interrelationships between the environmental parameters investigated in the study area. GraphPad Prism version 9.2.0 (283) (San Diego, California, USA) and Microsoft Excel were used to prepare the linear graphs. The scatter plots were made in Microsoft Excel with linear regression lines, and R-squared values were calculated for each plot to evaluate the linearity of the respective trend lines and their significance. SPSS version 21.0 (for Windows) (SPSS Inc., USA) was used for the statistical analyses.

Results

Data Analyses of Environmental Indicators and Emergent Trends

Surface Air Temperature

In the present study, the range of surface air temperature varies from 32.30 °C to 36.40 °C in the western sector, 32.10 °C to 36.20 °C in the central sector, and 32.20 °C to 35.70 °C in the eastern sector, as shown in Fig. 13.5. It is observed that the surface air temperature increases over time, irrespective of the sectors. A sudden increase and drop in surface air temperature can be observed in 2009 and 2020, respectively, for all three stations considered in this study.

Relatedly, the overall mean air temperature in the western, central, and eastern sectors are given to be 33.74 °C, 33.40 °C, and 33.44 °C, respectively, showing the highest value in the western sector compared to the central and eastern sectors of the

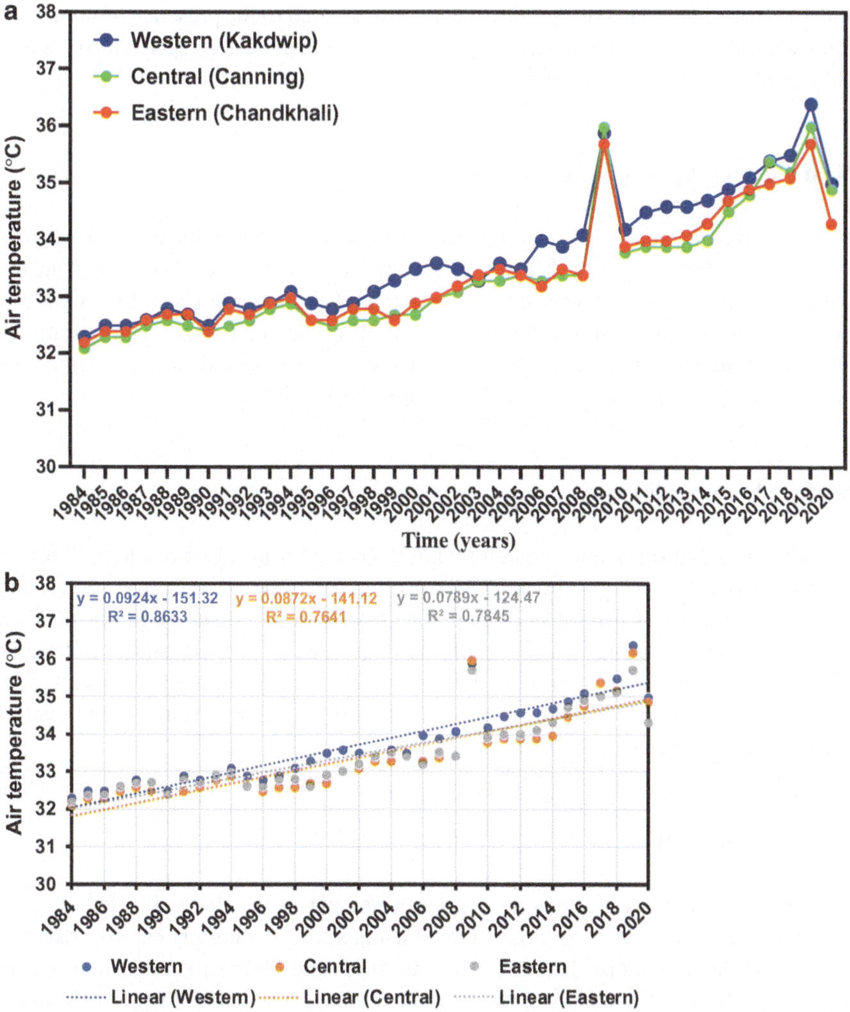

Fig. 13.5 Spatio-temporal variations in surface air temperature in the three Indian Sundarban sampling stations. **a** Sector-wise linear plot where each closed circle (blue, green, and red) represents the mean value of three seasons [pre-monsoon (summer season), monsoon (rainy season), and post-monsoon (winter season)] for every year for each sampling station. **b** Scatter plot with linear regression lines showing linearity of the trend lines

Indian Sundarbans. This dissimilarity may be attributed to the presence of numerous industries along the bank of the Hooghly estuary in the western sector of the study area (Mukhopadhyay et al. 2006).

Near-Surface Atmospheric Carbon Dioxide

It is observed that near-surface atmospheric CO_2 concentrations have increased from 348.38 to 414.97 ppm, 345.07 to 406.22 ppm, and 342.87 to 400.93 ppm in the western, central, and eastern sectors, respectively, as shown in Fig. 13.6. A sudden decrease in near-surface atmospheric CO_2 was observed in 2020 for all three stations. The overall mean atmospheric CO_2 levels in the western, central, and eastern sectors are 379.58 ppm, 374.90 ppm, and 372.17 ppm, respectively.

Surface Water Salinity

It is observed that surface water salinity has decreased from 11.65 psu to 3.27 psu in the western sector, increased from 20.02 psu to 26.86 psu in the central sector, and decreased from 19.41 psu to 9.12 psu in the eastern sector, as shown in Fig. 13.7. A sudden increase in surface water salinity was observed in 2009 for all three stations. The overall mean aquatic salinity in the western, central, and eastern sectors are 7.38 psu, 22.64 psu, and 15.91 psu, respectively, with the highest value in the central sector, followed by the eastern and western sectors.

Surface Water pH

The spatio-temporal variations of surface water pH in the three selected stations are presented in Fig. 13.8. A trend of decreasing surface water pH can be observed in all three stations until the year 2019. In 2009 and 2020, a sudden increase in aquatic pH values was observed for all three stations. The surface water pH ranges from 8.26 to 8.10, 8.34 to 8.24, and 8.30 to 8.20 in the western, central, and eastern sectors, respectively. The decline in aquatic pH values is more pronounced in the western sector (overall mean = 8.22), followed by the eastern sector (overall mean = 8.26) and central sector (overall mean = 8.31). The ANOVA result also confirms the spatio-temporal variation in pH values (Table 13.2 in the following section).

Analysis of Variance Showing Spatio-temporal Variations

Considerable variations are observed in the ANOVA results of all the environmental parameters between sectors (except for air temperature) and between years (except for aquatic salinity), as shown in Table 13.2.

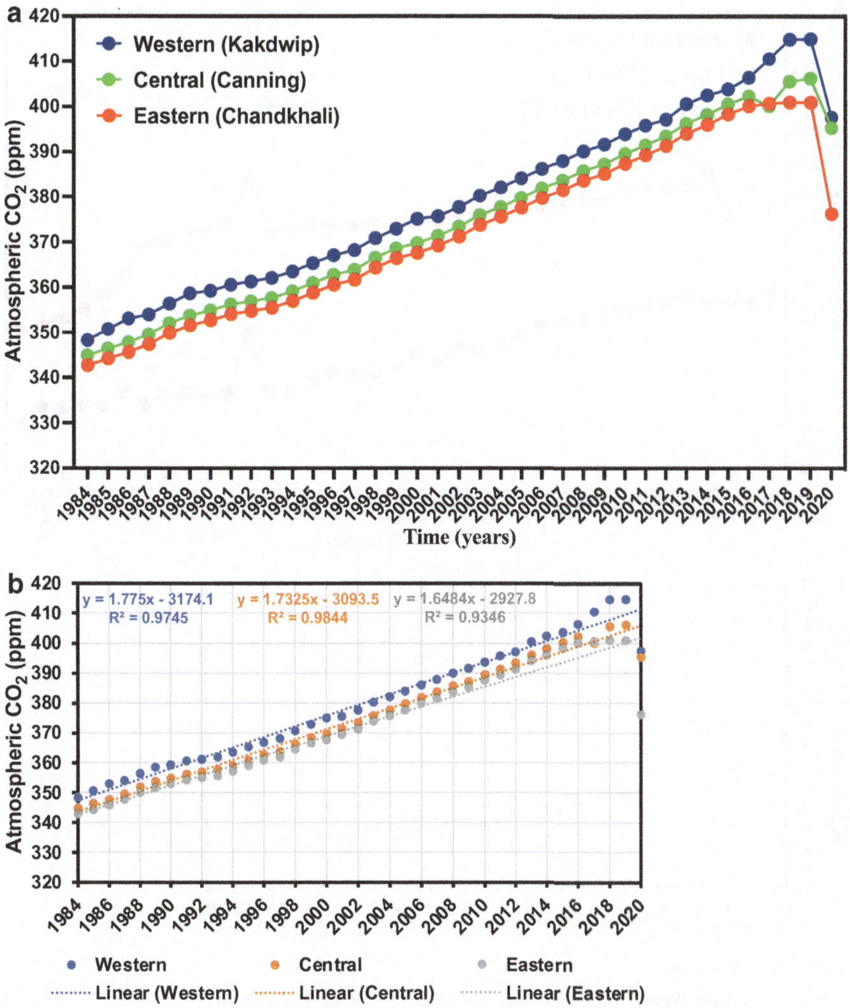

Fig. 13.6 Spatio-temporal variations in near-surface atmospheric carbon dioxide concentrations in the three Indian Sundarban sampling stations. **a** Sector-wise linear plot where each closed circle (blue, green, and red) represents the mean value of three seasons [pre-monsoon (summer season), monsoon (rainy season), and post-monsoon (winter season)] for every year for each sampling station. **b** Scatter plot with linear regression lines showing linearity of the trend lines

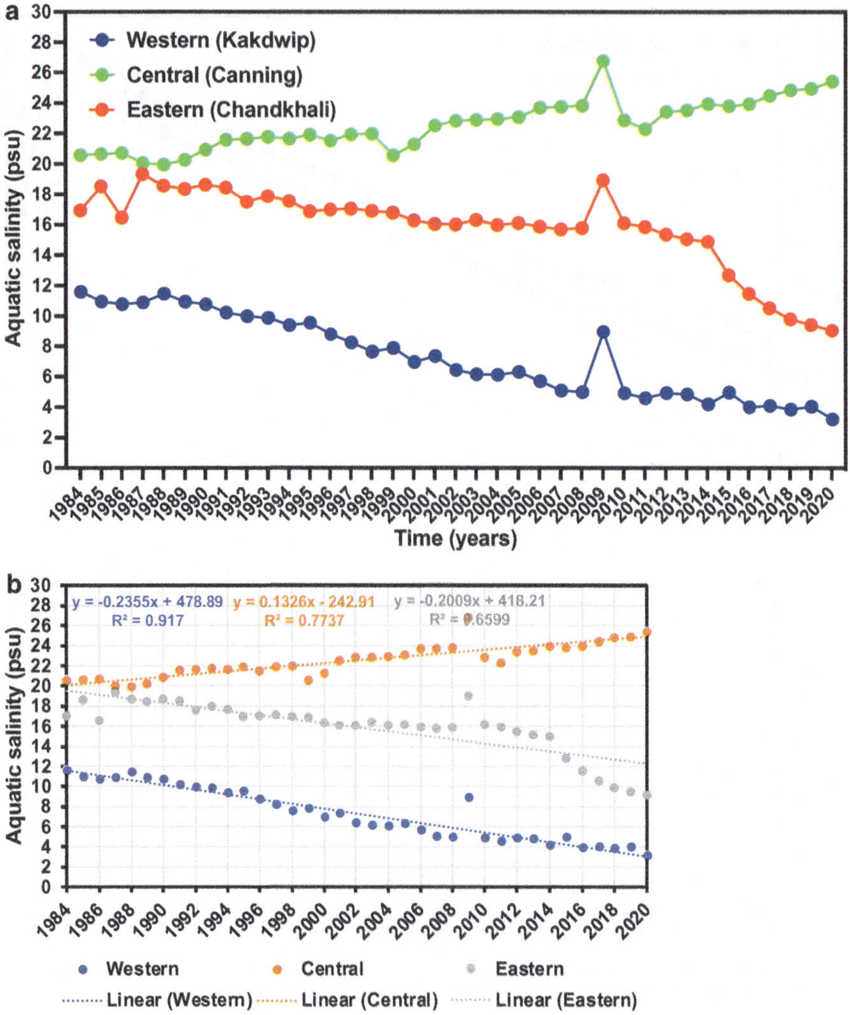

Fig. 13.7 Spatio-temporal variations in surface water salinity in the three Indian Sundarban sampling stations. **a** Sector-wise linear plot where each closed circle (blue, green, and red) represents the mean value of three seasons [pre-monsoon (summer season), monsoon (rainy season), and post-monsoon (winter season)] for every year for each sampling station. **b** Scatter plots with linear regression lines showing linearity of the trend lines

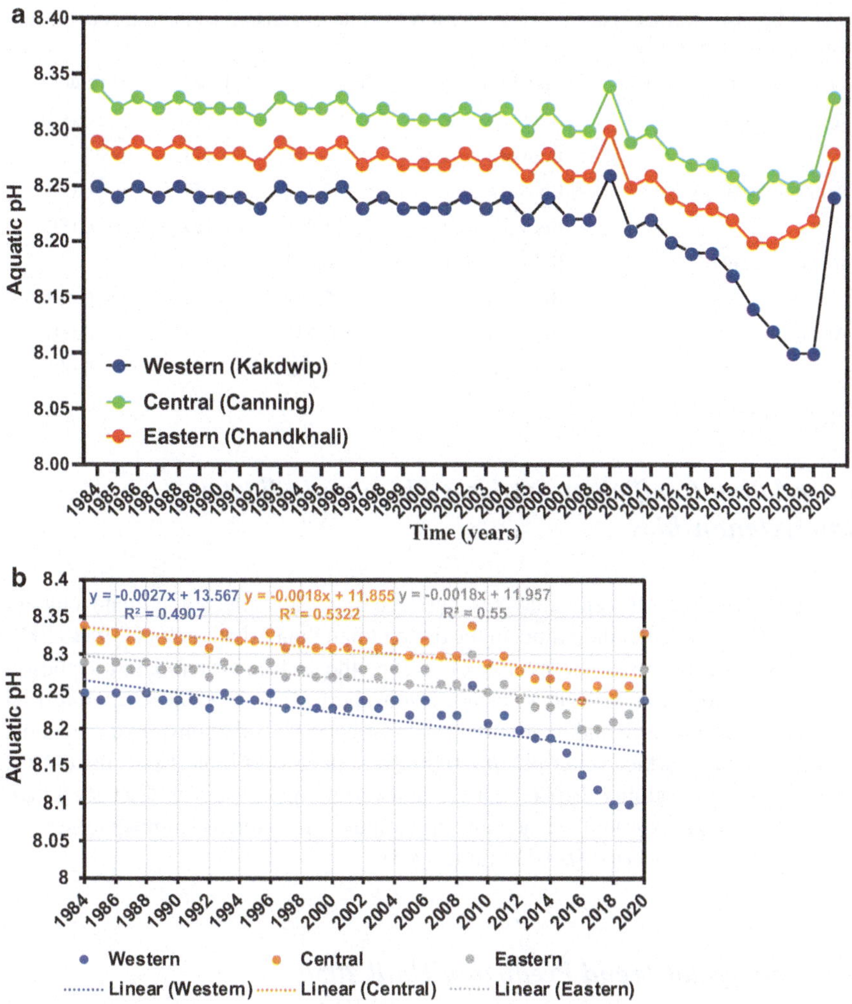

Fig. 13.8 Spatio-temporal variations in surface water pH in the three Indian Sundarban sampling stations. **a** Sector-wise linear plot where each closed circle (blue, green, and red) represents the mean value of three seasons [pre-monsoon (summer season), monsoon (rainy season), and post-monsoon (winter season)] for every year for each sampling station. **b** Scatter plots with linear regression lines showing linearity of the trend lines

Table 13.2 Two-way analysis of variance (ANOVA) of the environmental indicators in the three study sectors (1984–2020)

Environmental parameters	Variables	F_{cal}	F_{crit}	*P*-value
Air temperature	Between years	98.709	1.757	$p < 0.001$
	Between sectors	1.318	4.121	NS
Atmospheric CO_2	Between years	155.542	1.757	$p < 0.001$
	Between sectors	31.908	4.121	$p < 0.001$
Aquatic salinity	Between years	0.257	1.757	NS
	Between sectors	105.281	4.121	$p < 0.001$
Aquatic pH	Between years	205.632	1.757	$p < 0.001$
	Between sectors	4422.895	4.121	$p < 0.001$

NS Not significant

Correlation Coefficient Analyses Showing Indicator Interrelationships

The correlation coefficient analyses between the four selected climate change-oriented environmental indicators in the three Indian Sundarban sectors (1984–2020) are shown in Table 13.3. Irrespective of sectors, the surface air temperature showed a positive correlation with near-surface atmospheric CO_2 ($p < 0.001$), while both surface air temperature and near-surface atmospheric CO_2 demonstrated a negative correlation with surface water pH ($p < 0.001$ in both cases) (Table 13.3). In contrast, the surface water salinity displayed a sectoral variation in the correlation coefficient (positive or negative) with surface air temperature, near-surface atmospheric CO_2, and surface water pH ($p < 0.001$) (Table 13.3).

Environmental Trend Prediction Until 2050

As noted above, whereas data sets from 1984 to 2018 were used to train the NAR model, data sets from 2019 and 2020 were used to validate the trained model.

Prediction of Surface Air Temperature

A future trend of rising surface air temperatures is visualized in Fig. 13.9. The regression curves derived from regression analyses (training, validation, and test) are provided as Annexure 1.

Table 13.3 Correlation coefficients of the environmental indicators in the three study sectors (1984–2020)

Sl. No.	Correlation variables	Sectors	r value	*P*-value
1	Surface air temperature × Near-surface atmospheric CO_2	Western	0.9440	$p < 0.001$
		Central	0.8780	$p < 0.001$
		Eastern	0.8903	$p < 0.001$
2	Surface air temperature × Surface water salinity	Western	−0.8105	$p < 0.001$
		Central	0.8938	$p < 0.001$
		Eastern	−0.7229	$p < 0.001$
3	Surface air temperature × Surface water pH	Western	−0.7342	$p < 0.001$
		Central	−0.6313	$p < 0.001$
		Eastern	−0.7027	$p < 0.001$
4	Near-surface atmospheric CO_2 × Surface water salinity	Western	−0.9388	$p < 0.001$
		Central	0.8639	$p < 0.001$
		Eastern	−0.7366	$p < 0.001$
5	Near-surface atmospheric CO_2 × Surface water pH	Western	−0.7798	$p < 0.001$
		Central	−0.7834	$p < 0.001$
		Eastern	−0.8170	$p < 0.001$
6	Surface water salinity × Surface water pH	Western	0.6574	$p < 0.001$
		Central	−0.4931	$p < 0.001$
		Eastern	0.7536	$p < 0.001$

$p < 0.001$ = Statistically significant

Prediction of Near-Surface Atmospheric Carbon Dioxide

A future trend of increasing near-surface atmospheric CO_2 concentrations is observed in Fig. 13.10. The regression curves derived from regression analyses (training, validation, and test) are provided as Annexure 2.

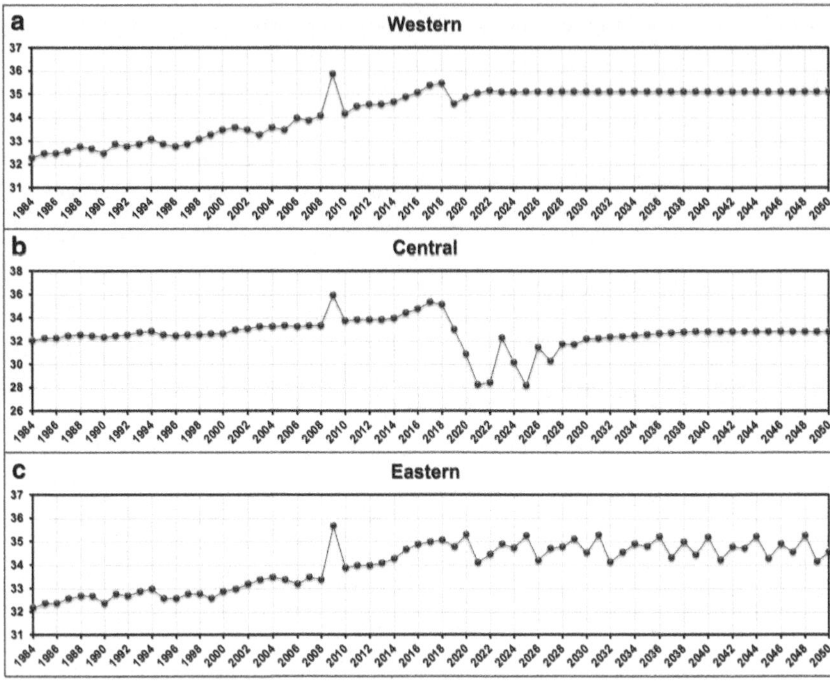

Fig. 13.9 Predicted surface air temperatures until 2050. A nonlinear autoregressive (NAR) model is used for trend prediction. The x-axis denotes time in years, while the y-axis denotes air temperature in °C

Prediction of Surface Water Salinity

A future trend of decreasing surface water salinity is observed in both the western and eastern sectors. In contrast, an increasing trend is observed in the central Indian Sundarbans due to siltation-linked freshwater cut-off (Fig. 13.11). The regression curves derived from regression analyses (training, validation, and test) are provided as Annexure 3.

Prediction of Surface Water pH

A future trend of decreasing surface water pH is observed in Fig. 13.12. The regression curves derived from regression analyses (training, validation, and test) are provided as Annexure 4.

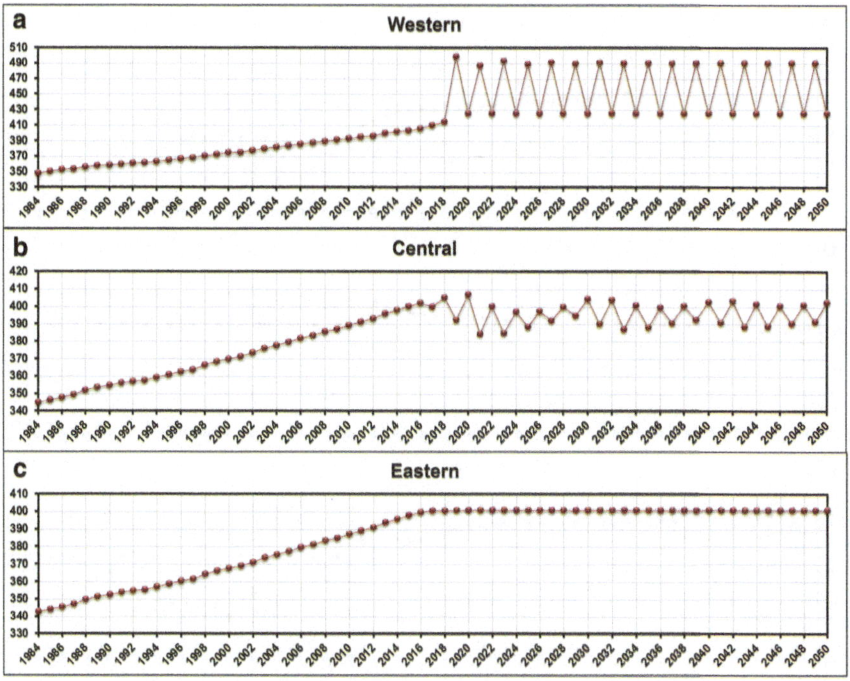

Fig. 13.10 Predicted near-surface atmospheric carbon dioxide (CO_2) concentrations until 2050. A nonlinear autoregressive (NAR) model is used for trend prediction. The x-axis denotes time in years, while the y-axis denotes atmospheric CO_2 in ppm

Discussion

Recent climate change impacts have negatively affected biodiversity, weather, rainfall, atmospheric temperature, water quality, sea levels, and even human health. For example, a recent study by Mukherjee et al. (2021) showed that climate change-induced rises in sea level and concomitant increases in salinity in the Matla estuary (located in the hypersaline central Indian Sundarbans) have had a deleterious impact on the commercially important variety of ichthyoplankton (fish eggs and larvae). The proxies used in our research (namely surface air temperature, near-surface atmospheric CO_2, surface water salinity, and surface water pH) are discussed here separately and consecutively to critically examine different climate change impacts affecting the present study area.

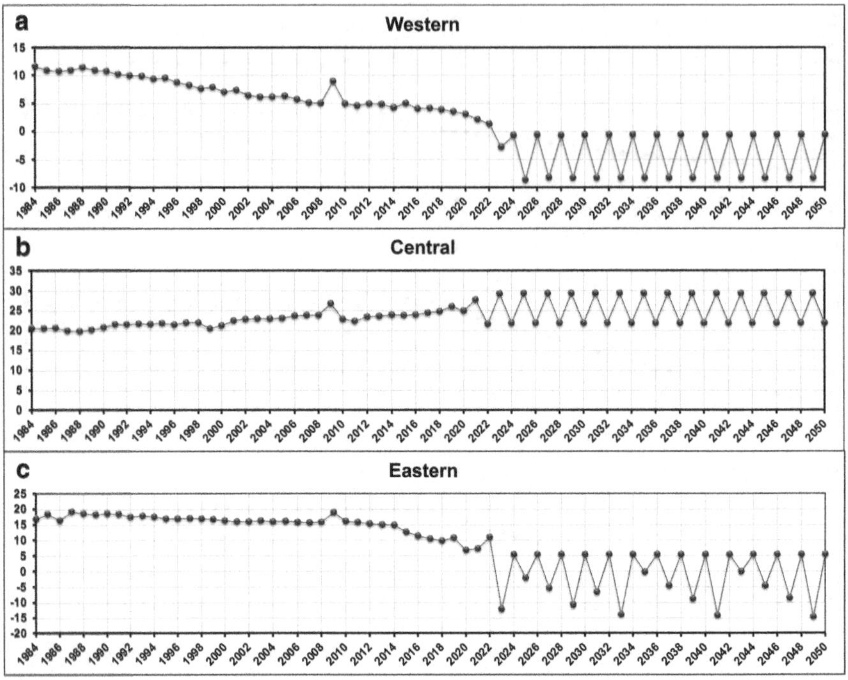

Fig. 13.11 Predicted surface water salinity until 2050. A nonlinear autoregressive (NAR) model is used for trend prediction. The x-axis denotes time in years, while the y-axis denotes aquatic salinity in psu

Surface Air Temperature

Air temperature serves as a proxy for climate change as it alters several variables related to regional weather, such as the rate of evaporation, relative humidity, types and patterns of precipitation, and wind direction or speed (IPCC 2007, 2022). As such, air temperature is a function of the intensity and amount of the sunlight directly heating the Earth's surface. Moreover, air temperature is regulated by atmospheric conditions like cloud cover and humidity, which trap incoming heat energy from the sun. Furthermore, it has been well established that GHGs being heavy, form a blanket over the Earth's crust, which absorbs and traps the sun's heat. Humans are contributing different GHGs to the atmosphere, and the effect of each gas on climate change is based on three primary factors, namely the amount (quantity) of GHG added, the period (extent) of GHG addition, and the capacity of the added GHG in terms of its warming potential. CO_2 is one of the major GHGs emitted by vehicles, industries, thermal power plants, etc., that absorbs the heat and traps it in the air (Earth Observatory 2018), thus raising air temperatures (Leal Filho et al. 2021a).

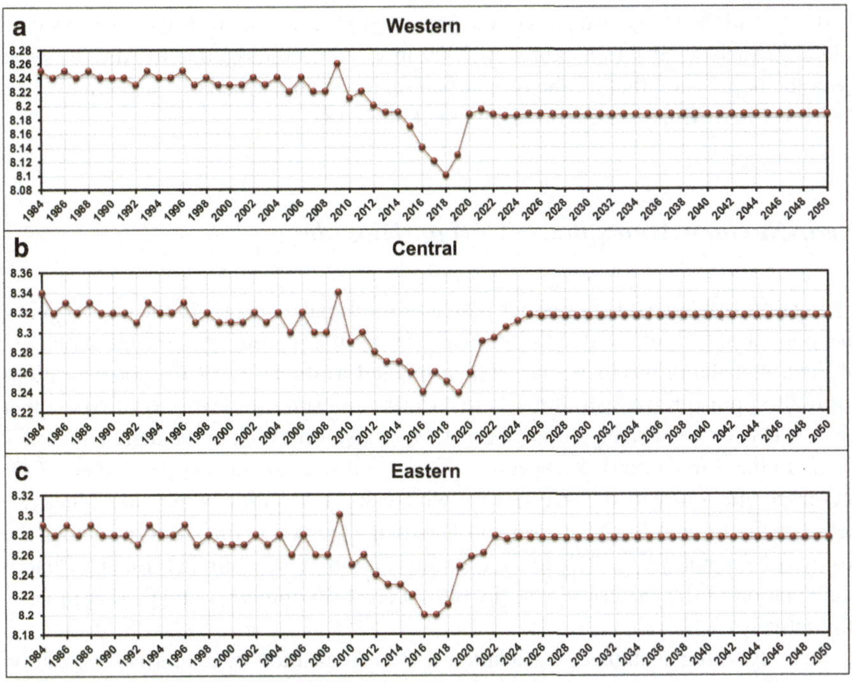

Fig. 13.12 Predicted surface water pH until 2050. A nonlinear autoregressive (NAR) model is used for trend prediction. The x-axis denotes time in years, while the y-axis denotes aquatic pH

While global average temperatures have already increased by more than 1 °C since 1850, it is noteworthy that global warming has appreciably accelerated since 1975, with global average temperatures rising by about 0.2 °C per decade since (Hansen et al. 2006; IPCC 2022). Given that water absorbs and releases heat more slowly (thermal inertia) than soil, warming over land tends to be more pronounced than warming over the oceans (Mitra et al. 2013).

The Indian Sundarban ecosystem is noted for its rich blue carbon propensity. Temporal variation of surface air temperature in the Sundarbans has been observed to range from 11.96 °C to 37.0 °C (Pitchaikani et al. 2017). This mangrove-dominated deltaic complex experiences maximum air temperatures during the pre-monsoon season, followed by the monsoon and post-monsoon seasons (Mitra and Zaman 2014, 2015, 2016).

In the current study, we found that irrespective of study sites, the surface air temperature increased uniformly in all three sectors, with a rate of increase of 0.073 °C yr^{-1}, 0.076 °C yr^{-1}, and 0.057 °C yr^{-1} in the western (Kakdwip), central (Canning), and eastern (Chandkhali) sectors, respectively. This uniform increase may be attributed to land-use changes, which are most likely caused by the massive-scale cutting of mangroves for shrimp culture establishment, industrial activities, urban developments, the proliferation of tourism, and the construction of fish landing stations in all the mangrove belts of the present study area. Interestingly, we observed a hike in surface air temperatures in 2009 in all three sectors due to Aila, a severe cyclonic depression that hit the Sundarban region with a speed of ~120 km hr^{-1} on 25 May 2009 (Mitra et al. 2011; Kelman et al. 2018).

Near-Surface Atmospheric Carbon Dioxide

Rising atmospheric CO_2 is another indicator of climate change. In the last 60 years, the amount of CO_2 emitted in the atmosphere was mainly owing to the increased use of coal, petroleum products, etc., which has led to the rise of CO_2 concentrations from 280 ppm (pre-industrial level) to 415.01 ppm (present level) as of November 2021.[2] This trend of rising atmospheric CO_2 is also observed in the deltaic complex of the Indian Sundarbans. Irrespective of study sites, near-surface atmospheric CO_2 increased uniformly in all three sectors, with a slight deviation in the rate of increase between the three selected stations. In Kakdwip (located in the western sector), the rate was 1.33 ppm yr^{-1}, while in the Canning station of the central sector and Chandkhali station of the eastern sector, the rates were 1.36 ppm yr^{-1} and 0.90 ppm yr^{-1}, respectively.

Strikingly, the sustained degradation and deforestation of mangrove vegetation for human settlement, industrial activities, tourism development, timber production, and shrimp culture farms (except in Chandkhali in the eastern sector, which is a restricted Reserve Forest zone), along with copious other human activities, have massively reduced the carbon sequestration capacity of the ecosystem while significantly diminishing its potential to act as a sink for future CO_2 emissions. This has been exacerbated by the release of GHGs from industries, brick kilns, and other sources that have cumulatively added substantial amounts of CO_2 to the atmosphere.

Surface Water Salinity

The Indian subcontinent is presently under the appreciable influence of climate change-induced sea-level rises, with India being listed among the 27 most vulnerable countries in the world (Mitra et al. 2013). In the Indian subcontinent, the footprints of climate change are felt across several ecosystems, including glaciers, estuaries, bays, and open oceans. The mangrove-dominated Indian Sundarban region at the apex of the Bay of Bengal in the lower Gangetic delta has been experiencing the impacts of climate change for a long time (Mitra et al. 2016). There are reports of rising sea levels in the Indian Sundarbans at a rate of 3.14 mm yr^{-1} (Hazra et al. 2002). Sea-level rise on this scale results in a significant increase in salinity due to saline water intrusion. However, this universal rule is not followed uniformly across

[2] https://www.co2.earth/annual-co2.

the Indian Sundarbans due to contrasting geomorphological features between the western, central, and eastern sectors. Long-term data on surface water salinity indicates a bell-shaped salinity profile in the Indian Sundarban region. While the central sector exhibits high levels of salinity, this area finds itself straddled between the two hyposaline sectors of western and eastern Indian Sundarbans (Trivedi et al. 2016).

The low-saline environment of the western Indian Sundarbans may be caused by freshwater inputs from the Farakka barrage situated upstream of the Ganga-Bhagirathi-Hooghly River system (Mukhopadhyay et al. 2006). A 10-year survey (1999–2008) on freshwater discharge from the Farakka barrage revealed a mean discharge of $(3.70 \pm 1.15) \times 10^3\ m^3\ s^{-1}$. Higher discharge values were recorded during the rainy season, with a mean of $(3.81 \pm 1.23) \times 10^3\ m^3\ s^{-1}$, and the highest discharge of 4,524 $m^3\ s^{-1}$ was observed during freshet (i.e., the rush of fresh water flowing into the sea because of a flood of a river from melted snow or heavy rainfall) (September). Significantly lower discharge values were recorded during the summer season, with a mean of $(1.18 \pm 0.08) \times 10^3\ m^3\ s^{-1}$, and the lowest discharge of 846 $m^3\ s^{-1}$ was observed during May. During the winter season, discharge values were moderate, with a mean of $(1.98 \pm 0.97) \times 10^3\ m^3\ s^{-1}$, as recorded by earlier research (Mitra et al. 2013).

The central sector represents a region of high salinity owing to the complete blockage of freshwater inflows from the upstream region because of siltation in the Bidyadhari River since the late fifteenth century (Chaudhuri and Choudhury 1994; Mitra et al. 2013; Sengupta et al. 2013). The Matla estuary in the central Indian Sundarbans is not ideally positioned given the lack of head-on discharge for dilution of the system with fresh water. The eastern sector of the Indian Sundarbans displays a low-saline aquatic phase, perhaps owing to interconnection with many creeks and channels from the tributaries of the Bangladeshi Sundarbans that arise from the Padma-Meghna River system. Such contrasting salinity within the same ecosystem confirms the impact of climate change-induced sea-level rise in the central sector. However, the effect is suppressed in the western sector due to barrage discharge.

In summary, aquatic salinity decreased in the western and eastern sectors, with a rate of 0.0054 psu yr^{-1} and 0.0116 psu yr^{-1}, respectively, due to an increase of dilution factor stemming from freshwater inputs from the Farakka barrage and trans-boundary effects from creeks, channels, and tributaries from rivers in Bangladesh (Padma and Meghna), respectively. Aquatic salinity is elevated in the central sector, with a rate of 0.0165 psu yr^{-1}, due to the clogging of the Bidyadhari River with a massive quantum of silt. A hike in aquatic salinity in 2009 may be due to salt-water intrusion from the adjacent bay caused by cyclone Aila (Mitra et al. 2011). In short, differences may be explained by divergent dilution factors and siltation-linked freshwater obstructions.

Surface Water pH

In the present study, the surface water pH decreased in the western, central, and eastern sectors, with a rate of around 0.00027 yr^{-1}, due to increases in CO_2 emissions from adjoining industries, brick kilns, fishing trawlers and vessels, tourist vehicles, etc. Once present in the ambient atmosphere, emitted CO_2 from these point sources mixes rapidly with surface estuarine water through the gaseous exchange, leading to estuarine acidification by forming carbonic acid (H_2CO_3), which eventually gives rise to hydrogen ions (H^+) that in turn lowers the surface water pH and makes the water acidic, as shown in Eq. 13.1 (IPCC 2005).

$$CO_{2\ (aqueous)} + H_2O \rightleftharpoons H_2CO_3 \rightleftharpoons CO_3^- + H^+ \rightleftharpoons CO_3^{2-} + 2H^+ \qquad (13.1)$$

where

$CO_{2\ (aqueous)}$ represents dissolved carbon dioxide;
H_2O represents estuarine water;
H_2CO_3 represents carbonic acid;
${CO_3}^{2-}$ and ${HCO_3}^-$ represent carbonate and bicarbonate ions, respectively;
H^+ represents hydrogen ion.

The sudden hike in aquatic pH values in 2009 was likely caused by cyclone Aila, which triggered the massive intrusion of saline water from the Bay of Bengal via increased precipitation and concomitant storm surge-induced sea-level rise (Mitra et al. 2009a, b).

Spatio-temporal Variations in the Selected Environmental Parameters

Two-way ANOVA, carried out with the selected environmental parameter data from the three study sites, exhibits significant variations between the stations and years in the case of near-surface atmospheric CO_2 and surface water pH ($p < 0.001$). In the case of surface air temperature, considerable variation was observed between years but not between stations. On the other hand, in the case of surface water salinity, we did not observe any statistically significant variations between years, although pronounced variations were observed between stations. Our results show that spatial variations in salinity are noted in the three sectors of the Indian Sundarbans; however, it is not wise to relate these variations to climate change given the limited temporal variations between years. The opposite holds for surface air temperature. The other two parameters, namely near-surface atmospheric CO_2 and surface water pH, exhibited pronounced spatio-temporal variations. Hence, they can be used as proxies for climate change in the present geographical locale.

Correlation Between Selected Environmental Indicators

As depicted in Table 13.3, surface air temperatures showed a positive correlation with near-surface atmospheric CO_2 concentrations. Since CO_2 is a GHG that traps heat energy from the sun during the day and prevents its escape from the Earth during the night, incremental increases in atmospheric CO_2 concentrations lead to incremental rises in air temperatures. As also shown in Table 13.3, the surface water pH negatively correlated with both surface air temperature and near-surface atmospheric CO_2. This negative correlation can be attributed to the gaseous exchange of atmospheric CO_2 with surface estuarine water that causes a lowering of surface water pH due to the generation of H^+ ions from H_2CO_3, which is formed upon the dissolution of atmospheric CO_2 in water.

Environmental Indicators and Future Trends

As noted above, the NAR model was trained using 35 years of data (1984–2018) for predicting future trends to 2050. Validation of the neural network was done by comparing the sector-wise predicted values of the selected environmental indicators with the real-time observed values for two consecutive years, 2019 and 2020, as tabulated in Table 13.4. In most cases, the sector-wise predicted values showed only minor deviations from the real-time observed values for both years (i.e., predicted and real-time observed values are very close); however, there are some exceptions, primarily for near-surface atmospheric CO_2, as shown in Table 13.4 with a single asterisk or double asterisks. Taken together, our analysis suggests that the model accuracy can be considered high and the model performance satisfactory.

From the regression analyses, it was found that the sector-wise correlation coefficient (R) values, as shown in the regression curves, range from 0.87461 (Chandkhali) to 0.92767 (Kakdwip) (for surface air temperature), 0.97549 (Kakdwip) to 0.99978 (Chandkhali) (for near-surface atmospheric CO_2), 0.67959 (Canning) to 0.92665 (Kakdwip) (for surface water salinity), and 0.8349 (Canning) to 0.9459 (Kakdwip) (for surface water pH). The overall high correlation coefficient suggests good model performance and satisfactory prediction accuracy. Therefore, the NAR model can be used for future predictions of environmental parameters.

The reasons for fluctuations observed in the future predicted NAR plots in the cases of surface air temperature (eastern sector), near-surface atmospheric CO_2 (western and central sectors), and surface water salinity (all three sectors) could be due to the randomness of natural phenomena and inherent chaos of the environmental parameters being taken into consideration.

Table 13.4 Validation of the nonlinear autoregressive (NAR) model by comparing the sector-wise predicted values with the real-time observed values of selected environmental indicators for the years 2019 and 2020

Environmental parameters	Sectors	Year 2019			Year 2020		
		Predicted value	Real-time observed value	% Deviation	Predicted value	Real-time observed value	% Deviation
Surface air temperature	Western	34.60	36.40	−4.945	34.90	35.00	−0.286
	Central	33.07	36.20	−8.646	30.91	34.90	−11.433
	Eastern	34.80	35.70	−2.521	35.33	34.30	3.003
Near-surface atmospheric CO_2	Western	499.00**	414.97**	20.250	425.00**	397.61**	6.889
	Central	393.00*	406.22*	−3.254	407.00*	395.33*	2.952
	Eastern	401.00	400.91	0.022	401.00**	376.23**	6.584
Surface water salinity	Western	3.55	4.11	−3.625	3.14	3.27	−3.976
	Central	26.07	25.02	4.197	25.00	25.49	−1.922
	Eastern	10.85	9.50	14.211	6.95	9.12	−23.794
Surface water pH	Western	8.13	8.10	0.370	8.19	8.24	−0.607
	Central	8.24	8.26	−0.242	8.26	8.33	−0.840
	Eastern	8.25	8.22	0.365	8.26	8.28	−0.242

*/** Noticeable difference between predicted and real-time observed values

Impacts of COVID-19 Lockdowns on Environmental Indicators

The outbreak of severe acute respiratory syndrome coronavirus 2 (SARS-CoV-2) causing coronavirus disease (COVID-19), which was initially detected in Wuhan, China, in late December 2019, swiftly spread worldwide.[3] The effects were first felt in India in late March 2020. Although the pandemic was a bane for the global populace as it claimed millions of lives, research also synthesised some positive side effects that could be considered a boon for the environment (Leal Filho et al. 2021b). The impacts of COVID-19 were also noticed in the data collections and analyses that informed the current study. The sharp dip in air temperatures and atmospheric CO_2 concentrations and the steep rise in aquatic pH levels in 2020 in all three stations under investigation are synchronous with the COVID-19 pandemic and lockdown periods, which underpinned widespread reductions in traffic and industrial activities. In short, decelerated human activities lowered the atmospheric carbon footprints and air temperatures and increased estuarine acidification (i.e., dissolution of atmospheric CO_2 in the estuarine or brackish water of the Indian Sundarbans). Correspondingly, the dip in atmospheric CO_2 concentrations and air temperatures, and the rise in

[3] https://www.who.int/health-topics/coronavirus.

aquatic pH levels in 2020–2021, as observed in the NAR plots, can be attributed to COVID-19-induced lockdown periods, especially in the western and central sectors.

Conclusions

This section synthesises six key findings that may be concluded from this research on selected environmental indicators. They combine to highlight a clear anthropogenic climate change signal that is unmistakably observed in all three Indian Sundarban study sites:

1. Measurements over time exhibit a clear trend of increasing surface air temperatures in all three study sites.
2. Levels of near-surface atmospheric CO_2 concentrations are synchronous with surface air temperature measurements and exhibit a gradual increase over time.
3. Rising air temperatures and increasing CO_2 concentrations in all three Indian Sundarban sectors correlate with the massive deforestation of mangroves (a major carbon sink) in the area. Reasons for the degradation and decimation of mangroves are manifold and include progressive human encroachment and settlement, urbanisation, industrial development, the proliferation of shrimp farms, fish landing stations, and growth in tourism, among other factors. Emissions from fishing trawlers and vessels that ply in and around the present study sites, coupled with brick kiln emissions, also contribute an appreciable amount of CO_2 to the atmosphere, directly impacting both air temperature and air quality.
4. Significant differences between study sites are observed in surface water salinity. While the western and eastern sectors exhibit a decreasing trend, the central sector exhibits a rising trend. Differences may be attributed to increased run-offs and dilution factors in the western and eastern sectors and siltation in the central sector, respectively.
5. In contrast to air temperature and CO_2, surface water pH data reflect a decreasing trend in all three sectors over time, which casts a negative correlation between surface air temperatures and near-surface atmospheric CO_2 concentrations. This is understood to be caused by the gaseous exchange of atmospheric CO_2 with surface water that results in lowering the surface water pH due to the formation of H_2CO_3, which subsequently generates H^+ ions.
6. Finally, analyses of environmental parameters using NAR networks point to some alarming future trends. These encompass surface air temperatures, near-surface atmospheric CO_2 concentrations, and surface water pH levels in all three study sites, and surface water salinity levels principally in the central study site. The prediction model for each of the parameters forecasts that surface air temperature, near-surface atmospheric CO_2, surface water salinity, and surface water pH may change to 34.21 °C, 410 ppm, 8.97 psu, 8.26, respectively, by 2050 for the entire Indian Sundarbans (mean of three sectors).

Recommendations for Climate Change Mitigation and Adaptation

Current trends in the Indian Sundarban region reflect increasing near-surface atmospheric CO_2 concentrations, rising surface air temperatures, surging surface water salinity, and decreasing surface water pH levels. These trends are poised to dramatically alter the biotic community (flora and fauna) in this unique mangrove ecosystem of the Indian subcontinent. This final section synthesises critical steps that may be implemented to counteract the adverse impacts of climate change as observed from data drawn from the Kakdwip, Canning, and Chandkhali study sites. The study advances six key recommendations:

1. To effectively reduce GHG emissions from industries, brick kilns, and fishing trawlers, which are the principal emission sources of CO_2 and N_2O in this mangrove-rich ecosystem, requires stringent emission controls. Hence there is an urgent need for a mitigation management action plan that may achieve, regulate, and sustain extensive GHG reductions over time.
2. Relatedly, there is an urgent need to develop a large-scale nursery of blue carbon (mangrove saplings) for plantation. Given the significant and manifold ecosystem services that mangroves provide as natural shelterbelts against cyclones and storm surges, as nursery beds for fishes and habitat for a plethora of species and biotic communities, and as absorbents for toxic heavy metals, in addition to their dual capacity to sequester GHGs and down-regulate temperatures, the benefits of mangroves can hardly be overstated. This makes the sustainable development of blue carbon nursery one of the most promising strategies for long-term environmental conservation and remediation. As such, mangrove reforestation can simultaneously play a significant role in both mitigating climate change and helping local communities adapt to its effects (see Fig. 13.13).
3. Furthermore, considering that mangroves are known to have very effective disaster mitigating propensities, there is an opportunity to connect mangrove reforestation initiatives to disaster awareness, preparedness, resilience, risk reduction, and locally relevant disaster risk education.
4. The introduction of low-carbon alternative livelihoods can concurrently uplift the economic profile of island dwellers without harming the ambient climate. This may include the preparation of snacks, cookies, jellies, and other food products from mangroves and mangrove associate species (provided a backup nursery is created to maintain the supply chain), seaweed culture, and oyster culture, among others.
5. The practice of mangrove-based shrimp culture using probiotics can minimise artificial feed that adversely impacts the aquatic phase through the generation of organic load, H_2S, NH_3, etc.
6. Finally, rainwater harvesting practices should be upscaled in nearshore agricultural areas (preferably in the central Indian Sundarbans) to circumvent the problem of increasing water salinity.

Fig. 13.13 The Indian Sundarbans. Mangroves and mangrove associate species provide a variety of essential ecosystem services (Mitra 2020). Besides acting as a powerful carbon sink, mangroves prevent soil erosion, serve as a buffer against cyclonic storms and wave actions, provide nursery beds for fishes, crabs, oysters, etc., and offer habitat for a plethora of species and biotic communities, including microbes. Mangroves are also noted for their disaster mitigating propensities (Luetz 2008; Menéndez et al. 2020), in addition to their dual capacity to actively sequester greenhouse gases (GHGs) and down-regulate temperatures. This profusion of benefits makes the sustainable development of blue carbon nursery one of the most promising strategies for long-term climate change mitigation and adaptation (Photos taken by authors during field research)

Figure 13.14 shows how alternative development pathways can cause different levels of GHG emissions. It also identifies mitigation and adaptation as the two essential response strategies to climate change, namely (1) the magnitude of temperature rises can be lessened by reducing GHG emissions, and (2) the negative impacts of climate change that are already occurring may be minimised through adaptation measures that may both enhance the coping capacities of local communities while concurrently lowering their vulnerabilities (IPCC 2001; Menéndez et al. 2020; Doust et al. 2021; Leal Filho et al. 2021a; Sultana and Luetz 2022).

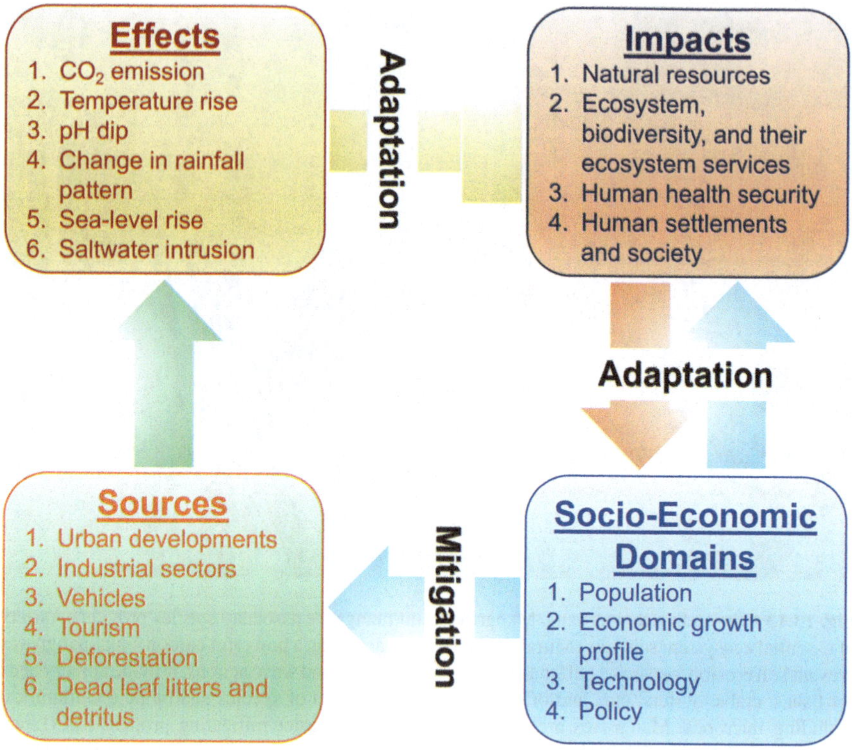

Fig. 13.14 An integrated framework of climate change. Sources, effects, impacts, and mitigation or adaptation of CO_2 emissions

Annexure 1

Regression analysis and sector-wise correlation coefficient values for surface air temperature using a nonlinear autoregressive (NAR) model.

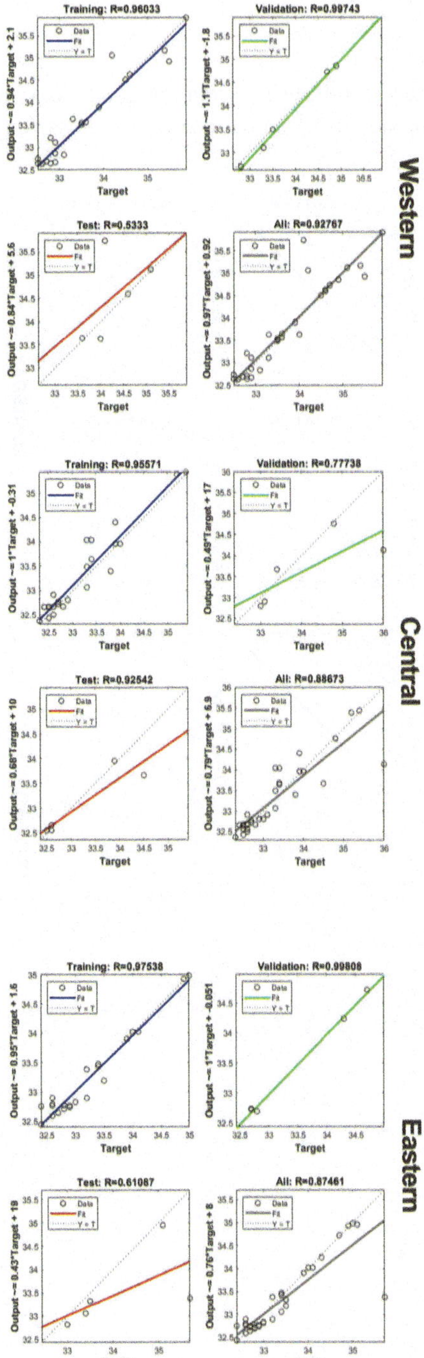
Training: R=0.96033
Output ~= 0.94*Target + 2.1
Validation: R=0.99743
Output ~= 1.1*Target + -1.8
Test: R=0.5333
Output ~= 0.84*Target + 5.6
All: R=0.92767
Output ~= 0.97*Target + 0.92
Western
Training: R=0.95571
Output ~= 1*Target + -0.31
Validation: R=0.77738
Output ~= 0.49*Target + 17
Test: R=0.92542
Output ~= 0.68*Target + 10
All: R=0.88673
Output ~= 0.79*Target + 6.9
Central
Training: R=0.97538
Output ~= 0.95*Target + 1.6
Validation: R=0.99808
Output ~= 1*Target + -0.051
Test: R=0.61087
Output ~= 0.43*Target + 19
All: R=0.87461
Output ~= 0.76*Target + 8
Eastern
Target
Data
Fit
Y = T

Annexure 2

Regression analysis and sector-wise correlation coefficient values for near-surface atmospheric carbon dioxide using a nonlinear autoregressive (NAR) model.

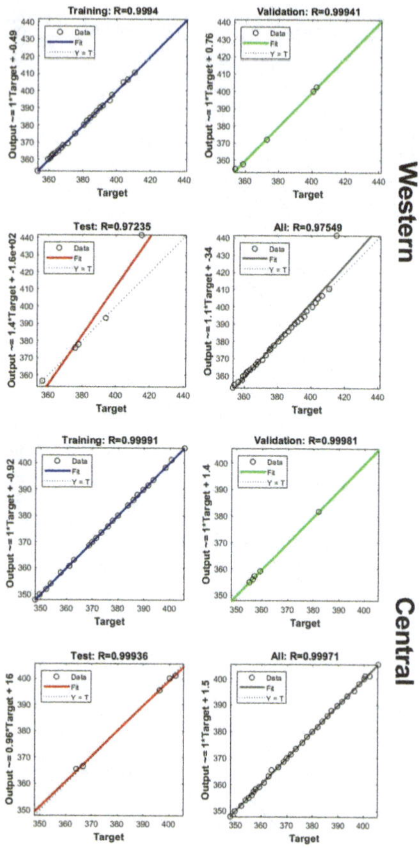

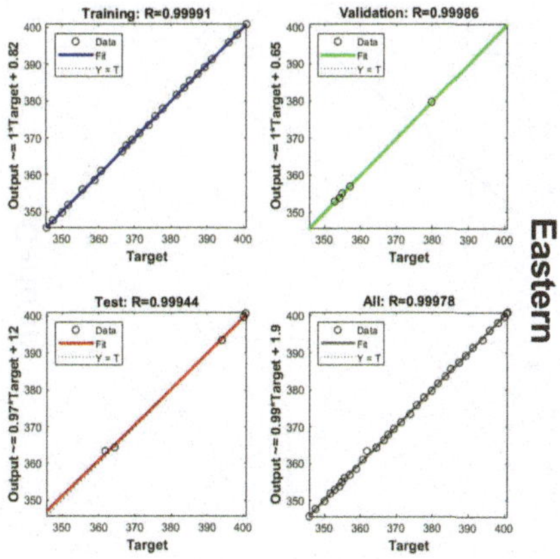

Annexure 3

Regression analysis and sector-wise correlation coefficient values for surface water salinity using a nonlinear autoregressive (NAR) model.

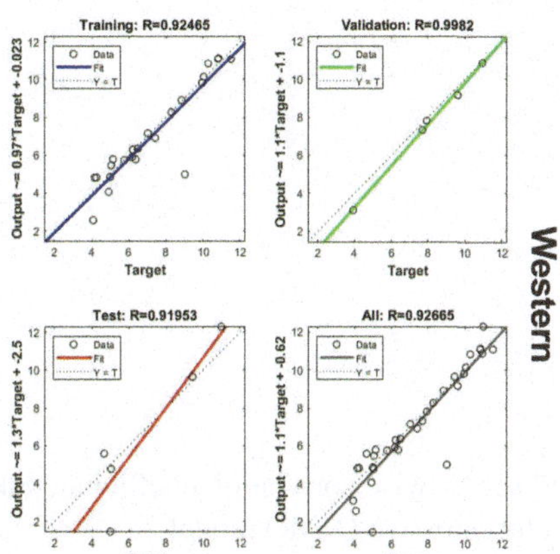

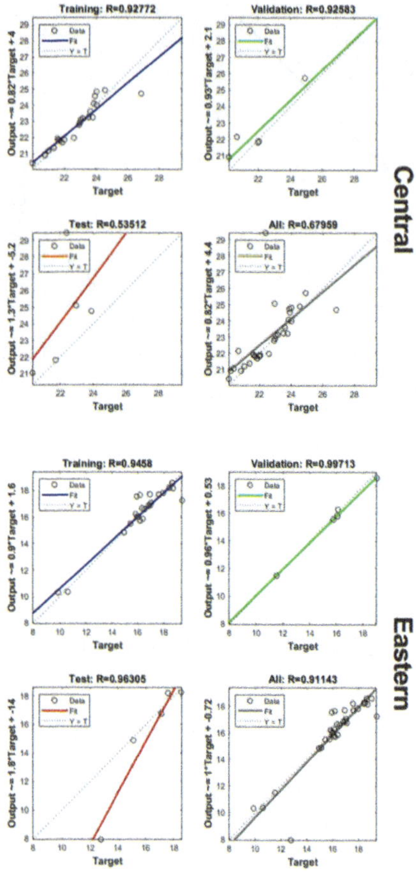

Annexure 4

Regression analysis and sector-wise correlation coefficient values for surface water pH using a nonlinear autoregressive (NAR) model.

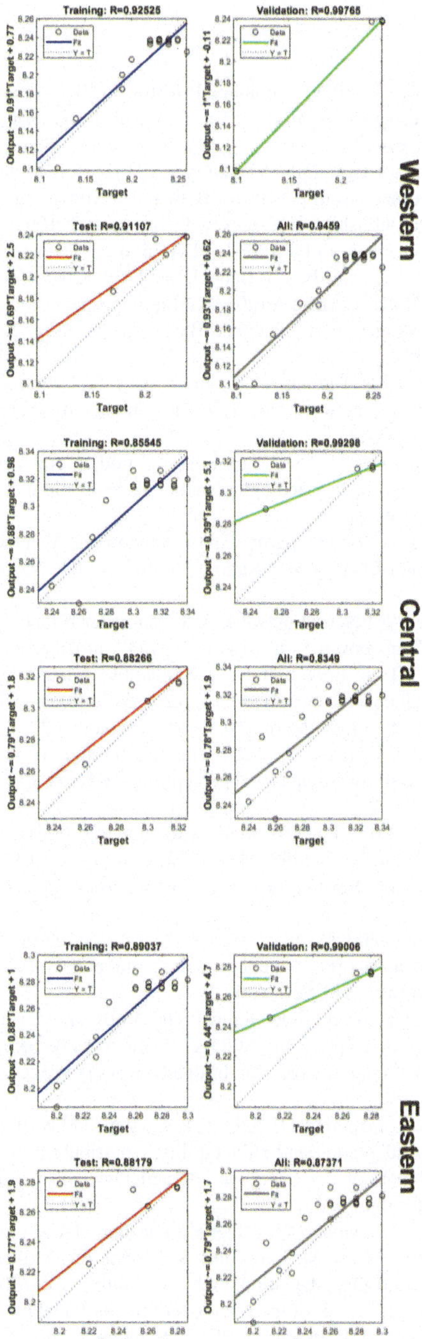
Training: R=0.92525
Validation: R=0.99765
Test: R=0.91107
All: R=0.9459
Western
Training: R=0.85545
Validation: R=0.99298
Test: R=0.88266
All: R=0.8349
Central
Training: R=0.89037
Validation: R=0.99006
Test: R=0.88179
All: R=0.87371
Eastern
Target

References

Agarwal S, Zaman S, Biswas S, Pal N, Pramanick P, Mitra A (2016) Spatial variation of mangrove seedling carbon with respect to salinity: A case study with *Bruguiera gymnorrhiza* seedling. Int J Adv Res Biol Sci 3(8):7–12

Banerjee K, Mitra A, Bhattacharyya DP, Choudhury A (2002) Role of nutrients on phytoplankton diversity in the North east coast of the Bay of Bengal. In: Kumar A (ed) Ecology and Ethology of Aquatic Biota. Daya Publishing House, New Delhi, pp 102–109

Banerjee K, Mitra A, Bhattacharyya DP (2003) Phytopigment level of the aquatic subsystem of Indian Sundarbans at the apex of Bay of Bengal. Sea Explorers 6:39–46

Banerjee K, Sengupta K, Raha A, Mitra A (2013) Salinity based allometric equations for biomass estimation of Sundarban mangroves. Biomass Bioenergy 56:382–391. https://doi.org/10.1016/j.biombioe.2013.05.010

Bhattacharya M, Sinha M (2021) Basin scale wind-wave prediction using empirical orthogonal function analysis and neural network models. Results in Geophys Sci 8:100032. https://doi.org/10.1016/j.ringps.2021.100032

Chakraborty SK, Choudhury A (1985) Distribution of fiddler crabs in Sundarbans mangrove estuarine complex, India. In: Proceedings of national symposium on biology, utilization and conservation of mangroves, pp 467–472

Chaudhuri AB, Choudhury A (1994) Mangroves of the Sundarbans. Volume one: India. International Union for Conservation of Nature and Natural Resources (IUCN). https://portals.iucn.org/library/node/6855

Chow TWS, Leung CT (1996) Neural network based short-term load forecasting using weather compensation. IEEE Trans Power Syst 11(4):1736–1742. https://doi.org/10.1109/59.544636

Doust K, Wejs A, Zhang T-T, Swan A, Sultana N, Braneon C, Luetz J, Casset L, Fatorić S (2021) Adaptation to climate change in coastal towns of between 10,000 and 50,000 inhabitants. Ocean Coast Manag 212:105790. https://doi.org/10.1016/j.ocecoaman.2021.105790

Ghosh A, Schmidt S, Fickert T, Nüsser M (2015) The Indian Sundarban mangrove forests: history, utilization, conservation strategies and local perception. Diversity 7(2):149–169. https://doi.org/10.3390/d7020149

Hansen J, Sato M, Ruedy R, Lo K, Lea DW, Medina-Elizade M (2006) Global temperature change. Proc Natl Acad Sci USA 103(39):14288–14293. https://doi.org/10.1073/pnas.0606291103

Hazra S, Ghosh T, DasGupta R, Sen G (2002) Sea level and associated changes in the Sundarbans. Sci Cult 68(9–12):309–321

IPCC (2001) Climate change 2001: synthesis report. A contribution of Working Groups I, II, and III to the third assessment report of the intergovernmental panel on climate change. Cambridge University Press, Cambridge and New York

IPCC (2005) In: Metz B, Davidson O, de Coninck HC, Loos M, Meyer LA (eds) IPCC Special Report on Carbon Dioxide Capture and Storage. Prepared by Working Group III of the intergovernmental panel on climate change. Cambridge University Press, Cambridge and New York, NY

IPCC (2007) In: Parry ML, Canziani OF, Palutikof JP, van der Linden PJ, Hanson CE (eds) Climate change 2007: impacts, adaptation and vulnerability. Contribution of Working Group II to the Fourth Assessment Report of the Intergovernmental Panel on Climate Change. Cambridge University Press, Cambridge

IPCC (2022) Summary for Policymakers [Pörtner H-O, Roberts DC, Poloczanska ES, Mintenbeck K, Tignor M, Alegría A, Craig M, Langsdorf S, Löschke S, Möller V, Okem A (eds)]. In: Climate change 2022: impacts, adaptation, and vulnerability. Contribution of Working Group II to the Sixth Assessment Report of the Intergovernmental Panel on Climate Change [Pörtner H-O, Roberts DC, Tignor M, Poloczanska ES, Mintenbeck K, Alegría A, Craig M, Langsdorf S, Löschke S, Möller V, Okem A, Rama B (eds)]. Cambridge University Press, Cambridge, UK and New York, NY, USA, pp 3–33. https://doi.org/10.1017/9781009325844.001

Kelman I, Ahmed B, Esraz-Ul-Zannat M, Saroar MM, Fordham M, Shamsudduha M (2018) Warning systems as social processes for Bangladesh cyclones. Disaster Prev Manag 27(4):370–379. https://doi.org/10.1108/DPM-12-2017-0318

Leal Filho W, Luetz JM, Ayal D (eds) (2021a) *Handbook of climate change management: research, leadership, transformation*. Springer, Cham.https://doi.org/10.1007/978-3-030-57281-5

Leal Filho W, Wall T, Alves F, Nagy GJ, Fernández Carril LR, Li C, Mucova S, Joost JP, Rayman-Bacchus L, Totin E, Ayal DY, Lütz JM, Azeiteiro UM, Garcia Vinuesa A, Minhas A (2021b) The impacts of the early outset of the COVID-19 pandemic on climate change research: Implications for policy-making. Environ Sci Policy 124:267–278. https://doi.org/10.1016/j.envsci.2021.06.008

Luetz JM (2008) Planet prepare: preparing coastal communities in Asia for future catastrophes. Los Angeles/Singapore: World Vision International. 1–126. https://www.wvi.org/asia-pacific/publication/planet-prepare

Luetz J (2018) Climate change and migration in Bangladesh: empirically derived lessons and opportunities for policy makers and practitioners. In: Leal Filho W, Nalau J (eds) Limits to climate change adaptation. Climate Change Management. Springer, Cham, pp 59–105. https://doi.org/10.1007/978-3-319-64599-5_5

Luetz JM (2020) Disaster-resistant schools for disaster-resilient education. In: Leal Filho W, Azul AM, Brandli L, Özuyar PG, Wall T (eds) Quality education: Encyclopedia of the UN Sustainable Development Goals. Springer, Cham, pp 1–17. https://doi.org/10.1007/978-3-319-69902-8_15-1

Luetz JM, Sultana N (2019) Disaster risk reduction begins at school: research in Bangladesh highlights education as a key success factor for building disaster ready and resilient communities—a manifesto for mainstreaming disaster risk education. In: Leal Filho W, Lackner B, McGhie H (eds) Addressing the challenges in communicating climate change across various audiences. Climate Change Management. Springer, Cham, pp 617–646. https://doi.org/10.1007/978-3-319-98294-6_37

Majra JP, Gur A (2009) Climate change and health: why should India be concerned? Indian J Occup Environ Med 13(1):11–16. https://doi.org/10.4103/0019-5278.50717

Menéndez P, Losada IJ, Torres-Ortega S, Narayan S, Beck MW (2020) The global flood protection benefits of mangroves. Sci Rep 10(4404):1–11. https://doi.org/10.1038/s41598-020-61136-6

Mitra A (2013) Sensitivity of mangrove ecosystem to changing climate. Springer, New Delhi

Mitra A (2020) Mangrove forests in India: exploring ecosystem services. Springer, Cham. https://doi.org/10.1007/978-3-030-20595-9

Mitra A, Choudhury A (1993) Heavy metal concentration in oyster Crassostrea cucullata of sagar Island, India. Indian J Environ Health 35(2):139–141

Mitra A, Sundaresan J (2016) How to study stored carbon in Mangroves. A start up manual. CSIR-National Institute of Science Communication and Information Resources (NISCAIR), New Delhi

Mitra A, Zaman S (2014) Carbon sequestration by coastal floral community: a ground zero observation on blue carbon. TERI Press, New Delhi, The Energy and Resources Institute

Mitra A, Zaman S (2015) Blue carbon reservoir of the blue planet. Springer, New Delhi

Mitra A, Zaman S (2016) Basics of marine and estuarine ecology. Springer, New Delhi

Mitra A, Zaman S (2021) Estuarine acidification: an overview. In: Estuarine acidification. Exploring the situation of mangrove dominated Indian Sundarban Estuaries. Springer, Cham, pp 1–48. https://doi.org/10.1007/978-3-030-84792-0_1

Mitra A, Ghosh PB, Choudhury A (1987) A marine bivalve Crassostrea cuculata can be used as an indicator species of marine pollution. National Seminar on Estuarine Management. Cochin, pp 177–180

Mitra A, Choudhury A, Zamaddar YA (1992) Effects of heavy metals on benthic molluscan communities in Hooghly estuary. Proceedings of Zoological Society 45:481–496

Mitra A, Halder P, Banerjee K (2011) Changes of selected hydrological parameters in Hooghly estuary in response to a severe tropical cyclone (Aila). Indian J Mar Sci 40(1):32–36

Mitra A, Zaman S, Bhattacharyya SB (2013) Heavy metal pollution in the lower Gangetic ecosystem. In: Abedin MA, Habiba U, Shaw R (eds) Water insecurity: a social dilemma. Published by Emerald Group Publishing Limited, ISBN: 978-1-78190-882-2

Mitra A, Banerjee K, Sengupta K, Gangopadhyay A (2009a) Pulse of climate change in Indian Sundarbans: a myth or reality? Natl Acad Sci Lett 32(1):1–7

Mitra A, Zaman S, Banerjee K (2009b) Effects of climate change on marine and estuarine ecosystems with special reference to Gangetic delta. Journal of Indian Ocean Studies 16(3):194–201

Mitra A, Pramanick P, Zaman S, Fazli P, Pal N, Mitra A (2015) Response of *Sonneratia apetala* to salinity in the frame work of Indian Sundarbans. Int J Innovative Stud Aquat Biol Fish 2(1):1–5

Mitra A, Rudra T, Guha A, Ray A, Pramanick P, Pal N, Zaman S (2016) Ecosystem service of Avicennia alba in terms of Carbon sequestration. J Environ Sci Comput Sci Eng Technol 5(1):155–160

Mitra A, Sundaresan J, Banerjee K, Agarwal SK (eds) (2017) Environmental coastguards-understanding mangrove ecosystem & carbon sequestration. Climate Change Series: 3. CSIR-National Institute of Science Communication and Information Resources (CSIR-NISCAIR), New Delhi

Mukherjee P, Mitra A, Zaman S, Mitra A (2021) Impact of climate-change-induced salinity alteration on Ichthyoplankton diversity of Indian Sundarbans. In: Leal Filho W, Luetz J, Ayal D (eds) Handbook of climate change management. Springer, Cham, pp 1–27. https://doi.org/10.1007/978-3-030-22759-3_276-2

Mukhopadhyay SK, Biswas H, De TK, Jana TK (2006) Fluxes of nutrients from the tropical River Hooghly at the land–ocean boundary of Sundarbans, NE Coast of Bay of Bengal, India. J Mar Syst 62(1–2):9–21. https://doi.org/10.1016/j.jmarsys.2006.03.004

Pal N, Gahul A, Zaman S, Biswas P, Mitra A (2016) Spatial variation of stored carbon in *Avicennia alba* seedlings of Indian Sundarbans. Int J Trend Res Dev 3(4):100–103

Pitchaikani JS, Sarma KS, Bhattacharyya S (2017) First time report on the weather patterns over the Sundarban mangrove forest, East Coast of India

Saha SB, Mitra A, Bhattacharyya SB, Choudhury A (1999) Heavy metal pollution in Jagannath canal, an important tidal waterbody of the north Sundarbans aquatic ecosystem of West Bengal. Indian J Environ Prot 19(11):801–804

Sengupta K, Roy Chowdhury M, Bhattacharya SB, Raha A, Zaman S, Mitra A (2013) Spatial variation of stored carbon in *Avicennia alba* of Indian Sundarbans. Discovery Nature 3(8):19–24

Strickland JDH, Parsons TR (1972) A practical handbook of seawater analysis, 2nd edn. Fisheries Research Board of Canada, Ottawa

Sultana N, Luetz JM (2022) Adopting the local knowledge of coastal communities for climate change adaptation: a case study from Bangladesh. Front Clim 4:823296. https://doi.org/10.3389/fclim.2022.823296

The Earth Observatory (2018) NASA Goddard Space Flight Center. https://earthobservatory.nasa.gov.

Trivedi S, Zaman S, Chaudhuri TR, Pramanick P, Fazli P, Amin G, Mitra A (2016) Inter-annual variation of salinity in Indian Sundarbans. Indian J Mar Sci 45(3):410–415

Zaman S, Biswas S, Pal N, Datta U, Biswas P, Mitra A (2016) Ecosystem service of *Heritiera fomes* seedlings in terms of carbon storage. Int J Trend Res Dev 3(4):183–185

Part II
Implementing Mitigation and Adaptation Measures

Chapter 14
Carbon Trading and Sustainable Development Goal 13: The Malaysia Perspectives

Zainorfarah Zainuddin and Tengku Adeline Adura Tengku Hamzah

Abstract Nations worldwide have made significant progress toward sustainable development due to climate change. As a result of the increased emphasis on achieving a wide range of economic, social, and environmental objectives, firms have begun to realign their operations toward cleaner production processes. Due to the carbon market's anticipated evolution over time, carbon trading development has experienced significant changes that require ongoing innovation and involvement to achieve Sustainable Development Goal 13 (SDG 13). Malaysia has implemented various strategies for mitigating climate change. However, the country's development needs to be aligned with the national economy. This research examines the scope of the SDG 13 framework on carbon trading implementation in Malaysia as well as past and ongoing clean development mechanism projects in the energy, forestry, and agriculture sectors. According to the analysis, the project's adaptability and growth are consistent with SDG 13's goal of lowering carbon emissions. This study exposes the project's challenges, opportunities, contributions and impacts.

Introduction

Global warming poses a serious threat to human economic progress and the natural environment. Many studies have shown that the human impact on climate change is vital (IPCC 2018). The concentration of carbon dioxide and other greenhouse gases has increased by 48% since its pre-industrial level in 1750 (EU 2021). Thus, new strategies and policies that lead to greater potential outcomes of aligning with solutions and mitigation actions and that respond to climate change's impacts are

Z. Zainuddin (✉) · T. A. A. T. Hamzah
Department of Geography, University of Malaya, 50603 Kuala Lumpur, Malaysia
e-mail: zainorfarah@um.edu.my

T. A. A. T. Hamzah
e-mail: adelineadura@um.edu.my

W. Leal Filho et al. (eds.), *Climate Change Strategies: Handling the Challenges of Adapting to a Changing Climate*, Climate Change Management,
https://doi.org/10.1007/978-3-031-28728-2_14

needed. Therefore, governments worldwide are taking steps to build resilient societies, environments and economies to minimise future risks and challenges (UN AR17 2017).

Combating climate change requires all sectors and communities to work collectively. The impact on business and social perspectives has encouraged rapid growth and development through carbon emissions project implementation. Despite disagreements about implementation and outcomes, there is a climate action initiative to raise awareness and develop resources and capabilities. In the World Economic Forum (2022), governments agreed at the United Nations of Climate Change Conference (COP26) to find ways to strengthen the climate action plans more frequently rather than every year as stipulated by the Paris Agreement. With multiple pressures, alongside the new pandemic crisis and climate change challenges, Malaysia has lagged behind the economic performance of Asian countries, as South Korea and Taiwan, which started at a lower level than Malaysia in the 70s, have begun their transformation. The problem exists within the context of investment and productivity, which have declined over the last 20 years. The government's role in creating policies and incentive frameworks needs to be redesigned. It has a lot of potentials to combine investments in large-scale programmes to cut emissions with market mechanisms that promote innovation in low-carbon technology in a cost-effective way (BUR 2021).

Malaysia has reduced its greenhouse gas (GHG) emission intensity to GDP by 29.4% by the end of 2016 (EPU 2021). However, the implementation of carbon trading and environmental sustainability could be solutions to mitigating climate change, increasing economic growth and green development (Hamzah et al. 2019). An incentive for businesses to shift to low-carbon practices and bring in more green investment can be achieved by utilising the carbon trading programme. According to the Ministry of Environment and Water (MEWA), carbon trading is one of the climate change strategies to encourage Malaysian companies to declare a goal of net-zero GHG emission to ensure the long-term viability of the company's operations (MEWA 2021). As the global carbon market downturn has affected carbon trading, the implementation has changed drastically since the Kyoto Protocol, and it has moved forward to a new kind of transformation that requires long-term strategies. Climate change issues affect the world, businesses, the environment and human well-being. Concerns about being involved in low carbon strategies and implementation of a more efficient and sustainable future in the energy, forestry, and agriculture sectors allow rapid progress on new systems and development policies. However, these require more action to move towards sustainable practices.

This chapter discusses the carbon trading project implementation and the Sustainable Development Goal 13 on climate action in Malaysia, which are expected to provide significant benefits and knowledge to businesses, policymakers and society. However, to justify the links between carbon trading and many other interconnected goals, such as renewable energy development, sustainable agriculture and poverty reduction (related to skills, advanced technologies and job creation), sustainable industrialisation must be considered as it contributes to long-term development goals and transformations. Health, poverty, equality, energy, economic growth and climate change are among the 17 complex and interwoven goals aimed to address the most difficult development concerns. The goals are set apart from target indicators, yet they cannot be viewed as separate entities. Understanding the complexities of accomplishing the SDGs requires a systematic approach as the economic, social and environmental systems are interconnected and impact each other.

Participating in environmental sustainability projects and carbon trading opens up a few new perspectives. Similarly, when considering sustainability from a broader perspective, there are several advantages and opportunities, as well as risks and possibilities to consider. One goal's progress is amplified when synergies occur, while another may also accelerate. For example, carbon trading has been linked to many other interconnected goals, including job and skills creation that could reduce poverty, encourage sustainable industrialisation, and adopt climate change mitigation strategies and renewable energy, all of which contribute to long-term development and progress. Thus, carbon trading progress efforts lead to poverty eradication by creating job opportunities (Goal 1), promoting sustainable agriculture (Goal 2), promoting a healthy lifestyle (Goal 3), renewable energy and bio-energy (Goal 7), economic growth (Goal 8) and climate change mitigation action (Goal 13).

Carbon Market in Malaysia

According to the Sustainable Development Report (2021), Malaysia's statistical performance index by the SDGs contributed to a country score of 70.9% that was aligned with the SDGs, including SDG 17 (Partnerships for the Goals). According to UNEP (2020), a country's ability to identify opportunities to reduce emissions and translate them into projects is indirectly reflected in the number of projects relative to the country's emissions. As a result, carbon trading schemes such as the Clean Development Mechanism (CDM) are in line with the potential for lowering emissions and with an effort to increase awareness among relevant sectors prior to the project's actual implementation. However, BUR (2019) reported little interest in the CDM implementation due to the low carbon price and the carbon market's uncertainty issues. According to the Nationally Determined Contribution (NDC), Malaysia intends to reduce its GDP emission intensity (GHG) by 45% by 2030 compared to GDP emission intensity (GHG) in 2005.

Businesses in Malaysia have gradually begun to implement environmental practices, projects and goods as a value-added approach to increase the corporate reputation and financial performance to maximise economic benefits. Many businesses have begun to incorporate sustainability projects into their business strategies to stay competitive. Malaysia's commitment and attempts to educate Malaysians about environmental issues, sustainable living and climate change, in particular, have been persistent (BUR 2015). The action of developing solutions necessitates a distinct approach to decision-making and a need to go beyond current and future viewpoints (Zainuddin et al. 2017). According to Amran et al. (2013), Malaysia has solid institutional arrangements, the stability to invest in climate change mitigation action and increasing awareness of CDM venture projects. Industries would benefit from developing environmental projects under CDM (Hamzah et al. 2019).

In keeping with its commitment to becoming one of the developed nations, Malaysia's government continues to pursue sustainable and resilient growth in its Twelfth Malaysia Plan (2021–2025). The government should improve the environment for green growth, implement sustainable consumption and production, conserve natural resources and increase resilience to climate change and natural disasters. Malaysia will benefit from these actions as they reduce its carbon footprint, ultimately meeting the SDGs. All the goals are a collection of parts linked to achieving a certain goal, which is sustainability. In pursuit of global development and growth, the United Nations operates collectively through a "win–win" process, bringing cooperation and significant gains to all countries (UN 2015).

A plan to change the trajectory of CO2 levels in the atmosphere by triggering long-term systemic shifts must be implemented to deal with climate emergencies and post-pandemic recovery (IPCC 2019). In recent years, the post-pandemic crisis has been tied to developing plans and strategic redesigns for a more sustainable future. Communities are actively seeking more green products, sustainability and carbon trading projects, and many firms are looking to adopt them (MIDA 2020). The development of low-carbon industries and green growth has led businesses to be proactive in developing and implementing green strategies in their practices.

Researchers and scholars frequently cite environmental policies and regulations. Legal compliance has been identified by Yang et al. (2019) and Paulraj (2009) as a major issue in developing strategies and planning climate change mitigation activities. Companies have become increasingly concerned about sustainability disclosures' determinants. Understanding the connections between international trade in carbon trading, competitiveness and global resources is necessary. Environmental projects need to be adopted and adapted, particularly in developing countries, to discover networking potential and further develop the sustainability pillars of economic development, social development and environmental protection. A company that manages environmental affairs will improve relations with consumers, regulators, vendors and other sectors that contribute to the advancement of a more sustainable value chain.

Challenges of Carbon Trading Implementation in Malaysia

Disaster management systems worldwide are confronted with enormous challenges caused by climate-related catastrophes. As was observed during the recent year-end flooding in Malaysia, which has affected many states and contributed to food-shortage and waterborne illnesses, it is apparent that maintaining a clean water supply and ensuring proper sewerage systems are extremely crucial. Other than flooding, other climate-related disasters which are equally challenging to be managed are heatwaves and forest fires. Despite the increasing sophistication of global climate change models, it remains unclear how temperature and climate changes will affect local communities and economies (Soltesova et al. 2014). Scientists still have a long way to go when limiting climate data to smaller regional units, including cities.

The Government of Malaysia submitted the Second National Communication Report (2011) to UNFCCC in 2011. The report shows that Malaysia has the ability to achieve a 40% reduction in GDP emissions through various actions in multiple economic sectors. Despite the possibility of these reductions in emissions, significant effort is required from all stakeholders; one of which is by implementing carbon trading projects. However, one huge hurdle which limits many mitigation actions in carbon trading projects is the cost and suitability of the technology involved. As was observed by Parikh and Parikh (2004), the generation of renewable energy is more expensive than conventional energy due to high cost of the technology involved. High technology costs have resulted in increased transaction costs (Parikh and Parikh 2004). Transaction costs were also mentioned by Parikh and Parikh (2004) as a major problem that makes it hard to start carbon trading projects.

Michaelowa (2015) noted that transaction costs are linked to the environmental project's formalisation, validation and monitoring implementation verification. Depending on their size, development projects require a large financial investment. In monitoring and evaluating carbon projects—the action to optimise environmental performance and reduce transaction costs are in conflict. Regulations may fail to meet the carbon reduction targets specified in the trading proposal, due to the pressure between low transaction costs and the environmental efficiency of cap and trade schemes (Donald 2015). As these changes and related outcomes might require some time to be resolved, it creates challenges in the transfer of technology, adaptation and adoption.

According to IISD (2008), some forest areas have been degraded by previous management effects regarding land usage, land use transition and forestry activities (LULUCF). From the forestry sector's perspective, the conservation and reforestation of these forests require high costs and support to fulfil the activities. Palm oil plantations' impact on deforestation creates concern among environmental activists and the European Union (EU). The EU condemned palm oil as it affects wildlife and promotes unsustainable logging (MPOB 2019). New skills and capabilities must be developed to understand the process, as carbon trading has never resulted in deforestation. The MPOB has implemented palm oil replanting successfully to sustain the industry and contribute to its sustainability (MARDI 2019; FELDA 2017; MPOB 2019). According to some scholars, global emissions are not currently being reduced enough to prevent dangerous climate change (Ronald et al. 2010).

Several scholars stated that it was already too late to prevent catastrophic climate change (Loarie et al. 2009). Therefore, one can argue that carbon trading is less effective since it is insufficient to avert dangerous climate change. Additionally, the limitation on clear and fair environmental policies on carbon trading has to bridge the competitiveness and uncertainty of carbon trading implementation to achieve the SDGs (Zainuddin 2021). Furthermore, political issues have also influenced carbon trading implementation's accuracy. A controversial carbon trading deal has been declared illegal by trading experts in Malaysia.

Carbon trading also poses challenges due to global resources owned by communities. Still, no person or entity has the right to acquire private property that excludes others from using those resources (Ott and Sachs 2000). As a general rule, governments are also expected to protect the common goods by protecting common resources, including preventing entities from using those resources for their benefit. A carbon cap and trade regime may result in pollution if property rights are assigned while governments are required to protect common resources. In the meantime, the ability to trade low-cost credits for emissions reductions reduces the pressure on high-emitting entities to change their behaviour. Additionally, carbon trading does not invest in sociological, political and historical analysis of achieving significant reductions in the required GHG emissions. Neither correlation nor conclusion is drawn as to whether the economic conditions are favourable for continuing the activity or if the absence of Certified Emission Reductions (CERs) or carbon credits reflects their absence at their inception.

This study aims to conduct an original research study rather than critically analyse previously published literature reviews. The data gathered is based on interviews conducted with CDM practitioners per the interview analysis used in the study. This research includes a discussion of the results' potential and implications, as well as suggestions for future research. This research has the potential to assist companies in understanding carbon trading schemes and enhance future climate policy in Malaysia to achieve SDG 13.

Research Method and Data Analysis

This study utilises the literature and interviews to answer the research questions and achieve convincing results. The study uses a thematic analysis with interview data collected from nine CDM practitioner companies in Malaysia that have implemented carbon trading. Malaysia is making progress on many SDGs. Despite the continuance's uncertainty, carbon trading projects in Malaysia have shown a progressive impact on meeting SDGs in the future. This section explains this study's fundamental structure, research design and data collection process. Figure 14.1 illustrates the conceptual framework and identified themes in this study.

The study underwent various phases to evolve and filter its content. Several variables and dimensions are determined in the first part of the literature review to represent the study's key concepts. This research includes interviews with nine carbon trading companies that have implemented carbon trading projects. The nine companies were selected based on the companies engaged in CDM and voluntary carbon market (VCM) projects in Malaysia. This study uses a qualitative approach to data collection with a field visit, focusing on staff at the managerial level who were familiar and knowledgeable with carbon trading implementation activities in their companies. The data collection process is taken seriously to lessen common method bias, which exclusively interviews within the managerial level to provide in-depth interview sessions, accuracy and reliable data.

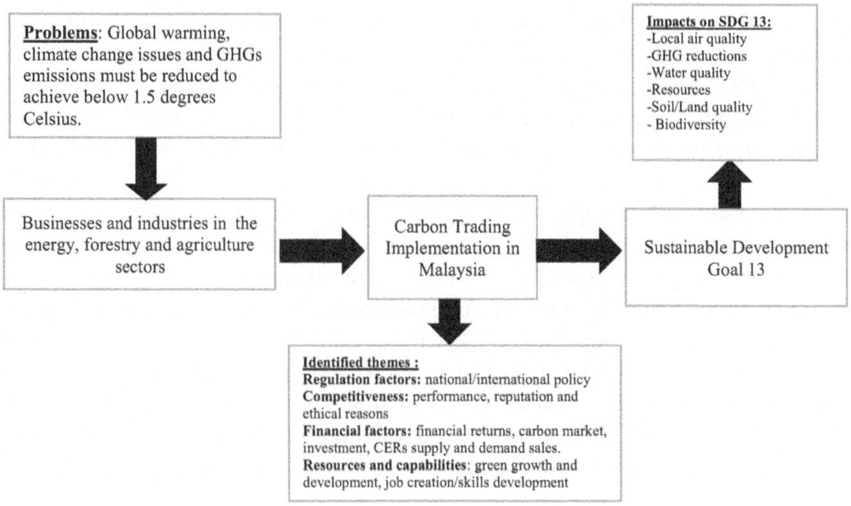

Fig. 14.1 Conceptual framework and identified themes

The study aimed to produce quantitative insights that were difficult to find in the quantitative method and to gain a deeper understanding of the phenomena's whys and hows (Davy and Valecillos 2010). Through semi-structured interviews, the researchers were able to gain a deeper understanding of the various aspects of carbon trading. The nine participating organisations, which ranged in size and industry were selected using a sampling technique called purpose convenient sampling. After getting verbal permission from the participant, the researcher conducted and digitally recorded the interview.

Boyatzis (1998) argues that thematic analysis is a means of communication amongst researchers by drawing on a variety of qualitative approaches. Using thematic data, the researcher can translate observations and use statistical analysis by identifying, analysing, and reporting patterns. Regulation forces, competitiveness, financial factors, resources and capabilities, and impacts on SDG 13 are the five primary categories explored in this analysis. Several themes were inductively produced under each heading using data collected from the subjects.

A combination of secondary and primary data was used in this study. Primary data is used to support the findings from secondary data in terms of the industry's nature and issues. In addition to the UNFCCC website, UNDP, Department of Statistics Malaysia (DOSM), journals and annual reports, secondary data was collected from different sources. In this study, the SDG 13 indicators have been measured following the CDM Handbook (2009) from Malaysia Green Technology Corporation (MGTC). The indicators are based on the project's impact on sustainable development and its contribution to sustainable development. Researchers Zainuddin et al. (2017), Michaelowa et al. (2019) and Paulraj (2009) have classified the determinants to be the strengths that affect how well a company reaches its objectives and the effect the carbon trading project has on a business. An interview process was used in this research, and coding and thematic analysis were performed. This research divided several factors into five themes and categories based on the anticipated assessment forms. The initial research themes consist of the construct focusing on the regulation factor, competitiveness, financial factors, resources and capabilities and impact on SDG 13.

In this study, carbon trading projects were examined in light of improved policies and procedures as well as the Sustainable Development Goals (SDGs) agenda for more sustainable production. Malaysia might be able to reduce regional temperatures by 1.5 °C with technology transfer and participation in carbon trading projects, as promised in the Paris Agreement. This paper aims to contribute to the research field of carbon trading project implementations and develop an understanding for practitioners and policymakers in carbon trading. Contributions towards achieving Climate Action Goal 13 will be made by supporting a sustainable future and creating opportunities for sustainable development and climate change mitigation.

Results and Discussions

This section discusses the findings on the factors affecting carbon trading implementation towards SDG 13. The discussions cover the factors on regulation factors, competitiveness, financial factors, resources and capabilities and the impacts on SDG 13. Table 14.1 summarises the instrument items and the results of this study.

Regulation Factors

Regulation factors such as the policy framework for carbon trading and strategic approaches must provide certainty and confidence for organisations to invest in and address climate actions. This study's results justified the positive outcomes in regulatory factors that lead to carbon trading implementation. Most companies agreed that carbon trading implementation is compatible with national regulations, engagement with international partnerships and strategy for additional carbon emission projects.

The increased focus on businesses in reducing carbon emissions, adopting environmental projects, improving resilience and adding values are integrated with regulative factors. As carbon trading is one of the mechanisms that could promote sustainability, engagement to provide suitable policies and approaches is the key focus to demanding new green growth in Malaysia. According to Delmas and Toffel (2003), when adopting environmental projects such as carbon trading, the authority or government is responsible for encouraging industries to engage in environmental practices. Supported by various studies and scholars, regulation factors are one of the motivations and drivers for encouraging businesses to adopt green practices (Hamzah et al. 2019; Amran et al. 2013; Bansal and Roth 2000; Paulraj 2009; Buysse and

Table 14.1 Interview Findings on Factors and Positive Sustainable Development Goal 13 Impacts

Categories	Sub-categories	Companies								
		A	B	C	D	E	F	G	H	I
Regulation Factors	Compatible with national regulations	*		*			*	*		*
	Engagement in international partnership/policy		*		*	*		*	*	
	Additionally, in emissions projects		*	*		*	*		*	
Competitiveness	Higher performance	*			*			*	*	*
	Good reputations			*	*		*			*
	Ethical reasons	*	*		*			*		*
Financial Factors	Confidence level in financial returns		*			*	*	*		
	The carbon market's progress	*	*				*			*
	Financial investment benefits			*		*		*	*	*
	Demand in CER sales		*		*		*			
Resources and Capabilities	Growth in green resources and clean technology infrastructures		*			*		*	*	
	A requirement for new technical skills and job creations	*		*			*		*	
Impacts on SDG 13	Local air quality	*	*		*		*	*		
	GHG reductions		*	*		*			*	*
	Water quality				*	*	*			*
	Resources	*	*				*	*	*	
	Soil/land quality	*			*			*		*
	Biodiversity			*	*	*		*		*

From: Authors (2021)

Verneke 2003). The data collected revealed that carbon trading implementation influences the outcomes in emission reduction and is one of the climate action strategies. Justified by most interviewees, the government's transformation plan on advancing green growth and enhancing energy sustainability is considered the main factor in moving toward low-carbon nations. However, a representative from Company D said that manipulations that happen with double-counting are to cause problems with reducing emissions and other strict additionality issues (Company D, 2019).

A representative from the Company I stated that there is a lack of information among industries on carbon trading and that most firms are unable to understand the current situation of the carbon market. Thus, the public and businesses need high-quality information to learn more about carbon trading, and environmental policies need to improve.

Competitiveness

Competitiveness plays an important role in driving companies to implement carbon trading. For a company to remain competitive in the market, environmental and business strategy elements must be balanced and co-aligned to achieve higher performance and sustainability (Tuuli and Rowlinson 2009). Companies believe they will achieve competitiveness, boost their performance and company image and gain a competitive advantage that demonstrates sustainability's importance.

Companies A, D, G, H and I considered that implementing carbon trading improved their image and increased their performance. As highlighted by representatives D and G, competitive advantage has turned into an opportunity to improve production and waste processes, shifting production and processes to be more eco-efficient. Thus, businesses in different fields that competed with each other started to think about protecting the environment as part of their business strategy and build good relationships with stakeholders.

Maintaining a competitive edge is crucial for companies to thrive in the market and increase profits over time (Paulraj 2009). Companies can boost their brand recognition and differentiate themselves from their competitors by engaging in carbon trading (Hart 1995). Because of this, businesses may be able to learn new skills, innovate and develop creative solutions that provide them with an advantage over the competition. Carbon trading might be motivated by perceived risks to competitiveness, employment and cost distribution (Decker and Wohar 2009).

Five out of nine companies pointed out that ethical reasons are a factor in earning stakeholders' confidence. Company G states that behaviours and attention are reflected in a firm's value systems and outcomes (Company G, 2019). Zainuddin (2021) and Klassen and Vachon (2003) stated that environmental intentions and behaviours have strong ties to the company's leader for economic and social reasons.

Financial Factors

According to Bohm and Dabhi (2008), the primary reason the industry is interested in implementing carbon trading projects is the financial incentives that are available to companies. With more opportunities and incentives, more firms will be able to influence their businesses' market value and financial success. Most of the companies agreed with the financial benefits. However, five out of nine were not confident about financial returns. Companies B and D mentioned that the carbon market's uncertainty affects the level of confidence in supporting carbon trading projects. According to the World Bank (2019), the financial pillar is key to engaging in the carbon market. Most firms realise that engagement in carbon trading projects is related to financial benefits and improved corporate performance (Li et al. 2019). The carbon credits and incentives have positive outcomes for increasing product value, reducing production costs and fully utilising waste, particularly in the palm oil industry, which could contribute to financial performance and economic development.

From the findings, most companies recognise the importance of long-term sustainability and will reduce carbon emissions by implementing more environmentally friendly projects. Consequently, waste could be reduced, increasing productivity while also enhancing good reputations and improving the company image.

Resources and Capabilities

This study focuses on the organisations within large and medium-scale projects, and most of the projects were from palm oil companies that deal with Palm Oil Mill Effluent (POME) and composting production. The findings in this study reveal that larger firms are expected to perform better in terms of resources and capabilities when engaged with carbon trading project implementations. As supported by Bowen (2003), larger firms are expected to have better availability and accessibility of resources and capabilities than smaller firms.

Firms' ability to finance and invest in environmental projects such as renewable energy, bio-energy, solar and hydro energy requires a high financial investment in equipment and facilities. Furthermore, skills and expertise to support the projects, not only the carbon trading projects but also the facilities, equipment, people and training, have to be considered to make the projects successful. Regarding resources and capabilities, Companies C, D and F suggested a high need to improve green technology with skills, knowledge and incentives for green technology projects. A representative from the Company I informed that most carbon trading projects no longer receive financial incentives on technology transfer and high capital implementation costs have been incurred. Thus, most carbon trading projects are generally carried out with the minimal risk of discontinuation.

Impacts on SDG 13

Carbon trading's main objective is to reduce carbon emissions and achieve sustainable development. The interactions with climate change adaptation and adoption are highly interlinked with other goals to support sustainability with the carbon trading project. In the context of Goal 13, investments in the agriculture sector increase the sector's resilience and adaptive capacity for climate change (ICS 2017). Since the goal of the CDM is to reduce greenhouse gas emissions, putting it into action could be part of the solution to improve adaptation strategies and help achieve the SDGs.

The impacts of SDG 13 from carbon trading implementation revealed high expectations amongst palm oil producers in Malaysia that are connected to GHG reduction and improved local air quality. Even though some issues are associated with the palm oil industry, such as deforestation, biodiversity loss and habitat destruction, Malaysia has established policies on sustainable palm oil plantations to minimise the risk of carbon emissions. The findings reveal that the interconnections and relationships between the SDGs and the agriculture sector significantly integrate with carbon trading projects.

Certainly, deforestation for agricultural purposes and expansion for development will impact mitigation efforts, increasing climate instability and extreme events such as landslides, Malaysia's second most significant natural disaster threat after flooding (Lawal et al. 2014). Dams may contribute to these trends by increasing greenhouse gas emissions (ICS 2017). Forest conversion for agricultural purposes may further undermine Goal 13 by exposing soils to diseases and contamination.

As an additional indication of the correlation between SDG 13 and carbon trading, findings reveal that certain environmental factors influence how firms respond to carbon trading implementation. Even though carbon trading is an investment in a strategy to deal with climate change, the community must be made aware of the need to adapt and use sustainable practices to fight climate change (Zainuddin 2015, 2021). To reduce the negative effects on the environment, measures to adapt to climate change, manage the risk of natural disasters, and protect natural resources should be paired with standards and laws that protect the environment all the time.

As a result of Malaysia's green growth strategy, good quality of growth will be achieved, as will food, water and energy security, while environmental and ecological risks will decrease. This will lead to a greater sense of well-being and quality of life for Malaysians. By reducing carbon dioxide emissions, we can preserve terrestrial and inland water, coastal and marine ecosystems. Buildings, transportation, products and services will become more energy-efficient and low-carbon as sustainable consumption and production practices increase.

As the world's energy consumption continues to increase, GHG emissions are also expected to increase over time. To ensure sustainable energy management, reduced GHG emissions and decreased growth through bio-energy projects, carbon trading will enhance economic development, increase national capabilities and capacity for innovation in green technology and ensure sustainable development for future generations (BUR 2019). Malaysia's ambitious plan for the SDGs to address climate change and GHG emissions could be achieved by developing adaptation strategies from the implementation of carbon trading projects. The energy sector constitutes an important mitigation factor in ensuring access to sustainable development.

This study has several limitations, which have been controlled and minimised to ensure the data's quality is subject to the respondents' perceptions, implying real data. Respondents are likely to rate their performance as favourable, as they want to present a good reputation and company image by reporting good performance (Zainuddin 2021). Thus, the researcher cannot objectively verify the interpretations and findings against the scenarios stated by the respondents. Data collection is time-consuming and requires categorisation, recoding and objective scales to obtain the needed information from respondents. Moreover, as this study was conducted only in the Malaysian context, its generalizability is limited, as its findings might not apply to other developing countries. Due to time and cost constraints, this study doesn't explore the sustainability performance between the Kyoto Protocol period and Paris Agreement that may affect other factors, such as political and international relations. However, the current study's limitations provide room for improvement by other scholars in the future.

Conclusion

To mitigate climate change and meet the SDGs, effective climate change-related planning and management must be provided, alongside support in the form of finance, technology, resources and capacity building for better mechanisms and approaches. Governments and private sectors around the world are adopting ambitious sustainable development goals. However, transformation must follow a certain pace and scale in the process of achieving major long-term changes to reduce GHG emissions.

In Malaysia, carbon trading development in areas where CDM or any VCM projects exist has gained significant ground and power. A growing body of research with ambitious climate action strategies and well-managed procedures will benefit sustainable development and contribute to Malaysia's economic growth, creating a sustainable future. Despite this, Malaysia's climate action remains limited. It must establish a comprehensive and ambitious climate action system for sustainability to meet its NDC targets. In conclusion, carbon trading requires stronger environmental policies, and natural resources and ecosystems must be managed responsibly. Consumers and businesses are confronted with their environmental costs with a carbon price. In that sense, economic policies must be reoriented, which is still possible and compatible with market functioning. Malaysian carbon trading should consciously use the market to help achieve sustainable growth and quality of life.

The SDGs integrate direct and indirect measures to determine carbon trading capacity. Regulators and governments must actively partner with businesses to promote more sustainable production that fosters innovation. Carbon trading and market-based instruments are important because they can foster competition, creativity and innovation among businesses to achieve environmental and social goals. Government support is needed to streamline policies and dismantle those that hinder sustainable development. Carbon trading project development will take time until the right framework is in place. This is because the process needs to be made less complicated and fit the local market. Changing how the market works helps people become more aware of and knowledgeable about a better policy framework for the good of society.

This research contributes to the body of literature in multiple ways. The study extends our knowledge of the issues relating to climate change and mitigation policies. Carbon trading has derived some lessons, and the industry has to use these lessons to plan the process for shared prosperity. There is extensive space and inputs to create carbon reduction opportunities. The value-added strategies that contribute to a company's sustainability have to go beyond business networking and should be more exposed. Thus, despite carbon trading's implementation within the geographical context to achieve the SDGs, a strategy such as domestic carbon trading implementation within states and sectors is one of the alternative strategies to increase economic development and create opportunities for introducing new technologies.

While the economy and the environment are interconnected, both through the SDGs and to one another, participation in carbon trading fosters a deeper understanding of achieving sustainable development, in which the processes and connection chains should be fulfilled and addressed. It is critical to carry out the project not only because it benefits the environment but also because it connects society, human development and economic growth—and this is only the tip of the iceberg: there is a responsibility to create long-term jobs, improve development, end poverty and hunger and, of course, stop pollution. As business continues to be the most powerful force for economic and wealth creation, carbon trading can be used to pursue economic goals while also contributing to social goals.

References

Amran A, Zainuddin Z, Zailani SH (2013) Carbon trading in Malaysia: review on policies and practices. Sustain Dev 21(3):183–192

Bansal P, Roth K (2000) Why companies go green: a model of ecological responsiveness. Acad Manag J 43(4):717–736

Bohm S, Dabhi S (2008) Upsetting the offset: the political economy of carbon markets—Part I: images from North and South. MayFlyBooks Publisher. Retrieved from https://www.semanticscholar.org/paper/Upsetting-the-Offset%3A-The-Political-Economy-of-Part-Boehm/be4488d04ce8d8123d187d3170989d0d53d62ba4

Bowen FE (2003) Organizational slack and corporate greening: broadening the debate. Br J Manag 13(4):305–316

Boyatzis RE (1998) Transforming qualitative information: thematic analysis and code development. Sage

BUR (2015) Malaysia: biennial report to the UNFCCC. Ministry of Energy, Science, Technology, Environment and Climate Change Malaysia (MESTECC). Available at: https://unfccc.int/sites/default/files/resource/MALBUR1.pdf

BUR (2019) Malaysia: 3rd National Communication and 2nd Biennial Report to the UNFCCC. Ministry of Environment and Water Malaysia (MEWA), Putrajaya

BUR (2021) Malaysia: 3rd National Communication and 3nd Biennial Report to the UNFCCC. Ministry of Environment and Water Malaysia (MEWA), Putrajaya

Buysse K, Verbeke A (2003) Proactive environmental strategies: a stakeholder management perspective. Strateg Manag J 24(5):453–470

Davy D, Valecillos C (2010) Qualitative research in technical communication. Routledge, pp 347–375

Decker CS, Wohar ME (2009) Do increase in petroleum product prices put the incumbent party at risk in US presidential elections? Appl Econ 39(6):727–737

Delmas M, Toffel M (2003) Institutional pressures and environmental management practices. Presented at the 11th International Conference of the Greening of Industry Network San Francisco, October 12–15, 2003

Donald BA (2015) Ethical issues raised by carbon trading. The Rock Ethics Institute. Climate Ethics, PennState College of the Liberal Arts

EPU (Economic Planning Unit) (2021) Twelfth Malaysia Plan 2021–2025. Ministry of Economic Affairs Malaysia Publications, Putrajaya

European Union (2021) Climate change: causes of climate change. European Commissions Publications, Brussels. Retrieved from https://ec.europa.eu/clima/eu-action/european-green-deal_en

FELDA (Federal Land Development Authority) (2017) Annual report 2017. FELDA Publications, Kuala Lumpur

Hamzah TA, Zainuddin Z, Yusoff M, Osman S, Abdullah A, Md SK, Sisun A (2019) The conundrum of carbon trading projects towards sustainable development: a review from the palm oil industry in Malaysia. Energies, MDPI 12(18):3530

Hart SL (1995) A natural-resource-based view of the firm. Acad Manag Rev 20(4):986–1014

ICS (International Council for Science) (2017) A guide to SDG interactions: from science to implementation. ICS Publications, Paris

IISD (International Institute for Sustainable Development) (2008) Climate change and food security in Pacific Island countries. IISD Publications, Canada. Retrieved from https://enb.iisd.org/publications-resources/sust_devt2008.htm

IPCC (Intergovernmental Panel of Climate Change) (2018) Summary for policymakers. In: Global warming of 1.5°C. An IPCC special report on the impacts of global warming of 1.5°C above pre-industrial levels and related global greenhouse gas emission pathways, in the context of strengthening the global response to the threat of climate change, sustainable development, and efforts to eradicate poverty. World Meteorological Organization, Geneva, Switzerland

IPCC (Intergovernmental Panel of Climate Change) (2019) Climate change and land: an IPCC special report on climate change, desertification, land degradation, sustainable land management, food security, and greenhouse gas fluxes in terrestrial ecosystems [Shukla PR, Skea J, Calvo Buendia E, Masson-Delmotte V, Pörtner HO, Roberts DC, Zhai P, Slade R, Connors S, Diemen RV, Ferrat M, Haughey E, Luz S, Neogi S, Pathak M, Petzold J, Portugal Pereira, Vyas P, Huntley E, Kissick K, Belkacemi M, Malley J, (eds)]. World Meteorological Organization, Geneva, Switzerland

Klassen RD, Vachon S (2003) Collaboration and evaluation in the supply chain: the impact on plant level environmental investment. Prod Oper Manage 12(3):336–352

Lawal DU, Matori AN, Yusof KW, Hashim AM, Balogun AL (2014) Analysis of the flood extent extraction model and the natural flood influencing factors: a GIS-based and remote sensing analysis. IOP Conf Ser Earth and Environ Sci 18(1):012059

Li L, Liu D, Hou J, Xu D, Chao W (2019) The study of the impact of carbon finance effect on carbon emissions in Beijing-Tianjin-Hebei Region—based on logarithmic mean divisia index decomposition analysis. Sustainability 11(5):1465

Loarie SR, Duffy PB, Hamilton H, Asner GP, Field CB, Ackerly DD (2009) The velocity of climate change. Nature 462(7276):1052–1055

MALSN2 (2011) Malaysia Second National Communication to the UNFCCC. Ministry of Natural Resources and Environment Malaysia Publications, Putrajaya. Retrieved from https://unfccc.int/resource/docs/natc/malnc2.pdf

MARDI (Malaysian Agricultural Research and Development Institute) (2019) Annual report 2019. MARDI Publications, Putrajaya. Retrieved from http://www.mardi.gov.my

MEWA (Ministry of Environment and Water) (2021) The tree, the sky, the sun: a pathway towards Malaysia's carbon-neutral future. MEWA Publications, Putrajaya

MGTC (Malaysia Green Technology Corporation) (2009) CDM Statistic Energy Project August 2010 Quarter 3. CDM Energy Secretariat. Ministry of Natural Resources and Environment Publications, Putrajaya

MIDA (Malaysian Investment Development Authority) (2020) Malaysia sets to maximise green industry, renewable energy potential in 2020. MIDA Publications, Putrajaya. Retrieved January 2020 from: http://www.mida.gov.my

MPOB (Malaysia Palm Oil Board) (2019) Biology and sustainability research division. MPOB Publications, Putrajaya. Retrieved at www.bio.mpob.gov.my

Michaelowa A, Hermwille L, Obergassel W, Butzengeiger S (2019) Additionality revisited: guarding the integrity of market mechanisms under the Paris Agreement. Clim Policy 19(10):1211–1224

Michaelowa A (2015) Opportunities for and alternatives to global climate regimes post-Kyoto. Ann Rev Environ Resour 40:395–417

Ott HE, Sachs W (2000) Standards and legislation for the carbon economy. In: Green technologies: concepts, methodologies, tools and applications, pp 1319–1323

Parikh J, Parikh K (2004) The Kyoto protocol: an Indian perspective. Int Rev Environ Strat 5(1):127–144

Paulraj A (2009) Environmental motivations: a classification scheme and its impact on environmental strategies and practices. Bus Strategy Environ 18(7):453–468

Ronald JE, Westra L, Bosselmann K (2010) In democracy, ecological integrity and international law. Cambridge Scholars Publishing, pp 376–383.

Soltesova K, Brown A, Dayal A, Dodman D (2014) Community participation in urban adaptation to climate change. Community Participation in Urban Adaptation to Climate Change 12:214–225

Tuuli MM, Rowlinson S (2009) Performance consequences of psychological empowerment. J Constr Eng Manag 1355(12)

UNEP (2020) CDM Projects. Centre on energy, climate and sustainable development. UNEP DTU Partnership. Available at: http://www.cdmpipeline.org/cdm-projects-type.htm

United Nations General Assembly (2015) Transforming our world: the 2030 agenda for sustainable development. United Nations Publications, New York

United Nations General Assembly (2017) Annual report 2017. United Nations Publications, New York

WEF (World Economic Forum) (2022) Global risks report 2022. WEF Publications, Geneva. Retrieved at https://www.weforum.org/reports/global-risks-report-2022/in-full

World Bank (2019) The World Bank Annual Report 2019: ending poverty, investing in opportunity (English). Washington DC. World Bank Group. Available at: http://documents.worldbank.org/curated/en/156691570147766895/The-World-Bank-Annual-Report-2019-Ending-Poverty-Investing-in-Opportunity

Yang F, Shi B, Xu M, Feng C (2019) Can reducing carbon emissions improve economic performance? Evidence from China. Economics Discussion Papers 2019(13)

Zainuddin Z (2015) Factors influencing environmental management practices towards clean development mechanism implementation. Master thesis, School of Management, Universiti Sains Malaysia, Penang

Zainuddin Z (2021) Carbon trading implementation towards sustainability in energy, forestry and agriculture sectors in Malaysia. Doctoral Thesis. Department of Geography, Faculty Arts and Social Science, University of Malaya, Kuala Lumpur

Zainuddin Z, Zailani S, Govindan K, Iranmanesh M, Amran A (2017) Determinants and outcome of a clean development mechanism in Malaysia. J Cleaner Prod 142:1979–1986

[illegible]

Chapter 15
Composite Water Resources Management: A Decentralized Approach for Climate Change Adaptation

Vadahanambi Ramachari Sowmithri, Parthasarathy Radhapriya, Rajeev Ahal, Jagdish Kumar Purohit, Rajendiran Nagarajan, and Raj Rengalakshmi

Abstract India is one among the ten most affected countries due to climate risks according to Global Climate Risk Index 2021. Water is the principal channel by which climate change influence ecosystem, community's well-being and livelihoods. Therefore, investments in water management has been received attention due its intrinsic ecological, economic and social values in the context of climate crisis. Equitable and efficient management of water resources requires science-led composite resources mapping, joint action and collective efforts among users, planners and policy makers in both planning and execution process. Besides, participation, ownership and local governance in water resources management are linked to bottom-up approach and decentralized planning. However, maintaining the balance between supply and demand of water has been a challenging and complex issue. Use of appropriate water management tool which captures both supply and demand by mapping the key water challenges in meeting demand and potential water actions for augmentation helps to address the issue. The Composite Water Resources Management plan (CWRM) is a tool supports in scientific analysis on the gap between supply and demand and identify potential ways for augmenting the supply was piloted in India with the partnership of Ministries of Rural Development and Jal Shakthi with GIZ's technical guidance at the lowest level. It has been tested in 1289 gram panchayats/villages in two districts of South India. The chapter explains the scientific approach adopted using climate, socio-economic, water and agriculture based non-spatial and natural resources related spatial datasets. The multi-dimensional vulnerabilities, climate, land-use, micro-watershed, profile of soil and catchment area, surface and groundwater resources, irrigation and greywater and demand for human, agriculture and livestock for water budgeting were analyzed and appropriate water actions/climate solutions were identified using both spatial and non-spatial datasets. The implementation process adopted participatory and convergence

V. R. Sowmithri · P. Radhapriya · R. Ahal · J. K. Purohit
Water Security and Climate Adaptation (WaSCA) in Rural India, NRM and Agroecology Deutsche Gesellschaft für Internationale Zusammenarbeit (GIZ) Gmbh, New Delhi, India

R. Nagarajan · R. Rengalakshmi (✉)
M.S.Swaminathan Research Foundation, Chennai, India
e-mail: rengalakshmi@mssrf.res.in

W. Leal Filho et al. (eds.), *Climate Change Strategies: Handling the Challenges of Adapting to a Changing Climate*, Climate Change Management,
https://doi.org/10.1007/978-3-031-28728-2_15

approach in which Mahatma Gandhi National Rural Employment Guarantee Scheme, a flagship developmental programme is the base converged with schemes of other line departments. The identified and implemented works are mapped with Sustainable Development Goals and India's Nationally Determined Contributions to understand its role in national policies and targets. The chapter describes the process adopted in planning, implementation and engaging different actors/policy makers and expected impacts.

Introduction

The sixth assessment report of the Intergovernmental Panel on Climate Change forewarned that climate continues to be warmer with extreme events and trends and difficult to predict (IPCC 2021). India is placed in 10th position on Global Climate Risk Index, 2021 which confirms the high degree of exposure and vulnerability to changing weather patterns. The national climate change assessment report of 2021 evidently shown the increased frequency and intensity of droughts between 1951 and 2016. Besides, nuanced climate analysis between 1989 and 2018 confirmed the changes in frequency of dry days, rainy days and heavy rainfall days, particularly the increasing frequency of dry days (MoEFCC 2021). These climate risks have direct impact on water resources and widening the gap between supply and demand. Such climate risks are obvious from the highly variable and unpredictable rainfall patterns, extreme seasonal droughts and flooding, sea level changes and coastal salinity. Water is rightly acknowledged as driver of several climate disasters. As water is crucial to adapt and mitigate the adverse impacts of climate change, integrating the importance of water for climate solutions in climate adaptation plans needs attention (Mauroner et al. 2021). As rightly highlighted by United Nations, water is the 'climate connector' among key targets for climate change—Paris Agreement, Sustainable Development Goals—2030 agenda and Disaster Risk Reduction as per the Sendai Framework (UNESCO, UN Water 2020). The Global Adaptation Commission (2019) highlighted the need for investment in water resources management as it is one of the five areas yield higher rate of economic returns, to the tune of more than 3:1 and 1.7 trillion net benefits apart from environmental and societal benefits.

Currently, India is experiencing stark water stress condition. It is evident from declining per capita water availability from 5178 M^3 in 1951 to 1544 M^3 in 2011, estimated to reduce further to 1174 M^3 by 2051. Also, India's water stress is evident from another global assessment study in which India is ranked 13 out of 17 extremely water stressed countries (WRI 2019). It means already more than 80% of the existing water in the country has been annually used. However, the severe water stress is primarily due to inadequate resource management measures between supply and demand rather than the physical shortage of water (Gulati et al. 2019). The "bottom-up" approaches in assessment have been evolved to address finer scale issues (Hegga et al. 2020). Since it is practical to understand local water systems, vulnerabilities, adaptive capacities and support in decisions involving relevant stakeholders. Besides, such decentralized, situation-specific planning is desirable to develop a composite analysis of water resources and plan for sustained outcomes and community participation (OECD 2015; UNESCO, UN Water 2020).

On this line, the India's National Water Policy, 2012 is envisaged to bring efficiency and local ownership by promoting decentralized planning and social inclusiveness through engaging panchayat raj institutions and other local institutions. Further, it has provisions to operationalize the Integrated Water Resources Management (IWRM) guidelines in planning, development and management of water resources. In particular it emphasizes to adopt processes governing climate along with economic and social dimensions at different spatial scales ranging from local, regional and state (NWP 2012). Besides, the policy is in-line with the 1992 Dublin Statement on Water and Sustainable Development by fostering multi-stakeholder participation, one of the four pillars of IWRM. In this backdrop, the interventions to ensure water security and climate adaptation through strengthening the rural water resources assumes importance and GIZ has developed a programme—Water Security and Climate Adaptation in Rural Areas (WASCA) in 2019 (GIZ 2019). The programme piloted the CWRM framework which is a science led methodology embedded with participatory approaches by promoting planning and decisions at the lowest units (administrative/hydrological) engaging stakeholders in planning and implementation of water works as envisaged in the Dublin Statement on Water and Sustainable Development (ICWE 1992) and IWRM principles (Global Water Partnerships 2000).

The chapter attempts to share the methods and processes of rural water resource management planning, and implementation to ensure water security and climate adaptation through science led bottom up planning.

Methodology

Identification of Climate Hotspots

The process started with the scoping study to examine the rural water security from the climate lens with 18 biophysical and socio-economic indicators at the district scale (Palanivelu et al. 2019). These indicators are organized into four dimensions namely climate (changes in maximum and minimum temperature, changes in rainfall, excess and deficit rainfall years), water (ground water extraction and recharge, surface water availability, water gap and contamination) agriculture (area under rainfed, cropping intensity, soil moisture and evapotranspiration) and socio-economic (rural proportion, multi-dimensional poverty index, source of drinking water within premises and proportion of marginal farmers) dimensions. It was further grouped into three areas: adaptive capacity, sensitivity and exposure indicators for the analysis. The data for each indicator is normalized, aggregated and prepared district-wise composite vulnerability index (CVI) and ranked to identify hotspot districts. Of the 38 districts, two districts with diverse agro-ecosystems were selected for piloting the CWRM planning tool (Palanivelu et al. 2019).

Assessment of Key Water Challenges and Climate Resilient Action Tool

The pilot adopted the CWRM planning tool (Purohit et al. 2020). The tool was prepared based on the principles of IWRM, climate science in identifying vulnerable hotspots, analysis of four waters namely surface water, ground water, soil moisture and rain water and socio-economic context. The tool is operationalized at the lowest administrative (gram panchayat) and hydrological unit (micro watershed) in a catchment/watershed or sub-basin to develop and implement scientific plans. The plan consists of four main components, 115 non-spatial and 15 spatial data based on the key components of IWRM and climate adaptation framework. In particular, the spatial data provided evidence to understand the issues pertaining to land use and land cover (LULC), waste land, salt and erosion affected lands, drainage lines, ground water potential, lineament, geomorphology and slope for science-based decision on water actions. Table 15.1 and Maps 15.1, 15.2, 15.3, 15.4, 15.5, 15.6, 15.7, 15.8, 15.9, and 15.10 (illustrative example of one GP is provided) provides the use of different spatial data to identify the key water challenges along with the non-spatial data. The analytical framework used both spatial and non-spatial data on land use, water, soil characters, forest, agriculture including livestock, drinking water and grey water for characterizing the landscape and natural resources, and estimated water budget based on supply and demand at the lowest unit level. Based on the ground level challenges, water actions/climate solutions pertaining to each of the land-use category was identified.

Maps 15.1–15.10: Spatial maps of Kilapakam Gram Panchayat, Vandavasi Block, Tiruvannamalai District, Tamil Nadu in India.

Water actions/Climate solutions: The comprehensive and holistic mapping of the key water challenges using the lens of eco-system approach enabled to identify innovative site-specific water action/works. It is categorized into public and common land (afforestation, soil and water conservation, improving the traditional water storage and catchment assets etc.), agriculture and allied sector (farm ponds, artificial recharge structures, on-farm plantation, irrigation methods, livestock—fodder development etc.) and rural infrastructure (safe drinking water and efficient handling of grey water). These water actions and climate solutions are identified in the field focusing on strengthening land use systems, rain water harvesting, storage and effective use of both surface and ground water, management of catchment areas to adapt/mitigate the extreme weather events including seasonal drought and flooding.

Table 15.1 Use of spatial data in identifying key water challenges in land use categories

Ten fold classification	Type of run-off	Reason for type of run off	Thematic area
1. Forest	Good (HR)	Degraded forests, slopes	LULC, Lineament map
2. Area under non agricultural use	Good (HR)	No area for infiltration (built up area, road, drainage lines, water bodies)	LULC
3. Barren un-cultivable land	Good (HR)	Poor Infiltration, slopes, degraded	Waste land, salt affected area
4. Grazing lands	Ave (AR)	Middle slopes, medium infiltration	LULC, Geomorphology, Ground water prospects
5. Tree crops	Ave (AR)	Low infiltration	LULC, Geomorphology
6. Culturable waste lands	Ave (AR)	Low infiltration	LULC, Geomorphology
7. Fallow lands	BR (LR)	Low infiltration, gentle slopes	LULC, Geomorphology/Ground water prospectus, Lineament map
8. Current fallow lands	BR (LR)	Medium infiltration, gentle slopes	LULC, Geomorphology/Ground water Prospectus
9. Unirrigated area (NSA)	BR (LR)	Soil moisture low, gentle slopes	LULC, Geomorphology, Ground water prospects, Lineament map
10. Irrigated area (NSA)	BR (LR)	Good infiltration, gentle and low slopes	LULC, Geomorphology, Ground water prospects

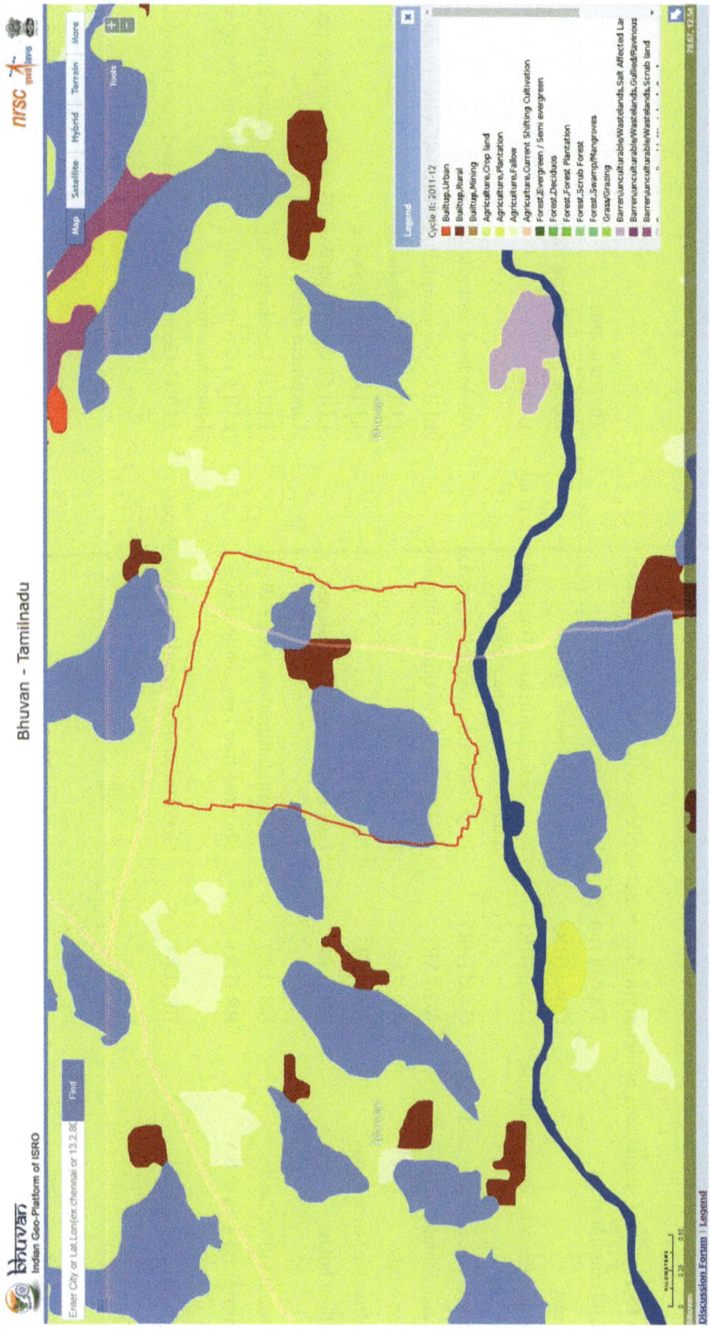

Map 15.1 Land use and land cover

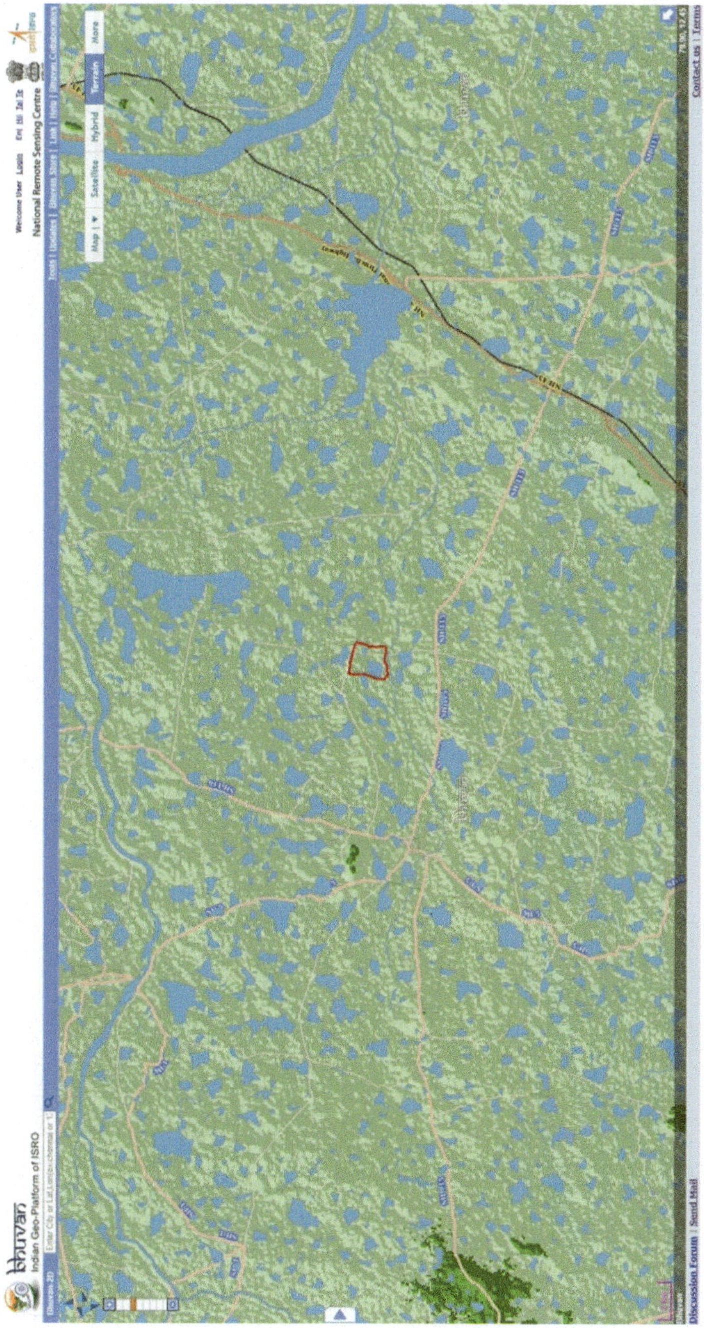

Map 15.2 Terrain

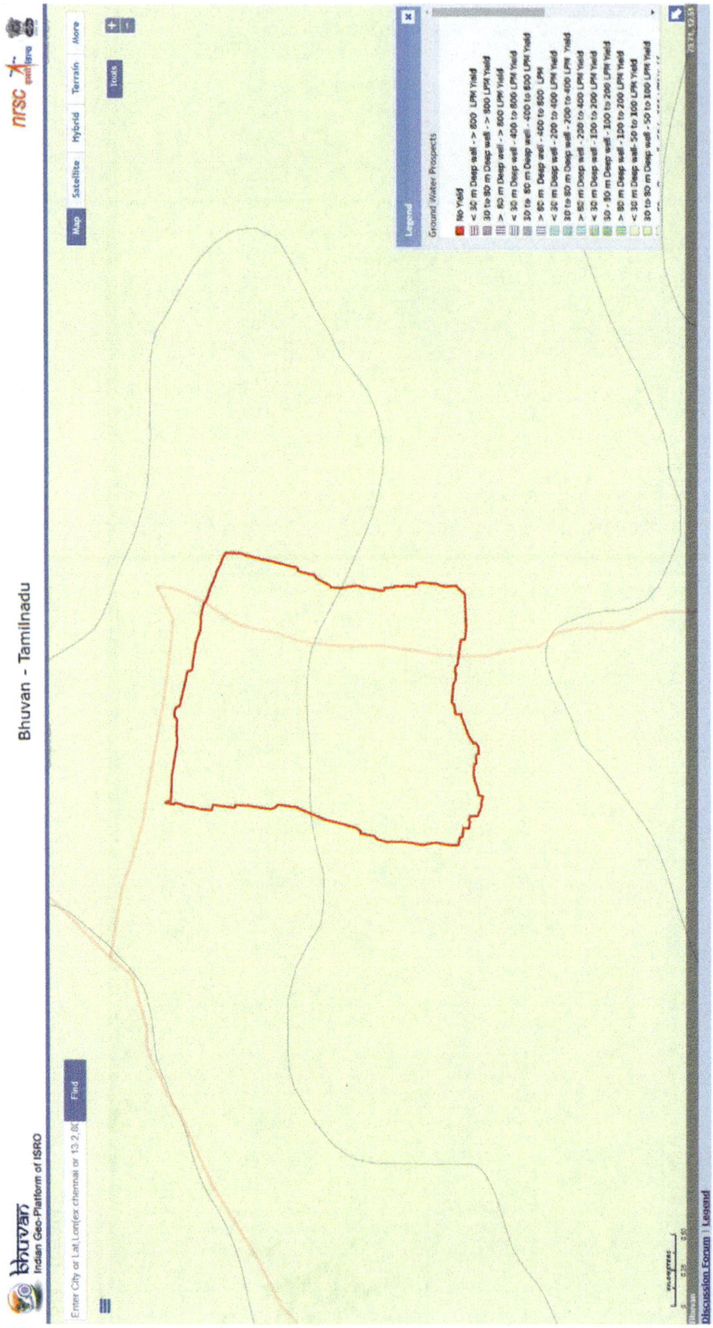

Map 15.3 Ground water prospects

Map 15.4 Lineament

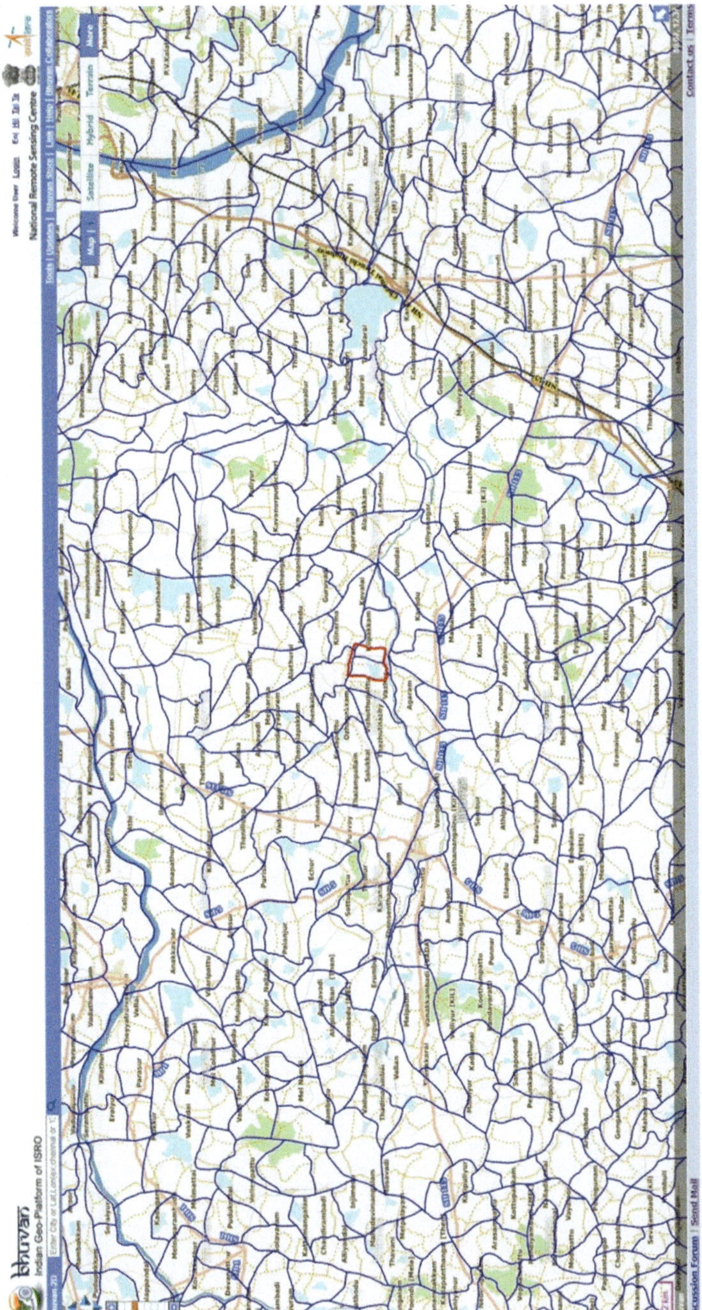

Map 15.5 Watershed

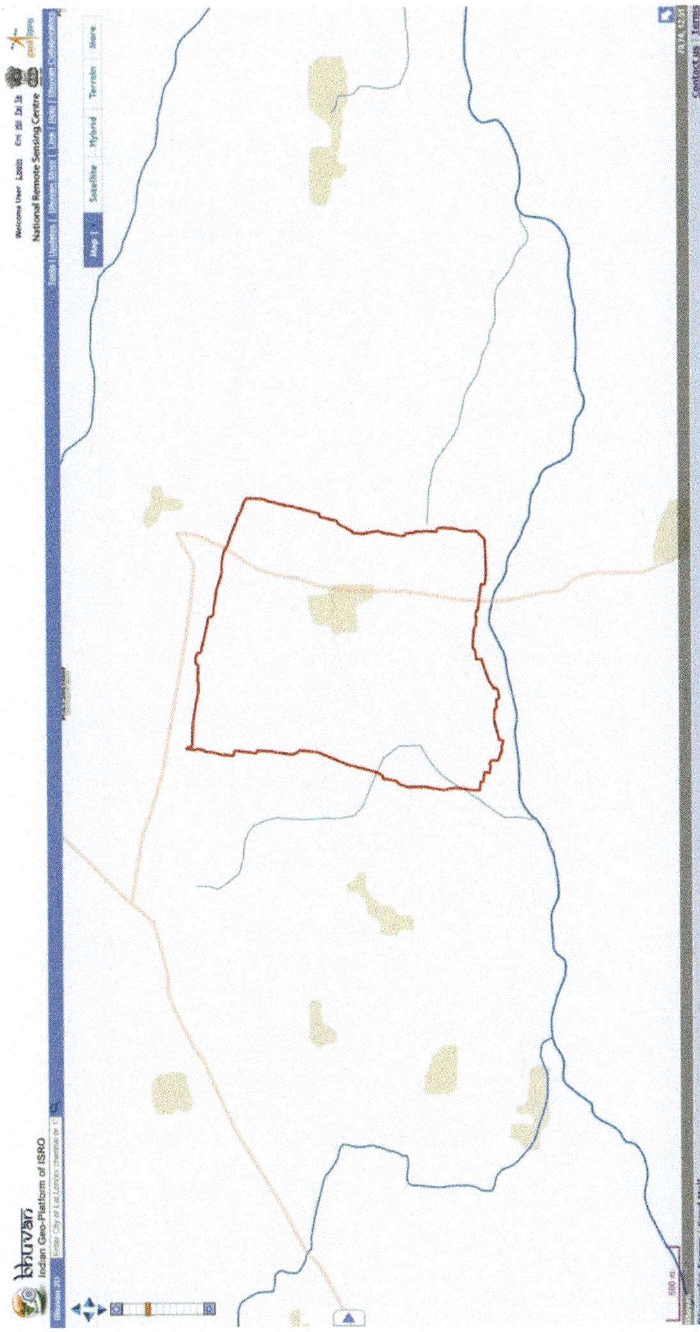

Map 15.6 Drainage and surface water bodies

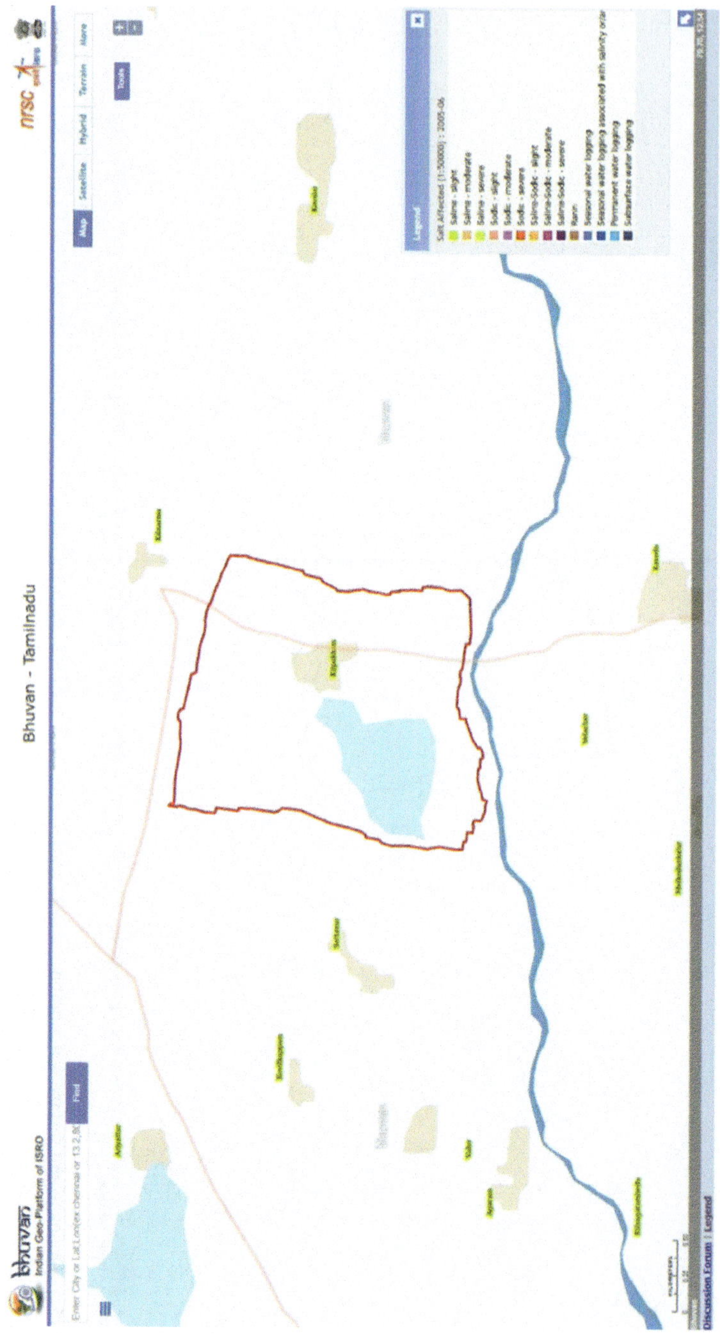

Map 15.7 Salt affected area

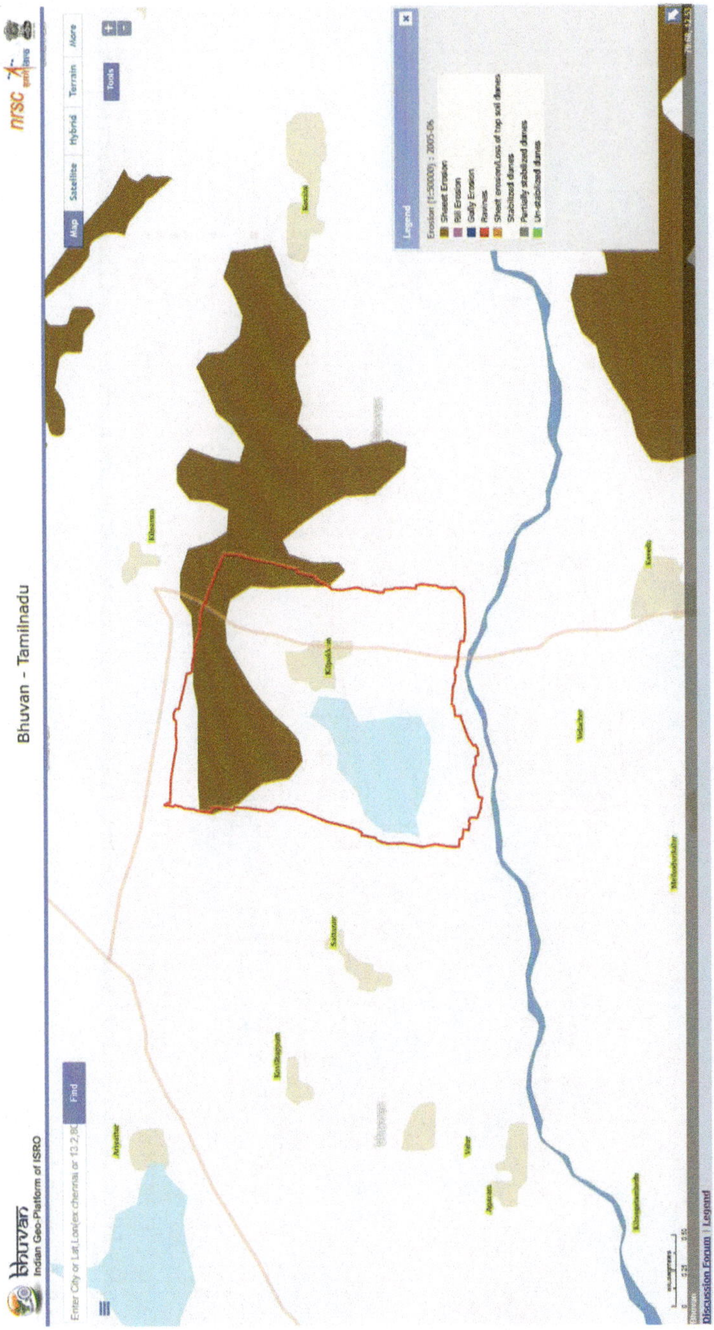

Map 15.8 Soil eroded areas

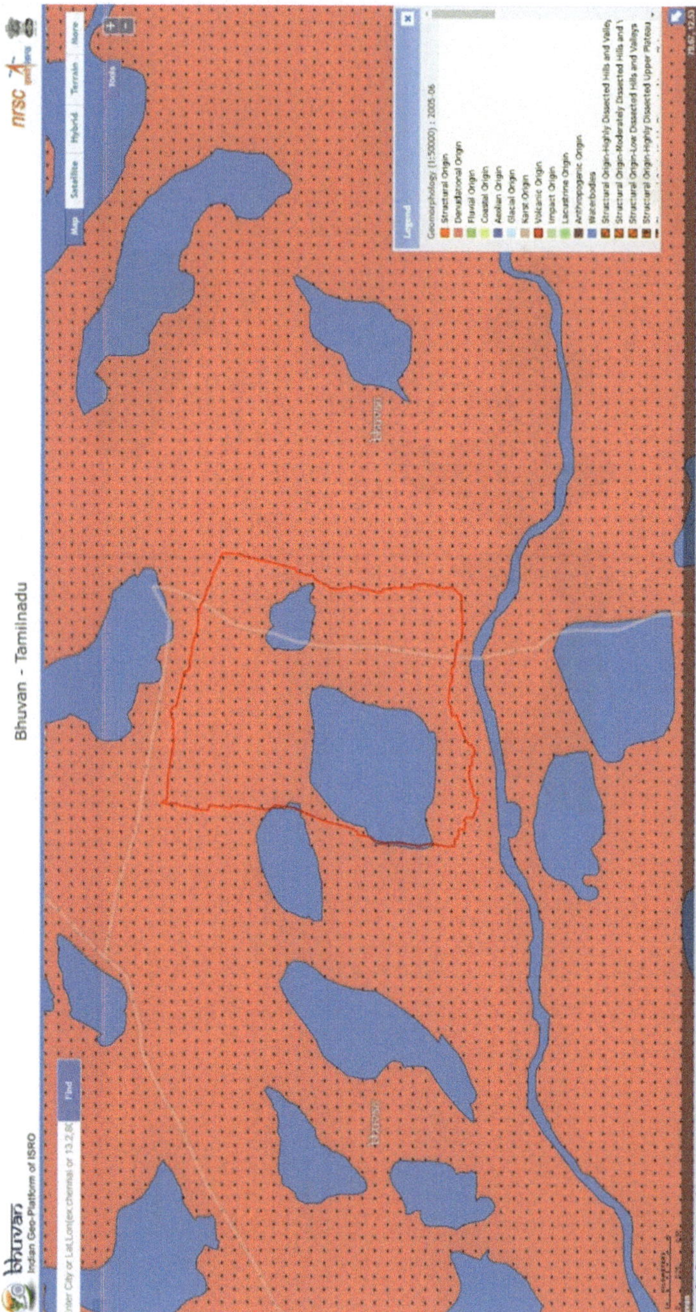

Map 15.9 Geomorphology

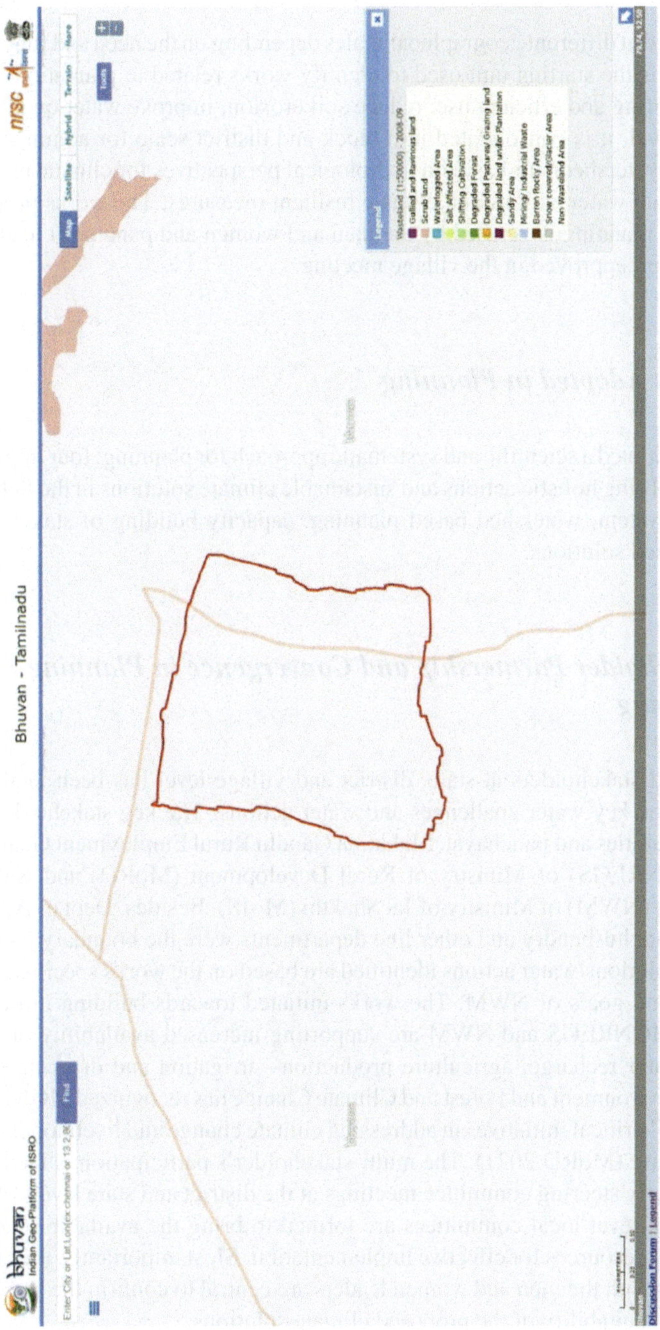

Map 15.10 Wasteland areas

Bottom-Up Planning for Water Resource Management

The plan is used at different geographical scales depending on the need and functional use. GP scale is the starting unit used to identify works related to plantation, water harvesting, storing and efficient use, reduce soil erosion, improve water quality etc. At the next level, it is consolidated into block and district scale for administrative purposes and watershed/sub-basin on hydrological perspectives for climate proofing with appropriate water actions and climate resilient measures. The works proposed using CWRM planning is verified by the men and women and panchayat leaders at the GP level and approved in the village meeting.

Approaches Adopted in Planning

The process adopted a scientific and systematic approach for planning; four important approaches to bring holistic actions and sustainable climate solutions in the field are: adopting ecosystem, watershed based planning, capacity building of stakeholders and nature based solutions.

Multi Stakeholder Partnership and Convergence in Planning and Financing

The associated stakeholders at state, district and village level has been facilitated in analyzing the key water challenges and water actions. The key stakeholders are village communities and panchayat, Mahatma Gandhi Rural Employment Guarantee Scheme (MGNREGS) of Ministry of Rural Development (MoRD) and National Water Mission (NWM) of Ministry of Jal Shakthi (MoJS). Besides, Dept of Agriculture and Animal husbandry and other line departments were the boundary partners. The climate solutions/water actions identified are based on the works specified in the MGNREGS and goals of NWM. The works initiated towards building infrastructure in both MGNREGS and NWM are supporting increased availability of water for ground water recharge, agriculture production—irrigation and drinking water. Ministry of Environment and Forest and Climate Change has recognized MGNREGS as one of the 24 critical initiatives in addressing climate change and livelihoods of the local community (MoRD 2021). The multi-stakeholder's participation is facilitated through quarterly steering committee meetings at the district and state levels. Along with this block level local committees are formed to bring the available schemes, funds and other resources for effective implementation. Most importantly, the village level meetings with the men and women leaders are central to confirm the key water challenges and suitability of the proposed climate solutions.

Study Sites

Ramanathapuram District: It is a coastal district (Map 15.11a) highly water stressed and vulnerable to rainfall variations and saline soil conditions. This is one of the 117 aspirational districts in the country and agro-climatically it falls under southern zone in the state. The district is administratively divided in to 429 Gram Panchayats and 11 blocks.

Tiruvannamalai district: The second selected district is Tiruvannamalai, which is second largest district in the state (Map 15.11b). It is a mix of hill and semi-arid agro-ecosystem, situated in the north eastern agro climatic zone. The district is divided in to 52 firkas, with 860 Grama Panchayats and 18 blocks.

Results and Discussion

The CWRM planning framework has analysed socio-economic, climate, water and agriculture vulnerabilities and mapped the identified the key water challenges at the GP level, using non-spatial and spatial datasets as mentioned in the methodology section. As an adaptation/mitigation measure to address the risks, appropriate water actions/climate solutions are identified scientifically to augment the water resources. The details are described at the district level.

Ramanathapuram District

Socio-Economic Vulnerability

The sex ratio of the district is 961 which is far below the state average. The population density is 331 per sq.km with a proportion of higher (70%) rural population and marginal farmers (93%). Of the total population 22.05% belongs to SC and ST population who are socio economically vulnerable, mostly landless with less assets. The multi-dimensional poverty index is 0.63 which is high among other districts in state, with wider variation intra district variations at block level. Only 5.6% of the total population has access to safe drinking water. Among farmers, 82% of them are marginal farmers holding less than one hectare of land, which is 43% of total land in the district. In addition, of the total holdings, only 13% are from SC and ST communities. In MGNREG scheme access, only 75% of total job card holders are active. Of the total drinking water demand, 86% is met through ground water resources and remaining 14% is met by surface water sources. Increasing salinization of ground water is an area of concern in the district. It is estimated that 17.21 MCM grey water is generated annually and decentralised reuse and recycle strategies not adequate to meet the requirement.

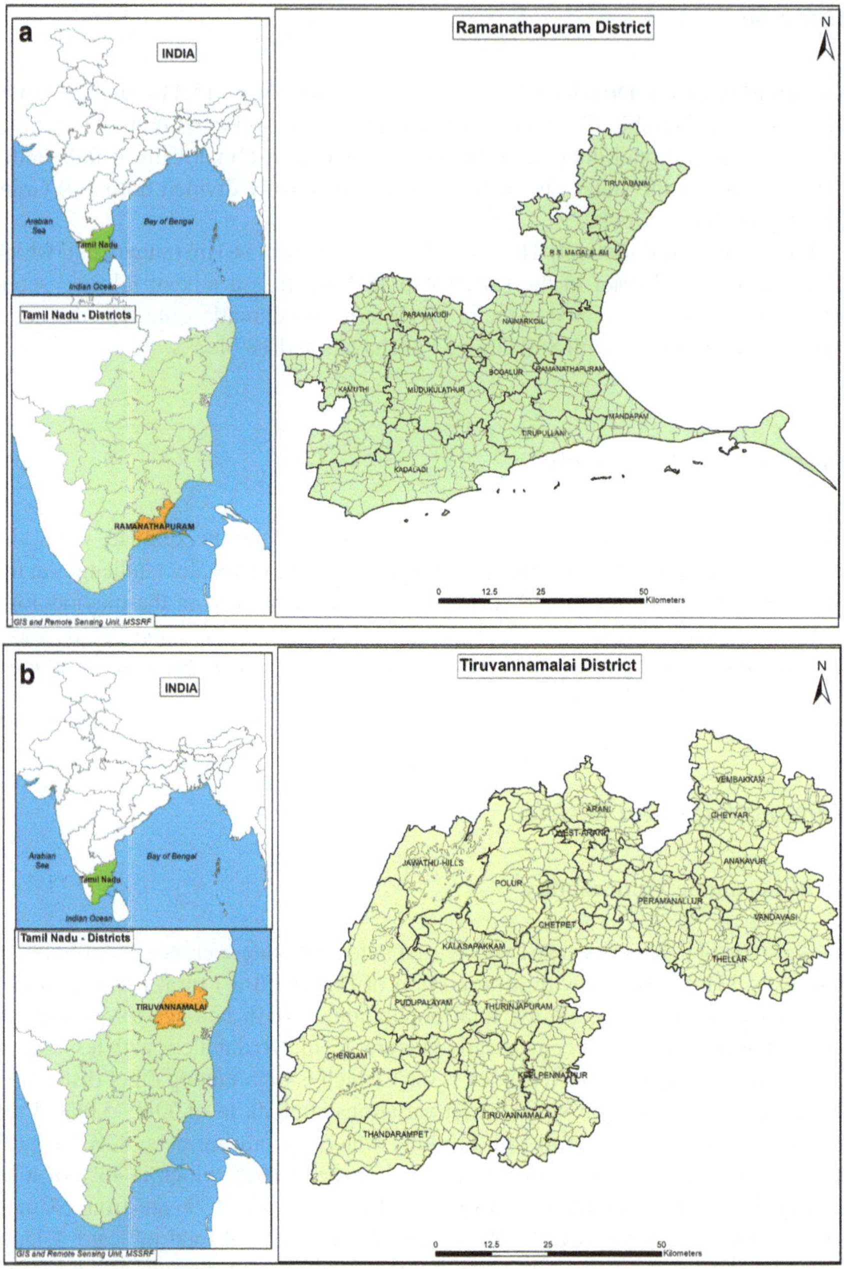

Map 15.11 Location of the study districts in Tamil Nadu, India

Climate Vulnerability

The average annual rainfall is 827 mm with high intraseasonal variation and North East Monsoon is the main rainy season, which depend more on cyclone and depressions in sea. The soil is less productive and low in moisture content coupled with salinity. The existing vulnerability will be further exacerbated under changing climate scenarios in both mid (2050) and end century (2080). It is projected that there will be an increase in annual rainfall by 2050s and in 2080s it will be +1.0% with respect to baseline (1970–2000) of 821 mm along with sharp increase in both minimum and maximum temperatures. The average maximum temperature range in the district is predicted to be 1.83–2.51 °C mid century and 2.71–3.73 °C in end century. Similarly, average minimum temperature in the district is predicted to be 1.56–2.20 °C mid century and increase would be of 2.39–3.23 °C (Palanivelu et al. 2019).

Water Resources Vulnerability

It has a distinctive ecosystem with a long coastline of about 280 km. There are seven watersheds and 914 micro watersheds covering an area of 5,54,570 Ha. These watersheds are spread across ten sub basins and three river catchments. Of the total 914 micro watersheds, 253 are coastal micro watersheds covering 1,75,200 Ha in the district. The forest cover of the district is only 1.09% of total geographical area. The district has 1,476 km total length of natural drainage lines. The key issues are reduced storage area, siltation, damaged surplus weirs, leaking tank outlet, destabilized bunds and disconnected links between water bodies in tank cascades. The district has 1,344 tanks and 3,883 *ooranis* (smaller water bodies) supporting 69,124 Ha of ayacut area for irrigation. There is 1,097 km length of main canal and 5,922 km length of minor canal systems. Further it has 423 km of distributaries and 1,576 km length of field channels in the distribution network. Most of the channels have issues of encroachment, siltation, reduced flow area, and lack of vegetation in the canal bunds.

With regard to water quality, the major issue is increasing total dissolved salt (TDS) in six out of the eleven blocks that are located in the coastal area. Although the ground water is in safe category in general, salinity is increasing, which is evident from the recent data that out of 39 firkhas eight are saline (revenue unit). The pre-monsoon groundwater quality indicates high concentration of electrical conductivity, TDS, sodium and chloride along the coastal region, possibly due to the seawater intrusion in aquifers. In addition, high concentration of calcium in the groundwater, possibly due to existence of calcium rich minerals such as gypsum, limestone, etc. The Water Quality Index indicates that only 9% of groundwater is excellent, 24% each comes under the good and medium categories, 10% comes under poor and 33% under very poor quality. This could be attributed to seawater intrusion, and agricultural intensification using chemical. Seawater mixing index shows 30% of the groundwater samples are affected by seawater intrusion.

In catchment types, district has 22.5% of total geographical area is under good catchment, 9.4% under average and 68.40% in bad catchments. While in surface runoff, bad catchment generates 45.24%, good at 37.64% and average at 17.12%. The total demand for water including human, agriculture and livestock is 27,362.97 HaM. Agriculture is the biggest user of water which is about 96%, only 2.4% for human and 1.6% for livestock. Of the sources, 79% is met through surface water while the remaining 21% is met by groundwater resources. The average annual volumetric soil moisture accounts to 52.15 TMC. The annual total evapotranspiration (ET) loss during 2018–2019 was 522 mm (monthly average of 43.5 mm). The average percentage area influences the water loss through ET in the district was 54% and the total annual losses due to ET alone 110 TMC in the district.

Agriculture Vulnerability

Of the total geographical area 31.35% of the land is under public and degraded land. With regard to ownership, 68.65% of land is under individual ownership, of which 21.85% is under fallow land other than current fallow and fallow land and only 46.80% of the total area is currently under cultivation. Of the cultivated area about 67% of the agriculture land is rainfed with low cropping intensity. 82% of the farmers are small farmers, predominantly cultivating paddy, chillies, millets, cotton and pulses. Under public and degraded land, the district has negligible area under permanent pastures, however have considerable number of small ruminants which are normally open grazed. Paddy is primary crop cultivated in 68% of total cultivated area, followed by chilli (8.81%), coconut (4.45%), other pulses (3.28%), Jowar (2.98%) and other crops in 8.81% of the area. Of the total crops, 42% is cultivated under irrigated condition and 58% is under rainfed cultivation. The total water demand for agriculture is 1,47,696 HaM. Paddy, being a predominantly cultivated, 41.06% of the area is under irrigation and remaining 58.94% is under rainfed cultivation. With reference to water requirement, of the total water needed for cultivation paddy consumes more than 82% of the total water used for irrigation, 82% is through surface water resources followed by remaining 18% through ground water resources.

It has diverse soil types namely clay, coastal alluvium, sandy loam, alluvium, sandy and red soil clay, and black cotton. Predominant soil type is clay occupying 45% of the total cultivated area of the district followed by coastal alluvial soil to an extend of 17% in the northern part of the district (17%). Remaining area is characterized by sandy loam in 15% of the total area. Among the major nutrients, nitrogen and phosphorus are very low to low category in the total number of soil samples tested, while potassium is medium to high. Also, the content of the organic carbon also ranges between very low to low category. This indicates that the soil fertility is very poor and further intensive practices make soil more vulnerable to soil erosion and land becomes degraded over a period of time. While micro nutrients such as zinc and boron are deficient in more than 58 to 85% of the soils tested and 70% of the soils are moderately acidic to moderately alkaline in nature.

The increasing salinity of soil is severely affecting the productivity of main crops and affecting the on-farm livelihoods of farmers and employment opportunities for landless labourers, especially women. The predominant type of irrigation is controlled flooding, and gravity is the main type in canalfed areas. The total area under irrigation in the district is 64,394 Ha, of which 82% is irrigated through surface water stored in the tanks/lakes while remaining 18% is through ground water using open/tube wells. The changing rainfall pattern, extreme events and long dry spells are adversely affecting the agriculture production along with the changes in the quality of land and soil.

Tiruvannamalai District

Socio-Economic Vulnerability

The sex ratio is 994 which is far below the state average. About 80% of the population live in rural areas and multi dimensional poverty index is 0.53 which is very high within the state. The population density is 398 per sq.km with a proportion of higher (70%) rural population. The share of scheduled caste is 23.00% with a minimum percentage of tribal population (3.70%) who are socio-economically marginalized and vulnerable, mostly they are landless and own very less assets. Among farmers, 81% of them are marginal farmers holding less than one Ha of land and hold only 47% of the total land in the district. In addition, of the total holdings only 12% are from SC and 2% ST communities. With reference to access to employment guarantee scheme, only 74.66% of the total job card holders are active in status. Of the total drinking water demand 85% is met through ground water resources and remaining 15% is met by surface water sources. However, the ground water sources are increasingly getting saline due to salt water intrusion. It is estimated that the households generate 39.95 MCM grey water annually and reuse and recycle structures are inadequate.

Climate Vulnerability

There has been changes in the maximum and minimum temperature as well as rainfall quantity compared the annual normal (1970–2000) of the district. The annual rainfall of the district is 1041 mm, the estimated projections for the period is there will be a decrease in 5 and 4% rainfall in both mid and end century. With regard to maximum temperature, it will increase upto 2.1 °C mid-century. For End-century, this increase would be of 3.2 °C. While the average minimum temperature in the district is predicted to 2.4 °C mid-century and 3.7 °C increase in end-century.

Water Resources Vulnerability

The district has 1,847 micro watersheds and 13 watersheds spread across 15 sub basins and three river catchments covering an area of 879,431 Ha. It has 3,787 ponds and 1,966 tanks with 767 km length of field channels for irrigation. The key issues are ownership and encroachments, siltation, repaired surplus weirs and denuded bunds with high degree of erosion. Similar to Ramanathapuram district, here also most of the supply channels are encroached and silted with reduced flow area and destabilized bunds. The ground water recharge is very low in the district, more than 71% of the blocks are under over exploited category and fluoride and nitrate contamination is high. The total runoff is 108,066 HaM. The total demand for water including human, agriculture and livestock is 257,647 HaM. Among the total water demand, agriculture is the biggest user of water which is about 98%, only 1.4% for human and 0.6% for livestock. With reference to the sources, on the total demand, 88% is met through ground water while the balance proportion of 12% is met by surface water resources. The average annual volumetric soil moisture stored in soil accounts to 84.93 TMC, which is almost equal to the amount of surface runoff. While the Annual total ET loss during 2018–2019 was 805 mm. The average percentage area influences the water loss through ET in the district was 44% and the total annual losses due to ET alone 160,654 HaM in the district.

Agriculture Vulnerability

In the district, 27.05% of the land is under public and degraded land. With regard to ownership, 72.90% of the land is under individual ownership, of which 30.7% is under fallow land other than current fallow and the fallow land and only 42.18% of the total area is currently under cultivation. The area under irrigation is 69% while remaining 31% of total cultivated area is under rainfed. The gross area under cultivation is 166,289 Ha. and the cropping intensity is 136%. Paddy and sugarcane, the high-water requiring crops are the primary crops followed by groundnut, vegetables and flowers. Paddy is the primary crop cultivated in 39% of total area cultivated followed sugarcane and vegetables. Of the total water used for irrigation, 88% is through surface water resources and 12% through ground water resources. Under public and degraded land, district has negligible area under permanent pastures, however considerable number of small ruminants which are normally open grazed are recorded. The soil is red loamy to silt and clay depending on the slope. The district has high number of small and marginal farmers of 94.7%. Predominant soil type is clay occupying 65% of the total cultivated area of the district followed by red and loamy soil. With reference to macro nutrients, nitrogen and phosphorus are very low to low category in the total number of soil samples tested, while potassium is medium to high. Also, the content of the organic carbon also ranges between very low to low category. This indicates that the soil fertility is very poor and further intensive practices make soil more vulnerable to soil erosion and land becomes degraded over a period of time. Similarly, important micro nutrients such as manganese, zinc and boron are deficient in more than 54–91% of the soils tested. With reference to the soil physical condition 70% of the soils are moderately acidic to moderately alkaline in nature. Controlled flooding is the predominant type of irrigation irrespective of the sources and ground water is the main source for agriculture through open and tube wells. Besides cropping, the main livestock species are goat, sheep, cow and poultry. The total water demand for livestock in the district is 7,647 HaM, of which 42% is through ground water and 58% through ground water.

Water Actions/Climate Solutions for Building Water Resilience

The deeper understanding on the multiple dimensions of the vulnerabilities in water resources management, enabled to identify contextualized water actions or climate solutions for augmenting the water resources. While identifying the water actions, watershed principles was adopted. These works proposed are divided into three categories: public and common land, agriculture and allied activities and common rural infrastructures for both the districts (Tables 15.2 and 15.3 and Map 15.12). These climate solutions are comprehensively identified addressing 'reduce, respond and restore' approach in climate risk management. These solutions include short-term to long-term benefits. Of the solutions, rainwater harvesting, storage and recharge supports in reducing risk and build resilience for water. In both the districts suitable soil and water conservation actions are identified such as drainage line treatments, restoration of water bodies and supply channels, land management technologies like continuous contour bunds, trenches, compartmental bunding with vegetation. Similarly, another nature based activities that are cost effective and sustainable in conservation of natural resources and dealing climate risks such as flooding/drought/extreme events are vegetation cover. The roots of these trees serve as sponges to recharge water in the below soil during precipitation and improve ground water and reduce surface runoff and flooding. Several innovative methods of improving the vegetation cover such as afforestation, silvi-pasture, farm bund plantation, agro-forestry, dry land horticulture etc. Tropical cyclones and high winds are addressed through shelter belt plantation with locally suitable tree species and restoring mangrove plantation with diverse species. Such nature-based solutions help to address issues related to sea water intrusion and sea erosion in coastal habitations.

The cumulative impacts of different water actions/climate solutions undertaken in reducing the vulnerability and building resilience of local communities are provided below. The GP based planning and integration at the block and watershed/sub-basin levels enable to adopt ecosystem approach in promoting nature-based solutions. These impacts are envisaged at the end of the three years of implementation by adopting saturation and convergence approach in mobilizing necessary finance, knowledge and technologies.

Table 15.2 List of climate solutions proposed in Ramanathapuram district and links with climate vulnerability, SDG and INDCs

S. No.	Climate solutions proposed	No. of works	Climate vulnerability area influenced	INDCs	SDGs
CWRM Water Action 1. Public and common land development					
1	Afforestation	11,217	Climate, Water Resource and socio-economic	1. To better adapt to climate change by enhancing investments in sectors vulnerable to climate change, particularly water resources 2. To create an additional carbon sink of 2.5 to 3 billion tonnes of CO_2 through additional forest and tree cover by 2030	SDG 1,2,6,13&15
2	Contour Continuous Bunds	43,894	Water Resource		SDG 1,2,6,13&15
3	Composting	11,931	Water Resources		SDG 1&6
4	Drainage Line Treatment	23,248	Water Resource		SDG 1&6
5	Silvi-pasture Development	191	Agriculture, Climate and Socio-economic		SGG 6,12&13
6	Linear Plantation	754	Climate, Water Resources and socio-Economic		SDG 1,2,6,12&13
7	Avenue plantation	1,801	Climate, Water Resource and socio-economic		SDG 1,6&13
8	Block Plantation	29,487	Climate, Water Resource and socio-economic		SDG 1,6&13
10	Restoration of water bodies: (a) Tanks	1,344	Water Resource, agriculture and Socio-economic		SDG 1,6&13
11	(b) Ooranis	3,883			SDG 1,6&13
12	Restoration of water bodies (c) Ponds	47	Water Resource, agriculture and Socio-economic		SDG 1,6&13
13	Artificial Recharge Structure	4,627	Water Resource, agriculture and Socio-economic		SDG 1,6&13
14	Canal Bund Plantation	2,618	Water Resource, agriculture and Socio-economic		SDG 1 &15
15	WC—Irrigation channels—Desilting	157,589	Water Resource, agriculture and Socio-economic		SDG 1,6&13
16	WC—Irrigation channels—canal side plantation	157,589	Climate, Water Resource and socio-economic		SDG 1,6&13

(continued)

Table 15.2 (continued)

S. No.	Climate solutions proposed	No. of works	Climate vulnerability area influenced	INDCs	SDGs
17	Agro forestry	1,614	Climate, Water Resource and social		SDG 1,6,13&15
18	Check dam	27	Climate and Socio-economic		SDG 1, 6,13&15
19	Mangrove plantations	135	Climate and Socio-economic		SDG 1,6,13&15
20	Fish Drying Yard	34	Socio Economic		SDG 1,6,13&15
21	Nursery development	12	Climate and Socio-economic		SDG 1,6,13&15
22	Shelter belts	16	Water Resources and Climate		SDG 1,6,13&15
23	Coastal wetland—Bund strengthening	22,579	Water Resources and Climate		SDG 1,6,13&15
24	Bund Plantation wet lands	10,964	Water Resources and Climate		SDG 1,6,13&15
25	Wetland plantation (inner)	133	Water Resources and Climate		SDG 1,6,13&15
26	Wetland Inlet improvement works	2,856	Socio Economic		SDG 1,6,13&15
Sub Total		**488,590**			
CWRM Water Action 2: Agricultural and allied Sector development					
1	Farm Bunding	32,926	Water Resources, Agriculture	For better adaptation to climate change by enhancing investments in sectors vulnerable to climate change, particularly agriculture and allied activities	SDG 1,2&6
2	Micro Irrigation	3,081	Water Resources, Agriculture and Climate		SDG 1,2&6
3	Construction of farm ponds	10,084	Water Resources, Agriculture and Climate		SDG 2&6
4	Land development	12,668	Agriculture and socio economic		SDG 2&6
5	Nursery Development—Plantation	9,769	Agriculture and Socio Economic		SDG 2&6

(continued)

Table 15.2 (continued)

S. No.	Climate solutions proposed	No. of works	Climate vulnerability area influenced	INDCs	SDGs
6	Cattle Shelters	2,329	Agriculture, Water Resources, Socio-Economic and Climate		SDG 2,6&13
7	Goat Sheep Shelters	19,385	Agriculture and Water Resources		SDG 2&6
8	Fodder development for cattle	2,329	Socio Economic and Climate		SDG 1,2&6
9	Azolla units	2,329	Socio Economic		SDG 1&2
10	Cattle trough	2,329	Socio Economic		SDG 1&2
11	Poultry shed	5,949	Agriculture and Socio Economic		SDG 1&2
12	Dry land Horticulture/Agro-forestry	16,460	Agriculture and Socio Economic		SDG 1&2
13	Vermi compost	2,329	Water Resources and Socio Economic		SDG 1&2
Sub Total		**121,966**			
CWRM Water Action 3: Rural Water Management					
1	Soak pits (Community)	3,895	Water Resources and Socio-Economic	To better adapt to climate change by enhancing investments in sectors vulnerable to climate change, particularly water resources	SDG 1,2&6
2	Soak pits (Individual)	38,985	Water and Socio-Economic		SDG 1,2&6
3	Roof-rain water harvesting in common buildings	858	Water Resources		SDG 2&6
4	Tanka—community	2	Water resources, socio-economic		SDG 2&6
Sub Total		**43,740**			
District Total		**654,296**			

Actions to Improve Socio-Economic Equality

Addressing the socio-economic vulnerability depends on how vulnerability issues are addressed in other three themes. As a first step, it is proposed to invest actions in asset creation for SC and ST households and female headed households with high priority. Also, priority action areas include access to safe drinking water, land

Table 15.3 List of climate solutions identified in Tiruvannamalai district and links with climate vulnerability area, SDG and INDCs

S. No.	Climate solutions proposed in Tiruvannamalai district	No. of works	Climate vulnerability area influenced	INDCs	SDGs
CWRM Water Action 1. Public and common land development					
1	Afforestation	18,771	Climate, Water Resource and socio-economic	1. To better adapt to climate change by enhancing investments in sectors vulnerable to climate change, particularly water resources 2. To create an additional carbon sink of 2.5–3 billion tonnes of CO_2 through addition al forest and tree cover by 2030	SDG 1,2,6,13&15
2	Contour Continuous Bunds	46,771	Water Resource		SDG 1,2,6,13&15
3	Composting	12,331	Water Resources		SDG 1&6
4	Drainage Line Treatment	13,071	Water Resource		SDG 1&6
5	Silvi-pasture Development	2,841	Agriculture, Climate and Socio-economic		SGG 6,12&13
6	Linear Plantation	60	Climate, Water Resources and socio-Economic		SDG 1,2,6,12&13
7	Avenue plantation	57	Climate, Water Resource and socio-economic		SDG 1,6&13
8	Block Plantation	8,233	Climate, Water Resource and socio-economic		SDG 1,6&13
9	Restoration of water bodies: (a) Tanks	1,966	Water Resource, agriculture and Socio-economic		SDG 1,6&13
10	(b) Ponds	3,787			SDG 1,6&13
11	Artificial Recharge Structure	26,113	Water Resource, agriculture and Socio-economic		SDG 1,6&13
12	Canal Bund Plantation	23,839	Water Resource, agriculture and Socio-economic		SDG 1,6&13
13	WC—Irrigation channels—Desilting	3,949	Water Resource, agriculture and Socio-economic		SDG 1&15
14	WC—Irrigation channels—canal side plantation (Mtr)	3,949	Water Resource, agriculture and Socio-economic		SDG 1&15
Sub Total		**165,738**			

(continued)

Table 15.3 (continued)

S. No.	Climate solutions proposed in Tiruvannamalai district	No. of works	Climate vulnerability area influenced	INDCs	SDGs
CWRM Water Action 2: Agricultural and allied Sector development					
1	Farm Bunding	14,099	Water Resources, Agriculture	For better adaptation to climate change by enhancing investments in sectors vulnerable to climate change, particularly agriculture and allied activities	SDG 1,2&6
2	Micro Irrigation	1,451	Water Resources, Agriculture and Climate		SDG 1,2&6
3	Construction of farm ponds	9,482	Water Resources, Agriculture and Climate		SDG 2&6
4	Land development	22,483	Agriculture &socio economic		SDG 2&6
5	Nursery Development	2,303	Agriculture and Socio Economic		SDG 2&6
6	Cattle Shelters	36,428	Agriculture, Water Resources, Socio-Economic and Climate		SDG 2,6&13
7	Goat/Sheep Shelters	17,649	Agriculture and Water Resources		SDG 2&6
8	Fodder development	27,091	Socio Economic and Climate		SDG 1,2&6
9	Azolla units	33,669	Socio Economic		SDG 1&2
10	Cattle Trough	30,453	Socio Economic		SDG 1&2
11	Poultry shed	26,006	Agriculture and Socio Economic		SDG 1&2
12	Dry land Horticulture/Agro-forestry	24,892	Agriculture and Socio Economic		SDG 1&2
13	Vermi compost	37,889	Water Resources and Socio Economic		SDG 1&2
14	Construction of new well	27,960	Agriculture, Water Resources, Socio-Economic and Climate		SDG 1,2&6
Sub Total		**311,855**			

(continued)

Table 15.3 (continued)

S. No.	Climate solutions proposed in Tiruvannamalai district	No. of works	Climate vulnerability area influenced	INDCs	SDGs
CWRM Water Action 3: Rural Water Management					
1	Soak pits (Community)	16,547	Water Resources and Socio-Economic	To better adapt to climate change by enhancing investments in sectors vulnerable to climate change, particularly water resources	SDG 1,2&6
2	Soak pits (Individual)	49,167	Water and Socio-Economic		SDG 1,2&6
3	Roof-rain water Harvesting in common buildings	4,640	Water Resources		SDG 2&6
Sub Total		**70,354**			
District Total		**547,947**			

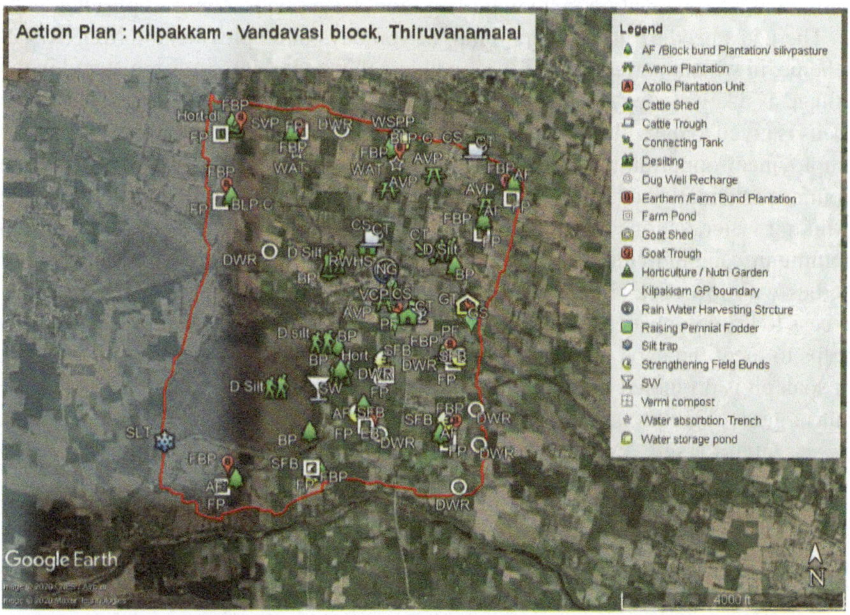

Fig. 15.12 Illustrative example of climate solutions identified for Kilapakkam GP in Tiruvannamalai district in Tamil Nadu, India

Table 15.4 Expected socio-economic impacts from the identified water actions

Sl. No.	Impact indicators	Ramanathapuram	Tiruvannamalai
1	Number of waste management systems	3,895 community and 38,985 individual level soak pits	16,547 community and 49,169 individual level soak pits
2	Roof rainwater harvesting and storagemeasures established	858 units	1640 units
3	No. of tankas planned	Two	Not suitable
4	Nutri gardens established	390,750 households	472,845 households
5	Number of vulnerable families benefitted through plantation	40,704 vulnerable families guaranteed employment for three consecutive years	45,477 vulnerable families guaranteed employment for three consecutive years

development, creating additional employment days, explore the possibility to bring skilled jobs especially for women. Besides, marginal farmers are targeted in the individual assets' creation including plantation, farm ponds, compost pits etc.

There is a need to ensure a minimum of 100 days of employment in MGNREGS scheme, in which increasing active job holders to the total job cards registered in the village as a key strategy to increase work participation rate in rural areas. Here also focus is given to individual assets creation which facilitate men and women to access employment opportunities. Also, by encouraging skilled works, unemployed rural youth's participation can be motivated in climate solutions. With regard access to safe drinking water, actions to improve availability by roof rainwater harvesting at both community and individual houses, models like tanka, restoring the traditional water bodies with low cost, simple water treatment plants etc. which further ensure their access to drinking water and also here convergence under Jal Jeevan Mission (JJM) helps to meet the gaps in the rural areas. For grey water management, the concept of soak pit is promoted both under community, and individual category. In addition, nutria-gardens are promoted to reuse grey water and improvement to existing drains are essential for safe disposal (Table 15.4).

Actions to Build Climate Resilience

The specific climate risks pertinent to local ecosystems are addressed through following innovative context-based climate resilient models in both the districts (Table 15.5) which are expected to provide specific impacts.

Ramanathapuram district:

1. Coastal watersheds—focus on wetland management, sand dune development, creek and other water bodies restoration, coastal plantation etc.
2. Community level tanka—safe rain water storage structure

Table 15.5 Number of vulnerable blocks and climate resilient measures proposed

Sl. No.	Impact indicators	Ramanathapuram	Tiruvannamalai
1	Number of vulnerable blocks (Area of Interest) identified in each district	11 blocks with selected GPs	18 blocks with selected GPs
2	Number of climate resilient measures identified	11 models	11 models

3. Plantations—Mini forest, silvi-pasture, agroforestry—dryland horticulture
4. Degraded land restoration—community level village based integrated farming system models with a focus to horticulture, millets, pulses etc.
5. Nurseries at block and regional level mini nurseries with locally suitable tree species
6. Avenue plantation on both the sides of the roads to reduce the soil erosion
7. Cascade of tanks—for effectively facilitating the excess water
8. Kottakariyar river rejuvenation—includes tank cascades, artificial recharge structures and check dams
9. River bank stabilization through plantation with diverse tree species to arrest breakage and flooding and
10. Coastal shelter belts with non-mangrove and mangrove species.

Tiruvannamalai district:

1. Greening of Hillocks to reduce land degradation and soil erosion
2. Agroforestry and Integrated farming systems for strengthening the recycling of water, improve soil and land management
3. Silvi-pasture development
4. Dry land agriculture/ horticulture for fallow land restoration
5. Nursery raising with locally relevant tree species for plantation
6. Cascade Tanks
7. River Rejuvenation for recharging the aquifers
8. Farm ponds for in-situ water harvesting and use
9. Spring sheds in the hilly areas
10. Bamboo cultivation in public lands and
11. Recharge Shaft—Borewell recharge structures.

Actions to Restore Water Resources

- Watershed profile including the natural drainage lines: The finer scale mapping and deeper understanding on the issues supported to practice suitable actions. The approach adopted ridge to valley treatment in watershed by analysing micro watersheds and drainage lines along with other land use.

Table 15.6 Expected impacts on water resources from the identified water actions

Sl. No.	Impact indicators	Ramanathapuram	Tiruvannamalai
1	Number of water bodies restored	7,561 number of traditional water bodies restored	5,264 number of traditional water bodies restored
2	Area under afforestation	40,704 Ha brought under tree cover	26.8% brought under tree cover
3	Area under linear plantation (tank bunds and avenue)	2,555 km length of linear plantation established	2,641 km length of linear plantation established
4	Percentage reduction in the annual surface runoff	219.38 MCM runoff harvested and stored	407 MCM runoff harvested and stored
5	Proportion of total geographical area treated under WASCA	19%	29%
6	Drainage line treatment	1476 km length treated	8,488 km length treated
7	Area under silvi-pasture development	191 Ha	2907 Ha

- Existing water recharge or storage structures and canal networks: The restoration of water storage structures includes deepening and de-silting of active storage areas, providing silt traps at inlets, bund strengthening and planting as well as weir repair and construction of sluices, restoring the connecting channels of tanks to establish cascade links are significant actions.
- Status of the ground water: Artificial recharge structures both at common and individual lands, check dams, check walls, percolation tanks, sunken bunds, contour bunds, water absorption trenches, compartmental bunds etc.
- Run-off management: The catchment profile-based planning is proposed by assessing the type of land and its current and past use pattern with the categorization of good, average and bad catchment.
- Water demand estimation: Sector wise water demand is worked out namely human, livestock and agriculture sectors
- Water budgeting: The water budget is estimated using surface runoff water, ground water, soil moisture and evapotranspiration with water demand for different sectors (Table 15.6).

Adaptation Measures in Agriculture

- Soil health management: Measures that improve soil fertility as well as conservation were proposed including composting, bund plantation with fast growing nitrogen fixing plants and mulching, Farm bund with trench cum bund to allow excess water flow out of the farmland, improve moisture conservation and have better drainage are few important illustrations

- Landscape based water actions: Actions for each of the land types—common and individual with a set of logics were applied to identify the potential areas for actions in each of the land use types. Through these measures 19% of the additional area has been proposed under WASCA with different soil and water conservation actions
- Crop cultivation: Diversification of cropping system with low water requirement crops and cropping systems, promoting crop intensification with inter/mixed crops and agro forestry etc. and increasing the water use efficiency within the field
- Livestock resources: Forage needs are crucial for livestock as both the districts have limited scope for irrigation to raise grasses under irrigated conditions: hence focus is given to actions such as silvi-pasture, agro-forestry with trees having forage value, azolla, promoting good rearing practices by ensuring infrastructures like sheds, troughs, composting units etc.
- Irrigation profile: Improve the conveyance efficiency by restoring the supply channels, promoting improved irrigation methods including micro irrigation, alternate wetting and drying in paddy etc. (Table 15.7)

Linkages with SDGs and INDCs

Climate change is of global issue and closely linked to lives and livelihoods of human and environment. The changes are quite dynamic requires constant adaptation and actions to mitigate the reduction of GHGs. The current level of actions to combat the climate risks are not sufficient to meet the increasing challenges such as from highly unpredictable rainfall, rising temperature and sea level, increasing tropical cyclones and low depression in oceans. Although adaptation measures are on-going, need more innovative frameworks that works at scale with the participation of associated stakeholders. Paris Agreement 2016 provided such framework to the countries at global level and reflect such climate solutions in Nationally Determined Contributions. At national and state level, National Action Plan on Climate Change and Tamil Nadu State Action Plan on Climate Change has been developed integrating the local context and advocated short, medium- and long-term actions. Different water actions developed to promote adaptation and mitigation strategies have been mapped and potential links with climate vulnerability, SDGs, and INDCs were carried out (Tables 15.2 and 15.3). The mapping helps to understand and prioritize the climate actions which has more weightage to address climate change issues and at same time ensure water resilience.

Convergence and Financing

The bottom up CWRM plan aims to achieve "Climate Resilience for Future Livelihoods" through the climate solutions by working closely with MGNREGS and

Table 15.7 Expected impacts in Agriculture and allied activities from identified water actions

Sl. No.	Impact indicators	Ramanathapuram	Tiruvannamalai
1	No. of structures established for on-farm (in-situ) water harvesting in dry lands	10,084 farm ponds established in dry lands	9,482 farm ponds established in dry lands
2	Reducing area under fallow lands	84,537 Ha of fallow land restored for cultivation (22% of total geographical area)	1,41,370 Ha of fallow land restored for cultivation (31% of total geographical area)
3	Improvement in soil health—with saturation approach	2329 vermi compost units 12,268 Ha land improved through silt application, 32,926 Ha area improved by farm bunds cum trench and planting, 16,149 Ha area covered under mulching and 12,668 ha of area improved under land development	37,889 vermi compost units, 22,483 Ha land improved through silt application, 14,099 Ha area improved by farm bunds cum trench and planting, 52,989 Ha area covered under mulching
4	Area covered under micro irrigation for improved water use efficiency	7,700 Ha area adopting micro irrigation practices	20,143 Ha area adopting micro irrigation practices
5	Number of vulnerable households benefitted through fodder and azolla cultivation	19,170 vulnerable households established fodder and 2,329 units of Azolla production	27,091 vulnerable households established fodder and 33,669 units of Azolla production
6	Area brought under agro-forestry—dry land (unirrigated) land use category	1,16,332 Ha	24,892 Ha

NWM, the key stakeholders of WASCA. Apart from these, the works under WASCA are closely linked with Ministry of Agriculture and Farmers Welfare and Ministry of Environment and Forest and Climate Change. The identified works under each of the categories are primarily mapped with MGNREGS at first level for implementation under its 100 days of guaranteed employment in rural areas. Following this, other potential schemes/projects of line departments are linked under convergence action keeping MoRD at the centre. Under MGNREGS a detailed link between listed works and climate change has been mapped to identify and explain how different works are promoting both adaptation and mitigation measures at individual and community level. The convergence among the line departments has been closely facilitated both at the state and district levels through state/district steering committees as well as structured meetings under the leadership of MoRD representative (Table 15.8). These convergence planning is facilitated for two areas: knowledge sharing and funding across different actors.

Table 15.8 Indicative areas of convergence for implementing the proposed works

S. No.	Name of the department	Areas of convergence
1	Public Works—Water Resources Organization and Central Ground Water Board	Traditional water harvesting structures restoration, Canal network improvements
2	Agriculture	Fallow land—individual development, IFS and Water management—Micro irrigation
3	Agriculture Engineering	Farm ponds and Micro watersheds
4	Horticulture	Dry land horticulture—fallow land dvt, crop diversification with reduced WR
5	Forestry	Plantations in public land
6	Animal Husbandry	Silvi pasture development in common lands, livelihood activity centres—technical support for fodder development in individual lands
7	Fisheries	Alternate livelihoods
8	NABARD	Horticulture Development and restoration of water bodies

Concluding Remarks

The CWRM framework and the process adopted in bottom-up planning for mapping vulnerabilities and relevant climate resilient actions are supportive to develop a plan keeping water at central stage. As the climate change issues are primarily revolving around water in rural areas especially on the impacts of water cycle and water related extreme events (Sadoff and Muller 2009), such an approach is significant and necessary at this point of time to make decisions at grassroots level. The experiences sofar in the field are indicating its significance that the CWRM plan is a better tool and provide platform for the development agents to converge and facilitate climate solutions by leveraging the implementation with government developmental schemes.

The identification of hotspots/areas of interest based on the vulnerability mapping and climate prediction at finer spatial scale and engagement with concerned stakeholders added value to the CWRM plan as it has more relevance for decision making at the point of implementation. Such a process based, science driven framework helped to link policy and practice in operationalizing the climate resilience and water security. The participatory process enabled the implementation of decentralized water governance from bottom up and paid more attention to building the capacities of local actors and systems to absorb the responsibilities and sustain it. The climate solutions proposed are having closely associated in contributing to several SDGs and address the commitments in Paris Agreement.

Recommendation

- The works proposed in CWRM plan is mapped for two important international policies such as Paris Agreement as well as SDGs. However, considering the unpredictable climate risks, linking the analysis with Sendai framework of disaster risk reduction will help to bring the dimension of preparedness, reduction and restoration over the climate adaptation solutions.
- Access to reliable and timely availability of spatial and non-spatial data for the lowest planning unit are to be readily accessible to the planners at the village level.
- Continuous capacity building and concerted actions to implement the proposed water actions/climate solutions under convergence approach is essential to make a positive impact on water budget.

Acknowledgements The authors sincerely thank the Department of Rural Development and Panchayath Raj, Tamil Nadu, the District Rural Development Agency Ramanathapuram and Tiruvannamalai, Ministry of Rural Development for their valuable inputs and their continuous support and Deutsche Gesellschaft für Internationale Zusammenarbeit (GIZ) Gmbh for their technical inputs and financial support.

References

Deutsche Gesellschaft für Internationale Zusammenarbeit (GIZ) GmbH (2019) Water security & climate change in rural India. Retrieved on 18 December from https://www.giz.de/en/downloads/giz2019-en-wasca-india.pdf.
Global Commission on Adaptation (2019) Adapt now: a global call for leadership on climate resilience. World Resources Institute, Washington, DC. © Global Commission on Adaptation. https://openknowledge.worldbank.org/handle/10986/32362 License: CC BY 4.0 International
Global Water Partnership (2000) Integrated water resources management. Global Water Partnership Technical Advisory Committee Background Papers, No. 4, 2000. https://www.gwp.org/globalassets/global/toolbox/publications/background-papers/04-integrated-water-resources-management-2000-english.pdf
Gulati A, Sharma B, Banerjee P, Mohan G (2019) Getting more from less: story of India's shrinking water resources, NABARD and ICRIER report, Indian Council for Research on International Economic Relations, New Delhi, p 170
Hegga S, Kunamwene I, Ziervogel G (2020) Local participation in decentralized water governance: insights from north-central Namibia. Reg Environ Change 20:105. https://doi.org/10.1007/s10113-020-01674-x)
ICWE (1992) The Dublin Statement on Water and Sustainable Development, International Conference on Water and the Environment, Dublin, 26–31 January 1992
IPCC (2021) Climate change 2021: the physical science basis. In: Masson-Delmotte V et al (eds) Contribution of Working Group I to the Sixth Assessment Report of the Intergovernmental Panel on Climate Change. Cambridge University Press, in Press. www.ipcc.ch/report/sixth-assessment-report-working-group-i/

Mauroner A, Timboe I, Matthews J, Taganova J, Mishra A (2021) Planning water resilience from the bottom-up to meet climate and development goals. UNESCO and AGWA, Paris, France and Corvallis, USA
MoEFCC (2021) Third Biennial Update Report to the United Nations Framework Convention on Climate Change. Ministry of Environment, Forest and Climate Change, Government of India. Retrieved on 12 December from https://unfccc.int/sites/default/files/resource/INDIA_%20BUR-3_20.02.2021_High.pdf
MoRD (2021) Annual master circular of Mahatma Gandhi National Rural Employment Guarantee Act 2005. Ministry of Rural Development, Government of India, Retrieved on 28 October from https://nrega.nic.in/Netnrega/WriteReaddata/Circulars/2419Annual_Master_Circular_2020-21_English.pdf
NWP (2012) National water policy. Retrieved on 12 December from http://nwm.gov.in/sites/default/files/national%20water%20policy%202012_0.pdf
OECD (2015) OECD principles on water governance. Adopted by the OECD Regional Development Policy Committee on 11 May 2015. Welcomed by Ministers at the OECD Ministerial Council Meeting on 4 June 2015. Retrieved on 12 December from https://www.oecd.org/cfe/regionaldevelopment/OECD-Principles-on-Water-Governance-en.pdf
Palanivelu K, Ramachandran A, Mudgal BV, Jeganathan A (2019) Scoping study report of Indo-German project: Water Security and Climate Adaptation in Rural India (WASCA) Tamil Nadu
Purohit JK, Tyagi K, Sowmithri VR (2020) GIZ India, Deutsche Gesellschaft für Internationale Zusammenarbeit (GIZ) GmbH. Water Security & Climate Change in Rural India
Sadoff C, Muller M (2009) Water management, water security and climate change adaptation: early impacts and essential responses. Global Water Partnership, Stockholm, Sweden
UNESCO, UN-Water (2020) United Nations world water development report 2020: water and climate change, UNESCO, Paris
WRI (2019) Aqueduct 3.0: updated decision-relevant global water risk indicators, World Resources Institute, USA. https://www.wri.org/data/aqueduct-30-country-rankings

Chapter 16
Innovative Ways to Mobilise Private Sector Capital in Climate Change Adaptation Investments in Developing Countries—Mechanisms and Forward-Looking Vision from Practitioners' Standpoint

Ayaka Fujiwara and Rajeev Mahajan

Abstract Developing countries are faced with a unique challenge. In addition to adapting to the extreme climate change events that are affecting their economies and citizens, they are also striving for the economic, industrial, and social development needed to lift their growing populations out of poverty. For this development to be aligned to the new climate reality, private sector investment needs to contribute to both goals. While many governments are prioritising climate investments and sustainable growth, their efforts to increase private investment are often constrained by investor perceptions of high-risk and the insufficiency of local technical and financial structuring capacity and know-how. Though investors are interested in the opportunities offered by climate investment in developing countries, the underdeveloped investment environment often hinders the mobilisation of private capital, especially in relation to climate change adaptation. Such challenges have become even more pronounced in the aftermath of COVID-19. International and multilateral agencies such as Green Climate Fund can play a pivotal role in alleviating the risk perceptions of the private sector. Their role can include supporting governments to develop adaptation plans, developing mitigation projects with co-adaptation benefits to enhance resilience, developing public–private partnerships in climate, demonstrating commercially viable investments through risk-sharing with a partner from the region, supporting a creation of a viable business model in climate adaptation projects. Another critical element during these critical times is preventing continued climate impacts. In this regard, supporting climate adaptation technologies through innovative blended finance structure and creating a conducive environment

A. Fujiwara (✉)
Climate Investment Specialist, Private Sector Facility, Green Climate Fund, 175 Art-Center Daero, Yeonsu-gu, 22004 Incheon, South Korea
e-mail: ayaka.fujiwara.1211@gmail.com; afujiwara@gcfund.org

R. Mahajan
Project Finance, Private Sector Facility, Green Climate Fund, 175 Art-Center Daero, Yeonsu-gu, 22004 Incheon, South Korea
e-mail: mahajan_rajeev@hotmail.com; rmahajan@gcfund.org

W. Leal Filho et al. (eds.), *Climate Change Strategies: Handling the Challenges of Adapting to a Changing Climate*, Climate Change Management,
https://doi.org/10.1007/978-3-031-28728-2_16

by familiarizing local financial institutions with lending to MSMEs will be crucial. Using relevant case-studies from the Green Climate Fund's portfolio, this chapter explores how such mechanisms can be put into place and provides a forward-looking vision of what more can be done to promote private sector climate change adaptation investments on a sustained basis.

Introduction

Climate change is a global challenge. But its impacts differ hugely between developing and developed countries.

Developing countries bear more damage from adverse climate impacts because of their lesser ability to cope with damage caused by severe climactic events, erratic weather patterns and rising sea levels. On the other hand, the developed countries are affected less, because of their higher capacity and available financing to manage any immediate climate events. In addition, the governments of developing countries are also faced with the challenge to deliver rapid economic and social development to respond to the needs of their growing populations. In short, developing countries today face unique challenges, because they need to adapt to the extreme climate change events that are affecting their economies and citizens, while working on overall economic development in the country. It is particularly challenging for developing countries to respond to these challenges by public investments alone. Thus, private investments are increasingly becoming critical for countries to adapt to the new climate reality.

It has been widely recognized—most recently at the 2021 COP26 in Glasgow—that the pledge to reach USD 100 billion in climate finance per year has not yet been met (Nature 2021). And the vast majority of climate finance is mobilized for mitigation goals, not for adaptation. The adaptation finance gap is still huge. While adaptation finance has gained momentum in 2019/2020, increasing 53% to an annual average of USD 46 billion from USD 30 million in 2017/2018, it still falls well short of investment in climate change mitigation, and is significantly below the estimated annual adaptation needs of USD 180 billion (Climate Policy Initiative 2021).

Much more needs to be done, and the contribution of the private sector will be key to closing the adaptation gap. Against this backdrop, it would make a lot of sense for leveraging public money to facilitate risk taking by private sector investors in developing credible business models related to climate adaptation.

Limitations of the Public Sector—Why Governments Alone Cannot Achieve Paris Agreement Targets on Their Own

As the world struggles to emerge out of successive waves of the COVID-19, public finances are becoming increasingly strained. With supply chains broken and tax collection becoming increasingly erratic, governments around the world are increasingly focusing on reviving their economies whilst shouldering the economic burden of pandemic induced healthcare costs.

In this context, climate investments can be the worrying victim of the pandemic, although a green, resilient recovery gives the best opportunity to deliver benefits for both economic development and climate change. To prevent a slide towards an unsustainable future from a climate perspective, there is a growing recognition that if the governments cannot themselves invest substantial amount of capital, then the private sector could potentially play a critical role in plugging in the missing investments, towards the USD 100 billion target.

Against this backdrop, governments will have to reimagine and recast their role by firstly recognising their fiscal constraints yet recognising their inherent capacity and capability to set the stage to enable private sector investments to come in.

There is much discussion about barriers to climate adaptation and resilience investments. The World Bank Group in a recent report has identified key barriers impeding investments in adaptation and resilience. Firstly, while there is a recognition on the criticality of resilience investments, there is inadequate climate risk and vulnerability data and a general lack of information services at a country level, which collectively impedes investment decision-making (World Bank Group 2021). Secondly, despite a growing recognition for significant investments in adaptation and resilience, there is limited clarity on where private investment is needed to fill in investment gaps that governments or the public sector is not able to fill in, due to their fiscal and capacity constraints.

Thirdly, critical from the private sector's perspective is the perception of low returns on investment, that inhibits their investment appetite in the sector (World Bank Group 2021).

The report also identifies ambiguous and restrictive regulatory frameworks as potential impediments to climate adaptation and resilience investments.

Furthermore, it must be recognised that attracting investments into the climate adaptation and resilience space is not merely a climate challenge but also more fundamentally an investment challenge, requiring an integrated approach.

Governments may require support in developing institutional capacity to address the barriers in a systematic and integrated fashion. These measures include adopting an interdisciplinary and integrated approach for building a suitable, conducive and predictable investment climate for attracting investments in the climate adaptation space.

Setting up inter-ministerial and inter-departmental forums to work in a coordinated fashion can not only address the three aforementioned barriers, but could also remove bottlenecks and establish a conducive regulatory and investment environment. It is

Table 16.1 Financing for adaptation and mitigation split by public and private sources (USD bn)

Sources	2019	2020	2019/2020 Average
Private	280	340	310
Adaptation	0	2	1
Mitigation	279	335	307
Multiple objectives	1	3	2
Public	343	300	321
Adaptation	42	48	45
Mitigation	288	240	264
Multiple objectives	14	12	13
Total	**623**	**640**	**632**

Source Reproduced with permission from Climate Policy Initiative. Original table from Climate Policy Initiative's "Global Landscape of Climate Finance 2021", directly quoted from Page 44

only then that private sector investments will flow into the climate adaptation and resilience space on a sustained basis.

The challenges surrounding private sector investments in the climate adaptation space are exemplified by the fact that in 2019–2020, only about 2 percent of adaptation funding came from corporations and institutional investors (Climate Policy Initiative 2021). The dominant actors in adaptation finance are from public sector, with the main actors being the multilateral Development Financial Institution (DFI)s, national DFIs, governments, bilateral DFIs, multilateral climate funds (Table 16.1).

A five-step approach has been identified for credible and sustained action to attract private sector investment in adaptation (World Bank Group 2021). This approach begins with support to develop long term adaptation strategies in coordination with international partners and agencies that have adaptation expertise, which would then translate into national adaptation investment plans, led by national level actors, based on their country priorities. This would then lead onto market assessment and pipeline screening in coordination with private sector investors, project preparation support followed by actual implementation and development of projects led by the private sector, often funded with public resources from multilateral and national development banks and institutions such as the Green Climate Fund.

Even in the last leg, the financing of adaptation projects can be facilitated by national governments by enabling regulation facilitating investment of long-term capital by pension funds and through the capital markets into this emerging asset class.

Ultimately, as an increasing number of projects develop credibly through this cycle, risk perceptions surrounding adaptation projects as an asset class are expected to reduce, setting in place the kind of virtuous cycle seen in quite a few developing countries in the renewable energy space over the last 15–20 years.

From the perspective of developing financing mechanisms, governments will have to think about ways in which financing for climate adaptation projects becomes a lot more predictable, which essentially implies that such funding is stable from a macroeconomic perspective. Given that developing countries are quite vulnerable to exchange rate movements, which inhibit the use of foreign currency solutions, there is a need to provide for a stable source of local currency financing. Governments could facilitate mechanisms in which domestic savings are put to productive use, without putting pressure on scarce foreign exchange resources.

The Role of Multilateral Funders, Including the Green Climate Fund (GCF)

With the large funding gap, there is an urgent need to scale up and further increase public adaptation finance both for direct investment and for overcoming barriers to private-sector adaptation (UN Environmental Programme 2021). Being the biggest investor in the climate adaptation finance with total investment amount of USD 16.1 billion in 2019/2020 (Climate Policy Initiative 2021), multilateral DFIs' investment can play a pivotal role in leveraging private sector investments and alleviating the risk perceptions in the market, consequently creating a conducive investment environment.

GCF, set up by the United Nations Framework Convention on Climate Change in 2010, is the biggest multilateral climate fund in the world with over USD 10 billion of overall committed portfolio in developing countries. Its goal is to help developing countries reduce their greenhouse gas emissions and enhance their ability to respond to climate change. After commencing operations in 2015, GCF has approved USD 3.4 billion in 39 private sector projects and programmes, which aim to mobilise USD 14.5 billion, as of December 2021.

In recognition of the adaptation finance gap, GCF's Updated Strategic Plan for 2020–2023 targets an equal balance in GCF's funding between mitigation and adaptation (Green Climate Fund 2020a). With its twin focus on unlocking private investment, and closing the adaptation gap, GCF is well placed to deliver on adaptation projects that model how private sector adaptation finance can be unlocked. GCF has some unique characteristics that can facilitate the development of well-designed and impactful climate projects. In particular, GCF financing is flexible and instrument agnostic, meaning it is adaptable to meet the unique financing needs of impactful projects. GCF's approach is also characterised by a high-risk tolerance and patient financing tenors. Together these features mean that GCF can be a catalytic force to mobilise private sector capital.

De-Risking and Unlocking Private Sector Capital for Adaptation—Case Studies from GCF's Private Sector Portfolio

In this section, we will elaborate on ways in which GCF has helped deliver the goal of enhancing adaptation financing in the private sector, using some case studies from the existing investment portfolio. We will highlight a number of ways in which GCF investment can unlock private sector capital for investment, by:

1. Supporting governments to develop adaptation plans
2. Developing mitigation projects with co-adaptation benefits to enhance resilience
3. Developing public–private partnerships in climate
4. Demonstrating commercially viable investments through risk-sharing with a partner from the region
5. Supporting a creation of a viable business model in climate adaptation projects
6. Preserving continued climate impacts in the aftermath of COVID-19
7. Supporting climate adaptation technologies through innovative blended finance structure
8. Creating a conducive environment by familiarizing local financial institutions with lending to MSMEs for adaptation measure implementation.

1. **Supporting governments to develop adaptation plans**

A lack of national planning is a critical impediment to the development of investment proposals in the adaptation space. In response, the GCF Readiness Programme provides support to developing countries to develop national adaptation plans.

Support for national adaptation plans will enable developing countries to strategise and attract financing at scale for a more resilient future. The GCF sees this planning process as key to countries' ongoing efforts to bolster adaptive capacities, coordinate actions of governments, public and private sectors and attract adaptation investments from a diversity of sources to make the countries more climate resilient.

So far, 69 proposals for developing national adaptation plans aggregating USD 162.4 million have been approved of which USD 68.5 million has been disbursed, including support for 34 SIDS and LDCs (Green Climate Fund 2021d).

From a private sector investment perspective, the GCF also focuses on the development of policy guidelines or regulations to remove barriers and incentivize adaptation investment. Support is also given for integrating private sector actors engaged in national, sectoral and/or sub-national adaptation planning, including planning for climate resilience of individual businesses and supply chains. Lastly, but equally critically, readiness funds can help match private financiers with solutions.

2. **Developing mitigation projects with co-adaptation benefits to enhance resilience**

As outlined above, most private sector climate investments are in the mitigation segment. While it is still relatively difficult for private sector to commit to investments in adaptation projects considering the aforementioned barriers, investments in mitigation projects with adaptation benefits are gaining more attention. The following two case studies will elaborate on an investment programme where GCF has successfully invested in a mitigation project with significant co-adaptation benefits.

Case Study: Arbaro Fund—Sustainable Forestry Fund (Arbaro Fund), Accredited Entity: MUFG Bank, Ltd: Category: Mitigation with Adaptation Benefits

Article 5 of the Paris Agreement emphasises the critical role that joint mitigation and adaptation approaches can play in sustainable management of forestry resources. Accordingly it is increasingly being recognised that policy approaches and incentives for reducing emissions from deforestation will play a significant role in battling climate change (UNFCCC 2015).

In 2020 March, the Arbaro Fund was approved by the GCF Board as a mitigation programme with adaptation co-benefits. The Arbaro Fund is an equity fund size with USD 140 million, out of which USD 25 million was invested by the GCF. GCF came in as an investor in the second close of the fund, with the objective of mobilising investments by private sector investors, following the first close, which raised USD 60 million mostly from public sector investors.

The fund focuses on sustainable forestry plantation investments in Latin America, the Caribbean, and Sub-Saharan Africa. Over its life, Arbaro Fund will aim to sequester 20 million tons of CO_2 through carbon sinks that are created from the sustainable forest plantation sites.

Arbaro fund supports sustainably managed and Forestry Stewardship Council (FSC) certified forest plantation and wood production, meaning that new forests will be planted in non-forest areas or long-deforested areas. The fund's approach is pioneering as FSC certified forestry management is still not prevalent in the targeted countries. The intervention of GCF helped facilitate the investments in forestry projects that are compliant with FSC standards, which is especially a gamechanger, in Latin America and African, where the ratio of forestry projects that comply with FSC is just 5% and 2%, respectively. While FSC certification is still not a common practice in the target region due to its rigorous standards, it is an innovative approach in the target region that could make the industry more sustainable eventually.

In line with FSC certification, Arbaro Fund will maintain and enhance the social and economic well-being of local communities and employees. Employment is an important adaptation co-benefit in Latin America and Sub-Saharan Africa where Arbaro Fund is investing, as the underemployment in the rural economy is a severe problem. As most of the population strongly depend on agriculture, the rural community is vulnerable to extreme climate events and climate change.

Employment in sustainable forestry improves livelihoods and increases resilience of the rural communities. The plantation of sustainable forestry will also contribute to improved resilience of the surrounding landscapes, as by following the strict environment requirements under FSC, the programme will seek to afforest or reforest degraded landscapes for the commercial, yet sustainable wood production (Green Climate Fund 2020a, b, c, d).

The programme is under implementation and has already received about USD 17.5 million of disbursement (Green Climate Fund 2021a, b, c, d, e, f).

Effectively, the Arbaro Fund has established a profitable mitigation case with significant adaptation benefits, which is a much-needed demonstration for private sector players. GCF's participation addressed the critical risk perceptions that private sector investors may have that lessens their appetite to invest in sustainable forestry investment.

3. **Developing public–private partnerships in climate**

In the adaptation sector where private sector actors find it too risky to participate, one effective approach is to form a public–private partnership. The following case study demonstrates how GCF's Private Sector Facility has successfully formed a public–private partnership by deploying different financial instruments.

Case Study: Global Fund for Coral Reefs, Accredited Entity: Pegasus Capital Advisors (PCA): Category: Adaptation

Coral reefs are biologically rich and productive ecosystems that provide valuable benefits to coastal communities. These ecosystems are important sources of food and income for local communities. They serve as nurseries for commercial fish species, generate income from tourism and equally importantly, provide protection from storms.

However, coral reef ecosystems are fragile. They are extremely vulnerable to the impacts of climate change. Warming seas have already caused widespread damage to coral reefs, thereby threatening coastal economic activity. Thus, it is imperative and essential to act with urgency.

Recognizing that healthy reefs and coastal ecosystems are the bedrock for sustainable livelihoods for reef-dependent communities, there is a need to increase investments aimed at enhancing the resilience of coral reef ecosystems. And much of these investments have to come from the private sector.

To stimulate investments in the sector, a unique public–private partnership model, Global Fund for Coral Reefs (GFCR), was conceptualised. The GFCR is comprised of both a grant window and an investment window.

The grant window has the presence of experienced technical partners and UN-agencies providing a mechanism for sectoral development, regulatory and government engagement, addressing policy or institutional barriers and facilitating early-stage project development activities such as technical and feasibility studies. The grant window deploys grants for technical assistance, financial structuring, baseline studies, site monitoring and impact assessment. The grant window is capitalised with contributions from European governments and also from philanthropic foundations.

The investment window on the other hand is structured as a private equity fund and would provide catalytic equity capital to private sector businesses and activities that reduce or eliminate existing local stressors in the most resilient reefs or climate refugia. The investments are intended to be made in three different sectors, namely (i) Sustainable ocean production (ii) Sustainable ecotourism and (iii) Sustainable infrastructure and waste management.

The investment window benefits from the early-stage groundwork already done by the grant window that creates a conducive environment for private sector investors to invest in the sector, facilitated by the catalytic intervention of the investment window.

The investment window aims to address critical financing and investment barriers which include (i) limited investor interest in investing in themes centred around the blue economy in general and coral reefs in particular; and difficulties in assessing investment opportunities, (ii) high risk perception as coral reef businesses are new-age businesses and relevant models are not well proven, (iii) policy, institutional and regulatory gaps and (iv) adverse impacts of COVID-19.

The grant window and the investment window are designed to work in a synergistic fashion. The two will enter into an implementation framework agreement with the UNDP as manager of the grant window. The agreement will set forth a framework under which the investment window and the grant window will collaborate to enhance the technical expertise of GFCR, including but not limited to pipeline screening, fundraising, communications, field engagements and reporting.

Furthermore, an advisory committee shall be set up to provide recommendations and advice to the grant window's executive board and the investment committee of the investment window. The committee will consist of scientists, representatives from national governments and public institutions, blue economy experts and other independent stakeholders, as appropriate and relevant (Green Climate Fund 2021a, b, c, d, e, f).

Effectively, the GFCR functions as a public private partnership to unlock hard-to-get investments in the climate adaptation space with the grant window paving the way for private sector investments to be unlocked.

The GCF board has approved an anchor investment in the investment window of the GFCR, as junior capital. This addresses the critical risk perceptions that private sector investors may have that lessens their appetite to invest in coral reef sector.

4. **Demonstrating commercially viable investments through risk-sharing with a partner from the region**

Due to the lack of track record in particular region, local financial intermediaries often find it difficult to finance innovative climate technologies at scale. The following case study shows an investment programme where GCF aims to alleviate perceived risks in the market by demonstrating commercially viable investments in climate mitigation and adaptation, in collaboration with an Accredited Entity from the targeted region. The case study will also elaborate on how GCF financing has supported to upscale of the financing towards vulnerable beneficiaries of the climate, who often do not get enough financing due to the perceived risks.

Case Study: Low Emissions and Climate Resilient Agriculture Risk Sharing Facility (Facility), Accredited Entity: Inter-American Development Bank (IDB): Category: Cross-Cutting

In November 2017, GCF Board approved an investment to finance a risk sharing facility in climate resilient agriculture, to unlock innovative and scalable financial instruments for MSMEs in Guatemala and Mexico.

In Mexico and Guatemala, there is a clear causal link between agro-forestry and climate impacts because the important drivers of deforestation are poor practices in commodity food production, such as coffee, cocoa, and avocados. Such unsustainable food production drives down productivity and forest conversion.

While Climate-Smart Agriculture (CSA) projects in the countries are much needed, they face the difficulty in securing financing due to the lack of appetite from the local financial intermediaries who do not offer tailored products in this sector. Given similar challenges that Latin America and the Caribbean (LAC) countries face, the project should provide a much-needed track record of a successful business model for peripheral countries in the same region.

The risk sharing facility is created by GCF and Inter-American Development Bank (IDB), the Accredited Entity, and it will support MSMEs that demonstrates environmentally sustainable practices in agro-forestry. The MSMEs will use the long-terms loans from the facility for CSA investments.

GCF's hypothesis is that with its pioneering anchor investments in early-stage companies, it would be able to demonstrate the viability of these innovative ideas thereby overcoming high-risk perceptions in CSA projects, ultimately building the confidence of national, regional and private sector financial intermediaries, capital markets, insurance providers to invest in the CSA sector on a continued basis.

GCF's equity investment will support conceptualisation and launch of a landscape restoration equity fund as well as development of a regional re-insurance business model focused on catastrophic climate risks such as droughts and storms. These insurance/re-insurance products are currently either not available in the market or when available are not cost effective, leaving MSMEs exposed to bear these risks on their own.

With GCF's participation in the guarantee component, perceived risks of financial intermediaries towards CSA investments, especially those stemming from lack of collateral and differing perspectives on collateral valuation, would be significantly alleviated, thereby allowing the intermediaries to scale up their funding in the sector.

In addition to investment via equity and guarantees, GCF will also invest through loans and grants. GCF's participation through different financial instruments will allow private sector investors to comfortably invest to the facility along with GCF and is expected to eventually result into a significant additional private capital being channelled into the sector of sustainable agriculture.

The project is currently being implemented, and USD 8.54 million out of the total of USD 20 million of GCF approved investment has been disbursed so far (Green Climate Fund 2017).

Case Study: Leveraging Energy Access Finance (LEAF) Framework, Accredited Entity: African Development Bank (AfDB): Category: Mitigation with Adaptation Co-Benefits

LEAF is USD 900+ million framework with the objective of unlocking local currency financing and overcome market barriers to support the growth of Distributed Renewable Energy (DRE) in Africa. Through its catalytic participation in the framework, GCF aims to enhance participation of local banks and financial institutions by de-risking their participation through guarantees and subordinated debt. Additionally, in addition to its financing participation, GCF's financing also aims to build and enhance institutional capacity within local banks and financial institutions to appraise and finance DRE businesses.

GCF's contribution to the framework is USD 170.9 million, out of which both loan and guarantees are USD 80 million each. The loan and guarantee would enable the deployment of commercial and institutional renewable solutions, solar home systems and mini grids at scale, which would provide households and small businesses with emissions-free access to energy. GCF is also contributing a grant of USD 10.9 million to the project.

The framework has significant adaptation co-benefits, as solar-powered assets would strengthen communities' adaptive capacity and resilience through economic, social, and environmental co-benefits. The electricity systems and appliances can result into income-generating activities among the local community and would make the community more climate resilient by allowing more access to formal and informal support networks and recover faster from impacts from disasters (Green Climate Fund 2021a, b, c, d, e, f).

Both aforementioned case studies present excellent opportunities for local banks and financial institutions to get a flavour of investing in the sector with risks assumed by the GCF, thereby providing them with a relatively low risk proposition of "learning by doing".

5. **Supporting a creation of a viable business model in climate adaptation projects**

Commercially viable business model is one of the biggest reasons why private sector actors are often hesitant to invest in climate adaptation project. It is a challenge to create a viable business model with climate adaptation beneficiaries who often not have enough collaterals or cash to finance climate adaptation practices, due to their decreased productivity by the adverse climate impacts. The following case study will show how GCF has supported a creation of viable business model in climate adaptation.

Case Study: Acumen Resilient Agriculture Fund (ARAF), Accredited Entity: Acumen Fund, Inc

With one seventh of the world's population, Africa is expected to bear nearly half of estimated global adaptation costs in health, water supply agriculture and forestry. Agriculture is a critical sector in Africa, where large populations are engaged in primitive agriculture. While the sector is heavily impacted by climate change, the development in the sector has not accelerated yet, particularly in East and West Africa.

Since advancement in agriculture across most countries in Africa are in a nascent stage, climate resilient investments could proceed smoothly without the costs of unwinding legacy investments. Yet, the investment from private sector investors in the climate resilient agriculture has been limited, due to the risk perceptions, the concerns in commercial viability well as the lack of track records.

Climate resilient agriculture is key to ensuring a long term sustainable increase in agricultural income and productivity for 2.4 billion of smallholder farmers across the world. In the process, it is essential that the climate resilience in agriculture is strengthened in alignment with ongoing climate change impacts. Recognizing the needs above, Acumen Resilient Agriculture Fund (ARAF), the world's first climate adaptation focused equity investment fund for smallholder farmers, was conceptualised to build climate resilience through investments in three different areas.

1. Agribusiness platform that will provide solutions to key needs of the farmer including inputs, knowledge, finance and buyers, so that the farmers could be integrated in into the markets;
2. Digital platforms that will enhance the supply chain efficiency through providing farmers access to information on knowledge, inputs, finance and markets; and
3. Financial services that will enable the payment, credit and insurance solutions to the farmers.

By offering the above services as a bundled package, ARAF aims to establish an ecosystem that allows farmers to raise their incomes and increase their resilience. The ecosystem will allow investees to build climate resilience, through developments of crop choices that are aligned with the climate forecasts and spreading adaptation tools and techniques, which will strengthen the stability of the farmers' income.

In 2018, GCF board approved an anchor investment to ARAF as first loss capital, to address the critical risk perceptions that private sector investors may have that lessens their appetite to invest in climate resilient agriculture sector. To achieve the effect, GCF has invested through grants and equity.

GCF provided USD 3 million of grant funding as part of USD 6 million for a technical assistance facility for impact measurement, gender integration through investee companies and to support operational and ESG capacity building among investee companies.

To attract capital from private sector investors, GCF provided the first loss equity capital of USD 23 million. Given that ARAF targeted USD 50 million, this is an anchor investment, which is essential in reducing risk to mobilise private sector capital to climate adaptation projects. With GCF's participation to the first loss tranche, private investors in the senior equity tranche would feel confident to participate to the fund investment. As a result of having GCF's first loss layer which catalysed the fund, ARAF was able to raise another USD 2 million in first loss and USD 33 million in senior equity from reputed investors such as FMO, PROPARCO and Soros Economic Development Fund. This resulted in a USD 58 million fund: USD 8 million above the initial target.

The fund is investing in portfolio companies with business models where farmers could make the best decisions despite the changing weather patterns. With no direct access to the buyers in the market, the smallholder farmers would offer their harvest to the middleman at 20–40% lower than market price and receive inputs from middleman on credits at an expensive mark-up rate, with very little information on climate resilience. ARAF invests in companies that are changing those dynamics while helping farmers increase their income and reduce their income volatility.

The solution provided by ARAF is innovative in allowing smallholder farmers to have access to buyers and access to inputs and insurance on credits. The system would remove the middleman that is lowering the income and productivity of the farmers, so that the farmers would not only be able to benefit from these services to improve the climate resilience, but also receive their income based on the market price in the off-take markets and improve the profitability at the same time (Green Climate Fund 2018b).

The project is currently under implementation. ARAF closed in June 2021 at USD 58 million, three months after the originally intended final close date due to fundraising delays occasioned by COVID-19. Out of USD 26 million of GCF approved investment, about 7 million has been disbursed so far (Green Climate Fund 2020a, b, c, d).

Through ARAF, there is an ambition to demonstrate the viability and scalability of new business models that provide climate adaptation benefits to the smallholder farmers and in turn the local community.

6. **Preserving continued climate impacts in the aftermath of COVID-19**

In 2019 and 2020, the climate investment sector has faced unique challenges due to COVID-19, which has lowered the liquidity available in already implemented climate investments. In such a critical time, concessional finance is needed not only to create new paradigm shifting investments but also to safeguard existing investments, that are at temporary risk through lack of liquidity. The following case study elaborates on how GCF aims to maintain the positive paradigm shift in the off-grid sector in Africa, which is a sector that is critical for climate adaptation.

Case Study: Energy Access Relief Facility (EARF), Accredited Entity: Acumen Fund, Inc: Category: Mitigation (In a Sector That Is Important for Climate Adaptation)

Energy access sector in Sub-Saharan Africa has experienced adverse impacts from COVID-19 pandemic. Many companies in the sector have experienced slower sales growth and reduced revenues due to the decreased funding for the sector. A significant number of the companies are struggling from low levels of liquidity, which constrains their current operations. Unless the funding gap is filled, preservation of the companies would not be possible, thereby potentially jeopardizing future growth of the sector.

The Energy Access Relief Facility (EARF) has a goal to ensure the solvency of companies operating in the energy access sector, helping them remain in business post-pandemic, so that they can continue to achieve positive climate and clean energy transformation.

In Sub-Saharan Africa, where EARF will invest, there are unique challenges in the fight to mitigate and adapt to climate change. The temperature rise in Africa will be more severe than the global temperature rise, and it will have severe impacts to multiple aspects in life such as health, agriculture and food security. Additionally, 548 million people still lack access to electricity and 900 million lack access to clean cooking fuels and technologies, which results into deforestation and high level of indoor air pollution. Africa faces an acute challenge to reduce emissions from energy sector and help vulnerable people adapt to climate change, while making electricity and clean fuels available to more people.

Until COVID-19, off-grid solar has contributed significantly to resolve this challenge and has already avoided 74 million metric tonnes of CO_2e emissions. However, progress has stagnated due to COVID-19. The operations of these companies have been endangered, which have limited cash reserves and are experiencing disrupted supply chains, as well as reduced sales and revenues.

Most governments acknowledge that a collaborative public–private partnership is needed to deliver nation-wide electrification and energy access. Especially for rural communities, private sector companies often play a pivotal role in filling the gaps where the unit economics do not make sense for grid extension. EARF's intervention to ensure the continuation of these private sector companies is critical for national climate change adaptation strategies and plans.

GCF's intervention was needed in order to maintain positive progress. It will also help companies retain over 10,000 qualified jobs in a sector that is critical for climate change adaptation in Africa. In order to mobilise other concessional finance providers, philanthropic institutions and private sector investors, GCF has invested in USD 30 million of equity into the fund as an anchor investor (Green Climate Fund 2020a, b, c, d).

The fund reached first close at USD 68 million in September 2021, meeting the expected target. It aims to support over 90 off-grid solar SMEs and protect energy access for over 20 million people. The loan tenor is about 3.5 years–5.5 years, by the end of which the negative effects from COVID-19 will have hopefully come to an end.

EARF is a timely idea that prevents small businesses from being adversely impacted by COVID-19 induced liquidity constraints in the financial sector, thereby saving the way for a continuation of mitigation and adaptation benefits.

7. **Supporting climate adaptation technologies through innovative blended finance structure**

Financial structure is critical when including more non-concessional investors to join the climate adaptation technology investment. The following case study outlines a programme where GCF aims to support the growth of adaptation technologies through its participation in concessional junior layer.

Case Study: CRAFT—Catalytic Capital for First Private Investment Fund for Adaptation Technologies in Developing Countries, Accredited Entity: Pegasus Capital Advisors (PCA): Category: Adaptation

Climate change affects water availability and quality, agriculture sector, energy production, infrastructure and human health. While adaptation has been considered traditionally as public good and service provided by government/public funds, mobilising private sector actors is becoming more and more critical to strengthen the resilience of these sectors from the adaptation perspective. The demand for adaptation technologies exists in all markets, and there is an urgent need to mobilise private sector capital and innovative technology to scale up adaptation investments in developing countries.

In October 2021, the GCF Board approved USD 100 million of catalytic equity investment to scale-up technologies for climate resilience and adaptation, in developing countries where such technologies are most needed. In the second close where GCF will be anchoring the investment, GCF is expected to additionally mobilise USD 136 million of senior non-concessional capital, which the fund was struggling to mobilise because of the effect of COVID-19. The total fund size is expected to be USD 400 million.

CRAFT mobilises technologies that reduces the exposure to climate hazard by resilient infrastructure, intelligence application and knowledge, as well as other risk-reducing assets. It will support investments in the following six categories of technologies: (1) Catastrophe Risk Modelling and Weather Forecasting, (2) Agricultural

Analytics (3) Supply Chain Analytics, (4) Geospatial Imaging and Mapping, (5) Water Harvesting and Efficiency and (6) Resilient Food Systems.

The intervention of CRAFT is critical in the growth of adaptation technologies in developing countries, where the adaptation intelligence is yet to be developed. The unique and innovative blended finance structure model is designed in a way that a junior concessional layer will mitigate downside risk for senior non-concessional investors, encouraging adaptation investment (Green Climate Fund 2021a, b, c, d, e, f).

CRAFT intends to facilitate technology transfer and mainstream cutting-edge technologies in developing countries, for enhancing and accelerating climate adaptation benefits.

8. **Creating a conducive environment by familiarizing local financial institutions with lending to MSMEs for climate adaptation**

While adaptation measures are most needed by vulnerable beneficiaries, the local financial system may not always offer suitable financial services for them. For instance, when MSMEs attempt to borrow money to improve their climate resilience through adaptation measures, there are very few options available in the market, especially in developing countries. The situation is often caused by the lack of familiarity by the local financial institutions in lending to not only to MSMEs, but also for adaptation measures. The following case study shows how GCF aims to create a conducive environment through an initiative to de-risk local financial institutions to lend to MSMEs for adaptation measures.

Case Study: Productive Investment Initiative for Adaptation to Climate Change (CAMBio II), Accredited Entity: Central American Bank for Economic Integration (CABEI): Category: Adaptation

Productive Investment Initiative for Adaptation to Climate Change (CAMBio II/Initiative) aims to increase climate resilience of MSMEs in Guatemala, El Salvador, Honduras, Nicaragua, Costa Rica, Panama and the Dominican Republic. Its objective is to remove barriers to access financial and non-financial services to adopt and implement climate adaptation measures.

Although agriculture, livestock and forestry activities are the foundations of the economy in the targeted countries, MSMEs from the sector struggle with limited financing options. In most Central American countries, credit to the whole agricultural sector accounts for only about 3% of total credit offered by financial systems, and the number is expected to be lower among MSMEs, and even lower for adaptation purposes. Given that the sector is vulnerable to climate change, production from the sector is likely to drop if climate resilience is not strengthened to combat aggravating climate factors.

This initiative has been established by CABEI and GCF with an understanding of the importance and challenges of the MSMEs in the target region and sector. The programme will provide USD 12.5 million of senior loans to provide more credit to MSMEs for adaptation solutions. GCF will also invest USD 3 million of grant for capacity building. The total size of the programme is USD 28 million.

By having local Intermediary Financial Institutions (IFIs) pass senior loans to beneficiary companies of the targeted countries, the initiative aims to increase awareness among financial institutions of potential market opportunities for lending to MSMEs that invest in adaptation measures to improve their climate resilience. With USD 25 million in available loans, it will reduce barriers to credit for 5,000 MSMEs targeting 69,700 beneficiaries.

With the USD 3 million of funds for capacity building, the programme will develop production models that are resilient to climate change. Capacity building includes workshops as well as financial rewards to MSMEs and IFIs for their successful implementation of adaptation activities (Green Climate Fund 2018a, b, c).

The programme was approved by the GCF Board in 2018 and so far, about USD 400,000 has been disbursed (Green Climate Fund 2021a, b, c, d, e, f).

The intention of CAMBio II is to mainstream adaptation financing business for both MSMEs and local IFIs, to in turn provide easy access to capital for new adaptation businesses thereby ensuring replicability and continuity.

Conclusion—Roadmap Going Forward

Case study	Key learnings
GCF Readiness Programme	Financial support for developing investment plans for adaptation is critical
Arbaro Fund	Developing mitigation projects with co-adaptation benefits can enhance climate resilience, provided industry best practices, certifications and safeguards are incorporated
Global Fund for Coral Reefs	Stimulating financing for adaptation projects through a unique public-private partnership model
Low Emissions and Climate Resilient Agriculture Risk Sharing Facility	Catalytic risk sharing facility to stimulate investments by banks in MSMEs engaged in climate resilient agriculture
Leveraging Energy Access Finance (LEAF) Framework	Financing of mitigation investments with significant adaptation co-benefits with the catalytic risk financing promoting learning by doing

(continued)

(continued)

Case study	Key learnings
Acumen Resilient Agriculture Fund (ARAF)	Equity financing through an investment fund structure finance adaptation and reliance investments among small holders
Energy Access Relief Facility (EARF)	Maintaining and preserving climate adaptation impacts of distributed energy projects through an emergency liquidity relief facility in the aftermath of COVID-19
CRAFT—Catalytic Capital for First Private Investment Fund for Adaptation Technologies in Developing Countries	Concessional and catalytic junior equity to stimulate adoption of climate adaptation technologies
CAMBio II	Mainstream adaptation financing business for both MSMEs and local IFIs, to in turn provide easy access to capital for new adaptation businesses

These case studies show practical solutions for the mobilisation of the private sector capital and deployment of adaptation finance. But further innovations will be needed beyond the ones outlined above. To conclude this chapter, here are a few ideas that could continue to drive an increase in private investment for adaptation in developing countries.

Bearing in mind that climate adaptation investments often lack an investment track record, one of the approaches that can be used is to finance adaptation projects in their developmental phases, through mechanisms such as GCF's Project Preparation Facility. This will allow private sector investors to only invest in projects when they have become commercially viable. While such approach is more time consuming, it would allow the market to accumulate track records for concessional finance in this nascent sector.

A novel approach adopted in FMO's Climate Investor One (Green Climate Fund 2018a, b, c) programme can be adapted for adaptation projects as well. This approach involves a multi-tiered investment fund comprising of sub-funds to address the specific needs of each phase of project development. The first level is a development fund, to provide scarce developmental capital at a project's high-risk developmental phase. The next tier involves a construction equity fund that can potentially finance the relatively high-risk construction phase, which many commercial banks may not be able to finance. Once the projects are up and running, revenue models are established, and viability is demonstrated, then they can be refinanced through a refinancing fund, co-opting the participation of private sector bank and local financial institutions (Convergence, 2021). In Climate Investor One, GCF supports both the development fund and the construction equity fund. Such an approach can be replicated for adaptation projects as well.

Other novel mechanisms of raising adaptation finance can also be explored. These could include the development of mechanisms that facilitate the participation of capital markets in financing these projects. Given that capital markets would typically seek a credit rating, mechanisms of credit enhancement can be contemplated. Another approach could be to create a portfolio of operational projects that are securitised in the capital markets and where securitisation proceeds are deployed to finance new and greenfield adaptation projects. This approach will unlock the participation of pension funds, that are uniquely placed to provide longer term liquidity when compared with banks that are constrained by Basel norms.

Institutions like GCF and other players in climate finance have a pivotal role to play in making this mechanism work. This would include convening relevant governmental regulatory and financial sector stakeholders on the same table to develop such mechanism. This could be complemented by credit enhancement for the first wave of capital markets issuances. This could be a potential game changer as this should unlock long-term saving for financing adaptation projects.

The challenge of adapting to climate change requires significant increases in private investment flows, alongside the use of public funds. GCF's experience is that there are multiple barriers to increasing the level of private sector investment in adaptation solutions within developing countries. There is no single solution, but rather a need to take a flexible and tailored approach to overcoming barriers. It is clear that without such solutions, and the innovative deployment of public funds, it will not be possible to catalyse the power of the private sector to support adaptation measures.

References

Climate Policy Initiative (2021) 'Global landscape of climate finance 2021' [online]. https://www.climatepolicyinitiative.org/publication/global-landscape-of-climate-finance-2021/

Convergence (2021) 'Climate investor one (CIO) case study' [online]. https://www.convergence.finance/resource/c98944c2-2391-43e2-bf3d-80611b0b4d4c/view?utm_source=social&utm_medium=post&utm_campaign=cio

Green Climate Fund (2017) 'FP048: climate-smart agriculture (CSA) risk sharing facility for MSMEs' [online]. https://www.greenclimate.fund/sites/default/files/document/funding-proposal-fp048-idb-guatemala-and-mexico.pdf

Green Climate Fund (2018a) 'FP097: productive investment initiative for adaptation to climate change (CAMBio II)' [online]. https://www.greenclimate.fund/sites/default/files/document/funding-proposal-fp097-cabei-guatemala-el-salvador-honduras-nicaragua-costa-rica-panama-and.pdf

Green Climate Fund (2018b) 'FP078: acumen resilient agriculture fund (ARAF)' [online]. https://www.greenclimate.fund/sites/default/files/document/funding-proposal-fp078-acumen-fund-inc-ghana-nigeria-and-uganda.pdf

Green Climate Fund (2018c) 'FP099: climate investor one' [online] https://www.greenclimate.fund/sites/default/files/document/funding-proposal-fp099-fmo-burundi-cameroon-djibouti-indonesia-uganda-kenya-malawi-madagascar.pdf

Green Climate Fund (2020a) 'Updated strategic plan for the green climate fund 2020–2023' [online]. https://www.greenclimate.fund/document/updated-strategic-plan-green-climate-fund-2020c-2023

Green Climate Fund (2020b) 'FP078: annual performance report 2020' [online]. https://www.greenclimate.fund/sites/default/files/document/fp048-annual-performance-report-cy2020-disclosable.pdf

Green Climate Fund (2020c) 'FP128: arbaro fund—sustainable forestry fund' [online]. https://www.greenclimate.fund/sites/default/files/document/funding-proposal-fp128.pdf
Green Climate Fund (2020d) 'FP148: participation in energy access relief facility ("EARF")' [online]. https://www.greenclimate.fund/sites/default/files/document/funding-proposal-fp148.pdf
Green Climate Fund (2021a) 'FP097: productive investment initiative for adaptation to climate change (CAMBio II)'. https://www.greenclimate.fund/project/fp097
Green Climate Fund (2021b) 'FP168: leveraging energy access finance ("LEAF") framework' [online]. https://www.greenclimate.fund/sites/default/files/document/funding-proposal-fp168.pdf
Green Climate Fund (2021c) 'FP180: global fund for coral reefs investment window' [online]. https://www.greenclimate.fund/sites/default/files/document/funding-proposal-fp180.pdf
Green Climate Fund (2021d) 'Country readiness' [online]. https://www.greenclimate.fund/readiness/naps
Green Climate Fund (2021e) 'Arbaro fund—sustainable forestry fund' [online]. https://www.greenclimate.fund/project/fp128
Green Climate Fund (2021f) 'FP181: CRAFT—Catalytic capital for first private investment fund for adaptation technologies in developing countries' [online]. https://www.greenclimate.fund/sites/default/files/document/funding-proposal-fp181.pdf
Nature (2021) 'The broken $100-billion promise of climate finance—and how to fix it' [online]. https://www.nature.com/articles/d41586-021-02846-3
UN Environment Programme (2021) 'Adaptation gap report 2021: gathering storm' [online]. https://www.unep.org/resources/adaptation-gap-report-2021
UNFCCC (2015) 'Paris agreement' [online]. https://unfccc.int/files/meetings/paris_nov_2015/application/pdf/paris_agreement_english_.pdf
World Bank Group (2021) 'Enabling private investment in climate adaptation and resilience' [online]. https://openknowledge.worldbank.org/bitstream/handle/10986/35203/Enabling-Private-Investment-in-Climate-Adaptation-and-Resilience-Current-Status-Barriers-to-Investment-and-Blueprint-for-Action.pdf?sequence=5&isAllowed=y

Chapter 17
Lessons in Adaptation and Innovation of Selected Local COVID-19 Responses in the Philippines

Maria Fe Villamejor-Mendoza

Abstract During pandemics, climate action seems to take a backseat as the priority action of governments seems to be in containing the emerging infectious disease and minimizing its adverse effects on public health, security and the economy. For COVID-19, for example, this crisis disrupted almost everything in our lives. It has drastically changed the ways of governing and governance. It has re-arranged policy priorities to the detriment or advancement of climate change mitigation and adaptation. This research looks at the interventions on climate action and health during the COVID-19 pandemic. It assesses the Philippines' national and local governments' COVID-19 Responses and explores their features, with or without climate action. For the local Response, it draws lessons in adaptation and innovation from selected local COVID-19 programs, which the Galing Pook (GP) Foundation deems as exemplars in its 2021 Awards. Specifically, it describes the local COVID-19 response programs of six cities and provinces and infers features that make them successful and better than the national COVID-19 response. Emphasis will be on the impact, participation, innovation, and sustainability of these programs on the communities they serve as well as their proactive climate action amidst the pandemic. Generally, the national COVID Response appears failing and with concentration on addressing the pandemic and recovering from the crisis. Despite strides, the 2021 Bloomberg COVID Resilience Index puts the Philippines at the bottom list as "it scored lowest in virus containment, the quality of healthcare, vaccination coverage, overall mortality and progress toward restarting travel and easing border curbs, among others" (Aguilar 2021). In contrast, local COVID Responses appear to effectively contain the virus, without sacrificing the economy, environment and people's lives. This study is mainly qualitative. It analyzes secondary materials such as studies and reports gathered from the GP Foundation and other sources. It theorizes that adaptation, innovation, participatory governance and strong leadership are key to successful (local) COVID response. Also, adding mitigation of climate risk escalation puts these programs a cut above the rest. These are crucial elements, which the national government could ponder on and emulate.

M. F. Villamejor-Mendoza (✉)
National College of Public Administration and Governance, University of the Philippines, UP Diliman Quezon City, Raul P. De Guzman St, 1101 Metro-Manila, Philippines
e-mail: mvmendoza@up.edu.ph

W. Leal Filho et al. (eds.), *Climate Change Strategies: Handling the Challenges of Adapting to a Changing Climate*, Climate Change Management,
https://doi.org/10.1007/978-3-031-28728-2_17

The Philippine National Response to the Challenges of COVID-19 Pandemic

The COVID-19 pandemic presents one of the greatest challenges in government responsiveness, resiliency and effectiveness. Although there are other players that may be tapped in addressing the multifaceted challenges of this pandemic, people look up to governments, both national and local, as the main agents to mitigate COVID-19's effects on public health and safety, economic security and recovery, and overall development.

The Philippines had at least four surges of COVID-19 variants since January 2020. As of December 31, 2021, total COVID-19 cases in the country have reached 2.8 million cases, some 98% of which have recovered. For the remaining 2%, some 11,772 are active cases and 51,373 are deaths (Department of Health 2021; please see Table 17.1).

Of the total active cases as of December 31, 2021, some 3% are considered critical; 15% are severe. The remaining 82% are moderate, mild or asymptomatic.

The Duterte administration was caught by surprise by this seemingly unprecedented global disruption. It nevertheless adopted a national COVID-19 Response which is appreciated by many as basically a composite of "(a) the World Health Organization (WHO)'s prescribed testing, tracing, isolation and treatment/vaccination protocols; (b) long hard community lockdowns that restricted mobility of people and non-operation/closure of schools, businesses and other work places, except those considered essential, e.g., supermarkets, pharmacies, etc.; (c) restrictions of public/mass gatherings; and (d) encouragement of non-pharmaceutical protocols of washing hands, wearing of face masks, and following social distancing" (Amit et al. 2021). Since the concentration of interventions was basically in containing the spread of the virus, other urgent policy actions like climate change mitigation and adaptation seem to take a backseat.

Table 17.1 Philippine COVID-19 cases as of December 31, 2021

Total cases	2,841,260	
Total Recovered	2,778,115	97.8%
Total Deaths	51,373	1.8%
Total New Cases	11,772	0.4%
Asymptomatic	577	4.9%
Mild	5,757	48.9%
Moderate	3,325	28.2%
Severe	1,786	15.2%
Critical	327	2.8%

Source Department of Health (DOH 2021)

For the long lockdowns from 2020 to early 2021, the following restrictions were in place in terms of population movement, public transport and work (Table 17.2).

Enhanced community quarantine (ECQ) is the highest level and strictest restriction. People are required to stay at home and there should be no movement within the ECQ area regardless of age and health status. In modified ECQ and general community quarantine (GCQ), there is limited movement within the containment area or buffer zone and mobility is allowed only in obtaining essential needs like food, groceries, and medicines. Essential workers like the medical and other frontliners, e.g. doctors, nurses, grocery personnel, others, are allowed to go out; non-essential workers may go out but only to buy essential necessities. The modified GCQ (MGCQ) is supposed to be the least restrictive and "a transition phase to the new normal" (Mypanoplies 2020), where socio-economic activities are permitted with certain health guidelines (EO 112 2020).

The long lockdowns led to business closures, displacement of workers, and mental anxiety in being 'caged' in one's home, among many. To address these adverse effects, the Duterte administration conceived of a national COVID-19 Response entitled the "Philippine Program for Recovery with Equity and Solidarity" (Fig. 17.1).

It includes the following pillars:

1. *Emergency support* for poor and low-income households, small business employees and other vulnerable groups through various assistance programs and wage subsidies amounting to some P590 billion or $11.8 billion;
2. *Marshaling resources to fight COVID-19,* particularly expanding medical resources to ensure the safety of front liners such as expanded health insurance coverage for all COVID-19 patients, special risk allowance, hazard pay, and personal protective equipment (PPE) for front line health workers; increased testing capacity, etc. under the Bayanihan Heal as One Act or RA 11469 (Bayanihan I);
3. *Monetary actions and other financing support* to keep the economy afloat and other financing support for emergency response and recovery initiatives; and
4. *An Economic Recovery Program* to create jobs and sustain growth, including provisions under the Bayanihan Recover as One or RA 11494 (Bayanihan II) (Department of Finance 2020).

The Problem with the National COVID-19 Response

This seemingly comprehensive response "has been marked by missteps and confusion, and even criticized as late and slow, with sluggish roll-out of testing, tracing and quarantine protocols, and weak leadership on the part of the incumbent President Duterte" (Robles and Robles 2021). The latter has also been described as initially complacent and dismissive towards COVID-19, consistent with his "strongman" persona. This led to the late closing of international borders in the first months of the pandemic when only two (2) confirmed positive cases were detected among Wuhan tourists who went to Boracay, a resort island in the Visayas. His response was also

Table 17.2 The Philippine community quarantine classification,[1] 2020–2021

Particulars	Quarantine classification/Restrictions	Quarantine classification/Restrictions	Quarantine classification/Restrictions	Quarantine classification/Restrictions
	Enhanced community quarantine (ECQ)	Modified ECQ	General community quarantine (GCQ)	Modified GCQ
Population/Movement	Everyone is not allowed to go out (with exceptions)	Everyone is not allowed to go out (with exceptions)	Everyone is not allowed to go out (with exceptions)	People may move between GCQ and MGCQ areas, except for leisure. Movement to and from stricter quarantine areas is not allowed
Public Transport	All public transport is suspended	All public transport is suspended	Public transport is allowed with social distancing observed	Public transport is allowed with social distancing observed
Work/Commerce	Only sectors that provide essential goods and services are allowed	More businesses are allowed, including manufacturing and processing plants	More businesses are allowed such as malls, hardware, TV and music productions	More businesses allowed to resume operations

Source EO 112 (2020)

[1] These classifications were changed to Alert Levels 1–5 from low risk to high risk, in the latter part of 2021.

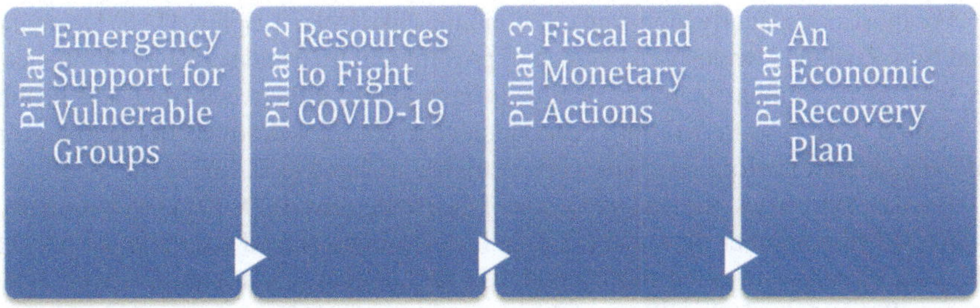

Fig. 17.1 The Duterte Administration's National COVID-19 Response: Philippine Program for Recovery with Equity and Solidarity (PH-PROGRESO) (*Source* Department of Finance 2020)

a militarization response to a primarily public health issue, with many military men manning the the inter-agency task force he created. This somehow marginalized credible health experts in key policy bodies (Rivera 2021).

In general, the national COVID-19 response was also seen as a "quick-fix" and non-sustainable response to complex problems similar to what was done in other development concerns in the country like illegal drugs and "tokhang", corruption, the Federalism project, the problem of political dynasties and the "oligarchy". His was also an unfocused and divisive leadership during the pandemic when it confronted the issue of press freedom, for example, especially after the Legislature did not grant the congressional franchise to a seeming unfriendly media giant, the ABS-CBN (Rivera 2021).

Lasco (2021) also alluded to the (mis)use of science and expertise in the 'one size fits all' solutions to the COVID-19 Response. Here, the voices of the health experts were muted in favor of the big voices of the close-in circle of Duterte's mostly military men. He also found unhealthy, the punitive, militaristic and disempowering regime based on compliance and discipline, which is opposite to the public health paradigm based on awareness, adherence and solidarity. Moreover, he found something wrong with "the 'Covidization' of health care, leading to various other health (and development) concerns, which are likewise a matter of life and death, being sidelined or not given equal attention to by government" (Lasco and San Pedro 2020).

In addition, the Philippines ranked last among 53 countries in the September 2021 Bloomberg's COVID Resilience Ranking Report. A monthly snapshot of the virus is being handled most effectively with the least social and economic disruption, the Index uses 12 data indicators that span from virus containment, overall mortality and progress toward restarting travel, among others. It captures which of the world's biggest 53 economies are responding best—and worst—to the same once-in-a-generation threat (Bloomberg News 2021).

The Philippine COVID Resilience Index was lowest "in virus containment, the quality of healthcare, vaccination coverage, overall mortality, severity of lockdowns and restrictions, progress toward restarting travel and easing border curbs, among others" (Aguilar 2021).

For those consistently high-ranked economies like Ireland, Spain, UAE, Denmark and New Zealand, "widespread degree of government trust and societal compliance, as well as investment in public health infrastructure, effective testing and health education were keys to avoiding economically crippling lockdowns in the first year of the pandemic, before vaccines were available" (Chang et al. 2021).

For the Philippines, the first two years of the pandemic (2020–2021) was similarly addressed in a number of ways: (1) long lockdowns remained for the most part of the year as surges ravaged the country; (2) vaccination is still sluggish with only 49.6 million of the 109 million total Filipinos or only 46% of the total population with complete 2 doses; 57 million with one dose; and 1.8 million with booster shots as of December 30, 2021 (DOH 2021); and (3) in the middle of it all, lawmakers investigated alleged anomalies involving billions of pesos in the Duterte administration's pandemic contracts. It was only after the latter part of 2021 when new daily COVID-19 cases were reported only in the hundreds that restrictions were eased and mandatory face shield policies were dropped—finally giving some kind of sense that perhaps, the Philippines was getting out of the COVID rut (Abad 2021).

The Philippine COVID-19 response has been reactive, ad hoc and inadequate. To reiterate, these general failings were a result "partly of weak leadership and lack of experience in dealing with a pandemic of this scale, as well as of the sorry state of the public health system in the country" (Quintos 2020).

It was a hit-and-miss, creating massive adverse effects on the economy, social relations, education and other aspects of the society. "The GDP decreased by 9.5% in 2020. When the tightest lockdown was implemented in the second quarter of 2020, GDP fell as low as –16.9%. Foreign debt increased by 26.7% to P10.3 trillion as of the end of January 2021 (Laforga 2021). Unemployment rate was 10.3% or 4.5 million unemployed in 2020. According to Moody's, the Philippine economic output will not return to pre-pandemic levels until the end of 2022. China, Taiwan, South Korea, and Vietnam, on the other hand, have already returned to their previous output levels, while Indonesia and Thailand are predicted to do so later in 2021" (FutureLearn 2021).

More importantly, because of the preoccupation at health and safety, the National COVID-19 Response seems to focus on preventing the spread of the virus by implementing the usual health protocols of Prevention, Detection, Isolation, Treatment and Reintegration (PDITR) strategy, including the albeit late but aggressive vaccination program, providing assistance to displaced and disadvantaged vulnerable sectors, providing resources to medical and other front liners, and slowly easing up travel and other restrictions to pave the way to economic recovery. All other policy concerns appear to take a lower priority, including urgent climate actions in a highly natural hazardvulnerable country like the Philippines.

Research Objectives and Methodology

Thus, many countries have slowly reclaimed normalcy in their lives and economies; but there are others like ours, which have remained stressed and seemingly unable to get out of this COVID-19 crisis. Although there have been positive developments, nationally, the country's COVID response seems failing. But locally, we can see gems that have swayed the tide towards effective containment of the virus, without sacrificing the economy, environment and people's lives.

This paper assesses both the national and local governments' COVID-19 Responses, with or without climate action. After the cursory assessment of the National Response above, it draws lessons in adaptation and innovation from selected local COVID-19 programs, which the Galing Pook (GP) Foundation deems as exemplars in its 2021 Awards. Specifically, it describes the local COVID-19 programs of selected cities and provinces and infers features that make them successful and better than the national COVID-19 response. Emphasis will be on the impact, participation, innovation, and sustainability of these programs on the communities they serve, as well as on the climate action they implemented amidst COVID-19.

This study is mainly qualitative. It analyzes and reviews secondary materials such as studies and reports gathered from the GP Foundation and other sources. It theorizes that adaptation, innovation, participatory governance and strong leadership are key to successful (local) COVID response. Also, adding climate action to mitigate natural hazard risk escalation puts these programs a cut above the rest. These are crucial elements, which the national government could ponder on and emulate.

The Case of COVID-19 Responses in Selected Local Governments

The Philippines is a unitary state, a system of political organization in which most or all of the governing power resides in a centralized government. The President acts both as the head of state and head of government. S/he exercises general supervision over local governments, which are territorial and political subdivisions of the Republic. As of September 2020, local government units (LGUs) consist of 81 provinces, 146 cities, 1,488 municipalities and 42,046 barangays (Department of Interior and Local Governments 2020 at https://www.dilg.gov.ph/facts-and-figures/Regional-and-Provincial-Summary-Number-of-Provinces-Cities-Municipalities-and-Barangays-as-of-30-September-2020/32). They enjoy local autonomy and have the power to create their own sources of revenue in addition to their equitable share in the national taxes. They also have the power to allocate resources in accordance with their own priorities. (Philippine Constitution of 1987, Article X). This does not, however, leave LGUs with unbridled discretion in the disbursement of public funds because they remain accountable to their constituency (Agra 2016).

LGUs are the closest to the people and may know the needs and requirements of their constituency better. Though they are 'merely' parts of the State and theoretically should implement COVID-19 programs similar to how the national government does it, they are nevertheless given leeway to address pressing development problems as they see fit. Because of their autonomy and the resources at their disposal, they may craft policies, programs and projects, which may be more responsive and suited to their needs. These may include climate action before, during and after the pandemic.

A 'one size-fits all' model may not be the usual way they may take to address pressing development problems like COVID-19. Thus, an evaluation of selected COVID-19 and related programs of LGU-applicants to the 2021 Galing Pook Awards may be in order to ascertain if their local COVID-19 Response is better, more creative and adaptive, less panic-driven and more successful in containing the virus without sacrificing lives, the economy and even the environment, particularly in mitigating climate risks while battling the pandemic.

The *Galing Pook* (literally translated as '*the excellent place*') Awards is a pioneering program that recognizes innovation and excellence in local governance. It is implemented by the Galing Pook Foundation (GPF), a leading resource institution that promotes innovation, sustainability, citizen empowerment, and excellence in local governance. It promotes the latter through recognition, sharing of information and support of efforts to replicate best practices at the local level (Galing Pook 2021b).

The GP Awards have for the past 28 years recognized programs that have become models of good governance and sustainable cities and municipalities in the country. These have also become exemplars for innovative solutions to common and shared problems in our planet, like the COVID-19 pandemic and climate change. The Awards, which may be likened to a 'mini-Nobel (Peace) Prize in local governance' started in October 21, 1993 under the joint initiative of the Local Government Academy-Department of the Interior and Local Government, the Ford Foundation, and other individual advocates of good governance from the academe, civil society and the government. The Asian Institute of Management carried on the awards program until 2001. Earlier in 1998, the Galing Pook Foundation was formed as a juridical institution to sustain the program (Galing Pook 2021b).

Since 1994, more than 250 programs have already won recognition. The Galing Pook winners are chosen each year from a wide array of programs from local governments after undergoing a rigorous multi-level screening process. The winning programs are selected based on, among others, whether the programs (a) led to positive results and impacts on the community they serve; (b) promoted people's participation and empowerment; (c) showcased innovation, transferability and sustainability; and (d) epitomized efficiency of program service delivery. The weights for these criteria are in the GP website (Galing Pook 2021b).

For 2021, of the total 206 entries, the following COVID-19 and related programs were highly evaluated:

1. San Juan City's Kalingap (Kaagapay, Lingkod at Gabay sa Pandemya).

2. The Roxas City's Unified Pandemic Operational Deployment (UPOD) Kita Program.
3. Mandaue City's COVID-19 Community Infectious Waste Management.
4. Bohol Province's Culture and the Arts for Healing and Recovery Program.
5. Negros Occidental Province's Network of Alliances for Coastal Wetlands Conservation.
6. Agusan del Sur Province's Upland Sustainable Agro-forestry and Development (USAD) Convergence Program.

Of these six, the programs of San Juan City, Negros Occidental and Agusan del Sur made it to the Finals of the 2021 Galing Pook Awards. They are geographically located in the National Capital Region/Metro-Manila, Visayas and Mindanao, respectively. The other three programs were semi-finalists and are still worth studying; thus, they are also included here to ascertain the features that make them a cut above the national pandemic response.[2]

San Juan City's Kalingap Program

There is a Filipino word, 'Kalinga' which means 'care' and with a 'p' could mean 'kalingang maagap' or ready care, now. However, San Juan's *Kalingap* program does not mean the latter. It is an acronym of the major components of the program, e.g., "Kaagapay, Lingkod at Gabay sa Pandemya", which literally means 'care, service and guidance during the pandemic.'

San Juan, a highly urbanized first class city has recorded the first COVID-19 local transmission in the country. The first COVID-19 local patient frequented the Muslim Prayer Hall in Greenhills Shopping Center, both as member of the Muslim flock and retailer in this shopping mall. The mall was temporary closed and sanitized thereafter, triggering the all-inclusive and comprehensive program, *KALINGAP* (San Juan GP Paper 2021).

The fear and anxiety of the San Juaneños[3] even at the very onset of COVID-19 in the country was overwhelming: the people started to lose jobs and livelihood as the country was placed on ECQ/hard lockdown and the trend of patients testing positive for COVID-19 one after another has dampened the spirits of the San Juaneños. Moreover, they have other needs and concerns such as (a) lack of isolation and quarantine facilities for COVID-19 patients who are asymptomatic or with mild symptoms. The San Juan Medical Center also lacks hospital facilities for patients who have moderate to severe symptoms; (b) economic issues for those who lost their jobs and means of livelihood; (c) the surge in the number of COVID-19 cases; (d) San Juaneños' inadequate knowledge and awareness on COVID-19 and the importance

[2] Permission was sought to use these Galing Pook (GP) cases for this chapter, which was granted by the GP Board on 2 September 2021. The author is a member of the National Selection Committee for the 2021 GP Awards.

[3] San Juanenos refer to the people of San Juan.

of vaccination; (e) lack of mechanisms to support the students from public schools in their online classes; and (f) unavailability of public transportation for the frontliners (San Juan GP Paper 2021).

Thus, *KALINGAP* is a holistic and comprehensive Response, offering the constituents of San Juan not just a strengthened city-version of the national government's Prevention, Detection, Isolation, Treatment and Reintegration (PDITR) strategy in combatting the spread of the virus but giving above and beyond to secure food, financial aid, mobility, commerce, education, connectivity and vaccination to its people. It has the following components:

1. Establishment of San Juan's own quarantine facilities and COVID-19 additional wards using container vans. The former are called Bahay Kalinga[4] (formerly Kalinga Kontra Korona)[5] Quarantine Vans that were constructed in partnership with the Xavier School Foundation, San Juan City National High School and the Department of Public Works and Highway. The latter are called Charlie Wards in San Juan Medical Center. The container vans were also turned into Emergency Room (ER) Vans that were used as Triage area in San Juan Medical Center and as quarantine facility near the San Juan Bureau of Fire Station.
2. Launch of TUPAD or *Tulong Panghanapbuhay sa ating Disadvantaged/Displaced Workers*, a temporary "cash for work" project for the indigents and jobless San Juaneños due to the pandemic. In coordination with various NGOs ad private companies, the city government also provided bicycles to indigent San Juaneños who were having a hard time commuting to and from their workplaces during the ECQ.
3. Provision of food packs/ayuda[6] to all the San Juaneños. This consisted of 1 food pack per family every week during the ECQ and modified ECQ (MECQ) or up to 16 waves of Ayuda.
4. Provision of temporary shelter located in San Juan Elementary School with food allocation; free Wi-Fi and financial assistance for medical front liners.
5. Financial assistance of PhP 3,000 was given to all the COVID-19 patients who were confined and completed the quarantine period in any of the San Juan City's quarantine facilities
6. Adoption of the E-Ayuda Scheme in disbursing the Social Amelioration Program's financial assistance from the national government, e.g., the Department of Social Work and Development (DSWD). The beneficiaries can get their cash assistance thru ATM machine in partnership with the Robinson's Bank.
7. In partnership with the Department of Information and Communications Technology (DICT), laptops, tablets and pocket Wi-Fi were given to public school students in time for the opening of online classes. This is on top of the Internet

[4] Literally means the House of Care.

[5] This means Care against COVID-19.

[6] Ayuda means support, financial, in kind and other forms.

satellites, which can be accessed for free in areas near schools, barangay halls and other public establishments. San Juan also partnered with Converge, an Internet service provider, for the installation of Fiber Optic Intranet and Learning Management System in every household of the public school students.

8. Provision of bicycles to indigent workers and medical front liners who are deprived of public commute due to ECQs and lockdowns. Pop up bike lanes[7] were also designated in all main streets of the city to make the city safe for bicycles, e-scooters and other forms of alternative transportation.
9. San Juan launched its Vaccination Program, starting with the online vaccination registration and on-ground registration for vaccination. The Filoil Flying V Arena and the Cinemas 1&2 of the Theatre Mall in Greenhills were the designated vaccination sites in San Juan City. In partnership with the DICT, the city's Vaccination Program used the Vaccination Information Management System Immunization Registry (VIMS-IR), which efficiently shortens the process time for the whole vaccination process, from registration, actual vaccination and monitoring.
10. The Vaccination Incentive Program (VIP) was launched to encourage the people to get fully vaccinated and be able to receive perks and discounts from the establishments in San Juan City. The Seal of 100% Vaccinated Establishment is also given to all the establishments whose owners and all workers have already been fully vaccinated (San Juan GP Paper 2021).

The Roxas City's Unified Pandemic Operational Deployment (UPOD) Kita Program

Roxas City is the seafood capital of the Philippines. It is a tourist haven because of its culture, crafts and arts. Its *UPOD Kita* Program is a collective of programs designed to address the impacts and effects of the COVID-19 Pandemic. The key pillar of UPOD Kita is the concept of participation, hence, the word '*upod*' which means, 'together'. The main objective of the program is to capacitate, empower, and operationalize the local government of Roxas City and its constituents in pandemic response. The entire program is hinged on three basic principles: impact mitigation, contagion containment and prevention, and recovery. In all these phases of the program, participation and partnership cut across implementation (Roxas City GP Paper 2021).

The objective for the first component of the program, which is impact mitigation, is to respond to and cushion the current impacts of the pandemic by carefully managed and calibrated programs. For contagion containment and prevention, the objective is to contain and further prevent the spread of the virus and protect the health and

[7] Pop up bike lanes are temporary cycling infrastructure that were turned into permanent fixtures by the city by installing bollards in wider roads and paint and signages in smaller roads. These created clear delineation of road spaces for cars and bikes and ensured the safety of bikers and other users of major thoroughfares (San Juan GP Paper 2021).

safety of the constituents; for recovery, the objective is to pro-actively address the remnant effects of the pandemic and assist the constituents in adapting to the new normal (Roxas City GP Paper 2021).

The first component of the program consists of the passage of legislation and policies strictly implementing minimum health protocols, deployment of palliative programs designed to address the immediate needs of the constituents like provision of food aid to affected individuals, alternative sources of income, and providing for health and medical intervention and assistance to affected individuals. In this component the city government has passed Ordinance no. 044-2020, and Executive Order No. 16, among others, which all seek to enhance and improve the implementation of health protocols at the local level. The city has also distributed rice to 32,000 families monthly as well as special food packs to families who are under home quarantine (Roxas City GP Paper 2021).

For the cultural sector, the City conducted the Healing Brushstrokes Painting Competition, which sought to provide income to local artists and promote their work online. To supplement this, the artists were provided an Art Village that served as venue to sell their crafts and artwork. Food packs were also distributed to artists-beneficiaries. Funding was also released to the private sector-Capiz Medical Society to assist them in their work as medical front liners during the pandemic. Additional isolation facilities were also constructed (Roxas City GP Paper 2021).

For the second component, the City implemented the following: (a) complied with the rigorous regulation of the Department of Health (DOH) and established its own Reverse Transcription-Polymerase Chain Reaction (RT-PCR) Diagnostic Laboratory and facility for moderate cases with six airflow cannula machines it also developed its localized QR Code contact tracing system; (b) E-Konsulta,[8] an online medical diagnostic and reporting system; (c) use of Ateneo's[9] Fasster Tool, which is an algorithmic prediction software that makes forecasts of future pandemic trends; and (d) the constant and regular inspection of tourism establishments for compliance with health protocols. Roxas City has also diverted school board funds to provide requirements, e.g., computer and other gadgets, Internet connectivity, etc., in the implementation of blended learning (Roxas City GP Paper 2021).

In the third component, which is recovery, the city has invested in continuing food aid beyond the pandemic, fast-tracked complementary infrastructure like new roads and road repairs, installed more city street lights, rehabilitated its the main trade center, empowered local businesses through webinars and consultations, in preparation for their renewed operation after the pandemic and during the new normal, and established the Agri Kita Urban Gardening Program with the Gerry Roxas Foundation. Vaccination is also currently in full swing (Roxas City GP Paper 2021).

[8] E-Konsulta means (medical) consultation by electronic means, e.g., by email exchange or 'face to face'-virtual exchange through various platforms, e.g., Viber or Messenger video call, Google Meet, etc.

[9] Ateneo is Ateneo University, a private academic institution tapped for their computer application for contact tracing, data analytics and the like.

Roxas City is a third-class component city with a population of 179,000, its main business is agro-fishery and tourism as it is known for its "abundant seafood and ecotourism destinations, heritage structures and cultural interests that could pack one's itinerary with unforgettable experiences" (Roxas City 2021). The pandemic has adversely affected these local businesses and enterprises and displaced workers during the lockdowns. Thus, the various programs above were implemented to ensure the virus is contained and the road to recovery is laid out, with the participation of the constituents, the private sector, academe, non-government organizations and the workers in art villages, fishponds and other stakeholders. Its health system was also upgraded with the construction of an RT-PCR diagnostic laboratory and a quarantine facility for moderate cases, whose services are for free.

Mandaue City's COVID-19 Community Infectious Waste Management

In addition to the regular PDITR plus vaccination strategy in combatting the spread of the virus, Mandaue City, a first class highly urbanized city focused on the pandemic's community infectious waste management (CIWM). The city has been at the forefront of maintaining a clean and safe environment through its Solid Waste Management Program (SWMP). With the pandemic, however, protocols for managing community infectious wastes like face masks, personal protective equipment and the like, have been a challenge as people are unaware how to dispose and manage these wastes (Mandaue City's GP Paper 2021).

The goal of the CIWM program is to protect Mandaue citizenry and prevent the spread of the virus through disposed wastes by integrating the Infectious Waste Management Program (IWMP) to the existing Solid Waste Management Plan of the city. Hence, the Infectious Waste Response Team was created in May 2020. The Mandaue City Environment and Natural Resources Office (MCENRO) through the order of the local government unit spearheaded the implementation of the infectious waste management program and protocol by engaging and equipping partners from the local communities, conducting information campaigns, and managing resources allocated by the city for this purpose (Mandaue City's GP Paper 2021).

The city conceived an infectious waste (IW) management process flow that engages, equips and empowers local communities from (a) collection of IW at the source, e.g., households in lockdown, testing/swabbing areas, city quarantine facility, city/national facility; to (b) disinfection, transport to the city transfer station; and finally (c) disposal and management of infectious waste (Mandaue City GP Paper 2021).

Bohol Province's Culture and the Arts for Healing and Recovery Program

The Provincial Government of Bohol highlights the "Reawakening of the Boholano[10] Culture" as one of three pillars for the province's strategic sustainable development. Towards its realization, the Bohol Arts and Cultural Heritage (BACH) Council plays a key role in coordinating and unifying efforts of individuals, groups, and institutions in pursuing programs to ensure the preservation and promotion of Bohol's cultural heritage and arts (Bohol GP Paper 2021).

A civil society organization with a multisectoral composition, the BACH Council was created in 2001 by virtue of Provincial Ordinance No. 2001-017. After a Strategic Planning Workshop on February 27–29, 2020, sets of prioritized programs and activities were ready for implementation when COVID-19 was declared a global pandemic, thereby, prompting a pivot to respond to the more crucial needs of the time. The provincial government implemented the usual PDITR programs to contain the virus, without compromising the health of the people, the economy and culture of the Boholanos. The BACH Council, meanwhile, synergized the Arts and Culture, Tourism and Youth sectors and kicked off the "Culture and Arts for Healing and Recovery Program" with the launch of DASIG-Bohol on April 1, 2020 (Bohol GP Paper 2021).

"*Dasig*" is a vernacular word that means "to enliven" or "to inspire" and which became the title of a digital streaming program aimed to lift the spirits of the Boholanos during the quarantine. The crisis triggered by the pandemic has serious and dire implications for the province; tourism being one of its top two sources of revenue. In an unprecedented blow to the industry, COVID-19 has drastically cut tourist arrivals causing a ripple effect that has heavily impacted on many other industries relying on tourism. These include not only the big players in the accommodation, recreation, events, food and beverage services, wellness, transportation, and travel services sectors but also the micro, small, and medium enterprises (MSMEs). The latter involves local artists, musicians, performers, photographers, crafts people, souvenir makers, and many other individuals who suddenly found themselves without any market and income for their services (Bohol GP Paper 2021).

The staging of DASIG-Bohol, thus, was conceptualized as an intervention to provide opportunity to local artists and craftspeople to showcase their skills and creativity and also correspondingly be paid for their art. A virtual show produced under strict safety protocols, DASIG-Bohol was also projected as an effective intervention for Bohol residents who were afforded free entertainment while adhering to "stay-at-home" restrictions to curb the spread of the virus. With school and business closures, the psychosocial effects on individuals could not be taken lightly (Bohol GP Paper 2021).

As the COVID-19 crisis continues, it also exposed the vulnerability of the youth and children in coping with loneliness, fear, violence, and uncertainty experienced

---

[10] Boholano means from or of Bohol. Boholanos are the people of Bohol province.

due to the pandemic. Cognizant of the vital role the arts had played in the healing and recovery of Boholanos after the 7.2 magnitude earthquake in 2013, the BACH Council spearheaded in harmonizing initiatives and rallying various sectors to respond to the mental health and other needs of the young people. LAUM-Bohol was created as the provincial government's multi-pronged approach utilizing culture and the arts, education, counseling, sports, and various training programs to reach out and capacitate the youth who are most in need of a lifeline in this troubled time. The Bohol Governor signed Executive Order No. 50 instituting LAUM-Bohol and the program was launched on October 15, 2020 (Bohol GP Paper 2021).

Thus, the "Culture and Arts for Healing and Recovery Program" has two major collaborative components, namely: DASIG-Bohol, with the Center for Culture and Arts Development (CCAD) as the lead office and LAUM-Bohol spearheaded by the Provincial Youth and Development Office (PYDO). All these initiatives are under the auspices of the Office of the Governor and implemented by the Bohol Arts and Culture Heritage Council (Bohol GP Paper 2021).

Negros Occidental Province's Network of Alliances for Coastal Wetlands Conservation

"*Abanse Negrense*", which is the vernacular for "Onward Negrense[11]!" is the rallying call of the administration of Negros Occidental to its heightened COVID-19 response to rebuild lives and recover from the health and economic crises as well as destruction due to various calamities (Guadalquiver 2021).

A cumulative report of the Provincial Incident Management Team showed that from March 27, 2020 to Dec. 30, 2021, Negros Occidental has logged 34,880 confirmed COVID-19 cases, of which 32,962 have recovered while 1,730 have died (Guadalquiver 2021). The province thus intensified its prevention, detection, isolation, treatment and reintegration programs by among others, closing its borders, testing probable cases, house to house information campaign and vaccination, and giving of financial support to those who were displaced and needed ayuda (ABS-CBN News 2020 at https://news.abs-cbn.com/news/03/16/20/negros-oriental-closes-all-points-of-entry-to-prevent-covid-19-spread).

In addition, its disaster response also included not only attending to COVID-19 but also to flooding and other natural disasters that wreck havoc to people's lives, crops, property, the economy and the environment. On December 16, 2021, Negros Occidental was hit by Typhoon Odette, leaving some PHP6-billion damage across all sectors. The onslaught left almost 40 Negrenses dead as well as 190,868 partially damaged houses and 65,871 destroyed dwellings. Last New Year's Day, massive floods hit several localities in northern Negros leaving thousands displaced, and homes and infrastructures destroyed. A week after, the second wave of flooding happened, affecting more residents in the same area (Guadalquiver 2021).

[11] Negrenses are the people of Negros Occidental.

Construction of flood control structures and implementation of various climate-proofing measures worth PHP1.2 billion has been funded in 2020 and 2021 for the affected areas. Priority sites for the implementation of socialized shelter projects as part of a long-term resilient and safe housing plan, particularly for informal settler families, were also identified (Guadalquiver 2021).

More importantly, the province continued the project, *Negros Occidental's Network of Alliances for Coastal Wetlands Conservation* amidst the pandemic, because it believed that only through alliance building and partnerships can sustainable coastal wetlands conservation and development be achieved (Negros Occidental GP Paper 2021).

Since the early 2000s, the Negros Occidental Coastal Wetlands Conservation Area (NOCWCA) has been managed by two inter-LGU alliances, e.g., KAHIL-ICAMC or Kabankalan City, Himamaylan, Ilog Integrated Coastal Area Management Council covering said LGUs; and CENECCORD or the Central Negros Council for Coastal Resource Development covering Bago City, Pulupandan, Valladolid, San Enrique, Ponteverda, Hinigaran and Binalbagan. In 2014, as part of enhancing the management of the wetlands through the initiatives of the LGUs led by the provincial government of Negros Occidental, and in cooperation with other national government agencies (NGAs) and non-government organizations (NGOs), the Negros Occidental Coastal Wetland Area Management Alliance (NOCWAMA) was formed. A Memorandum of Agreement among these stakeholders officially sealed this alliance. It aims to improve and wisely manage the coastal wetlands of international importance through the establishment of ten (10) Coastal Wetlands Local Conservation Areas (LCAs), thereby ensuring wetlands ecosystems can adapt and withstand the impacts of climate change with restored biodiversity that are sustainably managed and protected by the coastal communities. The 10 LCAs are known for their rich and bio-diverse coastal resources particularly the mangroves, shellfishes, waterbirds and Irrawaddy dolphins (Negros Occidental GP Paper 2021).

The Alliance specifically aims to (1) establish contiguous local coastal wetlands conservation areas (LCA) in Negros Occidental. This includes community mapping and surveying of wetlands in the 10 LGUs, profiling, capacity building for technical personnel in these LGUs on protected area management and the like, policy support and internal monitoring system; (2) enforce biodiversity regulations through the organization of community-based environmental law enforcers who will also work as Bantay Katunggan[12] Volunteers (BKVs) and Wildlife Enforcement Officers (WEOs). This will also consist of giving training programs on environmental laws, deputation of these BKVs and WEOs, and provision of equipment support to said officers; and (3) promote biodiversity-friendly practices and intensity information, education and communication (IEC) campaign drives on coastal wetlands management, coastal law enforcement, local conservation area establishment, mangrove rehabilitation and environmental thematic events related to CW conservation and protection in the

[12] Bantay Katunggan is the vernacular for Mangrove Forest Watchers.

province. These are consistent with the vision of the province, i.e., "Negros with a healthy environment where empowered citizenry enjoy sustainable economic growth based on equity" (Negros Occidental GP Paper 2021).

Its common goal is to resolve the pressing issues and problems on illegal activities that degrade the rich and abundant resources through harmonized conservation initiatives to wisely manage the coastal wetlands resources of the 10 LGUs. Specifically, it hopes to, among others, reduce the incidence of mangrove cutting and wildlife hunting; promote science-based project through the conduct of wetlands' characterization and updating of results; mainstream gender and development in the Alliance site and be included in the Ramsar List of Wetlands of International Importance (Negros Occidental GP Paper 2021).

Agusan Del Sur Province's Upland Sustainable Agro-Forestry and Development (USAD) Convergence Program

USAD is Agusan del Sur's building back better anti-poverty program after the province and the rest of Mindanao were decimated by the category 5 typhoon Pablo in December 2012. Determined to rise up from such devastation, the province conceptualized a disaster recovery and rehabilitation strategy to resolve the distress issues and uplift the economic condition of the marginalized Agusanons.[13] Also an environmental protection, people-centered and sustainable economic strategy, it continuously evolved to tailor-fit to the needs of the communities, especially the upland communities where poverty incidence is relatively higher (Agusan del Sur's GP Paper 2021).

USAD has been a flagship program of the province since 2013. And with the national lockdown imposed in the country in 2020, many of the Agusanons did not have food on their table because work has been suspended, leaving many without means of subsistence or support. Thus, the province deemed it fit to continue with the USAD to tide the people over, particularly those in the upland villages. USAD aims to (1) protect and conserve upland natural resources; (2) create/increase income through sustainable farming and livelihood; (3) reduce poverty incidence in the upland barangays; and (4) empower government through the creation of an USAD Coordinating Unit under the Office of the Provincial Governor through the spirit of volunteerism and strong community partnership based on shared governance, shared accountability and shared future (Agusan del Sur's GP Paper 2021).

In order to purge the dole-out mentality of the traditional beneficiaries of government programs and to practice the cohesive and collaborative efforts between the LGUs and the farmers, beneficiaries of the USAD Program were acknowledged as Farmer Enrollees (FEs). As FEs, they were provided farm inputs like seedlings for three high value commodities—rubber, cacao and cutflower, and technology-training programs that capacitated them. (Agusan del Sur's GP Paper 2021).

[13] Agusanons refer to the people of Agusan.

Analysis: The Defining Features of Local COVID-19 Programs

The defining features that make these local programs stand out vis-à-vis the national COVID-19 Response will be analyzed in this section, in order to draw lessons on adaptation and innovation. Emphasis will be on the impact, participation, innovation, and sustainability of these programs on the communities they serve, as well as the value added health and climate action they implemented during the pandemic.

The General Features and Value Added of Selected Local COVID-19 Responses

The main features of the six local COVID-19 programs presented above consist of adopting the general framework of the National COVID-19 Response of prevention, detection, isolation, treatment and reintegration + recovery, closing down of borders and imposing lockdowns; restriction of mass gatherings and encouragement of washing of hands, wearing of face masks and adopting physical distancing.

In addition, the local programs are perceived to be more comprehensive, responsive and adaptive to local conditions and needs, quicker in containing the spread of the virus, more medium to longer-term in perspective and engaged the participation of communities, the health sector, the private organizations, non-government organizations, the academe and other stakeholders in the planning and implementation of these programs. Their approach is not militaristic; they did not 'covidify health services' as other health issues, sectors, culture and the arts, environment, business and commerce were also attended to, amidst the pandemic. They also use modern technology and the health and other experts' contributions more. They exhibited more empathy to the peoples' plight; corollary, the people felt their local government was really helping them tide over the pandemic (Galing Pook 2021a).

Value Added Health and Climate Action

There is also a prevailing sentiment that the local programs value-added to the National Response. San Juan KALINGAP program enhanced the public health system by constructing additional wards and quarantine facilities using container vans, helped frontliners with additional incentives, conducted the testing and treatment of COVID patients for free[14] with other freebies, and incentivized the vaccination of people. It also secured the mobility of frontliners and other people during the long lockdowns where public transport was not allowed, with free bikes and dedicated bicycle lanes; providing ayuda to all constituents; provided cash for work for

[14] Intervention for COVID-19 testing from the national government is not always free.

displaced workers; secured the remote learning of public school children; and used technology to speed up the grant of national financial assistance. Although the free bikes and the dedicated bike lanes were borne out of necessity, they paved the way to citywide initiatives to reduce carbon footprints in San Juan.

The Roxas City's UPOD Kita enhanced the public health system by constructing an RT-PCR laboratory and quarantine facility for testing and treatment of COVID-19 patients for free; used technology for contact tracing, medical consultation and data analytics; provided ayuda (rice and other necessities) for people asked to stay at home; used arts, culture and technology to help workers and businesses to operate even during the pandemic; paved the way for the resumption of commerce by ensuring (tourism and other) establishments and public infrastructure are ready; also encouraged urban gardening to provide steady supply of vegetables in one's backyard.

Mandaue City's COVID-19 Community Infectious Waste Management focused on an important new threat to virus transmission through improper disposal of infectious waste at the community level, and at hospitals and quarantine facilities when they test, isolate and treat COVID-19 patients. The ever growing amount of infectious waste that is generated by protective personal lifesaving advances if not treated properly, may cause enormous suffering, pollution, unnecessary carbon emission, and waste of resources (No Harm Global 2022). Mandaue City in addressing this issue may have prevented large-scale problems in COVID-19 transmission, plastic pollution that may harm sea and other creatures, flooding and other environmental concerns.

Bohol Province's Culture and the Arts for Healing and Recovery Program used culture, arts, education, counseling, sports, and various training programs to address the mental health problems of the Boholanos who were asked to 'quarantine' at home during the lockdowns. They also used technology to promote arts and culture and keep these enterprises afloat amidst COVID-19.

Negros Occidental's Network of Alliances for Coastal Wetlands Conservation focused on enhancing local coastal wetlands management areas to provide food and livelihood to the people, and at the same time protect the wetlands and mangrove forests as mitigation measures against disasters and climate change. It also prevented illegal activities that degrade the rich and abundant resources in the province, as well as empowered the people to protect and conserve their mangroves, shellfishes, water birds and Irrawaddy dolphins. The Alliance also dealt with flooding and displacement of people after disastrous typhoons by constructing flood control projects and climate-proof shelter houses in the province.

Agusan del Sur's USAD Program focused on high-yielding crops to alleviate the poverty of the upland communities and on capacitating them for upland resource conservation, disaster prevention, and building back better. It also embedded environmental interventions and climate action in COVID-19 cum disaster response.

Community Participation, Innovation, Impacts and Sustainability

Community participation is evident when relevant stakeholders are engaged in program planning, implementation and other aspects of the program life. Innovation is present when a new idea, service or way of doing things is embedded and contextualized to local conditions. This also value-adds to solving the problem. Impacts are positive effects resulting from the program. Sustainability is evident when there are program documentation, legislation and commitment to continue with the program (Galing Pook 2021b).

a. ***Community Participation***: The National Response used the whole of government approach with stakeholders mainly from government participating in a dichotomized decision-making and implementation, i.e., the national government plans; the LGUs implement. Local governments, being closer to the people, used the whole of society approach in engaging various stakeholders from the government, community, health experts, private sector, non-government organizations and the academe, in planning, implementing and even monitoring the results of the interventions. In other words, the local governments employed collaborative governance more while the national response was more government-centric in its interventions.
b. ***Innovation***: The local interventions were more adaptive to the situation and innovative in solving the problem, which were absent in the command-control type in the national response. For example, when mobility was restricted and public transportation was not available for medical frontliners and other workers, San Juan provided bicycles and constructed pop up bike lanes. When COVID wards and quarantine facilities were inadequate, the cities of San Juan and Roxas built additional facilities using container vans. Roxas City also built an RT-PCR laboratory because the nearest one is one region away and costly. It also provided these services for free. This maybe because the national government usually provides the ('one size-fits all') template while the LGUs customize.

   When people were restricted and 'quarantined' at their homes, Bohol used the online platform to use culture (dances and songs) to address the anxiety and mental health issues, especially of the young. The artists were also helped through the supply of art materials and the selling of their paintings online rather than traditionally in their physical art shops. In Roxas City, since face to face-medical consultation for COVID and other cases was difficult during the lockdowns, the e-konsulta was introduced to provide the needed advice or service to those needing medical attention.

   The timely focus of Mandaue City on infectious waste management reduced the risk of COVID-19 transmission, as people were not aware of their proper disposal at home, in their communities or even at the testing, isolation and treatment facilities. This also averted environmental catastrophe at sea, especially as sea creatures may see anything lying on sea floors, including improperly disposed plastics, e.g., personal protective wastes, as food. San Juan's provision of bikes to front liners reduced the carbon footprints during the pandemic. Its aggressive awareness campaign, house-to-house registration, use of online platforms for scheduling, and incentivizing vaccination compliance raised the vaccination rate of the city to 70% of the total population in just six months (San Juan GP Paper 2021). This enabled the city to open its doors to vaccinating people from other LGUs.

c. ***Impact***: The enhancement of the partnership and alliances in the conservation and protection of Negros coastal wetlands improved the resiliency of these ecosystems. With more stakeholders and local governments involved in their conservation, bio-diversity regulation and development, the program also improved the level of bio-diversity in the area with the following resources and wildlife being protected: 46 macrobenthos[15]; 7 mangrove; and 3 species of egret. The habitat improvement also resulted in the increase of the fisherfolk's average income and catch of 4 kg/day of fish in 2017 to date. The Negros Wetlands was also included in the Ramsar List of Wetlands of International Importance in 2021. In addition, because of its vulnerable location prone to flooding and other devastation after strong typhoons, Negros has learned to climate-proof its relocation, shelther sites and other public works (Negros Occidental GP Paper 2021)
d. ***Sustainability***: In terms of sustainability, these interventions are all covered by local policy issuances and legislation that would ensure their continuous implementation beyond the pandemic. The USAD program of Agusan del Sur started as a recovery mechanism after the devastation by super typhoon Pablo. It has been carried out until now, under a new administration, ensuring that the upland communities would continue to harvest high yielding crops and earn their keep. The local programs also embedded COVID-19 response in their disaster risk mitigation and adaptation programs. This would ensure that these LGUs would be better prepared when the next pandemic and disaster strike again.

More importantly, the movers and shapers of these local programs are adaptive and innovative servant leaders who would strongly push good ideas and public interests in their policies and programs, beyond their personal interests.

Conclusions and Recommendations

The six local COVID-19 responses studied here are more successful than the national response because they provided a more comprehensive, innovative and integrative, medium-longer term perspective in addressing the pandemic. Being closest to the people, the LGUs used interventions that are more adaptive to disruptions, and included culture and the arts, smart technologies, and environmental concerns in working for solutions on the ground. They were not panic-driven and used collaborative and participatory governance in planning, implementing and even monitoring

[15] Macrobenthos is the dominant component of marine benthic assemblages. It is larger than 0.5 mm, and includes some of the most speciose groups, i.e., annelids, crustaceans, mollusks, etc. (Snelgrove 1999). Also cited in https://www.frontiersin.org/articles/10.3389/fmars.2021.756054/full.

the results of their intervention. They are also led by innovative and adaptive servant leaders whose vision and actions are more for the interest of their constituency.

The local COVID-19 Response did not sideline climate change risk mitigation and adaptation. They included climate actions in addressing the pandemic, e.g., bike lanes and bikes to reduce carbon footprints; infectious waste management to prevent disasters at landfills and at sea, especially for sea creatures who may be trapped or digest these toxic plastic wastes; wetlands conservation and climate-proofing infrastructure after disasters; upland sustainable agro-forestry and conservation. Being integrated to the COVID-19 Response, the disaster risk mitigation, adaptation and management framework of the LGUs studied here can be said to be more ready when the next pandemic occurs. Corollary, embedding COVID-19 response to local climate change risk reduction and management may have enhanced the climate change prevention, mitigation and adaptation of these LGUs.

The LGUs harnessed science, governance and economics to ensure more sustainable responses to disruptions and disasters. They also approached the solutions to the pandemic using the whole of society approach, rather than the top-down command-control and militaristic whole of government-approach of the national response. They also showed more empathy.

These are the defining features of the local COVID-19 response of the six GP 2021 Awards qualifiers and winners. These are lessons and best practices in health pandemic and disaster risk mitigation and adaptation, which the national government may emulate.

Future researches may build on the lessons from this study and interrogate and explain other factors that may contribute to better health emergency and disaster responses, both at the national and local government levels.

References

Abad M (2021) Pandemic in 2021: How the Philippines responded to COVID-19. Available via Rappler. https://www.rappler.com/newsbreak/iq/list-review-2021-covid-19-pandemic-philippines/. Accessed 20 Dec 2021

ABS-CBN News (2020) Negros Occidental closes all points of entry to prevent COVID-19 spread. Available via ABS-CBN. https://news.abs-cbn.com/news/03/16/20/negros-oriental-closes-all-points-of-entry-to-prevent-covid-19-spread. Accessed 3 Oct 2020

Agra AC (2016) Reviewer on local government law. Available via Agra Law. https://www.albertocagra.com/wp-content/uploads/2016/04/Reviewer.pdf. Accessed 2 Dec 2021

Aguilar K (2021). Among 53 countries, PH falls at the bottom in COVID-19 resilience report. Available via Inquirer News. https://globalnation.inquirer.net/199275/ph-last-among-53-countries-in-covid-19-resilience-report. Accessed 2 Oct 2021

Agusan del Sur's GP Paper (2021) Documents on Agusan del Sur's upland sustainable agro-forestry and development (USAD) convergence program entry to the 2021 Galing Pook Awards

Amit AML, Pepito VCF, Dayrit MM (2021) Early response to COVID-19 in the Philippines. Western Pac Surveill Response J 12(1):56–60. https://doi.org/10.5365/wpsar.2020.11.1.014

Bloomberg News (2021) Why the Philippines became the worst place to be in Covid. Available via Bloomberg. https://www.bloomberg.com/news/articles/2021-09-29/why-the-philippines-just-became-the-worst-place-to-be-in-covid. Accessed 2 Oct 2021

Bohol GP Paper (2021) Documents on the Bohol's culture and the arts for healing and recovery program entry to the 2021 Galing Pook Awards

Chang R, Varley K, Tam F, Munoz M (2021) The COVID resilience ranking: The best and worst places to be as covid reopening gathers pace. Available via Bloomberg News. https://www.bloomberg.com/graphics/covid-resilience-ranking/. Accessed 2 Dec 2021

Department of Finance (2020) The Duterte administration's 4-pillar socio-economic strategy against COVID-19. https://www.dof.gov.ph/the-4-pillar-socioeconomic-strategy-against-covid-19/. Accessed 2 Oct 2021

Department of Health (2021) COVID-19 situationer issue 613 December 31, 2021 https://www.doh.gov.ph. Accessed 3 Jan 2022

Department of Interior and Local Government (2020) LGU facts and figures as of September 2020. https://www.dilg.gov.ph/facts-and-figures/Regional-and-Provincial-Summary-Number-of-Provinces-Cities-Municipalities-and-Barangays-as-of-30-September-2020/32. Accessed 2 October 2021

EO 112 (2020) Imposing an enhanced community quarantine in high-risk geographical areas of the Philippines and a general community quarantine in the rest of the country from 01 to 15 May 2020, Adopting the omnibus guidelines on the implementation thereof, and for other purposes. Republic of the Philippines s 2020

FutureLearn (2021) The Philippine economy and the impact of COVID-19. Available via FutureLearn. https://www.futurelearn.com/info/futurelearn-international/philippines-economy-covid-19. Accessed 2 Oct 2021

Galing Pook (2021a) Minutes of the National Awards Selection Board's Validation Interviews with Stakeholders. A set of confidential materials, which the GPF granted access to the author/researcher.

Galing Pook (2021b) The Galing Pook Awards. Available via Galing Pook Awards. https://www.galingpook.org/what-we-do/awards/galing-pook-awards/. Accessed 2 Oct 2021

Guadalquiver N (2021) NegOcc surmounts trials brought by health crisis, disasters. Available via Philippine News Agency. https://www.pna.gov.ph/articles/1164195. Accessed 3 Jan 2022

Laforga BM (2021) NG debt rises to P10.3 trillion in Jan. Available via BusinessWorld. https://www.bworldonline.com/editors-picks/2021/03/03/347846/ng-debt-rises-to-p10-3-trillion-in-jan/. Accessed 4 Mar 2021

Lasco G (2021) COVID-19 in the Philippines: Pathologies, opportunities, and outlook. Paper presented at the Philippine public policy network webinar on "Moving to the next normal: Behavioral insights for public health policy" held on April 9, 2021

Lasco G, San Pedro J (2020) The 'covidization' of health care. Available via Philippine Daily Inquirer. https://opinion.inquirer.net/130544/the-covidization-of-health-care. Accessed 2 Oct 2021

Mandaue City's GP Paper (2021) Documents on the Mandaue City's COVID-19 community infectious waste management program entry to the 2021 Galing Pook Awards

Mypanoplies (2020) Available via Mypanoplies. https://mypanoplies.com/2020/07/16/ecq-mecq-gcq-mgcq-simplified/. Accessed 5 Jan 2022

Negros Occidental GP Paper (2021) Documents on Negros Occidental's Network of alliances for coastal wetlands conservation program entry to the 2021 Galing Pook Awards

No Harm Global (2022) Why health care waste management? Available via No Harm Global Organization. https://noharm-global.org/issues/global/why-health-care-waste-management. Accessed 18 May 2022

Philippines. Constitution of 1987, Article X. Local Governments

Quintos PL (2020) The Philippines' COVID-19 response: Symptoms of a deeper malaise in the Philippine health system. Policy Paper. National College of Public Administration and Governance, University of the Philippines

Rivera TC (2021) Five years of governance crises under Duterte: A summing up. Paper presented at the Center for People Empowerment and Governance (CenPEG). State of the Presidency Webinar, July 24, 2021

Robles A, Robles R (2021) Analysis: 'Late and slow motion': Where the Philippines' pandemic response went wrong. Available via South China Morning Post. https://www.scmp.com/week-asia/health-environment/article/3122257/late-and-slow-motion-where-philippines-pandemic. Accessed 29 Oct 2021

Roxas City (2021) Profile of Roxas City. Available via Roxas City Website. http://roxascity.gov.ph/. Accessed 2 Oct 2021

Roxas City GP Paper (2021) Documents on Roxas City's unified pandemic operational deployment (UPOD) Kita program entry to the 2021 Galing Pook Awards

San Juan GP Paper (2021) Documents on the San Juan City's Kalingap (Kaagapay, Lingkod at Gabay sa Pandemya) program entry to the 2021 Galing Pook Awards

Snelgrove P (1999). Getting to the bottom of marine biodiversity: Sedimentary habitats—Ocean bottoms are the most widespread habitat on earth and support high biodiversity and key ecosystem services. BioScience 49:129–130. https://doi.org/10.2307/1313538. Also cited in https://www.frontiersin.org/articles/10.3389/fmars.2021.756054/full. Accessed 3 Dec 2021

Chapter 18
Climate Change and Disaster Risk Management Policy Integration in Pacific Island Countries: Trajectories and Trends

Fernanda Del Lama Soares

Abstract This chapter provides a macro perspective on policy trajectories and trends of disaster risk management and climate change policy integration in Pacific Island Countries (PICs). It does so against the backdrop of key regional and global disaster risk and climate change governance milestones. The aim is to provide a contextualized overview of policy change in PICs, highlighting key points for further research. A desktop review and content analysis were conducted on national policies endorsed in the period from 1980 to 2020 for 12 PICs. The results show that: (i) policy integration in those countries is an innovation within regional and global disaster risk management and climate change governance contexts; (ii) integrated policies go beyond the integration of climate change adaptation and disaster risk management, and; (iii) there is a trend of integrated policies emphasizing climate change-related issues. These findings serve as new avenues for future research on disaster risk management and climate change policy integration in PICs, and as points for reflection on the development and practice of policy integration in the region.

Introduction

Climate change and disaster risk are governed through a complex network of institutional arrangements, policies, and actors at multiple levels of governance (Renn 2008; Bernauer and Schaffer 2010; Jones et al. 2015; Thompson 2021). Historically, climate change and disaster risk have been researched and managed separately, but the recognition of their cross-cutting and intertwined nature has brought these two fields into the policy integration debate (Ireland 2010; Hay and Mimura 2010; Djalante et al. 2013; Birkmann and Pardoe 2014; Kelman et al. 2016; Islam et al. 2020). Climate change refers to a long-term change in the state of the climate caused by natural variability and anthropogenic greenhouse gases emissions, which has the

F. Del Lama Soares (✉)
Centre for Urban Research, RMIT University, Swanston St. Building 8, Level 11, Melbourne City CampusMelbourne, VIC 3000, Australia
e-mail: fernandadellama@gmail.com

W. Leal Filho et al. (eds.), *Climate Change Strategies: Handling the Challenges of Adapting to a Changing Climate*, Climate Change Management,
https://doi.org/10.1007/978-3-031-28728-2_18

potential to create unprecedented extreme weather events (Intergovernmental Panel on Climate Change 2012, 2014). Disaster risk is the potential disruption of human systems due to hazardous event (s) caused by natural or human-induced hazards (United Nations 2016).

Disaster risk management (DRM) and climate change adaptation/mitigation are approaches to handle disaster risk and climate change. Whereas DRM aims to prevent the emergence of new disaster risks and to reduce and manage existing risk (United Nations 2016), climate change mitigation focuses on tackling the anthropogenic causes of climate change (e.g. greenhouse gases emissions), and climate change adaptation is centred on adjusting human systems to changes in the climate (Intergovernmental Panel on Climate Change 2014).

There are two main streams of policy integration research involving climate change and disaster risk. The first is concerned with the mainstreaming of climate change efforts (CC), be it for adaptation or mitigation, and/or disaster risk management into development planning (e.g. Carby 2018; Clar and Steurer 2019; Russel 2019). The second focuses on the integration of DRM and CC, with a particular focus on climate change adaptation (CCA) (e.g. Venton and La Trobe 2008; Schipper 2009; Birkmann and von Teichman 2010; Dias et al. 2021). This chapter falls into the second stream of integration studies.

Disaster risk management and climate change adaptation are discussed in the literature in terms of their synergies, as well as the challenges and barriers that hinder their integration (see Islam et al. 2020). Among the barriers commonly mentioned is the mismatch between integrated approaches and the separation of those policy domains and funding structures at the international level (Hay 2012; United Nations Office for Disaster Risk Reduction 2013; Handmer et al. 2014; Nalau et al. 2015). Nevertheless, such barriers did not prevent countries from pursuing and sustaining integrated policy approaches at the national level.

Pacific Island Countries (PICs) are among the pioneers in adopting DRM and CC/CCA integrated policy approaches. Several previous studies have analysed this experience (Gero et al. 2011; Hay 2012; Bijay et al. 2013; Secretariat of the Pacific Regional Environment Programme 2013a; United Nations Office for Disaster Risk Reduction 2013; Handmer et al. 2014; Oxfam 2016; Vachette 2017; Ronneberg and Leavai 2019; Natoli 2020; Hallwright and Handmer 2021). However, there is a lack of work that systematically reviews policy integration across multiple countries, while taking into consideration the international context in which such integration emerged and is sustained. This is the focus of this chapter.

Exploring countries' policy trajectories and trends within the context of global and regional international CC and DRM arrangements can shed light on the interactions across levels and serve as a steppingstone to understanding and theorizing internal and external factors of policy change, as well as its implications to countries' overall risk governance. Thus, such a macro perspective serves as a useful background to open avenues for further research on DRM and CC integration that transcend the identification of synergies, challenges, and barriers. It can also contribute to advance on the explanation of integrated approaches endurance at the national level in PICs, and the role such approaches can play in current times of unprecedented crisis.

Aims and Objectives

This chapter aims to provide a contextualized overview of DRM and CC policy integration trajectories and trends in Pacific Island Countries. In order to achieve this, this chapter analyses patterns of policy change in these countries against the backdrop of key global and regional disaster risk and climate change governance milestones, with the purpose to open avenues for further discussion and research in the field of DRM and CC integration in PICs.

Material and Methods

The present study is part of a larger research project. One of this project's objectives is to understand what DRM and CC integrated policies in Small Island Developing States (SIDS) in the Pacific are, and this has driven the selection of study cases analysed in this chapter. Due to the absence of consensus on a list of SIDS, study case countries were selected based on the following criteria: (i) membership in the United Nations,[1] (ii) membership in the Alliance of Small Island States,[2] and; (iii) membership in the Pacific Islands Forum.[3] Countries that were members of those three institutions were selected as case studies: Federated States of Micronesia, Fiji, Kiribati, Marshall Islands, Nauru, Palau, Papua New Guinea, Samoa, Solomon Islands, Tonga, Tuvalu, and Vanuatu.

A desktop review was conducted between August 2019 and August 2020 to map policy documents that addressed disasters, disaster risk, and climate change in the 12 selected PICs. The document collection was conducted in three rounds. The first round consisted of searching, country by country, for policy documents on the following online platforms: Prevention Web,[4] NAP Central,[5] and Voluntary National Reviews Database (Sustainable Development Goals—SDGs).[6] Those platforms were selected for the initial search as they provide an online space where countries' policy documents are publicly available. Policy documents identified in those platforms were selected when they: (i) were related to disaster/emergency

[1] https://www.un.org/about-us/member-states.

[2] https://www.aosis.org/about/member-states/.

[3] https://www.forumsec.org/who-we-arepacific-islands-forum/.

[4] https://www.preventionweb.net/knowledge-base/type-content/policy-plans?term_node_tid_depth=All&field_policy_type_target_id=All&items_per_page=8.

[5] https://unfccc.int/NAP-CENTRAL.

[6] https://sustainabledevelopment.un.org/vnrs/.

management, disaster risk reduction/management, and climate change efforts; and (ii) were issued by the national government. In the second round, these selected documents were reviewed in search of references to additional relevant policy documents. In this second round, countries' latest SDG Voluntary National Review Report, and Nationally Determined Contributions/National Communications available in the Voluntary National Reviews Database and the United Nations Framework Convention on Climate Change (UNFCCC)[7,8] websites, respectively, were consulted to ensure all relevant documents were mapped. The third round of document collection was conducted by searching the additional policy documents on countries' official websites and via the Google search engine. Whenever available, documents were downloaded for analysis. A total of 133 documents were identified, of which 83 were available for download (see Annexs 1 and 2 for the list of documents available and unavailable for download).

Following the identification of the 83 publicly available policy documents in the desktop review stage, a qualitative, deductive content analysis was then undertaken. This content analysis was based on a framework proposed by Candel and Biesbroek (2016). Candel and Biesbroek (2016) identify four dimensions of policy integration: policy frame, subsystem involvement, policy goals, and policy instruments. These four dimensions were used to 'code' different aspects of the 83 policy documents. The results discussed in this chapter focus exclusively on the dimensions (i.e. codes) related to 'policy frame' and 'policy goal'. The 'policy frame' dimension includes sub-codes related to types of hazards addressed in the policy, and the 'policy goal' dimension includes sub-codes related to the approaches policies include in their scope (e.g. disaster management, CCA, etc.). This content analysis underpins the policy modality classifications that are presented in the results section.

The key global and regional disaster risk and climate change governance milestones were identified via historical timelines provided by the United Nations Framework Convention on Climate Change (UNFCCC) (United Nations Framework Convention on Climate Change 2021a) and United Nations Office for Disaster Risk Reduction (UNDRR) (United Nations Office for Disaster Risk Reduction 2021a) websites. Identified official decision documents were then reviewed for references of CC and DRM integration. This non-exhaustive mapping exercise was conducted to provide an overview of the global context. An official historical timeline was not identified for the Pacific region. Thus, regional milestones were identified in grey and academic literature reviewed as part of the larger research project. In this case, governance milestones related to the formalization of international agreements for CC and/or DRM, and joint CC and DRM events were considered in this study. Reports and agreements related to those milestones were reviewed using the same protocol applied to the global milestones. The document collection and analysis were conducted in October and November 2021.

---

[7] https://unfccc.int/non-annex-I-NCs.

[8] https://www4.unfccc.int/sites/ndcstaging/Pages/Home.aspx.

Results

The results are presented in two subsections. The first subsection presents the findings of the content analysis in the form of a policy modality classification. The second subsection presents an analysis of the policy integration trajectories and trends and, in doing so, draws parallels with key global and regional disaster risk and climate change governance milestones.

Policy Modality Classification

In the context of this study, developing a policy modality classification is useful as it allows the differentiation between policy artefacts, thus enabling the analysis of patterns of policy change over time. Within the diversity of policy documents analysed, broad policy categories were identified based on the approaches they include (stand-alone or integrated) and types of hazards/issues addressed. These categories are described in Table 18.1.

National Policy Trajectories

According to the classification described above, PICs have adopted assorted types of policies to address disasters, disaster risk, and climate change over the period between 1980 and 2020 (Fig. 18.1). Often these policies are stand-alone and focus mainly on single issues but, in some cases, they integrate approaches for managing climate change, disasters, and disaster risk, and attempt to align objectives, priorities, and activities. According to the classification presented in Table 18.1, from the 12 countries analysed, eight presented at least one integrated policy: Kiribati, Marshall Islands, Federated States of Micronesia, Nauru, Palau, Tonga, Tuvalu, and Vanuatu.

As represented in Fig. 18.1, most of the countries considered in this review have initiated their policy trajectories by adopting disaster management instruments. This reflects the discourse, not only at the region but also at the global level, in which disasters are understood principally as a phenomenon to be managed rather than prevented (Jones et al. 2015; United Nations Office for Disaster Risk Reduction 2019). These policies mainly establish emergency protocols to prepare, respond and recover from disasters, and to mobilize resources domestically and internationally for relief.

Disaster risk management policies started to emerge in the 1990s, coinciding with the beginning of the International Decade of Natural Disaster Reduction (IDNDR), which was a global initiative established by the United Nations General Assembly in 1987 (United Nations 1987). This initiative aimed to raise awareness among member states regarding the need of efforts for risk reduction and sectoral mainstreaming.

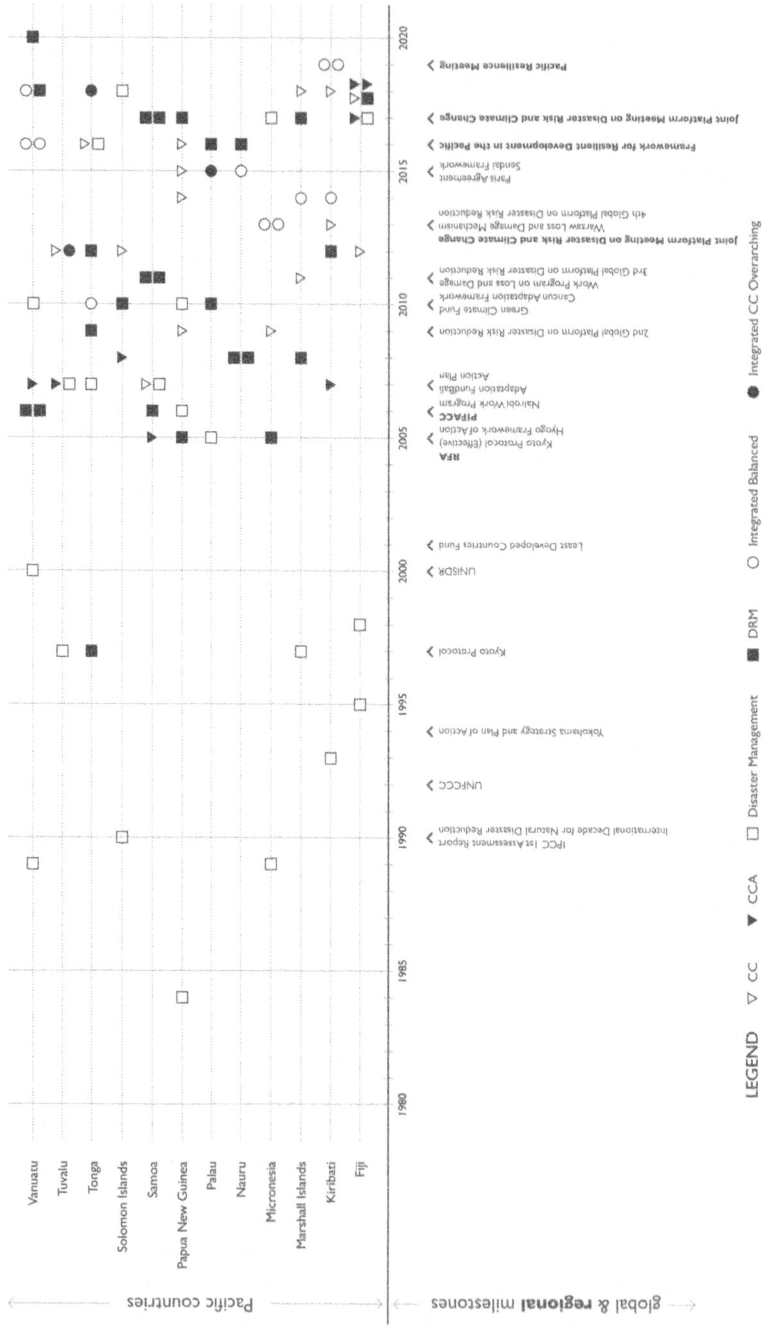

Fig. 18.1 Policies by country over time and key regional and global CC and DRM governance milestones

Table 18.1 Policy modalities classification

Stand-alone policy		
Scope	Disaster Management	The document addresses one or multiple hazards and mainly provides a strategy/activities for disaster/emergency response
	Disaster Risk Management (DRM)	The document addresses multiple hazards and mainly provides a strategy/activities for risk management/reduction (alone or in addition to disaster management arrangements), with or without reference to climate change impacts as one of the underlying factors of disaster
	Climate Change Adaptation (CCA)	The document addresses climate-related issues and mainly provides a strategy/activities for climate change adaptation
	Climate Change (CC)	The document addresses climate-related issues and mainly provides a strategy/activities for climate change adaptation and mitigation
Integrated policy		
Scope	Balanced	The document blends CC/CCA and DRM scopes in a balanced way (e.g. multi-hazards, rapid and slow onset impacts)
	CC overarching	The document blends CC/CCA and DRM scopes, but prioritizes the CC/CCA scope (e.g. neglecting or undermining non-climatic hazards/impacts)

In doing so, the initiative fostered a shift in countries' mentality towards a more preventive risk management approach. This initiative led to the development of the first global agreement for DRM, the Yokohama Strategy and Plan of Action for a Safer World (1994–2004), and later the creation of a secretariat in the United Nations System to serve as the focal point for DRM—the United Nations International Strategy for Disaster Reduction (UNISDR) (United Nations Office for Disaster Risk Reduction 2021a). These arrangements set the stage for the development of a DRM global governance system. Parallel to these developments in the DRM arena, the first IPCC Assessment Report on Climate Change was released in 1990 pushing for the establishment of the Intergovernmental Negotiating Committee for a Framework Convention on Climate Change. This led to the adoption of the United Nations Framework Convention on Climate Change (UNFCCC) in 1992 (United Nations Framework Convention on Climate Change 2021a).

Despite those global developments during the 1990s, it was only from 2005 that DRM policies became more prevalent among the countries in the Pacific region. Policies, particularly those from 2005 onwards, went a step further to consider the full cycle of disaster risk management, emphasizing the prevention and mitigation of risk by articulating more explicitly efforts for risk reduction and including references to climate change.

Climate change-related policies also started to emerge from 2005, starting with CCA policies. The emergence of CCA policies is mostly due to Least Developed Countries' international engagement in the development of National Adaptation Programs of Action (NAPAs) to access the Least Developed Countries Fund. Those NAPAs became the first climate change-related instrument to be developed at the national level in the region. After this wave of CCA policies, there was a proliferation of climate change-related policies, in which the mitigation component was also brought into the agenda.

The period around 2005 was also emblematic at the regional and the global levels. At the global level on the DRM front, the Hyogo Framework for Action: Building the Resilience of Nations and Communities to Disasters 2005–2015 (HFA) was adopted. This Framework articulates the relationships between climate change and disasters (United Nations Office for Disaster Risk Reduction 2005), reflecting earlier discussions on the synergies between DRM and CC held by the UNISDR Taskforce. This Taskforce later became the Global Platform for Disaster Risk Reduction (Inter-Agency Task Force on Disaster Reduction 2004). On the CC side, the UNFCCC's Kyoto protocol entered into force binding industrialized countries to reduce greenhouse gases emissions, and, in 2006, the Nairobi Work Program was created to support countries to integrate CCA into development programs, and to create methodologies and tools to assess climate impacts and vulnerabilities (United Nations Framework Convention on Climate Change 2006). Also in 2007, the Adaptation Fund was launched, six years after its establishment (United Nations Framework Convention on Climate Change 2021a).

In the Pacific, following the global approach, regional agreements were adopted specifically to address DRM and CC: the Pacific Islands Disaster Risk Reduction and Disaster Management Framework for Action 2005–2015 (RFA) and the Pacific Islands Framework for Action on Climate Change 2006–2015 (PIFACC). These agreements established parallel regional governance systems for DRM (aligned with the HFA) and for CC (aligned with the UNFCCC). These agreements seem to have particularly influenced the way national governments in the Pacific region framed and articulated their policies to address disaster, disaster risk, and climate change. The proliferation of DRM and CC policies from 2005 that mirror and refer to those global and regional agreements, is evidence of this idea. However, the shift towards integrating CC and DRM policies—which became widespread in countries in the region since 2010—seems to have taken an independent road, marking the point where those PICs adopted an approach distinct to global and regional arrangements and to their own previous trajectories.

Before the first national integrated policy (which was Tonga's Joint National Action Plan on Climate Change Adaptation and Disaster Risk Management of 2010), sporadic references to the synergies between DRM and CC (particularly CCA) were articulated at the global level. An example is the UNFCCC/COP13 Bali Action Plan (2007), where member states called for enhanced action on adaptation and to consider: "Risk management and risk reduction strategies, including risk sharing and transfer mechanisms such as insurance" and "Disaster reduction strategies and means to address loss and damage associated with climate change impacts in developing countries that are particularly vulnerable to the adverse effects of climate change" (United Nations Framework Convention on Climate Change 2007, p. 4). As another example, in 2009 the Chair's Summary of the 2nd Global Platform on Disaster Risk Reduction reported that "The overwhelming view of the Global Platform is that urgent action is required to harmonise and link the frameworks and policies for disaster risk reduction and climate change adaptation, and to do so within the broader context of poverty reduction and sustainable development (...)" (United Nations Office for Disaster Risk Reduction 2009, p. 2). At the regional level, the idea of a regional integrated approach had started to be discussed (e.g. Pacific Climate Change Roundtable 2009). Moreover, regional organizations, led by the Pacific Islands Applied Geosciences Commission (SOPAC), had started advocating the idea among PICs, as indicated by the Secretariat of the Pacific Regional Environment Programme (2013a) in the case of Tonga. However, it was at the national level that DRM and CC policy integration was first materialized, making this approach not only a policy innovation of the Pacific among SIDS in general[9] but also a bottom-up movement that provided the basis for the development of the first regional integrated framework in 2016, the Framework for Resilient Development in the Pacific (FRDP).

The Integration Era

There were two waves of policy change discernible regarding the integration of DRM and CC, the first starting in 2010 and the second in 2018. Until the first wave of integration, countries were in a policy trajectory that addressed disaster risk and climate change in parallel, with different levels of cross-reference among them. Countries, like Fiji, Solomon Islands, Papua New Guinea, and Samoa remained in this ***parallel trajectory*** in terms of national policy arrangements, while the other countries took a ***convergent trajectory*** towards integration, in which they brought together CC and DRM policies. It is noteworthy that, in the PICs studied, most of the integrated policies also include climate change mitigation in their scope (exceptions are Nauru's RONAdapt and Vanuatu's National Policy on Climate Change and Disaster-Induced Displacement). This suggests that integration in these countries goes beyond the attempt to build on DRM and CCA synergies (which has been the focus of the debate in the literature, as indicated in the introduction of this chapter). Moreover, half of the countries with integrated policies also embed disaster management arrangements into their integrated policy scope. This suggests a holistic approach towards risk in which they address all phases of the disaster risk management cycle (See Annex 1).

[9] It is important to note that the Maldives, was the only small island developing state out of the Pacific that adopted an integrated approach at the same period (December 2010), only a few months after Tonga endorsed its JNAP (July 2010).

The first wave of integrated policies started to emerge in the region in 2010, being Tonga the pioneer, followed by Tuvalu (2012), Micronesia (2013), Marshall Islands and Kiribati (2014), Palau and Nauru (2015), and Vanuatu (2016). Special cases are Nauru and Palau, which did not have a climate change policy prior to integration, and thus have "leap-frogged" into an integrated approach where CC-related objectives were articulated ***including*** DRM arrangements that were already in place. Most of these policies were characterized as having a balanced scope suggesting the countries' understanding and willingness to address risk from a multi-hazard perspective considering both climatic and non-climatic drivers of risk. Tuvalu and Palau are the exceptions, as they present a Climate Change Overarching Integrated Policy in which climate-related hazards are emphasized.

At the global level, the period between 2010 and 2015 was marked by several relevant milestones. First, the adoption of the Cancun Adaptation Framework, at the UNFCCC/COP16 in 2010, which called for action on "Enhancing climate change related disaster risk reduction strategies, taking into consideration the Hyogo Framework for Action, where appropriate, early warning systems, risk assessment and management, and sharing and transfer mechanisms such as insurance, at the local, national, subregional and regional levels, as appropriate" (United Nations Framework Convention on Climate Change 2010, p. 5). Then, the creation of the Green Climate Fund (GCF) in the same year. Another key milestone for the approximation of the DRM and CC agenda was the creation of a Work Program on Loss and Damage at the COP17 in Durban (2011), which later led to the establishment of the Warsaw Loss and Damage Mechanism in 2013 (United Nations Framework Convention on Climate Change 2021a). The 3rd and 4th Global Platforms on Disaster Risk Reduction, conducted in 2011 and 2013, also reiterated the need for an integrated approach to disaster risk reduction, climate change adaptation, and sustainable development. This is evident in paragraph 8.10 of the 3dr Global Platform Chair's Summary: "Avoid the inefficient use of existing resources by ensuring technologies for risk reduction are accessible as a means for adaptation and promoting integrated approaches to development that address climate change adaptation, disaster risk reduction and ecosystem management and restoration" (United Nations Office for Disaster Risk Reduction 2011, p. 3). At the Pacific regional level, a Joint Platform Meeting on Disaster Risk and Climate Change was carried out in 2013 bringing together, for the first time, the regional agreements' implementation and coordination mechanisms, the Pacific Platform for Disaster Risk Management (PPDRM), and Pacific Climate Change Round Table (PCCR). This event opened the space for DRM and CC communities of practice to establish a common ground of work and to gather support from member states to develop a regional integrated strategy, which would later become the FRDP (Secretariat of the Pacific Regional Environment Programme 2013b). These milestones reinforced PICs' integrated approach and opened the opportunity for more articulation between DRM and CC arrangements at regional and global levels, which in turn could result in a better harmonization with PICs integrated policies.

The second wave of integration is characterized by the review/update of the integrated policies. Only two of the policy artefacts representing this second wave were available for analysis: Kiribati's Joint Implementation Plan for Climate Change and

Disaster Risk Management 2019–2028 (KJIP2) and Tonga's Joint National Action Plan 2 on Climate Change and Disaster Risk Management 2018–2028 (JNAP2). Based on the analysis of these documents, it was possible to observe a shift in focus towards climate change. For instance, even though Kiribati's KJIP2 has a balanced scope in terms of hazards addressed, when presenting its linkages to other national frameworks it fails to include the National Disaster Risk Management Plan (NDRMP 2012) alongside the country's latest Climate Change Policy (2018) (see Government of Kiribati 2018, p. 62). This contrasts with the arrangements presented in the country's previous integrated policy (see Government of Kiribati 2014, p. 38). Moreover, the KJIP2 was developed alongside the Climate Change Policy (2018), bringing these two in close alignment, whereas the NDRMP was not updated concomitantly. Another interesting link to the climate change agenda in the case of Kiribati, is the fact that the KJIP2 was submitted to the UNFCCC as Kiribati's National Adaptation Plan in 2020 (United Nations Framework Convention on Climate Change 2021b).

In the case of Tonga, the JNAP2 (2018) places significantly more focus on climate-related impacts, restricting the reference to other hazards to a minimum (see Government of Tonga 2018, pp. 1, 17), in contrast to the comprehensive hazard profile description presented on the JNAP1. It also explicitly makes reference only to climate change impacts in its rationale (see Government of Tonga 2018, p. 23). Therefore, the JNAP2 is classified as a Climate Change Overarching Integrated Policy. Moreover, the JNAP2 does not refer to DRM policies and recurrently indicates its linkage and alignment with Tonga's Climate Change Policy—A resilient Tonga by 2035 (2016) (see, for example, Government of Tonga 2018, pp. 23, 26), as opposed to the JNAP1 which referred to both the National Emergency Management Plan (2009) and the Climate Chance Policy (2006). Furthermore, among the integrated policies analysed, this was the only document that received funds from the Green Climate Fund for its formulation.

Paradoxically, this second wave of national policy integration and its shift towards climate change emphasis happened at a time when a turn towards integration was being materialized at both the regional and the global levels. For instance, at the regional level, the FRDP superseded previous separated regional agreements (RFA and PIFACC) and established a regionally integrated governance system. At the global level, for the first time, agreements for DRM (Sendai Framework), CC (Paris Agreement/UNFCCC), and Sustainable Development (Agenda 2030 and the Sustainable Development Goals) were forged almost simultaneously allowing cross-pollination (United Nations Office for Disaster Risk Reduction 2019). Also, on the technical side, both the UNFCCC and the UNDRR have taken efforts to increase coherence between their processes and activities. This can be observed, for example, in the UNFCCC technical paper on the opportunities for integration of CCA into the Sendai Framework and Sustainable Development Goals (United Nations Framework Convention on Climate Change 2017) and the UNDRR supplementary guidelines for promoting better coherence between disaster risk reduction and the National Adaptation Plan process (United Nations Office for Disaster Risk Reduction 2021b). However, despite these examples signalizing a convergent trajectory towards integration at regional and global levels, it is noteworthy that the imbalance in international finance—with higher volumes to climate change (Hay 2013), and the escalation of the climate crisis along with its recognition as a security issue (Rasheed 2021) are arguably powerful forces for the prioritization of the climate change agenda.

Conclusion

This chapter analysed policy trajectories PICs have taken to address disasters, disaster risk, and climate change against key global and regional milestones for climate change and disaster risk governance. By doing so, this chapter demonstrated that PICs have followed a similar parallel pathway as global and regional arrangements, until the post-2010 period when they embarked upon CC and DRM policy integration. This indicates that DRM and CC policy integration in the Pacific region is a policy innovation for risk governance driven by PICs' national governments.

Secondly, it demonstrated that most of these integrated policy approaches initially presented a balanced scope, in that they addressed multiple hazards and took into consideration both climatic and non-climatic drivers of risk. Furthermore, it showed that some of these policies also integrated climate change mitigation and disaster management into their scope. This suggests a holistic approach to risk governance at the national level, which goes beyond building on disaster risk management and climate change adaptation synergies.

Finally, this chapter identified that the second generation of integrated policies tends to emphasize climate change-related issues, overshadowing other non-climate-related hazards and drivers of risk. Such a shift towards a Climate Change Overarching approach has happened in a period where, paradoxically, DRM and CC integration is getting traction in regional and global arenas, but also at a time where the climate crisis and the imbalance of international resources for CC and DRM are increasingly evident. Given that climate change is the major challenge PICs currently face, and where most of the international cooperation resources are being concentrated, it is understandable that there is a tendency in those countries to prioritize climate change adaptation efforts. However, framing the multiple problems PICs face this way, could end up undermining efforts towards risk reduction of other hazards to which those countries are also exposed (e.g. Tsunami, epidemics, earthquakes, volcano eruptions, etc.) (United Nations Economic and Social Commission for Asia and the Pacific 2021). This could result in an "Achilles heel" type of situation, which could take governments and societies by surprise and generate catastrophic impacts, as has been learned with the COVID-19 pandemic.

With the climate change crisis increasingly taking over the agenda of many of the small island developing countries in the world, and particularly in the Pacific, it is important that governments, as well as regional and global CC and DRM communities of practice, are attentive to broader risks to make sure integrated approaches can serve the purpose of being a holistic instrument for risk governance.

The findings of this chapter have implications for the research and practice of DRM and CC policy integration. The findings shed light on PICs' agency and leadership in policy integration practice; they point to the need to consider disaster management and climate change mitigation in the scope of holistic risk governance, and; they call attention to the potential effects of CC-biased integrated policies to PICs risk governance. In this way, this review opens avenues for practitioner reflection and research in the field of DRM and CC integration in small island developing countries in the Pacific.

Research Limitations

The findings presented are based on the analysis of documents publicly available and, therefore, can only provide a partial view of the trajectories of countries' policy frameworks for risk governance. For instance, the findings on trends related to the scope of integrated policies were based on only two cases and therefore would need to be reassessed in the future when more policies are revised and available to the public. Also, due to the high number of documents assessed, only specific aspects of the policies were analysed. This limited a deeper understanding of the roles and challenges of policy integration in the context of each PIC, and the meaning of such trends to those countries.

A further limitation is that this study did not assess the levels of implementation of those policies and, therefore, is unable to make inferences on the effects of such policy choices on countries' risk governance in practice.

Recommendations for Future Research

In light of the limitations of this study and its purpose to provide an overview of policy trajectories and trends, this chapter serves as a background for further discussion and research in the field of DRM and CC integration in PICs. As such, the analysis raises a few points that could be considered in future research. Firstly, on the DRM and CC integration as a policy innovation among small island developing states, countries' deviation from regional and global trends could be explored from the lenses of developing countries' agency and policy ownership in contrast to the siloed global and regional funding systems. Secondly, the inclusion of climate change mitigation and disaster management into integrated policies could be further explored to expand the understanding of holistic risk governance approaches and move the debate beyond disaster risk management and climate change adaptation synergies. Thirdly, the role of policy integration for the practice of DRM and CC and the challenges PICs face to integrate and sustain those policies could be further explored. Finally, the shift of integrated policies towards emphasizing climate change impacts could be further investigated to understand its implications for countries' risk governance.

Annex 1—Documents Available and Analysed ($n = 83$)

Country	Document	Classification
Fiji	Fiji Disaster Management Plan (1995)	DM
Fiji	Fiji Natural Disaster Management Act (1998)	DM
Fiji	National Climate Change Policy (2012)	CC
Fiji	Fiji Tsunami Response Plan (2017)	DM
Fiji	National Adaptation Plan Framework (2017)	CCA
Fiji	National Adaptation Plan (2018)	CCA
Fiji	Fiji's National Disaster Risk Reduction Policy (2018–2030)	DRM
Fiji	Planned Relocation Guidelines—A Framework to Undertake Climate Change Related Relocation (2018)	CCA
Fiji	Low Emission Development Strategy (2018)	CC
Kiribati	National Disaster Act (1993)	DM
Kiribati	National Adaptation Programme of Action—NAPA (2007)	CCA
Kiribati	Disaster Risk Management Plan (2012)	DRM
Kiribati	National Framework for Climate Change and Climate Change Adaptation (2013)	CC
Kiribati	Kiribati Climate Change Policy (2018)	CC
Kiribati	Joint Implementation Plan for Climate Change and Disaster Risk Management (2014–2023)*∞	Integrated-balanced
Kiribati	Joint Implementation Plan for Climate Change and Disaster Risk Management (2019–2028)*∞	Integrated-balanced
Kiribati	Disaster Risk Management and Climate Change Act (2019)*∞	Integrated-balanced
Marshall Islands	National Disaster Management Plan (1997)	DM
Marshall Islands	National Climate Change Policy Framework (2011)	CC
Marshall Islands	Republic of the Marshall Islands National Disaster Risk Management Arrangements (2017)	DRM
Marshall Islands	2050 Climate Strategy (2018)	CC
Marshall Islands	National Action Plan for Disaster Risk Management (2008–2018)	DRM
Marshall Islands	Joint National Action Plan for Climate Change Adaptation and Disaster Risk Management (2014–2018)*∞	Integrated-balanced

(continued)

(continued)

Country	Document	Classification
Micronesia	Disaster Relief Assistance Act (1989)	DM
Micronesia	Multi-State Multi-Hazard Mitigation Plan for FSM (2005)	DRM
Micronesia	Nationwide Climate Change Policy (2009)	CC
Micronesia	Nationwide Integrated Disaster Risk Management and Climate Change Policy (2013)$^{\infty}$	Integrated-balanced
Micronesia	Climate Change Act (2013)$^{\infty}$	Integrated-balanced
Micronesia	National Disaster Response Plan (2016)	DM
Nauru	Disaster Risk Management Act (2008)	DRM
Nauru	National Disaster Risk Management Plan (2008)	DRM
Nauru	Framework for Climate Change Adaptation and Disaster Risk Reduction (RONAdapt) (2015)	Integrated-balanced
Nauru	National Disaster Risk Management Act (2016)	DRM
Palau	Republic of Palau Pandemic Influenza Response Plan (2005)	DM
Palau	National Disaster Risk Management Framework (2010)	DRM
Palau	Climate Change Policy—for Climate and Disaster Resilient Low Emissions Development (2015)$^{\infty}$	Integrated-CC Overarching
Palau	National Disaster Risk Management Framework _ Amended (2016)	DRM
Papua New Guinea	Disaster Management Act (1984)	DM
Papua New Guinea	National Contingency Plan for Preparedness and Response for Influenza Pandemic Papua New Guinea (2006)	DM
Papua New Guinea	National Climate Compatible Development Management policy (2014)	CC
Papua New Guinea	Climate Change Management Act (2015)	CC
Papua New Guinea	Paris Agreement Implementation Act (2016)	CC
Papua New Guinea	National Disaster Mitigation Policy (2010)	DM
Papua New Guinea	Disaster Risk Deduction and Disaster Management National Framework for Action (2005–2015)	DRM
Papua New Guinea	Forestry and Climate Change Framework for Action (2009–2015)	CC
Papua New Guinea	Papua New Guinea National Disaster Risk Reduction Framework (2017–2030)	DRM

(continued)

(continued)

Country	Document	Classification
Samoa	National Adaptation Plan for Action—NAPA (2005)	CCA
Samoa	National Policy of Combating Climate Change (2007)	CC
Samoa	Disaster and Emergency Management Act (2007)	DM
Samoa	National Disaster Management Plan (2011)	DRM
Samoa	National Disaster Management Plan (2017)	DRM
Samoa	National Disaster Management Plan (2006)	DRM
Samoa	Samoa National Action Plan for Disaster Risk Management (2011–2016)	DRM
Samoa	Samoa National Action Plan for Disaster Risk Management (2017–2021)	DRM
Solomon Islands	National Disaster Council Act (1990)	DM
Solomon Islands	National Adaptation Plan for Action—NAPA (2008)	CCA
Solomon Islands	National Disaster Risk Management Plan (2010)	DRM
Solomon Islands	National Disaster Management Plan (2018)	DM
Solomon Islands	National Climate Change Policy (2012–2017)	CC
Tonga	National Disaster Plan and Emergency Procedure (1997)	DRM
Tonga	Emergency Management Act (2007)	DM
Tonga	National Emergency Management Plan (2009)	DRM
Tonga	Tsunami Plan (2012)	DRM
Tonga	Climate Change Policy—A Resilient Tonga by 2035 (2016)	CC
Tonga	Tropical Cyclone Winston Response Plan (2016)	DM
Tonga	Joint National Action Plan on Climate Change and Disaster Risk Management (2010–2015)$^{\infty}$	Integrated-balanced
Tonga	Joint National Action Plan 2 on Climate Change and Disaster Risk Management (2018–2028)$^{\infty}$	Integrated-CC Overarching
Tuvalu	National Disaster Management Plan (1997)	DM
Tuvalu	National Disaster Management Act (2007)	DM
Tuvalu	National Adaptation Programme of Action—NAPA (2007)	CCA

(continued)

(continued)

Country	Document	Classification
Tuvalu	National Strategic Action Plan for Climate Change and Disaster Risk Management (2012–2016)*$^{\infty}$	Integrated-CC Overarching
Tuvalu	Te Kaniva—Climate Change Policy (2012)	CC
Vanuatu	Meteorology Act (1989)	DM
Vanuatu	National Adaptation Programme for Action—NAPA (2007)	CCA
Vanuatu	National Disaster Plan (2010)	DM
Vanuatu	National Policy on Climate Change and Disaster-induced Displacement (2018)	Integrated-balanced
Vanuatu	National Disaster Risk Management Bill (2018)	DRM
Vanuatu	Recovery Strategy 2020–2023: TC Harold and COVID-19 (2020)	DRM
Vanuatu	National Disaster Act 2006 (2000)	DM
Vanuatu	Priorities and Action Agenda 2006–2015—Supplementary for Mainstreaming Disaster Risk Reduction and Disaster Management (2006)	DRM
Vanuatu	Disaster Risk Reduction and Disaster Management National Action Plan (2006–2016)	DRM
Vanuatu	Meteorology, Geo-hazards, and Climate Change Act (2016)$^{\infty}$	Integrated-balanced
Vanuatu	Climate Change and Disaster Risk Reduction Policy (2016–2030)*$^{\infty}$	Integrated-balanced

Note Integrated policy includes: *Disaster management, $^{\infty}$Climate Change Mitigation

Annex 2—Other Documents Mapped, But Unavailable for Analysis ($n = 50$)

Country	Document
Fiji	Fiji Drought Response Plan (2015)
Fiji	Draft Climate Change Adaptation Strategy (2011)
Fiji	National Climate Change Policy Framework (2007)
Fiji	National Humanitarian Policy for Disaster Risk Management (2017)
Fiji	Climate Change and Health Strategy Action Plan (2016–2020)

(continued)

(continued)

Country	Document
Fiji	Disaster Risk Management Agriculture Strategy (2010)
Fiji	Fiji National Influenza Pandemic Plan (2006)
Fiji	Emergency Manual (1979)
Fiji	Clean Development Mechanism (CDM) Policy Guideline (2008)
Fiji	National REDD + Policy (2011)
Kiribati	Climate Change Adaptation Policy and Strategy (2005)
Kiribati	Avian and Pandemic Influenza Preparedness and Response Plan (2008)
Kiribati	Draft National Disaster Management Plan (1995)
Kiribati	Draft National Disaster Plan (2010)
Marshall Islands	Joint National Action Plan for Climate Change Adaptation and Disaster Risk Management (2018)
Marshall Islands	National Disaster Risk Management Arrangements (2012)
Marshall Islands	Hazard Mitigation Plan (1994)
Marshall Islands	National Disaster Manual (1994)
Marshall Islands	National Adaptation Plan
Marshall Islands	Disaster Assistance Act (1987)
Marshall Islands	RMI National Emergency Response Plan (2010)
Marshall Islands	National Disaster Management Plan (1987)
Marshall Islands	Drought Disaster Plan (1996)
Marshall Islands	Standard Hazard Mitigation Plan (2005)
Marshall Islands	RMI Climate Change Roadmap (2010)
Micronesia	FSM Pandemic Influenza Plan (2008)
Micronesia	Disaster Preparedness Plan
Micronesia	FSM Climate Change and Health Action Plan (2011)
Nauru	Tsunami Support Plan (2011)
Palau	National Disaster Management Plan (1998)
Papua New Guinea	Climate Compatible Development Strategy
Papua New Guinea	Office of Climate Change and Environmental Sustainability Policy Framework
Papua New Guinea	REDD Framework (2007/2008)
Samoa	Disaster and Emergency Management Act (2006)
Samoa	National Disaster Management Plan and Emergency Procedures (2000)
Samoa	Samoa National Avian and Pandemic Influenza Preparedness Plan (2008)
Samoa	National Tropical Plan (2007)
Samoa	National Tsunami Plan (2008)
Samoa	Emergency Response Plan for Animal and Plant Pests (2005)
Samoa	National Fire Plan (2009)
Solomon Islands	National Disaster Plan (1987)

(continued)

(continued)

Country	Document
Solomon Islands	National Disaster Plan (1997)
Tonga	National Disaster Management Plan (1987)
Tonga	Climate Change Policy (2006–2015)
Tonga	Climate Change Policy (2015–2020)
Tonga	National Influenza Preparedness and Response Plan (2006)
Tuvalu	National Disaster Risk Management Arrangements (2012)
Tuvalu	National Disaster Management Plan (1978)
Vanuatu	National Disaster and Emergency Plan (1987)
Vanuatu	National Disaster Recovery Framework

Acknowledgements I am grateful to David Stevens and Prof. Ralph Horne for their insightful comments in early drafts of this chapter, and to the anonymous reviewers for their valuable feedback. This research was made possible thanks to funding provided by RMIT University.

References

Bernauer T, Schaffer LM (2010) Climate change governance. Center for Comparative and International Studies. Available via https://papers.ssrn.com/abstract=1661190. Accessed 21 Dec 2020

Bijay P, Leal Filho W, Schulte V (2013) Understanding the links between climate change and disaster management in Pacific Island countries. In: Leal Filho W (ed) Climate change and disaster risk management. Springer, Heidelberg, pp 55–69

Birkmann J, Pardoe J (2014) Climate change adaptation and disaster risk reduction: Fundamentals, synergies and mismatches. In: Glavovic BC, Smith GP (eds) Adapting to climate change: lessons from natural hazards planning. Springer, Dordrecht, pp 41–56

Birkmann J, von Teichman K (2010) Integrating disaster risk reduction and climate change adaptation: key challenges—Scales, knowledge, and norms. Sustain Sci 5:171–184. https://doi.org/10.1007/s11625-010-0108-y

Candel JJL, Biesbroek R (2016) Toward a processual understanding of policy integration. Policy Sci 49:211–231. https://doi.org/10.1007/s11077-016-9248-y

Carby B (2018) Integrating disaster risk reduction in national development planning: Experience and challenges of Jamaica. Environ Hazards 17:219–233. https://doi.org/10.1080/17477891.2017.1415864

Clar C, Steurer R (2019) Climate change adaptation at different levels of government: Characteristics and conditions of policy change. Nat Resour Forum 43:121–131. https://doi.org/10.1111/1477-8947.12168

Dias N, Amaratunga D, Haigh R, Clegg G, Malalgoda C (2021) Critical factors that hinder integration of CCA and DRR: Global perspective. In: Luetz JM, Ayal D (eds) Handbook of climate change management: Research, leadership, transformation. Springer International Publishing, Cham, pp 4107–4128

Djalante R, Holley C, Thomalla F, Carnegie M (2013) Pathways for adaptive and integrated disaster resilience. Nat Hazards 69:2105–2135. https://doi.org/10.1007/s11069-013-0797-5

Gero A, Méheux K, Dominey-Howes D (2011) Integrating disaster risk reduction and climate change adaptation in the Pacific. Clim Dev 3:310–327. https://doi.org/10.1080/17565529.2011.624791

Government of Kiribati (2014) Joint implementation plan for climate change and disaster risk management 2014–2023. Available via https://www.preventionweb.net/publication/kiribati-joint-implementation-plan-climate-change-and-disaster-risk-management-kjip. Accessed 7 Sep 2019

Government of Kiribati (2018) Joint implementation plan for climate change and disaster risk management 2019–2028. Available via https://www.preventionweb.net/publication/kiribati-joint-implementation-plan-climate-change-and-disaster-risk-management-2019. Accessed 4 Sep 2020

Government of Tonga (2018) Joint National Action Plan 2 on climate change and disaster risk management 2018–2028. Available via https://www.preventionweb.net/publication/tonga-joint-national-action-plan-2-climate-change-and-disaster-risk-management-jnap-2. Accessed 2 Jun 2020

Hallwright J, Handmer J (2021) Progressing the integration of climate change adaptation and disaster risk management in Vanuatu and beyond. Clim Risk Manag 31:100269. https://doi.org/10.1016/j.crm.2020.100269

Handmer J, Mustelin J, Belzer D, Dalesa M, Farmer N, Foster H, Greimel B, Harper M, Kauhiona H, Pearce S, Vines K (2014) Integrated adaptation and disaster risk reduction in practice. RMIT University, Griffith University and the National Climate Change Adaptation Research Facility. Available via https://www.nab.vu/sites/default/files/nab/documents/14/04/2014%20-%2014:20/handmer_etal_integratedadaptationdrrinpractice_1.pdf. Accessed 5 Dec 2019

Hay JE (2012) Disaster risk reduction & climate change adaptation in the Pacific: An institutional and policy analysis. United Nations Office for Disaster Risk Reduction. Available via https://www.preventionweb.net/files/26725_26725drrandccainthepacificaninstitu.pdf. Accessed 14 Oct 2019

Hay JE (2013) Roles of Pacific regional organizations in disaster risk management: Questions and answers. University of South Pacific. Available via https://www.brookings.edu/wp-content/uploads/2016/07/Brookings_Regional_Orgs_Pacific_July_2013_FINAL.pdf. Accessed 15 Jan 2020

Hay JE, Mimura N (2010) The changing nature of extreme weather and climate events: Risks to sustainable development. Geomat Nat Hazards Risk 1:3–18. https://doi.org/10.1080/19475701003643433

Inter-Agency Task Force on Disaster Reduction (2004) Working group on climate change and disaster risk reduction. https://www.unisdr.org/2005/task-force/tf-working-group-cc-drr-eng.htm. Accessed 17 Nov 2021

Intergovernmental Panel on Climate Change (2012) Managing the risks of extreme events and disasters to advance climate change adaptation: Special report of the intergovernmental panel on climate change. Cambridge University Press, Cambridge

Intergovernmental Panel on Climate Change (2014) Climate change 2014: Synthesis report. Contribution of working groups I, II and III to the fifth assessment report of the intergovernmental panel on climate change. Intergovernmental Panel on Climate Change, Geneva

Ireland P (2010) Climate change adaptation and disaster risk reduction: Contested spaces and emerging opportunities in development theory and practice. Clim Dev 2:332–345. https://doi.org/10.3763/cdev.2010.0053

Islam S, Chu C, Smart JCR, Liew L (2020) Integrating disaster risk reduction and climate change adaptation: A systematic literature review. Clim Dev 12:255–267. https://doi.org/10.1080/17565529.2019.1613217

Jones S, Manyena B, Walsh S (2015) Disaster risk governance: Evolution and influences. In: Shroder JF, Collins AE, Jones S et al (eds) Hazards, risks and disasters in society. Academic Press, Boston, pp 45–61

Kelman I, Gaillard JC, Lewis J, Mercer J (2016) Learning from the history of disaster vulnerability and resilience research and practice for climate change. Nat Hazards 82:129–143. https://doi.org/10.1007/s11069-016-2294-0

Nalau J, Handmer J, Dalesa M, Foster H, Edwards J, Kauhiona H, Yates L, Welegtabit S (2015) The practice of integrating adaptation and disaster risk reduction in the south-west Pacific. Clim Dev 8:365–375. https://doi.org/10.1080/17565529.2015.1064809

Natoli T (2020) Law and policies that protect the most vulnerable against climate-related disaster risks: Findings and lessons learned from Pacific island countries. International Federation of Red Cross and Red Crescent Societies. Available via https://www.preventionweb.net/publication/law-and-policies-protect-most-vulnerable-against-climate-related-disaster-risks. Accessed 17 Nov 2021

Oxfam (2016) Pathways and challenges towards integration and implementation of DRR and CCA in Tonga. Oxfam. Available via https://www.preventionweb.net/publications/view/51376. Accessed 12 Sep 2019

Pacific Climate Change Roundtable (2009) Report of the Pacific climate change roundtable meeting, 19–23 October 2009. Secretariat of the Pacific Regional Environment Programme. Available via http://archive.iwlearn.net/sprep.org/climate_change/PCCR/documents/PacificClimateChangeRoundtableMeetingREPORTFINAL.pdf. Accessed 17 Nov 2021

Rasheed DAA (2021) Island nations demand climate security at COP26. ANU Dept Pac Aff DPA Former State Soc Gov Melanes SSGM Program 32:1–2. https://doi.org/10.25911/9009-XB48

Renn O (2008) Risk governance: Coping with uncertainty in a complex world, 1st edn. Routledge, London

Ronneberg E, Leavai PD (2019) Status of climate change adaptation in the Pacific region. In: Alam M, Lee J, Sawhney P (eds) Status of climate change adaptation in Asia and the Pacific. Springer International Publishing, Cham, pp 97–123

Russel D (2019) Enabling conditions for the mainstreaming of adaptation policy and practice. In: Keskitalo ECH, Preston BL (eds) Research handbook on climate change adaptation policy. Edward Elgar Publishing, Cheltenham, pp 108–124

Schipper L (2009) Meeting at the crossroads? Exploring the linkages between climate change adaptation and disaster risk reduction. Clim Dev 1:16–30. https://doi.org/10.3763/cdev.2009.0004

Secretariat of the Pacific Regional Environment Programme (2013a) JNAP development and implementation in the Pacific, experiences, lessons and way forward. Secretariat of the Pacific Regional Environment Programme. Available via https://www.sprep.org/attachments/Publications/CC/JNAP.pdf. Accessed 15 Oct 2019

Secretariat of the Pacific Regional Environment Programme (2013b) Many positive lessons learnt from joint platform meeting on disaster risk and climate change. https://www.sprep.org/news/many-positive-lessons-learnt-joint-platform-meeting-disaster-risk-and-climate-change. Accessed 17 Nov 2021

Thompson DDP (2021) Disaster risk governance: Four cases from developing countries. Routledge Taylor & Francis Group, Milton

United Nations (1987) International decade of natural disaster reduction. Available via https://undocs.org/en/A/RES/42/169. Accessed 17 Nov 2021

United Nations (2016) Report of the open-ended intergovernmental expert working group on indicators and terminology relating to disaster risk reduction. Available via https://www.preventionweb.net/files/50683_oiewgreportenglish.pdf. Accessed 26 Sep 2021

United Nations Economic and Social Commission for Asia and the Pacific (2021) Asia-Pacific disaster report 2021: Resilience in a riskier world. United Nations, Bangkok

United Nations Framework Convention on Climate Change (2006) Report of the conference of the parties on its eleventh session, held at Montreal from 28 November to 10 December 2005 Addendum part two: Action taken by the conference of the parties at its eleventh session. Accessible via https://unfccc.int/documents/4249. Accessed 17 Nov 2021

United Nations Framework Convention on Climate Change (2007) Bali action plan. Accessible via https://unfccc.int/resource/docs/2007/cop13/eng/06a01.pdf. Accessed 17 Nov 2021

United Nations Framework Convention on Climate Change (2010) Cancun adaptation framework. Accessible via https://unfccc.int/resource/docs/2010/cop16/eng/07a01.pdf. Accessed 17 Nov 2021

United Nations Framework Convention on Climate Change (2017) Opportunities and options for integrating climate change adaptation with the Sustainable Development Goals and the Sendai Framework for Disaster Risk Reduction 2015–2030. United Nations Framework Convention on Climate Change. Accessible via https://unfccc.int/sites/default/files/resource/techpaper_adaptation.pdf. Accessed 18 Nov 2021

United Nations Framework Convention on Climate Change (2021a) Timeline: 25 years of effort and achievement. https://unfccc.int/timeline/. Accessed 4 Nov 2021a

United Nations Framework Convention on Climate Change (2021b) National adaptation plans. https://unfccc.int/topics/adaptation-and-resilience/workstreams/national-adaptation-plans. Accessed 30 Nov 2021b

United Nations Office for Disaster Risk Reduction (2005) Hyogo framework for action: Building the resilience of nations and communities to disasters 2005–2015. Available via https://www.preventionweb.net/files/1037_hyogoframeworkforactionenglish.pdf. Accessed 17 Nov 2021

United Nations Office for Disaster Risk Reduction (2009) Chair's summary of the second session global platform for disaster risk reduction. Available via https://www.preventionweb.net/files/10750_GP09ChairsSummary.pdf. Accessed 17 Nov 2021

United Nations Office for Disaster Risk Reduction (2011) Chair's summary third session of the global platform for disaster risk reduction and world reconstruction conference. Available via https://www.preventionweb.net/files/20102_gp2011chairssummary.pdf. Accessed 17 Nov 2021

United Nations Office for Disaster Risk Reduction (2013) The Pacific experience in developing policy and legislation on disaster risk reduction and climate change adaptation. United Nations Office for Disaster Risk Reduction. Available via https://www.preventionweb.net/files/34003_34003pacificexperienceonlegislation.pdf. Accessed 15 Oct 2019

United Nations Office for Disaster Risk Reduction (2019) Global assessment report 2019. United Nations Office for Disaster Risk Reduction, Geneva

United Nations Office for Disaster Risk Reduction (2021a) Milestones in the history of disaster risk reduction. https://www.undrr.org/about-undrr/history. Accessed 4 Nov 2021a

United Nations Office for Disaster Risk Reduction (2021b) Promoting synergy and alignment between climate change adaptation and disaster risk reduction in the context of national adaptation plans: A supplement to the UNFCCC NAP technical guidelines. United Nations Office for Disaster Risk Reduction. Available via https://www4.unfccc.int/sites/NAPC/Documents/Supplements/2021b_Promoting%20Synergy%20-%20Supplement%20to%20the%20NAP%20Technical%20Guidelines.pdf. Accessed 17 Nov 2021

Vachette A (2017) Integrating disaster risk reduction and climate change adaptation in Vanuatu: The art and practice of building resilience to hazards. In: Leal Filho W (ed) Climate change adaptation in Pacific countries: Fostering resilience and improving the quality of life. Springer International Publishing, Cham, pp 119–136

Venton P, La Trobe S (2008) Linking climate change adaptation and disaster risk reduction. Tearfund. Available via https://www.preventionweb.net/files/3007_CCAandDRRweb.pdf. Accessed 11 Sep 2019

Chapter 19
Mainstreaming Agri-Compatible Virtual Resource Flows in Agri-Food System Adaptation to Climate Change in the Caribbean

David Oscar Yawson and Michael Osei Adu

Abstract Climate change does not only threaten to amplify hydrometeorological disaster risks (to which the Caribbean countries are highly exposed and vulnerable) but also to unravel the agri-food systems of the Caribbean. Land and water are indispensable to food production but are scarce in the Caribbean Small Island Developing States where food imports play a major role in food security. Food trade indirectly circulates productive resources such as land and water and this has been captured in the virtual resource flow concept. However, Caribbean states have failed to realize and take strategic advantage of the virtual resource flow concept in a manner compatible with resource and food supply needs. Since food trade will play important role in future food security, it is worthy of consideration as part of a suite of adaptation measures. This chapter uses agri-compatible virtual land use associated with maize production and trade in Barbados to promote the need for mainstreaming virtual resource flows in the adaptation of agri-food systems to climate change in the Caribbean. Gaps and linkages in the production-trade-consumption nexus are explored in relation to agri-food system resilience based on the virtual resource flow concept in a changing climate. The chapter concludes that strategic use of agri-compatible virtual resource flows can be instrumental to the adaptation of Caribbean agri-food systems in the face of resource scarcity and climate change.

D. O. Yawson (✉)
Centre for Resource Management and Environmental Studies (CERMES), The University of the West Indies, Cave Hill Campus, Bridgetown, St. Michael BB11000, Barbados
e-mail: david.yawson@cavehill.uwi.edu

M. O. Adu
Department of Crop Science, University of Cape Coast, Cape Coast, Ghana

W. Leal Filho et al. (eds.), *Climate Change Strategies: Handling the Challenges of Adapting to a Changing Climate*, Climate Change Management,
https://doi.org/10.1007/978-3-031-28728-2_19

Introduction

Agri-food systems face manifold challenges: increased demand for diverse and nutritious food, dwindling productive resources, and climate change. Demand for food is expected to increase considerably in the next few decades. For example, global average meat consumption is projected to reach about 49 kg per person per year in the 2050s, or 91 kg per person per year in rich countries, with a corresponding increase in feed demand at 1.1 billion tonnes (Alexandratos and Bruinsma 2012). Land availability is fundamental to food production. Globally, agriculture accounts for the largest share of land use, with a significant share (about 25% of global land surface) used to support animal feed production (Foley et al. 2011). Feed use accounts for about 50% of total global crop production (Herrero et al. 2015) and 35% of total grain produced (Alexandratos and Bruinsma 2012). At the same time, the area of arable land is expected to decrease significantly due to pressures from urbanization, economic development, anthropogenic degradation, and climate change. As a result, land scarcity, a situation where more than 60% of a country's prime and good arable land is already committed to production or regular cultivation (Alexandratos and Bruinsma 2012), will be a major constraint to food production. Climate change threatens to escalate the biophysical stresses on biomass production and crop yields, enhance land degradation, and increase competition from other land use sectors. For example, climate mitigation measures or policies might unintentionally have adverse impacts on food supply (Hasegawa et al. 2018; Rowe et al. 2009; Yawson et al. 2017, 2020). The threat and impacts of climate change to agri-food systems are geographically disparate but particularly escalated in regions that are also noted to be highly vulnerable (Zimmermann et al. 2018; Yawson 2020; Huang et al. 2011). A combination of these factors would likely shape relative productive capacities, create new geographies of food surplus and deficits, and therefore demand a broader swath of adaptation measures for agri-food systems.

Food commodity trade provides a vital bridge between food-scarce and food-surplus regions and, thus, enables spatial decoupling of resource constraints and food supply. Trade would continue to play vital roles in future agri-food system resilience and food security (Zimmermann et al. 2018) in the face of the adverse impacts of land scarcity and climate change on agri-food systems. The added advantage of food trade is the classical economic notion that traded commodities embody and indirectly circulate productive resources consumed in their production (Huang et al. 2011; Ansink 2010). The utility of this idea for coupled resource-food security policy has been rationalized in the virtual resource concept (Yawson 2013). For a unit food commodity that is traded, the virtual land use transferred is a product of the virtual land use content and the quantity of the food commodity traded (Würtenberger et al. 2006; Yawson 2021). Virtual land use content refers to the area of land used in producing that unit of food commodity which is traded. While there is a copious body of literature on virtual water, few studies have focused on estimates of virtual land use (e.g., Kastner and Nonhebel 2010; Kissinger and Rees 2010; Fader et al. 2011; Fader et al. 2013; Qiang et al. 2013; Yawson 2021) or its utility for agri-food system adaptation to climate change (Yawson 2021).

Inherent in the virtual resource concept is the proposition that countries faced with food insecurity due to limited or scarce productive resources can import resource-intensive food commodities to overcome the challenge of both resource scarcity and food insecurity. So, nations faced with land scarcity, for example, can maintain food security by using food commodity trade to overcome the constraint of land scarcity. However, this promise and the virtual resource concept itself were met with sharp criticisms when first introduced via the virtual water concept by Allan (1998). The ensuing debate (see Neubert and Horlemann 2008; Yawson 2013), summarized here, remains active. On one hand, it was argued that the qualifier 'virtual' was unnecessary for something or a term that is already clearly named and understood, and was therefore misleading (Merrett 2003a, b). Indeed, it is the food commodity, not the embodied resource, that is traded. In response, Allan (2003) argued that the qualifier was a useful metaphor to capture both the intensive (resource use in production) and extensive (impact of food trade and consumption on resource balances, uses or policies in the trading nations) sides of the virtual resource concept. As a result, focusing on the intensive side alone constitutes an incomplete view. In the same vein, an overfocus on the resource alone, rather than a coupled resource-food security perspective, should be an incomplete view (Yawson 2013; Yawson et al. 2013). The extensive side provides an opportunity to analysing the potential impact of food commodity trade on overcoming domestic constraints to production due to resource limitations and maintenance of food security in the destination country. Another argument against the inherent promise of the virtual resource concept was the limitation of the practical utility of the resource savings (if at all) for policy and practice in the destination country (Yawson 2013). Here, the argument and research effort have been to generate evidence that resource scarcity underpins the structure, direction, and magnitude of virtual resource transfers (Yawson 2013).

Food commodity trade is underpinned by several factors, ranging from biophysical to socio-economic and political. It is therefore difficult to isolate a target resource scarcity as the singular driver of trade. Rather, food security imperatives might be a strong determinant of food imports. As a result, the utility of the virtual resource concept for policy, or the related debate, needs to be re-focused on a coupled resource-food security perspective rather than limiting it to only the resource. To this end and to address these issues, the idea of agri-compatible virtual resource flow was introduced. A detailed description of the idea of agri-compatibility can be found in Yawson (2013) and Yawson et al. (2013). In short, agri-compatibility suggests that, prior to analysing the value of food commodity trade for resource savings, two conditions should be met: (i) the production of the target food commodity in the importing nation should be limited by the target resource (a situation referred to as agri-compatible resource scarcity), and (ii) the food commodity import should address food insecurity resulting from gaps in production due to resource scarcity (i.e., resource-dependent food import). In other words, the magnitude of food import should, at least, correspond to or exceed the gap in food production introduced by the resource scarcity. By so doing, the partial or complete impact of virtual resource flows on addressing both food insecurity and resource scarcity can be ascertained, as well as its utility for policy and practice. Agri-compatibility lends further strength to the utility of the virtual resource concept for resource-food security policy.

The Sixth Assessment Report (AR6) of the Intergovernmental Panel on Climate Change (IPCC) indicates that anthropogenic warming of the atmosphere, ocean, and land has certainly occurred, with resultant increases in the frequency, intensity, and severity of impacts of extreme events such as droughts, heatwaves, heavy precipitation, and tropical cyclones (IPCC 2021). The report indicates that, without aggressive mitigation actions, increase in global mean temperature could reach 4 °C by the end of this century. As Tropical Small Island Developing States (SIDS), Caribbean countries are not only among the most vulnerable to climate change, but also face acute land and water scarcity (FAO and CDB 2019; Karmalkar et al. 2013). The Caribbean agricultural sector is dominated by smallholder, rainfed production systems that are highly exposed and vulnerable to climate change (FAO 2019). Previous climate-related disasters have had devastating impacts on Caribbean agriculture (FAO 2015). Climatic studies show that the Caribbean is becoming increasingly drier, and drought conditions and heavy precipitation events would become highly frequent and pervasive in the future (FAO 2015; Karmalkar et al. 2013). The destabilizing impacts of climate change on agri-food systems are already being felt in the Caribbean through the increased frequency and intensity of hydrometeorological disasters, including droughts, floods, and hurricanes (FAO 2015). Currently, Caribbean countries depend considerably on food import to ensure food security. For example, annual food import to the Caribbean Community (CARICOM) is estimated at about US$ 5 billion, with some countries importing more than 50% of their food needs (CARICOM Today 2021). Yet, Caribbean countries have not given due consideration to incorporating the virtual resource flow concept into the suite of measures for adapting agri-food systems to climate change. Agri-compatible virtual land use can serve as an analytical lens for food security contexts and support the design of a useful mix of adaptive production and trade strategies in response to land scarcity and climate change at varying administrative and temporal scales. The objective of this chapter is to use the virtual land use associated with maize production, trade, and consumption in Barbados to promote the need for mainstreaming agri-compatible virtual resource flows in agri-food system adaptation to climate change in the Caribbean and other regions.

Methods

Barbados (13° 10' N, 59° 30' W) is the Easternmost Island in the chain of the Eastern Caribbean Small Island States (see Fig. 19.1). It has an area of 430 square kilometres or 166 square miles and is divided into 11 parishes. It has a tropical, sub-humid climate with distinct wet season (June-November) and dry season (December-May). Barbados has a population of about 287,000 and is one of the most densely populated countries in the world as it has 663 persons per square kilometre (UNEP 2010). Agriculture used to be the main backbone of the economy but has recently been

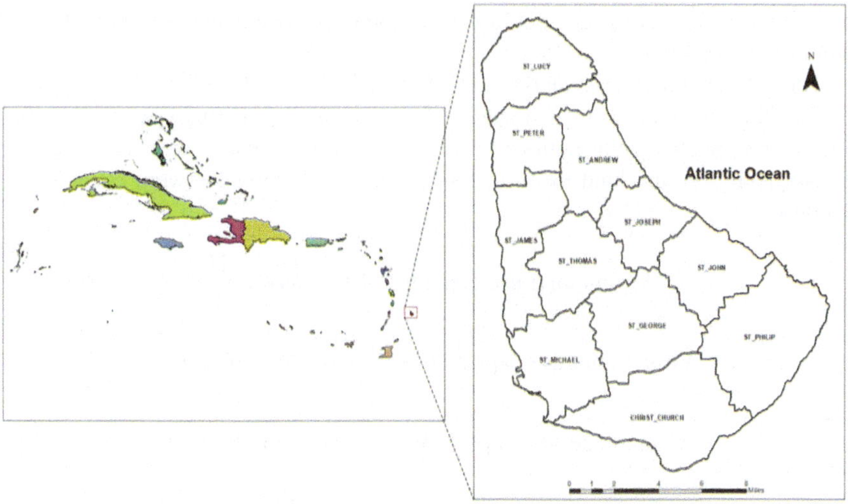

Fig. 19.1 The Caribbean Island States (left) and Barbados with its 11 parishes (right)

superseded by tourism and financial services. As a result, and as observed in other Caribbean countries, food import has increased dramatically. Currently, Barbados imports more than 50% of its food needs and food import bill is estimated to be more than US$500 million per annum.

Data were derived from the FAOSTAT database and the Food Balance Sheet (FBS). The FAOSTAT database provides data on food production, trade, and other information at national level. The database is globally and temporally consistent, permitting national comparisons of food production and trade over time. The FAO FBS (FAO 2001) provides an account of national food supply balances (from production, imports, and exports) and distribution of available supplies to different uses for a given reference year. It is therefore a useful tool that is widely used to assess national food shortages and supplies across the world. The FAOSTAT database was used to identify the top food commodity import to Barbados. Maize was identified to be the topmost food commodity imported to both the Caribbean and Barbados. Data on maize production for Barbados for the period 2010–2019 were then extracted from the database. Similarly, a detailed maize trade matrix for Barbados for the same period was generated. The trade matrix shows the import and export partner countries that Barbados traded maize with for the period of interest. The latest FBS of Barbados was used to identify the distribution of total maize supply to end uses (or consumption) in Barbados. From this, it was observed that feed use accounted for 86% of total quantity of maize supplied for domestic use, while direct consumption as food accounted for only 10%. Maize is largely used as energy source in monogastric feed formulation. As a result, the production, trade, and consumption of poultry was analysed from both the FAOSTAT database and the food balance sheet. Amongst all animal meat, poultry has the largest per capita food supply (46.78 kg) per annum in Barbados. For poultry meat, total import amounted to only 13% of domestic production.

The extracted data was analysed in Microsoft Excel. Charts on trends of maize production and trade were generated. The main trading partners were identified and their respective contributions to total trade with Barbados were decomposed. Subsequently, virtual land use transfers to and from Barbados were estimated as follows:

$$\text{Virtual land use content (VLUC, ha)} = \frac{\text{TP}}{Y}$$

$$\text{Virtual land use export (VLUE, ha)} = \text{VLUC} \times \text{Qe}$$

$$\text{Virtual land use import (VLUI, ha)} = \text{VLUC} \times \text{Qi}$$

$$\text{Net virtual land use (NVLU, ha)} = \text{VLUI} - \text{VLUE}$$

where TP denotes total production or quantity; *Y*, denotes yield; Qe denotes Quantity exported; and Qi denotes Quantity imported.

Subsequently, the equivalent land demand (total area of land, in ha, that would have been required to produce the total quantity of maize imported, given Barbados yield) was estimated. Then, residual land (i.e., total arable land less the cultivated area, which was assumed to be 70% of arable land area for each year studied as indicated in the FAO database) was calculated. Agri-compatible land scarcity was estimated as the difference between the equivalent land demand and the residual land. The corresponding maize deficit was calculated to represent agri-compatible import to fill the gap in supply created by land scarcity. The difference between the net virtual land use and the equivalent land demand was also computed to gauge land use efficiency in relation to the Barbadian maize trade.

Results

Production and Trade

Availability of arable land is an important indicator of food production potential. The area of arable land in Barbados has declined steadily from the year 2000 to 2017 (Fig. 19.2). For this period, arable land of Barbados ranged from 15,000 to 7,000 ha. While the area of arable land remained somewhat stable for the period 2004–2009 at 13,000 ha, it declined more steeply than the period before 2004. In the absence of productivity gains, Fig. 19.2 shows a declining food production potential for Barbados and increasing land scarcity.

For the period 2010–2019, total maize production in Barbados increased slightly from 260 tonnes in 2010 to 340 tonnes in 2013 (Fig. 19.3). It then declined sharply to 81 tonnes in 2014 and never recovered to 2019. Total maize production in 2019 was 42 tonnes and total maize production for the entire period (2010–2019) was 1,713 tonnes. The curve of total area harvested corresponded very well with the total production curve (Fig. 19.3), showing the dependence of the latter on the former, with little variation in yield. Total area harvested was highest in 2013, with a value of 120 ha, and least in 2019 (15 ha). Overall, Fig. 19.3 shows that current maize production in Barbados is very low.

For the period 2010–2019, total import of maize to Barbados from partner countries ranged from 3 to 37,501 tonnes, with a median value of 35 tonnes. Total annual import for the period studied ranged from 28,956 tonnes to 37,585 tonnes (Fig. 19.4).

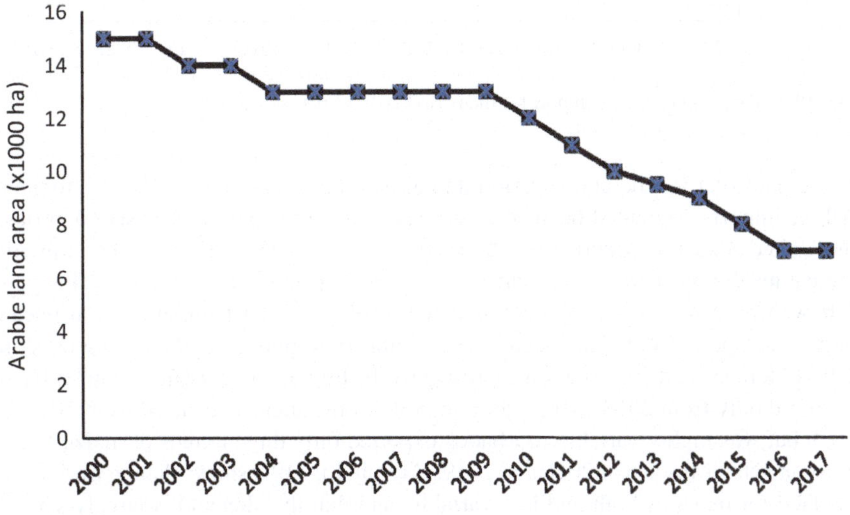

Fig. 19.2 Arable land area for the period 2000–2017

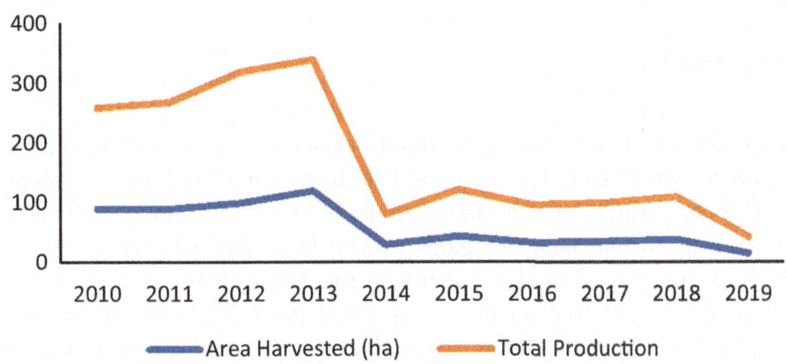

Fig. 19.3 Total maize production (metric tonnes) and area harvested (ha) for the period 2010–2019

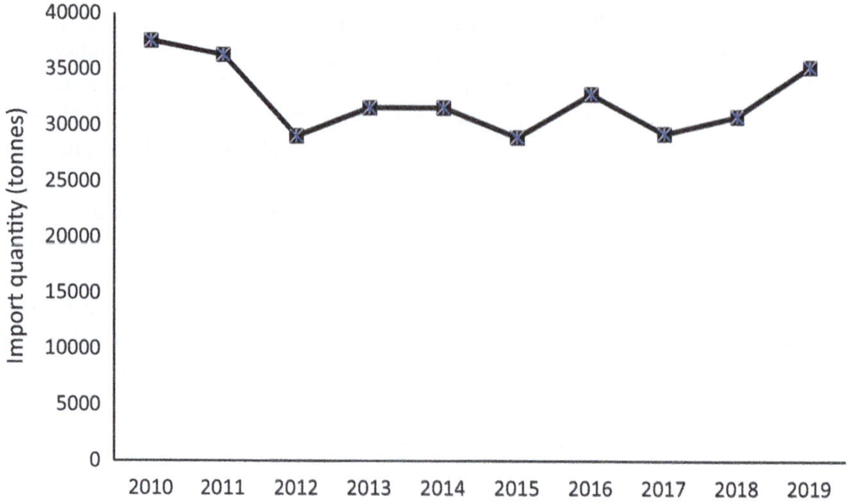

Fig. 19.4 Total annual maize import for the period 2010–2019

The grand total for the entire period and all source countries was 323,325 tonnes. All the imports originated from six countries (Fig. 19.5). Out of these six countries, the United States of America (USA) accounted for 99.75% of total maize import, indicating the most important and stable source of maize to Barbados. This was followed by Jamaica which accounted for 0.16% (the most important Caribbean partner). Imports from Canada and the Dominican Republic occurred only once in 2019 (3 tonnes) and 2011 (6 tonnes), respectively. Imports from Trinidad and Tobago occurred only from 2010–2012, while imports from Jamaica occurred from 2012 to 2019 but was highly variable. Barbados exported only three tonnes of maize to St. Kitts and Nevis in each of 2018 and 2019. Overall, Fig. 19.4 shows that, in Barbados, demand for maize is high and less variable, and that this demand is largely served by imports in contrast with very low export.

Virtual Land Use

Net virtual land use was calculated as import less export. As a result, positive values indicate net import or gain. For the period studied, virtual land use transfers (gains in this case) per partner country ranged from 0.32 ha (Canada) to 3,930 ha (USA). Total annual net virtual land use gains ranged from 2,670.86 ha to 3,961.37 ha (Fig. 19.6). While virtual land use gains declined marginally from 2010 to 2017, it seems to have picked up an increasing trend from 2017. Again, as would be expected, the USA contributed the largest share of total land use gains to Barbados via maize import. This was followed by Jamaica and Trinidad and Tobago as the

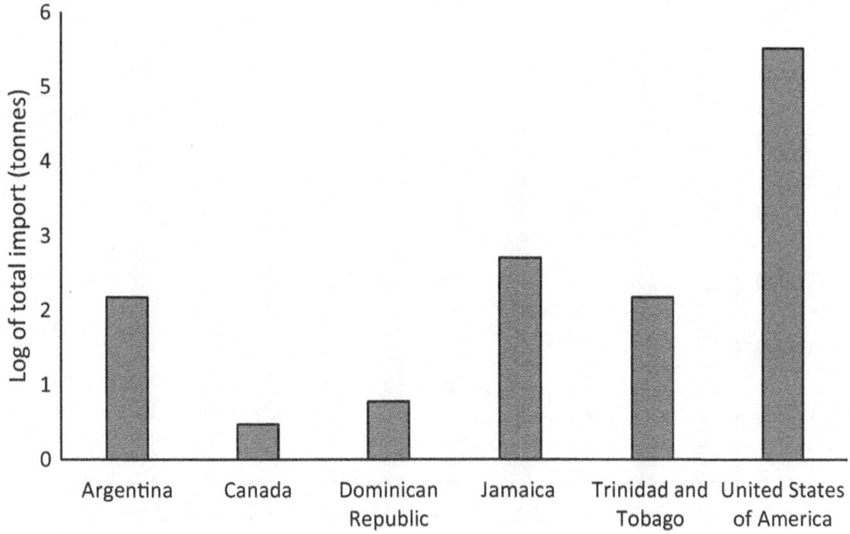

Fig. 19.5 Total import (log scale) and sources for the period 2010–2019

main Caribbean sources. Estimates of equivalent land demand can be a proxy for contribution to global land use efficiency, as well as the degree to which trade helps overcome local resource constraint to food production. The equivalent land demand in Barbados (i.e., given Barbados productivity, the total area of land that would have been required to produce the quantity of maize imported) ranged from 9,079.06 in 2012 to 13,010.19 ha (Fig. 19.6, primary vertical axis). Net land deficit represents the area of land that would have been required to produce the net imported maize if unused (residual) arable land was cultivated to maize (in other words, deducted from the equivalent land demand). In the current study, 30% of arable land was considered residual as the FAO database shows that about 70% of arable land in Barbados is committed to cultivation. Net land deficit, which represents agri-compatible land scarcity in this study, ranged from 5,779.06 ha to 10,196.07 ha (Fig. 19.7, left panel). The corresponding maize deficit (representing agri-compatible import) ranged from 18,493 tonnes to 28,549 tonnes (Fig. 19.7, right panel). These values are highest for the year 2019 due to higher import quantity and lower area of arable land. However, variations in import volumes per country over the years underpin the yearly variations in net virtual land use and equivalent land demand or land deficit. In all, Figs. 19.6 and 19.7 show that Barbados gains large area of virtual land use but this is actually a fraction of the total area of land that would have been required to produce the imported maize domestically.

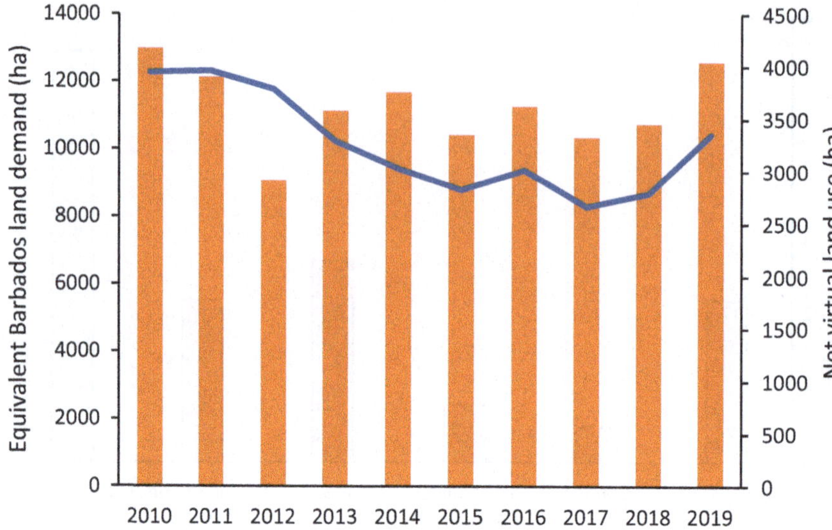

Fig. 19.6 Net virtual land use transfer (secondary vertical axis) and the equivalent land demand in Barbados (primary vertical axis)

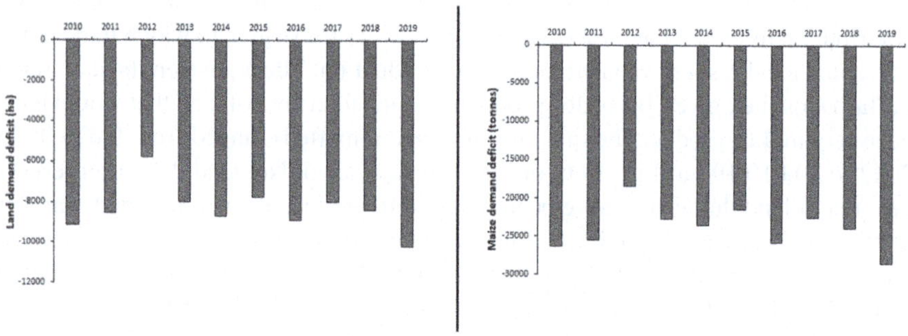

Fig. 19.7 Land deficit (ha) after accounting for 30% usable arable land (left panel) and resulting maize deficit (tonnes) (right panel)

Discussion

This study assessed the virtual land use gains of Barbados via maize trade to promote the need to mainstream agri-compatible virtual resource flows in the adaptation of agri-food systems to climate change in the Caribbean and elsewhere. As indicated earlier, agri-compatibility has two basic requirements: (i) agri-compatible resource scarcity and (ii) resource-dependent import. The results show that Barbados faces land scarcity, and its limited arable land has declined steadily from the year 2000 (Fig. 19.2). The area allocated to maize production was not just low but also declined sharply in the period studied. It is not clear what could account for the sharp decline in the area allocated to maize production. However, these, coupled with low yields, drove down total production. Imports remained quite steady over the period studied. This suggests that maize is important to Barbados and that land scarcity might contribute substantially to the observed scale of imports. The magnitude of net import observed in the current study cannot be supported by the total arable land of Barbados. The magnitudes of imports were more than 30 times the level of domestic production for each year in the period studied. Correspondingly, both the equivalent land demand and net land deficit were about twice the total available arable land. Additionally, the curve of the harvested area closely matched that of total production, suggesting a strong dependence of the latter on the former throughout the period studied. It can be deduced that while Barbados needs maize for its food security and is able to grow maize, land scarcity is or could be a major constraint to higher domestic production to significantly reduce the observed magnitude of import. Feed use accounts for the largest share of maize utilization in Barbados but this underpins poultry production for which imports are much lower compared to maize. Poultry meat has the largest per capita supply in Barbados and this underscores the importance of maize for Barbadian food security. It can therefore be said that the observed results in the current study considerably satisfy the requirements of agri-compatible virtual resource flows as land scarcity seemed to contribute substantially to the scale of maize imports.

The observed reduction in arable land area in Barbados is like most countries in the Eastern Caribbean. Globally, declines in the area of arable land per person have been observed and this would likely continue to the future (Alexandratos and Bruinsma 2012). Availability of land would be crucial for increasing total food production whether the biophysical limit of productivity gains is reached or not (Pollock 2011; Alexandratos and Bruinsma 2012; Yawson 2013). Regardless of the impact of climate change, it is unlikely that the land scarcity of Barbados could improve substantially to permit Barbados to meet its maize requirement via domestic production. For countries or regions that face land scarcity, for example, some gains in virtual land use would be helpful in conserving limited resources and maintaining food security. In the current study, the range of virtual land use gains or land savings (Fader et al. 2013; Qiang et al. 2013; Yawson 2021) was substantial (about half or more than half of the total available land for the years studied). This suggests that maize trade helps Barbados fill the gap in production or demand created by local land scarcity. Additionally, Barbados contributes to global land use efficiency because its equivalent land demand or net land deficit is higher than the net virtual land use (Yang et al. 2006; Fader et al. 2013; Yawson 2021). In other words, maize is imported from countries with higher land productivities than Barbados itself. Given the current low yields, it is improbable that Barbados might substantially increase yields or production in the face of climate change as this will require substantial long-term investment.

The instrumentality of food commodity trade for overcoming local food security challenges in the context of climate change has been highlighted (Zimmermann et al. 2018; Yawson 2020) but not so much about the possibility for addressing resource challenges. Trade enables spatial decoupling of gaps in local food production due to resource constraint from the supply of rich, diverse, and large quantities of food

(Fader et al. 2013). It is in this context that food trade is viewed as a crucial instrument for agri-food system adaptation to the uneven adverse impacts of climate change (Zimmerman et al. 2018; Yawson 2020) and points to the utility of the virtual resource concept for coupled resource-food security adaptation policy (Yawson et al. 2020). Obviously, food trade is underpinned by multiple factors and it is difficult to single out a resource as the main driver of trade. It is for this reason that agri-compatibility contends that analysis of the utility of virtual resource flows for policy should be based on coupled resource-food security goals. From such a perspective, the contribution of an agri-compatible resource scarcity to food import (and for that matter virtual resource flows) to serve food security needs can be realized. Agri-compatibility provides an opportunity for analyzing the feedback relationships between resource balances, production, and trade to support strategies or policies for production-trade mix that are harmonious with resource scarcity and food security goals. In the case of Barbados or the Caribbean, this can be applied in the context of adaptation planning or efforts at increasing domestic production for import substitution.

Barbados faces acute land and water scarcity. It is in a region that is highly vulnerable to the adverse impacts of climate change on food production. Barbados is currently a large, net food importer and this might not significantly change to the nearest future. Given these conditions, it is important for Barbados to seriously consider a broader agri-compatibility-based analysis of its food trade in relation to resource conservation and food security goals, and mainstreaming agri-compatible virtual resource flows in its suite of agri-food system adaptation plans. Such an integrated approach can help Barbados design adaptation policies or strategies that spread and minimize future trade risks associated with key food commodities, such as maize. In the current study, the USA supplies almost all the maize import to Barbados and this is risky. Mainstreaming agri-compatible virtual resource flows into adaptation plans or policies can enable Barbados to identify key food commodities and their production-trade thresholds to support sustainability of food security and local livelihoods. It can help Barbados identify source partner countries that engage in sustainable production practices to optimize virtual resource gains, and strategically replace those that face production penalties in the future due to climate change or other risks. Finally, it can foster a transparent approach to virtual resource accounting as an input to strategic resource (such as land or water) management or policy decisions for agri-food system resilience to climate change. Mainstreaming virtual resource flows in adaptation policy decisions implies taking an expanded and a more holistic view of the value of food commodity trade. It means understanding and quantifying the collateral benefits of food commodity trade in addressing not just food insecurity but also the conditions that drive food insecurity. By influencing the availability of productive resources, climate change will affect food production, and for that matter food availability, which is the foundation of food security. Mainstreaming agri-compatible virtual resource flows would be both a process and a state. As a process, mainstreaming here would require an explicit recognition of the dual value of agri-compatible virtual resource flows, and subsequent incorporation into adaptation policy goals, processes, and institutions. This process will, in turn, require an understanding of the impacts of climate change and resource scarcity on Caribbean food production and adoption of an expanded view of adaptation measures and policies that include agri-compatible virtual resource flows. Once the mainstreaming of agri-compatible virtual resource flows into policy or adaptation measures is achieved (state), trade strategies and practices that harmoniously balance domestic resource and food security goals should become conventional and routine. In all, agri-compatibility helps keep a focus on a trade strategy that is strongly premised on tightly-coupled resource-food security goals.

Limitations

This chapter was based on a single crop and country, albeit a clearly important food commodity for the food security of Barbados and the Caribbean. Such a single crop approach might not capture the full picture of the balance between resource scarcity and domestic food production in relation to trade. However, it helps bring a sharp focus to the issue and the value of an agri-compatible approach. Future studies can increase the number of food commodities and, potentially, countries involved. The approach used in this chapter appears simplistic but intuitive. The approach is at initial stages of methodological development, but its value is demonstrated. An improved quantitative approach would be a helpful step forward.

Conclusions

The adverse impacts of climate change on resource use and food production would be geographically uneven. Trade would continue to be crucial for global food security as it bridges regions with surpluses to those with deficits. Because virtual resource circulation is coupled to food trade flows, it can be strategically used as part of adaptation measures of agri-food systems to climate change. To this end, however, virtual resource flows need to be agri-compatible to be effective. The current chapter aimed to use maize-based virtual land use of Barbados to promote a need to mainstream agri-compatible virtual resource in agri-food system adaptation to climate change in Barbados and the Caribbean. Barbados was found to have limited arable land and faced land scarcity, especially in relation to maize production. Regardless of the reasons for low domestic production, maize seemed very important to Barbados and so large imports, corresponding to net land deficits, were sustained through the period studied. Barbados gains considerable virtual land use from food commodity trade. The virtual land use gains of Barbados, though huge, are about a third of the equivalent land that would be required for domestic production and, therefore, contribute to global land use efficiency. It is concluded that virtual land use gains to Barbados based on maize considerably satisfy the agri-compatibility requirements as arable land scarcity might make considerable contribution to the observed magnitudes of maize import. Given the unique characteristics of Barbados (and, by extension, the

Caribbean), such as large net food importer, high land and water scarcity, and high vulnerability to climate change, there is instrumental value in the strategic use of agri-compatible virtual resource flows as an integral component of measures or plans for agri-food system adaptation to climate change. This is necessary to designing adaptation strategies focused on production-trade mix for key food security commodities in balance with resource availability or scarcity. Mainstreaming of agri-compatible virtual resource flows in adaptation policies or strategies should be viewed as both a process and a state. The first step in the process would be a recognition of the dual value of food trade for addressing coupled resource-food security goals to permit consideration in adaptation policies and strategies or measures. As a state, trade practices and strategies that simultaneously enhance the security of both food and target resources should become verifiably the norm. The current study is limited in the number of crops and countries involved, but this was necessary to attempt a demonstration of the value of agri-compatibility. Future studies can expand the number of crops and/or countries and improve the quantitative aspect of the agri-compatibility approach.

References

Alexandratos N, Bruinsma J (2012) World Agriculture towards 2030/2050: The 2012 Revision. ESA Working Paper, No. 12-03. Food and Agriculture Organization, Rome, Italy, 2012

Allan JA (1998) Moving water to satisfy uneven global needs: trading water as an alternative to engineering it. ICID J 47(2):1–8

Allan JA (2003) Virtual water—the water, food, and trade nexus—useful concept or misleading metaphor? Water Int 28(1):106–113

Ansink E (2010) Refuting two claims about virtual water trade. Ecol Econ 69:2027–2032

CARICOM Today (2021) *'25 in 5' Plan to tackle CARICOM food import bill.* Accessed October 8, 2021, from https://today.caricom.org/2020/07/27/25-in-5-plan-to-tackle-caricom-food-import-bill/

Fader M, Gerten D, Krause M, Lucht W, Cramer W (2013) Spatial decoupling of agricultural production and consumption: quantifying dependences of countries on food imports due to domestic land and water constraints. Environ Res Lett 8(1):014046

Fader M, Gerten D, Thammer M, Heinke J, Lotze-Campen H, Lucht W, Cramer W (2011) Internal and external green-blue agricultural water footprints of nations, and related water and land savings through trade. Hydrol Earth Syst Sci 15:1641–1660

FAO (2001) Food balance sheets: a handbook. Economic and Social Development Department, Food and Agriculture Organization of the United Nations, Rome, Italy, 2001

FAO (2015) Addressing disaster risk management in Caribbean agriculture. Issue Brief #12, February 2015, FAO Sub-regional Office for the Caribbean

FAO (2019) Current status of agriculture in the Caribbean and the implications for agriculture policy strategy. 2030-food, agriculture, and rural development in Latin America and the Caribbean. Document No.14, Food and Agriculture Organization of the United Nations (FAO), Santiago de Chile, 28 pp

FAO and CDB (2019) Study on the state of agriculture in the Caribbean. Food and Agriculture Organization (FAO) of the United Nations and the Caribbean Development Bank (CDB), Rome, 212 pp

Foley JA, Ramankutty N, Brauman KA, Cassidy ES, Gerber JS, Johnston M et al (2011) Solutions for a cultivated planet. Nature 478(7369):337

Hasegawa T, Fujimoro S, Havlik P et al (2018) Risk of increased food insecurity under stringent global climate change mitigation policy. Nat Clim Chang 8:699–703

Herrero M, Wirsenius S, Henderson B, Rigolot C, Thornton P, Havlík P et al (2015) Livestock and the environment: what have we learned in the past decade? Annu Rev Environ Resour 40:177–202

Huang H, von Lampe M, van Tongeren F (2011) Climate change and trade in agriculture. Food Policy 36 (Suppl. 1), S9–S13.

IPCC (2021) Summary for policymakers. In: Masson-Delmotte V, Zhai P, Pirani A, Connors SL, Péan C, Berger S, Caud N, Chen Y, Goldfarb L, Gomis MI, Huang M, Leitzell K, Lonnoy E, Matthews JBR, Maycock TK, Waterfield T, Yelekçi O, Yu R, Zhou B (eds.), Climate change (2021): the physical science basis. Contribution of the Working Group I to the Sixth Assessment Report of the Intergovernmental Panel on Climate Change (in press). Cambridge University Press, Cambridge, United Kingdom and New York, NY, USA, pp. 3–32. https://doi.org/10.1017/9781009157896.001

Karmalkar AV, Taylor MA, Camp bell J, Stephenson T, New M, Cantella A, Benzanilla A, Charley J (2013) A review of observed and projected changes in climate for the islands in the Caribbean. Atmósfera 26(2):283–309

Kastner T, Nonhebel S (2010) Changes in land requirements for food in the Philippines: a historical analysis. Land Use Policy 27:853–863

Kissinger M, Rees WE (2010) Importing terrestrial biocapacity: the U.S. case and global implications. Land Use Policy 27:589–599

Merrett S (2003a) Virtual water and Occam's Razor. Water Int 28(1):103–105

Merrett S (2003b) Virtual water and the Kyoto Consensus: a water forum contribution. Water Int 28(4):540–542

Neubert S, Horlemann L (2008) Strategic virtual water trade – a critical analysis of the debate. In Scheumann W, Neubert S, Kipping M (eds), Water politics and development cooperation. Springer-Verlag, Berlin, Heidelberg, Germany, pp.123–145. ISBN 978-3-540-76707-7

Pollock C (2011) Food for thought: options for sustainable increases in agricultural production. In Foresight (2011). Foresight project on global food and farming futures. Regional case study R1: the UK in the context of North-west Europe. Government Office for Science, UK

Qiang W, Liu A, Chenga S, Kastner T, Xie G (2013) Agricultural trade and virtual land use: the case of China's crop trade. Land Use Policy 33:141–150

Rowe RL, Street NR, Taylor G (2009) Identifying potential environmental impacts of large-scale deployment of dedicated bioenergy crops in the UK. Renew Sustain Energy Rev 13:271–290

UNEP (2010) *National environmental summary: barbados*. Accessed December 28, 2020, from http://www.pnuma.org/publicaciones/FINAL%20Barbados%20NES%20Nov%202010-%20edited.pdf

Würtenberger L, Koellner T, Binder CR (2006) Virtual land use and agricultural trade: estimating environmental and socio-economic impacts. Ecol Econ 57:679–697

Yang H, Wang L, Abbaspour KC, Zehnder AJB (2006) Virtual water trade: an assessment of water use efficiency in the international food trade. Hydrol Earth Syst Sci 10(3):443–454. https://doi.org/10.5194/hess-10-443-2006

Yawson DO (2013) Climate change and virtual water: implications for UK food security. PhD thesis, University of Dundee, UK

Yawson DO (2020) Climate mitigation and hidden vulnerabilities: widening the food gap between the global north and south. Environ Justice 13(6):210–221

Yawson DO (2021) Estimating virtual land use under future conditions: application of a food balance approach using the UK. Land Use Policy 101(2021):105132. https://doi.org/10.1016/j.landusepol.2020.105132

Yawson DO, Mulholland B, Ball T, Mohan S, White P (2013) Food security in a water-scarce world: making virtual water compatible with crop water use and food trade. Sci Paper Ser Manag Econ Eng Agric Rural Dev 13(2):431–444

Yawson DO, Mulholland BJ, Ball T, Adu MO, Mohan S, White PJ (2017) Effect of climate and agricultural land use changes on UK feed barley production and food security to the 2050s. Land 2017(6):74

Yawson DO, Adu MO, Armah FA (2020) Impacts of climate change and mitigation policies on malt barley supplies and associated virtual water flows in the UK. Sci Rep 2020(10):376

Zimmermann A, Benda J, Webber H, Jafari Y (2018) Trade, food security and climate change: conceptual linkages and policy implications. FAO, Rome, p 48

Chapter 20
Climate Change Perceptions and Adaptation Strategies in Vulnerable and Rural Territories

Filipa Marques, Fátima Alves, and Paula Castro

Abstract The consequences of climate change are unavoidable, making adaptation actions more important than ever. The engagement of communities and civic participation is also essential in adaptation strategies, and part of that involves studying citizens' perceptions of climate change. This chapter aims to understand and discuss how people's perceptions of climate change, mainly those belonging to rural and vulnerable communities, evolved over time and are important in designing adaptation strategies tailored to their local vulnerabilities and territorial contexts that may effectively increase climate change resilience to climate change.

Climate Change: Consequences and Plan of Action

Climate change results from anthropogenic emissions of greenhouse gases, mainly carbon dioxide, methane, and nitrous oxide, causing the observed global warming since the mid-twentieth century (IPCC 2014). These conditions are expected to bring multiple severe environmental impacts in different world regions. Many of them are already being felt, such as changes in the water cycle, rainfall patterns, or rising sea levels (IPCC 2021).

Climate change is a global problem with local impacts that differ between regions and countries, influenced by their level of development, socioeconomic and demographic conditions, politics, culture, systemic vulnerabilities and adaptive capacity (McMichael 2003; Santos and Miranda 2006; Glaas et al. 2010; IPCC 2014, 2018; Loureiro et al. 2017).

However, the difficulties will be felt differently by developing countries, whose vulnerabilities are more pressing than those of developed countries (Santos and Miranda 2006; COM 2007, 2009; IPCC 2014; Dasgupta et al. 2014). In addition,

F. Marques (✉) · F. Alves · P. Castro
TERRA Associate Laboratory, Department of Life Sciences, Centre for Functional Ecology—Science for People and the Planet (CFE), University of Coimbra, Coimbra, Portugal
e-mail: filipajsmarques@gmail.com

F. Alves
Open University, Coimbra, Portugal

W. Leal Filho et al. (eds.), *Climate Change Strategies: Handling the Challenges of Adapting to a Changing Climate*, Climate Change Management,
https://doi.org/10.1007/978-3-031-28728-2_20

some populations are more vulnerable than others, such as communities whose livelihoods depend on agriculture and coastal resources (IPCC 2018). Other target groups include the elderly, children, women and people with disabilities or financial difficulties (MEA 2005; COM 2009; European Commission 2021a; Eriksen et al. 2021).

Southern Europe and regions like the Mediterranean, where Portugal is located, are also particularly susceptible to climate change (Diário da República 2019; Echave et al. 2019). Impacts such as sea-level rise, water scarcity, heat waves, droughts or forest fires will be of the most deeply felt (Santos and Miranda 2006; COM 2007; Bindi and Olesen 2010; European Commission 1995–2021a; Echave et al. 2019).

According to the Intergovernmental Panel on Climate Change, some of these changes, like the sea level rise, will be irreversible for the next hundreds to thousands of years, and unless "rapid and large-scale reductions" in the emissions of greenhouse gases are placed in effect, the goal of not increasing the average temperature by more than 1.5 °C will not be accomplished (IPCC 2021). With this scenario in mind, both mitigation and adaptation actions are paramount to tackle climate change (Pietrapertosa et al. 2018). Adaptation measures, especially, will hold a relevant role, considering that we are not likely to reach the target of 1.5 °C or even 2 °C, as the IPCC (2021) mentioned in their press release.

Mitigation refers to measures that have the goal of decreasing the emissions of greenhouse gases (IPCC 2014), while adaptation implies improving the capacity of people/communities to adjust and adapt to the changes in climate and its associated effects that will occur in different sectors of society and if possible, explore potential benefits and opportunities that come with those changes (Storbjork 2010; European Commission 2014; Alves et al. 2020). No country has the same capacity to adapt, especially developing countries that experience more constraints when compared to developed ones (Dasgupta et al. 2014; IPCC 2018). There are various limits to adaptation; they can be either physical, ecological, technological, economic, political, institutional, psychological or sociocultural (Vasconcelos et al. 2013; Shackleton et al. 2015; IPCC 2019; Mechler et al. 2020).

The data so far indicates that the cost of not adapting is higher than that of doing so (European Environment Agency 2007; European Commission 2014, 2021a, b, c, d; COACCH 2018). The European Commission (2014) estimates the costs of not adapting could reach up to 250 billion € by 2050. Nevertheless, it is fundamental to note that adaptation measures will only work if they help people achieve a "reasonable standard of living" (Roy et al. 2020). They must also consider "poverty and sustainable development"; inequalities derived from gender, ethnicity, age, social class or disability (Shackleton et al. 2015; IPCC 2018; Eriksen et al. 2021) and consider the past and current sociopolitics of the regions they are aimed at (Eriksen et al. 2021). Otherwise, adaptation measures can result in trade-offs or negative impacts in those areas and even increase the vulnerability of the populations they aim to help (Shackleton et al. 2015; IPCC 2018; Eriksen et al. 2021). An example of this is Gambella, in Ethiopia, where adaptation measures increased the tensions in the region due to their impact on human security (Milman and Arsano 2013).

The success of climate policies is dependent upon collaboration, being at an international, national, regional or local scale, between private and public institutions or between governments, local powers and citizens (Santos and Miranda 2006; COM 2007, 2009; Ahmad et al. 2012; Wolf and Moser 2011; European Commission 2014, 2021a, b; IPCC 2014; Guerra et al. 2015; IPCC 2018). The exchange of information and good practices between countries is encouraged by the European Union in their Strategy on Adaptation to Climate Change, including the support of actions at the regional and local level (European Commission 2021b).

Adaptation measures work better when they are suited to local contexts and their motivations, abilities and resources, including being informed by local knowledge and the involvement of communities (IPCC 2018). Some policies count on the behavioural response of individuals for their success (Wolf and Moser 2011). Public participation and their perception of climate change is hence crucial for the implementation and success of climate policies (IPCC 2014, 2018).

The Case of Europe and Portugal

So far, mitigation has been a priority for Europe. However, the urgency for territories to become more resilient to climate change (Pietrapertosa et al. 2018) has been driving adaptation actions and policies, which are currently being increasingly developed and implemented throughout the European Union (European Environment Agency 2018). Presently, 29 out of 33 European countries have a National Adaptation Strategy to Climate Change (European Environment Agency 2019). Portugal is a leading example as it was one of the first European countries to adopt in 2010 a National Climate Change Adaptation Strategy that was later revised in 2020. The plan of action is particularly concerned with what was considered to be the major vulnerabilities of the country: extreme weather events, forest fires, sea-level rise and coastal erosion (Diário da República 2019).

For the European Union, extreme climate events are also very concerning due to their economic and social impacts (European Commission 2021a). Economic losses related to these events account for over 12 million € per year, a value that may reach to 170 billion € in a scenario where global warming reaches 3 °C pre-industrial levels (European Commission 2021b, d).

As the impacts of climate change are unavoidable, the European Union seeks to reduce its emissions by 55% by 2030 (European Commission 2021a, b, c, d; European Commission 1995–2021a) and promote its climate resilience by 2050 (European Commission 2021b, d). To achieve this goal, the European Union plans to promote and finance adaptation actions at multiple levels, including helping local authorities to move from plan to action, especially those under the European Union's Covenant of Mayors for Climate and Energy (European Commission 2021b).

These objectives are not mere empty ambitions, as they are translated into laws as foreseen by the First European Climate Law (European Commission 2021a, b, c, d). Furthermore, its objective is to make information accessible to all citizens and not only to decision-makers (European Commission 2021b).

Local and Vulnerable Communities

Climate change impacts are context-specific, and each country's vulnerability level depends on many factors. The involvement of local authorities is fundamental for an adequate adaptation strategy (Gentle et al. 2018) since almost half of the world's population lives in small rural municipalities (Bausch and Koziol 2020). The European Environmental Agency highlights the need to involve municipalities in national strategies and share knowledge across all sectors (Campos et al. 2017), as implied by the European Union Strategy on Adaptation to Climate Change (European Commission 2021a). Despite this, research involving municipalities (local geographical scales) is usually neglected (Bausch and Koziol 2020).

Rural areas, in particular, are especially vulnerable to climate change (Houghton et al. 2017; Echave et al. 2019) due to their often ageing population (Echave et al. 2019) and migration to areas with better job opportunities and resources such as cities and the littoral, as it is the case in Portugal (Gouveia 2009; Echave et al. 2019; Oliveira and Penha-Lopes 2020). These circumstances lead to desertification and abandonment of the land, including agricultural (and forest) landscapes (Tonini et al. 2018; Oliveira and Penha-Lopes 2020), which entail several socioeconomic and ecological consequences (Oliveira and Penha-Lopes 2020).

Rural regions and small municipalities may not have the required planning capacity or resources to adapt to climate change (Vasconcelos et al. 2013; Bausch and Koziol 2020; Pattison et al. 2021). Many times, people such as farmers adapt in their own way to changing climate conditions (Kihila 2017; Kichamu et al. 2018), but their actions are mostly reactive and autonomous, lacking the proper understanding and financial resources to function well in the long-term (Kichamu et al. 2018; Rijal et al. 2021). In order to avoid such situations, local and national authorities must act appropriately (Kichamu et al. 2018; Lone et al. 2020).

Adaptation actions have been primarily implemented through a top-down approach which can lead to actions that do not address the real needs of vulnerable communities (Fenton et al. 2014; Reid et al. 2015; Eriksen et al. 2021). A top-down approach refers to policies created by the government with the assistance of technical expertise, which is then implemented at the local level, while a bottom-up approach tries to formulate and implement policies with the participation of the local community. A top-down approach is generally preferred when difficulties faced by bottom-up approaches are recognised, such as lack of information, incentives and resources, which are constraints to effective policy implementation (Fenton et al. 2014).

Many municipalities reveal a gap between the discussion of climate policies and their implementation. Their ability to adapt may be weak, especially in municipalities with less than 5,000 inhabitants, as is the case in many regions of Europe (Bausch and

Koziol 2020). Even though a top-down approach is usually the preferred type, bottom-up approaches have been advocated for and implemented since the 90s (Kirkby et al. 2018). They are viewed as an effective management method and a way of ensuring that local communities and marginalised groups in society have a say in climate policies (Harris 2014; Chaudhury et al. 2016).

Some countries, including in regions like the Middle East and North Africa, are evolving from a top-down approach to another one that endeavours to incorporate the perspective of local communities and increase their resilience (Froehlich and Al-Saidi 2018). Community-based adaptation (CBA) is an example of an approach that aims to increase the adaptive capacity of local and vulnerable communities to climate change (Allen 2006; Fenton et al. 2014; Chaudhury et al. 2016; Kirkby et al. 2018). In Europe, this is foreseen by the Covenant of Mayors for Climate and Energy, which aims to act at the local level to reduce CO_2 emissions by 40% until 2030. An effective climate policy at the local level is indispensable, not only due to the reasons mentioned above but because the lack thereof can be a driver of vulnerability and further increase social inequalities (Gentle et al. 2018).

Current research emphasises the need for communication and integration of local narratives and perceptions in adaptation strategies and overall climate policies (Reid et al. 2015; Gentle et al. 2018; Hugel and Davies 2020; Lai et al. 2021; Alizadeh et al. 2022). Many municipalities and governments follow a top-down approach with an absence of citizens' involvement in local policies (Carvalho et al. 2014; Bausch and Koziol 2020), which may explain the lack of awareness of its citizens about the local government initiatives towards climate change in some cases (Marques 2019). Nevertheless, in the European Union, citizens expect local policymakers and the administration to act and take responsibility (Bausch and Koziol 2020), as demonstrated in the Eurobarometer's results regarding climate change where national governments and the European Union are considered the main parties responsible for acting.

On another note, even community-based approaches can be problematic and not give much space for the participation of local communities and marginalised people (Eriksen et al. 2021). Many of these initiatives are still dominated by local governments or municipalities (Thaler et al. 2019). In these cases, real engagement of the local communities is lacking. In engaging with communities, Nkoana et al. (2018) consider good practices the use of their livelihoods, recognising their culture "do's and don'ts", the community leaders, the priorities of vulnerable stakeholders and communication about climate change that should be two-sided. Other studies preferred a combination of top-down and bottom-up approaches for a more effective floods mitigation strategy (Genovese and Thaler 2020) and water management (Girard et al. 2015).

Perceptions of Climate Change

Public perceptions of climate change have evolved. How people perceive global warming climate change is dependent on many variables: their country of origin, culture, gender, age, socioeconomic status, pre-existing worldviews and even the way climate change is communicated through media. For example, messages that elicit negative emotions such as fear and do not present solutions tend to disengage individuals and not promote action, whereas communications that inspire people are more likely to achieve positive outcomes (Wolf and Moser 2011; Salama and Aboukoura 2018; Ettinger et al. 2021). Incorporating a psychological perspective that considers the emotions behind the message and how people may react differently to climate change impacts could improve the effectiveness of the message itself being conveyed (Hugel and Davies 2020).

According to the systematic review of Capstick et al. (2015), between the late 1980s and early 1990s, climate change was already known by people, although usually associated with the hole in the ozone layer and air pollution. By the mid-1990s to mid-2000s there was an increase in concern that dropped by the late 2000s, where there was an upsurge in scepticism and polarized opinions, attributed by scholars to "climate fatigue, misleading media representations, the global financial crisis of 2008 and social attenuation of risk". Finally, by the late 2000s and early 2010s, public concern towards climate change seemed to increase again (Capstick et al. 2015). Climate activism is rising as well, so public participation in adaptation strategies is becoming more relevant (Hugel and Davies 2020).

When looking at the results of the last three Special Eurobarometer on Climate Change carried out by the European Commission, it is clear that climate change is becoming an increasingly worrying problem, both in Portugal and for European citizens in general (Fig. 20.1). In the Eurobarometer of 2021, for the first time, climate change was selected as the most severe problem the world faces (18% of responses), above other issues such as Poverty, Hunger and Lack of drinking water (European Commission 2021d).

The first national inquiry in Portugal regarding the environment was performed in 1997. However, the Eurobarometer surveys have mentioned climate change since the 1980s. So far, the studies about Portuguese citizens' perceptions of climate change characterize the population as showing high levels of concern despite having low levels of information and civic participation (Schmidt et al. 2011; Schmidt and Delicado 2014; European Commission 2017a, b). Portuguese citizens commonly perceive climate change as distant and a problem of the future (Lima and Schmidt 1996; Whitmarsh 2008; Brechin and Bhandari 2011; Adger et al. 2013). Climate change is considered less important than environmental problems such as pollution and extreme weather events (Lima and Schmidt 1996; Schmidt et al. 2011; Schmidt and Delicado 2014; Valente et al. 2017) or health and poverty issues (Lima and Schmidt 1996; European Commission 2017b; Alves et al. 2017).

The most common actions practised by Europeans in 2017 were recycling and saving water and energy (European Commission 2017a), but that changed in the results of 2019 and 2021, where the most common actions were recycling and reducing the consumption of disposable materials (European Commission 2019, 2021d). However, in the study of Marques (2019) in a rural municipality, in Portugal, those two actions (recycling and saving water and energy) are still the most practiced.

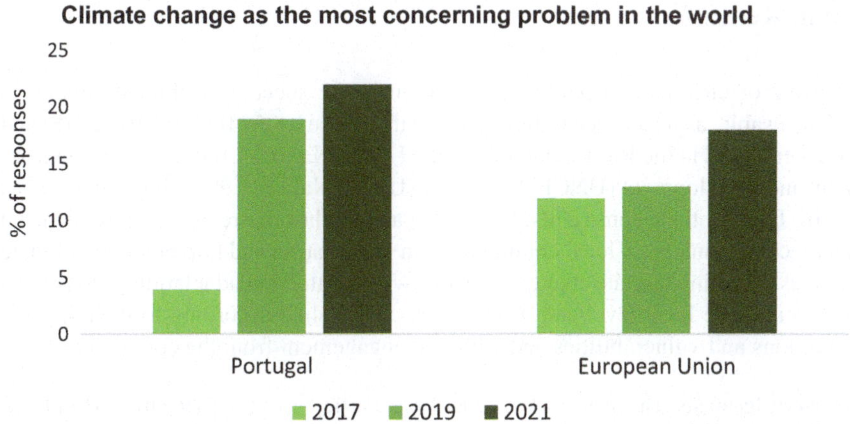

Fig. 20.1 Percentage of people that chose climate change as the most concerning problem in the world (*Data source* European Commission 2017b, 2019, 2021d. Figure elaborated by the authors with copyright permission—September 2017, September 2019, July 2021 by European Union, made available under Creative Commons Attribution 4.0 International: https://creativecommons.org/licenses/by/4.0/)

In 2021, respondents also mentioned for the first time "buy and eat more organic food" and "buy and eat less meat". The responsibility to combat climate change is imposed mainly upon the national governments and European Union, and majority of the respondents think more ambitious actions should be taken by both entities (European Commission 2021d).

Believing in climate change does not necessarily translate into active participation and climate-friendly actions (Wolf and Moser 2011; Hornsey et al. 2016; Marques 2019), as suggested by studies where respondents demonstrated low levels of participation (Schmidt et al. 2011; Schmidt and Delicado 2014; European Commission 2017a, b; Marques 2019). There are multiple explanations for citizens' inaction towards climate change. Some theories associate it with lack of information, time or money; unaccountability; scepticism or downright denial of climate change; lack of urgency; fatalism; feeling useless and unempowered against an overwhelming problem; lack of trust in information sources; faith in the role of technology to answer the problem; distancing themselves from climate change or simply the lack of willpower. Non-action from governments and companies and social norms are also seen as an additional barrier (Coelho et al. 2004; Cabecinhas et al. 2008; Wolf and Moser 2011; Morgado et al. 2017).

Despite these listed factors, most European citizens state they take personal action to tackle climate change and are now more likely than before to recognise that it is also their responsibility. As we can see from the previous results, concern towards climate change is increasing (European Commission 2017a, 2019, 2021d) and is especially prevalent in the younger generations (European Commission 2019).

Final Remarks

The role of citizens and public participation in the success of climate policies is indispensable, as it has been reiterated more than once by the IPCC (2014, 2018) and first mentioned in the Rio Declaration at the United Nations Conference on Environment and Development (UNCED), in 1992 (United Nations 1992; Hugel and Davies 2020). Despite their constraints, bottom-up approaches are recognized as adequate approaches to integrate local communities in formulating and implementing climate policies. Tackling climate change is a worldwide matter, but adaptation actions must be taken at the local level and should be accomplished according to their specific conditions and vulnerabilities and with real engagement from the community.

Acknowledgements The authors acknowledge ReNATURE project (CENTRO-01-0145-FEDER-000007) and F4F-Forest for the future (CENTRO-08-5864-FSE-000031, Programa Operacional Regional do Centro, Fundo Social Europeu). This work was carried out at the R&D Unit Centre for Functional Ecology–Science for People and the Planet (CFE), with reference UIDB/04004/2020, financed by FCT/MCTES through national funds (PIDDAC). P.C. contributed to the paper in the scope of the CULTIVAR project (CENTRO-01-0145-FEDER-000020), co-financed by the Regional Operational Programme Centro 2020, Portugal 2020, and the European Union, through the European Fund for Regional Development (ERDF).

References

Adger W, Barnett J, Brown K, Marshall N, O'Brien K (2013) Cultural dimensions of climate change impacts and adaptation. Nat Clim Chang 3(2):112–117. https://doi.org/10.1038/nclimate1666

Ahmad W, Soskolne C, Ahmed T (2012) Strategic thinking on sustainability: challenges and sectoral roles. Environ Dev Sustain 14:67–83. https://doi.org/10.1007/s10668-011-9309-5

Alizadeh MR, Adamowski J, Inam A (2022) Integrated assessment of localized SSP–RCP narratives for climate change adaptation in coupled human-water systems. Sci Total Environ 823:153660

Allen K (2006) Community-based disaster preparedness and climate adaptation: local capacity building in the Philippines. Disasters 30(1):81–101. https://doi.org/10.1111/j.1467-9523.2006.00308.x

Alves F, Leal C, Castro P, Pires S, Loureiro J, Santos L (2017) Inquéritos. In Loureiro J, Castro P, Alves F, Figueiredo A. (coord), *Plano Intermunicipal de Adaptação às Alterações Climáticas da CIM-RC* (pp. 1021–1080). Comunidade Intermunicipal da Região de Coimbra. https://www.cim-regiaodecoimbra.pt/wp-content/uploads/2018/10/PIAAC-CIM-RC-vers%C3%A3o-web.pdf

Alves F, Filho W, Casaleiro P, Nagy G, Diaz H, Al-Amin A, Guerra J, Hurlbert M, Farooq H, Klavins M, Saroar M, Lorencova E, Suresh J, Soares A, Morgado F, O'Hare P, Wolf F, Azeiteiro U (2020) Climate change policies and agendas: Facing implementation challenges and guiding responses. Environ Sci Policy 104:190–198. https://doi.org/10.1016/j.envsci.2019.12.001

Bausch T, Koziol K (2020) New policy approaches for increasing response to climate change in small rural municipalities. Sustainability 12(5):1894. https://doi.org/10.3390/su12051894

Brechin S, Bhandari M (2011) Perceptions of climate change worldwide. Wiley Interdiscipl Rev Climate Change 2(6):871–885. https://doi.org/10.1002/wcc.146

Bindi M, Olesen J (2010) The responses of agriculture in Europe to climate change. Reg Environ Change 11(1):151–158. https://doi.org/10.1007/s10113-010-0173-x

Cabecinhas R., Lázaro A, Carvalho A (2008) Media uses and social representations of climate change. In Carvalho A (ed), Communicating climate change: discourses, mediations and perceptions (pp. 170–189). Centro de Estudos de Comunicação e Sociedade, Universidade do Minho. http://repositorium.sdum.uminho.pt/handle/1822/29909

Campos I, Guerra J, Gomes J, Schmidt L, Alves F, Vizinho A, Lopes G (2017) Understanding climate change policy and action in Portuguese municipalities: A survey. Land Use Policy 62:68–78. https://doi.org/10.1016/j.landusepol.2016.12.015

Capstick S, Whitmarsh L, Poortinga W, Pidgeon N, Upham P (2015) International trends in public perceptions of climate change over the past quarter century. Wires Clim Change 6:35–61. https://doi.org/10.1002/wcc.321

Carvalho A, Schmidt L, Santos F, Delicado A (2014) Climate change research and policy in Portugal. Wires Clim Change 5(2):199–217. https://doi.org/10.1002/wcc.258

Chaudhury AS, Helfgott A, Thornton TF, Sova C (2016) Participatory adaptation planning and costing. Applications in agricultural adaptation in western Kenya. Mitig Adapt Strateg Glob Chang 21:301–322. https://doi.org/10.1007/s11027-014-9600-5

COACCH (2018) The economic cost of climate change in Europe: synthesis report on state of knowledge and key research gaps. COACCH: CO-designing the Assessment of Climate Change costs. https://www.bing.com/search?q=The+Economic+Cost+of+Climate+Change+in+Europe%3A+Synthesis+Report+on+State+of+Knowledge+and+Key+Research+Gaps&aqs=edge..69i57j69i59i450l8&FORM=ANCMS9&PC=U531

Coelho C, Valente S, Pinho L, Carvalho T, Ferreira A, Figueiredo F (2004) A Perceção Social das Alterações Climáticas e do Risco de Cheia. Paper presented at Actas do VII Congresso Nacional de Água, APRH, Lisboa. https://www.aprh.pt/congressoagua2004/PDF/64.PDF

COM (2007) Commission of the European Communities: Green Paper—From the commission to the council, the European parliament, the European economic and social commitee and the commitee of the regions—adapting to climate change in Europe—options for EU. Brussels, European Union. https://eur-lex.europa.eu/legal-content/EN/TXT/PDF/?uri=CELEX:52007DC0354&from=EN

COM (2009) Commission of the European Communities: White paper—Adapting to climate change: Towards a European framework for action. Brussels, European Union. https://ec.europa.eu/health/ph_threats/climate/docs/com_2009_147_en.pdf

Dasgupta P, Morton JF, Dodman D, Karapinar B, Meza F, Rivera-Ferre A, Sarr T, Vincent KE (2014). Rural areas. In Field CB, Barros VR, Dokken DJ, Mach KJ, Mastrandea MD, Bilir TE, Chatterjee M, Ebi KL, Estrada YO, Genova RC, Girma B, Kissel ES, Levy AN, MacCracken S, Mastrandea PR, White LL (eds), Climate Change 2014: Impacts, Adaptation, and Vulnerability. Part A: Global and Sectoral Aspects. Contribution of Working Group II to the Fifth Assessment Report of the Intergovernmental Panel on Climate Change (pp. 613–657). Cambridge University Press. https://www.ipcc.ch/site/assets/uploads/2018/02/WGIIAR5-Chap9_FINAL.pdf

Diário da República (2019) PRESIDÊNCIA DO CONSELHO DE MINISTROS – Resolução do Conselho de Ministros n.º 130/2019. Diário da República Eletrónico, 147, pp. 10–45 https://dre.pt/dre/detalhe/resolucao-conselho-ministros/130-2019-123666112

Echave C, Ceh D, Boulanger A, Shaw-Taberlet J (2019) An ecosystemic approach for energy transition in the Mediterranean Region. Paper presented at the 1st International Conference on Energy Transition in the Mediterranean Area (SyNERGY MED), Cagliari, Italy. https://doi.org/10.1109/SyNERGY-MED.2019.8764107

Ettinger J, Walton P, Painter J, DiBlasi T (2021) Climate of hope or doom and gloom? Testing the climate change hope vs. fear communications debate through online videos. Clim Change 164:19. https://doi.org/10.1007/s10584-021-02975-8

European Commission (2014) Adaptation to climate change. Brussels, European Union. https://doi.org/10.2834/849380

European Commission (2017a) Special Eurobarometer 459 Climate change. Brussels, European Union. https://ec.europa.eu/clima/system/files/2017a-09/report_2017_en.pdf

European Commission (2017b) Special Eurobarometer 459 Climate change – Portugal. Brussels, European Union. https://ec.europa.eu/clima/sites/clima/files/support/docs/pt_climate_2017_en.pdf
European Commission (2019) Special Eurobarometer 490 Climate Change. Brussels, European Union. https://ec.europa.eu/clima/system/files/2019-09/report_2019_en.pdf
European Commission (2021a) COMMISSION STAFF WORKING DOCUMENT – closing the climate protection gap – scoping policy and data gaps. Brussels, European Union. https://ec.europa.eu/clima/system/files/2021a-06/swd_2021_123_en.pdf
European Commission (2021b) Questions and answers: new EU strategy on adaptation to climate change. Brussels, European Union. https://ec.europa.eu/commission/presscorner/detail/en/qanda_21_664
European Commission (2021c) Communication from the commission to the European Parliament, the council, the European Economic and Social Committee and the Committee of the Regions – forging a climate-resilient Europe – the new EU Strategy on Adaptation to Climate Change. Brussels, European Union. Brussels, European Union. https://eur-lex.europa.eu/legal-content/EN/TXT/?uri=COM:2021:82:FIN
European Commission (2021d) Special Eurobarometer 513 Climate Change. Brussels, European Union. https://ec.europa.eu/clima/system/files/2021-07/report_2021_en.pdf
European Environment Agency (2007) Climate change: the cost of inaction and the cost of adaptation. Copenhagen, European Environment Agency. https://www.eea.europa.eu/publications/technical_report_2007_13
European Environment Agency (2018) Climate-ADAPT 10 case studies How Europe is adapting to climate change. Luxembourg, European Union. https://climate-adapt.eea.europa.eu/about/climate-adapt-10-case-studies-online.pdf
European Environment Agency (2019, December 3) Overview of national adaptation strategies and plans. European Environment Agency. https://www.eea.europa.eu/data-and-maps/figures/status-of-national-adaptation-policy-1
European Commission (1995–2021a) Climate change consequences. Brussels, European Union. https://ec.europa.eu/clima/climate-change/climate-change-consequences_en
European Commission (1995–2021b) European Green Deal: Commission adopts new proposals to stop deforestation, innovate sustainable waste management and make soils healthy for people, nature and climate. Brussels, European Union. https://ec.europa.eu/commission/presscorner/detail/en/ip_21_5916
Eriksen S, Schipper E, Scoville-Simonds M, Vincent K, Adam H, Brooks N, Harding B, Khatri D, Lenaerts D, Liverman D, Mills-Novoa M, Mosberg M, Movik S, Muok B, Nightingale A, Ojha H, Sygna L, Taylor M, Vogel C, West J (2021) Adaptation interventions and their effect on vulnerability in developing countries: Help, hindrance or irrelevance? World Dev 141:105383. https://doi.org/10.1016/j.worlddev.2020.105383
Fenton A, Gallagher D, Wright H, Huq S, Nyandiga C (2014) Up-scalling finance for community-based adaptation. Climate Dev 6(4):388–397. https://doi.org/10.1080/17565529.2014.953902
Froehlich P, Al-Saidi M (2018) Local community perception of climate change adaptation in Egypt. Paper presented at The 4th International Conference on Water Resource and Environment, Kaohsiung, Taiwan. https://doi.org/10.1088/1755-1315/191/1/012003
Genovese E, Thaler T (2020) The benefits of flood mitigation strategies: effectiveness of integrated protection measures. AIMS Geosci 6(4):459–472. https://doi.org/10.3934/geosci.2020025
Gentle P, Thwaites R, Race D, Alexander K, Maraseni T (2018) Household and community responses to impacts of climate change in the rural hills of Nepal. Clim Change 147:267–282. https://doi.org/10.1007/s10584-017-2124-8
Girard C, Pulido-Velazquez M, Rinaudo JD, Page C, Caballero Y (2015) Integrating top-down and bottom-up approaches to design global change adaptation at the river basin scale. Glob Environ Chang 34:132–146. https://doi.org/10.1016/j.gloenvcha.2015.07.002

Glaas E, Jonsson A, Hjerpe M, Andersson-Skold Y (2010) Managing climate change vulnerabilities: formal instittituions and knowledge use as determinants of adaptive capacity at the local level in Sweden. Local Environ 15(6):525–539. https://doi.org/10.1080/13549839.2010.487525

Gouveia V (2009) Contributo das Praias Fluviais para o Desenvolvimento Regional: A Rede das Praias Fluviais do Pinhal Interior Norte. [MS Thesis, Universidade Nova de Lisboa]. Repositório Universidade Nova. https://run.unl.pt/handle/10362/4025

Guerra J, Ferreira JG, Schimdt L, Campos IS, Penha-Lopes G, Vizinho A (2015) Alterações climáticas nos municípios portugueses – resultados de um inquérito. In Silva I, Pignatelli M, Viegas S (coord) Livro de Atas do 1º Congresso da Associação Internacional de Ciências Sociais e Humanas em Língua Portuguesa (pp. 10320–10337). Repositório Universidade de Lisboa. https://repositorio.ul.pt/bitstream/10451/22530/1/ICS_JGuerra_JGFerreira_LSchmidt_Alteracoes_A.pdf

Harris U (2014) Communicating climate change in the Pacific using a bottom-up approach. Pac Journal Rev 20(2):77–95. https://doi.org/10.24135/pjr.v20i2.167

Houghton A, Austin J, Beerman A, Horton C (2017) An approach to developing local climate change environmental public health indicators in a rural district. J Environ Public Health 2017:1–16. https://doi.org/10.1155/2017/3407325

Hornsey M, Harris E, Bain P, Fielding K (2016) Meta-analyses of the determinants and outcomes of belief in climate change. Nat Clim Chang 6:622–626. https://doi.org/10.1038/nclimate2943

Hugel S, Davies A (2020) Public participation, engagement, and climate change adaptation: a review of the research literature. Wires Clim Change 11(4):1–20. https://doi.org/10.1002/wcc.645

IPCC (2014) Climate Change 2014 – Synthesis Report. Contribution of Working Groups I, II and III to the Fifth Assessment Report of the Intergovernmental Panel on Climate Change. IPCC. https://www.ipcc.ch/site/assets/uploads/2018/05/SYR_AR5_FINAL_full_wcover.pdf

IPCC (2018) Global Warming of 1.5°C. An IPCC Special Report on the impacts of global warming of 1.5°C above pre-industrial levels and related global greenhouse gas emission pathways, in the context of strengthening the global response to the threat of climate change, sustainable development, and efforts to eradicate poverty. IPCC. https://www.ipcc.ch/site/assets/uploads/sites/2/2019/06/SR15_Full_Report_High_Res.pdf

IPCC (2019) Summary for policymakers. In Portner H, Roberts DC, Masson-Delmotte V, Zhai P, Tignor M, Poloczanska E, Mintenbeck K, Alegria A, Nicolai M, Okem A, Petzold J, Rama B, Weyers NM (eds), IPCC Special Report on the Ocean and Cryosphere in a Changing Climate (pp. 3–35). Intergovernmental Panel on Climate Change. https://www.ipcc.ch/srocc/

IPCC (2021, August 9) Climate change widespread, rapid, and intensifying – IPCC [Press release]. https://www.ipcc.ch/2021/08/09/ar6-wg1-20210809-pr/

Kichamu E, Ziro J, Palaniappan G, Ross H (2018) Climate change perceptions and adaptations of smallholder farmers in Eastern Kenya. Environ Dev Sustain 20:2663–2680. https://doi.org/10.1007/s10668-017-0010-1

Kihila J (2017) Indigenous coping and adaptation strategies to climate change of local communities in Tanzania: a review. Climate Dev 10(5):406–416. https://doi.org/10.1080/17565529.2017.1318739

Kirkby P, Williams C, Huq S (2018) Community-based adaptation (CBA): adding conceptual clarity to the approach, and establishing its principles and challenges. Climate Dev 10(7):577–589. https://doi.org/10.1080/17565529.2017.1372265

Lai CH, Liao PC, Chen SH, Wang YC, Cheng C, Wu CF (2021) Risk perception and adaptation of climate change: an assessment of community resilience in rural Taiwan. Sustainability 13(7):3651. https://doi.org/10.3390/su13073651

Lima A, Schmidt L (1996) Questões ambientais – conhecimentos, preocupações e sensibilidades. Repositório do Instituto Universitário de Lisboa. https://repositorio.iscte-iul.pt/bitstream/10071/14089/1/Quest%C3%B5es%20Ambientais.%20Conhecimentos%2C%20preocupa%C3%A7%C3%B5es%20e%20sensibilidades.pdf

Loureiro J, Castro P, Alves F, Figueiredo A (coord) (2017) Plano Intermunicipal de Adaptação às Alterações Climáticas da CIM-RC. Comunidade Intermunicipal Região de Coimbra. https://www.cim-regiaodecoimbra.pt/wp-content/uploads/2018/10/PIAAC-CIM-RC-vers%C3%A3o-web.pdf

Lone F, Maheen M, Shafiq M, Bhat M, Rather J (2020) Farmer's perception and adaptation strategies to changing climate in Kashmir Himalayas, India. *GeoJournal*. https://doi.org/10.1007/s10708-020-10330-0

Marques F (2019) Estudo sobre as perceções das alterações climáticas no município de Penela [MS Thesis, Universidade de Coimbra]. Repositório científico da UC. https://estudogeral.sib.uc.pt/handle/10316/88143

McMichael AJ (2003) Global climate change and health: an old story writ large. In McMichael AJ et al. (eds), Climate change and human health – Risks and Responses (pp. 1–15). World Health Organization. https://www.who.int/globalchange/publications/climatechangechap1.pdf

MEA (Millennium Ecosystem Assessment) (2005) Ecosystems and human well-being: synthesis. Island Press. https://www.millenniumassessment.org/documents/document.356.aspx.pdf

Mechler R, Singh C, Ebi K, Djalante R, Thomas A, James R, ... Revi A (2020) Loss and damage and limits to adaptation: recent IPCCinsights and implications for climate science and policy. Sustain Sci 15:1245–1251. https://doi.org/10.1007/s11625-020-00807-9

Milman A, Arsano Y (2013) Climate adaptation and development: contradictions for human security in Gambella, Ethiopia. Glob Environ Chang 29:349–359. https://doi.org/10.1016/j.gloenvcha.2013.11.017

Morgado F, Bacelar-Nicolau P, Osten J, Santos P, Bacelar-Nicolau L, Farooq H, Alves F, Soares A, Azeiteiro U (2017) Assessing university student perceptions and comprehension of climate change (Portugal, Mexico and Mozambique). Int J Climate Change Strategies Manag 9(3):316–336. https://doi.org/10.1108/IJCCSM-08-2016-0123

Nkoana E, Verbruggen A, Huge J (2018) Climate change adaptation tools at the community level: an integrated literature review. Sustainability 10(3):796. https://doi.org/10.3390/su10030796

Oliveira H, Penha-Lopes G (2020) Permaculture in Portugal: socio-ecological inventory of a re-ruralizing grassroots movement. Eur Countryside 12:30–52. https://doi.org/10.2478/euco-2020-0002

Pattison A, Henke C, Pumilio J (2021) Community-based climate action planning as an act of advocacy: a case study of liberal arts education in a rural community. Journal of Environmental Studies 11:183–193. https://doi.org/10.1007/s13412-020-00655-0

Pietrapertosa F, Khokhlov V, Salvia M, Cosmi C (2018) Climate change adaptation policies and plans: A survey in 11 South East European countries. Renew Sustain Energy Rev 81(2):3041–3050. https://doi.org/10.1016/j.rser.2017.06.116

Reid H, Swiderska K, King-Okumu C, Archer D (2015, November 15) Vulnerable communities: climate adaptation that works for the poor. IIED. https://www.jstor.org/stable/resrep17971?seq=1#metadata_info_tab_contents

Rijal S, Gentle P, Khanal U, Wilson C, Rimal B (2021) A systematic review of Nepalese farmers' climate change adaptation strategies. Climate Policy. https://doi.org/10.1080/14693062.2021.1977600

Roy D, Datta A, Kuwornu J, Zulfiqar F (2020) Comparing farmers' perceptions of climate change with meteorological trends and examining farm adaptation measures in hazard-prone districts of northwest Bangladesh. Environ Dev Sustain. https://doi.org/10.1007/s10668-020-00989-3

Salama S, Aboukoura K (2018) Role of emotions in climate change communication. In Filho W, Manolas E, Azul A, Azeiteiro U, McChie H (eds), Handbook of Climate Change Communication Vol. 1 Theory of Climate Change Communication (pp. 137–150). Springer. https://doi.org/10.1007/978-3-319-69838-0_9

Santos FD, Miranda P (ed) (2006) Alterações Climáticas em Portugal - Cenários, Impactos e Medidas de Adaptação - Projecto SIAM II. Gradiva Publicações. http://cciam.fc.ul.pt/prj/siam/

Schimdt L, Delicado A, Ferreira J, Fonseca S, Seixas J, Sousa D, Truninger M, Valente S (2011) O Ambiente em 25 Anos de Eurobarómetro. Repositório da Universidade de Lisboa. https://repositorio.ul.pt/bitstream/10451/20349/1/ICS_LSChmidt_Ambiente_RN.pdf

Schmidt L, Delicado A (coord) (2014) Ambiente, alterações climáticas, alimentação e energia: a opinião dos portugueses. Repositório da Universidade de Lisboa. https://repositorio.ul.pt/bitstream/10451/10852/1/ICS_LSchmidt_ADelicado_Ambiente_LEN.pdf
Shackleton S, Ziergovel G, Sallu S, Gill T, Tschakert P (2015) Why is socially-just climate change adaptation in sub-Saharan Africa so challenging? A review of barriers identified from empirical cases. Wires Clim Change 6:321–344. https://doi.org/10.1002/wcc.335
Storbjork S (2010) "It takes more to get a ship to change course": Barriers for organizational learning and local climate adaptation in Sweden. J Environ Planning Policy Manage 12(3):235–254. https://doi.org/10.1080/1523908X.2010.505414
Thaler T, Attems MS, Bonnefond M, Clarke D, Gatien-Tournat A, Gralepois M, Fournier M, Murphy C, Rauter M, Papathoma-Kohle M, Servain S, Fuchs S (2019) Drivers and barriers of adaptation initiatives – how societal transformation affects natural hazard management and risk mitigation in Europe. Sci Total Environ 650:1073–1082. https://doi.org/10.1016/j.scitotenv.2018.08.306
Tonini M, Parente J, Pereira M (2018) Global assessment of rural–urban interface in Portugal related to land cover changes. Nat Hazard 18:1647–1664. https://doi.org/10.5194/nhess-18-1647-2018
United Nations (1992, August 12) Report of the United Nations Conference on Environment and Development (Rio de Janeiro, 3–14 June 1992) [Conference presentation]. United Nations Conference on Environment and Development, Rio de Janeiro, Brasil. https://www.un.org/en/development/desa/population/migration/generalassembly/docs/globalcompact/A_CONF.151_26_Vol.I_Declaration.pdf
Valente S, Guerra J, Schmidt L (2017) Ambiente e sustentabilidade em tempo de crise. Research Gate. https://www.researchgate.net/publication/313447346_Ambiente_e_sustentabilidade_em_tempo_de_crise
Vasconcelos A, Bonati M, Schlindwein S, D'Agostini L, Homem L, Nelson R (2013) Landraces as an adaptation strategy to climate change for smallholders in Santa Catarina, Southern Brazil. Land Use Policy 34:250–254. https://doi.org/10.1016/j.landusepol.2013.03.017
Whitmarsh L (2008) Are flood victims more concerned about climate change than other people? The role of direct experience in risk perception and behavioural response. J Risk Res 11(3):351–374. https://doi.org/10.1080/13669870701552235
Wolf J, Moser S (2011) Individual understandings, perceptions, and engagement with climate change: insights from in-depth studies across the world. Wiley Interdisciplinary Reviews: Climate Change 2(4):547–569. https://doi.org/10.1002/wcc.120

Chapter 21
Human Mobility: The Invisible Issue in Climate Change Adaptation Policies: The Case of Morocco

Carla Sofia Ferreira Fernandes, Fátima Alves, and João Loureiro

Abstract Climate change has led countries to develop mitigation strategies and policies to support populations in their adaptation to its impacts. By analyzing the climate change adaptation policies it is possible to understand how human mobility is included in (or excluded from) those same policies. In the case of Morocco, for example, the National Plan Against Climate Change includes various actions that address strategic sectors. Nevertheless, in practice, most adaptation actions are limited to the water management and agriculture sectors, which are perceived to be priority sectors. Human mobility can be an adaptation strategy and/or a factor that further intensifies vulnerability to climate change. However, it is generally missing from the overall adaptation policies. Most frequently, mobility-related policies are defined in different fora that do not necessarily include climate change-related considerations and often focus on international movements, which are a minority of the overall migratory movements. This study aims to analyze how the various national and regional climate change adaptation policies address mobility as a possible adaptation strategy using Morocco as a case study.

Introduction

Climate change interacts with other environmental factors alongside economic, political, social and demographic triggers to influence the decision of migrating or not migrating at individual and household levels (Foresight 2011). The multiple triggers for human mobility along with the varied impacts of climate change make it difficult to establish a direct and exclusive link between the two. In practice, research shows that the economic factors are the preponderant trigger in the decision to migrate and that climate change contributes to a decline in the overall economic resilience of

C. S. Ferreira Fernandes (✉) · F. Alves
Universidade Aberta, 1269-001 Lisbon, Portugal
e-mail: 1600737@estudante.uab.pt

C. S. Ferreira Fernandes · F. Alves · J. Loureiro
Centre for Functional Ecology-Science for People & the Planet, Department of Life Sciences, Faculty of Sciences and Technology, University of Coimbra, 3000-456 Coimbra, Portugal

W. Leal Filho et al. (eds.), *Climate Change Strategies: Handling the Challenges of Adapting to a Changing Climate*, Climate Change Management,
https://doi.org/10.1007/978-3-031-28728-2_21

the communities most affected (Verner 2012; Stern 2014). Overall, there will be a reduction in the resources necessary to support the communities in the much-needed adaptation to climate change, thus intensifying their vulnerability (Stern 2014).

Human mobility is considered by the United Nations Framework Convention on Climate Change (UNFCCC) as being part of Losses and Damages so it is perceived as a failure to adapt. The Warsaw Mechanism for Loss and Damage includes both economic and non-economic losses and the latter includes, in addition to human mobility, areas such as cultural heritage, indigenous knowledge, society/cultural identity among others. Nevertheless, countries have traditionally focused more on mitigation and adaptation in their national plans, as well as in their international communications related to climate change; losses and damages are not systematically incorporated into the plans (Rai and Acharya 2020).

IOM (2016, p. 41) has analyzed the climate change and climate change adaptation policies concerning environmental migration specifically for Morocco and has concluded that only remarks are included and that "[m]igration of Moroccans does not yet figure prominently within Morocco's climate change policies". Nevertheless the documents analyzed were more limited in number and new policies have been published since the study by IOM (2016). On an international level, IOM (2018) has created a mapping of the inclusion of human mobility within climate change-related national policies and institutional frameworks. It has been concluded that most national adaptation plans and national communications include references to human mobility as a consequence of climate change and that a limited proportion of countries also include it in the Nationally Determined Contribution (NDC). For the countries that include human mobility within their national policies related to climate change only a fraction perceives it to be a possible adaptation strategy (IOM 2018).

This study aims to expand the number of documents analyzed by including climate change-related policies, as well as sustainable development-related documents and to identify how Moroccan authorities frame human mobility focusing on two main scenarios: human mobility as part of a wider analysis on Loss and Damage or as a possible strategy for adaptation.

Climate Change and Human Mobility in Morocco

Morocco is a country in the northwest of the African continent bordering both the Mediterranean Sea and the Atlantic Ocean. IPCC suggests that climate change will negatively impact Morocco, particularly in its water resources and agricultural productivity (Niang et al. 2014) with an expansion of its arid areas (IPCC 2021). Despite having a relatively low level of water consumption per capita, Morocco faces a situation of water stress due to a decline in the available water resources (Wodon et al. 2013), which is further intensified by the change in the irrigation techniques that have been privileging the unsustainable use of aquifers for the last decades (Lejars et al. 2017). The way this decline will impact the populations will be strongly correlated to the efficacy of the application of sound policies (de Haas 2011). Through its impact on water availability, climate change will also be the source of socioeconomic consequences due to the foreseen decline in agricultural productivity, which leads to a deterioration of the employment opportunities and to an increase in food prices to the end consumer (Taheripour et al. 2020).

Regarding human mobility, it is largely recognized in international agreements to be one among possible responses of communities affected by climate change. Nevertheless, it is difficult to establish a direct and exclusive link between both phenomena: human mobility and climate change. Such difficulty arises partially from the complexity of the decision to migrate, which is influenced by various factors in addition to the environment and climate namely, economic, political, social, cultural and demographic factors (UNFCCC 2016).

In terms of international awareness of the relevance of human mobility within the context of climate change, the first global recognition was in 2010 when UNFCCC adopted the Cancún Adaptation Framework at COP16, of which paragraph 14f clearly stated:

> Measures to enhance understanding, coordination and cooperation with regard to climate change induced displacement, migration and planned relocation, where appropriate, at the national, regional and international levels.

Previously, on a regional level, the Kampala Convention adopted by the African Union in 2009 had already made similar recommendations to its member states. Various other platforms focus specifically on environmental mobility, such as the Warsaw International Mechanism for Loss and Damage from UNFCCC, which includes a task force dedicated to human mobility that provides recommendations to avert and minimize displacement as a response to climate change impacts (WIM TFD 2018). The Global Compact for Safe, Orderly and Regular Migration includes considerations on the contribution of climate change to migratory movements. The 2030 Agenda for Sustainable Development also includes objectives that focus on climate change and human mobility. However, it does not formalize a link between both phenomena.

Human mobility within the context of climate change can either be perceived as an adaptive strategy or instead as a result of a failure to adapt (Warner 2010). In the former, migration can either function as a strategy that removes communities from areas that are exposed to various climate change impacts or as one that allows increasing the resources available to the community of origin for it to adapt to climate change in the form of financial, technological or cultural remittances (Joseph et al. 2014). When human mobility aims to remove populations from exposed areas, it might be initialized by the population itself or by the governing bodies through the application of preemptive or planned relocation. It is worth noting that even when populations have been resettled in a planned manner, they still face a deterioration of their socioeconomic level (De Wet 2006).

Clement et al. (2021) present three different emissions scenarios and the number of climate migrants by the year 2050 for each of the scenarios. The values range from 0.5 to 1.9 million people in Morocco alone. In general, regarding the intensity of migratory movements, it tends to be dominated by internal mobility rather than international (de Haas 2011).

In particular, for Morocco, there are various studies already available that focus specifically on environmental mobility. For example, Sobczak-Szelc (2008) identified a reshaping of the dynamics of socio-economic activities due to environmental change in a community of the Drâa Valley where the reduction of the water resources led to a drop in agricultural activities, while the expansion of the desertification caused an increment in tourism-related activities and both created new triggers for in- and out-migration to and from the community. Another phenomenon that is visible in the region of the Drâa Valley is a change in the habits related to transhumance as well as a sedentarisation of previously nomad groups (Ait Hamza et al. 2009). Mobility within Morocco, particularly directed towards urban centers, is also creating new challenges as the new settlements might not be sufficiently resilient to the impacts of climate changes (Tribak et al. 2019).

Methodology

The study aims to understand how national policies related to climate change address the issue of human mobility by undertaking a systematic and detailed content analysis of the various official documents, which requires the identification of different categories such as migration, displacement and planned relocation as being designations commonly used within the topic of human mobility (IOM 2018). The list of categories is created inductively to accommodate possible modes of referring to human mobility, which were not initially taken into consideration. The final list of categories is presented under the section Results.

The selection of the documents was based on the official information made available by the Ministry of Energy Transition and Sustainable Development and complemented with other documents related to the general national strategy for sustainable development, which was assumed to include information relevant to environmental sustainability.

At the time of writing, the National Plan of Climate Change Adaptation and various regional plans were under preparation, according to the official information made available by the Ministry of Energy Transition and Sustainable Development. Table 21.1 presents the list of policies that were analyzed regarding human mobility.

Results

The content analysis of the various policies allowed to infer general information and to identify how mobility within the context of climate change is addressed. The first section in this chapter presents an overview of the various documents analyzed, while

Table 21.1 List of documents analyzed

Document	Name of the original document (if not in English)	Date of publication
New Model of Development	*Le Nouveau Modèle de Développement*	2021
National Strategy for Sustainable Development	*Projet de Stratégie Nationale de Développement Durable 2030*	2017
Nationally Determined Contribution		2021
Climate Change Policy in Morocco	*Politique du Changement Climatique au Maroc*	2014
National Plan for the Climate (2020–2030)	*Plan Climat National*	N/A
Three National Communications to UNFCCC	*The text of the National Communications was more extensive in the French version, therefore that was the version analyzed in the scope of this study*	2001, 2009 and 2016

the second section addresses specifically how human mobility is contemplated in the same documents.

Climate Change and Sustainability Policies in Morocco

New Model of Development

The New Model of Development was initially based on a countrywide consultation of numerous stakeholders, ranging from the individual citizens belonging to different socioeconomic levels and geographies up to institutional partners. The consultation aimed to identify the strengths and weaknesses of the current development model in Morocco.

The New Model of Development considers that the progress done in terms of adaptation to climate change is very limited, which is measured by the lack of inclusion of environmental considerations in public projects and programs and also in the practically inexistent circular economy. Climate change is considered in more detail for the agricultural sector, with recommendations on climate change adaptation that include advocating the use of technology, improving the resilience in food production and, specifically, addressing a sustainable use of water resources.

National Strategy for Sustainable Development

Morocco has defined the National Strategy for Sustainable Development as part of a process that was initiated as a response to the Cancún Agreements (ClimaSouth 2018). The National Strategy for Sustainable Development is supported by the national legal framework, where Sustainable Development is indicated as a right. The vision of the strategy was to set the foundation for a green and inclusive economy until the year 2020 and it identified 7 specific challenges to be addressed to fulfill that vision: Governance, Green Economy, Biodiversity, Climate Change, Sensitive Territories, Social Cohesion and Culture.

Under the challenge Green Economy, one area of intervention is the agricultural sector, where the needs identified focus on the decrease of its vulnerability to climate change, as well as the need to reduce the pressure on the soil and water resources. To achieve that, it is suggested, among others, to monitor and map the vulnerabilities and to conduct research on agricultural practices in extreme situations. In particular for certain types of agricultural produce, the report encourages the constitution of cooperatives and the equitable distribution among the producers. The fishing sector is also specifically addressed in the Strategy for Sustainable Development, and it is emphasized that the resources risk declining dramatically by 2050 if no action is taken to preserve the biodiversity in the oceans. The actions point towards the management of the fishing quotas but also to an improvement in the conditions of the workers in the sector.

Regarding Biodiversity, the report criticizes the delay in the planning of crises management in case of flooding and drought, thus limiting the possibility to support vulnerable communities. The section on Climate Change is focusing specifically on governance, preparation of relevant national and regional plans, research and technology and climate finance. However, it does not provide detailed information on adaptation.

The report considers the Sensitive Territories to be the oases, deserts, coastal areas and mountains. In particular for the oases and desert areas, the report goes more in detail on what measures should be followed within an adaptation response to climate change.

Overall, the strategy for Sustainable Development is focusing primarily on the governance of the different key areas relevant to sustainable development. Some sectors, particularly the management of water resources, are the subject of focus in the majority of the report.

Nationally Determined Contribution

With its revised Nationally Determined Contribution, Morocco reinforces its commitment to the Paris Agreement, focusing clearly on mitigation measures and less on adaptation, which is in line with what was required of the NDCs (IOM 2018).

The main sectors responsible for the projected reduction of emissions are electricity production and industrial activities, namely phosphates and cement production. Overall, the document presents concrete measures for reaching its intended level of greenhouse gases emissions dividing them into two categories: conditional (dependent on international funds) and unconditional (independent of international financing).

In terms of adaptation, the report highlights the sectors that are most affected by climate change: agriculture, fishing industry, forestry, as well as habitat and health. The vulnerability of the population to those impacts is intensified by the lack of financial resources and expertise necessary to develop appropriate responses.

Climate Change Policy in Morocco

A precursor to the various documents analyzed in this study, the Climate Change Policy emphasizes the vulnerability of Morocco within a context of intensification of the climate change impacts. It introduces the need to develop national strategies that are in line with the commitment of Morocco with the Cancún Agreements and Rio+20. The need for international financing is strongly stated as being necessary to support the activities of mitigation and adaptation in Morocco.

National Plan for the Climate

The National Plan for the Climate focuses on governance, resilience, transition to a low-carbon economy and strengthening of technological and financial resources. Within governance, it is included the need to define a national legal framework that is aligned with the commitments towards sustainable development and mitigation and adaptation to climate change.

Furthermore, with this Plan, Morocco positions itself within a wider international community, aiming to become a reference of leadership for climate change management in the region.

In terms of adaptation, the Plan follows the same priorities stated in the NDC: water resources, agriculture, fishing and health, with detailed suggestions for adaptation and including the future negative impacts of climate change in the current activities. The Plan for the Climate also addresses the concern related to food security, which is threatened by climate change.

National Communications

Morocco has issued three National Communications to the Conference of the Parties of the United Nations Framework Convention on Climate Change. They were issued in 2001, 2009, and 2016, and a fourth one is being elaborated at the time of the preparation of this study. The National Communications provide an extensive overview of the country focusing on numerous topics: climate and environment and also geography, demography, society, politics and economy.

In terms of adaptation, the First National Communication is mainly providing a brief list of projects that focus mainly on water resources and the agricultural sector. Regarding mitigation, the First Communication contains several reflections on the methodologies used for calculating the emissions and a breakdown of the emissions by sector of activity as allowed by the limited knowledge available at the time, which is clearly stated in the document.

The Second Communication provides a more extensive overview of the socio-economic implications of climate change. For the agricultural sector, the document provides an extensive overview of the impacts for one specific region: Souss-Massa. In comparison to the First Communication, it provides a more intensive overview of the initiatives and committees implemented, which shows a clear progression in the national response to climate change. It also portrays in detail the challenges faced on a national level to implement the proposed action plans.

The Third Communication adds more detail to the challenges and also to the strategies put in place regarding mitigation and adaptation to climate change, but also for data monitoring. There is clear progress that is visible throughout the National Communications in terms of policies and institutions created in Morocco that target climate change mitigation and adaptation.

Human Mobility in the Climate Change Adaptation Policies in Morocco

Overall, human mobility is not thoroughly addressed in the policies in Morocco, though it is mentioned on specific occasions. A section on Loss and Damage is never clearly presented, except for economic losses as a result of the future impacts of climate change. Human mobility is also not consistently mentioned as a possible strategy of adaptation. There are however several references to migratory movements done throughout the texts, as exemplified under Table 21.2.

The New Model of Development[1] refers to human mobility within the context of climate change as being a factor that further contributes to forced migration from arid areas towards the urban centers, namely the coastal cities. Human mobility is therefore perceived as a negative decision that people feel forced to make, i.e. a failure to adapt. The New Model of Development and Climate Change Policy acknowledge human mobility as a result of climate change for migrants from other African countries to Morocco. These migratory movements are also mentioned in

[1] The term mobility (*mobilité* in French) is often used to describe issues related to transport, e.g. in the document New Model of Development (p. 138), it is stated "[i]n regard to mobility, the Commission recommends the confirmation of public transport as a basic public service and preferential means of transport" (authors' translation). These references to mobility as a general term to designate transportation were not included in Table 21.2 as they are not relevant to the object of study.

Table 21.2 References to human mobility

Document	References to human mobility (with number of references)	Examples of quotes from the text (authors' translation from French)
New Model of Development	Climate migration (1)	*"Increased immigration from Subsaharan Africa, within a context of climate change"* p. 45
	Displacement due to climate change (1)	*"Climate change can also increase the forced displacement of population from arid rural areas to urban areas"* p. 45
	Migration not related to climate change (3)	*"Increase in the out-migration of Moroccans to developed economies"* p. 45
National Strategy for Sustainable Development	Climate migration (1)	
	Migration not related to climate change (5)	*"Tourism, as a source of employment, leads to an internal migration towards coastal areas"* p. 91
Nationally Determined Contribution	Transhumance (1)	*"National programme to regulate the transhumance routes"* p. 30
	Measure to avert migration (1)	*"Planting 3 million palm trees to improve the productivity in the oases and to prevent desertification and out-migration from the younger generations from the rural areas"* p. 30
Climate Change Policy in Morocco	Climate migration (2)	
	Migration not related to climate change (1)	*"Morocco presented on the sidelines of this year's United Nations General Assembly the initiative of an African alliance for migration and development"* p. 5
National Plan for the Climate (2020–2030)	Migration supporting Adaptation (1)	*"Reinforce scientific cooperation in key areas of climate change—Implement support programmes for Moroccan researchers abroad"* p. 80

(continued)

Table 21.2 (continued)

Document	References to human mobility (with number of references)	Examples of quotes from the text (authors' translation from French)
	Transhumance (1)	
First National Communication to UNFCCC	Climate migration (1)	*"The intensification of the fluctuations in the agricultural yield, due to climate change, will lead to cycles of expansion–contraction in the rural households revenues, thus influencing their behaviour (consumption, savings, out-migration)" p. 51*
	Migration not related to climate change (2)	*"Out-migration from rural areas" pp. 9 & 25*
Second National Communication to UNFCCC	Climate migration (2)	*"Climate change could lead to economic downturn, social fractures and displacement of population" p. 101*
	Planned relocation (1)	*"Retreat—this strategy consists of progressively abandoning the land and structures threatened by seal level rise (…) and relocating its inhabitants" p. 116*
	Displacement (1)	*"Climate change (…) can be responsible for (…) displacement of populations" p. 120*
	Migration not related to climate change (2)	
	Migration as Failure to Adapt (1)	*"Degradation of soil and its fertility lead to lower yields, thus pushing farmers to reach for new lands" p. 102*
	Migration as Adaptation (1)	*Migration as a response option to extreme meteorological events. (p. 122)*
	Conflict as a result of climate change (2)	*"The impacts [of climate change] can also accentuate conflicts over water use during periods of droughts" p. 101*

(continued)

Table 21.2 (continued)

Document	References to human mobility (with number of references)	Examples of quotes from the text (authors' translation from French)
Third National Communication to UNFCCC	Climate migration (1)	*"Out-migration from rural areas is intensified during periods of drought" p. 43*
	Nomadism/Transhumance (1)	*"This project includes the management of pastoral lands and preservation of biodiversity by supporting the resuming of transhumance" p. 162*
	Measures to avert migration (1)	*"Implementation of specific projects (…) with the purpose to fixate the population" p. 257*

the Climate Change Policy in Morocco within the context of portraying the strong connections between Morocco and other African countries in general.

Morocco has submitted an updated Nationally Determined Contribution in June 2021, but it does not contain a reference to Loss and Damage in general nor to human mobility within its climate change adaptation policies, except in a measure that aims to avert migration. When 165 countries submitted their initial INDCs, only 20% contained references to human mobility in the context of climate change (IOM 2018), which means that Morocco is in line with the global tendency.

The general trend of considering human mobility as the exception is further reinforced by the National Plan for the Climate, which states that a productive agricultural sector is necessary to stabilize the rural population. The movements within the context of transhumance are also targeted by the Plan for the Climate that favors its regulation, even though IOM (2016) claims that nomadic populations in Morocco are more resilient to climate change, which is considered in the Third Communication by stating the importance of resuming the ancient practices of transhumance. The Third Communication also suggests that the plans put in place to revitalize the oases farming practices might lead younger demographics to return to abandoned oases, in what was the only reference done in the documents analyzed to policies that seek to actively influence human mobility.

The earliest document analyzed for this study, the First National Communication, has established a relationship between climate change and out-migration from the rural areas by stating that the former will contract the level of income of the households dependent on agriculture which, in turn, will lead to a change in their habits of consumption, savings and decision to out-migrate. The Second and Third Communications establish a clearer link between both phenomena by stating that the out-migration from rural areas is more intense in periods of drought.

Migration as a positive adaptation response is mentioned briefly in the Second National Communication when providing a theoretical overview of planned relocation as one of the possible responses to sea-level rise. Nevertheless, relocation is not mentioned in the actual initiatives presented in that same document to adapt to sea-level rise in Morocco. Another option of migration as a support to the adaptation efforts is mentioned in the Second National Communication when underlining the role of the technological transfer from the Moroccan diaspora, particularly researchers that are active in the field of climate change. The New Model of Development and the National Strategy for Sustainable Development also mention the importance of the financial remittances and technological transfer from the Moroccan diaspora but the text does not establish a relationship between those remittances and the adaptation to climate change in the communities of origin.

Discussion

Despite a general inclusion in the policies of key strategic sectors such as water, agriculture, fishing, forest, desertification, biodiversity, health, tourism and habitat, UNDP (2018) refers that most of the adaptation actions are focused on the water and agricultural sectors and that there are challenges in the participation of local and regional actors. The priority assigned to the water and agricultural sectors is also visible in the documents analyzed.

The only document analyzed that included a public consultation was the New Model of Development but its final considerations did not include the environment/climate or human mobility; instead, the focus was mostly directed at economic issues of wealth distribution and access to basic services. These results are in line with the latest pan-African survey on citizen priorities published by Afrobarometer which concluded that climate change is not at the top for none of the surveyed countries, which included Morocco (Coulibaly et al. 2018). The same report also mentions that usually there is a wide interest in climate change, but it does not constitute a priority.

Warner (2010) defends that the governance of human mobility within a context of climate change requires that relevant institutions recognize the potential of mobility as an adaptive response and, therefore, aim to improve the resilience of the migrant and not necessarily to control the actual movement. There is a growing tendency to integrate human mobility within the options of adaptation as being a strategy for increasing resilience in communities that face challenges to adapt (Gemenne and Blocher 2017). Nevertheless, in the case of the actual policies in Morocco, there are very limited actual attempts to explore the potential of human mobility as adaptation. The few references found in the documents analyzed are mostly confirming a narrative that people migrate due to a failure to adapt and that out-migration from affected areas needs to be averted. This tendency is in line with the studies analyzed by Liaeter and Durand-Delacre (2021), which confirm that the communities, in general, are seeking to remain in vulnerable locations and that they are resourceful in using a range of possible adaptive responses and not viewing mobility as a desired solution. Nevertheless, these studies focus on current and past situations, which can be subject to change once the impacts of climate change intensify. It remains, therefore, unknown if that trend will be confirmed in the future. In that sense, having policies that acknowledge a possible future trend might be more adequate to respond to the needs of the population affected.

There are various challenges to consider human mobility within the environment and climate-related discussions—it is often framed as a security issue and also free movement of people is usually not allowed in cross-border settings, thus creating other needs in terms of regulation and response by the national authorities. For example, the latest report of the White House (2021) on "The Impact of Climate Change on Migration" states that it can be a cause of greater instability and can eventually lead to waves of migration, thus confirming a generalized perception of migration trends as a security risk and as a potential source of conflict.

Furthermore, human mobility is a complex phenomenon where multiple factors influence the final decision of migrating, thus rendering the assignment of the main cause very difficult. Environmental factors interact with economic, political, demographic and social drivers that can then be further supported or not by the legal context and migration networks influencing the final decision to migrate (Foresight 2011). This complexity, however, can be addressed if active actions are taken to further understand the process of migrating and how it is influenced, which would be supported by having policies in place that allow to monitor and collect relevant data and evidence.

Finally, by transferring the onus of adaptation to the individual, there is a risk of placing excessive responsibility on the individual for the adaptation, which raises issues related to climate justice (Liaeter and Durand-Delacre 2021) since, more often than not, the individuals in a situation of vulnerability are not the ones that have contributed the most to climate change. Furthermore, the projections for 2050 of up to 1.9 million climate migrants in Morocco by Clement et al. (2021) raise the urgency of addressing migration as a part of collective responsibility.

Implications of the Absence of Migration in the Policies

Despite the recognition in the Paris Agreement of the limits of adaptation and that there will be inevitable Losses and Damages, Morocco does not consistently consider human mobility in particular and Loss and Damage are not systematically addressed in the related policies. The risk for Morocco is that potentially available funding will not be accessible if data on Loss and Damage is not consistently collected (Rai and Acharya 2020).

In the policies, mobility is also not consistently framed as a potential adaptation strategy, even though, depending on the context, it might be an important strategy for communities to improve their livelihoods and/or exit areas highly exposed to the impacts. Morocco is a country with significant migratory movements that either take place internally or across borders. However, when that migration is an unplanned and unexpected response to situations of risk, then there is a serious potential to affect social development (Clement et al. 2021). It is, therefore, crucial to include considerations on human mobility when addressing issues that are per se already deeply disruptive to social development, such as climate change.

Comparison with Other Countries

Other countries have already integrated human mobility within the context of climate change in their policies. For example, Colombia and Togo have considered human mobility in their national adaptation plans, Intended Nationally Determined Contributions (INDCs), and in their most recent National Communications, making these countries a good example of policy coherence at the national level (IOM 2018). Bangladesh and Ghana have already integrated collaboration between the actors that work either on human mobility or climate change, while Fiji includes the relocation of entire communities in their national plan of adaptation (IOM 2018).

Regarding Loss and Damage, nearly half of Small Island Developing States and one-third of the Least Economically Developed Countries referred to Loss and Damage in their INDCs (Rai and Acharya 2020).

Recommendations for Policymakers

Due to the absence of Loss and Damage considerations in the policies, it is recommended that a specific section is included that addresses human mobility and all the remaining items defined by the Warsaw International Mechanism for Loss and Damage that include both Economic and Non-Economic Losses. The former encompasses losses of income (business operations, agriculture production and tourism) and physical assets (infrastructure and property). Non-Economic Losses are mentioned at different levels: individual (life, health and human mobility), society (territory, cultural heritage, indigenous knowledge and societal/cultural identity) and environments (ecosystem services and biodiversity).

Policymakers are also advised to consider human mobility as part of a range of possibilities of adaptation; this will allow in the beginning to collect data and evidence that will be later used to inform actual policies (and possibly legislation) themselves in line with the context-specific needs (Gemenne and Blocher 2017). Such an effective response will require coordination and partnerships between authorities and actors that deal with human mobility and/or climate change (IOM 2018).

The United Nations High Commissioner for Human Rights (2018) recommends including considerations on gender issues within a wider concern for human rights when creating policies and strategies to govern human mobility, including when it is a response to climate change impacts.

Finally, it is recommended that the policymakers translate at the national level what was agreed upon in international commitments regarding human mobility within the context of climate change, such as the Cancún Adaptation Framework and The Global Compact for Safe, Orderly and Regular Migration (IOM 2018).

Limitations of the Study

The main limitation of the study is the scope of the documents analyzed. The policies included in this study are not extensive and other sectoral policies could be added to the list, such as Plan Maroc Vert and Génération Green 2020–2030, which focus specifically on the agricultural sector. The list could also include legal texts that focus on the environment. Additionally, the purpose of this study was to analyze how environment-related documents approach human mobility, but it could be relevant to expand the analysis to include human mobility-related policies and to study how they consider environmental factors.

Conclusion

The policies related to climate change and sustainable development in Morocco are not consistently considering human mobility, in particular, and Loss and Damage, in general, despite growing impacts of climate change and projections of over 1 million migrants due to climate change by 2050 (Clement et al. 2021).

Even though there is not a consensus on whether it is ideal to include policies that view human mobility as a possible adaptation strategy, it is generally acknowledged as being part of the Losses and Damages. By not including mobility in the policies, it will remain an invisible issue and no official data and evidence will be collected on a systematic basis, which would be needed to inform future policies and even legislation that is in line with the various local contexts.

Funding This work was carried out at the R&D Unit Centre for Functional Ecology–Science for People and the Planet (CFE), with reference UIDB/04004/2020, financed by FCT/MCTES through national funds (PIDDAC).

References

Ait Hamza, M., El Faskaoui, B. & A. Fermin (2009) Migration and environmental change in Morocco: The case of rural oasis villages in the Middle Drâa Valley. Case Study Report: EACH-FOR - Environmental Change and Forced Migration Scenarios

Clement V, Rigaud KK, de Sherbinin A, Jones B, Adamo S, Schewe J, Sadiq N, Shabahat E (2021) Groundswell part 2: acting on internal climate migration. The World Bank, Washington, DC

ClimaSouth (2018) Implementing Nationally Determined Contributions (NDCs) in the South Mediterranean region: perspectives on climate action from eight countries. ClimaSouth Policy Series, Paper N.4. Prepared by Andrea Rizzo & Pendo Maro. European Commission

Coulibaly, M., Ségorbah Silwé, K. & C. Logan (2018) Taking stock: Citizen priorities and assessments three years into the SDGs. Afrobarometer Policy Paper No. 51

de Haas H (2011) Mediterranean migration futures: Patterns, drivers and scenarios. Global Environ Change 21(suppl 1):S59–S69. https://doi.org/10.1016/j.gloenvcha.2011.09.003

De Wet CJ (ed) (2006) Development-induced displacement. Problems, policies, and people (Studies in forced migration, 18). Berghahn Books, New York

Foresight (2011) Migration and Global Environmental Change: Final Project Report. The Government Office for Science. London

Gemenne F, Blocher J (2017) How can migration serve adaptation to climate change? Challenges to fleshing out a policy ideal. Geogr J 183:336–347. https://doi.org/10.1111/geoj.12205

IOM (2016) Assessing the Evidence: Migration, Environment and Climate Change in Morocco. International Organisation for Migration – IOM, Geneva

IOM (2018) Mapping Human Mobility and Climate Change in Relevant National Policies and Institutional Frameworks. Warsaw International Mechanism for Loss and Damage associated with Climate Change Impacts. Task Force on Displacement – Activity I.1. International Organisation for Migration (IOM)

IPCC (2021) Summary for Policymakers. In: Climate Change 2021: The Physical Science Basis. Contribution of Working Group I to the Sixth Assessment Report of the Intergovernmental Panel on Climate Change [Masson-Delmotte, V., P. Zhai, A. Pirani, S.L. Connors, C. Péan, S. Berger, N. Caud, Y. Chen, L. Goldfarb, M.I. Gomis, M. Huang, K. Leitzell, E. Lonnoy, J. B. R. Matthews, T. K. Maycock, T. Waterfield, O. Yelekçi, R. Yu, and B. Zhou (eds.)]. Cambridge University Press, Cambridge, United Kingdom and New York, NY, USA, pp. 3–32. https://doi.org/10.1017/9781009157896.001.

Joseph G, Wodon Q, Blankespoor B (2014) Do remittances reach households living in unfavorable climate areas? Evidence from the Republic of Yemen, MPRA Paper 56939. University Library of Munich, Germany

Lejars C, Daoudi A, Amichi H (2017) The key role of supply chain actors in groundwater irrigation development in North Africa. Hydrogeol J 25:1593–1606. https://doi.org/10.1007/s10040-017-1571-

Liaeter S, Durand-Delacre D (2021) Situating 'migration as adaptation' discourse and appraising its relevance to Senegal's development sector. Environ Sci Policy J 126:11–21. https://doi.org/10.1016/j.envsci.2021.09.008

Niang I, Ruppel OC, Abdrabo MA, Essel A, Lennard C, Padgham J, Urquhart P (2014) Africa. In: Climate change 2014: impacts, adaptation, and vulnerability. part B: regional aspects. Contribution of Working Group II to the Fifth Assessment Report of the Intergovernmental Panel on Climate Change [Barros VR, Field CB, Dokken DJ, Mastrandrea MD, Mach KJ, Bilir TE, Chatterjee M, Ebi KL, Estrada YO, Genova RC, Girma B, Kissel ES, Levy AN, MacCracken S, Mastrandrea PR, White LL (eds)]. Cambridge University Press, Cambridge & New York

Rai SC, Acharya S (2020) Anchoring loss & damage in enhanced NDCs. World Wide Fund For Nature, Gland, Switzerland

Sobczak-Szelc K (2008) Changes in the environment and migration in Southern Morocco – example of the Mhamid Oasis. Miscellanea Geographica 13:239–250. https://doi.org/10.2478/mgrsd-2008-0024

Stern N (2014) The economics of climate change: the Stern review. Cambridge University Press, Cambridge. https://doi.org/10.1017/CBO9780511817434
Taheripour F, Tyner WE, Haqiqi I, Sajedinia E (2020) Water scarcity in Morocco: analysis of key water challenges. World Bank, Washington, DC
The White House (2021) Report on the impact of climate change on migration. Washington DC
Tribak A, Paradiso M, Azagouagh K (2019) Climate refugees, housing in risk areas, and vulnerability of the built environment in the Fez Urban Area of Morocco (Case of the Medina and Outlying Districts). In: Paradiso M (eds) Mediterranean mobilities. Springer, Cham
UNDP (2018) National adaptation plan process in focus: lessons from Morocco [online]
United Nations High Commissioner for Human Rights (2018) Summary of the panel discussion on human rights, climate change, migrants and persons displaced across international borders. Human Rights Council Thirty-seventh session. A/HRC/37/35
UNFCCC (2016) Technical Meeting – Action Area 6: Migration, Displacement and Human Mobility. Executive Committee of the Warsaw International Mechanism for Loss and Damage associated with Climate Change Impacts
Verner, D. (2012) Adaptation to a changing climate in the Arab countries: a case for adaptation governance in building climate resilience. The World Bank, Washington DC
Warner K (2010) Global environmental change and migration: governance challenges. Glob Environ Chang 20(3):402–413. https://doi.org/10.1016/j.gloenvcha.2009.12.001
WIM TFD (2018) Synthesizing the state of knowledge to better understand displacement related to slow onset events. Internal Displacement Monitoring Centre (IDMC) & Warsaw International Mechanism for Loss and Damage – Task Force on Displacement. [online]
Wodon Q, Burger N, Grant A, Joseph G, Liverani A, Tkacheva O (2013) Climate change, extreme weather events, and migration: review of the literature for five Arab countries. In: Piguet E, Laczko F (eds) People on the move in a changing climate. Global migration issues, vol 2. Springer, Dordrecht

Chapter 22
Written Press's Approach to Climate Change in the Autonomous Region of Madeira and the Autonomous Community of the Canary Islands

Ana Maria Bijóias Mendonça, Walter Leal Filho, and Fátima Alves

Abstract The Autonomous Region of Madeira and the Autonomous Community of the Canary Islands belong to Macaronesia and are ultra-peripheral regions, being particularly vulnerable to climate change (CC). Due to their isolation and idiosyncratic characteristics, the written press has a decisive role in the context of the insular territories and on tackling CC, hence, the purpose of this paper is to assess the written press' approach to CC in a ten-year interval. We used the descriptor 'climate change' to collect articles from two newspapers (*Canarias 7* and *Diário de Notícias da Madeira*), comprising July 2008 and January 2009, as well as July 2018 and January 2019, gathering 170 news articles (in total). For further analysis, all the news articles were coded and categorised according to five major categories (scope, entities involved, topics related to CC, discourse content, and position on CC). Results confirmed the prevalence of news articles in *Diário de Notícias da Madeira* in both periods of analysis, but this does not necessarily mean that the stakeholders are more receptive or prepared to act in Madeira. The scope was broad, given the prevalence of regional level news in the two newspapers. On the entities involved, despite the large number and diversity of social actors depicted, a preponderance of political and institutional actors stood out. A wide variety of topics related to CC were reported, predominantly vulnerability, CC consequences, and CC policies (overall mitigation). Regarding content, the centre of attention was on informative/institutional, expert opinions and critical discourses. Lay rationalities, knowledge, and practices

A. M. B. Mendonça (✉) · F. Alves
TERRA Associate Laboratory, Department of Life Sciences, Centre for Functional Ecology—Science for People and the Planet (CFE), University of Coimbra, Calçada Martins de Freitas, 3000-456 Coimbra, Portugal
e-mail: ana_bijoias@hotmail.com

F. Alves
e-mail: fatimaa@uab.pt

Open University (UAb), Rua da Escola Politécnica, Palácio Ceia, Lisbon, Portugal

W. Leal Filho
Research and Transfer Centre "Sustainability and Climate Change Management", Hamburg University of Applied Sciences (Haw Hamburg), Hamburg, Germany
e-mail: walter.leal2@haw-hamburg.de

W. Leal Filho et al. (eds.), *Climate Change Strategies: Handling the Challenges of Adapting to a Changing Climate*, Climate Change Management,
https://doi.org/10.1007/978-3-031-28728-2_22

were virtually absent. As for the positioning on CC, most of the news evidenced the acceptance that it is an unavoidable problem and that possible solutions require structural transformations in the development model of current societies and a paradigm shift. Other narratives coexisted with this perspective (mainly in *Diário de Notícias da Madeira*), denoting manipulative discourses and greenwashing, along with the view of CC as opportunity. We argue that the media should extend the debate on environmental issues to all social actors, including local communities, bridge the gap between scientists, technicians and policy makers and society in general, and go beyond the mainstream approach to CC and the sensationalist trends to contribute to knowledge improvement, awareness-raising, and solution building, mostly at the regional/local level.

Introduction

There is widespread scientific agreement that climate change (CC), resulting from the increase in global temperatures, is the most serious and dangerous environmental problem that threatens life on Earth and the most unexpected consequence of the industrial era, assuming an undeniable global and anthropogenic character (Santos 2012). The environmental crisis and CC, which is its most visible feature, represent a demanding challenge to governments, science, institutions, and societies in general (Aldeia and Alves 2019). Climate change affects the availability of natural resources, reflecting on the balance of ecosystems, on livelihoods, on social and cultural organisation, and on the health and well-being of current generations. It may also compromise future generations (IPCC 2007, 2013, 2014). In fact, scientific research shows that anthropogenic influence has warmed the climate to an extent unprecedented in the last 2000 years, with human activities affecting all the major climate system components (IPCC 2021). The Autonomous Region of Madeira and the Autonomous Community of the Canary Islands are part of Macaronesia and are outermost regions (Fig. 22.1) being particularly vulnerable to CC (Gobierno de Canarias 2013; Regional and da Madeira – Secretaria Regional do Ambiente, Recursos Naturais e Alterações Climáticas (SRARNAC) 2015; IPCC 2021; Institute for European Environmental Policy 2013).

In a globalised world, where an intricate multiplicity of networks and actors are interconnected and associated with each other (Latour 1999, 2005; Law 1992, 1999), the media are the main source of information and the main element that can influence people's awareness of CC and can contribute to framing the public agenda (Alves et al. 2020). Working as a space for the dissemination and discussion of ideas emanated by different social actors, they play a central part in the construction and reconstruction of social reality (Carvalho 2011). As a critical arena for the production, reproduction, and transformation of the interpretation of complex and multidimensional public problems, such as CC, the media influence the public's understanding of risks, responsibilities, and the performance of democratic politics (Carvalho 2010).

Fig. 22.1 The Canary Islands and Madeira archipelagos' location

In this regard, media coverage is an intermediary between expert and lay rationality, and it cooperates in the shaping of society's awareness about climate change (Areia et al. 2019). Being permeable to hegemonic knowledge and power relations that dictate the system of truth (Foucault 1977, 1992, 1998), it moulds discourse and action at the crossing point of climate science and policy (Boykoff and Roberts 2007).

According to Boykoff and Smith (2010), climate science and politics helped to frame media reporting, but the opposite is likewise true, since media coverage shapes science, governance, and policy decisions. At a social level, the way CC is presented by the media has a significant impact on the way the public perceives it and consequently on the way people respond to it (Areia et al. 2019; Carvalho 2010). In reality, journalistic production has served to map out and frame environmental issues for policy agents and the public, whilst relevant social actors such as individual authority figures, scientists and experts, governments, and consortia attempt to access media sources in order to mould public perceptions of environmental issues in line with their own perspectives and interests. This framing labour contributes to contextualising multidimensional and complex environmental processes, in addition to finding interpretative categories to explain and describe them (Boykoff and Smith 2010).

Media attention is a critical indicator for CC's social problem construction. It measures the amount of attention given to one issue in relation to the amount of attention given to other issues at the same time. Absolute numbers of newspaper articles on CC indicate distinct national attention levels but are also shaped by the size of a newspaper, differences in journalistic culture or financial limitations (Schmidt et al. 2013). In other words, the media and their discursive practices are at the same time producers of meaning and product of a determined socio-cultural context, having with it a mutually constitutive relationship (Carvalho 2011).

Following international and national tendencies, CC has increasingly received attention from the local media, although news in Portugal and Spain still favour political/technical discourses over civil society's discussions. Thus, Iberian media coverage on climate change reflects undisputable bias and may add to the public's low levels of acknowledgment and commitment (Areia et al. 2019). Studies show that countries from the Global North normally cover CC more frequently, and that countries from the Global South tend to focus on its societal dimension and impacts on humans (Hase et al. 2021). As a solution, the media should democratise the communication of CC, bringing the reality of the problem closer to the reality of individuals and society. Instead of the significant projection given to news that is related to political-scientific discussions in the international forum or to environmental catastrophes that occurred in distant latitudes, journalists should frame the issue of CC at the level of local communities and at the individual level. This means that a more proactive discourse on CC can be substantive, influencing the adoption of sustainable behaviours in the public and the implementation of environmental policies and laws by politicians. In fact, democratic communication on environmental issues focusing on community actions to tackle climate change—and not just government failures in relation to environmental policy or climate-related disasters—would improve the active role of the media in involving individuals and would help promote active societal responses to CC (Areia et al. 2019).

Iberian journalists are aware of their gaps in CC coverage, insomuch that in the aftermath of the international conference Media and Climate Change (*Medios de Comunicación y Cambio Climático*), held on 22–23 November 2012 and sponsored by the Laboratorio de Estudios en Comunicación from the University of Seville (Reig 2013) and other relevant events, the ECODES Foundation prepared a meeting between researchers of the Dialectical Mediation of Social Communication (MDCS, in Spanish) and relevant environmental communication professionals. The main objective was to develop a catalogue of recommendations to report on climate change. Hence, within the framework of the Change the Change Conference, that took place on 6–9 March 2018 in San Sabastián, the "Decalogue of Recommendations for Reporting on Climate Change" (*Decálogo de recomendaciones para informar sobre el cambio climático*) was presented, and by the end of 2019 more than eighty media outlets had already signed it (UNESCO 2021). The document acknowledged the discrepancy between the challenges of contemporary CC, its relevance in the scientific/institutional field, and the sparse information content reaching public audiences. Bearing that in mind, the signatories have committed themselves to improving journalistic exercise and to observing the social role of the media in relation to the phenomenon.

Either way, with respect to CC information and communication, the states, governments, scientists and the media have failed to reach out to heterogeneous audiences in different geographical regions (Leal Filho 2009). There is, additionally, a plurality of languages and definitions to communicate about the environment in general, also because the identification of an environmental problem and the search for a solution depends on knowledge, values and worldviews that differ in socio-cultural terms (Carvalho 2011). Besides, the discourses of social actors addressing CC, including the media's, are often covered in drama and appeal to fear and other negative emotions, leading people to usually reject them due to feelings of disbelief or apathy (Cabecinhas et al. 2008). At best, media reporting adds to CC communication, review and discussion but barely provides answers to it (Boykoff and Smith 2010).

At the regional and local level, media coverage brings forth a privileged insight into CC impacts on the territories and current policies. Nonetheless, in general, local coverage and local newspapers tend to disappear and/or decrease their quality owing to the rise of social media and other platforms, along with the consistent pressure of technological and economic factors (Ardia et al. 2020). The 2008 crisis precipitated the situation (Reyes 2013). One of the resulting consequences is the information vacuum derived from the spread of misinformation and disinformation. Another critical consequence accrued from the removal of a local news source from a market is the negative effect it is likely to have on community engagement and issue knowledge. Regulatory and policy solutions are required. Moreover, in possible solutions to contemporary "wicked problems" (Rittel and Webber 1973), such as CC, a multifaceted and multi-disciplinary approach is crucial to congregating experts and policymakers in an effort to leverage significant changes, including a new vocabulary to grant the legal, social, economic and journalistic principles involved and the impairment to a functioning democracy and citizenship if the status quo prevails (Ardia et al. 2020).

The written press has a decisive role in the context of the insular regions and on tackling CC due to their isolation, vulnerability, and idiosyncratic characteristics. However, our research confirmed that there are virtually no studies on the journalistic coverage of CC in the Autonomous Region of Madeira nor in the Autonomous Community of the Canary Islands, nor in the whole Macaronesia, from what we were able to ascertain.

Materials and Methods

The aim of the paper is to assess how the written press has been addressing CC in the Autonomous Region of Madeira and the Autonomous Community of the Canary Islands in a ten-year interval. The main objectives are to understand how the topic's approach has been evolving in both regions and to identify the informative trends and how they relate to multiple social actors and to international, national, and regional/local contexts against the background of environmental and climatic policies. With the purpose of providing information about policies, informative trends, and the position in relation to CC, we have analysed the articles from two newspapers: *Canarias 7* (entirely online) and *Diário de Notícias da Madeira* (via manual consultation in the Madeira Regional Archive, and online), choosing a summer and a winter month, namely July and January, in a ten-year interval. Thus, we focused on the news articles from July 2008 and January 2009, as well as July 2018 and January 2019. The descriptors 'environment' and 'climate change' were initially selected, but given the large quantity of data collected, that is, 152 news articles in the *Canarias 7* and 286 news articles in the *Diário de Notícias da Madeira,* totalling 438 news articles, we ended up selecting only those with the descriptor 'climate change'. As a result, we collected 62 news articles in the *Canarias 7* (sixteen in July 2008, five in January 2009, twenty-eight in July 2018 and thirteen in January of 2019), and 108 news articles in the *Diário de Notícias da Madeira* (twenty-seven in July 2008, twenty-eight in January 2009, thirty-one in July 2018 and twenty-two in January 2019), for a total of 170 news articles.

Altogether, we have:

Canarias 7: July 2008 (16) + January 2009 (5) + July 2018 (28) + January 2019 (13) = **62.**

Diário de Notícias da Madeira: July 2008 (27) + January 2009 (28) + July 2018(31) + January 2019 (22 news) = **108.**

Total: **170 news articles.**

In view of the need for further clarification and further analysis of all the data, the 170 news articles were coded and categorised. Based on the document analysis, the aim of the study, and the examination of the information in both newspapers, five major categories emerged:

- Scope;
- Entities involved;
- Topics related to CC;
- Discourse content;
- Position on CC.

A matrix was designed according to those main categories and other subsidiary categories that were progressively added during the reading and the organisation of the content analysis processes:

1. Scope:
    - 1.1. Regional
        - 1.1.1. Canary archipelago
        - 1.1.2. Madeira archipelago
    - 1.2. National
    - 1.3. International
    - 1.4. Without defined scope
2. Entities Involved:
    - 2.1. Politicians
    - 2.2. Economic agents (managers, economists, companies, and consortia)
    - 2.3. Activists/environmentalists
    - 2.4. Scientists/universities/institutes
    - 2.5. Population
    - 2.6. Artists

   2.7. Social movements
   2.8. Non-unique official entities (ministries, schools, departments, governments, etc.)
   2.9. Non-official entities (NGOs/NGOs, media, etc.).

3. Topics Related to CC:

   3.1. Risks and vulnerabilities
   3.2. Extreme events
      3.2.1. Natural disasters in general
      3.2.2. Fires
      3.2.3. Earthquakes/tremors
      3.2.4. Heat waves/drought/desertification
      3.2.5. Cold waves/snow
      3.2.6. Floods/torrential rain/maritime agitation
      3.2.7. Collapses and landslides
   3.3. Global warming/pollution
   3.4. Causes
   3.5. Effects
   3.6. Responses
   3.7. Policies of CC
      3.7.1. Mitigation (energy, reduction of GHG emissions, etc.)
      3.7.2. Adaptation (plans, strategies)
   3.8. Awareness/education/prevention

4. Discourse Content:

   4.1. Institutional
   4.2. Critical
   4.3. Prescriptive
   4.4. Promotional

5. Position on CC

   5.1. Sceptics/deniers
   5.2. Believers/paradigm shift
   5.3. Manipulative discourses/greenwashing
   5.4. Persistence of the mainstream/CC as opportunity

Results

The study confirmed the prevalence of news articles in *Diário de Notícias da Madeira* on CC in both periods of analysis (July 2008/January 2009, and July 2018/January 2019), specifically, fifty-five news articles in the first period (comparing with only twenty-one in *Canarias 7*), and fifty-three news articles in the second period (forty-one in *Canarias 7*).

Forthwith, we will proceed to a qualitative analysis of the information gathered, yet as a means to illustrate the discourse and facilitate comprehension and examination, we also provide a table with absolute numbers of the news articles in both newspapers according to major and subsidiary categories (Table 22.1).

Regarding the scope of the news articles in both newspapers during the appointed timeframe, it is possible to ascertain its broadness, albeit the prevalence of regional levels in the two territories. Additionally, there were no national level news articles in Canarias, in contrast to Madeira where they appear in third place, after international level news. *Diário de Notícias da Madeira* also published more mixed level news and with no defined scope when compared to *Canarias 7*.

The data point to two different outlines: on one hand, the gradual importance of environmental issues over time, especially CC, not only in the Iberian mainland but in their autonomous regions as well, and on the other hand, the evidence that CC effects are becoming increasingly evident in the insular territories, including the archipelago of the Canary Islands and the archipelago of Madeira.

With respect to the entities involved, some aspects did stand out in both regions in the two periods. Thus, despite the large number and diversity of social actors depicted, there was a considerable preponderance of political and institutional actors (more numerous in Madeira), in particular with politicians and official non-singular entities such as governments, departments, and ministries, as Areia et al. (2019) pointed out regarding media coverage of CC in Portugal and in Spain. Other relevant regional/local entities surfaced, such as economic operators and scientists/universities/institutes. This means that local media in the insular regions follow, to some extent, the trends of national (and international) media, guided by the mainstream rationality, knowledge and practices that shape environmental/climatic politics and the formal *modus operandi*. Non-official entities were mentioned in larger number in *Diário de Notícias da Madeira*. However, civil society, environmentalists and activists, social movements, associations, together with supplementary social/cultural actors, e.g., artists, were barely included. The discrepancy is evident and denotes the absence, even at small levels and scales, of a considerable part of society in the CC approach and search for solutions. Therefore, it was made clear that CC is still viewed as a scientific/political/ institutional/economic issue, and that the narratives outside of that realm, such as those bolstered by lay rationalities, non-formal knowledge, and everyday practices, are consistently ignored by the decision makers and by the media (Aldeia and Alves 2019; Alves et al. 2014, 2020).

News articles exhibited a wide variety of topics related to CC, which turned out to be more diverse and more present over time in the reporting space of both newspapers, particularly in *Canarias 7*. Understandably, one of the major topics mentioned is 'risks and vulnerabilities', mainly in *Canarias 7* (ten news articles in the second period of analysis, but only four in *Diário de Notícias da Madeira*). In the archipelago of the Canary Islands, these refer to an increase in average temperatures, affecting water availability and water supply, aggravating the desertification and the occurrence of wildfires. The interference of CC in the ocean and coastal areas cannot be ignored either. In the presence of this scenario, public authorities, especially the *Cabildo* of Gran Canaria and the Consejo Insular de Energía, in conjunction with the

Table 22.1 Absolute numbers of news articles in both newspapers according to major and subsidiary categories of analysis

Categories	Canarias 7	Diário de Notícias da Madeira
Scope	Regional level (48); international level (6); mixed level (4); no defined scope (4)	Regional level (45); national level (16); international level (22); mixed level (14); no defined scope (11)
Entities involved	Large number and diversity of actors, mainly politicians (30); economic operators (14); scientists/universities/institutes (13), and official non-singular entities (governments, departments, schools, etc.) (47). Non-official entities (NGO, media...) (5); activists/environmentalists (3); population (1); social movements (1)	Large number and diversity of actors, mainly politicians (57); economic operators (20); scientists/universities/institutes (22); non-official entities (NGO, media, etc.) (21), and official non-singular entities (governments, departments, schools, etc.) (72). Population (6); activists/environmentalists (3); social movements (2); artists (1)
Topics related to CC	Wide variety of topics: causes, effects, responses to CC, mostly risks and vulnerabilities (10) and extreme events (26), like wildfires, in July (13), earthquakes (2), heat waves/droughts/desertification (8), cold waves/snow (2), floods/torrential rain/sea disturbance (1) Policies of CC: mitigation (17), adaptation (1), awareness/education/prevention (8)	Wide variety of topics: causes, effects, responses to CC, mostly risks and vulnerabilities (4) and extreme events (55), like wildfires, in July (24), earthquakes (3, being 2 int. and 1 in Madeira), "bad weather" (3), heat waves/droughts/desertification (4), cold waves/snow (12: mainly about the Continent and European countries, only 1 in Madeira), floods/torrential rain/sea disturbance (5: only 1 about Madeira), downfalls/landslides (4: 3 in Madeira and 1 int.) Policies of CC: mitigation (14), adaptation (3), awareness/education/prevention (15)
Discourse content	Informative/institutional (18); expert opinion (14), critical (5), prescriptive (4), promotional (3), alarmist (1) negationist (1, expert)	Informative (31), informative/institutional (31), expert opinion (12), critical (10), alarmist (7), negationist (2)
Position on CC	Believers/paradigm shift (15), manipulative discourses/greenwashing (5), mainstream discourse/CC as opportunity (2), sceptics/negationists (1)	Believers/paradigm shift (15), manipulative discourses/greenwashing (9), mainstream discourse/CC as opportunity (8), sceptics/negationists (3)

AEMET (Agencia Estatal de Meteorología) and sponsored by the Fundación Biodiversidad, were preparing, in July 2018, the report "Diagnosis of Risks, Vulnerabilities and Adaptation to Climate Change in the Island of Gran Canaria" (*Diagnóstico de Riesgos, Vulnerabilidades y Adaptación al Cambio Climático de la Isla de Gran Canaria*), framing initiatives and actions aimed at reducing risks and vulnerabilities and increase resilience. The report, set under the *Pacto de los Alcaldes*, also aimed to outline municipal emergency plans against adverse meteorological phenomena and the studies of risks and vulnerabilities at the municipal level. In *Diário de Notícias da Madeira*, the risks and vulnerabilities approach emphasised predominantly wildfires as a risk factor.

Another major topic linked with the previous one addresses 'CC consequences' (more than 'CC causes') in global terms, but exceedingly in the two Macaronesian territories. As it happens, extreme events that were accounted for either as isolated occurrences or in relation to CC were massively mentioned in *Canarias 7* and in *Diário de Notícias da Madeira* in the first period of analysis, but to the greatest extent in the second period (July 2018/January 2019). Specifically, there were twenty-six news articles of extreme events in Canarias: wildfires, always in July (13), earthquakes (2), heat waves/droughts/desertification (8), cold waves/snow (2), floods/torrential rain/sea disturbance (1). In Madeira there were fifty-five news articles: wildfires, always in July (24), earthquakes (3: 2 int. and 1 in Madeira), 'bad weather' (3), heat waves/droughts/desertification (4), cold waves/snow (12: mainly referring to Continental Portugal and to European countries, only 1in Madeira), floods/torrential rain/sea disturbance (5: only 1 in Madeira), and downfalls/landslides (4: 3 in Madeira and 1 int.). We can infer that the growing number of newspaper reports referring to the consequences of CC will be due, altogether, to the awareness and visibility that the phenomenon has gained in global, national, and regional/local agendas and in the media, along with the accelerating visible effects in the insular regions and the need to act to tackle it. However, one should not forget the appetite of certain media and the public for impactful news, often linked to tragedies and calamities. The topic 'CC responses' focused almost exclusively on the diligences of public authorities, fire brigades, emergency, and rescue entities to comply with extreme events' impacts both in the archipelago of the Canary Islands and in Gran Canaria (e.g., public campaigns urging the owners of forested land to clean up woodlands, the effort to increase BRIF's capacity to intervene in wildfires' extinction in all Canary islands, activation of the Special Plan of Civil Protection and Emergency Attention—*Plan Especial de Protección Civil y Atención de Emergencias*—by the government in different situations, including wildfires), as well as in the archipelago of Madeira and in Madeira Island. Similarly to the situation in the Canary Islands, and in consonance with the news in *Canarias 7*, wildfires are also a big issue in Madeira during the summer. Thus, *Diário de Notícias da Madeira* reported occasions when fire-fighters had to attend to several locations. In wintertime, and in both regions, these emergency and rescue actors were described to intervene primarily in floods, cold waves/snow, and downfalls/landslides.

The third main topic related to CC addressed by *Canarias 7* and *Diário de Notícias da Madeira* concerns 'CC policies'. Again, the ensuing international guidelines were anchored in paramount events like the UN climate summits (COP), from which structuring documents emanated such as the Kyoto Protocol (UNFCCC 1997) and the Paris Agreement (United Nations 2015). The Roadmap 2050, an initiative of the European Climate Foundation (ECF), seeks to provide pathways to achieve a low-carbon economy in Europe by 2050. These initiatives and the resulting documents served as a framework for the supranational policies of the EU and each member state, accompanied by their autonomous regions. It was clear that CC mitigation was prevalent in the two newspapers, given the large number of news articles concerning energy transition process, i.e., the technology-based switch from fossil fuels to renewables and the reduction in greenhouse gas emissions. Thus, for example, the *Canarias 7* cites the Environmental Impact Statement for the El Hierro Hydroelectric Power Plant project and the construction of the Las Palmas-Telde power plant, as well as the possibility of storing CO_2 underground in the Canary Islands in the period of July 2008/January 2009. It is worth noting that, in the period July 2018/January 2019, the number of news articles related to the discourses and actions of the main political-institutional actors (Government, Cabildo, Ayuntamientos) and some prominent companies (e.g., Cepsa) towards decarbonisation grew considerably. The *Diário de Notícias da Madeira* reported in July 2008 that twenty-two people had already registered for home energy production that could be sold to the electric company in the archipelago (Empresa de Eletricidade da Madeira). In January 2019, the construction of the Pico da Urze reservoir was emphasised. Along with the investment in green energies, the political and institutional discourses focused on themes such as I&DT projects, blue economy strategies and regional development. Overall, *Canarias 7* published seventeen news articles on CC mitigation and only one on CC adaptation, whereas *Diário de Notícias da Madeira* published fourteen news articles on mitigation, and three on adaptation. It is important to underline that a great amount of attention was given in the two newspapers to awareness/education/prevention (eight news articles in *Canarias 7* and fifteen in *Diário de Notícias da Madeira*), focusing altogether on administrations' and researchers' efforts to increase public awareness, promote environmental good practices and/or advise on how to act to cope with CC effects in everyday life.

For all that has been earlier mentioned, it was not unexpected to find a wide range of narratives regarding the discourse content, the centre of attention being informative/institutional, expert opinions and critical narratives. Correctly speaking, and in liaison with the entities involved, we have collected, in *Canarias 7*, eighteen news articles of informative/institutional essence, fourteen on expert opinion, and five of critical nature, whereas in *Diário de Notícias da Madeira,* thirty-one news articles had a strictly informative basis, while the same amount incorporated an informative/institutional basis, twelve news articles focused on expert opinion and ten comprised critiques (to political and institutional actors). In accordance with the insight into contemporary environmental crisis and CC, showing that time is pressing, and imminent collapse could materialise, *Canarias 7* reported one piece of alarmist content, while *Diário de Notícias da Madeira* reported seven pieces, most of them in the period of July 2018/January 2019. Negationist narratives were almost non-existent, comprising only one piece in *Canarias 7* (signed by an expert), and two in *Diário de Notícias da Madeira*. Other narratives assumed a prescriptive (four news articles in *Canarias 7*) or promotional (three article news in *Canarias 7*) axis.

The topics related to CC and the content of the discourses allow for access to the subject's position in relation to the phenomenon over time. Thus, most of the news covered by *Canarias 7* and *Diário de Notícias da Madeira* evidence a growing concern with the topic, the acceptance that it is an unavoidable problem in global, national, and regional/local terms, and that possible solutions implicate a structural change in the development model of current societies and a paradigm shift (societal, political, economic). In line with this notion, *Canarias 7* reported, in July 2018, that climate change was already a reality, highlighting the need to work more effectively and articulately in the Canary Islands and internationally to deal with the problem, contrasting with the perspective (residual, but still worthy of note) of a physics professor at the Universidad de Las Palmas de Gran Canaria, in July 2008, who sustained that linking processes of desertification occurring in the islands to CC is alarmist, because CC is a natural event and has always affected the planet. In between, the mainstream approach emerged, and some political and economic actors went so far as to consider CC as an opportunity or have embarked on manipulative discourses to evidence greenwashing, especially where the energetic sector is concerned. We therefore have fifteen news articles denoting CC acceptance and the need for a paradigmatic transformation, five denoting manipulative discourses/greenwashing, two denoting mainstream discourse/CC as opportunity, and one denoting sceptic/negationist positioning. Comparatively, *Diário de Notícias da Madeira* also reported fifteen news articles denoting believability in CC (mainly of anthropogenic origin) and the urgency for a paradigm shift, yet at the same time it published nine news articles concerning manipulative discourses/greenwashing and eight concerning mainstream discourse/CC as opportunity (including an interview, in January 2009, to the coordinator in the Autonomous Region of Madeira of a benchmark Portuguese environmental NGO). Three news articles evidenced sceptic/negationist positioning (including one, in July 2008, regarding the president of the then Regional Government, and one regarding the president of a city council).

We can thus infer that socio-environmental and climate issues summon a heterogeneous multiplicity of fields, actors, associated topics, narratives, together with distinct and sometimes incompatible positions.

Conclusions

We have argued that CC is a critical societal problem with global causes and consequences that resonate in national and regional/local territories. Media coverage has the potential to increase societal awareness and knowledge of the issue, ideally contributing to conscious individual and collective actions. Regarding political decisions, the mass media represent a decisive setting for opinion formation and legitimation. The research outlines that regional/local newspapers bestow a significant part of reporting space to CC and that it has steadily grown in importance. Therefore, the media have indicated to their audience that the topic has gained relevance and have provided some possibilities for opinion formation.

We have highlighted several distinct aspects in the development and overall level of the written press's attention in the Autonomous Region of Madeira and the Autonomous Community of the Canary Islands, largely due to these territories' idiosyncrasies. Thus, the results confirmed the prevalence of news articles in *Diário de Notícias da Madeira* in both periods of analysis (108 vs 62). It must not be inferred, however, that the stakeholders (e.g., society, public and private authorities, economic agents, the media) are more receptive or prepared to act in Madeira. As mentioned, a summer month and a winter month were originally selected, intending to skip December on account of Christmas holidays, but Spain celebrates the *Día de los Reyes Magos* on January the 6th, hence there would be fewer articles regarding CC in the Canary Islands in January. In any case, the two newspapers reported more regional and local news than national or international ones. Moreover, the prevalence of political and institutional discourses is noteworthy, as are the residual number of news articles referencing further social actors, such as population groups, social movements, activists/environmentalists, and other non-official entities. The policies of CC were focused mostly on mitigation and awareness/education/prevention than on adaptation, so it is critical to develop robust and effective adaptation strategies in the Macaronesian region and to increase the adaptive capacity and resilience of the populations. The undeniable preponderance of the CC mainstream approach has led to incompatible discourses and positioning (e.g., CC believers/paradigm shift and CC as opportunity/greenwashing), leaving out of the debate other relevant contributions.

Recommendations

- It is most crucial that the media extend the debate on environmental problems to the entire society, including local communities.
- The media should truly bridge the gap between scientists, technicians and policy makers and society in general, assuming their role as a public service, independent informers, and social educators.
- Therefore, the media must go beyond the mainstream approach to CC and the sensationalist trends to take up the task of acknowledging different realities in specific contexts and learn from them, facilitating the populations' adaptation mechanisms and their capacity for resilience.

Limitations

For future studies, it would be advantageous to go beyond issue attention to provide a rationale for the role of media debates in shaping societal reactions to the challenges presented by CC, and extend the research to the whole Macaronesia, providing a broader comparative overview of how CC is perceived, managed, and reported in each of its territories and amongst them.

Funding This work was carried out at the R&D Unit Centre for Functional Ecology–Science for People and the Planet (CFE), with reference UIDB/04004/2020, financed by FCT/MCTES through national funds (PIDDAC).

References

Aldeia J, Alves F (2019) Against the environment. Problems in society/nature relations. Front Sociol 4:1–12. https://doi.org/10.3389/fsoc.2019.00029

Alves F, Caeiro S, Azeiteiro UM (2014) Lay rationalities of climate change (editorial material). Int J Climate Change Strategies Manag 6(1):1–5. https://doi.org/10.1108/IJCCSM-10-2013-0121

Alves F, Filho WL, Casaleiro P, Nagy GJ, Diaz H, Al-Amin AQ, Guerra JBSOA, Hurlbert M, Farooq H, Klavins M, Saroar M, Lorencova EK, Suresh J, Soares A, Morgado F, O'Hare P, Wolf F, Azeiteiro UM (2020) Climate change policies and agendas: facing implementation challenges and guiding responses. Environ Sci Policy 104:190–198. https://doi.org/10.1016/j.envsci.2019.12.001

Ardia D, Ringel E, Ekstrand VS, Fox A (2020) Adressing the decline of local news, rise of platforms, and the spread of mis- and disinformation online: a summary of current research and policy proposals. North Carolina: UNC Center for Media Law and Policy, University of North Carolina. Retrieved 24 Nov. 2021, from https://citap.unc.edu/wp-content/uploads/sites/20665/2020/12/Local-News-Platforms-and-Mis-Disinformation.pdf

Areia NP, Intrigliolo D, Tavares A, Mendes JM, Sequeira MD (2019) The role of media between expert and lay knowledge: a study of Iberian media coverage on climate change. Sci Total Environ 682:291–300. https://doi.org/10.1016/j.scitotenv.2019.05.191

Boykoff MT, Roberts JT (2007) Media coverage of climate change: current trends, strengths, weaknesses. Human Development Report 2007/8 – United Nations Development Programme Occasional paper. Human Development Report Office. New York, NY, pp 1–53. Retrieved 23 Nov. 2021, from http://hdr.undp.org/sites/default/files/boykoff_maxwell_and_roberts_j._timmons.pdf

Boykoff M, Smith J (2010) Media presentations of climate change. In: Lever-Tracy C (ed) Routledge handbook of climate change and society. Routlede International Handbooks, Abingdon, pp 210–218

Cabecinhas R, Lázaro A, Carvalho A (2008) Media uses and social representations of climate change. In: Carvalho A (ed) Communicating climate change: discourses, mediations and perceptions. Centro de Estudos de Comunicação e Sociedade, Universidade do Minho, Braga, pp 170–189

Carvalho A (2010) Media(ted) discourses and climate change: a focus on political subjectivity and (dis)engagement. Wires Clim Change 1(2):172–179. https://doi.org/10.1002/wcc.13

Carvalho A (2011) Introdução. In Anabela Carvalho (org.) As Alterações Climáticas, os Media e os Cidadãos. Grácio Editor, Coimbra, pp 9–21

Foucault M (1977) Pouvoir et Savoir. Dits et Ecrits 1976–1979, Tomo III, 1994, [2001]. Gallimard, Paris

Foucault M (1992) Genealogía 1 – Erudición y Saberes Sujetos, primera lección 7 enero de 1976. In Michel Foucault Genealogia del Racismo. Caronte Ensayos – Editorial Altamira, La Plata, Argentina, pp 13–24

Foucault M (1998) Microfísica do Poder, 13ª. Graal, Rio de Janeiro

Gobierno de Canarias (2013) Plan de Acción de la Estrategia Europa 2020 en Canarias. Retrieved 20 Nov. 2021, from https://ec.europa.eu/regional_policy/sources/policy/themes/outermost-regions/pdf/canarias_es.pdf

Hase V, Mahl D, Schäfer MS, Keller TR (2021) Climate change in news media across the globe: an automated analysis of issue attention and themes in climate change coverage in 10 countries (2006–2018). Glob Environ Chang 70:1–12. https://doi.org/10.1016/j.gloenvcha.2021.102353

IPCC (Intergovernmental Panel on Climate Change) (2007) Climate change 2007: the physical science basis. Contribution of Working Group I to the [Fourth Assessment Report of the Marquis, K.B. Averyt, M. Tignor and H.L. Miller (eds)]. Cambridge University Press, Cambridge, United Kingdom and New York, NY, USA. Retrieved 12 Nov. 2021, from http://www.ipcc.ch/pdf/assessment-report/ar4/wg1/ar4_wg1_full_report.pdf

IPCC (Intergovernmental Panel on Climate Change) (2013) Climate Change 2013: The Physical Science Basis. Contribution of Working Group I to the Fifth Assessment Report of the Intergovernmental Panel on Climate Change [Stocker TF, Qin D, Plattner G-K, Tignor M, Allen SK, Boschung J, Nauels A, Xia Y, Bex V, Midgley PM (eds)]. Cambridge University Press, Cambridge, United Kingdom and New York, NY, USA, 1535 pp. Retrieved 12 Nov. 2021, from https://www.ipcc.ch/site/assets/uploads/2018/02/WG1AR5_all_final.pdf

IPCC (Intergovernmental Panel on Climate Change) (2014) Climate Change 2014: Synthesis Report. Contribution of Working Groups I, II and III to the Fifth Assessment Report of the Intergovernmental Panel on Climate Change. Geneva: IPCC. Retrieved 11 Nov. 2021, from https://www.ipcc.ch/site/assets/uploads/2018/02/SYR_AR5_FINAL_full.pdf

IPCC (International Panel on Climate Change) (2021) Summary for Policymakers. In: Climate Change 2021: The Physical Science Basis. Contribution of Working Group I to the Sixth Assessment Report of the Intergovernmental Panel on Climate Change [Masson-Delmotte V, Zhai P, Pirani A, Connors SL, Péan C, Berger S, Caud N, Chen Y, Goldfarb L, Gomis MI, Huang M, Leitzell K, Lonnoy E, Matthews JBR, Maycock TK, Waterfield T, Yelekçi O, Yu R, Zhou B (eds)]. Cambridge University Press. In Press. Retrieved 25 Jan. 2022, from https://www.ipcc.ch/report/ar6/wg1/downloads/report/IPCC_AR6_WGI_Full_Report_smaller.pdf

Institute for European Environmental Policy (2013) Impacts of Climate Change on all European Islands: final report. IEEP, London, Brussels. Retrieved 10 Nov. 2021, from https://ieep.eu/uploads/articles/attachments/72712cb5-7d9b-4730-966b-860e0a02c914/Final_report_EP_CC_impacts_on_islands_FINAL_clean.pdf?v=63664509835

Latour B (1999) On recalling ANT. In: Law J, Hassard J (eds) Actor network theory and after. Blackwell Publishers, Oxford, pp 15–25

Latour B (2005) Reassembling the social: an introduction to actor-network theory. Oxford University Press, Oxford

Law J (1992) Notes on the theory of the actor network: ordering, strategy and heterogeneity. Syst Pract 5(4):379–393. https://doi.org/10.1007/BF01059830

Law J (1999) After ANT: complexity, naming and topology. In John Law e John Hassard (eds) Actor network theory and after. Blackwell Publishers, Oxford, pp 1–14

Leal-Filho W (2009) Communicating climate change: challenges ahead and action needed. Int J Climate Change Strategies Manag 1(1):6–18. https://doi.org/10.1108/17568690910934363

Regional and da Madeira – Secretaria Regional do Ambiente, Recursos Naturais e Alterações Climáticas (SRARNAC) (2015) Estratégia CLIMA-Madeira: Estratégia de Adaptação às Alterações Climática da Região Autónoma da Madeira. In Ana G, Avelar D, Santos F, Costa H, Garrett P (eds) Funchal: Governo Regional da RAM, Secretaria Regional do Ambiente e Recursos Naturais, OCCIAM; Intervir+

Reig R (2013) Presentación. In Rogelio Fernández Reyes (dir.) and Rosalba Mancinas Chávez (coord.) Medios de Comunicación y Cambio Climático. Sevilla: Fénix Editora, pp 7–9

Reyes RF (2013) Introducción. In Rogelio Fernández Reyes (dir.) and Rosalba Mancinas Chávez (coord.) Medios de Comunicación y Cambio Climático. Fénix Editora, Sevilla, pp 11–15

Rittel H, Webber M (1973) Dilemmas in a general theory of planning. Policy Sci 4(2):155–169. https://doi.org/10.1007/BF01405730

Santos FD (2012) Alterações Globais: Os desafios e os riscos presentes e futuros. Fundação Francisco Manuel dos Santos, Lisboa

Schmidt A, Ivanova A, Schäfer MS (2013) Media attention for climate change around the world: a comparative analysis of newspaper coverage in 27 countries. Glob Environ Chang 23(5):1233–1248. https://doi.org/10.1016/j.gloenvcha.2013.07.020

UNESCO – Cátedra UNESCO de Investigación en Comunicación Universidad Rey Juan Carlos (2021) Decalogue of Recommendations. Retrieved 26 Nov. 2021, from https://www.comunesco.com/decalogo-de-recomendaciones/?lang=en

UNFCCC (1997) Kyoto protocol to the United Nations framework convention on climate change. Retrieved 01 Dec. 2021, from https://treaties.un.org/doc/Treaties/1998/09/19980921%2004-41%20PM/Ch_XXVII_07_ap.pdf

United Nations (2015) Paris agreement. Retrieved 01 Dec. 2021, from https://unfccc.int/files/essential_background/convention/application/pdf/english_paris_agreement.pdf

Chapter 23
Indigenous Knowledge of Artisanal Fisherfolks on Climate Change Adaptation in Ondo State, Nigeria

Mosunmola Lydia Adeleke and Johannes M. Luetz

Abstract This research investigated the Indigenous knowledge of artisanal fisherfolks in Ondo State, Nigeria. A multistage sampling technique was used. Two coastal communities (Aiyetoro and Igbobini) and two riverine inland communities (Ogbese and Owena) were purposefully selected because of extensive fishing activities in the region. Fifty respondents were randomly chosen from each fishing community and interviewed using a structured questionnaire and interview schedule. A total of two hundred respondents provided information on matters related to climate change adaptation. All respondents (85% male; 63% married) were literate and answered questions based on their geographical locations and personal experiences. The results revealed a high awareness of climate change among the respondents. The results also revealed that the methods of adapting to climate change vary across the communities and are informed by local Indigenous knowledge and practices. The impacts of climate change were most acutely felt by the communities via adverse effects on their fishing business. Some fisherfolk indicated using "spiritual approaches" and "prayers" to adapt to the effects of climate change. Outlining unique Indigenous perceptions and perspectives, this mixed methods study presents Indigenous practices of adapting to climate change, along with arising opportunities for further scientific research.

M. L. Adeleke
Department of Fisheries and Aquaculture Technology, The Federal University of Technology, Akure, Nigeria
e-mail: mladeleke@futa.edu.ng

J. M. Luetz (✉)
Graduate Research School, Alphacrucis University College, Brisbane, QLD, Australia
e-mail: johannes.luetz@ac.edu.au; j.luetz@unsw.edu.au; jluetz@usc.edu.au

School of Social Science, University of New South Wales, Sydney, NSW, Australia

School of Law and Society, University of the Sunshine Coast, Maroochydore, QLD, Australia

W. Leal Filho et al. (eds.), *Climate Change Strategies: Handling the Challenges of Adapting to a Changing Climate*, Climate Change Management,
https://doi.org/10.1007/978-3-031-28728-2_23

Introduction

Indigenous knowledge has been used in many parts of sub-Saharan Africa (SSA) to conserve the environment and promote higher yields in agriculture. It has also helped in developing a variety of measures to adapt to climate change, such as growing drought-tolerant and early maturing Indigenous crops, gathering wild fruits and vegetables, cultivating wetlands, and diversifying and trading livestock (Abate and Kronk 2013; Ford et al. 2020). Despite these benefits, the importance of Indigenous knowledge in responding to local climate change challenges is sometimes not adequately recognized or leveraged for effective adaptation to climate change (Allison et al. 2009; Akinola et al. 2019, 2020; Hosen et al. 2020; Leal Filho et al. 2021a). This can limit adaptive capacity in the face of progressive climate change impacts, including droughts, storms, flooding of crop fields, loss of fertile soil, and wind-blown damage to vulnerable crops, among others (IPCC 2007, 2022; Leal Filho et al. 2021b, c).

Research has emphasized the benefits of so-called "reversals of learning" whereby Indigenous communities, which are sometimes denigrated simply as being poor, rather "teach the profligate and so-called 'developed' rich about the interwoven nature of frugality, modesty, contentedness, spirituality and sustainability" (Luetz et al. 2019, p. 132). Given that Indigenous peoples have lived sustainably for thousands of years, the time is ripe to solicit their knowledge and worldview understandings for climate change adaptation purposes (Ellis et al. 2021; Fischer et al. 2022; Walshe and Nunn 2012; Leal Filho et al. 2021a; Sultana and Luetz 2022; Yunkaporta 2019). Crucially, Indigenous knowledge can make a vital contribution to climate forecasting (Balehegn et al. 2019; Chisadza et al. 2015). The opportunity costs of not harnessing Indigenous knowledge can limit local food security and negatively impinge on biodiversity, freshwater resources, natural ecosystems, and human health (Ford et al. 2020; Adeleke and Omoboyeje 2016; McKenzie et al. 2021; Sadiku and Sadiku 2011).

Indigenous knowledge is local grassroots expertise that has been built upon and passed on from generation to generation via word of mouth (UNESCO 2021; Osunade 1994). Many rural communities' decision-making is based on this expertise. Indigenous knowledge is lived experience that is informed by past generations' observations and experiments, supports local weather forecasting, and creates a natural link to people's surroundings and environments (Balehegn et al. 2019; Iticha and Husen 2019; Adesina et al. 1999). Moreover, Indigenous knowledge is valuable not only to the communities in which it develops but also to scientists and planners working to improve rural conditions (Makondo and Thomas 2018; Mundy and Compton 1991).

Climate change has a significant impact on African fisheries and threatens the lives of over 200 million people who rely on fishing and aquaculture for their livelihoods (Fregene and Ogunika 2013; Adeleke and Wolff 2016). This is evidenced by impacts on artisanal fishing, which is crucial to the livelihoods and food security of many communities in developing countries like Nigeria. Fishing is undertaken primarily for subsistence, local and small markets and generally employs traditional techniques and small boats (Béné 2006; Adeleke and Fagbenro 2013). Set against this background context, it is beneficial to research the Indigenous knowledge utilized by artisanal fisherfolks in adapting to progressive climate change and its adverse consequences.

The field research sites for this study were chosen because of the high level of fishing activity in these areas; these locations supported excellent data collection possibilities. The primary aim of this research is to better understand the Indigenous knowledge on climate change adaptation in artisanal fishing in Ondo State, Nigeria. Its specific objectives are to determine the:

- Fisherfolks' perception of climate change;
- Local communities' awareness of climate change;
- Socio-economic characteristics of the respondents in each local government area.

This chapter is organised as follows. First, we introduce the study's theoretical framework. Next, we discuss the research methodology, field study locations, and sampling techniques. Thereafter, we present the study's results, which we analyse in terms of their significance for local climate change adaptation. Data analysis will also explore the linkages between Indigenous knowledge, spirituality, and the capacity of local communities to adapt to climate change. Finally, we recapitulate the study's main findings and discuss key lessons and future opportunities.

Theoretical Framework

Agriculture is one of the most important economic sectors in Nigeria, accounting for 23.4% of the country's GDP (World Bank 2022). Over recent years, the sector has been negatively affected by climate change, which has resulted from a range of human activities, including the burning of fossil fuels, wood, solid wastes, and wood products; combustion of solid wastes and fossil fuels in industrial and agricultural activities; livestock raising and organic waste decomposition in solid waste landfills; bush burning; and deforestation (Idowu et al. 2011; Apata et al. 2009). The Intergovernmental Panel on Climate Change (IPCC 2007, 2022) has recurrently warned of the adverse impacts of climate change on ecosystems, including fisheries (Bindoff et al. 2019). Pertinent challenges include impacts on fish production, the supply of fishmeal and fish oils, and future aquaculture production, as well as arising knock-on effects. Such secondary impacts may be experienced via diminishing fishmeal or fish oils, particularly if stocks of species used in fishmeal production are negatively affected by climate change and live-fish production. Furthermore, according to Adeleke and Wolff (2016), climate change can significantly impact the productivity and distribution of fishery resources in both freshwater and marine environments. The confluence of intersecting climate change issues makes small to medium coastal towns and riverine communities fertile case study locations for grassroots-informed data collection and analysis (Doust et al. 2021; Sultana and Luetz 2022).

Research Methodology: Study Locations and Sampling Techniques

This study was carried out in Ondo State, Southwest Nigeria, where the mangrove and rainforest zones converge. Mean annual rainfall ranges from 2,000 to 3,000 mm, and the climate is characterized by significant seasonal variation in humidity. Temperatures range from 17.5 to 27 °C. Ondo State has a geographical area of 14,793 km^2 and a population of about 3,460,877[1] (National Bureau of Statistics 2011). According to Oseni (2010), agricultural livelihoods include fish farming/ aquaculture, food crops (e.g., cocoyam, maize, pepper, plantain, sweet potato, and tomatoes), and cash crops (e.g., timber and cocoa).

The four study locations for this field research are depicted in Figs. 23.1, 23.2, 23.3, and 23.4 (Local Government Areas are in parentheses) and include Aiyetoro (Ilaje), Igbobini (Ese-Odo), Ogbese (Akure North), and Owena (Oriade). Aiyetoro (Fig. 23.1) and Igbobini (Fig. 23.2) are located at or near the coast, and Ogbese (Fig. 23.3) and Owena (Fig. 23.4) are riverine water areas that are located further inland. Ogbese is located about 10 km east of Akure, the Ondo State capital. Notably, Owena (7°24′11.3″N 5°00′52.5″E) has a distinctive feature – this village is dissected by the Ondo State border and has a small part of the community in Osun State. Apart from this coincidental incongruity, all research sites are located in Ondo State, Nigeria. This includes the river in Owena, which is located entirely in Ondo State (Fig. 23.5).

The Ogbese River has its source at Ayede-Ekiti in Ekiti State and flows via Ogbese, Ondo State, and Edo State (Fig. 23.6). Ilaje and Ese-Odo are two of the State's eighteen local government units. They are bounded on the north by the Ikale Local Government Area (LGA) and on the south by the Atlantic Ocean, with an 80-km-long coastline. Because people in Aiyetoro (Fig. 23.7) have easy access to the sea, fishing is the principal source of income. Aiyetoro and its neighbouring communities are considered among the most important fishing locations along the coast, with rich biodiversity that includes shellfish (e.g., crabs, lobster, shrimp, and gastropods), fish, reptiles, and other species.

A mixed methods approach weighted in favour of quantitative analysis was used for this study (Babbie 2021; Bryman 2016). In Part A of the study, a structured questionnaire was disseminated among fisherfolks from the four fishing communities to

[1] Most recent official figures available. Projections by the Ondo State Bureau of Statistics put the estimated State population at approximate 5.2 million. http://www.ondostatistics.org/ondo_profile.php.

Fig. 23.1 Aiyetoro (Coordinates: 6.105100, 4.777900) https://goo.gl/maps/jY8SZVmkjGfrVud3A

collect primary survey data. In Part B of the study, interviews, focus group discussions, and physical observations were used to supplement the survey data and check for accuracy and reliability (Hesse-Biber 2017; Punch 2014). Fifty respondents were purposely recruited from each community and given a structured questionnaire to complete, resulting in a total of 200 respondents from all the communities. The questionnaire also formed the basis for the interviews and focus group discussions, which typically lasted about 2–3 hours, depending on the availability and engagement of the participants (Figs. 23.8 and 23.9). These methodological approaches were employed because they are suited to Indigenous settings and because some discussions digressed from the questions in the questionnaire (Kovach 2018; Usman

Fig. 23.2 Igbobini (Coordinates: 6.510500, 4.820000) https://goo.gl/maps/KdEWxq51YD9V9xC69

2010). Data collection took about 4–5 months (about one month in each community) and occurred during the period November 2020–March 2021. This duration allowed for proper familiarization and interaction with the communities. Data collection was supported by an extension worker affiliated with The Federal University of Technology Akure (FUTA). Data gathered from the respondents were analysed using descriptive statistics and content analysis and produced insightful grassroots perspectives (Babbie 2021; Creswell and Creswell 2018).

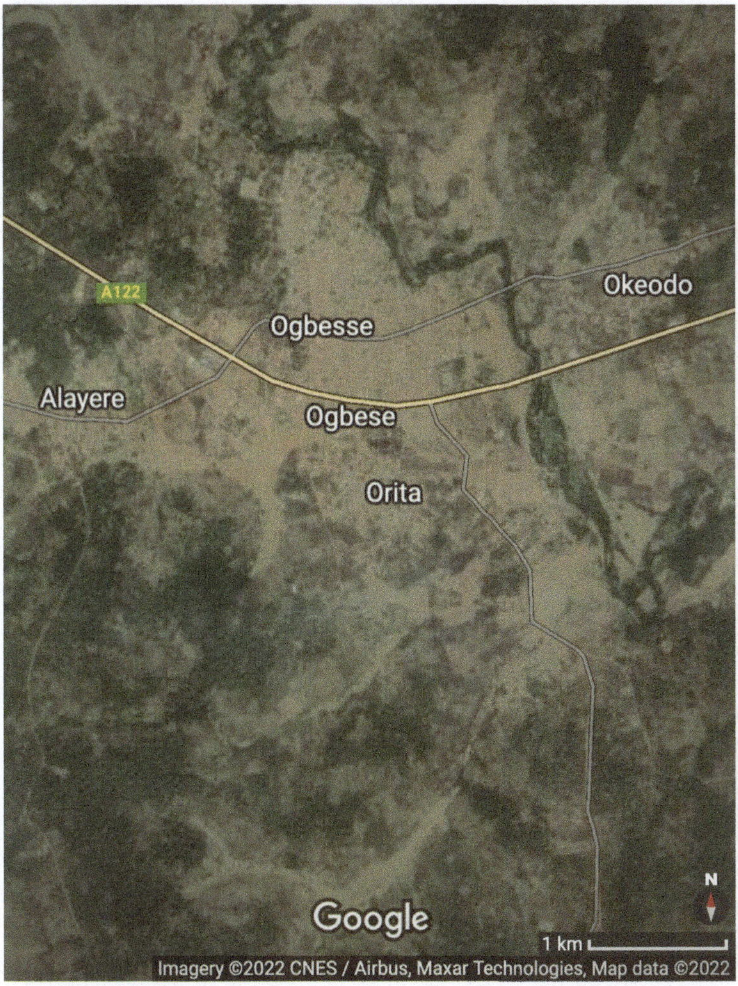

Fig. 23.3 Ogbese (Coordinates: 7.259086, 5.372463) https://goo.gl/maps/8y958yhdsd5mQvxt5

The study was approved by the Ethical Committee for Human Research and Laboratory Animals of the School of Agriculture and Agricultural Technology, The Federal University of Technology, Akure, Nigeria (Ref. No. FUTA/SAAT/07-2020-21/023). The field survey was conducted according to the human and animal use and care regulations of the Canadian Council on Animal Care Guidelines and Protocol Review (Canadian Council on Animal Care 1993/2020).

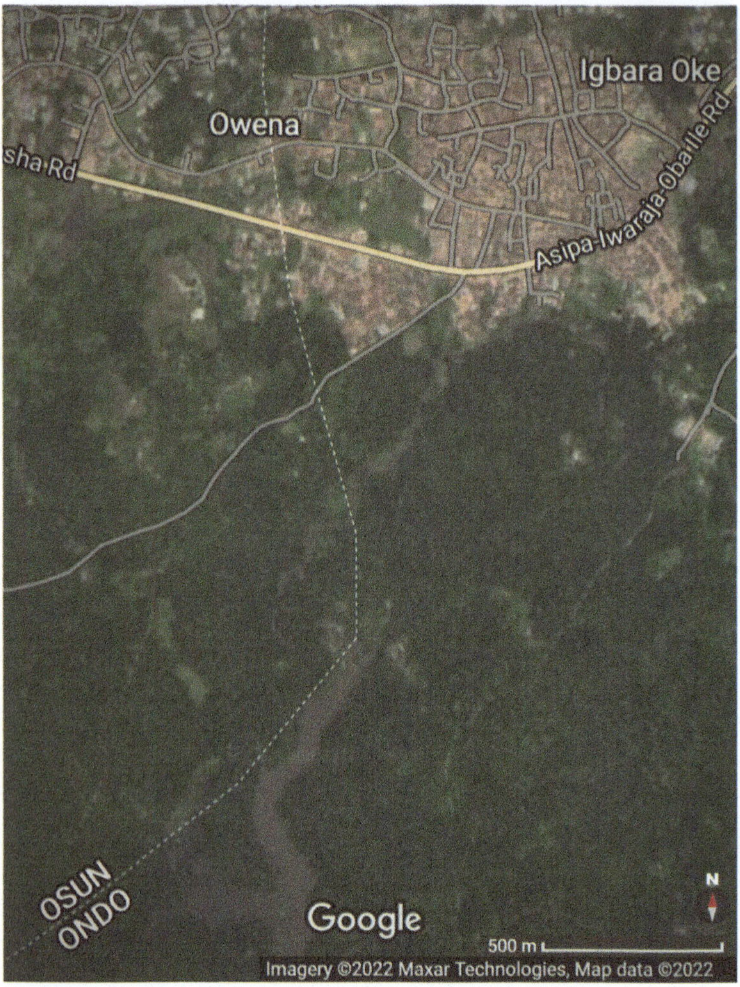

Fig. 23.4 Owena (Coordinates: 7.403135, 5.014589) https://goo.gl/maps/2PCLDP5VoEuMp5cj8

Fig. 23.5 Owena riverbank; 2021 Field Survey (*Photo* Fawale Iyanu)

Fig. 23.6 Ogbese river; 2021 Field Survey (*Photo* Fawale Iyanu)

Fig. 23.7 Freshwater body in Aiyetoro; 2021 Field Survey (*Photo* Fawale Iyanu)

Results and Key Findings

Demographic and Socio-Economic Respondent Characteristics

Table 23.1 reflects the distribution of respondent ages, gender, religion, marital status, occupations, and education in both coastal areas (Aiyetoro and Igbobini) and inland water areas (Ogbese and Owena). Most respondents were younger (18–29 years) in Igbobini (62%) and Obese (56%) and slightly older (30–49 years) in Aiyetoro (54%) and Owena (58%). Across all communities, elderly respondents (50–59 years) had the lowest representation; Aiyetoro (6%), Igbobini (8%), Ogbese (10%), and Owena (8%). Most of the respondents in the four communities were males; Aiyetoro (84%), Igbobini (76%), Ogbese (88%), and Owena (92%). In respect of religion, most respondents were Christian in the coastal areas of Aiyetoro (100%) and Igbobini (68%), and most respondents were Muslim in the inland riverine areas of Ogbese (96%) and Owena (94%). Out of all study areas, there were only two respondents in Igbobini (4%) who professed an exclusively "traditionalist" belief. Notwithstanding,

Fig. 23.8 Igobini community members responding to field enumerator and filling in questionnaires; 2021 Field Survey (*Photo* Ogunyemi Kola)

elements of syncretism (blended beliefs consisting of amalgamated spiritual traditions and Indigenous practices) cannot be ruled out in the communities. In Aiyetoro, Igbobini, Ogbese, and Owena, 54%, 40%, 80%, and 78% of the respondents were married, while the remainder was single (46%, 56%, 20%, and 22%) or divorced (4% in Igbobini).

As expected, fishing was the dominant livelihood in all field research areas, although Igbobini showed a slightly more diversified labour force. In Aiyetoro, Igbobini, Ogbese, and Owena, 76%, 40%, 84%, and 90% of the respondents were artisanal fisherfolks; and 18%, 32%, 16%, and 10% were students. A minority of respondents were traders in Aiyetoro (6%) and Igbobini (12%), and only 8% of respondents pursued other occupations in Igbobini. In respect of the highest level of education, most respondents in both the coastal (Aiyetoro and Igbobini) and inland riverine areas (Ogbese and Owena) had undertaken secondary or tertiary education.

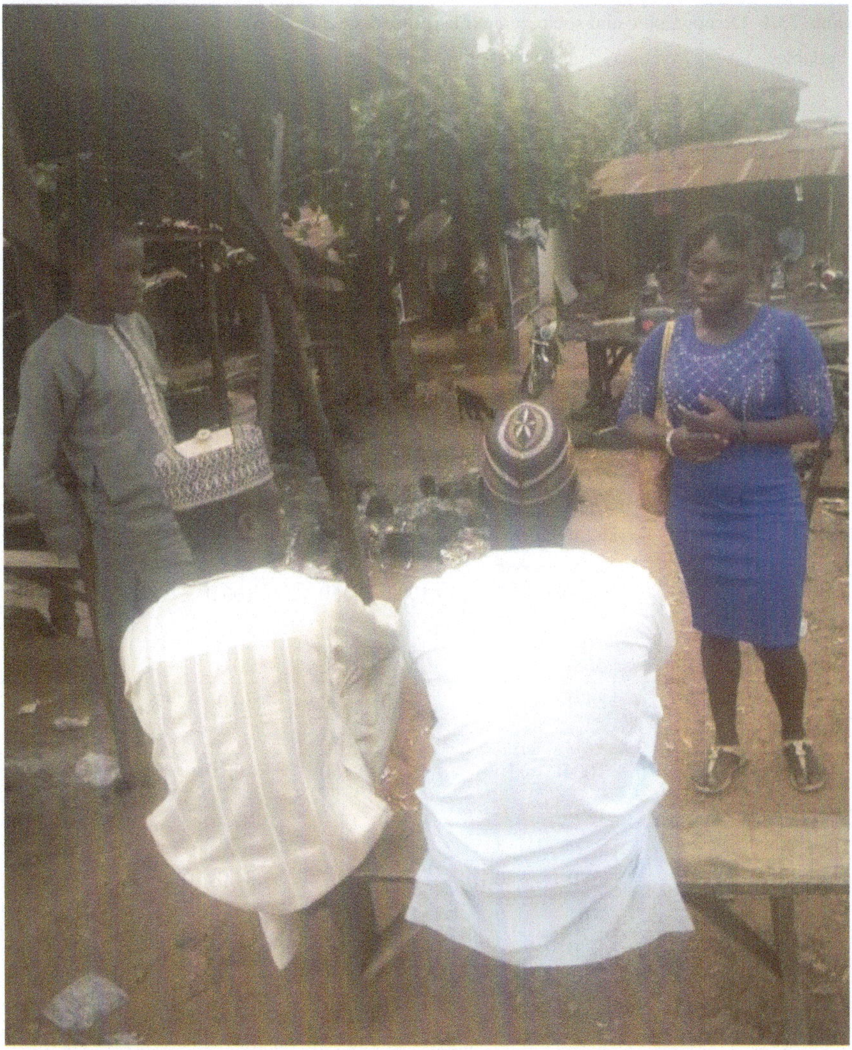

Fig. 23.9 Respondents from Owena; 2021 Field Survey (*Photo* Ogunyemi Kola)

Table 23.1 Demographic and socio-economic respondent characteristics

Variables	Aiyetoro		Igbobini		Ogbese		Owena	
	Freq.	%	Freq.	%	Freq.	%	Freq.	%
Age (years)								
18–29	20	40.0	31	62.0	28	56.0	17	34.0
30–49	27	54.0	15	30.0	17	34.0	29	58.0
50–59	3	6.0	4	8.0	5	10.0	4	8.0
Total	50	100.0	50	100.0	50	100.0	50	100.0
Gender								
Male	42	84.0	38	76.0	44	88.0	46	92.0
Female	8	16.0	12	24.0	6	12.0	4	8.0
Total	50	100.0	50	100.0	50	100.0	50	100.0
Religion								
Christianity	50	100.0	34	68.0	2	4.0	3	6.0
Islam	–	–	14	28.0	48	96.0	47	94.0
Traditionalist	–	–	2	4.0	–	–	–	–
Total	50	100.0	50	100.0	50	100.0	50	100.0
Marital status								
Single	23	46.0	28	56.0	10	20.0	11	22.0
Married	27	54.0	20	40.0	40	80.0	39	78.0
Divorce	–	–	2	4.0	–	–	–	–
Total	50	100.0	50	100.0	50	100.0	50	100.0
Occupation								
Government	–	–	4	8.0	–	–	–	–
Artisanal	38	76.0	20	40.0	42	84.0	45	90.0
Trader	3	6.0	6	12.0	–		–	–
Student	9	18.0	16	32.0	8	16.0	5	10.0
Others	–	–	4	8.0	–	–	–	–
Total	50	100.0	50	100.0	50	100.0	50	100.0
Education								
Primary	–	–	6	12.0	8	16.0	6	12.0
Secondary	15	30.0	24	48.0	33	66.0	34	68.0
Tertiary	35	70.0	20	40.0	9	18.0	10	20.0
Total	50	100.0	50	100.0	50	100.0	50	100.0

Key Frequencies (Freq.) and Percentages (%)
Data Source 2021 Field Survey

Table 23.2 Sources of climate change awareness

Variables	Aiyetoro		Igbobini		Ogbese		Owena		Pooled	
	Freq.	%	Freq.	%	Freq.	%	Freq.	%	Freq.	%
Television	30	60.0	21	42.0	17	34.0	17	34.0	85	42.5
Radio	4	8.0	14	28.0	24	48.0	22	44.0	64	32.0
Newspaper	6	12.0	5	10.0	–	–	2	4.0	13	6.5
Internet	10	20.0	5	10.0	7	14.0	6	12.0	28	14.0
Others	–	–	5	10.0	2	4.0	3	6.0	10	5.0
Total	50	100.0	50	100.0	50	100.0	50	100.0	200	100.0

Key Frequencies (Freq.) and Percentages (%)
Data Source 2021 Field Survey

Sources of Climate Change Awareness

Table 23.2 reflects the media sources through which respondents first became aware of climate change. In Aiyetoro, Igbobini, Ogbese, and Owena, respondents learned about climate change through television (60%, 42%, 34%, and 34%), radio (8%, 28%, 48%, and 44%), newspapers (12%, 10%, 0%, and 4%), the internet (20%, 10%, 14%, and 12%), and other media sources (0%, 10%, 4%, and 6%). Informal connections with friends and at school were reported as additional sources of climate change information by a small percentage of the respondents.

Time of Climate Change Awareness

Table 23.3 reflects the point in time when respondents first became aware of climate change. In Aiyetoro, Igbobini, Ogbese, and Owena, respondents started observing climate change in the past 30 years (16%, 8%, 6% and 0%), in the past 20 years (40%, 14%, 8%, and 6%), in the past 10 years (14%, 52%, 32%, and 24%), in the past 5 years (24%, 24%, 38%, and 38%), and in 'Other' time periods (6%, 2%, 16%, and 32%), which was understood to indicate a timeframe during the last couple of years. The results show that most of the respondents in the coastal areas have been aware of climate change for more extended time periods (10–30 years), whereas most of the respondents in the inland water areas have become aware of climate change more recently (1–10 years). This finding is consistent with the anecdotal accounts of coastal dwellers that sea level rises have caused the sea to move closer to their coastal communities. The data also suggest that climate change awareness has grown over recent years, which may be attributed to both recent impacts and relevant media coverage, as noted above.

Table 23.3 Time of climate change awareness in the study areas

Variables	Aiyetoro		Igbobini		Ogbese		Owena		Pooled	
	Freq.	%	Freq.	%	Freq.	%	Freq.	%	Freq.	%
30 years	8	16.0	4	8.0	3	6.0	–	–	15	7.5
20 years	20	40.0	7	14.0	4	8.0	3	6.0	34	17.0
10 years	7	14.0	26	52.0	16	32.0	12	24.0	61	30.5
5 years	12	24.0	12	24.0	19	38.0	19	38.0	62	31.0
Others	3	6.0	1	2.0	8	16.0	16	32.0	28	14.0
Total	50	100.0	50	100.0	50	100.0	50	100.0	200	100.0

Key Frequencies (Freq.) and Percentages (%)
Data Source 2021 Field Survey

Climate Change and Indigenous and Local Knowledge

Most respondents in Aiyetoro believed that climate change is real and is already affecting them through impacts on their fish farming and day-to-day activities. They also explained that the sea used to be 30 km away but has moved closer to their community because of climate change. Relatedly, fisherfolks in Aiyetoro stated that the sea had infiltrated the community's freshwater bodies, altering the flavour of some freshwater fish (Fig. 23.7). Some people stated that climate change is primarily due to the rising levels of atmospheric carbon dioxide caused by the usage of fossil fuels and that it impacts on both the biotic and abiotic factors of the environment, as well as the transportation of products and people. Some respondents also linked climate change to rough seas, which they said can hamper their fishing activities.

Most respondents in Igbobini said global warming was caused by human-induced emissions of greenhouse gases and had produced large-scale shifts in weather patterns. Some respondents reported that climate change is already affecting both farming productivity and health conditions. Others reported that climate change was mainly evidenced by warming temperatures and changes in precipitation. Notably, some respondents expressed the belief that climate change may be caused by a lack of traditional sacrifices. In their view, failure to perform annual sacrifices to appeal to their deities in nature may trigger possible punishment. Consequently, some respondents interpreted the adverse effects of climate change as a kind of chastisement or discipline.

In Ogbese, respondents expressed disparate perspectives on the causes and effects of climate change. Many stated the belief that climate change was caused by God and that the consequences could not be averted. Some fishermen believed that, because of future climate change, fish might even be more likely to come out to be caught. Others reported connections through increasing or decreasing sales, while yet others expressed the belief that climate change might even have an overall positive future impact on their business.

Table 23.4 Indigenous knowledge and climate change

Variables	Aiyetoro		Igbobini		Ogbese		Owena		Pooled	
	Freq.	%	Freq.	%	Freq.	%	Freq.	%	Freq.	%
Yes	–	–	13	26.0	8	16.0	8	16.0	29	14.5
No	50	100.0	37	74.0	42	84.0	42	84.0	171	85.5
Total	50	100.0	50	100.0	50	100.0	50	100.0	200	100.0

Key Frequencies (Freq.) and Percentages (%)
Data Source 2021 Field Survey

In Owena, some fisherfolks felt that climate change was raising the water levels and the size of water bodies and that it varied by season and time; others reported that it had both positive and negative impacts on their fishing business. Based on some of the responses, there was a sense that some interviewees may not have been clearly distinguishing in their responses between climate and climate change. Overall, there was a sense that while local awareness of climate change was generally high, substantive proficiency about climate science was limited and informed largely by individual experiences and levels of education.

Indigenous Knowledge and Climate Change

As shown in Table 23.4, all communities were hesitant to admit any immediate linkages between Indigenous knowledge and climate change. This sense was most pronounced in Aiyetoro, where all respondents expressed such caution (100%). In Igbobini (74%), Ogbese (84%), and Owena (84%), sentiments were similarly cautious but more nuanced. Our analysis suggests that community members are not straightforwardly associating traditional Indigenous knowledge with climate change issues. Even so, discussions revealed that many Indigenous practices are nevertheless aimed at adapting to climate change.

Indigenous Practices and Climate Change

Table 23.5 reflects Aiyetoro as the only community that did not admit to engaging in any Indigenous practices that are unequivocally linked to climate change adaptation. Notwithstanding this apparent reticence in Aiyetoro (0%), respondents in Igbobini (88%), Ogbese (86%), and Owena (94%) overwhelmingly admitted using a range of Indigenous practices to adapt to climate change. Our analysis suggests that each community adapts to climate change in its own unique way based on Indigenous traditions, worldviews, and belief systems; these may include reverence for ancestors and respect for water spirits, especially the deity *Omalokun* (the god of the Atlantic).

Table 23.5 Indigenous practices and climate change

Variables	Aiyetoro		Igbobini		Ogbese		Owena		Pooled	
	Freq.	%	Freq.	%	Freq.	%	Freq.	%	Freq.	%
Yes	–	–	44	88.0	43	86.0	47	94.0	90	45.0
No	50	100.0	6	12.0	7	14.0	3	6.0	110	55.0
Total	50	100.0	50	100.0	50	100.0	50	100.0	200	100.0

Key Frequencies (Freq.) and Percentages (%)
Data Source 2021 Field Survey

Analysis: Indigenous Practices Linked to Climate Change

As noted above, Aiyetoro is a community with a majority Christian worldview orientation. Most community members stated that they believe in only one God, the God of the Bible, to whom alone they address their prayers. Hence respondents from this community reported that because they identify as Christians, they do not observe any rituals in the community, and the only Indigenous practice they are aware of is praying to God. Contrastingly, in Igbobini (the other community in coastal proximity that also exhibits a majority Christian worldview orientation), some people also worship a sea goddess known as *Malokun* and *Oboroweh* (cf. Omojeje and Adu-Peters 2021). Furthermore, members of this community expressed the belief that water spirits are like humans in that they have personal strengths and weaknesses and that humans live among water spirits before birth. In Igbobini, the duty of prayer sanctions living in the good graces of the water spirits (Table 23.6). Moreover, in the literature, Omojeje and Adu-Peters (2021) report Indigenous worship of *Malokun* elsewhere along the Ilaje coast: "whenever the people are hungry they usually beckon on *Malokun* to come to their rescue and this is usually done through songs" (p. 227).

In Ogbese and Owena, respondents indicated observing similar Indigenous practices. Both communities are Hausas and pursue very similar fishing activities. Some of their Indigenous practices are summarized below (Table 23.7), including calabash use for fishing (Fig. 23.10).

Indigenous Practices and Climate Change on Water Bodies

As reflected in Table 23.8, respondents in Aiyetoro (100%) indicated that there were no discernible connections between climate change, local water bodies and Indigenous practices in their community. Contrastingly, respondents in Igbobini (8%), Ogbese (16%), and Owena (40%) reported that there were at least some connections between climate change and their water bodies that could be mitigated by Indigenous practices. Changes were mainly observed via excessive rainfall or changes in the colouration of the water (e.g., from freshwater to dirty water or from white to black water), including changes in water levels (Figs. 23.5 and 23.6).

Table 23.6 Indigenous practices in Aiyetoro and Igbobini

Indigenous practices	Indigenous beliefs and meanings
Malokun	People call on this spirit and believe that it helps them to catch an abundance of fish
Tuke (water lettuce)	People offer prayers to their gods while placing water lettuces on top of water bodies. People believe that this practice refreshes the water, allowing them to harvest different types of fish species
Oboroweh	This practice involves all the spirits in a water body. Before appeasing *Boroweh*, people summon the *Malokun* spirit. People believe that by appeasing the gods, the spirits will offer them guidance
Boabo	During this festival in December people are not allowed to fight. They make use of palm fronts, dance in the streets, and offer songs and prayers to their gods at the riverbank. After doing this, people believe that the spirits will come and rejoice with them
Ogunbebe	People believe that *Ogunbebe* is a spirit summoned from the water through the beat of a drum. The spirit will appear in human form and then dance on land without resting for three days. After the third day, the spirit will return to the water body. During the dance period, no one is allowed to go fishing. Then, after the spirit has returned, there will be peace in the land and an abundance of fish to catch
Traps	Traps are placed on the riverbank at night and stay in place until the following day when the fisherfolks will check whether the traps have caught any fish

Data Source 2021 Field Survey

Table 23.7 Indigenous practices in Ogbese and Owena communities

Indigenous practices	Indigenous beliefs and meanings
Calabash	The calabash is placed on the water body. The practice helps to collect 'trophies', whereby fish that are caught are kept inside the calabash, which also supports transportation. This practice helps to catch fish
Different hooks and lines	The selection of different hooks and lines depends mainly on the season
Different net sizes	Different net sizes are used to catch different sizes of fish

Source 2021 Field Survey

Festival Practices and Climate Change

A similar picture emerged in respect of the linkages between festival practices and climate change (Table 23.9). Once again, respondents in Aiyetoro (100%) indicated that there was no discernible connection between climate change and festival practices. On the other hand, respondents in Igbobini (16%), Ogbese (18%), and Owena (18%) reported that, in their view, there were at least some connections between climate change and their festival practices. More specifically, respondents in Igbobini

Fig. 23.10 Calabash use for fishing in Ogbese; 2021 Field Survey (*Photo* Fawale Iyanu)

Table 23.8 Indigenous practices and climate change on water bodies

Variables	Aiyetoro		Igbobini		Ogbese		Owena		Pooled	
	Freq.	%	Freq.	%	Freq.	%	Freq.	%	Freq.	%
Yes	–	–	4	8.0	8	16.0	20	40.0	32	16.0
No	50	100.0	46	92.0	42	84.0	30	60.0	168	84.0
Total	50	100.0	50	100.0	50	100.0	50	100.0	200	100.0

Key Frequencies (Freq.) and Percentages (%)
Data Source 2021 Field Survey

Table 23.9 Festival practices and climate change

Variables	Aiyetoro		Igbobini		Ogbese		Owena		Pooled	
	Freq.	%	Freq.	%	Freq.	%	Freq.	%	Freq.	%
Yes	–	–	8	16.0	9	18.0	9	18.0	26	13.0
No	50	100.0	42	84.0	41	82.0	41	82.0	174	87.0
Total	50	100.0	50	100.0	50	100.0	50	100.0	200	100.0

Key Frequencies (Freq.) and Percentages (%)
Data Source 2021 Field Survey

reported that some of their festival practices were linked to climate and climate change through rainfall during festival periods. Respondents seemingly interpreted this nexus as answers to prayers related to and arising from festival periods and practices.

Festival Frequencies

Table 23.10 reflects the frequencies of festivals in the communities. The results show that all important festival events occur on an annual basis. In Aiyetoro, the community observes an annual church anniversary on January 12, in addition to celebrating twin festival events in July. Festival observance involves community members gathering in a big cathedral and singing worship and praise songs to God. In Igbobini, the community celebrates a different festival every year in December (*Boabo*; see Table 23.6). Given that Ogbese and Owena are located in Muslim-majority areas, respondents in these communities reported observing only the Sallah Festival.

Table 23.10 Festival frequencies

Variables	Aiyetoro		Igbobini		Ogbese		Owena		Pooled	
	Freq.	%	Freq.	%	Freq.	%	Freq.	%	Freq.	%
Annually	50	100.0	50	100.0	50	100.0	50	100.0	200	100.0
Biannually	–	–	–	–	–	–	–	–	–	–
Monthly	–	–	–	–	–	–	–	–	–	–
Others	–	–	–	–	–	–	–	–	–	–
Totals	50	100.0	50	100.0	50	100.0	50	100.0	200	100.0

Key Frequencies (Freq.) and Percentages (%)
Data Source 2021 Field Survey

Table 23.11 Performing rituals to catch fish

Variables	Aiyetoro		Igbobini		Ogbese		Owena		Pooled	
	Freq.	%	Freq.	%	Freq.	%	Freq.	%	Freq.	%
Yes	44	88.0	38	76.0	35	70.0	40	80.0	157	78.5
No	6	12.0	12	24.0	15	30.0	10	20.0	43	21.5
Total	50	100.0	50	100.0	50	100.0	50	100.0	200	100.0

Key Frequencies (Freq.) and Percentages (%)
Data Source 2021 Field Survey

Performing Rituals to Catch Fish

Although the nexus between festivals and climate change appeared to be somewhat tenuous, as shown above, respondents overwhelmingly admitted performing a range of rituals to catch more fish. Table 23.11 reflects that a clear majority of respondents in all communities perform rituals intended to ameliorate their fishing operations; Aiyetoro (88%), Igbobini (76%), Ogbese (70%), and Owena (80%). In all communities, respondents admitted performing rituals through prayers. In Igbobini, some respondents additionally conceded performing rituals and ceremonies by offering sacrifices to their gods. In summary, fisherfolks recurrently and openly reported employing a range of "spiritual approaches" and offering "prayers" to adapt to the effects of climate change. Most respondents concurred with sentiments reported in other research, stating: "these changes are brought about by God and [we] can only pray for mercies" (Nzeadibe et al. 2011, p. 21).

New Fish Species and Climate Change

A considerable proportion of respondents indicated that, based on their observations, climate change is altering the composition of fish species in their environment. Table 23.12 reflects that in all communities, respondents thought climate change was contributing to the availability of at least some new fish species in their water bodies, including in Aiyetoro (20%), Igbobini (10%), Ogbese (4%), and Owena (22%). Even so, respondents agreed that overall, the availability of fish species was mainly declining because of climate change and that some species had already become extinct. The findings from this study agree with other research that aquatic organisms can be highly vulnerable to climate change (Bindoff et al. 2019). Notably, the vulnerabilities of fish species to climate change have been attributed to the average temperatures of both air and water changing in tandem, thus putting fish species under progressive stress and reducing their reproductive potential (Adeleke and Fagbenro 2013; Tedesco et al. 2013).

Table 23.12 New fish species and climate change

Variables	Aiyetoro		Igbobini		Ogbese		Owena		Pooled	
	Freq.	%	Freq.	%	Freq.	%	Freq.	%	Freq.	%
Yes	10	20.0	5	10.0	2	4.0	11	22.0	28	14.0
No	40	80.0	45	90.0	48	96.0	39	78.0	172	86.0
Total	50	100.0	50	100.0	50	100.0	50	100.0	200	100.0

Key Frequencies (Freq.) and Percentages (%)
Data Source 2021 Field Survey

Discussion and Conclusion

This empirical study found that fisherfolks in Ondo State have a high overall awareness of climate change. This was confirmed through mixed methods research that engaged 200 participants in four communities, two coastal communities (Aiyetoro and Igbobini) and two inland riverine communities (Ogbese and Owena). Conducting on-site research over a period of 4–5 months from November 2020 to March 2021 and interviewing 50 participants in each community, 200 in total, this study found that respondents perceive the effects of climate change principally through how it affects them through their fishing operations. Many seemingly believe that observable changes in the climate and environment are brought about by God. Such sentiments were recurrently expressed through statements such as, "all we can do now is pray for mercy". Accordingly, to adapt to the effects of climate change, many fisherfolks use a mix of Indigenous knowledge, "spiritual approaches", and "prayers". Their adaptation practices are informed by each community's own unique traditions, worldviews, and belief systems. Importantly, although climate change awareness was found to be generally high in all communities, local understandings of climatic change did not seem to have a strong basis in climate science. Therefore, by induction, it seems helpful to complement Indigenous knowledge surrounding risk reduction methods with climate scientific information, which could be disseminated to the local communities (Chisadza et al. 2015; Walshe and Nunn 2012). On the other hand, it is essential to harness the benefits of Indigenous knowledge through so-called "reversals of learning" so that Indigenous knowledge, values, and practices may also "teach the profligate and so-called 'developed' rich [elsewhere] about the interwoven nature of frugality, modesty, contentedness, spirituality and sustainability" (Luetz et al. 2019, p. 132). Hence there is an argument that Indigenous perspectives, values, and worldview understandings have an essential and growing role to play within the evolving global climate change adaptation agenda (Balehegn et al. 2019; Nalau et al. 2018; Nunn and Luetz 2021).

Over recent years there have been repeated calls for more holistic approaches to climate change adaptation that better integrate the domains of science, spirituality, and sustainability. These calls have been underpinned chiefly by the argument that holistic strategies will make sustainable development and climate change adaptation more inclusive, wholesome, and effective (Gupta and Agrawal 2017; Luetz and Nunn 2020, 2021). Given that the literature presents people's worldviews as being intricately entangled with development outcomes, climate change adaptation strategies should aim for integrated approaches that conjoin different knowledges, including in relation to spirituality, sustainability, and science (Fair 2018; Makondo and Thomas 2018; Nunn et al. 2016; Luetz and Leo 2021).

Sadly, Indigenous knowledge and spirituality continue to be widely eclipsed and underappreciated according to the technocratic norms of the modern global economic system (Fischer et al. 2022; Luetz and Walid 2019; Stein 2019; Telleria 2017). Given that Indigenous communities have sustainably subsisted for thousands of years, it is high time to research and (re)discover the worldview-sustainability-adaptation nexus from their perspectives (Ellis et al. 2021; Fischer et al. 2022; Leal Filho et al. 2021a, b; Schramm et al. 2020). There is a solid evidence base that Indigenous knowledge and spirituality can helpfully underpin sustainability and climate change adaptation and therefore deserve more attention from both policy and practice (Gupta and Agrawal 2017; Nunn et al. 2016; Yunkaporta 2019). As such, this research recommends holistic and inclusive adaptation approaches that conjoin and integrate elements of Indigenous and scientific inputs (Makondo and Thomas 2018). Integrated and spiritually attuned strategies will stand a greater chance of being socioculturally meaningful, contextually appropriate, locally practical, and sustainable over time (Balehegn et al. 2019; Chisadza et al. 2015; Luetz and Nunn 2020, 2021).

This research is subject to some limitations. Given that this mixed methods research was weighted methodologically in favour of quantitative data analysis implies that there are opportunities for further in-depth qualitative inquiry. More specifically, it would be interesting to examine areas of syncretism between traditional Indigenous spirituality and major world religions. Future studies could investigate how Indigenous spiritual beliefs may be nurtured and preserved, even when these have acquiesced to introduced religious traditions. Having more nuanced information in this area could additionally strengthen strategies that target using Indigenous local knowledge for effective and sustainable climate change adaptation practice.

Acknowledgements The authors wish to thank Samuel O. Akande (Ph.D.), Centre for Space Research and Applications (CESRA), The Federal University of Technology Akure (FUTA), Nigeria, for assistance with visualisations, and Miss Fawale Iyanu for research assistance in the communities. Finally, the authors wish to thank the community members in Aiyetoro, Igbobini, Ogbese, and Owena for generously sharing their experiences and perspectives.

Institutional Review Board Statement

This field research was approved by the Ethical Committee for Human Research and Laboratory Animals of the School of Agriculture and Agricultural Technology, The Federal University of Technology, Akure, Nigeria (Ref. No. FUTA/SAAT/07-2020-21/023). The field survey was conducted according to the human and animal use and care regulations of the Canadian Council on Animal Care Guidelines and Protocol Review (Canadian Council on Animal Care 1993/2020).

References

Abate RS, Kronk EA (2013) Commonality among unique indigenous communities: an introduction to climate change and its impacts on indigenous peoples. Tulane Environ Law J 26:179–195

Adeleke ML, Fagbenro OA (2013) Livelihood diversification and operational techniques of the artisanal fisherfolks in the coastal region of Ondo State, Nigeria. Int J Innov Res Dev 2:262–273

Adeleke ML, Omoboyeje VO (2016) Effects of climate change on aquaculture production and management in Akure Metropolis, Ondo State. Nigeria Niger J Fish Aquac 4(1):50–58

Adeleke ML, Wolff M (2016) Adaptation of the artisanal fisher folks to climate change in the coastal region of Ondo state, Nigeria. In: Leal Filho W (eds) Innovation in climate change adaptation. Climate change management. Springer, Cham. https://doi.org/10.1007/978-3-319-25814-0_13

Adesina FO, Siyambola WO, Oketola FO, Pelemo DA, Ojo LO, Adegbugbe AO (1999) Potentials of agroforestry for climate change mitigation in Nigeria: some preliminary estimates. Glob Ecol Biogeogr 8:163–173

Akinola JO, Olawusi-Peters, OO, Akpambang VOE (2019) Ecological hazards of total petroleum hydrocarbon in Brackish water white shrimp *Nematopalaemon hastatus* (AURIVILLUS 1898). *Egypt J Aquat Res* 45(3):205–210. https://doi.org/10.1016/j.ejar.2019.07.004

Akinola JO, Olawusi-Peters OO, Akpambang VOE (2020) Human health risk assessment of TPHs in Brackish water prawn (*Nematopalaemon hastatus*, AURIVILLUS 1898). Heliyon 6(1):1–6. https://doi.org/10.1016/j.heliyon.2020.e03234

Allison EH, Perry AL, Badjeck MC, Adger WN, Brown K, Conway D (2009) Climate change and fisheries: a comparative analysis of the relative vulnerability of 132 countries. Fish Fisheries 10(2):173–196

Apata TG, Samuel KD, Adeola AO (2009) Analysis of climate change perception and adaptation among arable food crop farmers in Southwestern Nigeria. Paper presented at the International Association of Agricultural Economists conference, Beijing, China, 16–22, 2009

Babbie E (2021) The practice of social research, 15th edn. Cengage Learning

Balehegn M, Balehey S, Fu C et al (2019) Indigenous weather and climate forecasting knowledge among Afar pastoralists of north eastern Ethiopia: role in adaptation to weather and climate variability. Pastoralism 9:8. https://doi.org/10.1186/s13570-019-0143-y

Béné C (2006) Small-scale fisheries: assessing their contribution to rural livelihoods in developing countries. FAO Fisheries Circular. No. 1008. Food and Agriculture Organization of the UN (FAO), 46pp. https://www.fao.org/3/j7551e/j7551e.pdf

Bindoff NL, Cheung WWL, Kairo JG, Arístegui J, Guinder VA, Hallberg R, Hilmi N, Jiao N, Karim MS, Levin L, O'Donoghue S, Purca Cuicapusa SR, Rinkevich B, Suga T, Tagliabue A, Williamson P (2019) Changing ocean, marine ecosystems, and dependent communities. In: Pörtner H-O, Roberts DC, Masson-Delmotte V, Zhai P, Tignor M, Poloczanska E, Mintenbeck K, Alegría A, Nicolai M, Okem A, Petzold J, Rama B, Weyer NM (eds) IPCC special report on the ocean and cryosphere in a changing climate. Cambridge University Press, pp 447–587. https://doi.org/10.1017/9781009157964.007

Bryman A (2016) Social research methods, 5th edn. Oxford University Press

CCAC—Canadian Council on Animal Care (1993/2020) Guide to the care and use of experimental animals, vol 1, 2nd edn. Publication Date: 1993; Revision Date: April 2020. https://ccac.ca/Documents/Standards/Guidelines/Experimental_Animals_Vol1.pdf

Chisadza B, Tumbare MJ, Nyabeze WR, Nhapi I (2015) Linkages between local knowledge drought forecasting indicators and scientific drought forecasting parameters in the Limpopo River Basin in Southern Africa. Int J Disaster Risk Reduct 12:226–233. https://doi.org/10.1016/j.ijdrr.2015.01.007

Creswell JW, Creswell JD (2018) Research design: qualitative, quantitative, and mixed methods approaches, 5th edn. SAGE

Doust K, Wejs A, Zhang T-T, Swan A, Sultana N, Braneon C, Luetz JM, Casset L, Fatorić S (2021) Adaptation to climate change in coastal towns of between 10,000 and 50,000 inhabitants. Ocean Coast Manag 212.https://doi.org/10.1016/j.ocecoaman.2021.105790

Ellis EC et al (2021) People have shaped most of terrestrial nature for at least 12,000 years. PNAS 118(17):e2023483118. https://doi.org/10.1073/pnas.2023483118

Fair H (2018) Three stories of Noah: navigating religious climate change narratives in the Pacific island region. Geo Geogr Environ 5(2). https://doi.org/10.1002/geo2.68

Fischer M, Maxwell K, Nuunoq, Perdersen H, Greeno D, Jingwas N, Blair JG, Hugo S, Mustonen T, Murtomäki E, Mustonen K (2022) Empowering her guardians to nurture our Ocean's future. Rev Fish Biol Fisheries 32:271–296.https://doi.org/10.1007/s11160-021-09679-3

Ford JD, King N, Galappaththi EK, Pearce T, Mcdowell G, Harper SL (2020) The resilience of Indigenous peoples to environmental change. One Earth 2:532–543

Fregene BT, Ogunika OF (2013) Perception of fish farmers to climate change and adaptation strategies in Oyo State, Nigeria. In: Ndimele PE (ed) Proceedings of the 28th annual conference of the fisheries society of Nigeria (FISON), 25–29 Nov 2013, Abuja, Nigeria, pp 408–411

Gupta K, Agrawal R (2017) Sustainable development and spirituality: a critical analysis of GNH index. Int J Soc Econ 44(12):1919–1939. https://doi.org/10.1108/IJSE-10-2015-0283

Hesse-Biber SN (2017) The practice of qualitative research, 3rd ed. Sage

Hosen N, Nakamura H, Hamzah A (2020) Adaptation to climate change: does traditional ecological knowledge hold the key? Sustainability 12(2):676. https://doi.org/10.3390/su12020676

Idowu AA, Ayoola SO, Opele AI, Ikenweiwe NB (2011) Impact of climate change in Nigeria. Iran J Energy Environ 2(2):145–152

IPCC (2007) In: Parry ML, Canziani OF, Palutikof JP, van der Linden PJ, Hanson CE (eds) Climate change 2007: impacts, adaptation and vulnerability. Contribution of working group II to the fourth assessment report of the intergovernmental panel on climate change. Cambridge University Press, Cambridge

IPCC (2022) Summary for policymakers. In: Pörtner H-O, Roberts DC, Poloczanska ES, Mintenbeck K, Tignor M, Alegría A, Craig M, Langsdorf S, Löschke S, Möller V, Okem A (eds) Climate change 2022: impacts, adaptation, and vulnerability. Contribution of working group II to the sixth assessment report of the intergovernmental panel on climate change. Cambridge University Press, Cambridge

Iticha B, Husen A (2019) Adaptation to climate change using indigenous weather forecasting systems in Borana pastoralists of southern Ethiopia. Clim Dev 11(7):564–573. https://doi.org/10.1080/17565529.2018.1507896

Kovach M (2018) Doing indigenous methodologies: a letter to a research class. In: Denzin NK, Lincoln YS (eds) The Sage handbook of qualitative research, 5th edn. Sage

Leal Filho W, Matandirotya NR, Lütz JM et al (2021a) Impacts of climate change to African indigenous communities and examples of adaptation responses. Nat Commun 12:6224. https://doi.org/10.1038/s41467-021-26540-0

Leal Filho W et al (2021b) African handbook of climate change adaptation. Springer Nature

Leal Filho W, Luetz JM, Ayal D (2021c) Handbook of climate change management—research, leadership, transformation. Springer Nature. https://doi.org/10.1007/978-3-030-57281-5

Luetz JM, Bergsma C, Hills K (2019) The poor just might be the educators we need for global sustainability—a manifesto for consulting the unconsulted. In: Leal Filho W, Consorte McCrea A (eds) Sustainability and the humanities. Springer Nature, pp 115–140. https://doi.org/10.1007/978-3-319-95336-6_7

Luetz JM, Nunn PD (2020) Climate change adaptation in the Pacific islands: a review of faith-engaged approaches and opportunities. In: Leal Filho W (ed) Managing climate change adaptation in the Pacific region. Springer Nature, pp 293–311. https://doi.org/10.1007/978-3-030-40552-6_15

Luetz JM, Leo RG (2021) Christianity, creation, and the climate crisis—ecotheological paradigms and perspectives. In: Luetz JM, Nunn PD (eds.) Beyond belief—opportunities for faith-engaged approaches to climate-change adaptation in the Pacific islands. Springer, pp 345–375. https://doi.org/10.1007/978-3-030-67602-5_18

Luetz JM, Nunn PD (2021) Beyond belief: opportunities for faith-engaged approaches to climate-change adaptation in the Pacific islands. Springer. https://doi.org/10.1007/978-3-030-67602-5

Luetz JM, Walid M (2019) Social responsibility versus sustainable development in United Nations policy documents: a meta-analytical review of key terms in human development reports. In: Leal Filho W (ed) Social responsibility and sustainability. Springer, pp 301–334. https://doi.org/10.1007/978-3-030-03562-4_16

Makondo CC, Thomas DSG (2018) Climate change adaptation: linking indigenous knowledge with western science for effective adaptation. Environ Sci Policy 88:83–91. https://doi.org/10.1016/j.envsci.2018.06.014

McKenzie DJ, Geffroy B, Farrell AP (2021) Effects of global warming on fishes and fisheries. J Fish Biol 98:1489–1492. https://doi.org/10.1111/jfb.14762

Mundy P, Compton L (1991) Indigenous communication and Indigenous knowledge. Dev Commun Rep 74(3):1–3

Nalau J, Becken S, Schliephack J, Parsons M, Brown C, Mackey B (2018) The role of indigenous and traditional knowledge in ecosystem-based adaptation. Weather Clim Soc 10(4):851–865. https://doi.org/10.1175/WCAS-D-18-0032.1

National Bureau of Statistics (NBS) (2011) Annual abstract of statistics. [Online]. Available. www.nigerianstat.gov.ng. Accessed 12 Apr 2021

Nzeadibe TC, Chukwudumebi LE, Nnaemaba AC, Victoria CA (2011) Climate change awareness and adaptation in the Nigeria delta region of Nigeria. African technology policy study network working paper series/no. 57. ISBN 978-9966-1552-6-9

Nunn PD, Mulgrew K, Scott-Parker B, Hine DW, Marks ADG, Mahar D, Maebuta J (2016) Spirituality and attitudes towards nature in the Pacific islands: insights for enabling climate-change adaptation. Clim Change 136(3–4):477–493. https://doi.org/10.1007/s10584-016-1646-9

Nunn PD, Luetz JM (2021) Beyond belief. In: Luetz JM, Nunn PD (eds) Beyond belief: opportunities for faith-engaged approaches to climate-change adaptation in the Pacific islands. Springer, pp 1–14. https://doi.org/10.1007/978-3-030-67602-5_1

Omojeje A, Adu-Peters RO (2021) Malokun festival and practices among the Mahin on the Ilaje Coast. Yoruba Stud Rev 2(1). https://doi.org/10.32473/ysr.v2i1.129855

Oseni JO (2010) Effects of deregulation policy on cocoa marketing in Ondo State, Nigeria. PhD. Dissertation, Department of Agricultural Economics and Extension, Federal University of Technology, Akure, Nigeria

Osunade MA (1994) Indigenous climate knowledge and agricultural practices in Southwestern Nigeria. Malays J Trop Geogr 1:21–28

Punch KF (2014) Introduction to social research: quantitative & qualitative approaches, 3rd edn. Sage

Sadiku NA, Sadiku IBS (2011) Indigenous efforts by African farmers in ensuring sustainability in agricultural productivity in the face of changing climate. Proceedings of the environmental management conference, Federal University of Agriculture, Abeokuta, Nigeria, 2011

Schramm PJ et al (2020) How indigenous communities are adapting to climate change: insights from the climate-ready tribes initiative. Health Aff 39:2153–2159. https://doi.org/10.1377/hlthaff.2020.00997

Stein S (2019) The ethical and ecological limits of sustainability: a decolonial approach to climate change in higher education. Aust J Environ Educ 35(3):198–212. https://doi.org/10.1017/aee.2019.17

Sultana N, Luetz JM (2022) Adopting the local knowledge of coastal communities for climate change adaptation—a case study from Bangladesh. Front Clim 4:823296. https://doi.org/10.3389/fclim.2022.823296

Tedesco PA, Oberdorff T, Cornu JF, Beauchard O, Brosse S, Dürr HH, Grenouillet G, Leprieur F, Tisseuil C, Zaiss R, Hugueny B (2013) A scenario for impacts of water availability loss due to climate change on riverine fish extinction rates. J Appl Ecol 50:1105–1115. https://doi.org/10.1111/1365-2664.12125

Telleria J (2017) Power relations? What power relations? The depoliticising conceptualisation of development of the UNDP. Third World Q 38(9):2143–2158. https://doi.org/10.1080/01436597.2017.1298437

UNESCO (2021) United Nations Educational, Scientific and Cultural Organization. Local and Indigenous Knowledge Systems (Links). https://en.unesco.org/links
Usman LM (2010) The indigenous knowledge system of female pastoral fulani of Northern Nigeria. In: Kapoor D, Shizha E (eds) Indigenous knowledge and learning in Asia/Pacific and Africa. Palgrave Macmillan. https://doi.org/10.1057/9780230111813_14
Walshe RA, Nunn PD (2012) Integration of indigenous knowledge and disaster risk reduction: a case study from Baie Martelli, Pentecost Island, Vanuatu. Int J Disaster Risk Sci 3:185–194. https://doi.org/10.1007/s13753-012-0019-x
World Bank (2022) Agriculture, forestry, and fishing, value added (% of GDP) – Nigeria. https://data.worldbank.org/indicator/NV.AGR.TOTL.ZS?locations=NG. Accessed 17 Oct 2022
Yunkaporta T (2019) Sand talk: how Indigenous thinking can save the world. Text Publishing

Chapter 24
Clarifying Local Government Policymakers' Needs on Climate Change Science and Technologies: Experiences of Science and Policy Deliberation at Co-Design Workshops in Japan

Kenshi Baba and Mitsuru Tanaka

Abstract This chapter clarifies the needs of local government officials for climate science and technologies and draws on future developments by science and policy deliberation in co-design workshops. To this end, first, the documents of local governments' climate change adaptation plans were scrutinized, and it was examined how the scientific data were used for the plans. Second, interviews were conducted with local governments. Simultaneously, science and policy deliberation was conducted at a series of co-design workshops that involved over a hundred scientists and policymakers, who exchanged information on needs and resources. The main results demonstrate the following: (i) From the results of the scrutiny of the documents of local governments' climate change adaptation plans, a higher need for scientific data was observed in irrigated rice cultivation, fruit cultivation, heat stroke, natural forest, and mudflow disasters for a shorter term and specific areas. (ii) From the results of the interviews, as one of the challenges, although there was no resistance to using the climate change projection data to formulate adaptation plans, at least in environmental bureaus, other bureaus responded differently. In addition, although population projections have been widely adopted, climate change projections are not at that stage. This may be due to the degree of uncertainty. (iii) From the results of the questionnaire surveys conducted after a series of nationwide co-design workshops, one of the most important lessons learned was that mutual understanding of the use of projection data between scientists and policymakers has not necessarily been promoted over time, although the degree of satisfaction with the workshops themselves has increased. This implies that further improvement in scientific literacy for policymakers and a deeper understanding of policy for scientists are required.

K. Baba (✉)
Faculty of Environmental Studies, Tokyo City University, 3-3-1 Ushikubonishi, Tsuzuki-Ku, Yokohama 224-8551, Kanagawa, Japan
e-mail: kbaba@tcu.ac.jp

M. Tanaka
Emeritus, Faculty of Social Sciences, Hosei University, Tokyo, Japan

W. Leal Filho et al. (eds.), *Climate Change Strategies: Handling the Challenges of Adapting to a Changing Climate*, Climate Change Management,
https://doi.org/10.1007/978-3-031-28728-2_24

Introduction

Since the establishment of the Law Concerning the Promotion of the Measures to Deal with Global Warming in Japan, many local governments have begun enacting global warming countermeasure implementation plans centered around mitigation measures, positioning, and implementing various institutions and projects under various ordinances and administrative plans to promote these countermeasures. Local governments' actions in implementing adaptation measures have been limited. However, in November 2015, the government's National Plan of Climate Change Adaptation was adopted by the cabinet, and internationally, the Paris Agreement was finalized, which stipulated that the international community focus on "holding the increase in the global average temperature to below 1.5 to 2 degrees Celsius above pre-industrial levels" and called for policies which aggressively position adaptation measures. Consequently, at the end of 2016, many local governments enacted climate change adaptation plans. In June 2018, the Climate Change Adaptation Act obliged local governments to make efforts to enact local climate change adaptation plans. Therefore, this number is expected to rise.

Climate change adaptation plans have been implemented in countries and cities worldwide. Previous studies on the formulation of adaptation plans include surveys examining the formulation of adaptation plans in large cities across the world, such as the United Kingdom, Europe, and large cities in the United States (Araos et al. 2016; Heidrich et al. 2013; Reckien et al. 2018; Woodruff and Stults 2016). Carmin et al. (2012) and Aylett (2015) conducted questionnaires for local governments that are members of the International Council for Local Environmental Initiatives (ICLEI), a network organization of local governments around the world regarding environmental policies, and clarified challenges with the adaptation plan formulation, such as budget acquisition, communication between departments, and national commitment. Bierbaum et al. (2012) surveyed documents including the adaptation plans of American local governments, federal governments, and non-governmental organizations. They identified issues such as uncertainties in information and decision-making, lack of resources, vertical decision-making between departments, institutional restrictions, and lack of leadership. Furthermore, Gero et al. (2012) listed the operational (limited access to scientific findings), policy (inconsistent rules between departments), financial (difficulty prioritizing), and cognitive (low awareness of risk in communities) aspects of the adaptation plan inhibiting factors of local governments based on their experiences inside and outside Australia. In addition, some studies have clarified the factors of climate change adaptation policy adoption among countries and local governments in terms of policy diffusion. Massey et al. (2014) indicate adoption of climate change adaptation policy among EU countries is largely being driven by internal factors such as extreme weather events and increased public awareness adaptation rather than external factors such as efforts by international organizations. Musah-Surugu et al. (2018) note that local governments' adoption of climate change adaptation policies in Sub-Saharan Africa is a function of contextual, agency, structural, and science-heaviness factors.

However, they tend to lack to know how to handle scientific knowledge, such as the output of climate models and the various impacts of climate change.

In Japan, according to the results of a questionnaire survey conducted on the environmental bureaus of prefectures, ordinance-designated cities, designated mid-level cities, and so on throughout Japan in February and March 2016 (155 distributed, 123 returned, for a response rate of 79.4%), more than half of all local governments responded that the presumed challenges to studying and implementing climate change adaptation plans in local governments were "insufficient accumulation of experience and specialization in administrative organs" and had "discrepancies in awareness of division of tasks and priorities between administrative bureaus." Additionally, 20% of these local governments pointed out "a mismatch between scientific knowledge and administrative needs" (Baba et al. 2017). Although climate change projections extending into the near future or to the end of this century and impact assessments in each sector have begun to be provided, many local governments are not in a position to obtain these with ease.

Therefore, establishing a place for deliberation between policymakers and scientists is essential for promoting local climate change adaptation plans. Efforts in co-design and co-production have been made in the context of climate change adaptation. Thus, Baba et al. (2021) and Gonçalves et al. (2022) introduced transdisciplinary approaches to local communities. There are a lot of such local case studies. On the other hand, Becsi et al. (2020) introduced the Austrian nationwide case of developing regional climate change impact maps in which a co-design effort was made together with stakeholders in adaptation planning and climate change communication experts. This rare case is also important in determining the direction of the future development of projections.

The purpose of this chapter is to clarify the potential needs of local government policymakers when formulating a local climate change adaptation plan. Furthermore, this chapter attempts to draw lessons from nationwide co-design workshops that aim to match the scientific knowledge of climate change impacts and adaptation measures with the needs of policymakers.

Methodology

The following two methods were employed. First, documents of local governments' climate change adaptation plans were scrutinized, and the scientific data used for the plans were examined. Simultaneously, interviews were conducted with local governments. Using these methods, the challenges and needs related to scientific data were clarified. Second, based on the above results, science and policy deliberation was designed and conducted at co-design workshops in which over a hundred scientists and policymakers participated, and their needs and resources were exchanged.

Clarifying the Potential Needs of Local Governments Policymakers for Scientific Knowledge

The use of scientific data in local climate change adaptation plans can be understood through a literature survey. The items on climate change impact provided for listing were based on subitems in the national government's climate change adaptation plan. To express the suitability of the used scientific data to the region, all impact items were counted by adding a weighting score: (a) global scale (quoted from IPCC Assessment Report, etc.) = 1 point; (b) Japan scale (quoted from the national government's adaptation plan) = 2 points; (c) broad region scale (quoted from the Japan Meteorological Agency, "global warming projection information, etc.") = 4 points; (d) one's own prefecture or city scale (quoted from past documents, etc.) = 8 points; and (e) projection data in smaller area scale = 16 points (provided by themselves with local universities and research institutes). This provides an overview of what kinds of climate change impact items are cited using what kinds of scientific data and clarifies the latent need for technology development of climate models and impact assessment, which could be utilized for future local climate change adaptation plans. The results of applying this method to the local climate change adaptation plans enacted until March 2017 are presented in the next section.

Simultaneously, interviews were conducted with local governments. The main questions were as follows: state of enactment of local climate change adaptation plans, motives for enactment, and organization of bureaus to examine the plan and challenges, characteristics of the plan, scientific data used, and challenges. This provides a background for the latent need for the technological development of climate models and impact assessments, which could be utilized for future local climate change adaptation plans. The results of interviews with policymakers of environmental bureaus in 22 prefectures and cities conducted from June to September 2016 are presented in a later section.

Science and Policy Deliberation at Co-Design Workshops

In addition, science and policy deliberation was conducted at a series of co-design workshops in which scientists and policymakers participated, and their needs and resources were exchanged. Holding a direct dialogue gives them a mutual understanding that will lead to technology development based on specific administrative needs and the utilization of scientific knowledge and data in the formulation of local climate change adaptation plans. Evidence-based policymaking (EBPM) will progress in terms of climate change adaptation. "The Adaptation Local Government Forum" was held once a year as an all-day co-design workshop during 2016–2019. The workshop, in which the latest developments in climate change adaptation technology, policies, and views were directly exchanged, was attended by local government officials from relevant departments, researchers from local environmental research institutes, and scientists from nationwide universities and national research institutes. The results of the series of co-design workshops are presented in a later section.

Results

Clarifying the Potential Needs of Local Governments Policymakers for Scientific Knowledge

Scrutiny of the Local Governments' Climate Change Adaptation Plans

The results of aggregating climate change impact items by scientific data that were used in local climate change adaptation plans by adding the weighting score mentioned above are shown in Table 24.1. This was aggregated by the seven sectors adopted by the government's climate change adaptation plan.

As shown in Table 24.1, the most cited impact items were in the agriculture, forestry, and fishery sectors, including paddy rice and orchards. The next is the natural disaster sector, which includes debris flows, landslides, and flooding. The third and fourth sectors are health and natural ecosystems, which include natural forests, second-growth forests, and heat stroke, respectively. In terms of the geographic distribution, even among these items, many more detailed studies have been done in eastern and northern Japan. Elsewhere, although it cannot be said to be clear, the items such as flooding, coastal erosion, debris flows, and landslides tend to be reported in eastern Japan. However, the items related to the risk of death, such as stock farming, insect pests or weeds, and aquaculture, tend to be studied to a certain degree in western Japan.

Among the scientific data used to assess impacts in the local climate change adaptation plans reviewed above, category (e) is extremely rare, and categories (a)–(d) are frequent. Therefore, it is necessary to study future technology development, while keeping in mind that there could be a latent need for such impact items when revising future plans.

Interviews with Some Remarkable Local Governments

Table 24.2 outlines the results of the interviews with local governments, focusing on the characteristics of planning and the challenges to be addressed by the internal system for examination. The table was not prepared to show information for all local governments, but only the results for local governments from which thorough information was obtained.

Table 24.1 Impact items used in climate change adaptation plans of local government by scientific data (weighted score)

	Agriculture, forestry, and fishery	Aqueous environment/water resources	Natural ecosystem	Natural disaster	Health	Industry/economic activity	National life/city life
a	4	6	5	9	4	1	2
b	63	26	38	53	38	7	19
c	2	0	0	3	0	0	2
d	106	19	59	60	52	12	27
e	4	2	2	4	3	0	0
Totals	179	53	95	129	97	20	50

Table 24.2 Challenges concerning climate change adaptation plans in local governments

	Typical responses
Characteristics of planning	• Most climate change adaptation plans are incorporated along with mitigation plans • The existing climate change adaptation plans will be examined when the impact assessment by the national government will be renewed • Evaluation items of climate change impact with medium confidence or above in the national government's climate change adaptation plan are selected in principle and are judged anew qualitatively in terms of the region • The local government officials conduct a literature survey of climate change impact assessments and look for impact examples in the city • Existing potential climate change adaptation measures in various areas are introduced in the climate adaptation plan • Enact a new climate change ordinance as the grounds • Enter numerical targets in the climate change adaptation plan
Challenges to formulate climate change adaptation plans	• The existing cross-bureau framework set to examine climate change mitigation measures or a new organization suitable for the broader issue of adaptation measures by adding bureaus is applied to adaptation planning • Awareness of the priority of policies differs between bureaus and each bureau tends to recognize that the existing measures have no relation to climate change, so there is a need for mutual understanding and coordination between related bureaus • When implementing climate change adaptation measures, it becomes a problem that the budgets are beyond the environmental bureaus' jurisdiction • Adjusting the climate change adaptation plan to the comprehensive administrative plan has not been done • It is difficult to formulate long-term climate change adaptation plans with scientific data including high uncertainty

One of the characteristics of planning is that most climate change adaptation plans are incorporated with mitigation plans. The existing climate change adaptation plans will then be examined when the impact assessment by the national government is renewed. In addition, in some local governments, the evaluation items of climate change impact with medium confidence or above in the national government's climate change adaptation plan were selected in principle and judged anew qualitatively in terms of the region. Existing potential climate change adaptation measures in various sectors have been introduced into the climate adaptation plans of some local governments. Very few local governments enacted new climate change ordinances, entered numerical targets in the climate change adaptation plan, and very few local government officials conducted a literature survey of impact assessments and looked for impact examples in the city.

Regarding the local governments' challenges to be addressed in the internal system for examining, in almost all cases, the existing cross-bureau framework set to examine climate change mitigation measures, or a new organization suitable for the broader issue of adaptation measures by adding bureaus, is applied to adaptation planning. However, in many cases, awareness of the priority of climate change adaptation measures differs between bureaus, and each bureau tends to recognize that the existing measures are not related to climate change. Therefore, it is necessary for environmental bureaus to mutually understand and coordinate between related bureaus. As a result, when implementing climate change adaptation measures, it becomes a problem that the budgets are beyond the environmental bureaus' jurisdiction. In addition, the climate change adaptation plan has not been adjusted to a comprehensive administrative plan, and it is difficult to formulate long-term climate change adaptation plans with projection data that include high uncertainty in all local governments.

Table 24.3 shows the results of interviews with local governments, focusing on the challenges and needs related to scientific data from local governments. The scientific data used in all local governments are listed in the table, all referred to past observation data such as "Meteorological agency observation data (Reports on climate change)," but they used projection data in two distinct patterns: (i) very limited local governments independently got climate change impact assessment of some items in detail in collaboration with local universities and research institutes, or independently got customized projection data provided by the Ministry of Environment S8 climate change impact/adaptation research project team. When unable to obtain this, (ii) most local governments referred to wide-area projection data that were more widely open and anyone could obtain, such as meteorological agency observation data (reports on climate change), and Meteorological Agency global warming projection information Vol. 8. The former approach was limited to a few local governments that were able to obtain a budget as a model project of the Ministry of the Environment and find experts in the corresponding area. In the former cases, each responsible bureau mainly extracted current measures that could be regarded as latent climate change adaptation measures, or rarely examined those measures that have never been seen before that could be regarded as supplementary climate change adaptation measures. However, many of the latter cases were only at the stage where they investigated worrisome phenomena qualitatively and examined sectors that should be emphasized to extract current measures, which could be regarded as latent climate change adaptation measures.

Regarding challenges and needs related to scientific data, it has been pointed out that as the problem is on a temporal scale; the target year (about 2030) of the next plan is vital, but projection data for 2050 or 2100 are rarely required. As for the problem of spatial scale, it is difficult for local government officials to answer citizens' frequent questions about the future state of specific locations based on projection data. This implies that the scale should be sufficiently wide to permit an understanding of the impact of several spheres within the jurisdiction of the local government. However, if 1 km grid size data were disclosed to the public, it would be worrying that people have (only) a false impression of the data. Perception gaps were also observed among the bureaus. Environmental bureaus cannot freely access data under the jurisdiction of other bureaus, whereas agriculture, forestry, fisheries, and disaster prevention bureaus only write in line with the guidelines of the ministry with jurisdiction. Although using projection data does not make the environmental bureau feel uncomfortable, in principle, agriculture, forestry, and fisheries bureaus do not use projection data in their climate change adaptation measures. This may

Table 24.3 Challenges and needs related to scientific data in local governments

	Typical responses
Scientific data used	• Meteorological agency observation data (reports on climate change) • Meteorological Agency global warming projection information Vol. 8 • Japan's climate at the end of the twenty-first century: projection calculations including uncertainty data • Ministry of Land, Infrastructure, Transport, and Tourism "A report concerning ideal climate change adaptation measures in the flood disaster area" (Result of research by the National Institute for Land and Infrastructure Management [NILIM]) • Original research paper • Customized projection data provided by the Ministry of Environment S8 climate change impact/adaptation research project team • Climate change impact assessment of some items in detail in collaboration with local universities and research institutes

(continued)

Table 24.3 (continued)

	Typical responses
Challenges and needs related to scientific data	• As for a temporal scale, the target year (about 2030) of the next plan is vital, but a projection of data in 2050 or 2100 is rarely required • It is difficult for local government officials to answer the citizens' frequent questions about the future state of specific locations based on projection data • While 1 km grid size data are disclosed to the public, it is worrying that people have (only) a false impression of the data • It is difficult to collect observation data because of a lack of the Automated Meteorological Data Acquisition System (AMeDAS) • Environmental bureaus cannot freely access data under the jurisdiction of other bureaus • Using projection data never makes the environmental bureau feel uncomfortable • Agriculture, forestry, fisheries, and disaster prevention bureaus only write in line with the guidelines of the Ministry with jurisdiction • Agriculture, forestry, and fisheries bureaus do not use projection data in their climate change adaptation measures in principle • The city government does not exercise complete control during emergencies, such as a disaster on a river that is within the jurisdiction of the prefecture or national government • It is difficult for local government officials to survey research papers, deal with climate models, and explain everything about scientific data • Even if the probability is shown, it is difficult to make policy decisions based only on numerical values • It is necessary to clearly indicate that there is uncertainty, and climate change adaptation measures should be taken based on it • If there are no scientific data, the local government officials cannot judge whether or not climate adaptation measures are necessary • Credibility from the perspective of third parties is important; it should be credible regardless of who looks at the scientific data

be because of the degree of uncertainty. The fact that population projections have been widely adopted, for example, while climate change projections are not at this stage, indicates that this may be due to the degree of uncertainty. Nevertheless, it is important to determine whether credibility exists from the perspective of third parties; it should be credible regardless of who looks at scientific data. It is often pointed out that credibility is particularly high when it is used for planning by a responsible government ministry or agency, and that this stage can be actively referred to. It is necessary for experts and scientists to keep in mind the need for basic scientific data by a local government when enacting a climate change adaptation plan to apply them to future technological development.

Science and Policy Deliberation at Co-Design Workshops

Overview

Table 24.4 presents an outline of the workshops. The number of participants has increased over the years, and the forum is attended not only by the Ministry of Education, Culture, Sports, Science, and Technology (MEXT), the Ministry of Environment (MOE), the Ministry of Land, Infrastructure, Transport, and Tourism (MLIT), the Ministry of Agriculture, Forestry, and Fisheries (MAFF), the Japan Meteorological Agency (JMA), and various bureaus of local governments.

The workshop was divided into small groups on the environment, disaster prevention, agriculture, and heat stroke; a group was formed with three to six policymakers related to each field, several members from local environmental research institutes, and scientists from national research institutes and universities who were developing climate models and impact assessment in addition to facilitators. The remaining participants listened to the audience members. As shown in the table, the discussion topics in the workshop changed during each session.

The purpose of the first session was to "clarify the needs of local governments for technological seeds related to climate change projection and impact assessment and to provide an opportunity to share issues for implementation." Although the topics of the small groups had some differences, they generally focused on aspects such as 'What kind of technological development is useful for local governments to plan for adaptation?' "What are you not sure of have problems with, or do not understand, adaptation policies?" Discussions were conducted with the material of "Procedures to examine adaptation policies and the reality of the implementation/Frequently raised issues on the ground," which was presented at the plenary session. Since the event was held 6 months after the formulation of the national adaptation plan, and many local governments were still working on their adaptation plans through trial and error, the main focus was to share basic information on what data and technologies were needed to formulate adaptation plans.

The second session aimed to "clarify the provision of climate change data for the development of adaptation measures and their use in administrative practice."

Table 24.4 Outline of the co-design workshops

	1st	2nd	3rd	4th
Date	Aug 31st, 2016	Aug 30th, 2017	Aug 28th, 2018	Aug 28th, 2018
Participants	Ministry of Education, Culture, Sports, Science, and Technology (MEXT), Ministry of Environment (MOE), local environmental research institutes, local govts, and scientists; 76 in total	MEXT, MOE, Ministry of Land, Infrastructure and Transport (MLIT), local environmental research institutes, local govts, and scientists; 109 in total	MEXT, MOE, MLIT, Ministry of Agriculture, Forestry and Fisheries (MAFF), Meteorological Agency (JMA), consultants, local environmental research institutes, local govts, and scientists; 150 in total	MEXT, MOE, MLIT, MAFF, JMA, consultants, local environmental research institutes, local govts, and scientists; 140 in total
Agenda	Plenary session	Plenary session	Plenary session	Plenary session
	• Technological development trend • Local govts' needs trend • MOE's policies • Local govts' policies	• Technological development trend • Local govts' needs trend • MOE's policies • Local govts' policies	• Technological development trend • Local govts' needs trend • MOE's policies • Local govts' policies	• Technological development trend • Local govts' needs trend • MOE's policies • Local govts' policies
	Small group workshop (Environment, disaster, and agriculture)	Small group workshop (Environment, disaster, agriculture and heatwave)	Small group workshop (Environment, disaster, agriculture and heatwave)	Small group workshop (Environment, disaster, agriculture and heatwave)
	• Topics: "What kind of technological development is useful for local governments to plan for adaptation?" and "What are you not sure of, have problem with, or do not understand about adaptation policies"	• Topics: "What climate change adaptation technologies did you find interesting in the presentations?", "What are the administrative practices where these technologies could be useful?", "How to communicate the information of the developed adaptation plans to citizens and stakeholders?"	• Topics: "mutual understanding of current impacts, needs, and seeds", "assumption of impacts and examination of issues when assuming a temperature increase (mid-century)", and "examination of hypothetical adaptation measures (plans)"	• Topics: "issues of prediction and assessment of climate change impacts", "issues of setting adaptation targets and progress management", "issues of establishment and maintenance of regional adaptation centers, and division of roles between national and local governments"

In the workshop, although there were some differences in topics among the small groups, discussions were conducted in response to questions such as, "What climate change adaptation technologies did you find interesting in the presentations?" "What are the administrative practices where these technologies are useful?" and "How can we communicate the information of the developed adaptation plans to citizens and stakeholders?" The discussion proceeded with the use of the catalog of climate change adaptation technologies within the ongoing research project as reference material. By this time, many local governments had begun to consider adaptation plans, and their specific needs became clearer; thus, the focus was on discussing the use of data and technology in line with the planning steps. Nevertheless, more time was allocated to topic presentations.

In the third session, the objective was, more specifically, to "verify the process of implementation of climate science and technology and data into local government administration and to clarify the issues." In the workshop, although there were some differences in topics among the small groups, discussions were generally held on "mutual understanding of current impacts, needs, and seeds," "assumption of impacts and examination of issues when assuming a temperature increase (mid-century)," and "examination of hypothetical adaptation measures (plans)," while introducing experiences and examples of social implementation by some model municipalities. As the Climate Change Adaptation Act was about to come into effect, local governments would be required to make efforts to develop adaptation plans and consider regional adaptation centers. In other subcommittees, more time was allocated to the workshop so that the experiences of some local governments could be closely shared during the workshop. The number of participants significantly increased in the third session, and the range of organizations to which they belonged significantly expanded.

The fourth session was conducted under the circumstances in which the Climate Change Adaptation Act was enforced, municipal adaptation plans were positioned or updated based on laws and ordinances, and regional adaptation centers were gradually established. The purpose of this session was to "examine the process of implementing climate science and data into local government administration and to identify issues to be addressed," and to discuss the findings of the third session in more detail. In the workshop, although there were some differences in topics among the small groups, they generally discussed "issues of projection and assessment of climate change impacts," "issues of setting adaptation targets and progress management," and "issues of establishment and maintenance of regional adaptation centers, and division of roles between national and local governments," while introducing experiences and cases of implementation by the model local authorities.

Changes in Post-Event Evaluation

The following shows the results of a brief questionnaire sent to the participants after the event.

Regarding the level of satisfaction with the plenary session in the first half of the event, 80%–90% of the evaluations were positive in all four sessions, and no major improvements were suggested, except for insufficient time. The level of satisfaction with the second half of the small-group workshops was even higher in all four sessions, and the fact that few other opportunities for direct and close exchange of opinions between experts and policymakers are available is thought to enhance positive evaluations (Figs. 24.1 and 24.2).

In stark contrast, however, positive evaluations of progress in the mutual understanding of the use of scientific data and administrative practices have become less frequent. This trend is also generally true for the evaluation of the potential of using scientific data in administrative practices (Figs. 24.3 and 24.4).

The following challenges emerged from the free-text column. The general trend was that, although the participants had a better understanding of the issues, they needed to improve their understanding and ability to read and understand the data to take the information back to their bureaus and result in the formulation of a plan, and it was difficult to link the results of projection and impact assessment to specific measures.

- While research on basic data is progressing, the overall impression is that the receiving system of the local government that uses the data is not yet in place.
- As individuals, the participants felt that they had achieved a better understanding of the content of scientific data, but that a more detailed explanation was needed to apply it in their work and to have others understand it.
- Many different types of scientific data and fields exist; thus, the more information one obtains, the less one knows what to do with it.
- It is difficult to utilize scientific data at the prefectural level because this would require a budget.

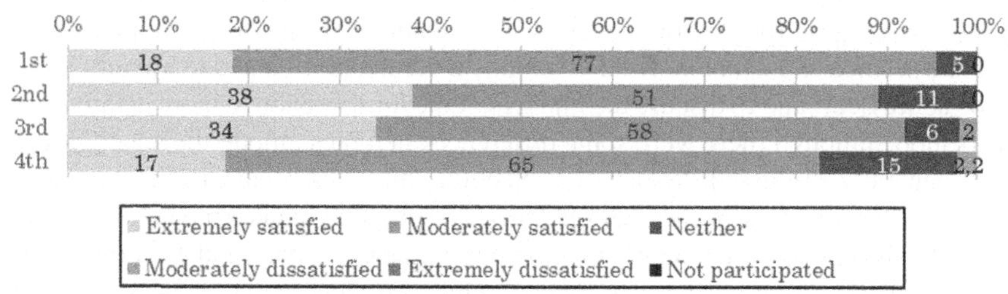

Fig. 24.1 Post-event evaluation: satisfaction of the plenary session

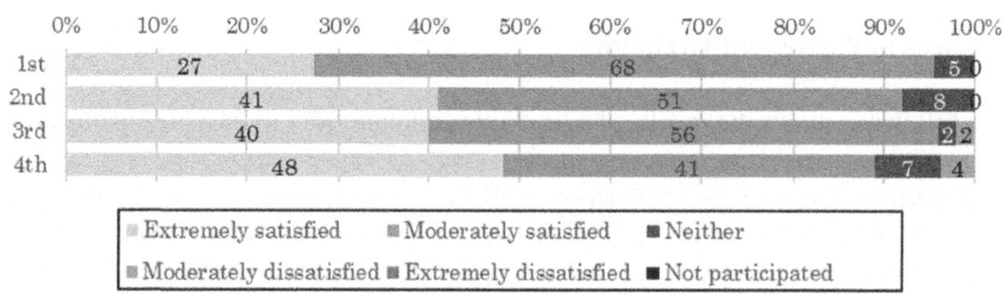

Fig. 24.2 Post-event evaluation: satisfaction of small group workshop (Altered from Tanaka and Baba 2021)

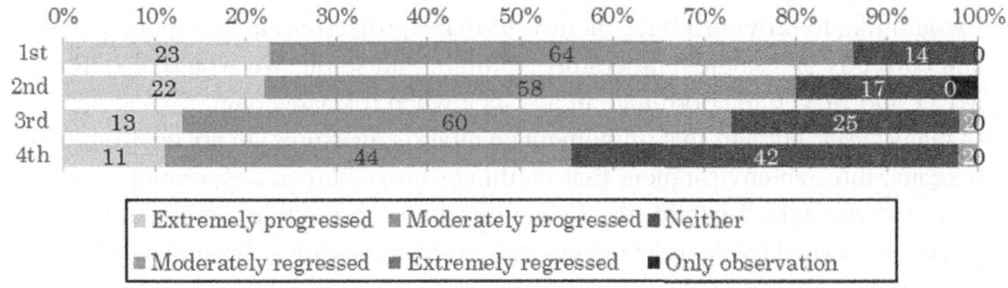

Fig. 24.3 Post-event evaluation: progress in mutual understanding of the use of scientific data and administrative practices (Altered from Tanaka and Baba 2021)

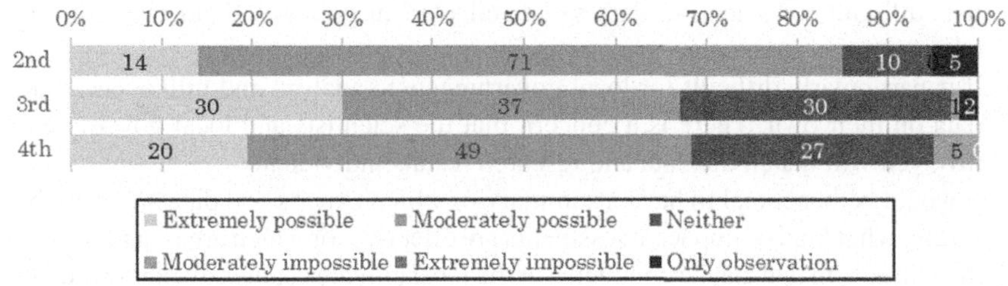

Fig. 24.4 Post-event evaluation: possibility of using scientific data in administrative practices

- Instead, it seems that more issues have arisen. In the face of uncertainty, local governments will accept policies that are based on projections.
- While I would like to actively use scientific data, it was difficult to come up with something concrete that the government could do with the data other than promoting awareness, and I sensed the difficulty in implementing adaptation measures.
- The administrative practice seeks error-free projection by downscaling, but since projection has its limitations, I felt that we should not rely too much on projection, but on vulnerability assessment and management through PDCA.

Policymakers also expressed the following views on the possibility of using scientific data: The use of uncertain data for planning and the gap between bureaus within the local government office remains a significant issue. This may be the reason for the split in evaluations.

- A methodology for translating the research findings (impact assessment) into policies must be developed. Currently, there is a disconnect between the scientific approach and how local governments work.
- As probabilistic thinking, which has become mainstream in climate change, does not fit into local government decision-making, it makes little sense to provide datasets created by scientists as they are. Is it best to create a small number of (socioeconomic) scenarios and have local governments endorse them?
- A system that can easily visualize various data would make it easier for local government policymakers and others to obtain a clearer picture.

- I would like to provide advice on how to make projections at the municipal level and correctly assess the temperature, rainfall, and so on. We would also like the MLIT and MAFF to coordinate in advance when releasing data.
- In many cases, bureaus that implement adaptation measures are not environmental bureaus; thus, an environment that facilitates cross-bureau cooperation is necessary (for example, I would like to see top-down instructions, notifications, and guidelines issued by the relevant central government ministries and agencies.)
- As a local environmental research institute, we can act as an intermediary; however, it is not easy to reflect on the latest scientific findings on local governance. It is important to make local government policymakers grasp scientific knowledge, but since government officials change every 2–3 years due to personnel turnover, it is difficult for scientific data to be reflected in policy without changing the administrative structure.
- It is particularly difficult for local governments to analyze and utilize necessary data on their own. There is a concern that the scientists and local governments involved will match the data and research results individually.
- It would be beneficial to have a summary of which experts and data are available where, what kind of impact assessments are effective for which areas, and provide them in a form that can be conducted at the local climate change adaptation center.
- No administrative plan can be formulated without scientific evidence. We believe that this method is necessary.
- It is necessary to present the usage of the system in a manner that is appropriate for a wide range of fields and people involved. It is also necessary to have human resources that can connect experts with the government and the public and private sectors. Prefectural residents: Data and pictures are sufficient to show that this is probably the way things go. Industry: Enough data to make the risk visible to mitigate, for example, the amount of damage. Prefectural government: Accurate and precise data on the prioritization of measures.

In addition, the following comments were received from scientists and consultants showing their awareness of responding to local government needs and their willingness to coordinate matching needs and resources through this kind of forum.

- In the future, with the establishment of the Climate Change Adaptation Act, local governments will make specific adaptation plans. Looking at group work, I felt that many specific needs arose in this process and that I could mediate by collecting them.
- It would be effective if there were more opportunities for dialogue between the creators/providers of scientific data and the users and to learn and understand each other's efforts, situations, and issues in detail.
- We would like to provide information that better meets the needs of the local governments.

Discussion

As mentioned, various challenges are expected from the examination and enactment of local climate change adaptation plans. It is important to examine whether the mismatch between scientific knowledge and local government needs can be resolved through these deliberations. Verifying whether scientific knowledge can be applied to future local climate change adaptation plans is important for explaining the possibilities of evidence-based policymaking (EBPM), that is, planning based on a precautionary principle.

As Cairney (2016) points out, EBPM has the following barriers: (i) Scientific findings that are reliable in terms of policy challenges and solutions are lacking. Additionally, they are not packaged in a way that allows policymakers to understand, become interested in, and hope for policy changes. (ii) Despite scientific findings, policymakers have not paid adequate attention because a gap exists in the timing of journal publication of scientific findings and policy formulation. Scientists also lack an understanding of policy agenda priorities. (iii) What policymakers want to do first, and then search for suitable or bending scientific findings to support policy decisions. The background of this is that national and local governments tend not to view climate change risks as long-term policy issues. Politicians do not try to understand data, and policymakers have a biased understanding of the preeminent challenges. It would be extremely difficult to make changes regarding these issues.

The following are the potential solutions: As for the problem (i), the quality of the scientific findings provided will likely continue to improve through a national research project. As an important function of local climate change adaptation centers, information and data must be provided to local stakeholders in a simplified manner. In terms of (ii) and (iii), a cultural difference exists between academia and policymaking, and not only in terms of terminology, while scientists pursue the accuracy of scientific findings (including uncertainties), policymakers demand certainty, clear solutions, and value alignment of interests. These tasks show a notable cultural gap, and filling this gap is crucial. To that end, it is important to create a setting where scientists, policymakers, and stakeholders can directly discuss co-design workshops sharing needs and resources. Analyzing the minutes of co-design workshops by text mining, local environmental research institutes can be an interface between policymakers and scientists, possibly playing an important role in the formulation of adaptation plans that utilize projection data and impact assessments (Iwami et al. 2020). As such, improvements to local climate change adaptation centers that break through the barriers to EBPM are urgently required.

Conclusions and Further Development

This chapter clarified local government officials' needs for climate change science and technologies and drew future developments through science and policy deliberation in co-design workshops. The principal results of this study are as follows.

First, from the results of the scrutiny of the local governments' climate change adaptation plans, the most cited impact items were in the agriculture, forestry, and fisheries sectors, including paddy rice and orchards. The next is the natural disaster sector, which includes debris flows, landslides, and flooding. The third and fourth sectors are health and natural ecosystems, which include natural forests, second-growth forests, and heat stroke, respectively. Among the scientific data used to assess impacts, the category "e) projection data in smaller area scale" is extremely rare and "a) global scale (quoted from IPCC Assessment Report, etc.)" to "d) one's own prefecture or city scale (quoted from past documents, etc.)" are frequent. Therefore, it is necessary to study future technology development while keeping in mind that there could be a latent need for such impact items when revising future plans.

Second, from the results of interviews conducted with some remarkable local governments, they used projection data in two distinct patterns: (i) very limited local governments independently received detailed climate change impact assessments of some items in collaboration with local universities and research institutes or independently obtained customized projection data provided by the Ministry of Environment Climate Change Impact/Adaptation Research Project Team. When unable to obtain this, (ii) most local governments referred to wide-area projection data that were more widely available and anyone could obtain. As one of the challenges and needs related to projection data, the problems of temporal and spatial scale, credibility, and uncertainty were raised. In particular, although there was no resistance to using climate change projection data to formulate plans, at least in environmental bureaus, other bureaus responded differently. In addition, although population projections have been widely adopted, climate change projections are not at that stage. This may be due to the degree of uncertainty. It is necessary for experts and scientists to keep in mind the need for basic scientific projection data by a local government when enacting a climate change adaptation plan to apply them to future technological development.

Third, one of the most important lessons learned from a series of nationwide co-design workshops based on the results of the questionnaire surveys conducted afterward was that a mutual understanding of the use of projection data between scientists and policymakers has not necessarily been promoted over time, although the degree of satisfaction with the workshops themselves has increased. The challenges are that although the participants had a better understanding of the issues, they needed to improve their understanding and ability to read and understand the data to take the information back to their bureaus, resulting in the formulation of a plan, and that it was difficult to link the results of projection and impact assessment to specific measures. Again, the difference in the use of uncertain projection data for planning between bureaus within the local government office remains a significant issue. These may be the reasons for the split in evaluations.

This seems to be a limitation of this co-design workshop; however, in terms of the original purpose of determining the direction of the future development of projections, it provides a valuable opportunity to learn about it. However, simultaneously, further improvement in scientific literacy for policymakers and a deeper understanding of policy for scientists are needed. To overcome these barriers, the role of local climate change adaptation centers is important. Many local climate change adaptation centers are expected to work as an interface between the local government and scientists; thus, they play an important role in applying projections to the formulation of adaptation plans. It is also important to take on the role of sharing needs and resources in co-design workshops, where scientists, policymakers, and stakeholders come together to hold direct discussions. The importance of such a forum for dialogue between science and policy is growing, not just for climate change issues. In the post-coronavirus era, spatial and geographical barriers have been lowered, and it is important to establish a forum for dialogue between science and policy using mechanisms such as online deliberation.

Acknowledgements This study was conducted with the support of MEXT, the Social Implementation Program on Climate Change Adaptation Technology (SI-CAT, JPMXD0715667201), JST RISTEX Grant Number JPMJRX20B5, and a Grant-in-Aid for Scientific Research (B) (JSPS 21H03675), Japan.

References

Araos M, Berrang-Forda L, Forda JD, Austina SE, Biesbroek R, Lesnikowskia A (2016) Climate change adaptation planning in large cities: a systematic global assessment. Environ Sci Policy 66:375–382

Aylett A (2015) Institutionalizing the urban governance of climate change adaptation: results of an international survey. Urban Climate 14:4–16

Baba K, Doi M, Tanaka M (2021) Developing future scenarios for climate change adaptation policy: case study of farming community in Japan. In: Leal Filho W, Luetz J, Ayal D (eds) Handbook of climate change management. Springer International Publishing AG, Cham. https://doi.org/10.1007/978-3-030-22759-3_280-1

Baba K, Matsuura M, Kudo T, Wtanabe S, Kawakubo S, Chujo A, Tanaka H, Tanaka M (2017) Climate change adaptation strategies of local governments in Japan. Oxford Encyclopedia of Climate Science, pp 1–26. https://oxfordre.com/climatescience/display/10.1093/acrefore/9780190228620.001.0001/acrefore-9780190228620-e-597?rskey=BpLfol&%20result=4

Becsi B et al (2020) Towards better informed adaptation strategies: co-designing climate change impact maps for Austrian regions. Clim Change 158(3–4):393–411

Bierbaum R, Smith JB, Lee A, Blair M, Carter L, Chapin FS III, Fleming P, Ruffo S, Stults M, McNeeley S, Wasley E, Verduzco L (2012) A comprehensive review of climate adaptation in the United States: more than before, but less than needed. Mitig Adapt Strat Glob Change 18(3):361–406

Cairney P (2016) The politics of evidence-based policy making. Palgrave Macmillan Publishers, London

Carmin J, Nadkarni N, Rhie C (2012) Progress and challenges in urban climate adaptation planning: results of a global survey. http://resilientcities.iclei.org/fileadmin/sites/resilient-cities/files/Resilient_Cities_2012/Urban_Adaptation_Report_23May2012.pdf

Gero A, Kuruppu N, Mukheibir P (2012) Cross-scale barriers to climate change adaptation in local government. Australia Background Report. https://nccarf.edu.au/cross-scale-barriers-climate-change-adaptation-local-government-australia/
Gonçalves C et al (2022) On the development of a regional climate change adaptation plan: Integrating model-assisted projections and stakeholders' perceptions. Sci Total Environ 805:150320
Heidrich O, Dawson RJ, Reckien D, Walsh CL (2013) Assessment of the climate preparedness of 30 urban areas in the UK, 2013. Clim Change 120:771–784
Iwami A, Matsui T, Kimura M, Baba K, Tanaka M (2020) Organizing the challenges faced by municipalities while formulating climate change adaptation plans. Sustainability 12(3):1203–1203
Massey E et al (2014) Climate policy innovation: the adoption and diffusion of adaptation policies across Europe. Glob Environ Change 29:434–443
Musah-Surugu IJ et al (2018) Diffusion of climate change adaptation policies among local governments in Sub-Saharan Africa: conceptual review. In: Fátima A, Leal Filho W, Azeiteiro U (eds) Theory and practice of climate adaptation, climate change management. Springer International Publishing AG, Cham, pp 65–85. https://link.springer.com/chapter/10.1007/978-3-319-72874-2_4
Reckien et al (2018) How are cities planning to respond to climate change? Assessment of local climate plans from 885 cities in the EU-28. J Clean Prod 191:207–219
Tanaka M, Baba K (eds) (2021) Local policy and social implementation towards climate change adaptation, Gihodo Shuppan. Tokyo. (in Japanese)
Woodruff SC, Stults M (2016) Numerous strategies but limited implementation guidance in US local adaptation plans. Nat Clim Change 6:796–802

Chapter 25
Climate Change Adaptation: An Overview of Contextual Factors Constraining Adaptation Responses of Smallholder Agricultural Producers

Shehu Folaranmi Gbolahan Yusuf
and Oluwaseun Oluwabunmi Popoola

Abstract Climate change remains a vital concern, with its impact over time threatening many sectors, including agriculture. One of the central issues is the inconsistency in climate variables, with extreme climate events likely to have cataclysmic consequences that will affect agriculture and every other sector in the global economy. A broad range of empirical and non-empirical research has been conducted on the effects of climate change on global agricultural productivity, farmers' adaptation actions, and factors influencing and constraining such adaptations. This paper reviews the implications of changing climate conditions for the agricultural productivity of smallholder farmers, the concept of adaptation, the nexus between perception and adaptation, and various theoretical underpinnings that may explain determinants and constraints to adaptation. The theories applied are protection motivation theory, (PMT), conservation of resources theory (COR), rational choice theory (RCT), the knowledge gap theory and the theory of perceived attributes. An attempt is made to understand the factors that constrain smallholders' adaptation responses through a review of the scholarly literature on the topic. This is followed by suggestions, based on the literature, for mitigating the challenges and enhancing the adaptive capacities of smallholder producers.

Introduction

Climate change as a subject matter is complex, requiring vast and detailed analysis from diverse disciplines and a convergent resolve to comprehend the complexities in pursuit of solutions to its impacts (Hertel and Rosch 2010). Various definitions of the concept have been generated over time by numerous researchers, institutions and bodies through analysis of empirical findings. These varied definitions agree on

S. F. G. Yusuf (✉) · O. O. Popoola
Department of Agricultural Economics and Extension, University of Fort Hare, Alice, South Africa
e-mail: fyusuf@ufh.ac.za

W. Leal Filho et al. (eds.), *Climate Change Strategies: Handling the Challenges of Adapting to a Changing Climate*, Climate Change Management,
https://doi.org/10.1007/978-3-031-28728-2_25

two basic aspects: (i) alteration occursinthe elements, properties and composition of climate (i.e., temperature, precipitation and humidity, amongst other aspects), and (ii) these alterations are and have been occurring over an extendedperiod of time (Hegerl et al. 2007: International Strategy for Disaster Reduction (ISDR) 2008 IFAD). The literature maintains that as temperaturescontinue to undergoa rising trend, extreme weather events such as temperature upsurge, droughts, floods and other natural disasters will keep occurring with increasing frequency and intensity, with presumed global consequences (United Nations Framework Convention on Climate Change (UNFCCC) 2007; ISDR 2008; United Nations Environment Programme (UNEP) 2010; Intergovernmental Panel on Climate Change (IPCC) 2013; International Fund for Agricultural Development (IFAD) 2014; Geological Society of America (GSA) 2015). Although projections for the magnitude, degree and regional patterns of climate variability are imprecise, most researchers agree that its consequences will affect generations (Maponya and Mpandeli 2012). These extreme weather events are alreadycausing damage tofarmlands, properties and even the human population (IFAD 2014). Theimpact over time threatens global welfare (Rasul 2021) on an immense scale.

Climate change affects many sectors of the global economy (UNFCCC 2007), with the agricultural sector one of the most susceptible to climatic variations (Ahmed et al. 2013; Fujisawa et al. 2015). Agriculture is the mainstay of many developing economies and the primary source of livelihoods for a number of their social classes (Smith et al. 2014). It is the backbone of most economies, providing employment for 70% or more of some populations and contributing a large proportion of national gross domestic product (GDP) (IFAD 2014).

As climate change impacts intensify, many agricultural regions will experience loss and degradation of critical agricultural soil and water assets that will continue to challenge both rain-fed and irrigated agriculture unless innovative conservation methods are implemented (Bathke et al. 2014). Decreasing crop and livestock production is projected for many regions, with consequences for global food security as crop yields, food prices, food processing, storage, transportation and retailing all bear the brunt of climate change (Bathke et al. 2014). The agricultural sector is therefore under immense pressure as a result of climate change, with its intensity threatening to exceed mankind's capacity to adapt. As global population grows, there will be increased demand for food, which may become increasingly difficult to meet if adequate and appropriate adaptation responses are not put in place. Given these realities, it is essential that efforts at innovating and implementing adaptation measures are relentless.

The drive to reinforce smallholders' adaptation to climate change in order to facilitate sustainable food production is weak in many regions (Ingutia 2021). The majority of smallholder farmers rely on what they produce for food security; however, such-securitycan rarely be guaranteed based on farmers' own production, as individual-production levels are low and few smallholders are able to generate revenues from the sale of surplus products (Marongwe et al. 2011). This increases their vulnerability, making them tremendously reliant (particularly during drought) on external aids for seeds, fertilisers and even food (Marongwe et al. 2011). Smallholder farming communities and associatedstakeholders (with whom farmersinteract and on whom they often depend) need,as a matter of urgency, to harness ways of adapting to longer-term climate changes.

Smallholder farming families are constrained by several factors that limit their access to climate-smart farming. 'Farmers' adaptation to climate change is modulated by their exposure to multiple and interacting stressors and their decision making is often largely affected by these various non-climatic stimuli' (Fujisawa et al. 2015, p. 2). Studies mention social, economic, environmental and institutional constraints as factors affecting the adaptation decisions of most farming populations (Sacramento et al. 2012; Komba and Muchapondwa 2012; Berman et al. 2013; Juana et al. 2013; Terdoo and Adekola 2014; Balew et al. 2014; Taruvinga et al. 2016). Many concur that resource-constrained smallholder producers may not have the financial capacity to adopt climate-smart technologies (Lin 2011). In the agricultural industry, adaptation to climate change islargely framed as a 'grassroots' challenge (Montmasson-Clair and Zwane 2016). There are limited coping and adaptation response options for many grassroots farming communities, with the majority implementing temporary measures to curtail climate change impacts on their productivity (Wiid and Ziervogel 2012; Otieno and Muchapondwa 2016). Unfortunately, climate events will continue to affect the smallholder agricultural sector if urgent attention is not given to implementing sound and long-term adaptation responses and policies (Turpie and Visser 2013). Smallholder adaptation responses remain significantly weakened (Ingutia 2021). This will have to be addressed across the board, as the success of any adaptation or intervention scheme largely depends on smallholder sector inclusion (Ingutia 2021).

The struggle to adapt to climate variations has intensified in recent years (Mitchell and Tanner 2006). Adapting to climate change means

> adjustment in natural or human systems in response to actual or expected climatic stimuli or their effect, which moderates harm or exploits beneficial opportunities. Adaptive strategies are longer-term (beyond a single season) strategies that allow people to respond to a new set of evolving conditions (biophysical, social and economic) that they have not previously experienced. (UNEP) United Nations Convention to Combat Desertification (UNCCD) and United Nations Development Programme (UNDP) 2009, p. 12)

It can be deduced from the literature that adaptation strategies are long-term measures that are planned and executed beyond a specific period in an effort to respond to climate change threats to enhance the adaptation capacities of a population (Abeka et al. 2012; Hizza 2013; Balew et al. 2014).

Humans have long had to develop various means of dealing with inconsistencies in climate (Mitchell and Tanner 2006); in recent decades, agricultural producers' vulnerability to climate change has intensified, motivating them to find ways of adapting to its effects (Fujisawa et al. 2015). Studies (Lynn et al. 2011; Berman et al. 2013) have observed that while many rural communities are adopting measures to lessen the effects of climate change on their productivity, these measures are largely coping mechanisms; they are generally short-term or temporary strategies (Abeka et al. 2012; Tesso et al. 2012; Hizza 2013; Mpandeli et al. 2015; Pandey et al. 2018). Studies suggest that indigenous adaptations are vital complementary measures that maybe used to moderate the deleterious impacts of climate change (Taruvinga et al. 2016), describing the multiple means that smallholders have devised to cope (Juana et al. 2013). However, it would be incorrect to presume that all farmers affected by climate change have taken considerable measures to combat the situation; some studies report 'no adaptation' as an adaptation strategy, 'yet such scenarios are very popular [common], which suggests the desperate nature often faced by rural households when exposed to climate change and variability shocks' (Taruvinga et al. 2016, p. 687).

In addition, existing climate stresses have been found occasionally to exceed individual capacity to cope, a situation which is likely to deteriorate if such capacity is not improved upon (Mitchel and Tanner 2006).

Making decisions at the farm level usually occurs within a short time frame; such decisions are influenced by social and economic factors, amongst others (Tesso et al. 2012). A crucial concern is whether these short-term measures are suitable for mitigating the increasing scale and frequency of risks associated with climate change in the distant future. The sustainability of communal livelihoods rests upon the implementation of long-term strategies (Hizza 2013), which is why it is critical to focus on the feasibility of developing appropriate and cost-effective long-term strategies. The intensity of the effects of climate change on livelihoods largely depends on the scope of adaptation responses (Gbetibouo 2009); also, increases in production efficiencies are achievable using appropriate adaptation strategies (The International Labor Foundation for Sustainable Development (ILFSD) 2011).

For these reasons, assessing existing adaptation measures and constraints to adaptation is a critical step to improving agricultural producers' current strategies and introducing more suitable adaptation systems (Kassie et al. 2013).

Purpose of Paper

The primary purpose of this review paper is to contextualise the constraints to climate change adaptation by smallholder agricultural producers. Under this overarching purpose, the review seeks to:

- highlight the implications of changing climate conditions for agricultural productivity;
- understand the nexus between perception and adaptation;
- explore theories that may explain determinants and constraints to adaptation;
- uncover existing barriers to adaptation through the lens of the literature; and
- make suggestions for mitigating climate change for smallholder producers.

Methodology

Using online academic search engines and data portals such Google Scholar, Science Direct, JSTOR, Web of Science, and the University of Fort Hare's digital library, among others, the authors carried out a wide ranging search of scholarly literature on climate change adaptation and related themes. Delimitations of the search were (i) Limiting the search to literature published in English; (ii) Focusing on literature centered on climate change risks to agricultural production, adaptation responses, determinants of and barriers to adaptation, and theoretical perspectives that may influence adaptation; (iii) Using key search words and phrases such as climate change, risks/threats, perception, adaptation responses, strategies, measures, barriers, determinants or constraints to adaptation, crop, livestock, theories, and other closely related strings of words; and (iv) Extracting relevant materials from predominantly peer-reviewed journals and grey literature with information of value to the investigation.

The authors narrowed down information sources to a combined total of one hundred and six scientific papers and non-traditional publications consisting of published research papers in peer-reviewed journals, books, reports, government publications, working papers and conference proceedings. As the primary aim of this review was to investigate and describe contextual factors that constrain smallholder producers' adaptation to climate change, the review was guided by key questions: (a) What are the effects of climate change on agricultural production? (b) How does people's perception of climate change influence their responses, in terms of either adapting or failing to take any steps to respond to the situation? (c) How can we use existing theories to better explain the factors influencing farmers' adaptation responses? and (d) What major constraints are mentioned in the literature as affecting adaptation responses, and what has been suggested to help mitigate these barriers?

Results and Discussion

Implications for Agricultural Productivity

Challenges arising from climate change have been elucidated by numerous research papers which point out its severe implications for agriculture (Mitchell and Tanner 2006; Trevors 2010; Juana et al. 2013; Bathke et al. 2014; Popoola et al. 2018, 2019a). Studies show increasing impacts of recurrent inconsistences in temperature, rainfall and other extreme climate events on crop and livestock production (Maponya and Mpandeli 2012; Nkomwa et al. 2014; Shongwe et al. 2014; Li et al. 2015; Popoola et al. 2018, 2019a, b). These impacts vary from an increase in pests and diseases to heat stress, altered soil conditions, loss of vegetation and a host of other drastic consequences affecting yields at both the commercial and smallholder level of production.

Some of the adverse effects include: (i) the depletion of production land, (ii) shortening of the growing season, (iii) drying out of soil moisture and nutrient content, affecting crop growth and yields, (iv) increasing infestation/infections of parasites and pathogens, (v) escalating heat stress on plants and animals, (vi) increasing infernos, which reduce grasslands for foraging, consequently affecting livestock/poultry weight, meat, milk and egg production, (vii) unhealthiness and even death, and (viii) loss of farm income earnings, amongst many other impacts (Eitzinger et al. 2010; Hatfield et al. 2011; Department of Economic Development and Environmental Affairs, DEDEA 2011; Maponya and Mpandeli 2012; Newton et al. 2011; Turpie and Visser 2013; Kassie et al. 2013; Alade and Ademola 2013; Megersa et al. 2014; Mpandeli et al. 2015; Shongwe et al. 2014; Nkomwa et al. 2014; Terdoo and Adekola 2014; Li et al. 2015; Popoola et al. 2018, 2019a, b).

The smallholder sector is already a vulnerable group (Lin 2011; Maponya and Mpandeli 2012; Obi 2013), and becomes especially so in the face of substantial increases in climate variability (Benhin 2008; Turpie and Visser 2013). The sector's vulnerability is likely to affect future revenue generation, heightening impoverishment amongst smallholders (Turpie and Visser 2013). For this reason, it is essential that smallholder farmers are assisted to adopt suitable adaptation strategies as quickly as possible.

The Nexus Between Perception and Adaptation

Psychology research papers show that adapting to climate change often depends less on what is learnt about it than on how, and from whom, it is learnt (Clayton et al. 2015). Perception of climate change is influenced by multiple factors, including information media, personal experience and/or other individuals (Clayton et al. 2015). Secondary sources of information may not be as effective in influencing a person's outlook or behaviour towards climate change as having a direct and personal experience of it (Spence et al. 2011; Rudman et al. 2013; Clayton et al. 2015). People's ability to perceive and interpret climate change informs their responses in terms of protecting their wellbeing and future (Ofuoku 2011).

Perception is therefore significant to farmers' adaptation responses (Gbetibouo 2009; Bello et al. 2013; Molua 2014). Contradictions between climate change predictions and the perception of farmers could, for instance, result in farmers adopting deficient or counterproductive adaptation measures (Kassie et al. 2013). Studies have shown that farmers perceptions of climate change on productivity are an important factor informing their decisions on whether and what coping and/or adaptation actions to take (Belay et al. 2017; Fahad and Wang 2018; Khanal et al. 2018; Popoola et al. 2018, 2019a, b). Understanding farmers' perceptions is therefore fundamental in assisting them to develop effective adaptation responses, and for the sustainable growth of communities (Clayton et al. 2015; Belachew and Zuberi 2015).

A study on protection motivation theory (PMT) by Le Dang et al. (2014) states that perception is the first part of the PMT cognitive mediating process. PMT is one of the psychological models applied to climate change adaptation (Reser and Swim 2011). The four major components (threat appraisal, coping appraisal, mal-adaptive coping, and protection motivation) in the thought/cognitive mediating processes of protection motivation theory have been effectively linked to climate change. In their new context, the components are referred to as: (i) climate change risk appraisal or risk perception of climate change, (ii) adaptation appraisal or adaptation assessment, (iii) avoidant maladaptation (iv) adaptation intention (Le Dang et al., 2014).

Individuals tend to evaluate climate change threats by themselves in two ways (Le Dang et al. 2014):

1. **Perceived probability**: This is the anticipation or likelihood that they may be exposed to climate change threats.
2. **Perceived severity**: This is the extent to which they may be affected if eventually exposed to climate change threatening conditions.

It is the perception of these potential threats (the threat assessment) that leads to the next step in the cognitive process, which is adaptation assessment. This involves more thought processes:

1. **Perceived self-efficacy**: This refers to the perception of individuals of their ability to adopt adaptation strategies in response to climate change threats.
2. **Perceived adaptation efficacy**: This is a form of perceived credence in the efficiency of adaptation measures. A number of socioeconomic, cultural, political and institutional factors influence this.
3. **Perceived adaptation cost**: This denotes the individual's perception of what resources, labour and time will be required to adopt and implement adaptation measures.

One of two possible cognitive processes is then activated; individuals may choose to deny or become passive about the threats (maladaptation) or face the threat headon by establishing the intent to take up adaptation measures or strategies (adaptation intention).

Understanding Factors that May Influence/Constrain Climate Change Adaptation

In responding to weather irregularities, smallholder producers experience and deal with multiple stresses arising from hostile environmental conditions caused by recurrent variations in climate, all of which escalate their vulnerability as production become increasingly constrained. These stresses tend to affect the psychological state of farmers, particularly those in remote communities who lack sufficient capacity to adapt to climate variations. Such psychological strains arise from their economic losses and the increased burden of eking out a living by any means possible. It is therefore crucial to understand the concept of stress.

Stress may be defined as a substantial disproportion existing between environmental demands and the response capacity of the principal organism in that environment (McGrath, 1970). Psychological stress is the association between an environment and persons living within the confines of that environment, in which such persons evaluate the environmental conditions as challenging, demanding, or surpassing their resources and threatening their state of wellbeing (Lazarus and Folkman 1984). Much of the literature builds on Cannon's (1929, 1932) conceptualisation of the diverse ways people react to psychological stress (Hobfoll 1988; Arun 2004; Turton and Campbell 2005; Chrousos et al. 2013). In illustrating Cannon's 'fight-or-flight' concept of stress reaction, Hobfoll (1988) stated that the fight-or-flight response is the marshalling of every available bodily resource to directly counter or evade a threat. The author elucidated that the fight-or-flight response can be described using a dual-axis stress response matrix which indicates the different reaction modes—action or inaction, flight or fight—of individuals to psychological stresses. According to this matrix, individuals might choose to acknowledge the reality of the situation and attempt to take certain actions to deal with it, or, alternatively, switch to denial mode and decline to take necessary actions to pre-empt adversity.

In the case of climate change threats, farmers who choose to recognise recurrent extremes of climate and who accept their likely consequences tend to take immediate actions to cope with or adapt to such extremes, while farmers who recoil into a denial state tend to do little or nothing to respond to such conditions. Fundamentally, how people choose to react when faced with challenging circumstances such as unstable weather conditions plays a pivotal role in determining their adaptation capacities.

The theories below help to conceptualise individual thought processes and evaluations of situations, which in turn lead either to being driven to respond or failing to respond and alleviate the stressed state.

Climate Change Responses through the Lens of Theories

The Conservation of Resources Theory (COR)

The principal assumption of the COR theory is that there is an inherent desire in individuals to conserve their resources (both quantity and quality) and to restrain any condition that might threaten the safety of these resources (Hobfoll 1988, 2001). Hobfoll's (1988, 2001) proposition suggests that individuals highly value their resources and have a natural instinct to protect them. Threats to resources could take three forms (Hobfoll 1988; Zamani et al. 2006):

i. **The threat of a net loss of resources**: This supposes that individuals are threatened or faced with a 'potential' and not 'actual' loss of their resources.
ii. **The net loss of resources**: Individuals in this position have to deal with the actual loss of resources.

iii. **The lack of resources gained following investment of resources**: This form applies to individuals who may at some point have invested existing resources in the hope of acquiring additional (turnover of) resources, but who fail to do so.

The actual or potential loss of resources jeopardises people's prized possessions or their very identities (Hobfoll 1988; Zamani et al. 2006); in the case of farmers, they could lose their resources and livelihoods owing to climate change events. They are therefore faced with the critical decision of whether to attempt to conserve their resources or to do nothing about the situation. In order for farmers to succeed at conserving their resources for sustainable production, they need to continuously evolve ways to adapt to changing climate conditions, which they can do only by further investing more resources. Individuals may attempt to reduce their resource losses when faced with potential or actual loss of resources, but even that is done at a cost (Hobfoll 1988). This option proves challenging to smallholder farmers, since most are already resource poor. Smallholder farmers tend to find the requirements of climate readiness beyond their abilities to meet; the necessary technologies that would enable them to adapt to changing climate conditions are extremely expensive. This poses an obvious obstacle for those who recognise the threats and are eager to invest in climate change adaptation strategies, but are unable to do so (Lin 2011).

The Rational Choice Theory (RCT)

RCT theory suggests that all actions are essentially rational in nature and that people estimate the possible costs and benefits of any action before making a decision on what course of action to take (Browning et al. 1999). The theory supposes that individuals make their own choices and that these choices are dependent on two factors:

i. their preferences; and
ii. the constraints that they face under given conditions.

Applying this theory's suppositions to farmers' responses to climate change, we find that farmers select their adaptation strategies based on their personal preference and the surrounding constraints. Any action taken by an individual is not based exclusively on his 'intentions' but also on constraints, which are influenced by at least two independent factors (Friedman and Hechter 1988):

i. **Shortage of resources**: Access to resources varies greatly amongst individuals; achieving certain goals is naturally easier for individuals with good access to resources but challenging or totally elusive for those with limited or no access to resources.
ii. **Social institutions**: Individuals live under laws, rules, ordinances, norms, values, etc., imposed by various societal institutions. These can systematically affect their choice of actions, either increasing or reducing the net benefit of any anticipated course of action.

Smallholder farmers are consistently faced with the challenge of selecting adaptation practices based not only on their preferences but also on other factors, some of which can be negating. This explains some of the determinants and barriers to adopting adaptation strategies. These determinants include age, education, gender, marital status, wealth, changes in agricultural policies, labour conditions, cost of inputs and market prices, availability of extension services, access to credit facilities and access to information, amongst other factors, as observed by numerous studies (Fujisawa et al. 2015; Uddin et al. 2014; Balew et al. 2014; Terdoo and Adekola 2014; Komba and Muchapondwa 2012; Berman et al. 2013; Juana et al. 2013; Sacramento et al. 2012). When farmers make adaptation choices, rational choice theory holds that they anticipate the consequences of taking various courses of action and determine the best action to opt for, based on their preferences and their constraints; as rational beings, they choose a course of action that is most likely to satisfy them and fit within their constraints (Heath 1976; Browning et al. 1999). For instance, the assumption is that farming households will choose a particular adaptation strategy only if the expected benefit is greater than that which might be expected from all other options (Balew et al. 2014). The goal is always to maximise the expected benefit at the end of the production period.

Knowledge Gap Theory

Lack of access to agricultural information impedes productivity (Yaseen et al. 2016). There has been 'concern for segments of the population facing disparities associated with differential patterns of information dissemination and underlying social inequities that influence how information is used' (Eastin et al. 2015, p. 416). This statement exemplifies the knowledge gap theory. It posits that.

> as the infusion of mass media information into a social system increases, segments of the population with higher socioeconomic status tend to acquire this information at a faster rate than the lower status segments, so that the gap in knowledge between these segments tends to increase rather than decrease. (Tichenor et al. 1970, pp. 159–160)

The theory stresses the inequality in information sharing amongst a given population, largely owing to differences in their socioeconomic status. Underlying factors influencing the knowledge gap include education, number of social contacts, time, and the type of media used in information dissemination (Tichenor et al. 1970; Eveland and Scheufele 2000; Eastin et al. 2015). Access to information on climate change, especially amongst smallholder producers, is critical to determining their level of awareness of current climate events and enabling them to adopt suitable adaptive measures. Many of these small producers do not have sufficient access to such information, or informationon adaptation choices and appropriate adoption (Girvetz et al. 2009; Onyeneke and Madukwe 2010; Acquah and Onumah 2011; Gandure et al. 2013; Abid et al. 2015; Popoola et al. 2020a, 2020b). Information and communication technologies (ICT) such as television, radio and cellular phones, among other new ICT media, are relevant information tools that could serve the purpose of massive dissemination of climate change information to a wide target population (Popoola et al. 2020a, b). This could conceivably bridge existing knowledge gaps in smallholder producing communities, where adaptation information is largely needed for sustainable production of crops and livestock.

Theory of Perceived Attributes

This theory posits that people adopt technologies if they are perceived to have the following characteristics (Botha and Atkins 2005):

i. has distinct benefits in relation to existing technologies;
ii. is well suited to communal ethics and practices;
iii. is not too complex in application;
iv. has trial-ability—meaning the technology can be put to test for a period of time without actual adoption; and
v. offers observable results.

This theory is applicable with respect to smallholder farmers adopting modern technology-driven adaptation to climate change. It explains why some farmers are apprehensive about adopting certain technologies, even where there is awareness that benefits of its use supersede indigenous coping strategies. Technologies will be accepted based on their performance, and rejected if they fail to make an impression; there is, however, no guarantee that the innovation, once adopted, will be used continually, as there is a possibility of rejection after adoption (Botha and Atkins 2005). This suggests that whatever new concept, technology or practice is developed to combat climate change must have all five listed attributes for farmers to use it over the long term.

Climate Change Responses through the Lens of Studies

There are various existing and emerging constraints to adaptation (Dapilah and Nielsen 2020). These barriers can prove challenging, especially to smallholder farmers (Masud et al. 2017). Many factors—socioeconomic, environmental and institutional—significantly influence the adaptation choices of farmers, particularly in relation to the accessibility, availability and affordability of specific adaptation measures (Komba and Muchapondwa 2012). These factors modulate the capacity of farmers to adapt to climate change (Fujisawa et al. 2015). Highlighting these factors is therefore critical to understanding farmers' vulnerability to climate change (Singh 2020). Table 25.1 lists these factors and the authors who raise them.

Accessibility to information on climate change is important to adaptation (Pandve et al. 2011; Semenza et al. 2011), helping farmers to make informed, operative and comparative decisions. This in turn enables them to adopt effective adaptation measures (Balew et al. 2014; Ndamani and Watanabe 2015; Belay et al. 2017).

Table 25.1 Determinants of/barriers to climate change adaptation

Factors	Sources
Age	Terdoo and Adekola (2014), Masud et al. (2017), Popoola et al. (2020a)
Education	Sacramento et al. (2012), Terdoo and Adekola (2014), Belay et al. (2017), Masud et al. (2017), Fagariba et al. (2018), Singh (2020), Popoola et al. (2020a)
Income	Kim et al. (2017), Alemayehu and Bewket (2017), Masud et al. (2017), Belay et al. (2017); Pandey et al. (2018), Dapilah and Nielsen (2020), Wang et al. (2020), Marie et al. (2020), Singh (2020), Talanow et al. (2021), Fahad and Wang (2018)
Farm labour/experience	Terdoo and Adekola (2014), Masud et al. (2017), Fagariba et al. (2018), Singh (2020), Fahad and Wang (2018), Belay et al. (2017)
Farm size/property rights/land tenure	Juana et al. (2013), Belay et al. (2017), Masud et al. (2017), Singh (2020), Popoola et al. (2020a), Khanal et al. (2018), Trinh et al (2018), Fahad and Wang (2018)
Access to agricultural markets (input/output)	Terdoo and Adekola (2014), Sacramento et al. (2012), Balew et al. (2014), Masud et al. (2017), Popoola et al. (2020a)
Access to credit facilities	Masud et al. (2017), Kim et al. (2017), Fagariba et al. (2018), Khanal et al. (2018), Fahad and Wang (2018), Popoola et al. (2020a), Singh (2020)
Availability/cost of farm inputs	Terdoo and Adekola (2014), Masud et al. (2017), Belay et al. (2017), Kim et al. (2017)
Institutional constraints (access to government support services and non-government support groups)	Masud et al. (2017), Kim et al. (2017), Belay et al. (2017), Fagariba et al. (2018), Khanal et al. (2018), Fahad and Wang (2018), Wang et al. (2020), Marie et al. (2020), Singh (2020), Popoola et al. (2020a), Talanow et al (2021)
Access to climate change information	Sacramento et al. (2012), Girvetz et al. (2009), Onyeneke and Madukwe (2010), Acquah and Onumah (2011), Gandure et al. (2013), Abid et al. (2015), Masud et al. (2017), Belay et al. (2017), Alemayehu and Bewket (2017), Kim et al. (2017), Pandey et al. (2018), Fagariba et al. (2018), Mutunga et al. (2018), Khanal et al. (2018), Fahad and Wang (2018), Wang et al. (2020), Singh (2020), Popoola et al. (2020a), Talanow et al. (2021)

(continued)

Table 25.1 (continued)

Factors	Sources
Climate change awareness/knowledge	Alemayehu and Bewket (2017), Kim et al. (2017), Trinh et al. (2018), Pandey et al. (2018), Wang et al. (2020), Singh (2020)
Access to contemporary adaptation technologies/potentials for use of technology	Masud et al. (2017), Alemayehu and Bewket (2017), Belay et al. (2017), Wang et al. (2020), Singh (2020), Popoola et al. (2020a)
Biophysical factors (farm topography, soil composition/fertility)	Juana et al. (2013), Terdoo and Adekola (2014), Masud et al. (2017), Singh (2020), Talanow et al (2021)
Availability of adaptation options/cost of implementing specific adaptation measures	Juana et al. (2013), Singh (2020)
Degree of climate change impact on productivity	Khanal et al. (2018), Trinh et al (2018), Fagariba et al. (2018)

Source Authors, 2021

Farmers with better access to information on climate change are more likelyto implement a diversity of adaptation measures (Belay et al. 2017; Khanal et al. 2018; Singh 2020). This may be attributed to their increased awareness, knowledge and perception of changing climate patterns and the varieties of existing response measures (Fagariba et al. 2018; Marie et al. 2020). The reasoning is that access to information is a critical determining factor for intensifying the awareness and knowledge of farmers, which potentially enhances their climate change adaptive capacities (Nkeme and Ndaeyo 2013; Kassie et al. 2013; Ajayi 2014; Balew et al. 2014; Shongwe et al. 2014). As Popoola et al. (2020b, p. 5) state, 'informed knowledge on climate change and resilient strategies is critical to sustaining small-scale agricultural farming systems.'

However, factors in addition to the availability of information also play an important role. Access to government support services is considered critical for strengthening the adaptive capacities of farmers, who are highly vulnerable to climate change impacts in the absence or near absence of government support (Fagariba et al. 2018). The extension arm of any national government is an important public support system for agricultural producers, particularly those in core rural communities (Adekunle 2013). Since adaptation is especially critical for grassroots communities (Montmasson-Clair and Zwane 2016), extension service support is key; it sensitises and educates through training programmes at the grassroots level in order to assist farmers to understand climate change events and how to adapt to them (Popoola et al. 2020b). Many studies have buttressed the positive effects of accessibility to extension and advisory services on farmers' use of diverse adaptation measures (Juana et al. 2013; Belay et al. 2017). As such, farmers who have the most interaction with extension field officers have higher awareness, knowledge and skill when it comes to adopting adaptation measures (Juana et al. 2013).

There are many support measures that national governments can implement to lessen the adverse effects of climate change on producers. Some of these support structures are still lacking in many communities. Wang et al.'s (2020) study, for instance, provides empirical evidence suggesting that farmers do not receive adequate government incentives, and lack support with respect to receiving information on climate change events, adaptation choices, subsidies, and technological and infrastructural aids. Popoola et al.'s (2020b) research also shows that there is a lack of government support services in relation to training on conservation, irrigation and water techniques. Farmers trained in climate change programmes have a higher likelihood of adopting various adaptation techniques (Trinh et al. 2018), while climate change policy regulations and incentives similarly play a very important role in farmers' adaptation decision-making (Wang et al. 2020). National governments therefore ought to provide resource and infrastructural support, promote intervention schemes, and run training events on the use of climate-smart technologies (Popoola et al. 2020a).

Adapting to climate change also requires access to resources, particularly capital resources. The extremely low adaptive response capacity of most smallholder farmers could be a consequence of their poor economic status. Farmers are less likely to make adaptation decisions when they have insufficient resources to do so (Balew et al. 2014; Marie et al. 2020; Singh 2020). For farmers who manage to take adaptation actions, lack of financial resources remains their major issue post adoption (Dapilah and Nielsen 2020). Smallholder farmers with low annual farm incomes are severely constrained when it comes to investing in capital-intensive adaptation technologies (Lin 2011), even assuming that there is awareness of such technologies. According to the conservation of resources theory, every individual has an innate desire to preserve their resources from any form of threat (Hobfoll 1988, 2001); thus it seems likely that where knowledge of climate change adaptation options exists, it is only financial incapacity that prevents smallholder farmers from adopting such. The rational choice theory then applies, as farmers are left with no other option than to adopt adaptation methods based not only on their preference but on their surrounding constraints. This is relatively dependent on the availability or scarcity of resources (Friedman and Hechter 1988). Perceived adaptation costs (capital, time and effort expended in the course of implementing the adaptation response or measure) are a critical cognitive consideration according to protection motivation theory, and also play a key role in determining whether an individual forms an adaptation intention or not (Le Dang et al. 2014).

All of these theories involve the issue of constraints, usually financial; the availability of resources for smallholder farmers is therefore a critical matter requiring prompt attention to enable farmers efficiently adapt to climate change threats.

Farmers require adequate capital in order to afford the expenses associated with certain adaptation measures, making access to affordable credit facilities an essential component of the climate change readiness of smallholder farmers. Access to credit has been hypothesised as a major factor influencing the adoption or lack of it when it comes to adaptation measures (Balew et al. 2014; Singh 2020).

Education, too, is of great importance, as knowledge allows farmers to easily understand available information on climate change as well as the various measures available for adaptation (Balew et al. 2014; Wang et al. 2020). The more educated farmers are, the greater the chances that they will adopt various adaptation measures in their efforts to respond to climate change threats (Trinh et al. 2018; Khanal et al. 2018; Singh 2020). In addition, farmers with advanced literacy levels tend to adopt strategies that are more complex (Fagariba et al. 2018).

Property rights is another critical factor that constrains adaptation. Farmers who do not have sole rights to their farmland may fear losing the land and therefore refrain from investing in the management of such land (Fagariba et al. 2018). Farm size also tends to affect adaptation. The larger the farm, the greater the chances of farm owners taking adaptation actions such as integrating and diversifying crop and livestock production (Belay et al. 2017).

Size and intensity of the labour force is another factor vital to the adoption of strategies. Farming households that have access to more labour are perceived to have greater capacity to adopt adaptation measures than households with limited access to labour (Balew et al. 2014; Singh 2020). In this regard, family size is a vital variable. Larger farming households have a higher probability of taking adaptation decisions that are more effective (Fagariba et al. 2018) or have better chances of engaging in non-farming activities to generate other income sources (Balew et al. 2014).

Market access also affects farm-level adaptation. The more access farmers have to various markets, the better their chances of adaptation. Access to input markets, for instance, affords farmers the opportunity to purchase appropriate agricultural inputs that could help boost their resistance to climate change impacts (Balew et al. 2014; Belay et al. 2017), although the high cost of inputs may well impede farmers' efforts to adapt (Fagariba et al. 2018). Farmers who also have access to output markets are able to increase their farm earnings and gain the financial capability to implement adaptation measures (Balew et al. 2014).

Biophysical factors such assoil composition/fertility, farmland topography, and farmers' perception of precipitation and temperature unpredictability are critical determinants of farmers' adaptation decisions (Alemayehu and Bewket 2017; Fagariba et al. 2018; Singh 2020; Talanow et al. 2021). The degree of climate change impact on farming households' productivity also influences their preparedness and their adaptation decisions. Households facing increasing damage to their production are more likely to respond by adopting adaptation strategies to mitigate the impacts (Khanal et al. 2018; Trinh et al. 2018). For example, farmers who perceive the effect of increased temperature on their soil moisture content tend to swiftly employ adaptation measures to improve soil moisture levels (Fagariba et al. 2018).

Conclusion and Recommendations

The consensus across studies appears to be that its complexity notwithstanding, climate change comes with grave consequences in the form of extreme weather events and resultant impacts, affecting major global economic sectors, including and especially agriculture. Studies have projected a drastic decline in agricultural production owing to climate variability. This threat to agriculture has massive implications for farmers, especially resource-poor smallholder producers who constitute one of the most vulnerable groups to climate change. The literature draws attention to the fact that smallholder farmers are already a constrained and vulnerable group, and are at greater economic risk from climate change than many other groups. As a result, they have become a significant population of interest.

In general, the literature reviewed in this paper shows that even though smallholder farmers are responding to climate change stresses to an extent, many factors hugely affect their adaptation capacities. Amongst these are age, farming experience, education, farm size, property rights, income, availability/cost of farm labour and inputs, access to agricultural markets, information, and credit facilities and institutional constraints, such as access to state and non-state support systems. Other climate change-related factors said to constrain adaptation include: knowledge and awareness levels of farmers about climate change, level of climate change impact on productivity, biophysical factors such as farmland terrains and soil composition, availability of and access to modern adaptation technologies, the potential to use such technologies, availability of adaptation options, and the cost of implementing such options.

Theories are described as logical explanations for why something is as it is or acts as it does. This paper has examined the theories that appear to best explain why farmers act the way they do in relation to climate change adaptation. The *theory of perceived attributes*, for example, explains that the features of whatever technology being introduced to farmers influences the adoption rate of such technology. If a climate smart technology is not simple to use or does not offer a better result when implemented than indigenous methods do, the likelihood of its adoption will be minimal. The *knowledge gap theory* stresses the bridging of information gaps between different social classes; if smallholder producers had the same access to climate change information as their counterparts (commercial producers) do, it could help inform their decisions with respect to an effective response to climate change stresses.

Adapting to climate change conditions requires access to resources, particularly capital resources. Resource-poor smallholder farmers tend to have extremely low response capacity across the board, considering their low farm income levels. The rationale is that they are severely constrained in investing in capital-intensive adaptation technologies, even where there is high awareness of such technologies. According to the *conservation of resources theory* (COR), individuals have an inborn desire to preserve their resources from any form of threat. This raises concerns for smallholder farmers, as the strong desire to protect their production resources is constrained by their financial incapacities. *Rational choice theory* (RCT) is thus forced into play, with farmers having no alternative than to take up adaptation choices based not just on preference but on constraints, making them dependent on the availability or scarcity of resources. Perceived adaptation cost (capital, time and effort incurred in implementing the adaptation response or measure) is a critical cognitive process in *protection motivation theory* (PMT), and plays a key role in determining whether an individual takes up an adaptation intention or not.

For all of these reasons, there is a need to roll out and implement, consistently, government policies to address a number of these constraints. There is a corresponding need to increase funding and skills in local authorities in order to implement and teach a variety of appropriate and cost-effective climate change adaptation strategies.

Limitation of Paper

There are existing and new barriers or constraints to climate change adaptation that this review article may not have covered. In addition, the theories presented in this paper to explain the influence of certain factors on adaptation are not necessarily the only theories applicable to the subject matter. This is a review article and not a research paper; it does not present any original survey findings on the factors constraining adaptation. As such, its content is limited to the findings of existing scientific and grey literature.

References

Abeka S, Anwer S, Huamaní RB, Bhatt V, Bii S, Muasya BP, Rozario AR, Senisse HR, Soría GV (2012) Women farmers adapting to climate change. Four examples from three continents of women's use of local knowledge in climate change adaptation. Dialogue 09. (Online) https://www.unscn.org/uploads/web/news/Women-farmers-adapting-to-Climate-Change.pdf. Accessed 18 Aug 2021

Abid M, Scheffran J, Schneider UA, Ashfaq MJESD (2015) Farmers' perceptions of and adaptation strategies to climate change and their determinants: the case of Punjab province. Pakistan. Earth System Dynamics 6(1):225–243

Acquah H, Onumah EE (2011) Farmers perception and adaptation to climate change: an estimation of willingness to pay. Agris on-line Papers in Economics and Informatics 3(665–2016–44813):31–39

Adekunle OO (2013) Analysis of effectiveness of agricultural extension service in among rural women: case study of Odeda Local Government, Ogun State, Nigeria. J Agric Sci 5(12):65

Ahmed M, Asif M, Sajad M, Khattak JZK, Ijaz W, Wasaya A, Chun JA (2013) Could agricultural system be adapted to climate change?: a review. Aust J Crop Sci 7(11):1642–1653

Ajayi JO (2014) Awareness of climate change and implications for attaining the Millennium Development Goals (MDGs) in Niger Delta Region of Nigeria. Agris on-line Papers in Econ Inform 6(665–2016–45004):3–11

Alade OA, Ademola AO (2013) Perceived effect of climate variation on poultry production in Oke Ogun area of Oyo State. J Agric Sci 5(10):176–182

Alemayehu A, Bewket W (2017) Smallholder farmers' coping and adaptation strategies to climate change and variability in the central highlands of Ethiopia. Local Environ 22(7):825–839

Arun CP (2004) Fight or flight, forbearance and fortitude: the spectrum of actions of the catecholamines and their cousins. Ann N Y Acad Sci 1018(1):137–140

Balew S, Agwata J, Anyango S (2014) Determinants of adoption choices of climate change adaptation strategies in crop production by small scale farmers in some regions of central Ethiopia. Journal of Natural Sciences Research 4(4):78–93

Bathke DJ, Oglesby R, Rowe C, Wilhite DA (2014) Understanding and assessing climate change: implications for Nebraska. A synthesis report tosSupport decision making and natural resource management in a changing climate. (Online) https://digitalcommons.unl.edu/geosciencefacpub/439/. Accessed 18 Aug 2021

Belachew O, Zuberi MI (2015) Perception of climate change and livelihood of a farming community of Maruf Kebele, Central Oromia. Ethiopia. American Journal of Climate Change 4(03):269–281

Belay A, Recha JW, Woldeamanuel T, Morton JF (2017) Smallholder farmers' adaptation to climate change and determinants of their adaptation decisions in the Central Rift Valley of Ethiopia. Agriculture & Food Security 6(1):1–13

Bello M, Salau ES, Galadima OE, Ali I (2013) Knowledge, perception and adaptation strategies to climate change among farmers of Central State Nigeria. Sustainable Agriculture Research 2(3):107–117

Benhin JK (2008) South African crop farming and climate change: an economic assessment of impacts. Glob Environ Chang 18(4):666–678

Berman RJ, Quinn CH, Paavola J (2013) Identifying drivers of household coping strategies to multiple climatic hazards in Western Uganda: implications for adapting to future. Climate change. Centre for Climate Change Economics and Policy Working Paper No. 149, Sustainability Research Institute Paper No 51, pp.1–33

Botha N, Atkins K (2005) An assessment of five different theoretical frameworks to study the uptake of innovations. Paper presented at the 2005 NZARES Conference Tahuna Conference Centre—Nelson, New Zealand: New Zealand Agricultural and Resource Economics Society Inc. Conference, August 26–27, 2005

Browning G, Halcli A, Webster F (1999) Understanding contemporary society: theories of the present. Sage Publications, London, Thousand Oaks and New Delhi

Cannon WB (1929) Bodily changes in pain, fear, hunger, and rage. An account of recent researches into the function of emotional excitement, New York: Appleton

Cannon WB (1932) The wisdom of the body, New York: NY Norton

Chrousos GP, Loriaux DL, Gold PW (2013) Mechanisms of physical and emotional stress (Vol. 245). Springer Science & Business Media

Clayton S, Devine-Wright P, Stern PC, Whitmarsh L, Carrico A, Steg L, Swim J, Bonnes M (2015) Psychological research and global climate change. Nat Clim Chang 5(7):640–646

Dapilah F, Nielsen JØ (2020) Climate change extremes and barriers to successful adaptation outcomes: disentangling a paradox in the semi-arid savanna zone of northern Ghana. Ambio 49(8):1437–1449

DEDEA (Department of Economic Development and Environmental Affairs) (2011) Eastern cape climate change response strategy. (Online) https://www.cityenergy.org.za/uploads/resource_182.pdf. Accessed 18 Aug 2021

Eastin MS, Cicchirillo V, Mabry A (2015) Extending the digital divide conversation: examining the knowledge gap through media expectancies. J Broadcast Electron Media 59(3):416–437

Eitzinger J, Orlandini S, Stefanski R, Naylor REL (2010) Climate change and agriculture: introductory Editorial. J Agric Sci 148(05):499–500

Eveland WP Jr, Scheufele DA (2000) Connecting news media use with gaps in knowledge and participation. Polit Commun 17(3):215–237

Fagariba CJ, Song S, Soule Baoro SKG (2018) Climate change adaptation strategies and constraints in Northern Ghana: evidence of farmers in Sissala West District. Sustainability 10(5):1484

Fahad S, Wang J (2018) Farmers' risk perception, vulnerability, and adaptation to climate change in rural Pakistan. Land Use Policy 79:301–309

Friedman D, Hechter M (1988) The contribution of rational choice theory to macrosociological research. Sociol Theory 6(2):201–218

Fujisawa M, Kobayashi K, Johnston P, New M (2015) What drives farmers to make top-down or bottom-up adaptation to climate change and fluctuations? A comparative study on 3 cases of apple farming in Japan and South Africa. PLoS ONE 10(3):e0120563

Gandure S, Walker S, Botha JJ (2013) Farmers' perceptions of adaptation to climate change and water stress in a South African rural community. Environmental Development 5(2013):39–53
Gbetibouo GA, (2009) Understanding farmers' perceptions and adaptations to climate change and variability: the case of the Limpopo Basin, South Africa. International Food Policy Reearchs Institute (IFPRI) Discussion Paper 00849, pp.1–36
Girvetz EH, Zganjar C, Raber GT, Maurer EP, Kareiva P, Lawler JJ (2009) Applied climate-change analysis: the climate wizard tool. PLoS ONE 4(12):e8320, 1–19
Hatfield JL, Boote KJ, Kimball BA, Ziska LH, Izaurralde RC, Ort D, Thomson AM, Wolfe D (2011) Climate impacts on agriculture: implications for crop production. Agron J 103(2):351–370
Heath A (1976) Rational choice and social exchange: a critique of exchange theory. Am J Sociol 82(6):1364–1366
Hegerl GC, Zwiers FW, Braconnot P, Gillett NP, Luo Y, Marengo Orsini JA, Nicholls N, Penner JE, Stott PA (2007) Understanding and attributing climate change. In: Solomon S, Qin D, Manning M, Chen Z, Marquis M, Averyt KB, Tignor M, Miller HL. (eds) Climate Change 2007: the physical science basis. Contribution of working group I to the fourth assessment report of the intergovernmental panel on climate change.Cambridge University Press, Cambridge, United Kingdom and New York, NY, USA
Hertel TW, Rosch SD (2010) Climate change, agriculture, and poverty. Appl Econ Perspect Policy 32(3):355–385
Hizza EL (2013) Coping and adaptation strategies to climate change impacts: the case of rural community in Vuga Village-Lushoto District. Moshi University College of Cooperative and Business Studies (MUCCoBS) Working Paper, No.3/2013
Hobfoll SE (1988) The ecology of stress. Hemisphere Publishing, New York
Hobfoll SE (2001) The influence of culture, community, and the nested-self in the stress process: advancing conservation of resources theory. Appl Psychol 50(3):337–421
IFAD (International Fund for Agricultural Development) (2014) The adaptation advantage: the economic benefits of preparing small-scale farmers for climate change. (Online) https://www.ifad.org/documents/38714170/40321175/adaptation_farmers.pdf/6dd54c56-6a70-40eb-bb56-d8fad167c472?t=1555405059000. Accessed 18 Aug 2021
ILFSD (International Labor Foundation for Sustainable Development) (2011) The social dimensions of climate change. ITC-ILO / ACTRAV TU training workshop on climate change policies, green jobs & decent work Bangkok, 21–25 Feb 2011
Ingutia R (2021) The impacts of COVID-19 and climate change on smallholders through the lens of SDGs; and ways to keep smallholders on 2030 agenda. Int J Sust Dev World 28(8):693–708
ISDR (International Strategy for Disaster Reduction) (2008) Climate change and disaster risk Reduction. Geneva, September 2008. Briefing Note 01. (Online) http://eird.org/publicaciones/Climate-Change-DRR.pdf. Accessed 18 Mar 2022
Juana JS, Kahaka Z, Okurut FN (2013) Farmers' perceptions and adaptations to climate change in Sub-Sahara Africa: a synthesis of empirical studies and implications for public policy in african agriculture. J Agric Sci 5(4):121–135
Kassie BT, Hengsdijk H, Rötter R, Kahiluoto H, Asseng S, Van Ittersum M (2013) Adapting to climate variability and change: experiences from cereal-based farming in the Central Rift and Kobo Valleys, Ethiopia. Environ Manag 52(5):1115–1131
Khanal U, Wilson C, Hoang VN, Lee B (2018) Farmers' adaptation to climate change, its determinants and impacts on rice yield in Nepal. Ecol Econ 144:139–147
Kim I, Elisha I, Lawrence E, Moses M (2017) Farmers adaptation strategies to the effect of climate variation on rice production: insight from Benue State, Nigeria. Environ Ecol Res 5(4):289–301
Komba C, Muchapondwa E (2012) Adaptation to climate change by smallholder farmers in Tanzania. Economic Research Southern Africa (ERSA) working paper, 299:1–32
Lazarus RS, Folkman S (1984) Stress, appraisal, and coping. New York Springer Publishing Company

Le Dang H, Li E, Nuberg I, Bruwer J (2014) Understanding farmers' adaptation intention to climate change: a structural equation modelling study in the Mekong Delta, Vietnam. Environ Sci Policy 41:11–22

Li S, An P, Pan Z, Wang F, Li X, Liu Y (2015) Farmers' initiative on adaptation to climate change in the Northern Agro-pastoral Ecotone. International Journal of Disaster Risk Reduction 12(2015):278–284

Lin BB (2011) Resilience in agriculture through crop diversification: adaptive management for environmental change. Bioscience 61(3):183–193

Lynn, K., MacKendrick, K. and Donoghue, E.M., 2011. Social vulnerability and climate change: synthesis of literature. Gen. Tech. Rep. PNW-GTR-838. Portland, OR: US Department of Agriculture, Forest Service, Pacific Northwest Research Station, pp.1–70. (Online) https://www.fs.fed.us/pnw/pubs/pnw_gtr838.pdf. Accessed 18 Aug 2021

Maponya P, Mpandeli S (2012) Climate change adaptation strategies used by Limpopo Province farmers in South Africa. J Agric Sci 4(12):39–47

Marongwe LS, Kwazira K, Jenrich M, Thierfelder C, Kassam A, Friedrich T (2011) An African Success: the Case of Conservation Agriculture in Zimbabwe. Int J Agric Sustain 9(1):153–161

Marie M, Yirga F, Haile M, Tquabo F (2020) Farmers' choices and factors affecting adoption of climate change adaptation strategies: evidence from northwestern Ethiopia. Heliyon 6(4):e03867

Masud MM, Azam MN, Mohiuddin M, Banna H, Akhtar R, Alam AF, Begum H (2017) Adaptation barriers and strategies towards climate change: challenges in the agricultural sector. J Clean Prod 156:698–706

McGrath JE (1970) A conceptual formulation for research on stress. Social and Psychological Factors in Stress, New York: Holt, Rinehart & Winston

Megersa B, Markemann A, Angassa A, Ogutu JO, Piepho HP, Zárate AV (2014) Livestock diversification: an adaptive strategy to climate and rangeland ecosystem changes in Southern Ethiopia. Hum Ecol 42(4):509–520

Mitchell T, Tanner T(2006) Adapting to climate change: challenges and opportunities for the development community. Institute of Development Studies and Tearfund, Teddington, UK

Molua EL (2014) Climate change perception and farmers' adoption of sustainable land management for robust adaptation in Cameroon. J Agric Sci 6(12):202–214

Montmasson-Clair G, Zwane, M (2016) Climate change adaptation and agriculture in South Africa: a policy assessment. Report compiled for World Wide Fund—South Africa (WWF-SA). (Online) http://awsassets.wwf.org.za/downloads/wwf_pfu_policy_brief__lowres_.pdf. Accessed 18 Aug 2021

Mpandeli S, Nesamvuni E, Maponya P (2015) Adapting to the impacts of drought by smallholder farmers in Sekhukhune District in Limpopo Province, South Africa. J Agric Sci 7(2):115

Mutunga EJ, Ndungu CK, Muendo P (2018) Factors influencing smallholder farmers' adaptation to climate variability in Kitui County, Kenya. Int J Environ Sci Nat Res 8(5):01–07

Ndamani F, Watanabe T (2015) Farmers' perceptions about adaptation practices to climate change and barriers to adaptation: a Micro-Level Study in Ghana. Water 7(9):4593–4604

Newton AC, Johnson SN, Gregory PJ (2011) Implications of climate change for diseases, crop yields and food security. Euphytica 179(1):3–18

Nkeme KK, Ndaeyo NU (2013) Climate change and coping strategies among peasant farmers in AkwaIbom State, Nigeria. International Journal of Basic & Applied Sciences 2(1):24–28

Nkomwa EC, Joshua MK, Ngongondo C, Monjerezi M, Chipungu F (2014) Assessing indigenous knowledge systems and climate change adaptation strategies in agriculture: a case study of Chagaka village, Chikhwawa, Southern Malawi. Phys Chem Earth 67(2014):164–172

Obi A (2013) Integration of crops and livestock in the smallholder farming system of the former Homelands of South Africa. J Agric Sci 5(10):183–198

Ofuoku, A.U., 2011. Rural farmers' perception of climate change in central agricultural zone of Delta State, Nigeria. *Indonesian Journal of Agricultural Science*, *12*(2), pp.63–69.

Otieno, J. and Muchapondwa, E., 2016. Agriculture and adaptation to climate change: the role of wildlife ranching in South Africa. *Economic Research Southern Africa (ERSA) Working Paper No.*579, pp.1–28.
Onyeneke RU, Madukwe DK (2010) Adaptation measures by crop farmers in the southeast rainforest zone of Nigeria to climate change. Science World Journal 5(1):32–34
Pandey R, Kumar P, Archie KM, Gupta AK, Joshi PK, Valente D, Petrosillo I (2018) Climate change adaptation in the western-Himalayas: household level perspectives on impacts and barriers. Ecol Ind 84:27–37
Pandve HT, Chawla PS, Fernandez K, Singru SA, Khismatrao D, Pawar S (2011) Assessment of awareness regarding climate change in an urban community. Indian Journal of Occupational and Environmental Medicine 15(3):109–112
Popoola OO, Monde N, Yusuf SFG (2018) Perceptions of climate change impacts and adaptation measures used by crop smallholder farmers in Amathole district municipality, Eastern Cape province, South Africa. GeoJournal 83(6):1205–1221
Popoola OO, Monde N, Yusuf SFG (2019a) Perception and adaptation responses to climate change: an assessment of smallholder livestock farmers in Amathole District Municipality, Eastern Cape Province. South African Journal of Agricultural Extension 47(2):46–57
Popoola OO, Monde N, Yusuf SFG (2019b) Climate change: perception and adaptation responses of poultry smallholder farmers in Amathole District Municipality, Eastern Cape Province of South Africa. South African Journal of Agricultural Extension 47(3):108–119
Popoola OO, Yusuf SFG, Monde N (2020a) Information sources and constraints to climate change adaptation amongst smallholder farmers in Amathole District Municipality, Eastern Cape Province, South Africa. Sustainability 12(14):5846
Popoola OO, Yusuf SFG, Monde N (2020b) South African national climate change response policy sensitization: an assessment of smallholder farmers in Amathole District Municipality, Eastern Cape Province. Sustainability 12(7):2616
Rasul G (2021) A framework for addressing the twin challenges of COVID-19 and climate change for sustainable agriculture and food security in South Asia. Front. Sustain. Food Syst. 5:679037
Reser JP, Swim JK (2011) Adapting to and coping with the threat and impacts of climate change. Am Psychol 66(4):277–317
Rudman LA, McLean MC, Bunzl M (2013) When truth is personally inconvenient, attitudes change: the impact of extreme weather on implicit support for green politicians and explicit climate-change beliefs. Psychol Sci 24(2013):2290–2296
Sacramento A, Matavel A, Basılio M, Bila S (2012) Climate change impacts and coping strategies in Chicualacuala District, Gaza Province, Mozambique. (Online) http://www.unep.org/climatechange/adaptation/Portals/133/documents/Chicualacuala_ReportClimate_Change_Impacts_n_Coping_Strateg%20ies.%20pdf. Accessed 18 Aug 2021
Semenza JC, Ploubidis GB, George LA (2011) Climate change and climate variability: personal motivation for adaptation and mitigation. Environ Health 10(1):46–53
Shongwe P, Masuku MB, Manyatsi AM (2014) Cost benefit analysis of climate change adaptation strategies on crop production systems: a case of Mpolonjeni Area Development Programme (ADP) in Swaziland. Sustainable Agriculture Research 30(1):37–49
Singh S (2020) Farmers' perception of climate change and adaptation decisions: a micro-level evidence from Bundelkhand Region, India. Ecol Indic 116:106475
Smith P, Bustamente M, Ahammad H, Clark H, Dong H, Elsiddig E, Tubiello F (2014). Agriculture, forestry and other land use (AFOLU). In: Edenhofer O, Pichs-Madruga R, Sokona Y, Farahani E, Kadner S, Seyboth K, Adler A, Baum I, Brunner S, Eickemeier P, Kriemann B, Savolainen J, Schlömer S, Von Stechow C, Zwickel T, Minx J (eds) ClimatecChange 2014: mitigation of climate change. Contribution of Working Group III to the Fifth Assessment Report of the Intergovernmental Panel on Climate Change. Cambridge University Press, Cambridge, United Kingdom and New York, NY, USA, pp. 1–179
Spence A, Poortinga W, Butler C, Pidgeon NF (2011) Perceptions of climate change and willingness to save energy related to flood experience. Nat Clim Chang 1(1):46–49

Talanow K, Topp EN, Loos J, Martín-López B (2021) Farmers' perceptions of climate change and adaptation strategies in South Africa's Western Cape. J Rural Stud 81:203–219

Taruvinga A, Visser M, Zhou L (2016) Determinants of rural farmers' adoption of climate change adaptation strategies: evidence from the Amathole District Municipality, Eastern Cape Province, South Africa. International Journal of Environmental Science and Development 7(9):687–692

Terdoo F, Adekola O (2014) Perceptions, knowledge, adaptation and socio-economic cost of climate change in Northern Nigeria. J Agric Sci 6(8):60–71

Tesso G, Emana B, Ketema M (2012) Econometric analysis of local level perception, adaptation and coping strategies to climate change induced shocks in North Shewa, Ethiopia. International Research Journal of Agricultural Science and Soil Science 2(8):347–363

Tichenor PJ, Donohue GA, Olien CN (1970) Mass media flow and differential growth in knowledge. Public Opin Q 34:159–170

Trevors J (2010) Climate change: agriculture and hunger. An International Journal of Environmental Pollution: Water Air Soil Pollution205(Supplement 1), pp. 105.

Trinh TQ, Rañola RF Jr, Camacho LD, Simelton E (2018) Determinants of farmers' adaptation to climate change in agricultural production in the central region of Vietnam. Land Use Policy 70:224–231

Turpie J, Visser M (2013) Chapter 4: the impact of climate change on South Africa's rural areas. (Online) www.ffc.co.za/.../300-chapter-4-impact-of-climate-change-on-south-africas-rural-area. Accessed 18 Aug 2021

Turton S, Campbell C (2005) Tend and befriend versus fight or flight: gender differences in behavioral response to stress among university students. J Appl Biobehav Res 10(4):209–232

Uddin MN, Bokelmann W, Entsminger JS (2014) Factors affecting farmers' adaptation strategies to environmental degradation and climate change effects: a farm level study in bangladesh. Climate 2(4):223–241

UNEP (United Nations Environment Programme), UNCCD (United Nations Convention to Combat Desertification), UNDP (United Nations Development Programme) (2009) Climate change in the African Dry lands: options and opportunities for adaptation and mitigation. (Online)https://www.droughtmanagement.info/literature/UNCCD_climate_change_african_drylands_2009.pdf. Accessed 18 Aug 2021

UNEP (United Nations Environment Programme) (2010) Climate change: an overview—Vision for 2010. (Online) http://www.unep.org/climatechange. Accessed 18 Mar 2022

UNFCCC (United Nations Framework Convention on Climate Change) (2007) Climate change: impacts, vulnerabilities and adaptation in developing countries. (Online) https://unfccc.int/resource/docs/publications/impacts.pdf. Accessed 18 Aug 2021

Wang W, Zhao X, Cao J, Li H, Zhang Q (2020) Barriers and requirements to climate change adaptation of mountainous rural communities in developing countries: the case of the eastern Qinghai-Tibetan Plateau of China. Land Use Policy 95:104354

Wiid N, Ziervogel G (2012) Adapting to climate change in south africa: commercial farmers' perception of and response to changing climate. South African Geogr J 94(2):152–173

Yaseen M, Xu S, Yu W, Hassan S (2016) Farmers' access to agricultural information sources: evidences from rural Pakistan. J Agricultural Chemistry and Environ 5(01):12–19

Zamani GH, Gorgievski-Duijvesteijn MJ, Zarafshani K (2006) Coping with drought: towards a multilevel understanding based on conservation of resources theory. Hum Ecol 34(5):677–692

Chapter 26
Spatial Distribution Modeling of Odonata in the New Aquitaine Region (France): A Tool to Target Refuge Areas Under Climate Change

Anouk Glad and Fanny Mallard

Abstract Odonata (dragonflies and damselflies) are good indicators of climate change effects due to their fast response to climatic variables such as temperature, humidity and amount of rainfall. This study aims to investigate the effect of three scenario of climate change at a regional scale (New Aquitaine region, France) on 59 odonata species distribution using species distribution modeling methods. Those results allow to identify species that will be the most impacted by climate change but also to evaluate changes in odonata diversity across the study area, through the calculation of diversity indices for each climate scenario. 24–33% of the species are predicted loss between 75 and 100% of suitable habitat by 2100 under two scenarios. Predicted distribution map can be use by managers, and stakeholders to target areas to be protect in priority. Different approaches can be pursued: protections of areas that are suitable or will be suitable in the future for rare species and/or target areas that will be suitable for high number of species leading to a higher diversity. By protecting wetland suitable for diverse odonata species, other wetland affiliated species such as amphibians, birds, and plants might benefits from those actions.

Introduction

Climate change affects species and their ecosystems at different rates and magnitudes. However, wetlands are known to be particularly sensitive to climate change due to potential temperature and hydrological variations (drought, rainfall periodicity, and rainfall amount) that can affect directly the presence of such ecosystems (Erwin 2008). Numerous species are depending on such ecosystems to survive, such as birds, insects, amphibians, or reptiles. Among those species, Odonata (dragonflies

A. Glad (✉) · F. Mallard
Cistude Nature, Chemin du Moulinat, 33185 Le Haillan, France
e-mail: gladanouk@gmail.com

F. Mallard
UMR PASSAGES the French National Center for Scientific Research (Centre National de La Recherche Scientifique, CNRS), Maison Des Suds 12, Esplanade Des Antilles, 33607 Pessac Cedex, France

W. Leal Filho et al. (eds.), *Climate Change Strategies: Handling the Challenges of Adapting to a Changing Climate*, Climate Change Management,
https://doi.org/10.1007/978-3-031-28728-2_26

and damselflies) are particularly sensitive to climate change effects due to their fast response to climatic variables such as temperature, humidity, and amount of rainfall (Hassall and Thompson 2008). Odonata are carnivorous insects which have an aquatic larval stage that can last 1–3 years, making populations highly dependent on water availability, quality, and temperature at a long-term scale. The vegetation associated with the ecosystem is also important as it allows females to lay their eggs and the emergence of adults. Thus, those species are making good indicators of climate change effects as first impacts on species phenology have already been reported (Dingemanse and Kalkman 2008). Understanding and monitoring the effects of climate change on such species is essential to guide conservation decisions and management actions on wetlands (Mallard and Couderchet 2019). Two major interrogations can be highlighted concerning species conservation:

- Where should the management actions take place?
- What kind of action should be applied?

The first step would be to spatially prioritize areas that need effective protection. To provide meaningful results for managers and decisions makers, a local scale is relevant. In France, the regional scale is a pertinent scale for the implementation of such actions (Mallard and Couderchet 2019). The New-Aquitaine region, located in the south-west of France, encompasses a large variety of ecosystems from the Atlantic coast, the Pyrenees mountains ranges, to the "Loire" margins, the Limousin plateau of the "Massif Central", and the northernmost part of the Poitou–Charentes sector, with intensive agricultural areas. This variety of ecosystems makes the New-Aquitaine region an area exposed to multiple effects of climate change. Furthermore, this region will be exposed to major climatic changes regarding other French regions (Le Treut 2020). One of the challenges thus will be to take into consideration this diversity in terms of ecosystems, macro-climate, and species presence to obtain reliable results for conservation planning.

To take into account the different Odonata species' ecological needs, species distribution models (SDMs) can be used. Such models, also known as habitat suitability models, are widely use to provide insights into climate change impacts on future species distribution (Elith et al. 2010) and to guide management planning and decisions (Elith and Leathwick 2009; Franklin 2009). Species distribution models link species occurrence to a selected set of environmental predictors. The results can be used to predict the potential distribution of the target species in the study area by showing the environment suitability variations regarding the predictors used for modeling. The predictions can also be extrapolated to other areas where the presence of the species is unknown or to different climate evolution scenarios (Heikkinen et al. 2006; Iturbide et al. 2018). The choice of the statistical methods is essential as it is a source of uncertainties regarding results obtained from different algorithms (Hijmans and Graham 2006). To solve this issue, ensemble modeling methods have been developed, allowing the use of multiple algorithms to produce average predicted maps along with uncertainty measurements (Thuiller et al. 2009). Nevertheless, the use of SDMs under climate change is challenging (Heikkinen et al. 2006), and future climate model scenario uncertainties lead to variations in the predicted results.

Furthermore, the realized niche is mainly approximated by the model, sometimes leading to incomplete responses to environmental predictors and, consequently, to non-pertinent results, in particular when the full range of climatic predictors is not observed in the current climate (Thuiller et al. 2004). Despite those constraints and limitations, SDM predictions can be used by managers and stakeholders to target priority areas for protection.

In the case of Odonata, such models have been tested. Using a large number of species, SDMs have been performed in both north America (Hassall 2012) and Europe (Keil et al. 2008). Both studies predicted only the actual distribution of the species and the overall Odonata assemblage diversity and richness by using water-energy and climate variables. However, those studies were conducted at a coarse scale (55 and 220 km^2, respectively), impeding the use of the results at a local scale. In both studies, species diversity was explained by water-energy variables, highlighting that climate change can greatly influence species assemblage. This hypothesis was confirmed by Cerini et al. (2020), as this study showed that large changes are ongoing in countries such as Morocco or Sweden. If some species were benefiting from the changes (generalist species), specialist species were more likely subject to local population extinction. Odonata communities affiliated with a lentic habitat were more likely to suffer important changes than communities affiliated with a lotic habitat. Finally, numerous species were found to have migrated toward north over the last 50 years. One study investigated the potential future distribution for some selected species over Europe (Jaeschke et al. 2013) and showed that some species were predicted to lose from −68% to −73% of their distribution range (*Coenagrion ornatum, Coenagrion mercurial, Ophiogomphus Cecilia, Leucorrhinia albifrons and Leucorrhina caudalis*). In particular, *Leucorrhinia albifrons* and *Leucorrhina caudalis* are predicted to disappear from the New-Aquitaine region (France).

These studies show the importance of climate for Odonata, although the potential future distribution of a large number of species at a local scale has not been investigated so far. Thus, this study investigates the effects of three climate change scenarios at a regional scale (New Aquitaine region, France) on the potential distribution of 53 Odonata species, using species distribution modeling methods. Based on the results, indices such as protected species index, heritage species index, or diversity index are calculated and used to provide a climatic sanctuary index, intending to transfer meaningful information to managers and decision makers. A similarity index was also calculated to identify areas with higher change rates.

Material and Methods

Study Area

New Aquitaine is a French administrative region created by the territorial reform of 2015 and containing 12 departments. It forms the largest region in France (84,000 km^2) and the 3rd most populated one, with 6 million inhabitants (source: Demographic National Institute INED January 1, 2021). It extends to the south-west of France from the Pyrenean Mountain landscapes to the "Loire" margins in the north; and from 723-km Atlantic coast in the west to the Limousin plateau of the "Massif Central" to the east. The northernmost part of the Poitou–Charentes sector extends over agricultural and wine-growing plateaus. In the southwest quarter, there is the "Landes de Gascogne", with a forest mainly containing pines and covering an area of 1 million hectares. A dense and diverse network of wetlands and streams is present in the region (Fig. 26.1). Large wetland areas are found along the Gironde estuary, the Dordogne River, the Seudre River, and the Charente River. The Poitevin swamp ('Marais Poitevin"), located in the north-west of the region, is the second largest wetland in France after the Camargue wetland, with a surface of 1,000 km^2, overlapping two regions (New-Aquitaine and Loire's lands (Pays de la Loire). The Deux-Sèvres department, located in the north of the region, hosts a dense network of permanent and intermittent streams, as well as the Limousin area, the Landes department, and the marge of the Pyrenees in the south.

Odonata Presence Dataset

Data on the presence of species were obtained from the "Observatoire de la Faune Sauvage de Nouvelle-Aquitaine", called FAUNA (2020). The dataset contains observations from observers affiliated to organizations (associations, universities, parks, reserves, departments), independent experts, and from people with non-specified affiliations. To avoid unreliable models, species with less than 20 observations were not included in this study. A first analysis of the results of future modeling was carried out to rule out the species whose results did not correspond to the known distribution of the species according to naturalist field knowledge of the region. Among the remaining species, 53 were selected. To limit the effects of sampling bias and spatial autocorrelation (Kramer-Schadt et al. 2013), the presence data were aggregated by a 1-km^2 mesh.

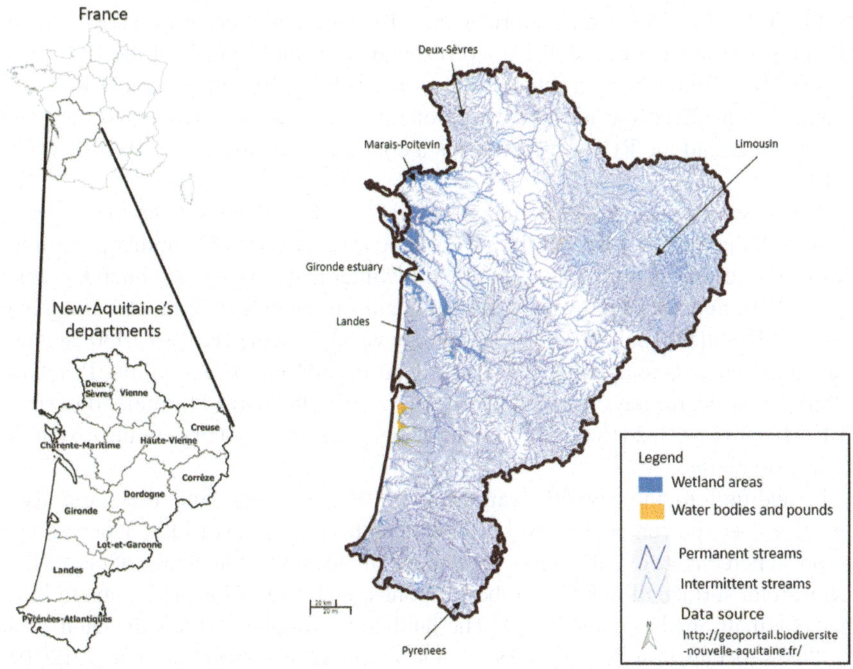

Fig. 26.1 Map of the landscaped sectors of the New Aquitaine region (South-West of France), Authors 2022

Climate Datasets

We combined climatic data with land cover (vegetation) and topographic data, as the distribution of species is not only influenced by the climate.

The climatic variables were Aladin (Limited area dynamic development adaption international) 52 CNRM (National center of meteorological research) 2014 climatic simulations, downloaded via the "Météo France" access to the French Regional climatic scenario for Impact and Adaptation of Society (DRIAS) website (Drias 2020; Lémond et al. 2011), from which indices were calculated. The raw data was at a resolution of 8 km^2, which was divided into a 1-km^2 mesh. Daily predictions were summarized into averages calculated. The present reference was defined by the years from 1991 to 2020, and the future periods or horizons were defined as follows: Horizon 1 (H1) = 2021–2050, Horizon 2 (H2): 2051–2070, and Horizon 3 (H3): 2071–2100. These horizons were determined according to the recommendations of the official service for meteorology and climatology in France, namely "Météo-France", which recommends a duration of around 30 years to smooth out the "noises" resulting from climate simulations (Ouzeau et al. 2014). The simulations in the period known as the present were derived for the years 2006–2020 from the predicted data of the Representative Concentration Pathway (RCP) 8.5 scenario, and the previous data from 1991 to 2005 were historical data. Three climatic scenarios were retained. The most optimistic one, RCP 2.6, corresponds to a stabilization of the CO_2 rates before 2100, followed by a return to the current level. The intermediate RCP 4.5 scenario, considered the most probable one, is a continuous increase to the level of 4.5W/m2, and the RCP 8.5 represents a continuous increase to 8.5W/m^2 (IPCC 2014).

A first selection of the variables was based on knowledge of the biology and ecology of the species. Five variables were kept: average relative humidity in summer (MoyHR_ete), mean temperature in autumn/winter and in spring/summer(Jaeschke et al. 2013), number of rainy days in spring and summer (NjP_print_ete) (Jaeschke et al. 2013), and the sum of the degree-days above 30 °C of the year (TotDJ30_annee), indicating extreme temperatures that can lead to mortality (Leggott and Pritchard 1986). These degree days were calculated from the mathematical formulas described by F. Mallard (2017, 2018a, 2019), linearly approximating the temperature evolution of the growth rate.

In addition to the climatic variables, those describing land use were used (expressed as a percentage of cover on a pixel with a resolution of 1 km^2): percentage of forest, percentage of cultivation, percentage of meadows, percentage of moors and lawns, total surface of ponds, number of ponds, and sum of linear streams in kilometers (small, medium, and large). The land cover variables were calculated from the "CES Occupation des sols (OSO)" and produced from satellite images (2019) with a resolution of 25 m. The 23 nomenclatures were merged into 11 classes. Elevation was selected as a topographic variable (IGN data). Among those variables, some were correlated (Spearman's correlation test ($p = 0.05$) but were nevertheless used in modeling as the correlation relationship may be different in future scenarios (Braunisch et al. 2013).

Modeling Methods

The analyses were carried out using the R software (version 3.6.3, R Core Team 2020). The species distribution models were performed with the BIOMOD2 package, version 3.4.6 (Thuiller et al. 2009), facilitating the creation of ensemble models on the basis of different algorithms. For this study, the algorithms generalized linear models (GLM), RandomForest, and MaxEnt (Phillips et al. 2006) were selected, and for each algorithm, three repetitions were performed to evaluate the performance (80% of the data for the training of the model and 20% for the evaluation). The background points were generated randomly over the New-Aquitaine study area ($n = 10{,}000$). The threshold for the creation of the overall models was the 0.5 quantile of the True Skill Statistic (TSS) values, obtained from the 24 models (Allouche et al. 2006). Furthermore, if all individual models had a TSS value less than 0.4; then, no overall model was created. Overall models represent the average of the predictions of the individual models weighted by the performance of each one. Presence/absence maps were created with the TSS threshold (rangesize function of the BIOMOD2 package). These maps were used to calculate the number of meshes going from "present" to "absent" (loss) and from "absent" to "present" (gain).

Indices

To explain the evolution of diversity profiles, regional indices were created in collaboration with the managers of natural environments from the "CEN" New-Aquitaine territory, with the aim to make the easier to understand and to respond to the manager's needs (Bailleux et al. 2017). These indices take into account the management issues based on protected species (species targeted by a national action planning) (IPi index) or heritage species (IAi index), including species targeted by a regional action planning (local actions taken to preserve protected, threatened (red list), rare species of naturalistic interest) (Bailleux et al. 2017; Mallard 2021b). The diversity index (IDi) represents the number of species in each mesh divided by the number of species studied to identify the areas at stake which have significant diversity and need to be preserved. Based on these indices, we calculated the climatic refuge index, which represents the mean of the three indices presented: protection index, heritage index, and diversity index. This index allows the identification of areas with conservation challenges containing protected and heritage species as well as a significant diversity. Finally, the similarity index was calculated, which is a comparison of the species' presence between the present and the future climatic scenarios, allowing to anticipate the possibilities of change in the population assemblage (number of species in common in the mesh divided by the total number of species). These areas should be protected from the destruction and degradation of natural habitats. In addition, natural habitats could be restored in current or future favorable climatic zones to provide high-quality refuge areas for multiple species. The indices were calculated using the following formulas:

IPi protection index in mesh *i*
npi number of protected species in mesh *i*.
NPi number of protected species in total = 8

$$\text{IPi} = \left(\frac{npi}{NPi}\right)x100 \quad (27.1)$$

IAi heritage index in mesh *i*
nai number of heritage species in mesh *i*
NAi number of heritage species in total = 21

$$\text{IAi} = \left(\frac{nai}{NAi}\right)x100 \quad (27.2)$$

IDi index of diversity in mesh *i*
ndi number of species in mesh *i*.
NDi number of species in total = 100

$$\text{IDi} = \left(\frac{ndi}{NDi}\right)x100 \quad (27.3)$$

ISi similarity index in mesh *i*
nsi number of common species present in mesh *i*
NSi number of species in total in the mesh

$$\text{ISi} = \left(\frac{nsi}{NSi}\right)x100 \quad (27.4)$$

ICi climatic sanctuary index

$$\text{ICi} = \left(\frac{IPi + IAi + IDi}{3}\right) \quad (27.5)$$

Results

Species Level

The species distribution models showed contrasting results regarding the scenarios and the species. Scenario RCP 2.6 had the lowest impact on the potential future distribution of Odonata in the New-Aquitaine region, with the majority of the species predicted to have a habitat loss from 0 to 25% (14 species H3) and a habitat gain from 0 to 50% (7 species in horizon H3 2071–2100). However, the impact on species distribution was stronger for scenarios RCP 4.5 and 8.5. Regarding scenario RCP 4.5, 24% of the species targeted in this study may lose between 75 and 100% of their habitat at horizon H3 (13 species). Concerning scenario RCP 8.5, 32% of the species are expected to lose between 75 and 100% of their habitat at horizon H3 (17 species). On the contrary, other species may benefit from the climatic changes in New-Aquitaine: two species in RCP 4.5 and four species in RCP 8.5 may gain between 500 and 1000% of their actual suitable habitat range. An additional 11 and 12 species, respectively, are predicted to gain between 100 and 500% of suitable habitat (Fig. 26.2).

As an example, three case species are presented below. *Somatochlora metallica* is predicted to be present mainly at the marge of the Central Massif plateau and occasionally in the "Landes triangle" and in the Poitou–Charentes area. This species is predicted to lose large parts of its habitat in scenarios RCP 4.5 and RCP 8.5. It may gain habitat at suitable climatic conditions at higher elevations, but over time, the

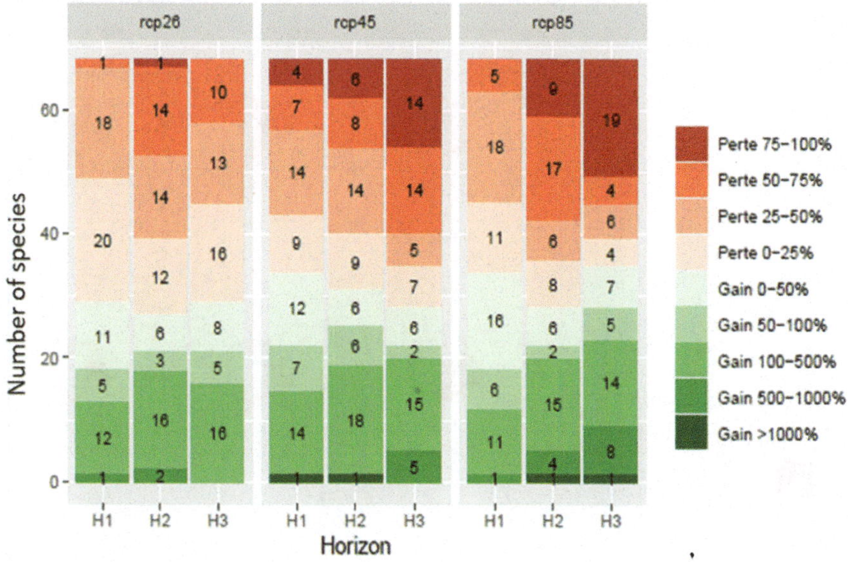

Fig. 26.2 Number of species per category of loss and gain of potential suitable habitat (H1 = 2021–2050, H2 = 2040–2070, H3 = 2071–2100)

proportion of suitable habitat in the New-Aquitaine region will decreased (Fig. 26.3). In the case of scenario RCP 8.5, in the long term, almost all remaining suitable areas may be lost for this species.

The species *Oxygastra curtisii* is predicted to be present along streams, according to our knowledge of the species' habitat preference. Under scenario RCP 2.6, the species is predicted to be still present in the region, gaining suitable habitat in the north and at higher elevations and still remaining in the Pyrenees-Altantique area (south). However, large parts of the suitable areas located in the center of the region are predicted to be lost under this scenario. In scenario RCP 4.5, the species does not find suitable climatic conditions in the area where it is currently present. However, it will gain suitable habitats at higher elevations (Limousin, Pyrenees) and along the southern part of the Atlantic coast at horizons H1 and H2. At horizon H3, only areas at higher elevations will remain suitable for this species. Scenario RCP 8.5 will result in the progressive disappearance of the optimal climatic conditions for *Oxygastra curtisii* in the region, leading to a potential extinction of the species in the study area (Fig. 26.4).

For the species *Platycnemis latipes,* scenario RCP 2.6 will result in the expansion of suitable climatic conditions in the northern part of the region (Poitou–Charentes and Limousin) over time. This pattern would also be observed for scenarios RCP 4.5 and 8.5, with a gain observed in the north and at higher elevations and a loss in the lower-elevation areas in the south. The predicted gains are lower for scenario RCP 8.5 H3 than for the intermediate scenario RCP 4.5 (Fig. 26.5).

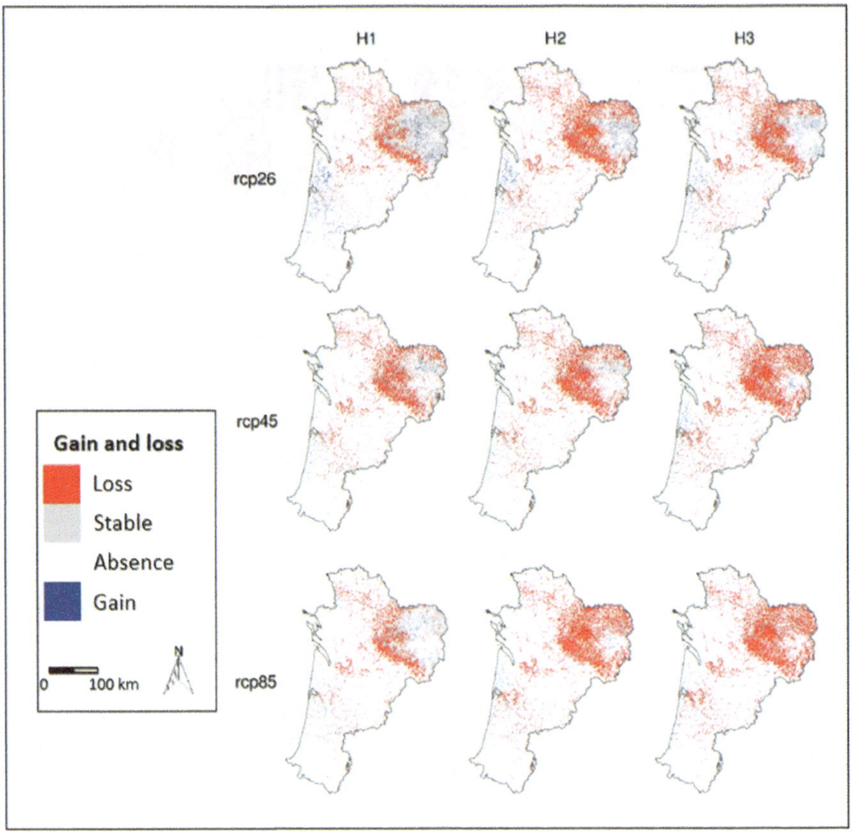

Fig. 26.3 Potential habitat gains and losses for *Somatochlora metallica*

Multi-Species Level: The Use of Indices

The protection index shows that in the New-Aquitaine region, those species are mainly affiliated with stream type (Fig. 26.6). Among the 53 species, only 2 are protected: *Oxygastra curtisii* and *Gomphus gasilinii.* Both species are known to be present in streams, and future scenarios show that they may be highly impacted by climate change. The number of protected species will decline over time for scenarios RCP 4.5 and RCP 8.5.

The heritage index shows that the species are mainly located in the "Landes triangle" on the Atlantic coast (Fig. 26.7). Of the 53 species targeted, 9 are categorized as heritage species in New-Aquitaine (*Aeshna juncea, Brachytron pratense, Erythromma najas, Gomphus graslinii, Lestes dryas, Lestes virens, Oxygastra curtisii, Somatochlora flavomaculata, Sympetrum danae*). This area is known for its pond network, associated with pine forest plantations. Future climatic changes are predicted to impact the number of heritage species that will find suitable habitats

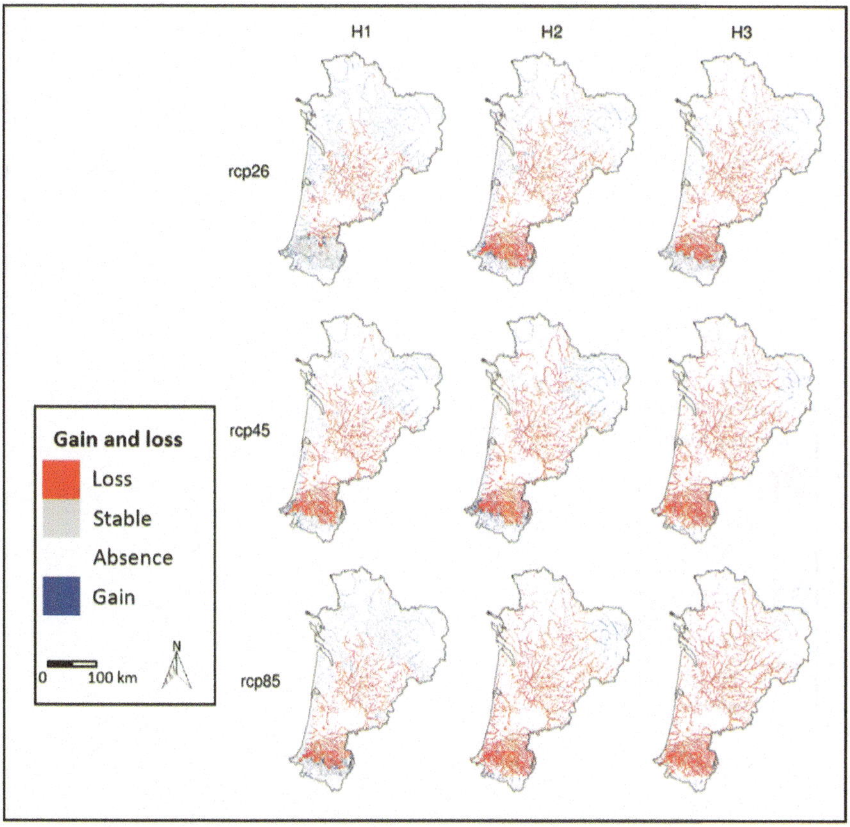

Fig. 26.4 Potential habitat gains and losses for *Oxygastra curtisii*

in the New-Aquitaine region. For scenarios RCP 4.5 and RCP 8.5, the number of species is predicted to decrease. In addition, the most suitable habitats are shifting from inland areas to coastal areas over time.

The diversity calculated for each mesh over the New-Aquitaine region shows that climate change will affect both diversity level and distribution, in particular for the two scenarios RCP 4.5 and RCP 8.5 (Fig. 26.8). For the present period, a higher diversity is found in the Limousin area and in the "Landais triangle" area, mainly around streams. Over time, for both RCP 4.5 and RCP 8.5, the diversity will decrease in some areas in the region. In particular, the Limousin area will show a large decrease in diversity, whereas the coastal area is predicted to have a higher diversity compared to the current period. Overall, the diversity is predicted to reach an intermediate level due to the predicted expansion of several species such as *Erythromma viridulum, Ischnura pumilio,* and *Sympetrum meridionale.*

The climatic sanctuary index shows important variations between the present and the different scenarios and horizons (Fig. 26.9). In the current period, the areas identified as refuges are located in the Landais triangle in the Limousin area and

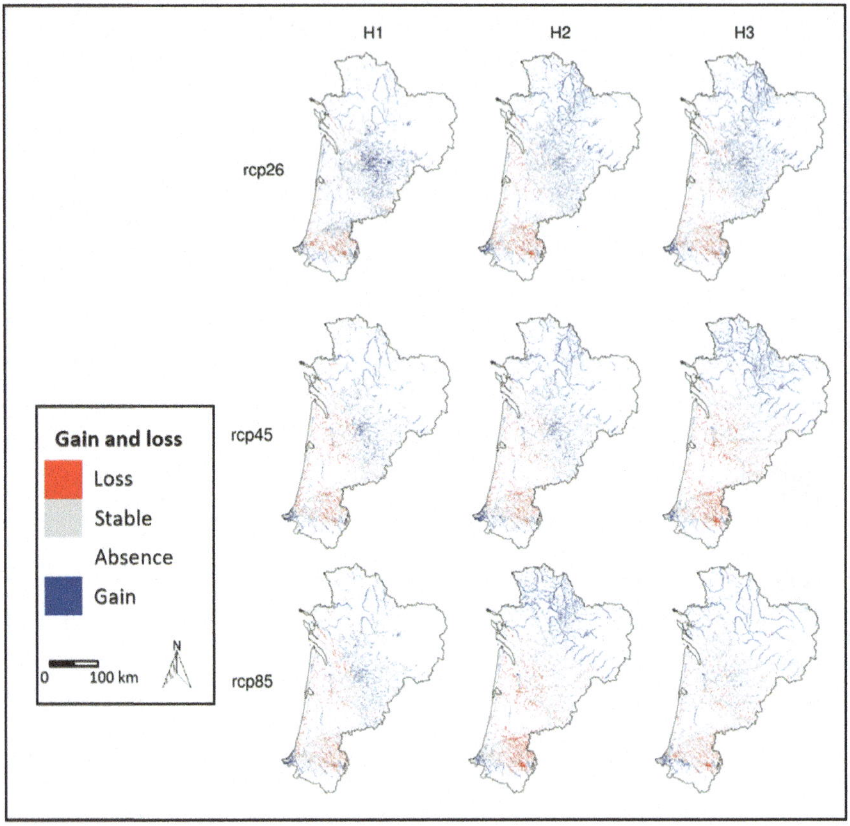

Fig. 26.5 Potential habitat gains and losses for *Platycnemis latipes*

in the Pyrénées-Atlantiques area. Those locations host a high odonatan diversity, along with protected and heritage species. Regarding scenario RCP 2.6, few changes are observed, mainly in the Limousin area at horizon H3, with a decrease in the proportion of the mesh that can be identified as sanctuaries at lower elevations. In scenario RCP 4.5, areas that may become climatic refuges in the future are located along the Atlantic coast; the refuge area in the Limousin no longer provides suitable climatic conditions for the target species. Scenario RCP 8.5 is similar concerning the location of the area at stake. The RCP 8.5 horizon H2 is similar to the RCP 4.5 H3 described earlier. However, regarding the scenario RCP 8.5 H3, the importance of the area located near the Atlantic coast is highlighted as a climatic refuge regarding the north of the region and the Limousin.

The similarity index compares the species' potential presence from models for the current period to models for future climate scenarios (Fig. 26.10). The results show that the number of similar species will decrease over time for scenarios RCP 4.5 and

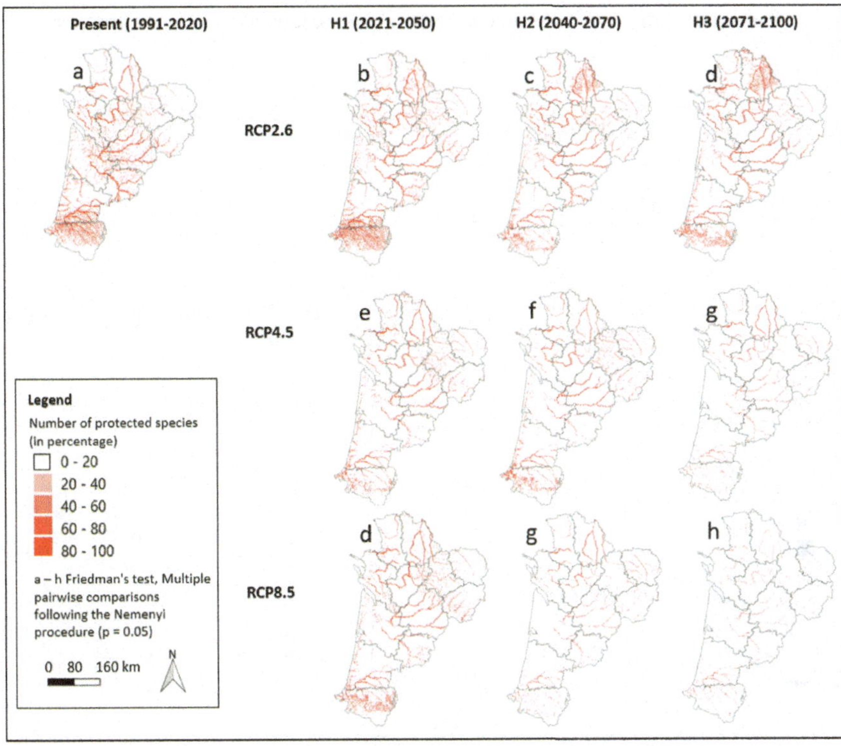

Fig. 26.6 Protection index values calculated for Odonata in the New-Aquitaine region

RCP 8.5. Thus, high changes in the Odonata species assemblage are expected in the region due to the disappearance of certain species, including protected and heritage species.

Discussion

We used species distribution models to target species and areas that will be threatened by climatic changes, with the aim to provide managers and decision makers with a tool to protect the biodiversity in the New-Aquitaine region (France). The use of Odonata species living in wetlands, such as ponds or streams, based on their sensitivity to climatic changes, allowed an insight of the impacts on the species' potential presence in the study area. To our knowledge, such information on odonatan species in the regional New-Aquitaine area is lacking. The results of this study allow to identify species that will be the most impacted by climate change but also to evaluate changes in Odonata diversity across the study area via the calculation of indices for each climate scenario.

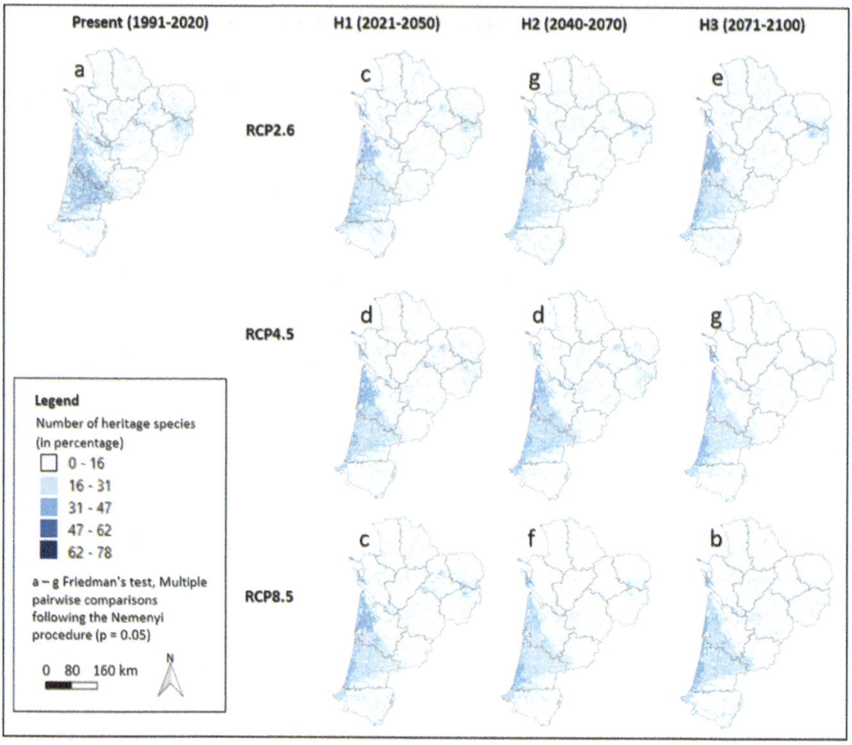

Fig. 26.7 Heritage index calculated for Odonata in the New-Aquitaine region

In the first part, species distribution models allowed us to predict potential suitable habitat for species individually, facilitating the identification of species that may have important changes in their future distributions. Our results showed that a large number of species will experience a reduction in their suitable habitats. Some species may even not find a suitable habitat in the New-Aquitaine region anymore in the case of scenarios RCP 4.5 and 8.5 at the long-term horizon H3 (2071–2100), which is the case for *Somatochlora metallica*, *Oxygastra curtisii*, or *Aeshna grandis*. These species, although they have different ecological needs, are still negatively impacted by climate change under the two most pessimistic scenarios. *Oxygastra curtisii* occurs in streams and is widely distributed in New-Aquitaine. It mainly occurs in the Mediterranean basin (Bailleux et al. 2017) but is nevertheless predicted to have no suitable habitat remaining in the RCP 8.5 H3. *Somatochlora metallica* and *Aeshna grandis* are found at higher elevations in the Central Massif (Limousin) and are predicted to be negatively impacted by climate change, with no suitable habitat remaining in the Limousin by 2071–2100 (RCP 8.5). Some other species will mostly benefit from the changes, with the range of suitable areas increasing over time by up to 1,000% of their current potential distribution range. Those species are

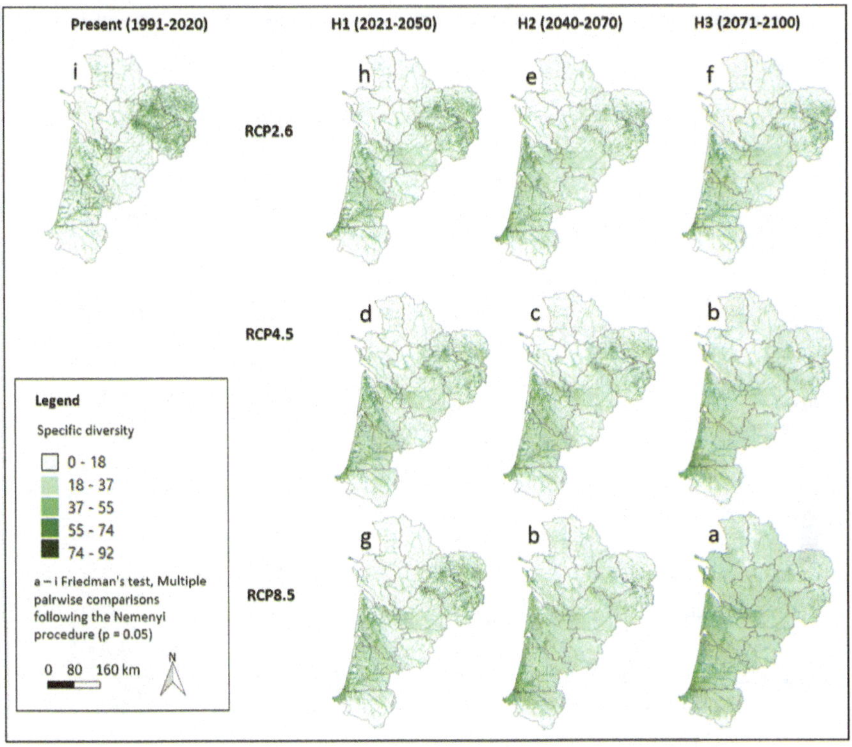

Fig. 26.8 Specific diversity index values calculated for Odonata in the New-Aquitaine region

thermophile species, and some of them area affiliated with Mediterranean ecosystems (Bailleux et al. 2017). The expansion of their range under climate change is thus expected and confirmed by the SDMs and the predictions of experts, particular for species such as *Anax imperator, Ishnura elegans,* and *Anax parthenope*, known either for their history of colonization or their preference for higher temperatures (Ott 2001). *Platycnemis latipes* is as species affiliated to the Mediterranean climate, and models for this species showed that an expansion towards the north and higher latitudes is expected. Yet, the intermediate scenario RCP 4.5 seems to be the most suitable one for this species. Nevertheless, these findings are limited. For example, by selecting only species that are already present in New-Aquitaine, our study does not take into account the arrival of new species both from the Mediterranean basin and from southern countries (Spain, North Africa) (Morghad et al. 2019).

Calculated from the species' individual predictions, the use of indices can help to estimate the amplitude of the impacts and to identify areas which may need further protection (Mallard 2020, 2021a). Among the various species impacted by climate change, protected and heritage species are largely negatively impacted. Indeed, the two protected species are likely to disappear by 2100, confirming their status in the New-Aquitaine region and the necessity to protect them in the area where populations

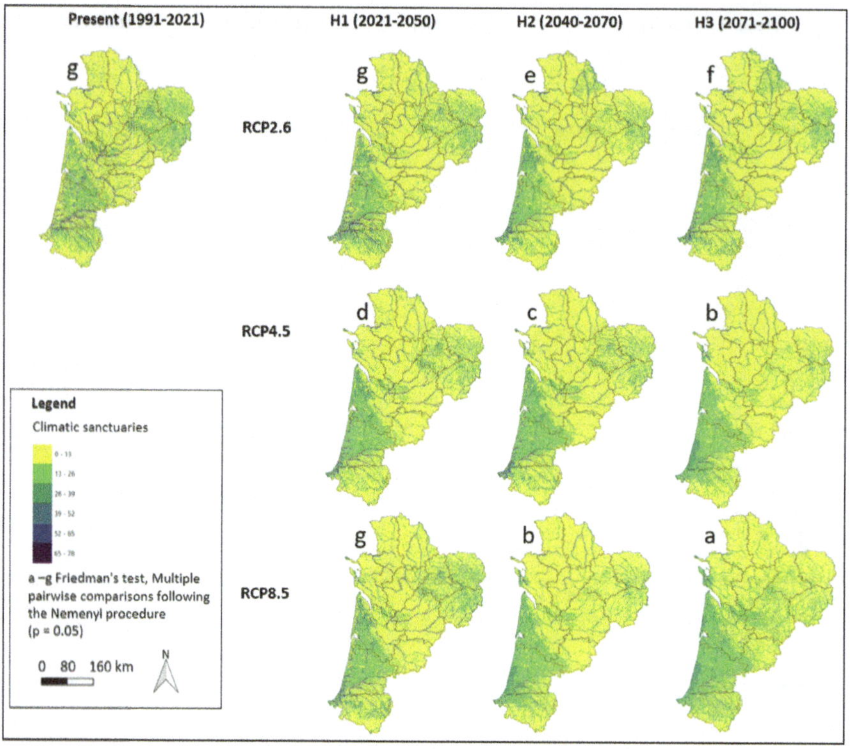

Fig. 26.9 Climatic sanctuary index values calculated for Odonata in the New-Aquitaine region

area still present. Heritage species may find suitable climatic conditions along the Atlantic coast as this region will be colder and will experience a higher relative humidity compared to inland regions. Thus, maintaining suitable wetlands in those areas seems to be crucial to allow heritage species to find refuge under climate change.

Nevertheless, those multiple impacts on species habitat suitability will lead to major changes in diversity in the study area. This is in agreement with a previous study Keil et al. (2008) highlighting that the New-Aquitaine region will be subject to a decrease of odonatan species richness if the temperature increases. However, in contrast, Termaat et al. (2019) reported a potential increase in the number of species under climate change in France. Yet, our study does not include species from southern areas. However, the use of species distribution models over a local study area might overestimate species local extinction as the whole distribution is not taken into account (Barbet-Massin et al. 2010). Nevertheless, our results are at a finer scale, allowing us to differentiate local climatic evolution. They also show that areas at higher elevations will negatively be impacted by diversity changes, making them high-priority areas for protection. If some species are not likely to still be present under scenario RCP 8.5, providing highly suitable habitat along with micro-climatic

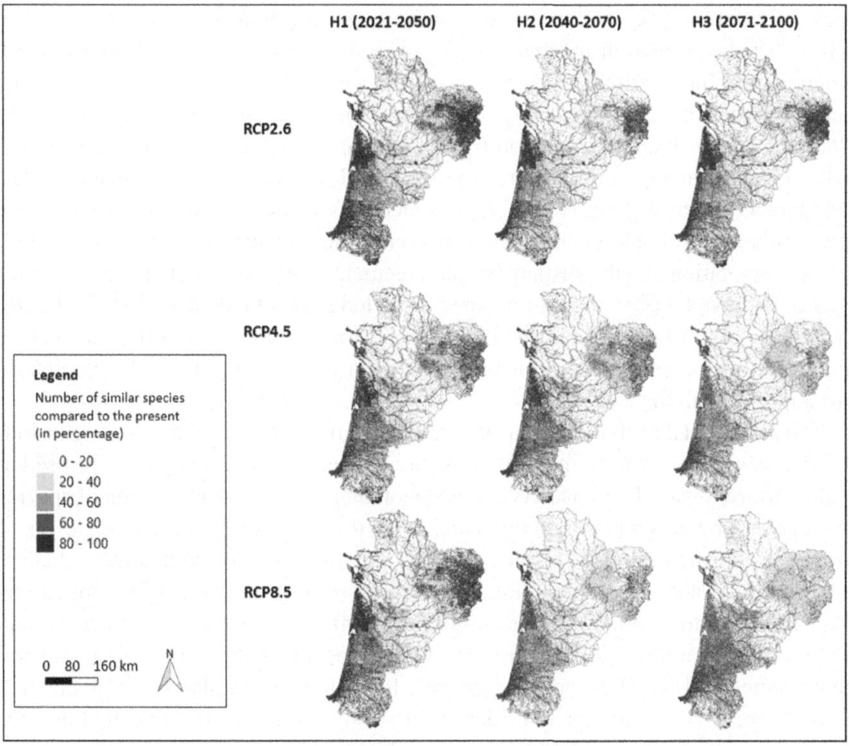

Fig. 26.10 Similarity index values calculated for Odonata in the New-Aquitaine region

refuges (vegetation) may allow some species to maintain their population in the area. The similarity index most significantly captures the potential changes for the New-Aquitaine region. Areas at higher elevations, such as the Central Massif and the Pyrenees, seem to be most impacted by species turnover, leading to a loss in the sanctuary status of those areas over time. Species affiliated with such colder areas are likely to move up north to find a suitable climate. Simultaneously, species associated with lower elevations and a higher temperature range may move to those higher-elevation areas, largely changing the assemblages of the species present (Hassall and Thompson 2008).

Another area is predicted to experience important changes, becoming a climatic refuge area for odonatan species under all scenarios tested. This area is located along the Atlantic coast and is highlighted by the climatic sanctuary index, which allows the identification of areas that may host the highest diversity of species, along with target species of interest (protected and heritage species). The conservation of wetlands along the coast is thus essential to allow a high diversity of species to maintain viable populations in the New-Aquitaine region. This index indicates the potential evolution of the odonatan assemblage and can therefore be used by managers and decision maker to spatially prioritize actions in the New-Aquitaine region. Different types of conservation actions can be pursued: *In-situ* conservation is based on the protection and maintenance of biodiversity in the natural environment, whereas non-interventionist protection can be achieved through the acquisition of land in natural areas by legal instruments and the awareness of various audiences (Mallard 2020). Interventionist protection is characterized by direct action on the natural environment, namely the restoration of a degraded state and management to maintain a reference state. The future presence of odonatan species is determined by the presence of suitable habitats, such as ponds and streams in the area, along with aquatic vegetation and the sustainable management of hydrological resources (water quality and availability), based on expert knowledge (Bailleux et al. 2017). The use of indices taking into account the national and local status of the different species allows a better communication between modelers and managers as the results are more in line with their needs.

The predicted distributions map and index calculations allow various insights into the changes that may occur in the New-Aquitaine region. These results can be used in many different ways by managers and decision makers. The identification of species that may not be present in the future can lead to decisions on the protection levels of those species, increasing the chance for the populations to cope with climate change. In addition, along with experts, the close monitoring of the remaining populations may also help to provide more accurate information on the species' presence and population dynamics. Furthermore, applying a special protection status to several new species and targeting areas of interest for Odonata can also be implemented. Two different views may be considered: First, the protection of areas that are, for the current period, either highly diverse or of importance for certain target species. Juridical protections of those areas may allow Odonata populations to be in better shape to cope with climate change (adaptation, migration) (Mallard 2020). A second option is to target areas that will become refuge areas in the future. Anticipating potential future species migrations will help to provide suitable habitast that will be used by species when the suitable climatic conditions are reached (Mallard 2020). The combination of the two likely greatly improves the conservation of the Odonata species in the New-Aquitaine region, allowing the species that can adapt to climate change to remain in a suitable habitat but providing, at the same time, suitable habitat as refuge for other species. Nevertheless, the migration capacity of each species is a key parameter that was not taken into account in the species distribution modeling. Jaeschke et al. (2013) showed that the addition of an estimate of the migration capacity for each species largely diminished the potential future distribution area. Migration capacity is not only affected by the physiology of the species but also by the presence of suitable habitats, allowing the species to migrate over time (Heller and Zavaleta 2009). For Odonata, the presence of a network of ponds and streams of high quality is essential; such a network should especially be reinforced towards the Atlantic coast and the higher-elevation areas.

Despite the advantages of using species distribution models, some limitations must be highlighted. First, only some of the species occurring in the region were used; in particular, species with a very low occurrence were excluded due to poor performance (Wisz et al. 2008). Many species with low occurrences are protected species, thus leading to a bias in the estimation of the protection index. In addition, species that may arrive from other areas (south, Mediterranean basin) were not taken into account. Yet, estimation of the diversity must take into account all species occurring to prevent underestimation of the diversity in some areas. Furthermore, the exclusion of species can sometimes lead to differences in priority area targeting (Kujala et al. 2018). Finally, the uncertainties inherent to the modeling process must be kept in mind. The results presented here depend on the quality of the data used. Spatial sampling bias in species presence can lead to biased results toward well-sampled areas (Kramer-Schadt et al. 2013). Although the climatic variables were at a fine scale (8 km^2), they may still not represent the future micro-climatic conditions in ecosystems as particular as wetlands or streams (Austin and Van Niel 2011). In addition, if the evolution of the climate was taken into account, the evolution of the land use (agriculture, forest composition, urban expansion) was not available for our study. Such predictions are a complex mix of climate change effects, economics, and political choices over time. Taking into account such evolutions might be the next step to improve model prediction across the study area.

Conclusions

The use of Odonata, which are species affiliated to wetlands such as ponds or streams and which are sensitive to climate change, provided an insight into the future impacts on those habitats in the New-Aquitaine region (France). The use of species distribution models can lead to the identification of species with future conservation issues, but it also facilitates the evaluation of changes in Odonata diversity across the study area through the calculation of diversity indices for each climate scenario. The results showed that 24–33% of the species are predicted to lose between 75 and 100% of suitable habitat by 2100 under two scenarios. Other species might benefit from climate change, and an extension of suitable habitat availability was predicted. The use of indices showed that the areas located along the Atlantic coast and at higher elevations (Central Massif) are key areas for current and future protection and management measures. Thus, the protection of Odonata habitats might be concentrated along the coast, as coastal regions might be the last refuges for numerous species, including protected and heritage species. Those indices, based on both national and local species protection criteria, ensure the comprehension of the results by mangers and local experts. By protecting wetlands suitable for a diverse Odonata assemblage, other wetland-affiliated organisms, such as amphibians, birds, and plants, might benefit from those actions. Nevertheless, numerous limitations, such as sampling bias, the climatic variable coarse scale, or the use of only species with a sufficient number of observations, may have influenced our results. In the future, the use of finer-scale variables and the inclusion of southern odonatan species might result in a better understanding of the future assemblages.

Recommendations

- Species threatened by climate change should benefit from active protection as soon as possible to help the remaining populations to cope with future changes.
- Areas with a high Odonata diversity should be protected, and management actions should be taken to preserve remaining wetlands in those areas.
- Areas identified as refuges under the different climate change scenarios should be protected, and management actions should be taken to preserve remaining wetlands in those areas.
- To allow the species to migrate towards climatically suitable habitats, the conservation of a fine network of ponds and streams is essential.

Acknowledgements This work would not have been possible without support from the European Union (the European Regional Development Fund—Feder), the French region of New Aquitaine, and the French departments of "Gironde" and "Pyrénées-Atlantiques." We thank these organizations for their support and funding from 2016 to 2021, as well as our technical partners Météo France and Conservatory of Natural Area of Aquitaine (CEN). We also thank the members of the 2016–2019 Scientific Council of the program, including Hervé Le Treut, Honorary President of the Scientific Council, Professor at Pierre and Marie Curie University, for their opinions, analyses, advice, and validation of the methods, protocols, models, and results. Finally, we thank Akaren Goudiaby, Pierre-Yves Gourvil, and Gilles Bailleux for sharing their knowledge on Odonata and their useful remarks on the model results.

References

Allouche O, Tsoar A, Kadmon R (2006) Assessing the accuracy of species distribution models: Prevalence, kappa and the true skill statistic (TSS). J Appl Ecol 43(6):1223–1232. https://doi.org/10.1111/j.1365-2664.2006.01214.x

Austin MP, Van Niel KP (2011) Improving species distribution models for climate change studies: Variable selection and scale. J Biogeogr 38(1):1–8. https://doi.org/10.1111/j.1365-2699.2010.02416.x

Bailleux G, Couanon V, Gourvil P-Y, Soulet D (2017) Pré-atlas des odonates d'Aquitaine—Synthèse des connaissances 1972—2014. CEN Aquitaine, LPO Aquitaine, p 117

Barbet-Massin M, Thuiller W, Jiguet F (2010) How much do we overestimate future local extinction rates when restricting the range of occurrence data in climate suitability models? Ecography 33(5):878–886. https://doi.org/10.1111/j.1600-0587.2010.06181.x

Braunisch V, Coppes J, Arlettaz R, Suchant R, Schmid H, Bollmann K (2013) Selecting from correlated climate variables : A major source of uncertainty for predicting species distributions under climate change. Ecography 36(9):971–983

Cerini F, Stellati L, Luiselli L, Vignoli L (2020) Long-term shifts in the communities of odonata : Effect of chance or climate change? North-Western Journal of Zoology 16(1):1–6

Dingemanse NJ, Kalkman VJ (2008) Changing temperature regimes have advanced the phenology of Odonata in the Netherlands. Ecological Entomology 33(3):394–402. https://doi.org/10.1111/j.1365-2311.2007.00982.x

Drias. (2020). Drias, Les futurs du climat—Accueil. http://www.drias-climat.fr/

Elith J, Leathwick JR (2009) Species distribution models : Ecological explanation and prediction across space and time. Annu Rev Ecol Evol Syst 40(1):677

Elith J, Kearney M, Phillips S (2010) The art of modelling range-shifting species. Methods Ecol Evol 1(4):330–342. https://doi.org/10.1111/j.2041-210X.2010.00036.x

Erwin KL (2008) Wetlands and global climate change : The role of wetland restoration in a changing world. Wetlands Ecol Manage 17(1):71. https://doi.org/10.1007/s11273-008-9119-1

FAUNA (2020) FAUNA—Accueil. https://observatoire-fauna.fr/

Franklin J (2009) Moving beyond static species distribution models in support of conservation biogeography. Divers Distrib 16(3):321–330

Hassall C (2012) Predicting the distributions of under-recorded Odonata using species distribution models. Insect Conservation and Diversity 5(3):192–201. https://doi.org/10.1111/j.1752-4598.2011.00150.x

Hassall C, Thompson DJ (2008) The effects of environmental warming on Odonata : A review. International Journal of Odonatology 11(2):131–153

Heikkinen RK, Luoto M, Araújo MB, Virkkala R, Thuiller W, Sykes MT (2006) Methods and uncertainties in bioclimatic envelope modelling under climate change. Prog Phys Geogr 30(6):751–777

Heller NE, Zavaleta ES (2009) Biodiversity management in the face of climate change : A review of 22 years of recommendations. Biol Cons 142(1):14–32. https://doi.org/10.1016/j.biocon.2008.10.006

Hijmans RJ, Graham C (2006) The ability of climate envelope models to predict the effect of climate change on species distributions. Glob Change Biol 12(12):2272–2281. https://doi.org/10.1111/j.1365-2486.2006.01256.x

IPCC (2014) Climate Change 2014 : Impacts, Adaptation, and Vulnerability. Working Group II Contribution to the IPCC 5th Assessment Report. Cambridge University Press, Cambridge, UK and New York, USA

Iturbide M, Bedia J, Gutiérrez JM (2018) Background sampling and transferability of species distribution model ensembles under climate change. Global Planet Change 166:19–29. https://doi.org/10.1016/j.gloplacha.2018.03.008

Jaeschke A, Bittner T, Reineking B, Beierkuhnlein C (2013) Can they keep up with climate change?–Integrating specific dispersal abilities of protected Odonata in species distribution modelling. Insect Conserv Divers 6(1):93–103

Keil P, Simova I, Hawkins BA (2008) Water-energy and the geographical species richness pattern of European and North African dragonflies (Odonata). Insect Conservation and Diversity 1(3):142–150. https://doi.org/10.1111/j.1752-4598.2008.00019.x

Kramer-Schadt S, Niedballa J, Pilgrim JD, Schröder B, Lindenborn J, Reinfelder V, Stillfried M, Heckmann I, Scharf AK, Augeri DM (2013) The importance of correcting for sampling bias in MaxEnt species distribution models. Divers Distrib 19(11):1366–1379

Kujala H, Moilanen A, Gordon A (2018) Spatial characteristics of species distributions as drivers in conservation prioritization. Methods Ecol Evol 9(4):1121–1132. https://doi.org/10.1111/2041-210X.12939

Le Treut, H (2020) Anticipating climate change in Nouvelle-Aquitaine. To guide policy at local level—Executive report. Éditions AcclimaTerra 2020, p 96

Leggott M, Pritchard G (1986) Thermal preference and activity thresholds in populations of Argia vivida (Odonata : Coenagrionidae) from habitats with different thermal regimes. Hydrobiologia 140(1):85–92

Lémond J, Dandin P, Planton S, Vautard R, Pagé C, Déqué M, Franchistéguy L, Geindre S, Kerdoncuff M, Li L (2011) DRIAS: a step toward Climate Services in France. Adv Sci Res 6(1):179–186

Mallard F (coord) (2017) Programme les sentinelles du climat. Tome IV : Ajustement des protocoles d'échantillonnage et analyses exploratoires des indicateurs des effets du changement climatique sur la biodiversité en Nouvelle-Aquitaine. C. Nature. https://hal.archives-ouvertes.fr/hal-02022916

Mallard F (coord) (2018) Programme les sentinelles du climat. Tome VI : Résultats exploratoires des indicateurs des effets du changement climatique sur la biodiversité en Nouvelle-Aquitaine. C. Nature. https://hal.archives-ouvertes.fr/hal-02060363

Mallard F (coord) (2019) Programme les sentinelles du climat. Tome VIII : Écologie du changement climatique en région Nouvelle-Aquitaine. C. Nature. https://hal.archives-ouvertes.fr/hal-02495935

Mallard F (coord) (2020) Programme les sentinelles du climat. Tome IX : Connaitre et comprendre pour protéger les espèces animales et végétales face au changement climatique, C. Nature : Le Haillan, Gironde, 822p. https://hal.science/hal-03130349

Mallard F (2021a) Climate Sentinels program : Meeting the Challenge of regional biodiversity conservation adaptation to climate change. In: Leal Filho W, Luetz J, Ayal D (eds) Handbook of climate change management. Springer, Cham, pp 1–39. https://doi.org/10.1007/978-3-030-22759-3_193-1.

Mallard F (coord) (2021b) Programme les sentinelles du climat—Tome X: Réponses des espèces animales et végétales face au changement climatique et pistes d'actions de conservation de la biodiversité en région Nouvelle-Aquitaine. C. Nature, Le haillan (Gironde). https://hal.inria.fr/hal-03647259/

Mallard F, Couderchet L (2019) Climate sentinels research program : Developing indicators of the effects of climate change on biodiversity in the region of New Aquitaine (south west, France). In: Leal Filho W, Barbir J, Preziosi R (eds) Handbook of climate change and biodiversity. Climate change management. Springer, Cham, pp 223–241 https://doi.org/10.1007/978-3-319-98681-4_14.

Morghad F, Samraoui F, Touati L, Samraoui B (2019) The times they are a changin' : Impact of land-use shift and climate warming on the odonate community of a Mediterranean stream over a 25-year period. Vie Et Milieu-Life and enVironment 69(1):25–33

Ott J (2001) Expansion of Mediterranean Odonata in Germany and Europe—Consequences of climatic changes. In "Fingerprints" of climate change. Springer, New York, pp 89–111

Ouzeau G, Déqué, M, Jouini, M, Planton, S, Vautard, R (2014) Le climat de la France au XXI siècle Volume 4, scénarios régionalisés: Édition 2014 pour la métropole et les régions d'outre-mer (No 4). Ministère de l'écologie et du développement durable, pp 1–64

Phillips SJ, Anderson RP, Schapire RE (2006) Maximum entropy modeling of species geographic distributions. Ecol Model 190(3–4):231–259. https://doi.org/10.1016/j.ecolmodel.2005.03.026

R Core Team (2020) R: a language and environment for statistical computing. R Foundation for Statistical Computing. https://www.R-project.org/

Termaat T, van Strien AJ, van Grunsven RHA, De Knijf G, Bjelke U, Burbach K, Conze K-J, Goffart P, Hepper D, Kalkman VJ, Motte G, Prins MD, Prunier F, Sparrow D, van den Top GG, Vanappelghem C, Winterholler M, WallisDeVries MF (2019) Distribution trends of European dragonflies under climate change. Divers Distrib 25(6):936–950. https://doi.org/10.1111/ddi.12913

Thuiller W, Brotons L, Araújo MB, Lavorel S (2004) Effects of restricting environmental range of data to project current and future species distributions. Ecography 27(2):165–172

Thuiller W, Lafourcade B, Engler R, Araújo MB (2009) BIOMOD–a platform for ensemble forecasting of species distributions. Ecography 32(3):369–373

Wisz MS, Hijmans R, Li J, Peterson AT, Graham C, Guisan A, NCEAS Predicting Species Distributions Working Group (2008) Effects of sample size on the performance of species distribution models. Divers Distrib 14(5):763–773

Chapter 27
Climate Change, Soil Saturation, and Risk of Yield Penalties to Key Cereal Crops: A Neglected Issue in Agri-Food System Adaptation

David Oscar Yawson and Michael Osei Adu

Abstract Soil water or moisture balance drives essential rhizospheric and physiological processes that culminate in yield formation in crop plants. Climate change threatens to alter temperature and precipitation patterns and thereby the hydrological processes underpinning soil moisture balance and related stresses such as saturation or deficit. While considerable volume of literature exists on soil moisture deficits under the climate change, the same cannot be said of soil saturation (the transitory water content of the soil when nearly all its pore spaces are filled with water and the water potential is close to zero). Few crop plants can survive soil saturation exceeding few days. This chapter demonstrates the potential effects of soil saturation on crop yield penalties under projected climate change. Cereals are the main dietary energy source in human diets and animal feed formulations. Hence, future yield penalties in cereals can be detrimental to global food security. Cereals can be highly sensitive to abiotic stresses such as heat stress and soil water deficit or saturation at specific growth stages. The FAO AquaCrop model was used to simulate biomass production and yield of the barley genotype Westminster for all UK administrative regions in the 2050s. The simulation was done for spring conditions, under the low and high climate change emission scenarios (LES and HES), respectively. Yield penalty was expressed as the mean percentage of all yearly yield reductions in the 30-year time slice relative to the 10th percentile yield for the entire time slice. The results showed that yield penalties were more extensive under the HES than the LES. The ranges of yield penalties were 5–76% (HES) and 5–41% (LES) under varied durations and severities of soil saturation. It is concluded that, due to climate change, soil saturation can potentially cause substantial yield penalties, especially in highly sensitive crops and is, therefore, worthy of research, policy and practical attention. Although soil saturation stresses could be less frequent compared to those of heat and soil moisture

D. O. Yawson (✉)
Centre for Resource Management and Environmental Studies (CERMES), The University of the West Indies, Cave Hill Campus, P. O. Box 64, Bridgetown, St. Michael BB11000, Barbados
e-mail: david.yawson@cavehill.uwi.edu

M. O. Adu
Department of Crop Science, School of Agriculture, College of Agriculture and Natural Sciences, University of Cape Coast, Cape Coast, Ghana

W. Leal Filho et al. (eds.), *Climate Change Strategies: Handling the Challenges of Adapting to a Changing Climate*, Climate Change Management,
https://doi.org/10.1007/978-3-031-28728-2_27

deficits, yield penalties under the former could, in some cases, be larger than the latter and could occur as a surprise. It is recommended that further studies on potential effects of soil saturation on crop yield penalties be integrated in the suite of research and measures for adaptation of key crops such as cereals, which are the main source of energy in human diets and animal feed.

Introduction

Soil is a three-phase system comprising solid, liquid, and gaseous phases. The solid phase includes minerals and or organic matter, which serves as the solid matrix or backbone of the soil and provides physical anchorage and nutrients to crops. The spaces between the solid particles are referred to as soil pores. The pores can be large (macropores) or small (micropores). The liquid phase is the soil solution (water with dissolved organic and inorganic substances). The gaseous phase is a mixture of gases that can be referred to as soil air. Typically, the solid fraction accounts for about 50% of well-drained soils that support upland crop production. In comparison, the liquid phase (which usually resides in the micropores) and the soil air (which generally resides in the macropores) account for 25% each of a given soil volume.

Soil moisture (used interchangeably with soil water in this chapter) refers to the water content of the unsaturated (vadose) zone of the soil profile (Lal and Shukla 2004). A soil is said to be saturated when all its pores are filled with water, and the force of gravity dominates the movement of water down the profile. If the macropores are substantially drained, and only the micropores contain water, the soil is said to be unsaturated, a condition preferred by most plants. Soil saturation, then, is the transitory water content when nearly all the soil pore spaces are filled with water, and the water potential is close to zero. In rain-fed agroecosystems, soil saturation occurs typically after heavy rainfall. Normally, due to drainage, soil saturation does not last for long, but few plants can withstand saturated conditions for a period exceeding a few days. Soil saturation is influenced by the hydraulic properties of the soil, precipitation, groundwater table, and capillary rise.

Soil moisture content plays crucial physiological roles in the growth of crop plants. The spatial and temporal dynamics of soil moisture content determine water availability to plant roots throughout the growing season. Water is indispensable in photosynthesis, uptake and use of nutrients from the soil, maintenance of cellular hydration and turgidity, and translocation of assimilates (Pinheiro and Chaves 2011; Barnabàs et al. 2008). Carbon dioxide and water are the two main ingredients required for photosynthesis and biomass production. Primarily, plant stomatal behaviour is affected by plant water status. Optimal plant water status is critical for improved photosynthesis, presumably because stomatal resistance to CO_2 absorption is lowered (Lambers et al. 2008). Together with light and nutrients, soil moisture availability regulates yield formation in crops (Rajala et al. 2011). For cereals, soil moisture availability is crucially important as it dominates the regulation of yield formation (Barnabàs et al. 2008; Rajala et al. 2011). Indeed, the indispensability of soil moisture availability for several critical physiological functions that regulate biomass production and yield formation has been recognized by researchers. As a result, substantial effort has been devoted to studying the effects of soil moisture deficits on crop plants under both present and climate change conditions (e.g., Bhattacharya et al. 2021; Gebre and Earl 2021; Han et al. 2018). In addition, because soil moisture deficit is a common occurrence in almost all agroecosystems, breeding has considerably focused on enhancing crop tolerance of soil moisture deficits.

However, soil saturation, though infrequent in agro-ecosystems, can unleash considerable yield reductions or penalties in crop plants. In saturated soils, nutrient and oxygen availability and uptake are the most significant limiting factors for crop yield formation (Lynch et al. 2012). Soil redox potential decreases with increasing saturation and negative redox potential can affect nutrient solubility and availability (Neumann and Römheld 2012). For example, in saturated soils with low pH, ample supply of root-borne carbohydrates can trigger excessive microbial activity, resulting in decreased redox potential in the rhizosphere. This condition is favourable to increased Manganese solubility and toxicity in plants. Anoxic conditions in saturated soils limit root respiration, can increase nitrogen losses via denitrification, and can cause mortality of plants (Neumann and Römheld 2012). Root development, growth, establishment, and nodule function are all inhibited under saturated conditions (Lynch et al. 2012). These effects can, in turn, make crop plants more susceptible to drought stress following saturated conditions, root infections, and lodging. Intense, heavy precipitation events can lead to fluctuations in groundwater table and render pore pressure highly dynamic. Sudden variations in pore pressures, due to saturation, might cause suction loss and disturb aggregate stability (Clarke et al. 2006; Lee and Jones 2004). Regardless of the potential adverse impacts of soil saturation on crop yields, it has received relatively little research attention, especially in climate change studies, compared to soil moisture deficits. The potential effects of climate change on soil saturation and, consequently, crop yields are not far-fetched.

Globally, cereals are the largest dietary energy source in human diets and animal feed (Yawson et al. 2016). Cereals are the most widely grown crops and the most traded food commodity. While cereal crops might have some tolerance to soil moisture deficit conditions (González and Ayerbe 2011), they might not be able to tolerate soil saturation to the same extent. Cereals might poorly compensate for losses in biomass production due to soil saturation at an early growth stage. Barley, for example, can tolerate and recover from short term soil water deficit but is sensitive to soil saturation over comparable periods. Barley crop is the fourth most important cereal crop in terms of area cultivated and quantity harvested (Newton et al. 2011). It is the most prominent coarse grain used in animal feed (Newton et al. 2011) and directly consumed as food. Barley production underpins the global malting industry. This chapter aims to promote the study of the effect of soil saturation on yields of key crops, such as cereals, under climate change to inform adaptation. The specific objective is to highlight the sensitivity of spring barley crop yields to soil saturation under projected climate change using the UK as an example.

Climate Change, Soil Saturation, and Crop Production

Recent reports from the Intergovernmental Panel on Climate Change (IPCC) indicate that climate change is unambiguous and that the effects of anthropogenic warming of the climate system are already evident (IPCC 2018, 2021). Without significant mitigation measures, global mean temperature could increase by 4 °C at the turn of the next century, with more frequent, intense, and severe hydrometeorological events such as heavy precipitation, tropical cyclones or hurricanes, and droughts (IPCC 2021). These reports highlight the enormity of the challenges associated with climate change management due to the cross-cutting or cascading effects of both direct and indirect impacts on biophysical processes underlying ecosystems. For example, recent observed warming could continue to the next decade due to prior commitment of greenhouse gases to the atmosphere. Aggressive mitigation measures are critical to meaningfully limit the catastrophic consequences of climate change (IPCC 2021). Currently, hydrometeorological hazards, mainly droughts and floods, account for over 80% of all damages and losses associated with disasters in the agricultural sector (FAO 2018). It is certain that climate change will alter temperature and precipitation patterns and, therefore, hydrological processes that underpin soil moisture balance (IPCC 2018; Daba et al. 2018). This implies climate change threatens to alter the fundamental properties of abiotic stresses to crop plants, change the geographic coverage and range of sensitive or vulnerable crops to abiotic stresses, and amplify the combined effects of multiple stresses in agroecosystems. Because projections of changes in precipitation show greater uncertainty (IPCC 2018; Daba et al. 2018; Murphy et al. 2009), the direct and indirect effects of future changes in precipitation patterns on crop production are equally imperilled with uncertainties. Although climate change is a global phenomenon, its impacts would be geographically varied. As a result, the complexity of climate change and its adverse impacts on soil moisture balance and crop production make adaptive responses and management complicated and challenging.

Variations in precipitation characteristics largely account for soil moisture balance (Wang et al. 2019). A critical potential risk of heavy precipitation events, under climate change, is soil saturation which creates anoxic conditions that hamper physiological processes underpinning biomass production and yield formation in crop plants (Neumann and Römheld 2012; Lynch et al. 2012). Here, soil saturation in agro-ecosystems can potentially persist over durations that are known to be tolerable to most crops. Soil saturation, due to climate change, can therefore cause significant yield penalties in crops depending on the temporal distribution of these climatological events in relation to crop phenology, duration of saturated conditions, sensitivity of crops, and effectiveness of management responses. Already, the challenge of adaptive management of climate change due to heavy precipitation events has been noted (Daba et al. 2018) but less so in relation to soil saturation. Due to the projected trajectories of climate change, it is possible that heavy precipitation events can follow severe drought events, and thereby subject the soil to rapid infiltration and variations in pore pressure due to cracks formed during the drought (Vardon 2015; Robinson and Vahedifard 2016). The sensitivity of crop plants and the effectiveness of their adaptive responses will be a key determinant of the scale of yield penalties to be incurred. These suggest a need to devote some research efforts to the potential effect of soil saturation on crop yields under climate change.

Materials and Methods

This study used the FAO AquaCrop simulation model to simulate barley yields under projected climate change for the 2050s. The AquaCrop model was explicitly developed for simulating crop responses to temporal variations in soil moisture content. The AquaCrop algorithms or components and their structural relationships in the modelling framework are provided in Raes et al. (2009). AquaCrop comprises soil, crop, climate, and management sub-models. The soil sub-model permits up to five user-defined soil layers with different textures and depths. This dispersed system approach allows the model to closely mimic the layered depletion of soil moisture to root development and soil moisture availability. The critical input in this sub-model is soil type and hydraulic properties.

AquaCrop uses a daily soil water balance simulation to estimate the available water in the root zone. The engine of the crop-growth sub-model is based on a water productivity function or coefficient, normalized for reference evapotranspiration and atmospheric carbon dioxide concentration, with the development of canopy cover. This underpins biomass production and harvest index and the model's applicability to different geographic locations or climates and seasons. AquaCrop decomposes the yield into biomass production and harvest index in simulating crop growth, permitting the user to assess the distinctive effects of environmental variables on biomass or harvest index (Geerts et al. 2009; Raes et al. 2009). AquaCrop uses three conservative parameters to capture the effect of water stress on the crop: decreased rate of canopy expansion, stomatal closure, and increased canopy senescence. These parameters are used to control the water productivity and harvest index, which are also affected by soil saturation conditions. The onset and intensity of these water-related stresses largely depend on timing, soil conditions, climate, and management (Raes et al. 2009). The climate sub-model inputs key climatic variables: evapotranspiration, temperature, and precipitation. These are used as input in the crop-growth sub-model to track the dynamics of soil water balance, canopy development and heat stress.

The details of the sources and handling of datasets and simulation can be found in Yawson (2013) and Yawson et al. (2016). A summary is provided here. In this study, the simulation was done for spring barley under rainfed conditions, with no additional management or fertility stress considered. Initial soil water content was set to field capacity. The crop information required in AquaCrop was based on the barley genotype Westminster, which is a high-yielding spring/winter genotype widely grown in the UK. The data was obtained from a field experiment used to calibrate and validate AquaCrop to simulate future yields. Downscaled to the UK administrative regional level, the projected climate data were obtained from the UKCP09 (Murphy et al. 2009). Each time slice in this dataset covered 30 years. An embedded Weather Generator in the UKCP09 (Jones et al. 2009) was used to obtain future climate variables on a daily resolution. In this chapter, the results of two emission scenarios for the 2050s time slice are reported, namely, the high emission scenario (HES) and the low emission scenario (LES), corresponding to the A1FI and B1 in the SRES scenarios, respectively (Nakićenović and Swart 2000). Soil data were obtained from the Crop Growth Monitoring System (CGMS) database in the SINFO (Soil Information System) of the European Union (Baruth et al. 2006). From this, soil hydraulic parameters required by AquaCrop were generated.

Once all the required datasets had been obtained or generated and processed to AquaCrop compatible formats, they were imported into the model, and the simulations were run. Multiple run project files were created and executed in the AquaCrop plug-in program (Raes et al. 2009) to capture the various UK regions and emission scenarios. The outputs were processed and analyzed in Microsoft Excel. Mean values of all model variants for each year in the time slice, UK region, and emission scenario were generated to represent the 30-year time slice. Subsequently, yield percentiles were computed, and the data was inspected for the occurrence of yield values below the 10th percentile. Yield value below the 10th percentile value in any given year in the time slice was flagged as yield penalty. All the yield penalty values were then inspected in the AquaCrop graphical outputs showing the effect of temperature and water stresses on biomass production and yield (see Fig. 27.1). Here, yield reductions due predominantly to soil saturation were selected and, in some instances, compared with the magnitude of yield reductions due to heat stress for other years in the same emission scenario and region. For each region and emission scenario, the mean yield penalty was calculated and then expressed as a percentage of the respective 10th percentile yield for the time slice to obtain the percent yield penalty (representing the intensity or severity of the yield penalty).

Results

Generally, within the studied time slice (the 2050s), yield values (mean of 100 model variants for each year in the 30-year time slice) under the LES were less variable than under the HES. The 10th percentile yields (based on the yearly values of the 30 years in the time slice) of the UK regions in the 2050s are shown in Fig. 27.2. Yields under the HES were higher than under the LES. Per region, substantial differences in the 10th percentile yields for the two emission scenarios were observed for East England (EE), East Midlands (EM), Northwest Scotland (NWS), Southeast England (SEE), West Midlands (WM) and Yorkshire and the Humber (YH). Particularly for Wales (WA), Northeast Scotland (NES), and Northwest England (NWE), the percentile

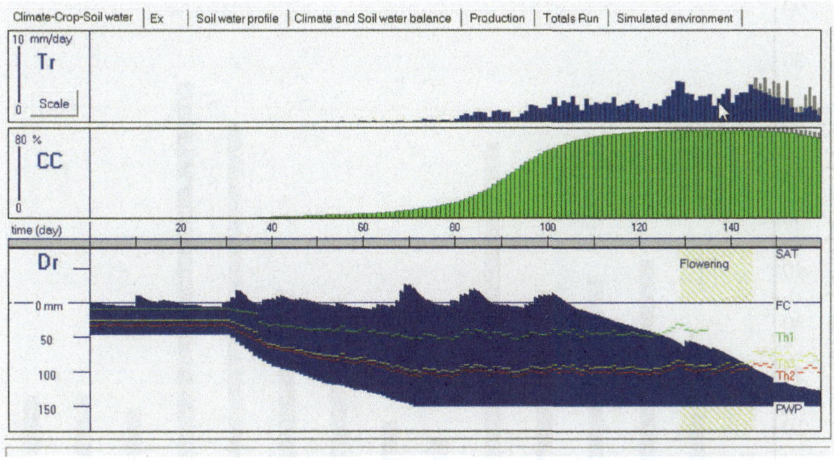

Fig. 27.1 Sample graphical output from AquaCrop showing soil water balance (lower panel), canopy development (middle panel) and transpiration (top panel) over the growing season. The FC denotes field capacity, and SAT represents soil saturation. Vertical grey bars indicate reductions in either transpiration rates (top panel) or canopy development (middle panel)

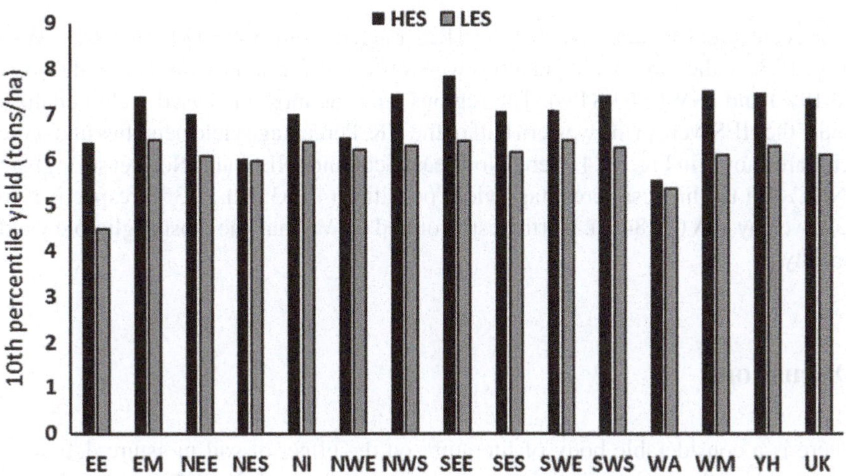

Fig. 27.2 The 10th percentile yields of UK regions under the high emission scenario (HES) and low emission scenario (LES)

yields for the two emission scenarios were quite similar. The 10th percentile yields for the HES ranged from 5.55 tonnes/ha (WA) to 7.84 tonnes/ha (SEE). The values ranged from 4.42 (EE) to 6.44 tonnes/ha (EM) for the LES.

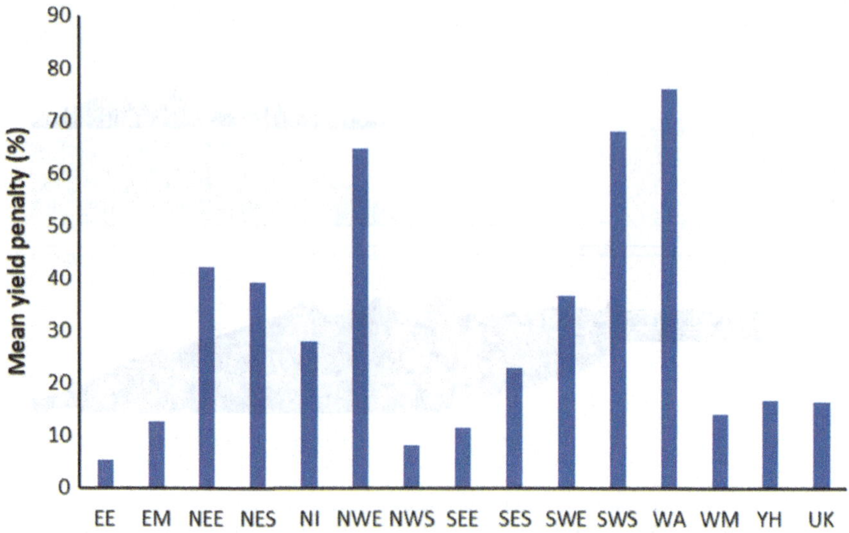

Fig. 27.3 Percentage yield penalty for UK regions under the HES. The mean of all regions represents the UK

Percent yield penalties under the HES ranged from 5.4% (EE) to 76% (WA) (Fig. 27.3). Other high yield penalties were observed for Southwest Scotland (SWS, 68.19%) and (NWE, 64.84%). The regions with the most increased yield penalties under the HES were in the western half of the UK. Percentage yield penalties under the LES are shown in Fig. 27.4. Here, Northeast Scotland (NES) and Northeast England (NEE) had the highest percentage yield penalties (40.63 and 25.3%, respectively), followed by WA (21.84%). Northwest Scotland (NWS) had the most negligible yield penalty.

Discussion

There is a considerable body of literature on the effect of soil moisture deficit on crop yields under projected climate change (e.g., Daba et al. 2018; Wang et al. 2019). However, the impact of soil saturation on yields under projected climate change is understudied. This chapter uses a simple case study to illustrate and draw research attention to the potential effect of soil saturation on barley yields (and by extension key cereal crops). Key results on actual yields and the statistical parameters are reported in Yawson et al. (2016) in which the baseline yields showed an upward trend. The results in the current chapter show that barley yields can be sensitive to soil saturation and that soil saturation under projected climate change can contribute to considerable yield reductions in spring barley. The yields become more variable under climate change. The results in the current chapter suggest that soil saturation could be

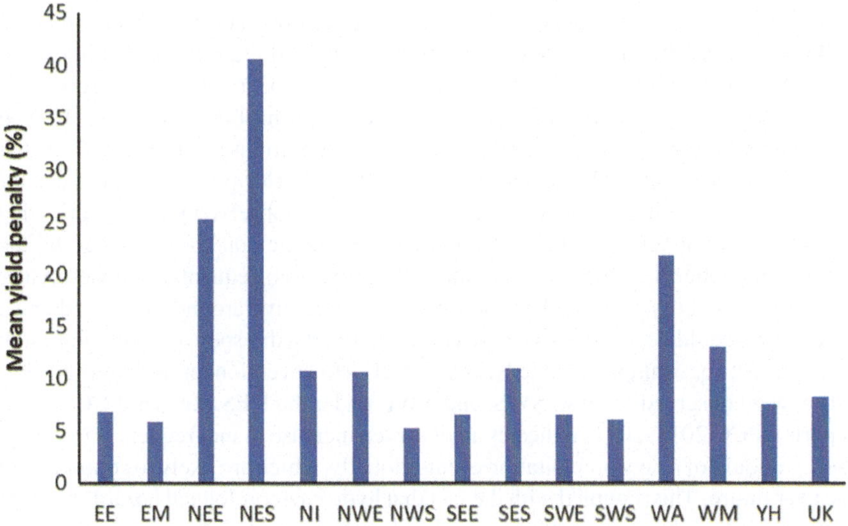

Fig. 27.4 Percentage yield penalty for UK regions under the LES. The mean of all regions represents the UK

a significant contributor to the yield variability, especially yield dips, observed earlier. In the UK, precipitation increases from south to north and east to west (Jenkins et al. 2008). The reverse is true for temperature. This geographical pattern would likely remain unchanged in the spring agricultural season (Yawson et al. 2016; Murphy et al. 2009). However, future changes in precipitation under projected climate change show greater uncertainty than for temperature (Murphy et al. 2009). The results in the current chapter show that, under the HES, where rainfall variability could be higher, western regions in the UK suffer the larger magnitudes of yield penalties. The UKCP09 report indicated that, under the HES, eastern (and southern) UK regions would likely experience more significant reductions in precipitation than the western regions. Thus, the result suggests that barley yields in the western regions, which are currently wetter than the eastern regions, could be more susceptible to penalties from increased soil saturation under highly variable precipitation regimes such as observed under the HES. In the simulation results, it was observed that soil saturation was less frequent compared to heat stress or soil water deficit, but more frequent in western regions than in the eastern or south-eastern regions. However, the magnitude of the yield penalties associated with soil saturation was higher in some cases, especially when it occurred around the establishment and the reproductive phases. There is the added danger of uncertainty here. In simulations such as reported in the current chapter, the years in time slices do not correspond to actual years, implying any observed event in the time slice can occur in any actual year. This means any yield penalty observed under any of the emission scenarios could occur in any given year in the 2050s, depending on the trajectory of climate change.

Like most cereals, barley is sensitive to drastic variations in soil moisture content and temperature around the establishment, anthesis, and grain filling (Anjum et al. 2011; Semenov and Shewry 2011). While barley can moderately tolerate soil moisture deficits through osmotic adjustment and recover from short-term water stress, it is sensitive to soil saturation as this creates anoxic conditions (González and Ayerbe 2011; González et al. 1999; Shone and Flood 1983). Barley poorly compensates for reduced tillering during the early stages of growth. The observed yield penalties due to soil saturation could be due to the duration of anoxic conditions created by the saturation. Under the HES, soil saturation occurred less frequently but had greater intensity or severity than the LES and was more effective around the reproductive phase. This could be due to successive combination of a dry spell and soil saturation, as observed, for example, for the NES and NEE. However, longer durations of saturation were observed for WA, SWS, and NWE under the HES. Recent IPCC special reports (IPCC 2018, 2021) indicates an observed increase in the frequency, intensity, and or amount of heavy precipitation events globally, which are likely to continue into the near future. This, coupled with the fact that hydrometeorological hazards account for over 80% of damages and losses incurred in the agricultural sector (FAO 2018), should direct research attention to soil saturation and its impacts on crop yields in the climate, agricultural, and disaster risk management communities. In the future, it is possible that crucial cereal crops could suffer severe yield penalties as precipitation becomes more variable and heavy precipitation events become more frequent and intense. Wet agroecosystems, under current climates, might be at higher risks of soil saturation and yield penalties, but the alternation of dry spells and saturation could also create yield penalties even in semi-dry agroecosystems. Further studies are required to explore these issues.

Limitations of the Study

The current study, though uses spatially varied regions in the UK, it is based on a single country and a single crop. Regardless of how typical the UK climate could represent northern temperate environments, more studies are required around the world using combinations of different types of soil, climate, and crop genotypes. Studies of this nature use datasets from varied sources, each of which has its own uncertainties. Projected climate data have their own inherent accuracies that limit their utility for extrapolation. Although this study used projected climate change dataset from the UK MET Office, it is limited to the accuracy of the dataset at the time of production and the study. Finally, uncertainties associated with model architecture, calibration and validation impose further limitations on the current study. In all, multiple studies are needed to improve understanding of the effects of climate change on soil saturation and, consequently, yield penalties in different agro-ecosystems.

Conclusions and Future Directions

Climate change would likely increase the frequency, intensity, and frequency of heavy precipitation events and tropical cyclones. Correspondingly, this could increase the frequency, duration, and severity of soil saturation conditions, with varying effects on crop yields. The findings in this chapter show that soil saturation, under climate change, can contribute to substantial yield penalties in barley and, therefore, constitutes a significant risk to future crop production and food security. Yield penalties, due to soil saturation, showed a strong dependence on emission scenario than the frequency of occurrence. Agro-ecosystems that currently have wet climates could potentially be at higher risks of yield penalties related to soil saturation under more varied or heavier precipitation events. While the effect of soil moisture deficit on crops has been widely studied, the same cannot be said of the impact of soil saturation. Poor understanding of the effects of soil saturation, either singularly or in combination with other stresses, under both current and future climates, could constitute an impediment to effective adaptation planning. Findings in the current chapter demonstrate that the effect of soil saturation on crop yields under climate change should no longer be neglected. It is time to pay research, policy, and practical attention to understanding the scale and scope of the risks of yield penalties from soil saturation, under climate change, to support adaptation planning. It is important to combine both experimental and simulation studies to understand the responses of key food security crop plants to varying frequencies, severities, and durations of soil saturation and the impacts on yields under both current and projected climates. Outcomes of such studies could inform plant selection and breeding strategies, as well as agronomic practices. However, resource allocation to such studies, as part of investment in adaptive capacity, ought to be prioritized from national to global scales. Again, efforts should be directed at quantifying the cascading impacts of yield penalties from soil saturation on national and transnational food security, including multi-model comparisons.

References

Anjum SA, Xie X, Wang LC, Saleem MF (2011) Morphological, physiological, and biochemical responses of plants to drought stress. Afr J Agric Res 6(9):2026–2032

Baruth B, Genovese G, Montanarella L (2006) New soil information for the MARS Crop Yield Forecasting System. European Commission Directorate General, Joint Research Centre, Ispra, Italy

Barnabás B, Jäger K, Fehér A (2008) The effect of drought and heat stress on reproductive processes in cereals. Plant, Cell Environ 31(1):11–38

Bhattacharya A (2021) Effect of soil water deficit on growth and development of plants: a review. In: Soil water deficit and physiological issues in plants. Springer, Singapore, pp 393–488. https://doi.org/10.1007/978-981-33-6276-5_5

Clarke GRT, Hughes DAB, Barbour SL, Sivakumar V (2006) The implications of predicted climate changes on the stability of highway geotechnical infrastructure: a case study of field monitoring of pore water response. EIC Climate Change Technology, 1–10
Daba MH, Bazi Z, Belay A (2018) Effects of climate change on soil and water resources: a review. J Environ Earth Sci 8(7):71–80
FAO (2018) The impact of disasters and crises on agriculture and food security. http://www.fao.org/emergencies/resources/documents/resources-detail/en/c/1106859/
Gebre MG, Earl HJ (2021) Soil water deficit and fertilizer placement effects on root biomass distribution, soil water extraction, water use, yield, and yield components of soybean (*Glycine max* (L)) grown in 1-m rooting columns. Front Plant Sci 12:581127. https://doi.org/10.3389/fpls.2021.581127
Geerts S, Raes D, Gracia M, Miranda R, Cusicanqui JA, Taboada C, Mendoza J, Huanca R, Mamani A, Condori O, Mamani J, Morales B, Osco V, Steduto P (2009) Simulating yield response of Quinoa to water availability with AquaCrop. Agron J 101:499–508
González A, Ayerbe L (2011) Response of coleoptiles to water deficit: growth, turgor maintenance and osmotic adjustment in barley plants (*Hordeum vulgare* L.). Agric Sci 2(3):159–166
González A, Martin I, Ayerbe L (1999) Barley yield in water-stress conditions: the influence of precocity, osmotic adjustment and stomatal conductance. Field Crop Res 62(1):23–34
Han M, Zhang H, Chavez JL, Ma L, Trout TJ, DeJonge KC (2018) Improved soil water deficit estimation through the integration of canopy temperature measurement into a soil water balance model. Irrig Sci 36:187–201
IPCC (2021) Summary for policymakers In: Masson-Delmotte V, Zhai P, Pirani A, Connors SL, Péan C, Berger S, Caud N, Chen Y, Goldfar BL, Gomis MI, Huang M, Leitzell K, Lonnoy E, Matthews JBR, Maycock TK, Waterfield T, Yelekçi O, Yu R, Zhou B (eds) Climate change (2021): the physical science basis contribution of the working group I to the sixth assessment report of the intergovernmental panel on climate change (In press)
IPCC (2018) Global warming of 15°C an IPCC special report on the impacts of global warming of 15°C above pre-industrial levels and related global greenhouse gas emission pathways in the context of strengthening the global response to the threat of climate change sustainable development and efforts to eradicate poverty. In: Masson-Delmotte VP, Zhai H-O, Pörtner D, Roberts J, Skea PR, Shukla A, Pirani W, Moufouma-Okia C, Péan Pidcock S, Connors JBR, Matthews Y, Chen X, Zhou MI, Gomis E, Lonnoy T, Maycock M, Tignor, Waterfield T (eds) (In press)
Jenkins GJ, Perry MC, Prior MJ (2008) The climate of the United Kingdom and recent trends. Met Office Hadley Centre Exeter, UK
Jones PD, Kilsby CG, Harpham C, Glenis V, Burton A (2009) UK climate projections science report: projections of future daily climate for the UK from the weather generator. University of Newcastle, UK
Lal R, Shukla MK (2004) Principles of soil physics. Marcel Dekker, New York
Lambers H, Chapin III FS, Pons TL (2008) Plant physiological ecology (2nd edn) Springer, New York, pp 16–22
Lee EM, Jones DKC (2004) Landslide risk assessment. Thomas Telford, London
Lynch J, Marschner P, Rengel Z (2012) Effect of internal and external factors on root growth and development. In: Marschner's mineral nutrition of higher plants. Academic Press, London, pp 331–346
Murphy JM, Sexton DMH, Jenkins GJ et al. (2009) UK climate projections science report: climate change projections. Met Office Hadley Centre, Exeter
Nakićenović N, Swart R (Eds) (2000) Special report on emissions scenarios: A special report of working group III of the intergovernmental panel on climate change. Cambridge University Press, Cambridge
Neumann G, Römheld V (2012) Rhizosphere chemistry in relation to plant nutrition. In: Marschner's mineral nutrition of higher plants. Academic Press, London, pp 347–368

Newton AC., Flavell AJ, George TS et al. (2011) Crops that feed the world 4. Barley: a resilient crop? Strengths and weaknesses in the context of food security. Food Secur 3(2):141–178

Pinheiro C, Chaves MM (2011) Photosynthesis and drought: can we make metabolic connections from available data? Journal of Experimental Botany 62(3):869–882. https://doi.org/10.1093/jxb/erq340

Raes D, Steduto P, Hsiao TC, Fereres E (2009) AquaCrop—the FAO crop model to simulate yield response to water: II. Main algorithms and software description. Agron J 101:438–447

Rajala A, Hakala K, Mäkelä P, Peltonen-Sainio P (2011) Drought effect on grain number and grain weight at spike and spikelet level in six-row spring barley. J. Agron. Crop Sci 197(2):103–112

Robinson JD, Vahedifard F (2016) Weakening mechanisms imposed on California's levees under multiyear extreme drought. Climatic Chang 137(1–2):1–14

Semenov MA, Shewry PR (2011) Modelling predicts that heat stress, not drought, will increase vulnerability of wheat in Europe. Sci Rep 1(66):5. https://doi.org/10.1038/srep00066

Shone MGT, Flood AV (1983) Effects of periods of localized water stress on subsequent nutrient uptake by barley roots and their adaptation by osmotic adjustment. New Phytol 94(4):561–572

Vardon PJ (2015) Climatic influence on geotechnical infrastructure: a review. Environ Geotech 2(3):166–174

Wang Y, Yang J, Chen Y, Fang G, Duan W, Li Y, De Meyer P (2019) Quantifying the effects of climate and vegetation on soil moisture in an arid area. China. Water 11:767. https://doi.org/10.3390/w11040767

Yawson DO, Ball T, Adu MO, Mohan S, Mulholland BJ, White PJ (2016) Simulated regional yields of spring barley in the United Kingdom under projected climate change. Climate 2016(4):54. https://doi.org/10.3390/cli4040054

Yawson DO (2013) Climate change and virtual water: implications for UK food security, PhD thesis. University of Dundee, Dundee, UK. https://discovery.dundee.ac.uk/en/studentTheses/climate-change-and-virtualwater.

Chapter 28
Soil Fertility Recovery at the Kara River Basin (Togo, West Africa): Local Solutions at the Interface of Climate and Land Use Change

M'koumfida Bagbohouna, Meine van Noordwijk, Badabaté Diwediga, and Sidat Yaffa

Abstract Degrading soils reduce the tolerance to a variable and changing climate; managing soil fertility is key to climate change adaptation. Rural communities of the Kara River Basin in Togo see soil and climate issues as closely linked. To identify local solutions to adapt to climate and land-use-change-induced soil fertility decline, this study employed semi-structured questionnaire interviews with 436 respondents (farming households) from 38 selected villages in the river basin, focus group discussions, and key informant interviews coupled with transect-walks in the landscape. Data were analysed using descriptive statistics through IBM IPSS version 25 and MS EXCEL 2016, and triangulation methods to verify results. The local solutions farmers mentioned beyond chemical fertilizers, were dominated by natural solutions such as: fallowing (53%), mixed farming (50%), composting (45%), contour cropping (43%), mulching (28%), agroforestry (8%), and (other) biological fertilizers (7%). All these solutions were rated 8.5 and 9 on a scale of 10 by farmers in terms of performance in improving soil fertility and increasing yields, respectively. The study concluded that local nature-based solutions as such are preferred by farmers and present huge potential to combat the soil fertility loss in existing land use in the basin that is aggravated by climate change. Such solutions should be prioritized for

M. Bagbohouna (✉) · S. Yaffa
West Africa Graduate School on Climate Change and Education, University of The Gambia, Kanifing Campus, MDI Road, Kanifing, Serrekunda 3530 PB, Gambia
e-mail: bagbohouna-wascal@utg.edu.gm

S. Yaffa
e-mail: syaffa@utg.edu.gm

M. van Noordwijk
Centre for International Forestry Research and World Agroforestry (CIFOR-ICRAF), Bogor 16155, Indonesia
e-mail: M.vanNoordwijk@cifor-icraf.org

B. Diwediga
Laboratory of Botany, Université de Lomé, Lomé, Togo

W. Leal Filho et al. (eds.), *Climate Change Strategies: Handling the Challenges of Adapting to a Changing Climate*, Climate Change Management,
https://doi.org/10.1007/978-3-031-28728-2_28

testing if rural households are to increase resilience vis-a-vis global environmental changes.

Introduction

Land degradation constitutes a critical global issue that threatens food security and rural development. The Intergovernmental Science-Policy Platform on Biodiversity and Ecosystem Services (IPBES) defined land degradation as the "many human-caused processes that drive the decline or loss in biodiversity, ecosystem functions or ecosystem services in any terrestrial and associated aquatic ecosystems" (IPBES 2018a). The same panel estimates land degradation in Africa of about 20% of the continent land surface (660 million ha) (IPBES 2018b). With projections in the global population to rise to 11 billion people by the end of this century (United Nations, 2015a), the food demand is anticipated to rise by 60–80% (Depenbusch and Klasen 2019), inducing land-use changes and competition for land (Runyan and Stehm 2019). Many studies highlighted that land degradation implies food insecurity (GCEC 2014; Lal 2004; West and Williams 2014), vulnerability to climate change (GCEC 2014; Reynolds et al. 2007; Lambin and Meyfroidt 2011; Foley et al. 2005; Gerber et al. 2014) and poverty (Gerber et al. 2014; Nachtergaele et al. 2010; Nkonya et al. 2011; Bai et al. 2008). Over 1.5 billion people experience land degradation (Bai et al. 2008), including large numbers of the rural poor (Gerber et al. 2014; Nachtergaele et al. 2010; Nkonya et al. 2011). From 2000 to 2010, in developing nations, the amount of degrading agricultural land grew by 14% in relation to rural population growth (Barbier and Hochard 2016). The same study indicated the highest rate (17–18%) of degradation in Sub-Saharan Africa (Barbier and Hochard 2016). In general, land cover changes can result in, among others, the loss in biodiversity, increase in soil erosion and sediment transport, river siltation, alteration in hydrological processes, reduction in forage production, soil fertility decline, and reduced yields in the long run. Several authors regard low soil fertility as the most significant limitation to improve agricultural production, as it limits utilization of, also scarce, available water (Critchley et al. 1992).

Circumventing land degradation and restoring degraded lands are critical for achieving the Sustainable Development Goals: SDGs (IPBES 2018). At the cross-road of carbon neutrality, SDGs and climate action, indigenous knowledge has been widely recognized at the frontline to enable a better food system, healthy landscapes and inclusive development (van Noordwijk et al. 2020; IPBES 2019; United Nations 2015b). There still is a need to document such knowledge to efficiently improve efforts towards building resilience and strengthening adaptive capacity of the rural communities to global changes (climate and environmental changes).

Multifunctional landscapes in the West Africa region have so far supported the major part of rural livelihoods. In Togo, for example, many landscapes are utilized for multiple purposes, including a vast majority of communities' products coming from agriculture (farming, livestock, etc.). One of these landscapes, the Kara basin currently witnesses a land cover transition that has left only a little space untouched by human influence (Badjana 2015). The core socio-economic activity in the Kara basin remains farming which first of all supports subsistence with limited market integration (Djagni 2003). A handful of studies (Badjana 2015; Badjana et al. 2016) documented an increase of agricultural lands at the cost of savannah, the main vegetation type of the basin. This situation is predominantly linked to an increase in human population during recent decades (Badjana 2015). As a matter of fact, agriculture in there is extensive relying on slash-and-burn land clearing, swiddens and fallows. All over the basin, land degradation, including soil erosion is prevalent. As land in the basin is being exploited without proper compensation of soil nutrients, this implies a decrease of soil fertility and subsequent failure in crop yields (Djagni 2003). Generally, land degradation concerns at least 85% of arable land in Togo, and nearly one-third of protected land areas have been irrevocably lost in recent years, or are too expensive to rehabilitate (GFDRR 2018). Soil fertility decline, a key limitation for both crop production and natural vegetation makes fertility management an issue of wide concern (Critchley et al. 1992). Local farming communities have embarked on an extensive range of strategies to offset the negative effects of soil fertility loss.

Beside the vegetation loss linked to human activities, climate variability and change influences farming. Recent decades have been characterized by a high spatial–temporal variation of rainfall, causing a shortening of the wet season and a rise in temperature (MERF 2009; Badjana et al. 2011), leading to a reduction of the seasonal vegetation period and hampering vegetation development. Adaptation approaches are important to handle and adjust with extreme weather (Shelar et al. 2022) and land use and land cover change impacts. Multifunctional landscapes are associated with climate resilience and mitigation (Duncan et al. 2020; Harvey et al. 2014; Scherr and McNeely 2012; Minang et al. 2015) and biodiversity conservation (Scherr and McNeely 2008). Their inhabitants possess traditional knowledge used over centuries, retaining what worked in local context. Considering the reputation of local ecological knowledge in successful management of multifunctionality in the landscapes (Carpenter et al. 2009; Potschin and Haines-Young 2013), it is vital to revive and document the often-underrated local knowledge in the face of weather vagaries and changes of land use systems.

Most of projects, so far, that fight poverty at rural level have largely employed scientific and policy knowledge—imported methods or approaches to halt soil fertility loss in the Global South; they often neglect local barriers to their implementation. The overall knowledge base seems patchy (Cradock-Henry et al. 2019), including challenges in establishing other forms of knowledge exterior to peer-reviewed literature, for instance the lived experiences of the Global South and Indigenous Knowledge (Parsons et al. 2016). Projects to reclaim soil fertility have resulted into failure throughout, with very little impact in changing conditions. However, barriers to scientific and policy ecological knowledge documented in the literature (including cost, knowledge, technical know-how, accessibility, replicability and adaptability) seem to be well handled by African local communities (Isaac

2015). Researchers tend to prefer small case studies somewhat because of the now deep-rooted postulation about the localness of adaptation (Nalau et al. 2015; Shi et al. 2016) and handling effectively climate change challenges. Effectiveness and adequacy of climate change management including adaptation have been discussed by researchers, it comes strongly out that effectiveness is variously seen at many scales and in diverse contexts (Dilling et al. 2019; Eriksen et al. 2011; Ford et al. 2013; Preston et al. 2013). Notwithstanding such challenges, stocktaking adaptation advancement remains key to reduce vulnerability to climate change and enlightening adaptation funding and priorities (Berrang-Ford 2019; Morgan et al. 2019).

This paper seeks to identify the set of local or traditional ecological knowledge used to address soil fertility decline in the Kara River Basin to inform climate and land management strategies, actions or policies. It applies a diagnostic study to appraise current issues on land use as perceived by local respondents.

Materials and Methods

Study Area

The multifunctional landscapes of the Kara River Basin cover a total area of 4594 km^2 in northern Togo (plus some area in neighbouring Benin; Badjana et al. 2017b). The study area lies between longitudes 0°30′ and 1°38′E and latitudes 9°15′ and 10°01′N (Maps 28.1 and 28.2). It is drained away from the Ocean by the north-flowing Oti River, a sub-watershed of the well-known Volta Basin in West Africa. The area has a tropical sudanian climate with two seasons: a wet season (April to October) with high rainfall variability, and a dry season (November to March). The average annual rainfall was 1320 mm over the period 1960–2012, and the mean annual temperature at Kara synoptic station was 27.3 °C over the period 1977–2011. The minimum and maximum temperatures were 15 °C and 41 °C, respectively (Badjana 2015). Climate variability in the region during the last decades has been characterized by a great inter-annual variability with a succession of dry and humid periods and rising temperature (Badjana et al. 2017a). The area is exposed to high climatic variability.

The Kara River basin is predominantly located in the administrative Kara Region which rural population's main socio-economic activity consists of agriculture and livestock. The livestock is made up of small ruminants (sheep, goats, cattle, pigs) and poultry (UNEP-GEF 2012). About 96.16% of rural households are involved in agricultural-based livelihoods with 81% headed by men and 19% by women (MAEP 2003). Land degradation happens in the basin (Brabant et al. 1996), leading consequently to increased soil erosion and declining crop yield (UNEP-GEF 2012).

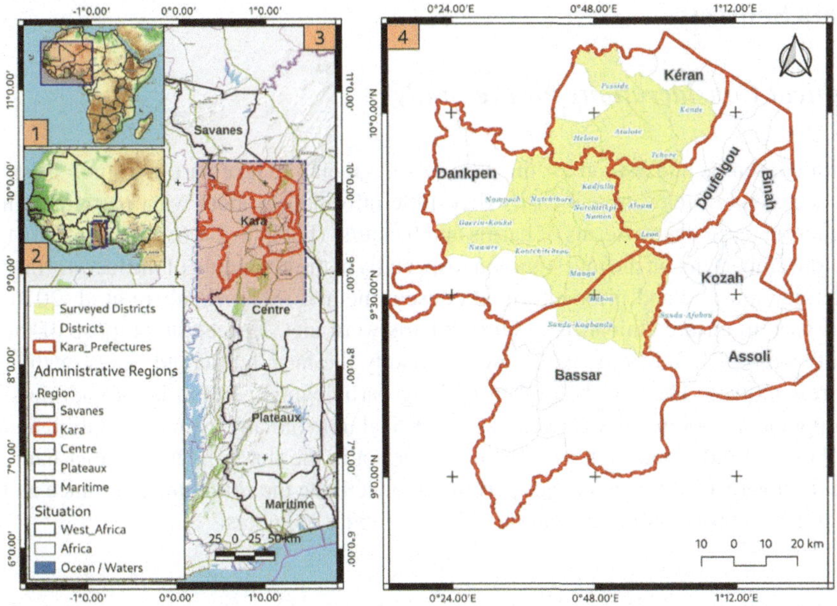

Map 28.1 Study area with surveyed districts and communities (*Source* Authors' compilation 2022)

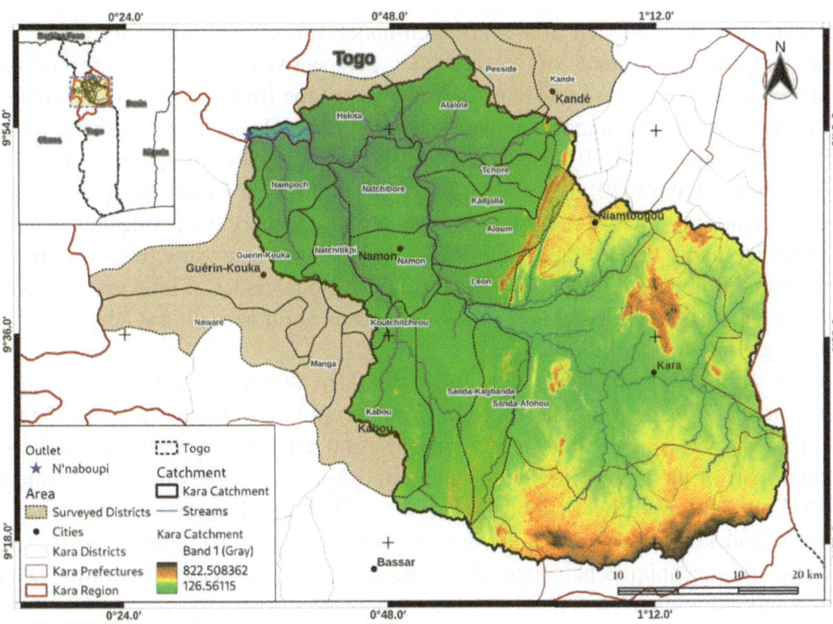

Map 28.2 The Kara River basin (*Source* Authors 2022)

Data Sources

Ethical Considerations for the Study

Prior consent is understood as 'the process of agreeing to take part in a study based on access to all relevant and easily digestible information about what participation means, in particular, in terms of harms and benefits' (Parahoo 2006; Hardicre 2014). Before embarking on the investigation (especially at the beginning of the face-to-face interview), a free and prior informed consent method (Schreckenberg et al. 2016) was used to ensure ethical research standards were met. Hence, the purpose of the study was disclosed to the interviewees, mainly the farm household heads or their representatives. This was helpful in seeking participants' consent before administrating the questionnaire (as this phase is critical to avoid any misinterpretation and violation of individual rights to participate or not in the survey). On the other hand, the disclosure of study objectives permitted to facilitate the community entrance and guarantee a proper communication with community members.

Data Acquisition

The participatory approach in this research underscores the worth of local knowledge and the importance of participant's perspectives. Such data generation approaches seem more suitable to gathering information from multiple perspectives, at different scales, and integrating socio-economic, biophysical, and climate information (Mwongera et al. 2017). Hence, this paper draws upon participatory fieldwork as in Duncan et al. (2020) to generate thorough information on how people interact with the landscape. This study predominantly used primary data collected from a field survey using face-to-face interviews. The tool is a semi-structured questionnaire containing the following aspects: local conditions, issues of land use and land cover change, soil fertility decline and local strategies for agricultural adaptation to changing climatic conditions.

From April 2021 to June 2021, 436 smallholder farmers from 38 communities were surveyed in the agropole of Kara. The agropole covers an area of 165,000 hectares and will be dedicated for agrobusiness and agricultural development in the basin as part of the nationwide large-scale Togo Agri-Food Transformation Project (PTA-Togo). followed by 19 focus group discussions (FGDs) on average of 8–12 peoples, and key informants' interviews techniques with 4 agricultural institutes. Triangulation techniques helped to revisit the interviews and verify the results. Data were analysed without individual identifiers, to protect privacy.

The study used a stratified sampling approach. At the first level, all the four (4) districts of the Kara Agropole were selected. At the second level, all the 19 cantons of the study area were surveyed. At the third stage, 38 villages or communities were randomly selected with an average of 2 per canton. In each village, about 10–50 smallholder farmers were randomly selected to contribute in the survey. Female-headed households were included as respondents where they were selected in the randomization (Table 28.1).

Field observations to collect factual information were conducted through transect-walk across the landscapes of the study area. This was useful in observing facts of real physical soil degradation (e.g., soil erosion, lateritic crusting, etc. and other landscape features relevant to better apprehend land degradation issues in the area). Field activities engaged in this research have been employed in other rural landscapes to comprehend how people utilize ecosystem services (Malmborg et al. 2018; Sinare et al. 2016).

Table 28.1 Surveyed communities

Districts	Cantons of the "Agrople of Kara"	Number of respondents
Dankpen		**136**
	Naware	36
	Guerin Kouka	11
	Koutchitcheou	24
	Natchitikpi	10
	Natchiboré	12
	Namon	20
	Nampoch	13
Bassar		**110**
	Manga	50
	Kabou	25
	Sanda Kagbanda	25
	Sanda Afohou	10
Keran		**90**
	Atalotè	28
	Hélota	20
	Kande	15
	Péssidè	27
Doufelgou		**100**
	Tchoré	25
	Alloum	22
	Léon	25
	Kadjalla	28
		N 436

Data Analysis

First, collected data were inputted and subjected to a clear and rigorous process of data cleaning and quality control. Next, the cleaned data were analysed using descriptive statistics through IBM IPSS 25. Microsoft EXCEL 2016 was used to report the results in the form of graphs.

Results

Socio-Demography and Diagnostic of Climate and Land Management Issues

Socio-demographic features of the respondents (Table 28.2) showed that the majority were male-headed household (89%); female-headed household (11%) were under-represented in our survey relative to the 20% of female-headed household in the national agricultural surveys (MAEP 2013). The fraction of respondents that were married (89%) also exceeds that (79.7%) in the national agriculture census of 2013 (MAEP 2013). Almost 60% of respondents have ever attended school while 40% did not. Such result matches with MAEP (2013) census data that found a decrease in the proportion of illiterate household heads from 62 to 46% thanks to efforts of the State and its partners in the education reflected in sectoral and national plans for education from 2010 to 2020 (Republique Togolaise 2010). These efforts include free of charge schooling and the establishment of a school feeding programme for primary schools since 2008. In fact, this has helped to keep children in school for a long period. 70% and more are above 40 years of age meaning that agriculture activity lies in the hand of the active group even though little is handled by the youths (less than 30%). The majority tends to be in a monogamous marriage (69% against 31% in polygamous) with most of the ethnic groups dominated by Kabye (61%) and Konkomba (16%) being autochthons. In terms of household size, most respondents are from a household of a size comprised between [3–7] followed by those above 7 members. The majority of respondents (82%) have over 20 years of experience in farming (Table 28.3). Beyond family labour, 65% of households makes use of at least some hired labour, while 63% indicated to be part of mutual aid, labour exchange for major tasks such as land clearing from fallow vegetation. Most of the households own lands (97%), that can be derived through heritage (60%) or land clearing (27%); some 10% became owners after borrowing land for many years (over 30 years).

Three types of terrain in the landscape are flat terrain, slopes, and swampy areas. Most of the farmers own more than 2 ha of land which are flat, swamp or sloping or a combination of two or three of these land types. A majority of farmers indicated low fertility of their farmlands (69%) while farm fertility was seen as decreasing in almost all farmlands of the respondents over the last 3-year period particularly. Most

Table 28.2 Demographic characteristics of surveyed farm household heads

Variables	Percentage of respondents (%)	Variables	Percentage of respondents (%)
Gender		**Age**	
Female	11	<30 years	4
Male	89	30–40 years	10
Marital status		41–50 years	29
Single	2	51–60 years	40
Married	89	>60 years	17
Divorced	3	**Educational level**	
Widowed	6	Primary	34
Household size		Secondary	20
<3	9	Tertiary	5
[3–7]	70	Koranic	1
>7	21	Never attended school	40
Marriage type		**Ethnic groups**	
Monogamy	69	Kabye	61
Polygamy	31	Konkomba	16
Origin of respondents		Lamba	5
Autochthonous	92	Tem	3
Migrants to Kara basin	8	Tchokossi	5
Before 1987	0	Bassar	5
1987–1990	20	Fulhani	2
1991–2000	20	Nawda	2
2001–2010	36	Ewe	1
After 2010	24		

Source Field work 2021

respondents stated that land is becoming scarce in the area. Also, chemical fertilizers are widely used by most farmers as land is becoming scarce and land needs to feed a growing population increase (Table 28.4).

Deforestation and Climate Change

Besides agriculture (crop-based farming and livestock), the majority (78%) of the farmers reported to also draw their livelihoods from forests with almost close to the half (47%) perceiving their dependency on forest related livelihoods as moderate to

Table 28.3 Socio-economic profile of the respondents

Variable	%	Variable	%
Farming experience		**Belonging to a structured group**	
<10 years	6	Yes	37
10–20 years	12	No	63
21–30 years	41		
>30 years	41		
Land tenure		**Land ownership types**	
Freehold	97	Inherited	87%
Leasehold	3	Rented	13%
Farmland topography		<2 ha	>2 ha
Flat	81	40%	60%
Swamp	59	46%	54%
Sloping	53	44%	56%
Perceived Farm Soil fertility		**Change in soil fertility over the last 3-years**	
Low	69	Decrease	94
Moderate	28	Unchanged	5
High	3	Increase	1
Available land size *vs* land need		**Is land becoming scarce?**	
Little	3	Scarce	91
Small	58	Unchanged	7
Average	36	Abundant	2
Sufficient	3		
Farm labour type[a]		**Permanent workforce** (pers.)	
Household	90	<3	23
Hired	65	3–7	53
Mutual aid	63	>7	14
		Seasonal workforce (pers.)	
		<3	26
		3–7	40
		>7	34
Farmer properties[a]		**Sources of cooking energy**[a]	
Radio	85	Fuel wood	97
TV	42	Charcoal	45
Phone	57	Crop residues Manure Others (specify)	1 1 1
Motorbike	34	**Time spent to collect source of energy**	

(continued)

Table 28.3 (continued)

Variable	%	Variable	%
Bike	10	<1 h	61
Car	1	>1 h	39
Access to power supply	22	**Distance to collect a source of energy**	
Grinding/milling machine	6	0-2 km	58
		3-5 km	34
		>5 km	18

[a]Multiple answers possible

Table 28.4 Other livelihoods, road network and commercialization, and farming inputs

Variable	%	Variable	%
Other source of livelihoods, beyond crops		**Distance to the nearest main road**	
Livestock	72	<5 km	50
Petty trading	21	5–10 km	22
Horticulture	1	11–15 km	16
Sale of petroleum products	2	>15 km	12
Taxi moto	3	**Distance to the nearest secondary road**	
Others	1	<5 km	66
Chemical fertilizer usage		5–10 km	26
<100 kg	20	11–15 km	18
100–500 kg	64	**Distance to the nearest water body**	
501–1000 kg	11	<5 km	60
>1000 kg	5	5–10 km	26
Fertilizer expenditure		11–15 km	12
<$100	55	>15 km	2
$(100–200)	25	**Distance to the nearest village market**	
$(201–400)	15	<5 km	72
>$400	5	5–10 km	25
		11–15 km	2
		>15 km	1
Did your farming revenues cover family		**Distance to the nearest main market**	
needs during the past 5 years?		<5 km	57
Yes	17	5–10 km	20
No	83	11–15 km	13
		>15 km	10

important. The Kara River basin landscape is multifunctional; thus, provides multiple purposes to its inhabitants. The relative low dependency on forest-based livelihoods in the area (53%) originates in the high interest farmers place on non-timber products for their subsistence, rather than timber or wood. Wood products are mostly used by wood exploiters for commercialization; and very few by local farmers.

Fuel wood and charcoal remain the main source of energy by farmers in the area; denotating the dependence vis-à-vis forest resources (or at least trees) in the landscapes. In general, water and energy source are relatively far from the respondents (with the average affirming that such facilities are less than 5 km of distance and less than 1 h of reach. Other socio-economic activities like livestock keeping stand out as the main activity after crop-based farming. Overall, farming revenues were not sufficient to cover family needs in the last five years; revenue decline is probably on the rise (83%).

Late arrival and early termination of the raining period was mentioned by 77%, insufficient rains by 63%, a rise in surface temperature by 57%, variability in the quantity of rains by 48%, strong winds by 47% and unpredictable seasons by 33%. Respondents attributed climate change primarily to deforestation (94% agreed). Attribution to agriculture (32%) was the second factor, with only 5 and 4%, respectively, attributing climate change to 'gods or ancestors' or 'bushfires'. Farmers who see climate change as a punishment of god(s) or ancestors mention the absence of, or negligence in the protection of their sacred woods/forests which sometimes are intruded or cleared as land is becoming scarce in the area.

The vast majority of respondents (93%) agreed that the climate change crisis is 'very important' to farmers, and only 7% found it to be 'less important'. The consequences for farmers of climate change, however, are diverse (Fig. 28.1).

Reasons pointed out by farming households to explain the seriousness of climate change issues are increase of heat (26%), reduction of agricultural livelihoods' production (20%), reduction of agricultural revenues (19%), abundance of

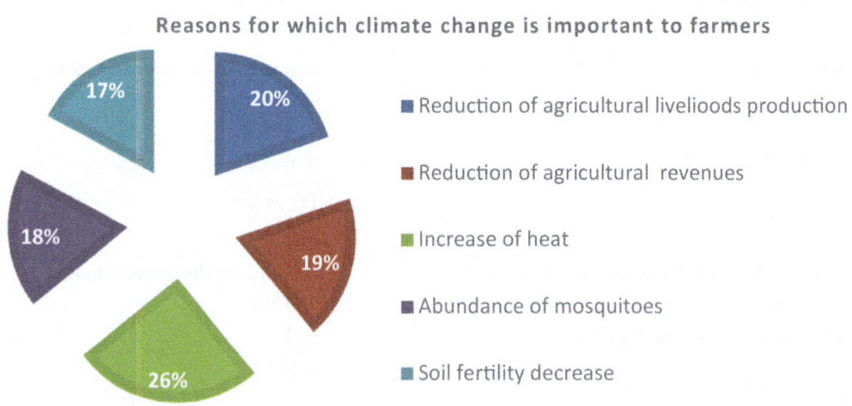

Fig. 28.1 Reasons explaining the seriousness of climate change to local peoples (*Source* Field work 2021)

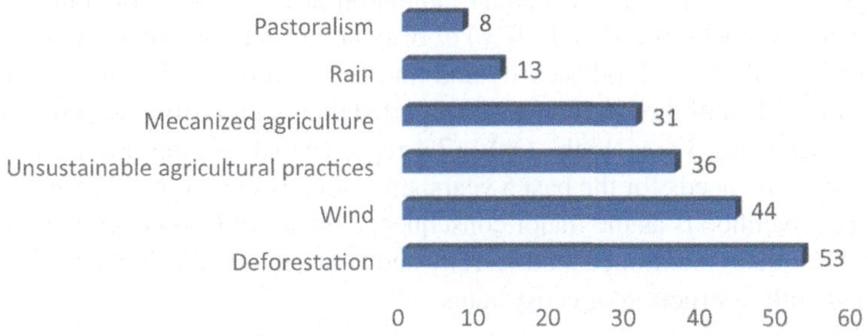

Fig. 28.2 Causes of soil erosion and soil fertility loss in the basin—multiple answers possible (*Source* Field work 2021)

mosquitoes and soil fertility decline or decrease (17%). During the FGDs, households have recognized that one manifestation of climate change is the rise in the surface temperature.

Land Degradation Issues in the River Basin

The majority (62%) of farmers saw signs of erosion on their farmland, 38% not. Farmers had mixed opinions about the role of livestock in land degradation: 42% saw such relationship, 11% said livestock played no role, and 47% wasn't sure about this. Opinions on causes of soil erosion were varied ranging from deforestation, wind, agriculture to rains (Fig. 28.2). Only 13% of respondents, however, reported erosion to be 'severe', 24% 'moderate' and 63% 'minor'.

Only 1% of respondents still reported fallow periods of 3–7 years, for 35% fallow periods were less than 3 years and 64% relied on permanent cropping, without fallow periods.

The area has seen a decrease in the availability of forest-based products that serve as livelihoods for the communities. Hence, non-timber forest products represent the scarcest products to be found in the area (62%), while access to charcoal, firewood and construction wood had become problematic for 2, 12 and 14% of respondents, respectively.

Local Farmers' Perspectives on the Causes of Land Conversion

Farmer perceptions on the causes of land conversion referred in only 1% of respondents to population growth; 7 and 8%, respectively, referred to logging and charcoal production; 24% to increase in human habitation and 60% to agricultural expansion and development. Nearly all (97%) of respondents reported reduced food security and 55% a threat to food security and nutrition, but only 3% saw an increased vulnerability. Most of the farmers (69%) reported their soil fertility as 'low', 28% as 'moderate' and only 3% as 'high'. Only 17% reported that farming revenues covered family livelihood needs for the past 5 years, 83% said this was not the case. Farmers saw loss of livelihoods as the major consequence (65%) of land cover and land use change, 41% (also) mentioned land scarcity and land degradation, and only 10% was concerned with destruction of ecosystems.

Local Knowledge of Soil Erosion Induced Soil Fertility Loss Biophysical Indicators

Farmers reported multiple biophysical indicators of decreasing soil fertility: proliferation of *Striga* (70%), increasingly hard soil (74%) and cracks in the soil (80%). The four practices that reduce or prevent soil fertility loss most frequently mentioned by farmers were:

- o Contour cropping (47%)
- o Keeping trees in farmlands (41%)
- o Terraced cultivation (38%)
- o Use of ridges in row cropping (37%).

Integrated local soil and water management practices (ILSWMP) used were diverse (Fig. 28.3).

Farmers perceived a number of benefits as the importance of these ILSWMP (Fig. 28.4). Their effectiveness in increasing yields was scored as 9 out of 10, their effect on improving soil fertility as 8.5 out of 10.

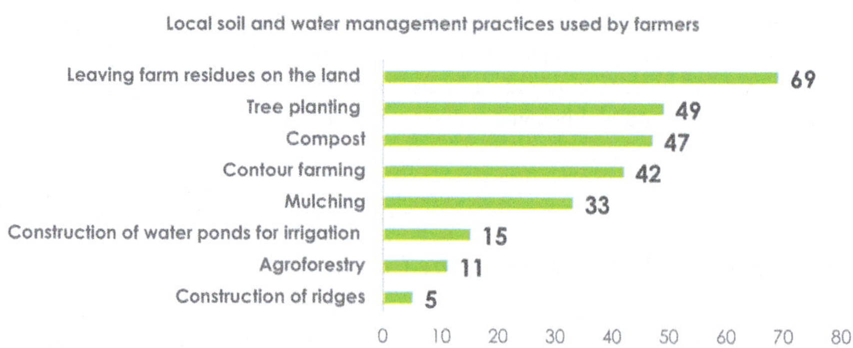

Fig. 28.3 Integrated local soil and water management practices (ILSWMP)—multiple answers possible (*Source* Fieldwork 2021)

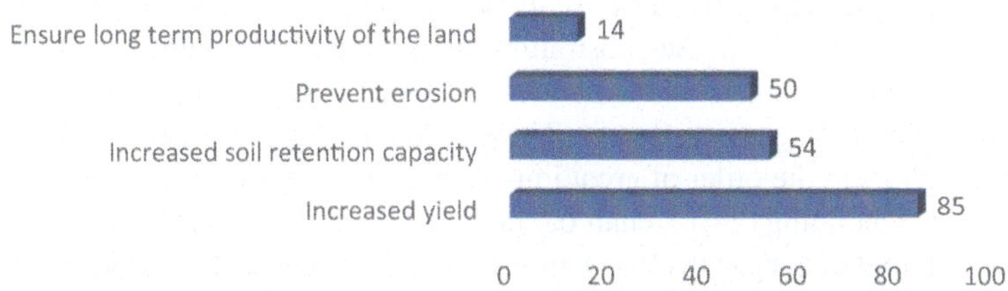

Fig. 28.4 Farmers' perceived effects of ILSWMP on farming activities—multiple responses (*Source* Fieldwork 2021)

Discussion

In the local knowledge of the study area, climate change is understood and characterized as late commencement and early cessation of rains, insufficient rains, temperature increase, variability in rainfall patterns, strong winds and unpredictable seasons. These findings are in agreement with observed data from meteorological stations in the study area and other previous studies in the Kara River Basin (Badjana et al. 2017a), which indicated an interannual variability of rainfall and a decrease in average annual rainfall in the basin. In few words, the empirical observations of farmers tend to corroborate with scientific records of rainfalls. Also, there was a consensus among respondents vis-à-vis certain changes in weather and frequency of extreme events.

In the quest to understand how serious global changes, including climate change, are to the farmers, we noticed that the majority of them (93%) acknowledged the climate crisis as very important for reasons in Fig. 28.1 as it poses threats to their means of survival (e.g., agriculture, livestock even forest-based products). Such information is in line with climatic change profile in the Kara river basin which predicted increase in surface temperature (Republique Togolaise 2015; Badjana 2015) and in the overall Oti Basin (Badjana et al. 2011). Such scenario undoubtedly may have led to burdens seen in climate sensitive livelihoods activities (reduction of agricultural livelihoods' production and revenues). Further, the rise in temperature coupled with increase in rainfall duration (Badjana et al. 2017a) tends to increase number of mosquitoes. One factor extensively discussed in the literature is the likely influence of regional warming linked to climate change on the development of malaria mosquitoes (Hay et al. 2002). Paaijmans et al (2010) showed that increased water temperature could affect mosquito development cycle. This finding corroborates with the outcome of a study by Yao et al. (2018) which revealed increased malaria cases all through the wet season in farming communities in Bole District of Northern Ghana, a sub-Saharan neighbouring country of Togo.

Farmlands within the multifunctional landscapes seem to be very much under pressure regarding the few available lands (91% of households declared that land is scarce). The few available dedicated to fallowing has in majority a short fallow duration (less than 3 years). Then, it will be reasonable to expect more willingness of farmers to adopt farming practices to counter land scarcity induced by land cover and land use changes. In the multifunctionality of landscapes, land plays a central role as it represents the substratum upon which the system lies. Therefore, farmers' perceptions on the land dynamics show that land transition has taken place as a result of anthropogenic activities. These activities found their root causes in the following factors in the order of great importance: agriculture expansion and development (60%), building (24%), charcoal fabrication (8%), timber logging (7%); and accessorily population growth (1%) acting as an indirect cause. Interestingly, farmers hardly mention population growth as a factor in increasing land scarcity and reduced fallow duration, while data in the national census (MAEP 2013) clearly point in this direction.

These perceived causes of farmers explain empirically well the interrelated nature of land-forest-food web with land and forest being the core sectors in this nexus. Population growth lags behind and seems not to act as a direct source of land conversion in the area. However, it constitutes a pressure for land conversion in many other studies (Molotoks et al. 2018; Maja and Ayano 2021; Jiang and Hardee 2011). As a matter of fact, based on the DPSIR (driver, pressure, system, impact, responses) analysis, population dynamics increases pressure on natural resources (land, forest, water, etc.). Migration and population dynamics that prevailed in the last decades: from 1987–2010 and 2010 to date (see Table 28.2) influence land use changes in the basin. This pressurises natural resources by reducing natural forests, triggering loss of habitat and biological diversity (Zegeye 2017; Molotoks et al. 2018).

Land cover changes exert an influence on forests and land resources in general. Within the Kara River basin, changes in land cover and land use induced by anthropogenic actions (agriculture, building, logging, to name a few) affect the solar net radiation on the ground and increase surface heat as well as soil erosion. Less vegetation cover is susceptible to an important risk of wind and water erosion coupled with increased pollution. This has been indicated by the local populace who perceived land use and land cover change as a threat to soil fertility (69%), reduction in livelihood provision (65%), land degradation and scarcity (41%), water pollutions and ecosystems destruction. An eroded soil losses nutrient (through physical degradation); thus, reducing soil fertility. Further, locals revealed an increased risk of soil fertility decline with continuous climate variability, changes and windstorms as the area faces environmental change. In the overall, this explains their farming revenues in the past 5 years not able to cover their livelihood needs. Thus, the need to go in for other alternative livelihoods (logging, etc.).

Farmers possess knowledge on soil erosion which induces loss of soil fertility. In fact, soil erosion causes top soil to be removed; thus, loss of soil nutrients content. However, locals are knowledgeable about the indicators of soil erosion.

A wide range of local, almost traditional measures are being implemented by farmers to prevent their farm soils from erosion and maintain their fertility. Besides strategies to prevent soil fertility loss, other blended measures are put in place by these farmers that integrate both soil and water management; as the area experiences monthly rainfall variability and rise in temperature (Badjana 2015).

Integrating soil and water management by locals have been identified as part of the local solutions inspired by nature that farmers mentioned to be of great reliability to improve soil fertility. In reality, the integration of soil and water management practices add more to the conservation efforts of soil fertility decline as water and soil are intrinsically connected and may not be handled separately as per sustainable agriculture and climate-smart agriculture require. These are leaving farm residues on the land (69%), tree planting (49%), compost (47%), contour farming (42%), mulching (33%), construction of water ponds for irrigation (15%), agroforestry (11%) and construction of ridges. Our findings here agree with Adji (2021) study which stated that compost making from vegetable or animal waste/residues through traditional practices is a prevalent technique in the country, particularly in the dry northern area of which Kara River basin is part. On the other hand, agroforestry systems (still in small proportion) introduced by farmers are gradually gaining popularity among women and men in the northern part of the country.

In terms of importance, the ILSWMP are well perceived in the area to be beneficial in increasing yields (85%), augmenting soil retention capacity (54%), preventing soil erosion (50%) and ensuring long term productivity of the farmland (14%).

The utilisation of local solutions by farmers to handle climate and land use-induced land degradation is seen by the majority of farmers to perform well rating on a scale of 1 to 10, 8.5 for improving soil fertility and 9 for increasing yields with respect to their performance. These local solutions used to cope and/or adapt to land degradation and soil fertility have the desired impacts. These findings may suppose that farmers willingly accept to adopt such LSWMP which performance are widely recognised.

Conclusions and Recommendations

In all, this paper provides empirical evidence that land degradation and climate change interact and pose enormous threats to local livelihoods in the Kara River basin. Farmers face soil erosion, consequence of land use and land cover changes aggravated by climate change. Local farmers perceived the existing connection between land cover changes and climate change and its effects on agricultural outputs. To handle the decline in soil fertility generated by land use, land cover change and climate change, farming households in the Kara River Basin have managed to adapt by employing locally proven solutions that they own and seem to work out well to augment food production and limit food insecurity. Such local solutions present huge potential to combat soil fertility loss aggravated by climate change, and land use and land cover change increase in the basin; which should be prioritized in other communities in the basin or other parts in Africa. These solutions may help to address the well documented barriers or constraints to adaptation in developing countries in scientific literature which chiefly derive from prevalent poverty, unequal land distribution, over reliance on rain-fed agriculture, restricted access to capital and technology, poor long-term weather forecasts and inadequate research and extension, and inadequate public infrastructure (e.g., such as roads).

On the other hand, in the light of the Paris Agreement requirements for countries to contribute periodically to stock takes, including revising the *adequacy* and *effectiveness* of adaptation (Singh et al. 2021a, b; Craft and Fisher 2018; Tompkins et al. 2018), the above findings would be key for drawing best practices and lessons learnt for handling climate change locally.

The following can be as part of the recommendations to policymakers, development actors and scientists intervening in the area for properly managing challenges posed by climate change on soil fertility:

- Upscale these local solutions to other smallholders in the region and other regions that share similar geophysical and climatic conditions with the river basin;
- Train smallholders on and promote soft, locally effective and low-cost soil fertility management technologies to support their adaptive capacity and building resilience;
- Leverage on the locally preferred and available soil management strategies in the region for projects or programmes targeting to secure livelihoods and combatting climate change;
- Initiate farmer-centred learning and exchange platforms to build smallholders resilience at local level.

Limitations of the Research

This study faces some shortcomings related to understanding gender disaggregated strategies to handling climate change and land use/land cover impacts induced soil fertility loss. Further studies could investigate farming households' willingness to adopt the investigated strategies as well as significant specific factors that hinder or promote the utilisation of the identified strategies to combat climate change and land use and land cover impacts in the basin.

Acknowledgements We recognize the support from the Agriculture Regional Directorate of the Kara Region, agricultural extension officers during fieldwork and community outreaches as well as farming households who played a key role for participating in the survey. This study is part of the doctoral thesis of the first author (M.B).

Conflicts of Interest

All the authors have read and approved the final manuscript and declare no conflict of interest.

Authors' Contribution M.B designed the methodology, conducted the analysis, and drafted the manuscript; M.v.N advised on research design and improved the draft manuscript with inputs, suggestions, and proofreading; B.D improved the draft manuscript with inputs and suggestions; S.Y contributed with comments.

Funding We appreciate the German Ministry of Education and Research (BMBF) for their financial support for this Doctoral research programme Scholarship and research grant (Grant nb. Batch 4) at the University of The Gambia through the West African Science Service Centre on Climate Change and Adapted Land Use (WASCAL) Initiative. This work was also produced with the financial aid of the Prince Albert II of Monaco Foundation through the IPCC Scholarship Programme (2021–2023). The work is solely the liability of the authors and under no circumstances may be considered as a reflection of the position of the Prince Albert II of Monaco Foundation and/or the IPCC.

References

Adji KM, Egbendewe AYG, Lokonon BOK (2021) Potential impacts of sustainable agricultural practices on smallholders' behavior in developing countries: Evidence from Togo. Nat Res Forum 1–15:77. https://doi.org/10.1111/1477-8947.12243

Badjana HM (2015) River basins assessment and hydrologic processes modeling for integrated land and water resources management (ILWRM) in West Africa. Doctoral thesis

Badjana HM, Batawila K, Wala K, Akpagana K (2011) Evolution Des Paramètres Climatiques Dans La Plaine De L'oti (Nord-Togo): Analyse Statistique. Perceptions Locales Et Mesures Endogènes D'adaptation 15(2):77–95

Badjana HM, Helmschrot J, Selsam P, Wala K, Flügel WA, Afouda A, Akpagana K (2016) Land cover changes assessment using object-based image analysis in the Binah River watershed (Togo and Benin). Earth and Space Science 3:46–67. https://doi.org/10.1002/2014EA000083

Badjana HM, Renard B, Helmschrot J, Edjamé KS, Afouda A, Wala K (2017a) Bayesian trend analysis in annual rainfall total, duration and maximum in the Kara River basin (West Africa). J Hydrol Region Stud 13:255–273. https://doi.org/10.1016/j.ejrh.2017.08.009

Badjana HM, Olofsson P, Woodcock CE, Helmschrot J, Wala K, Akpagana K (2017b) Mapping and estimating land change between 2001 and 2013 in a heterogeneous landscape in West Africa: Loss of forestlands and capacity building opportunities. Int J Appl Earth Obs Geoinf 63:15–23

Bai ZG, Dent DL, Olsson L, Schaepman ME (2008) Proxy global assessment of land degradation. Soil Use Manag 24:223–234

Barbier EB, Hochard JP (2016) Does land degradation increase poverty in developing countries? PLoS ONE 11(5):13–15. https://doi.org/10.1371/journal.pone.0152973

Berrang-Ford L (2019) Tracking global climate change adaptation among governments. Nat Clim Chang 9(6):440–449. https://doi.org/10.1038/s41558-019-0490-0

Brabant P, Darracq S, Égué K, Simonneaux V (1996) Togo: état de dégradation des terres résultant des activités humaines, carte des indices de dégradation, Paris: Éd. de l'ORSTOM, ISBN 2709913488 9782709913485

Carpenter SR, Mooney HA, Agard J, Capistrano D, DeFries RS, Díaz S, Dietz T, Duraiappah AK, Oteng-Yeboah A, Pereira HM (2009) Science for managing ecosystem services: beyond the Millennium Ecosystem Assessment. Proc Natl Acad Sci 106:1305–1312

Cradock-Henry NAF, Buelow S, Flood P, Blackett A, Wreford (2019) Towards a heuristic for assessing adaptation knowledge: impacts, implications, decisions and actions. Environ Res Lett 14. https://doi.org/10.1088/1748-9326/ab370c

Craft B, Fisher S (2018) Measuring the adaptation goal in the global stocktake of the Paris agreement. Climate Policy 18(9):1203–1209. https://doi.org/10.1080/14693062.2018.1485546

Critchley WRS, Reij CP, Turner SD (1992) Soil and Water Conservation in Sub-Saharan Africa: towards Sustainable Production by the rural poor. https://doi.org/10.1016/0143-6228(92)90018-i

de Jong R, de Bruin S, Schaepman M, Dent D (2011) Quantitative mapping of global land degradation using Earth observations. Int J Remote Sens 32(6823–6853):13

Depenbusch L, Klasen S (2019) The effect of bigger human bodies on the future global calorie requirements. PLoS ONE 14:1–15. https://doi.org/10.1371/journal.pone.0223188

Dilling L, Prakash A, Ahmad ZZ, Singh N, de Wit S, Nalau J, Daly M, Bowmman K, (2019) Is adaptation success a flawed concept? Nat Clim Chang 9:572–574. https://doi.org/10.1038/s41558-019-0539-0

Djagni KK (2003) L'agriculture togolaise face à des mutations environnementales multiples, Nécessité d'un ensemble d'innovations techniques et organisationnelles cohérentes. In: Jamin JY et al. (eds) Prasac, Savanes africaines : des espaces en mutation, des acteurs face à de nouveaux défis. N'Djamena, Tchad, Cirad, Montpellier, France

Duncan JMA, Haworth B, Boruff B, Wales N, Biggs EM, Bruce E (2020) Managing multifunctional landscapes: Local insights from a Pacific Island Country context. J Environ Manage 260:109692. https://doi.org/10.1016/j.jenvman.2019.109692

Eriksen S, Aldunce P, Bahinipati CS, Martins RDA, Molefe JI, Nhemachena C, O'Brien K, Olorunfemi F, Park J, Sygna K, Ulsrud K (2011) When not every response to climate change is a good one: Identifying principles for sustainable adaptation. Climate Dev 3(1):7–20. https://doi.org/10.3763/cdev.2010.0060

Foley JA, De Fries R, Asner GP, Barford C, Bonan G, Carpenter SR, Chapin FS, Coe MT, Daily GC, Gibbs HK (2005) Global consequences of land use. Science 309:570–574. https://doi.org/10.1126/science.1111772

Ford JD, Berrang-Ford L, Lesnikowski A, Barrera M, Heymann SJ (2013) How to track adaptation to climate change: A typology of approaches for national-level application. Ecol Soc 18(4):art40. https://doi.org/10.5751/ES-05732-180340

GCEC (2014) Better growth, better climate: The new climate economy report. The Global Commission of the Economy and Climate, New Climate Economy, Washington, DC, USA

Gerber N, Nkonya E, von Braun J (2014) Land degradation, poverty and marginality. In: von Braun J, Gatzweiler FW (eds) Marginality: addressing the nexus of poverty, exclusion and ecology. Springer, Berlin, pp 181–202

GFDRR (2018) Enhancing disaster preparedness in togo. https://doi.org/10.1016/c2018-0-02311-7

Hardicre J (2014) Valid informed consent in research: an introduction. Br J Nursing 4(4):17–18

Harvey CA, Chacon M, Donatti CI, Garen E, Hannah L, Andrade A, Bede L, Brown D, Calle A, Chara J (2014) Climate-smart landscapes: opportunities and challenges for integrating adaptation and mitigation In tropical agriculture. Conserv Lett 7:77–90

Hay SI, Simba M, Busolo M, Noor AM, Guyatt HL, Ochola SA, Snow RW (2002) Climate change and the resurgence of malaria in the East African highlands. Nature 415:905–909. https://doi.org/10.1038/415905a

Intergovernmental Science-Policy Platform on Biodiversity and Ecosystem Services (IPBES) (2018) Land degradation and restoration: Companion to environmental studies. https://doi.org/10.4324/9781315640051-105

Intergovernmental Science-Policy Platform on Biodiversity and Ecosystem Services (2019) The global assessment report on of the IPBES global asessment report on biodiversity and ecosystem services: summury on policymakers. https://ipbes.net/system/tdf/ipbes_global_assessment_report_summary_for_policymakers.pdf?file=1&type=node&id=35329

Isaac D (2015) Soil and water conservation technologies in the West African Sudan Savanna: cropping system options to address variability of crop yield and impacts of climate change. http://hss.ulb.uni-bonn.de/2015/4221/4221.pdf

Jiang L, Hardee K (2011) How do recent population trends matter to climate change? Popul Res Policy Rev 30:287–312

Lal R (2004) Soil Carbon Sequestration Impacts on Global Climate Change and Food Security. Science 304(5677):1623–1627. https://doi.org/10.1126/science.1097396

Lambin EF, Meyfroidt P (2011) Global land use change, economic globalization, and the looming land scarcity. Proc Natl Acad Sci 108:3465–3472

MAEP (2013) Aperçu de l'agriculture togolaise.

Maitima J, Reid RS, Gachimbi LN, Majule A, Lyaruu H, Pomery D, Mugatha S, Mathai S, Mugisha S (2004) Regional synthesis paper: the linkages between land use change, land degradation and biodiversity across East Africa. LUCID Working Paper 42:1–51

Maja MM, Ayano SF (2021) The impact of population growth on natural resources and farmers' capacity to adapt to climate change in low-income countries. Earth Syst Environ 5(2):271–283. https://doi.org/10.1007/s41748-021-00209-6

Malmborg K, Sinare H, Kautsky EE, Ouedraogo I, Gordon J (2018) Mapping regional livelihood benefits from local ecosystem services assessments in rural Sahel, pp 1–20.
Ministere de l'Environnement et des Ressources Forestieres du Togo (MERF) (2009) Plan d'Action National d'Adaptation aux changements climatiques—PANA-Togo
Minang PA, van Noordwijk, M, Freeman OE, Mbow C, de Leeuw J, Catacutan D (eds) (2015) Climate-smart landscapes: multifunctionality in practice. World Agroforestry Centre (ICRAF), Nairobi (Kenya)
Molotoks A, Stehfest E, Doelman J, Albanito F, Fitton N, Dawson TP, Smith P (2018) Global projections of future cropland expansion to 2050 and direct impacts on biodiversity and carbon storage. Glob Change Biol 24:5895–5908
Morgan EA, Nalau J, Mackey B (2019) Assessing the alignment of national-level adaptation plans to the Paris agreement. Environ Sci Policy 93:208–220. https://doi.org/10.1016/j.envsci.2018.10.012
Mwongera C, Shikuku KM, Twyman J, L€ aderach P, Ampaire E, Van Asten P, Twomlo, S, Winowiecki LA, (2017) Climate smart agriculture rapid appraisal (CSA-RA): a tool for prioritizing context-specific climate smart agriculture technologies. Agric Syst 151:192–203
Nachtergaele F, Petri M, Biancalani R, Van Lynden G, Van Velthuizen H, Bloise M et al. (2010) Global land degradation information system (GLADIS). Version 1.0. An information database for land degradation assessment at global level (Land Degradation Assessment in Drylands Technical Report No. 17). FAO, Rome, Italy
Nalau J, Preston BL, Maloney MC (2015) Is adaptation a local responsibility? Environ Sci Policy 48:89–98. https://doi.org/10.1016/j.envsci.2014.12.011
Nkonya E, Gerber N, Baumgartner P, von Braun J, De Pinto A, Graw E, et al (2011) The economics of desertification, land degradation, and drought (IFPRI Discussion Paper 01086) International Food Policy Research Institute, Washington, DC
Paaijmans KP, Imbahale SS, Thomas MB, Takken W (2010) Relevant microclimate for determining the development rate of malaria mosquitoes and possible implications of climate change. Malar J 9(1):1–8. https://doi.org/10.1186/1475-2875-9-196
Parahoo K (2006) Nursing Research: Principles, Process and Issues, 2nd edn. Palgrave Macmillan, Basingstoke
Parsons M, Fisher K, Nalau J (2016) Alternative approaches to co-design: insights from indigenous/academic research collaborations. Curr Opin Environ Sustain 20:99–105
Potschin M, Haines-Young R (2013) Landscapes, sustainability and the place-based analysis of ecosystem services. Landsc Ecol 28:1053–1065
Preston BL, Mustelin J, Maloney MC (2013) Climate adaptation heuristics and the science/policy divide. Mitig Adapt Strat Glob Change 20(3):467–497. https://doi.org/10.1007/s11027-013-9503-x
Republique Togolaise (2015) troisieme communication nationale sur les changements cllimatiques
Reynolds J, Fernando F, Maestre T, Paul R, Kemp D, Mark Stafford-Smith, Lambin E (2007) Natural and human dimensions of land degradation in drylands: Causes and consequences. Terres Ecosyst Chang World 247–57. https://doi.org/10.1007/978-3-540-32730-1_20.s
Runyan CW, Stehm J (2019) Land use change, deforestation and competition for land due to food production. In: Ferranti P, Elliot M, Berry JRA (eds) Encyclopedia of food security and sustainability. Elsevier, pp 21–26. https://doi.org/10.1016/B978-0-08-100596-5.21995-1.
Schreckenberg K, Franks P, Martin A, Lang B (2016) unpacking equity for protected area conservation, 22 November
Scherr SJ, McNeely JA (2008) Biodiversity conservation and agricultural sustainability: towards a new paradigm of 'ecoagriculture' landscapes. Philos Trans R Soc Biol Sci. 363:477–494
Shelar R, Singh AK, Maji, S (2022) Constraints in adapting the climate change in Konkan Region of Maharashtra. Ind J Exten Edu 58(1):169–171. https://doi.org/10.48165/ijee.2022.58132
Shi C, Chu E, Anguelovski I, Aylett A, Debats J, Goh KT, Schenk TKC, Seto KC, Dodaman D, Roberts D, Roberts JT, Van Deveer SD (2016) Roadmap towards justice in urban climate adaptation research. Nat Clim Change 6:131–137. https://doi.org/10.1038/nclimate2841

Sinare H, Gordon LJ, Kautsky EE (2016) Assessment of ecosystem services and bene fi ts in village landscapes: a case study from Burkina Faso. Ecosyst Serv 21:141–152. https://doi.org/10.1016/j.ecoser.2016.08.004

Singh C, Iyer S, New MG, Few R, Kuchimanchi B, Segnon AC, Morchain D (2021a) Interrogating 'effectiveness' in climate change adaptation: 11 guiding principles for adaptation research and practice. Climate Dev 1–15. https://doi.org/10.1080/17565529.2021.1964937

Togolaise R (2010) Plan Sectoriel De L'Education, 2015

Tompkins EL., Vincent K, Nicholls RJ, Suckall N (2018) Documenting the state of adaptation for the global stocktake of the Paris agreement. WIRES Climate Change, 9(5), 1–9. https://doi.org/https://doi.org/10.1002/wcc.545

UN (2015a) Sendai framework for disaster risk reduction 2015a–2030. United Nations Office for Disaster Risk Reduction, Geneva, Switzerland https://www.unisdr.org/we/inform/publications/43291

UN (2015b) Transforming our world: The 2030 agenda for sustainable development (A/RES/70/1, 25 September 2015b). United Nations, New York, NY https://sustainabledevelopment.un.org/?menu=1300

UNEP-GEF Volta Project (2012) Volta basin transboundary diagnostic analysis. Unep/gef/volta/rr 4:153

United Nations (2015) Paris agreement

van Noordwijk M, Gitz V, Minang PA, Dewi S, Leimona B, Duguma L MeybeckA (2020) People-centric nature-based land restoration through agroforestry: a typology. Land 9(8). https://doi.org/10.3390/LAND9080251

West SE, Williams RC (2004) Estimates from a consumer demand system: implications for the incidence of environmental taxes. J Environ Econ Manag 47(3):535–558. https://doi.org/10.1016/j.jeem.11.004

Yao KMA, Obeng F, Ntajal J, Tounou AK, Kone B (2018) Vulnerability of farming communities to malaria in the Bole district. Ghana. Parasite Epidemiology and Control 3(4):e00073. https://doi.org/10.1016/j.parepi.2018.e00073

Zegeye H (2017) Major drivers and consequences of deforestation in Ethiopia: implications for forest conservation. Asian J Sci Technol 8:5166–5175

Chapter 29
Can Biostimulants Mitigate the Negative Impact of Climate Change on Oliviculture?

Maria Celeste Dias, Rui Figueiras, Marta Sousa, Márcia Araújo, and Conceição Santos

Abstract The increase of extreme climate events in the Mediterranean region represents a threat to oliviculture, one of the most important agricultural sectors in this region. Therefore, it is urgent to develop sustainable strategies that can help to mitigate the impact of climate change on olive production. The application of eco-friendly compounds, such as biostimulants, promote plant growth and yield, and seems to alleviate stress negative effects. This study aims to elucidate the effects of a biostimulant pretreatment on the water status and antioxidant battery of *Olea europaea* plants under well-watered and drought stress conditions. Potted young olive trees were randomly divided in two groups, one was sprayed with a biostimulant based on *Ascophyllum nodosum* extract and the other with water. Then, both groups were subdivided in two and exposed to drought and well-watered conditions, respectively. Drought stress treatment reduce water availability, carbohydrates levels and total flavonoids. However, biostimulant application under drought conditions helps to maintain the leaf water content and to accumulate more antioxidants, such as total phenols and flavonoids. Total soluble sugar levels were not affected by the biostimulant application under drought conditions. Starch was more responsive to the

M. C. Dias (✉) · R. Figueiras · M. Sousa · M. Araújo
Department of Life Sciences, Centre for Functional Ecology - TERRA Associate Laboratory, Calçada Martim de Freitas, University of Coimbra, 3000-456 Coimbra, Portugal
e-mail: celeste.dias@uc.pt

M. Araújo
e-mail: marciaaraujo@fc.up.pt

R. Figueiras · M. Araújo · C. Santos
IB2Lab, Department of Biology & LAQV/REQUIMTE, Faculty of Sciences, University of Porto, Rua Do Campo Alegre, 4169-007 Porto, Portugal
e-mail: csantos@fc.up.pt

M. Araújo
Centre for the Research and Technology of Agro-Environmental and Biological Sciences (CITAB), University of Trás-Os-Montes and Alto Douro, 5001-801 Vila Real, Portugal

W. Leal Filho et al. (eds.), *Climate Change Strategies: Handling the Challenges of Adapting to a Changing Climate*, Climate Change Management,
https://doi.org/10.1007/978-3-031-28728-2_29

biostimulant, increasing in both treatments, well-watered and drought. These results show that this biostimulant can induce stress tolerance and alleviate drought negative effects on olive.

Introduction

Olive culture has a strong socio-economic impact in several countries of the Mediterranean basin since there is produced around 95% of the world olive oil (Fraga et al. 2021). Spain is the world leader in olive oil production, being Greece, Italy, Portugal, Tunisia, Syria, Turkey and Morocco also important countries in the oliviculture sector, contributing in large part for the world olive oil marketing (Sales et al. 2020; Fraga et al. 2021). The awareness of the benefits of olive oil to the human health boosted the consumption and production of olive oil, and the expansion of olive culture outside the Mediterranean basin to countries like China, Australia and Brazil (Dias et al. 2019). Furthermore, to feat the increasing demand of olives and the need of higher profitability, the traditional dry-farming/rainfed orchards in the Mediterranean region are being replaced by intensive and super intensive orchard growing-system. These systems are characterized by high-density orchards that use usually specific varieties well adapted to the region climate and to the mechanical harvesting and pruning, but with high needs of irrigation and agrochemicals (Fraga et al. 2021). Nowadays it is estimated that around 70% of worldwide olive orchards use irrigation (Romero-Trigueros et al. 2019), but despite the advantages (e.g., increase of olive fruit yields and size) of this agricultural practice, the current scenario of climate change and future persistence of extreme climatic events impose a need to develop sustainable practices towards both improving olive resilience to climate change challenges, while rationalizing the use of water in olive groves (Fotia et al. 2021).

The increase of frequency of extreme weather events (characterized by long periods with low or absent precipitation accompanied by high temperature and UV episodes) due to climate change exerts a dramatic challenge to the agriculture sector. The Mediterranean region is particularly susceptible to this scenario of climate change, and negative impacts on typical crops of this region are already visible, mostly due to changes in precipitation patterns causing severe drought conditions (Brito et al. 2019; del Pozo et al. 2019). For instance, the olive tree is well adapted to the harsh environmental conditions of the Mediterranean region, but the increasing of the frequency and fierceness of extreme weather events represents a threat to this culture (Brito et al. 2019; Dias et al. 2018 and 2020). Recent data support the susceptibility of this species to the already ongoing extreme climate events, appointed to, changes in olives and olive oil quality, alterations in the phenological times, and enhance in the vulnerability to pest and diseases (Brito et al. 2019; Fraga et al. 2021). Olive plants developed some morphophysiological mechanisms to cope with drought and heat conditions, such as a good capacity of stomatal control, osmotic adjustment, and high root water uptake (Brito et al. 2019). However, severe stress conditions can simultaneously expose plants to extreme heat, drought and excessive PAR/UV-radiation, reducing olive water availability, impairing photosynthesis, decreasing stomatal conductance, inducing oxidative stress, and shifting multiple secondary metabolic pathways of the plant (Silva et al. 2018; Araújo et al. 2019 and 2021). Under such conditions, the light energy absorbed by the plants usually exceeds the one necessary for photosynthesis, leading to an uncontrolled increase of reactive oxygen species (ROS) formation, which cause oxidative damages (Silva et al. 2019). Plants have several mechanisms to control the level of ROS. For instance, the antioxidant system, comprising the enzymatic and non-enzymatic antioxidants, can help to scavenger free radicals/ROS, decreasing the level of oxidative stress (Dias et al. 2014). Several antioxidant enzymes (guaiacol peroxidase, ascorbate peroxidase, glutathione reductase, superoxide dismutase and catalase) and non-enzymatic antioxidants, like phenolic compounds (e.g. the flavonoids luteolin-7-*O*-glucoside, quercetin-3-*O*-rutinoside, apigenin-7-*O*-glucoside and the secoiridoid oleuropein) (Araújo et al. 2021; Dias et al. 2019, 2020, 2021), ascorbate and glutathione were described to have an important role in olive stress protection.

Olive fruit's development and quality are also strongly influenced by climatic events (Valente et al. 2020; Fraga et al. 2021). For instance, water-deficit might negatively affect the fruit yield and oil accumulation, but accelerates the maturation of olives (Brito et al. 2019). Concerning the fruit and olive oil, the observed trend is to improve the quality due to the increase of the levels of some important polyphenols, changes in the quantitative profile of fatty acids and minor effects on free acidity, peroxide value, and specific absorption coefficients (K_{232}, K_{270}, ΔK) (Machado et al. 2013; Caruso et al. 2014; Brito et al. 2019). Thus, in the balance of the impact of the climate change on productivity, both yield versus quality must be considered, and if the yield loss remains within a sustainable threshold, it may be compensated by the gain in quality.

Considering the challenges posed by climate change, particularly the water shortage, it urges the necessity to implement sustainable agriculture management practices with a more reasonable use of irrigation water and able to satisfy the growing demand of food. In the last years, new eco-friendly compounds, like biostimulants, have been emerged, opening perspectives to make agriculture more sustainable and resilient. The biostimulants are natural substances (e.g., algae and amino acids) or microorganisms that can increase plant yield and fruit quality under both optimal and stress growth conditions (Shukla et al. 2019). Biostimulants' application in plants (foliar or in the soil, near the root zone) influence various metabolic processes and signalling cascades leading to the improvement of plant water content, nutrient uptake, photosynthesis, and chlorophyll content (Shukla et al. 2019). Another advantage of the use of biostimulants is that they are biodegradables and not toxic to humans and environment.

Biostimulants can be obtained from several natural sources, and the ones prepared from seaweeds extracts are very successful on the market (Staykov et al. 2021). Within the seaweed biostimulants, those based on the brown algae *Ascophyllum nodosum* are very used in agriculture (Staykov et al. 2021). The seaweed *A. nodosum* is rich in carbohydrates, and also vitamins, lipids, hormones, minerals, proteins and phenolic compounds (Pereira et al. 2020). The biostimulants based on extracts of this algae have been reported to act as priming agents, inducting molecular and physiological defense mechanisms that lead to enhanced tolerance to stress (Kerchev et al. 2020) (Fig. 29.1). *A. nodosum* extracts act in the upregulation of several genes involved in photosynthesis and ROS scavenging (Shukla et al. 2019). In model species (e.g. Arabidopsis and lettuce), several crops (e.g. tomato, pepper, soyabean, spinach and bean) and fruiting species (e.g. limon and orange tree) priming with *A. nodosum* extracts have been shown to improve plant water status via the modulation of ABA signal and induction of partial stomata closure, increase nutrient uptake, reduce ROS production and membrane damages through the enhance of the plant antioxidant system under drought stress conditions (Shukla et al. 2019; Baltazar et al. 2020). In olive, research on biostimulants is scarce. Chouliaras et al. (2009) demonstrated the beneficial effects of the application of *A. nodosum* in an olive rainfed orchard, improving productivity and oil nutritional value (increase of oleic and linolenic acids levels). Almadi et al. (2020) demonstrated that in young potted and well-watered olive trees, the application of a biostimulant based on protein hydrolysates increased tree biomass, photosynthesis and stomatal conductance, but did not change fruit yield and oil content. Nevertheless, the effects of biostimulants on plant response to stress depends on several factors, that include between others, species/genotype, type of stress, and biostimulants dose, time and application mode (Baltazar et al. 2020; Staykov et al. 2021).

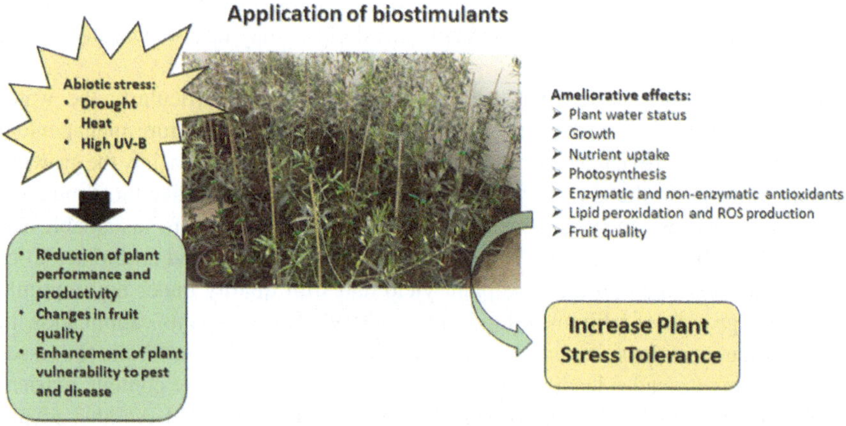

Fig. 29.1 Benefits of *A. nodosum* extract application in plants exposed to abiotic stresses

Aims and Objectives

The Mediterranean region is the main contributor for the world's oil production, and, in the last 15 years, olive oil world consumption increased substantially. To face this world's increased demand of olives and olive oil, the traditional dry-farming/rainfed orchards have being replaced by high-density orchards with more necessities of irrigation and agrochemicals, but with higher profitability. Unfortunately, these intensive practices lead to land overuse, soil salinization, desertification, deforestation, and, ultimately, to environmental degradation of the ecosystems. This challenge is aggravated by the ongoing increase of extreme episodes of drought and/or heat associated to climate change that are already threatening olive groves in the Mediterranean basin. Thus, it is urgent to implement more sustainable olive management practices to improve the rates of productivity under climate change scenarios, namely under a global water scarcity. The main objective of this study was to unveil the effect of the application of a commercial biostimulant based on *A. nodosum* extract on the tolerance of olive plants to water-deficit stress. In particular, we aimed to quantify the benefits of this biostimulant on plants of Arbequina, a cultivar highly used in intensive irrigated groves, and assess if the biostimulant protected plants under prolonged drought stress regarding water status, and by modulating the carbohydrate metabolism, and the secondary metabolism of phenols, to counteract the putative oxidative stress generated by the drought stress conditions.

Material and Methods

Olea europaea L. (cultivar Arbequina) plants with three years old were kindly provided by Viveiros Miguel Vaz (Semide, Coimbra, Portugal). Plants grew in black pots with 3L of turf at the Botanical Garden of the University of Coimbra (Portugal). Olive plants were randomly distributed in two groups. One group was assigned as biostimulant treatment and was treated (at the days 17/03/2020, 30/06/2020, 07/07/2020 and 14/07/2020) with an extract of *Ascophyllum nodosum* (Fitoalgas Green©, Tradecorp). The other group was assigned as control and was treated with distilled water. For the biostimulant treatment, each plant was sprayed with 200 mL (3 mL of biostimulant/L water). In the last day of the biostimulant treatment, all plants were irrigated at 100% field capacity (FC). The group assigned as biostimulant was subdivided in two groups—treated with biostimulant and well-watered at 100% FC (B-WW), and—treated with biostimulant and watered only at 50% FC, representing a water-deficit condition (B-WD). The group assigned as control was also subdivided in two groups: - without biostimulant treatment and well-watered at 100% FC (WW), and—without biostimulant and exposed to water-deficit, watered only at 50% FC (WD). The field capacity in the pots was evaluated two times per week and the water replaced when necessary. After 69 days under these treatments, leaves were collected for leaf water status, carbohydrates, and antioxidants determination.

Leaves were collected, immediately weighted for the determination of the fresh weight. Then, the leaves were dried at 80 ºC and the dry weight determined. Plant water content was calculated as:

$$\mathrm{WC}\ (\%) = (\text{fresh weight} - \text{dry weight})/(\text{fresh weight}) \times 100$$

For total soluble sugars (TSS) determination, leaves were homogenized with ethanol at 80% (v/v) and incubated for one hour at 80 °C (Dias et al. 2019). After centrifugation (6000 *g* for 10 min at 3 °C) the supernatant was mixed with an anthrone solution (40 mg anthrone dissolved in 1 mL dH_2O: 20 mL H_2SO_4) and placed in a bath at 100 °C for 10 min, as described in Irigoyen et al. (1992). After cooling in ice, the samples were centrifuged as described before. The absorbance of the supernatant was measured at 625 nm in a Multilabel Reader (EnSpire™, PerkinElmer, Norwalk, USA). TSS content was calculated using a calibration curve with glucose.

The level of starch was determined according to Osaki et al. (1991). The pellet resulted from the TSS extraction was used for starch extraction. After washing the pellet three times with ultrapure water, perchloric acid (30%, v/v) was added and the mixture incubated for one hour at 60 °C. Then, the mixture was centrifuged for 10 min at 10,000 *g* and 3 °C. The supernatant was collected and mixed with an anthrone solution (described above for TSS). After an incubation at 100 °C for 10 min samples were centrifugated for 10 min at 5000 *g* and 3 °C. One aliquot was collected and the absorbance was read at 625 nm in a Multilabel Reader (EnSpire™, PerkinElmer, Norwalk, USA). Starch level was calculated using a calibration curve with glucose.

For the determination of total phenols, orthodiphenols and flavonoids, leaves (100 mg) were homogenized with 1.25 ml of methanol and incubated for 30 min at 4 °C. Then the homogenate was centrifuged for 5 min at 4000 *g* at 5 °C, and the supernatant was used for analysis. Total phenols content was determined according to López-Orenes et al. (2018). An aliquot of the leaf extract (5 μL) was homogenized with a mixture of 405 μL of Folin-Ciocalteu and 75 μL Na_2CO_2 at 20%. The samples were incubated for 30 min at 37 °C. The absorvance of the mixture was read at 765 nm in a Multilabel Reader (EnSpire™, PerkinElmer, Norwalk, USA). Total phenols were determined using a calibration curve of gallic acid. Orthodiphenols were determined using the molybdate assay according to Giertych et al. (1999). An aliquot of the leaf extract (5 μL) was mixed with 10 μL of sodium molybdate (5% w/v in 50% methanol) and after incubation for 15 min at 20 °C the absorbance was measured at 370 nm in a Multilabel Reader (EnSpire™, PerkinElmer, Norwalk, USA). Orthodiphenols were calculated using a calibration curve of gallic acid. Flavonoids were determined according to López-Orenes et al. (2018). An aliquot of the leaf extract (5μL) was mixed with 37.5 μL of methanol, 75μL of $NaNO_2$ at 5%, and 75 μL $AlCl_3$ at 10%. After homogenization the samples were incubated for 6 min in dark. 125 μL of NaOH (1 M) was added and the absorbance measure at 510 nm in a Multilabel Reader (EnSpire™, PerkinElmer, Norwalk, USA). Flavonoids content was calculated using a calibration curve of rutin.

A total of 33 plants were used in this experiment, 8–9 per treatment. The data obtained were analyzed with the program SigmaPlot for Windows version 4.0. Two-way ANOVA was used to assess the effects of the factors watering (W) and biostimulant (B), and their interaction (WxB), on olive physiological performance and antioxidant profile. Pairwise differences among groups were compared by the post-hoc test Holm–Šidák, and the significance level of 5%. Data are presented as mean ± standard deviation. The treatments WD and B-WW were not compared in this statistical analysis.

Plant Water Status and Carbohydrate Responses to Biostimulant Treatment

Climate change events persistence in the Mediterranean region represent a threat for agricultural production, particularly for many crops typically of this region, as the case of *Olea europaea*. Several studies have demonstrated that beside the high tolerance of this species to abiotic stresses (e.g., Dias et al. 2020 and 2021; Araújo et al. 2019 and 2021), physiological performance and yield can be compromised by climate extreme events (Brito et al. 2019).

The watering and biostimulant factors tested here induced several changes in the physiological response of olive plants (Fig. 29.2). This species presents several strategies to conserve the water status, such as good control of the stomata aperture or osmotic adjustment, under drought conditions (Brito et al. 2019). However, a decrease of the leaf water status under severe drought was reported by Silva et al. (2018) and Araújo et al. (2019) in young olive plants exposed to water withholding for one month. Also, in the present work an effect of the factor watering was observed (Fig. 29.2a). The water-deficit condition applied for a long period (69 days at 50% of the FC) reduced the leaf water content in olive plants of the water deficit treatments (WD and B-WD) compared to their controls (WW and B-WW, respectively). These results highlight that water-deficit stress can affect negatively olive water status. However, the factor biostimulant was also significant. The application of biostimulant (B-WW and B-WD treatments) increased the availability of the leaf water content (Fig. 29.2a). A beneficial effect of biostimulants' application was also observed in Arabidopsis, hazelnut, sweet cherry and soybean species, where *A. nodosum* helped to maintain a high-water content under drought conditions (Santaniello et al. 2017; Shukla et al. 2018; Cabo et al. 2020; Correia et al. 2020). For some species, *A. nodosum* positive effects are related to stomata closure, which prevent excessive water loss due to transpiration (Santaniello et al. 2017). Under drought conditions *A. nodosum* induce ABA biosynthesis through the upregulation of 9-cis-epoxycarotenoid dioxygenase (NCED3) gene expression resulting in a reduction of the stomatal aperture and can also downregulate the expression of *MYB60* gene that participate in the regulation of stomata aperture movements (Shukla et al. 2018 and 2019; Staykov et al. 2021).

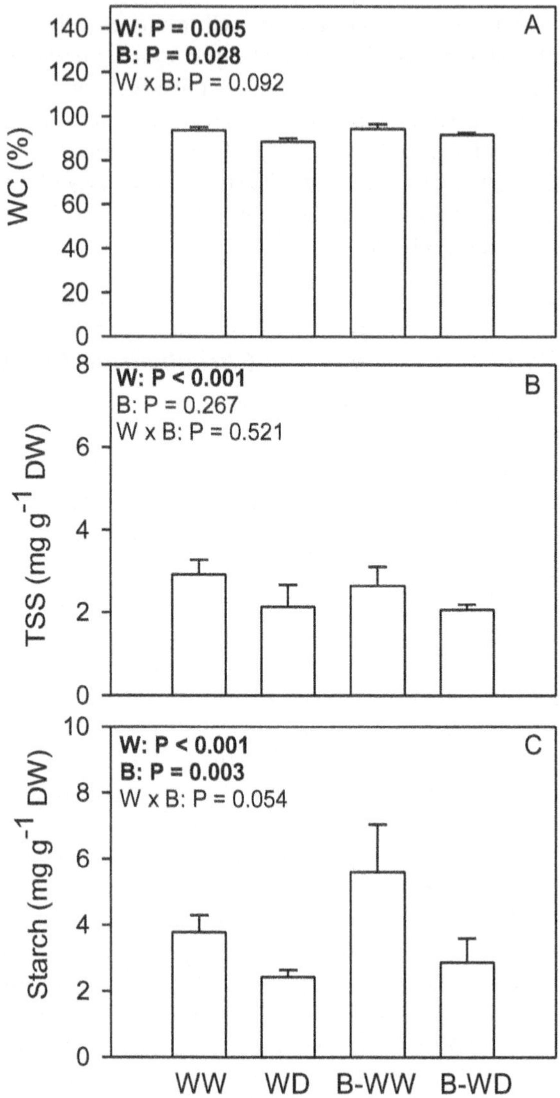

Fig. 29.2 Leaf water content (A), total soluble sugars (B) and starch (C) contents in *Olea europaea* plants exposed to well-watered (WW), water-deficit (WD), biostimulant + well-watered (B-WW), and biostimulant + water-deficit (B-WD) treatments. Data are presented as mean ± standard deviation ($n = 6$–8). W—actor watering, B—factor biostimulants, W × B—interaction between the factor watering and biostimulant.

One of the end products of photosynthesis are the carbohydrates, that provide the necessary energy for plant growth, development, and survival (Camisón et al. 2020). In olive, TSS was only affected by the factor watering, but the starch was affected by the factors watering and biostimulant (Figs. 30.2b and c). Water-deficit reduced the levels of TSS (WD and B-WD treatments). Also, the lowest levels of starch were observed in plants exposed to water-deficit (WD and B-WD), but the biostimulant increased the pool of this carbohydrate (B-WW and B-WD treatments). Under stress conditions, as the case of water-deficit, a higher use of soluble sugars may occur due to a greater need of these carbohydrates to maintain growth (Thalmann and Santelia 2017). Also, stress can increase the remobilization of starch to restore soluble sugars pools when photosynthesis is potentially limited by stress (Thalmann and Santelia 2017), or to produce other derived metabolites that help to sustain growth and to mitigate the stress negative effects (e.g. osmoprotectors) (Krasenky and Jonak 2012). In olive, the biostimulant improved the accumulation of starch (particularly under well-watered conditions), possibly due to a stimulation of the carbon metabolism or the activity of the enzymes related to the synthesis of this carbohydrate, as reported by Wadas et al. (2020). This increase of starch represents a higher availability of reserve sugars in olive plants. Different results were obtained in maize, sweet cherry, and grapevine plants were the biostimulant *A. nodosum* application under drought condition stimulated the levels of TSS, but starch levels were not affected (Baltazar et al. 2020; Correia et al. 2020).

Antioxidant Responses to Biostimulant Treatment

Biostimulants are considered as substances which, besides enhancing growth and yield of plants, also improve antioxidant defense (Baltazar et al. 2020). The application of biostimulants (based on seaweed extracts) in several species was able to alleviate the symptoms of oxidative stress under drought conditions, by inducing the accumulation of non-enzymatic antioxidants, such as phenolic compounds, and activating the enzymatic antioxidant system (Shukla et al. 2019; Baltazar et al. 2020). In olive plants we analysed the levels of non-enzymatic antioxidants and an interaction between the factors, biostimulant and watering, was observed for total phenols and orthodiphenols (Figs. 30.3a and b). Olive plants treated with biostimulant and exposed to water-deficit (B-WD) showed the highest levels ($p < 0.05$) of these antioxidants. The level of total flavonoids was also affected by the factor biostimulant and by the water regime separately (Fig. 29.3c). The water-deficit treatment decreased the levels of total flavonoids in olive plants, while the biostimulant treatment increased the pool of these antioxidants (B-WW and B-WD). In general, our results indicated that biostimulant can stimulate the levels of some antioxidants, particularly under water-deficit conditions. Also, these results corroborate the work of Fan et al. (2013), Elansary et al. (2016a, b) and Shukla and Prithiviraj (2021) which demonstrated that *A. nodosum* treatment under drought stress induced the stimulation of non-enzymatic and also enzymatic antioxidants. This stimulation can be in part related to an upregulation of important genes of the secondary metabolism associated with stress protection, such as flavonol synthase 1, dihydroflavonol-4-reductase and chalcone synthase, or genes involved in oxidative stress control, such as L-ascorbate peroxidase 2, and CAT1, 2 and 3 (Shukla et al. 2019).

The role of biostimulants in the stimulation of the phenylpropanoid pathway under stress condition was also highlighted by Koleška et al. (2017). This biosynthetic pathway is responsible for the synthesis of several phenolic compounds with antioxidant properties, such as flavanols and flavones (Dias et al. 2021). Moreover, a positive

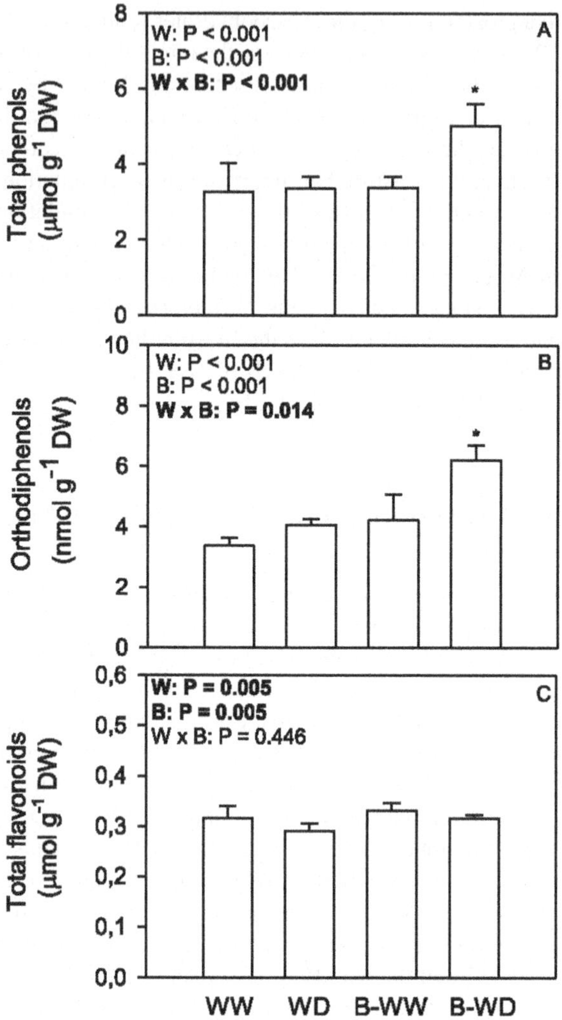

Fig. 29.3 Total phenols (A), orthodiphenols (B), and total flavonoids (C) of *Olea europaea* plants exposed to well-watered (WW), water-deficit (WD), biostimulant + well-watered (B-WW), and biostimulant + water-deficit (B-WD) treatments. Data are presented as mean ± standard deviation ($n = 6$–8). W—factor watering, B—factor biostimulants, W × B—interaction between the factor watering and biostimulant. Asterisk (*) means significant differences ($p < 0.05$) between treatments.

correlation between polyphenols synthesis and the activities of glucose 6-phosphate dehydrogenase responsible for sugars mobilization was also demonstrated in tomato plants (Koleška et al. 2017). In olive, the increase in the pool of total phenols and orthodiphenols in response to biostimulant under water deficit (B-WD in Fig. 29.3A and B) together with the reduction of soluble sugars (B-WD in Fig. 29.2b), can also suggest a mobilization of these carbohydrates to the phenylpropanoid pathway (Randhir and Shetty 2007).

Conclusions and Recommendations

Extreme climate events are increasing in frequency, extension and strength in the Mediterranean region representing a threat to olive culture, one of the main agriculture sector of this region. Our results also corroborate that drought stress can reduce olive performance, decreasing olive water and carbohydrates availability. However, *A. nodosum* extract use can help to reduce these negative effects, increasing leaf water content and the availability of reserve sugars in olive plants not only exposed to drought stress, but also under well-watered conditions. In addition, this biostimulant stimulate the antioxidant system, particularly the levels of total phenols and orthodiphenols, to counteract the putative oxidative stress generated by the drought stress conditions. These results highlight the potential of biostimulants based on *A. nodosum* extract to mitigate the negative impacts of stresses related to climate change, increasing olive performance and stress tolerance. Moreover, even under optimal growth conditions the use of this biostimulant is a valuable sustainable strategy to improve plant performance.

These results, obtained under controlled conditions, support the use of this biostimulant as environmentally sustainable alternatives to agrochemicals, aligned with the Farm to Fork European strategy. Interestingly, the environmental impact of adding biostimulants to the field remains poorly studied, as stressed by Yakhin et al. (2017). Also, before their wider use in oliviculture, it is important to transfer our results, obtained with potted young plants, to adult trees at field conditions. Monitoring for several years the effects of biostimulants in the field, where multiple biotic/abiotic stressors act simultaneously in combination, will provide more accurate information on the real benefits of the biostimulants on modulating the olive trees' water status, carbohydrates, and antioxidant battery. These field assays will also give insight on the need to adjust the biostimulant's formulation/doses to the orchard characteristics (soil, cultivars, and cultural practices). These assays will also provide the needed information on the impact of the biostimulants on the orchard's productivity, and fruit and oil quality. A final evaluation of the costs-benefits by the producers is also crucial to assess the potential acceptance of this strategy by the producers.

Acknowledgements This work was sponsored by Foundation for Science and Technology (FCT) and the Ministry of Science, Technology and Higher Education through national funds and co-funding by FEDER, within the PT2020 Partnership-Agreement, and COMPETE_2010, project: UI0183–UID/BIA/04004/2020. MCDias (SFRH/BPD/100865/2014) was funded by national funds (OE), through FCT, I.P., in the scope of the contract foreseen in the nº:4-6 of article-23, of the Decree-Law 57/2016, August 29, changed by Law-57/2017, July 19. FCT also supported the doctoral fellowships of M Araújo (SFRH/BD/116801/2016 and COVID/BD/151706/2021) through POCH/FSE. The authors acknowledge the Viveiros Miguel Vaz (Semide, Portugal).

References

Almadi L, Paoletti A, Cinosi N, Daher E, Rosati A, Di Vaio C, Famiani FA (2020) Biostimulant based on protein hydrolysates promotes the growth of young olive trees. Agriculture 10(12):618. https://doi.org/10.3390/agriculture10120618

Araújo M, Prada J, Mariz-Ponte N, Santos C, Pereira JÁ, Pinto DCGA, Silva MAS, Dias MC (2021) Antioxidant adjustments of olive trees (*Olea europaea*) under field stress conditions. Plants 10:684. https://doi.org/10.3390/plants10040684

Araújo M, de Oliveira F, Oliveira JMP, Santos C, Correia C, Dias MC (2019) Responses of olive plants exposed to different irrigation treatments in combination with heat shock: physiological and molecular mechanisms during exposure and recovery. Planta 249(5):1583–1598. https://doi.org/10.1007/s00425-019-03109-2

Baltazar M, Correia S, Guinan KJ, Sujeeth N, Bragança R, Gonçalves B (2020) Recent advances in the molecular effects of biostimulants in plants: an overview. Biomolecules 11(8):1096. https://doi.org/10.3390/biom11081096

Brito C, Dinis LT, Moutinho-Pereira J, Correia CM (2019) Drought stress effects and olive tree acclimation under a changing climate. Plants 8(7):232. https://doi.org/10.3390/plants8070232

Cabo S, Morais MC, Aires A, Carvalho R, Pascual-Seva N, Silva AP, Gonçalves B (2020) Kaolin and seaweed-based extracts can be used as middle and long-term strategy to mitigate negative effects of climate change in physiological performance of hazelnut tree. J Agron Crop Sci 206(1):28–42. https://doi.org/10.1111/jac.12369

Camisón Á, Ángela Martín M, Dorado FJ et al (2020) Changes in carbohydrates induced by drought and waterlogging in *Castanea sativa*. Trees 34:579–591. https://doi.org/10.1007/s00468-019-01939-x

Caruso G, Guccia R, Urbani S, Esposto S, Taticchi A, di Maio I, Selvaggini R, Maurizio Servili M (2014) Effect of different irrigation volumes during fruit development on quality of virgin olive oil of cv. Frantoio. Agric. Water Manag 134:94–103. https://doi.org/10.1016/j.agwat.2013.12.003

Chouliaras V, Tasioula M, Chatzissavvidis C, Theriosa I, Tsabolatidou E (2009) The effects of a seaweed extract in addition to nitrogen and boron fertilization on productivity, fruit maturation, leaf nutritional status and oil quality of the olive (*Olea europaea* L.) cultivar Koroneiki. J Sci Food Agric 89(6): 984–988. https://doi.org/10.1002/jsfa.3543

Correia S, Queirós F, Ferreira H, Morais MC, Afonso S, Silva AP, Gonçalves B (2020) Foliar application of calcium and growth regulators modulate sweet cherry (*Prunus avium* L.) tree performance. Plants 9(4): 410. https://doi.org/10.3390/plants9040410

del Pozo A, Brunel-Saldias N, Engler A, Ortega-Farias S, Acevedo-Opazo C, Lobos GA, Jara RR, Molina-Montenegro MA (2019) Climate change impacts and adaptation strategies of agriculture in mediterranean-climate regions (MCRs). Sustainability 11(10):2769. https://doi.org/10.3390/su11102769

Dias MC, Azevedo C, Costa M, Pinto G, Santos C (2014) *Melia azedarach* plants show tolerance properties to water shortage treatment: an ecophysiological study. Plant Physiol Biochem 75:123–127. https://doi.org/10.1016/j.plaphy.2013.12.014

Dias MC, Pinto DCGA, Figueiredo C, Santos C, Silva MAS (2021) Phenolic and lipophilic metabolite adjustments in *Olea europaea* (olive) trees during drought stress and recovery. Phytochemistry 185:112695. https://doi.org/10.1016/j.phytochem.2021.112695

Dias MC, Pinto DCGA, Correia C, Silva AMS, Santos C (2018) UV-B radiation modulates physiology and lipophilic metabolite profile in *Olea europaea*. J Plant Physiol 222:39–50. https://doi.org/10.1016/j.jplph.2018.01.004

Dias MC, Pinto DCGA, Freitas H, Santos C, Silva AMS (2020) The antioxidant system in *Olea europaea* to enhanced UV-B radiation also depends on flavonoids and secoiridoids. Phytochemistry 170:112199. https://doi.org/10.1016/j.phytochem.2019.112199

Dias MC, Santos C, Pinto G, Silva AMS, Silva S (2019) Titanium dioxide nanoparticles impaired both photochemical and non-photochemical phases of photosynthesis in wheat. Protoplasma 256(1):69–78. https://doi.org/10.1007/s00709-018-1281-6

Elansary HO, Norrie J, Hayssam M, Ali HM, Salem MZM, Mahmoud EA, Yessoufou K (2016a) Enhancement of *Calibrachoa* growth, secondary metabolites and bioactivity using seaweed extracts. BMC Comp Med Ther 16:341. https://doi.org/10.1186/s12906-016-1332-5

Elansary HO, Skalicka-Wozniak K, King IW (2016b) Enhancing stress growth traits as well as phytochemical and antioxidant contents of Spiraea and Pittosporum under seaweed extract treatments. Plant Physiol Bioch 105:310–320. https://doi.org/10.1016/j.plaphy.2016.05.024

Fan D, Hodges DM, Critchley AT, Prithiviraj B (2013) A commercial extract of brown macroalga (*Ascophyllum nodosum*) affects yield and the nutritional quality of spinach *in vitro*. Commun Soil Sci Plant Anal 44(12):1873–1884. https://doi.org/10.1080/00103624.2013.790404

Fotia K, Mehmeti A, Tsirogiannis I, Nanos G, Mamolos AP, Malamos N, Barouchas P, Todorovic M (2021) LCA-based environmental performance of olive cultivation in northwestern Greece: from rainfed to irrigated through conventional and smart crop management practices. Water 13(14):1954. https://doi.org/10.3390/w13141954

Fraga H, Moriondo M, Leolini L, Santos JA (2021) Mediterranean olive orchards under climate change: a review of future impacts and adaptation strategies. Agronomy 11(1):56. https://doi.org/10.3390/agronomy11010056

Giertych MJ, Karolewski P, De Temmerman LO (1999) Foliage and pollution alter content of phenolic compounds and chemical elements in *Pinus nigra* needles. Water Air Soil Pollut 110(3–4):363–377. https://doi.org/10.1023/A:1005009214988

Irigoyen JJ, Einerich DW, Sánchez-Díaz M (1992) Water stress induced changes in concentrations of proline and total soluble sugars in nodulated alfalfa (*Medicago sativa*) plants. Physiol Plant 84:55–60. https://doi.org/10.1111/j.1399-3054.1992.tb08764.x

Kerchev P, Meer TVD, Sujeeth N, Verlee A, Stevens CV, Breusegem FV, Gechev T (2020) Molecular priming as an approach to induce tolerance against abiotic and oxidative stresses in crop plants. Biotechnol Adv 40:107503. https://doi.org/10.1016/j.biotechadv.2019.107503

Koleška I, Hasanagić D, Todorović V, Murtić S, Klokić I, Parađiković N, Kukavica B (2017) Biostimulant prevents yield loss and reduces oxidative damage in tomato plants grown on reduced NPK nutrition. J Plant Interact 12(1):209–218. https://doi.org/10.1080/17429145.2017.1319503

Krasenky J, Jonak C (2012) Drought, salt, and temperature stress-induced metabolic rearrangements and regulatory networks. J Exp Bot 63(4):1593–1608. https://doi.org/10.1093/jxb/err460. PMID: 22291134; PMCID: PMC4359903

López-Orenes A, Dias MC, Ferrer MA, Calderón A, Moutinho-Pereira J, Correia C, Santos C (2018) Different mechanisms of the metalliferous *Zygophyllum fabago* shoots and roots to cope with Pb toxicity. Environ Sci Poll Res 25:1319–1330. https://doi.org/10.1007/s11356-017-0505-1

Machado M, Felizardo C, Fernandes-Silva AA, Nunes FM, Barros A (2013) Polyphenolic compounds, antioxidant activity and l-phenylalanine ammonia-lyase activity during ripening of olive cv. "Cobrançosa" under different irrigation regimes. Food Res Int 51(1): 412–421. https://doi.org/10.1016/j.foodres.2012.12.056

Osaki M, Shinano T, Tadano T (1991) Redistribution of carbon and nitrogen compounds from the shoot to the harvesting organs during maturation in field crops. Soil Sci Plant Nutr 37:117–128. https://doi.org/10.1080/00380768.1991.10415017

Pereira L, Morrison L, Shukla PS, Critchey AT (2020) A concise review of the brown macroalga *Ascophyllum nodosum* (Linnaeus) Le Jolis. J Appl Phycol 32:3561–3584. https://doi.org/10.1007/s10811-020-02246-6

Randhir R, Shetty K (2007) Elicitation of the proline-linked pentose phosphate pathway metabolites and antioxidant enzyme response by ascorbic acid in dark germinated fava bean sprouts. J Food Biochem 31:485–508. https://doi.org/10.1111/j.1745-4514.2007.00126.x

Romero-Trigueros C, Vivaldi GA, Nicolás E, Paduano A, Salcedo FP, Camposeo S (2019) Ripening indices, olive yield and oil quality in response to irrigation with saline reclaimed water and deficit strategies. Front Plant Sci 10:1243. https://doi.org/10.3389/fpls.2019.01243

Sales H, Šatovic Z, Alves ML, Fevereiro P, Nunes J, Vaz Patto MC (2021) Accessing ancestral origin and diversity evolution by net divergence of an ongoing domestication Mediterranean olive tree variety. Front Plant Sci 12: 688214. https://doi.org/10.3389/fpls.2021.688214

Santaniello A, Scartazza A, Gresta F, Loreti E, Biasone A, Di Tommaso D, Piaggesi A, Perata P (2017) *Ascophyllum nodosum* seaweed extract alleviates drought stress in *Arabidopsis* by affecting photosynthetic performance and related gene expression. Front Plant Sci 8:1362. https:/ /doi.org/10.3389/fpls.2017.01362

Shukla PS, Mantin EG, Adil M, Bajpai S, Critchley AT, Prithiviraj B (2019) *Ascophyllum nodosum*-based biostimulants: Sustainable applications in management. Front Plant Sci 10:655. https:// doi.org/10.3389/fpls.2019.00655

Shukla PS, Prithiviraj B (2021) *Ascophyllum nodosum* biostimulant improves the growth of *Zea mays* grown under phosphorus impoverished conditions. Front Plant Sci 11:601843. https://doi. org/10.3389/fpls.2020.601843. PMID: 33488647; PMCID: PMC7820112

Shukla PS, Shotton K, Norman E, Neily W, Critchley AT, Prithiviraj B (2018) Seaweed extract improve drought tolerance of soybean by regulating stress-response genes. AoB Plants 10(1): plx051. https://doi.org/10.1093/aobpla/plx051

Silva S, Oliveira JMPM, Dias MC, Silva AMS, Santos C (2019) Antioxidant mechanisms to counteract TiO_2-nanoparticles toxicity in wheat leaves and roots are organ dependent. J Hazard Mater 380:120889. https://doi.org/10.1016/j.jhazmat.2019.120889

Silva S, Santos C, Serôdio J, Silva AMS, Dias MC (2018) Physiological performance of drought-stressed olive plants when exposed to a combined heat-UV-B shock and after stress relief. Funct Plant Biol 45(12):1233–1240. https://doi.org/10.1071/FP18026

Staykov NS, Angelov M, Petrov V, Minkov P, Kanojia A, Guinan KJ, Alseekh S, Fernie AR, Sujeeth N, Gechev TS (2021) An *Ascophyllum nodosum*-derived biostimulant protects model and crop plants from oxidative stress. Metabolites 11(1):24. https://doi.org/10.3390/metabo11010024

Thalmann M, Santelia D (2017) Starch as a determinant of plant fitness under abiotic stress. New Phytol 214(3):943–951. https://doi.org/10.1111/nph.14491

Valente S, Machado B, Pinto DCGA, Silva AMS (2020) Modulation of phenolic and lipophilic compounds of olive fruits in response to combined drought and heat. Food Chem 329:127191. https://doi.org/10.1016/j.foodchem.2020.127191

Wadas W, Dziugieł T (2020) Quality of new potatoes (*Solanum tuberosum* L.) in response to plant biostimulants application. Agriculture 10(7): 265. https://doi.org/10.3390/agriculture10070265

Yakhin OI, Lubyanov AA, Yakhin IA, Brown PH (2017) Biostimulants in plant science: A global perspective. Front Plant Sci 7:2049. https://doi.org/10.3389/fpls.2016.02049

Chapter 30
The Vulnerability of Small-Scale Fisheries-Based Livelihoods to Climatic and Non-Climatic Stressors in Kani Ward, Binga, Zimbabwe

Douglas Nyathi, Joram Ndlovu, Thulani Dube, Prince Mathe, and Bakani Mathe

Abstract Fisheries are a leading livelihood portfolio supporting food and income security in developing countries. This study interrogated the susceptibility of fisheries to climate change and their contribution as a livelihood strategy for communities living in Kani ward, Binga. The study sought to evaluate the contribution of fisheries towards the sustainability of household livelihoods and the impact of climate change and multi-stressors on these fisheries-based livelihoods. We adopted a primarily qualitative approach. The results show that fishing is an important livelihood strategy for the Kani ward community. Fisheries were found to be capacitating households with various assets that include livestock, houses, refrigerators, cars and fishing rigs. Furthermore, fishing was found to be contributing to increased food production, household incomes, poverty reduction, better access to education, nutrition and health. Despite the contribution of fisheries as a livelihood strategy, it was found to be susceptible to climate variability and other multi-stressors. Multiple stressors included the cost of fishing permits and hawkers' licenses, poor infrastructure and COVID-19. The dynamic interaction of these stressors with climate change contributed to the vulnerability of livelihoods based on fisheries. The study concludes that small-scale fishing can buffer marginalised people against shocks since they can

D. Nyathi (✉) · J. Ndlovu
School of Social Sciences, University of KwaZulu Natal, Howard College, Durban 4001, South Africa
e-mail: douglasnyathi08@gmail.com

J. Ndlovu
e-mail: ndlovuj1@ukzn.ac.za

T. Dube
Centre for Evaluation Science, Faculty of Humanities and Social Sciences, Lupane State University, Off Victoria Falls Rd, P.O Box 170, Bulawayo, Zimbabwe

P. Mathe · B. Mathe
Department of Development Studies, Faculty of Humanities and Social Sciences, Lupane State University, Off Victoria Falls Rd, P.O Box 170, Bulawayo, Zimbabwe

W. Leal Filho et al. (eds.), *Climate Change Strategies: Handling the Challenges of Adapting to a Changing Climate*, Climate Change Management,
https://doi.org/10.1007/978-3-031-28728-2_30

enter and exit at a reasonably low cost. In order to address the challenges of climate change, the government needs to develop a climate change adaptation plan specific to fisheries, invest in data on small-scale fisheries, align fishing payment policies and improve household access to natural resources to ensure sustainable use of fisheries as an environmentally viable livelihood strategy.

Abbreviations and Acronyms

DFID	Department for International Development
FAO	Food and Agriculture
HIV	Human Immuno-Deficiency Virus
IPCC	Inter-governmental Panel on Climate Change
SSF	Small-Scale Fisheries
STI	Sexually Transmitted Diseases
UNSD	United Nations Statistical Department
ZPWMA	Zimbabwe Parks and Wildlife Management Authority

Introduction

Fisheries are a leading livelihood portfolio supporting food and income security in developing countries. Over half a billion people globally are supported by fisheries for their livelihoods (UNSD 2022), which also ensures employment, and other income generation activities (Muringai et al. 2020b; Funge-Smith and Bennett 2019; Utete et al. 2019). It is estimated that 52.8 million people engage in small-scale fisheries globally (Nico and Christiaensen 2022). Fisheries employ approximately 37 million people, around 90% of whom are from Asia (FAO 2020, 2021). Two-thirds of the global fish production are split between China and the rest of Asia with the Americas, Europe and Africa providing 14%, 10% and 7% of global fish supply, respectively (Belton 2021). In sub-Saharan Africa, it is estimated that up to 10 million people are employed in small-scale fisheries, with 3000 people being employed directly in fisheries in Zimbabwe (Muringai et al. 2020a). Amongst poor communities in Africa, small-scale fisheries have long been observed to provide a safety net against extreme poverty and food insecurity (Gandiwa et al. 2012; Béné and Merten 2008). Marginalised and disenfranchised communities tend to rely on ecosystem services such as fishing, hunting and gathering (Dube et al. 2021). In several countries, inland small-scale fisheries provide various services to local communities which include food security, poverty alleviation and revenue generation (Limuwa et al. 2018; Utete et al. 2018; Ogutu-Ohwayo et al. 2016). Fisheries are particularly suitable as a livelihood strategy in poor communities because they require relatively fewer resources to exploit (Shava and Gunhidzirai 2017; Smith et al. 2005).

A survey of the literature shows that very limited attention has been directed towards the influence of climate variability on inland fisheries ecosystems considering that millions of people in developing countries are dependent on inland small-scale fisheries value chains (Allison et al. 2009). People who work in the fisheries value chain include amongst others, fishers, fish traders, processors and support staff in the value chain (Allison et al. 2009). In the context of long-term shifts in weather patterns and increasing temperatures, it is important to appreciate how communities respond to multiple stressors in small-scale fisheries in order to develop effective adaptation strategies. Multiple stressors can be defined as socio-economic and environmental changes at local and global scales which make communities vulnerable. Current studies have mostly examined the linear interaction between climate change and fisheries (Isaacs et al. 2020; Freduah et al. 2017) and have neglected other multiple stressors to fisheries-based livelihoods. Researchers agree that there is limited knowledge available about the effects of climate change on fisheries-based livelihoods in the Zambezi valley, particularly around Lake Kariba.

Definition of Small-Scale Fisheries (SSFs)

SSF are categorized based on their intensity in terms of labour when harvesting fish, they use reasonably smaller vessels for fishing, they happen closer to the shore, they achieve relatively low catch per boat with limited capital injection. As a result of these characteristics, the returns from SSFs are limited and the communities that engage in this livelihood are amongst the poorest in developing countries (Muringai et al. 2020b). In some literature, SSF are defined as 'artisanal fishing' that is typically characterised by low level technology, limited capital investments and is mostly undertaken by individuals and households and not companies (Ndhlovu et al. 2017).

The Climate Change and Small-Scale Fisheries Nexus

Global evidence shows that as the climate changes, this will also be accompanied by human population, ecosystems and species changes. This is expected to have profound consequences for fisheries (Allison et al. 2009). To be able to effectively predict the effects of climate change on SSF, it is important to understand the characteristics of the fishing communities and the factors that affect their capacity to adapt. Small-scale fishing has a safety net function since it can absorb unskilled excess labour by supporting pluralism in different occupations through a combination of farming (livestock and crops) and fishing (Nico and Christiaensen 2022). The vulnerability of small-scale fishing communities is a construct of several factors including the severity of extreme climate events, the socio-economic conditions of the communities as well as the adaptation strategies that they pursue (Allison et al. 2009). The impact of climate change on fisheries varies by the type of fishery and geographical location. In coastal areas, some of the major effects of climate change include the salination of water due to thermal expansion and sea-level rises.

The research conducted in the Lake Kariba area shows that rainfall has been declining over the years, the water temperature has warmed by about 2 °C and evaporation rates have also significantly increased (Muringai et al. 2020a). Scientists have underscored the point that because fish are cold-blooded (poikilothermic) they are significantly impacted by warming temperatures. Warming temperatures are specifically expected to affect their rate of growth, metabolism, reproduction, size and distribution amongst other things (Muringai et al. 2020a). Projections show that in Africa, temperature increases induced by climate change will progress faster than in the rest of the world. It is estimated that temperatures will rise by between 3 °C and 6 °C by 2050. This will be accompanied by a decrease in rainfall in the drier tropics (Muringai et al. 2020a). Fisheries are highly sensitive to changes in climatic conditions due to their dependency on local ecologies (Muringai et al. 2020a). In view of these projections, their resilience will partly depend on their adaptability to emerging climate change impacts and other stressors (Freduah et al. 2017).

Some researchers have warned that overlooking the effect of climate change on the ecosystems in developing countries could lead to devastating consequences for the fishing industry, particularly in Zambia and Zimbabwe (Ndebele-Murisa et al. 2011a). A study conducted by Ndebele-Murisa and Mashonjowa (2011b) in Lake Kariba showed that since 1963, rainfall has been declining at an annual rate of 0.63 mm per year. The rate of evaporation has increased by approximately 31% over the same period. The reduction in the amount of Kapenta fish caught in Lake Kariba has largely been attributed to climate change (Ndebele-Murisa et al. 2011b). The impact of climate change at Lake Kariba is evidenced by the continued decline in fish catches (Ndhlovu et al. 2017). According to Utete et al. (2018), fish catches at Lake Manyame in Harare have been declining since 2006 and they continue to show a downward trend. The declining trend is associated with extreme changes in weather conditions including stronger winds and increased temperatures (Utete et al. 2018). The increasing frequency of extreme weather changes and a decrease in rainfall have altered the distribution and productivity of different fish species.

Other Multiple Stressors

In order for the fisheries sector to adapt to climate change, the impact of fisheries at a global and local scales needs to be understood and their interaction with the global environmental change and other social and economic factors in which they operate (Bennett et al. 2015). Climate change-related stressors that have been identified globally include rising sea levels, salination of coastal water sources, worsening storm events and erosion. Other observed environmental factors also include water pollution, overfishing and landslides (Bennett et al. 2015). Some non-climate stressors observed elsewhere include competition with other users of water and neglect by government institutions (Freduah et al. 2017). Competing water uses include municipal water supplies, hydro-power generation and irrigation. When combined with climate change, inland pollution activities such as contamination, littering, water pollution and wastewater runoff, have a significant effect on the sustainability of fisheries (Funge-Smith and Bennett 2019). In Malawi, the non-climatic stressors that have been observed around Lake Victoria include, amongst others, the growth of land demarcated for agriculture, population growth and settlements, intensive fishing and the effects of exotic species introduced to the Lake. Because of the convergence of these factors with climatic effects, it is argued that Lake Victoria has experienced a significant reduction in fish catches over the years (Hecky et al. 2010). In Malawi, some fisheries have collapsed due to multiple stressors that include weak governance structures, overfishing and climate change-related effects.

Study Objectives

The main objective was to explore the effect of climate and non-climate stressors on SSF in order to establish possible pathways for strengthening local livelihood portfolios. The following sub-objectives underpinned the study:

i. To examine the main livelihood portfolios used by small-scale fishers in Kani Ward;
ii. To discuss the impact of climate and non-climate stressors on fisheries-based livelihoods; and
iii. To determine how local communities, adapt to climate change in the advent of dwindling fish catches in Lake Kariba.

Case Study, Materials and Method

Study Site

Kani Ward is located on the lakeside of Lake Kariba. On the Zimbabwean side, Lake Kariba is split into five hydrological and geomorphic zones that are demarcated by subsidiary rivers. The focus of the study was mainly on the fishing basin in Kani ward. The Kani ward basin is one of the five identified hydrological zones. Because of the marginalised state of the area, data about the communities being studied is limited. The majority of households in the areas surrounding Lake Kariba depend to a varying extent on small-scale fisheries for livelihoods. The fishing method which is mostly practiced in gillnet fishing can be described as fishing using nets. This is a low-cost fishing method often employed by poor communities (Muringai et al. 2020b). The Zambezi River is the fourth largest river in Africa with a length of 2574 km and it passes through eight countries in Africa (Tweddle et al. 2015). Lake

Kariba is found on the Zambezi River which forms the border between Zambia and Zimbabwe. The lake was constructed between the years 1958 and 1963 by putting a dam wall on the Zambezi River. Lake Kariba is an artificial lake that is the second largest in the world. It has a capacity to store 185 million cubic meters of water. It stretches for a distance of approximately 280 km in length and it covers a surface area of 5580 km^2 (Ndhlovu et al. 2017). The lake is primarily regulated by the Zimbabwe Parks and Wildlife Management Authority (ZPWMA) and the relevant rural district local councils (Ndhlovu et al. 2017). The Statutory Instrument 362 of 1990 under ZPWMA prohibits crop farming and livestock rearing in the proximity of the lake (Ndhlovu et al. 2017). There are over 35 fishing villages dotted around the lake. These villages mostly comprise households who were moved from the lake area during its construction. Kani ward village was chosen due to its accessibility and proximity to Lake Kariba. The village is unique considering the number of community members who rely on fishing to sustain their livelihoods. Households are allowed to harvest fish as a source of income (for commercial purposes) and sustenance using primitive boats and nets. The village has been negatively affected by climate change and poor rainfall resulting in the scarcity of fish. Fishing being almost the only viable livelihood option in the area, fish populations have been depleted due to overfishing. As can be seen from Fig. 30.1, Kani ward village is close to Sinakoma, singalenge, Manjolo and Simatelele. Generally, Binga is surrounded by Nyaminyami, Gokwe, Lupane, and Gokwe. Zambezi river separates Zimbabwe from Zambia.

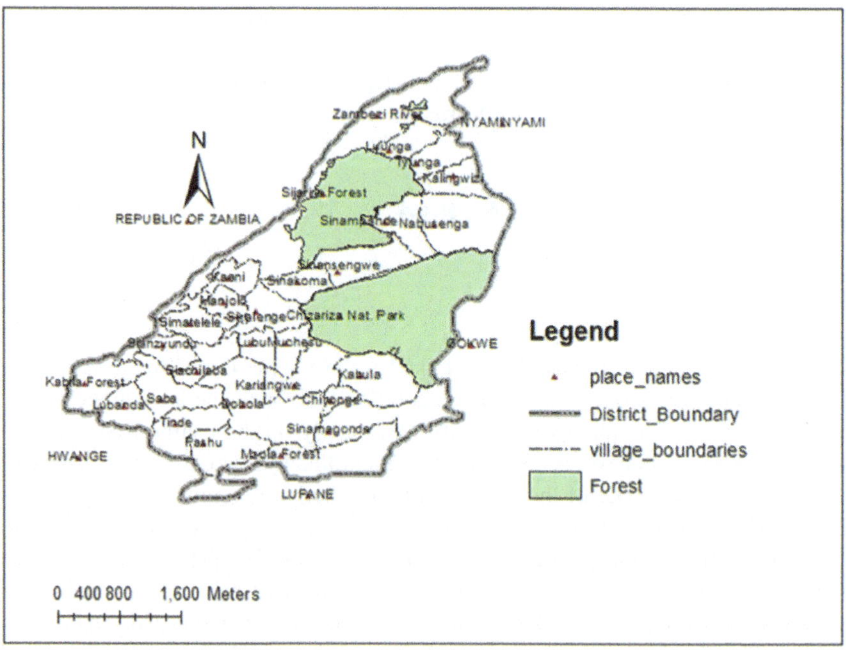

Fig. 30.1 The map showing the Villages in Binga (*Source* Authors)

Figure 30.1 shows the villages that are closer to the Zambezi river and depend on fishing for livelihood.

Research Approach and Tools

A case study approach was chosen for the study to enable the researchers to gain insight into livelihood options available to fisheries dependent households. A case study approach was used because the data is most often examined within the context in which it is used (Yin 2018), that is, examining the activity within its natural setting. A case study allows the use of multiple evidence or data needing to converge in a triangulation fashion (Yin 2018). By using a case study design, researchers wanted to capture the diverse perceptions of participants in order to illuminate the influence of different meanings on the topic being studied. Field work included visits to the study area so as to familiarize the researchers with the context in which the study was being done. Specifically, pre-data collection visits assisted the researchers in understanding the social, cultural, economic and environmental characteristics of the study areas. We also took advantage of such visits to pre-test our data collection tools. The purpose of pretesting tools was to ensure that the tools answer the research question and to see to it that the participants struggle to answer the questions. Purposive sampling was adopted for the identification and selection of study participants. This sampling approach allowed us to select knowledgeable participants about the phenomenon under investigation (Creswell and Plano-Clark 2011). Data collection was done through document analysis, one focus group discussion (6 key informants), and thirty in-depth interviews (30). In-depth interviews were used in this study to gain insight into the dynamics of fisheries-based livelihoods in the context of climate change.

The majority of the questions were open-ended to allow the participants to express their own views (Lucas et al. 2018). An interview schedule asked about the main livelihood portfolios used by small-scale fishers in Kani Ward. The data collection tool was carefully analysed and the additional key areas were identified and explored in detail with key informants through a focus group discussion (Lucas et al. 2018). The focus group was comprised of key informants who were a mix of selected individual household members, government officials and Non-Governmental Organisations. Focus group discussion with key informants sought to understand the impact of climate change and multi-stressors on fisheries-based livelihoods and the community's adaptation strategies to climatic and non-climatic stressors. The data was collected in Kani village for 2 months from 1 December 2020 to 31 January 2021. Snowball sampling was used in recruiting the key informants. The duration of the interview ranged from 30 to 45 min. The sample was determined after reaching saturation. During interviews, the interviewer asked for participants' consent to record the proceedings and took additional notes where necessary. District Development reports from the Rural District Council, National Parks and the Zimbabwe Vulnerability Assessment Committee (ZimVac) were analysed so as to have a comprehensive understanding of fisheries and other livelihood options pursued in the study area.

Document analysis involved the repeated review, examination, and interpretation of the secondary in order to gain meaning and empirical knowledge of fisheries-based livelihoods (Gross 2018). Data were analyzed using a summative content analysis approach. Typically, a study using a summative approach to qualitative content analysis starts with identifying and quantifying certain words or content in text with the purpose of understanding the contextual use of the words or content (Hsieh and Shannon 2005). This data analysis approach was used to determine the presence of certain words, themes or concepts within some given qualitative data. By using summative content analysis, researchers were able to count and make comparisons of key words in order to determine the meanings of words underpinning the context. The choice of summative content analysis was influenced by the researchers' considerable theoretical interest in the problem (Hsieh and Shannon 2005). Hence, the approach enabled the researchers to engage easily with the subject and interpret the underlying context.

Study Findings and Discussion

Profiling Households that Depend on Fisheries

The study revealed that most of the people that engage in fisheries are youths aged 35 and below. Most of these youths are school dropouts. Deliberations with participants indicated that youths engage in fisheries as a result of a lack of formal employment opportunities within the district and thus fisheries become the source of employment and income. Most of the land in Kani ward was found to be owned by men as a result of the patriarchal architecture which emanates from cultural beliefs. This arrangement poses constraints to women's access to and control of land and other productive resources. For instance, one of the female participants from the field visits alluded that:

> *Women do not have the power to own land in the Tonga culture as the land they are staying on as a family is called under the name of the father (husband) and he controls the day to day running of the home. The father is the head and decision maker as the home is called under him.* (Female participant)

Based on the above view, it is evident that women rarely own land. They also do not have the power to make decisions about the land. The inequalities in land ownership pose a threat to the empowerment of women and the achievement of gender equality. These research findings resonate with the findings of Akuffo and Quagrainie (2019) about similar traditional communities in Northern Ghana who also believe that women do not get the same opportunities in accessing and controlling land. Women in the fisheries-based households in Kani rarely own land because of customary ways of land distribution.

The study established that most fisheries households have large families with most of them having a minimum of six children. It was noted that at least 17 respondents had at least six children during the interviews. Children were viewed as a source of cheap labour for households, farming and fishing duties. Respondents indicated that some families fail to send their children to school as boys were quickly assimilated in fishing and livestock heading while the girls were taken into selling fishery products, sewing nets and doing other domestic household chores. For example, one respondent from the focus group discussion indicated that:

> *Tonga speaking households believe in large families and having many children in the past was seen as prestigious and children were considered as social capital. However, in the twenty first century households in fisheries have large families as a result of poor access to family planning. But they also bare more children because they want them to be a source of labour in household chores in the absence of the father.* (Government Official)

The division of labour within the household and market set-up is done according to gender considerations. Men and women are traditionally allocated different roles. Discussions showed that most women engage in the drying of fish (Kapeta fish), selling, packaging, sewing nets, farming and reproductive roles in the home set-up. The World Fish Centre (2010) concluded that more commonly, a woman's role in fishing is to manage household chores and engage in subsistence pre and post-harvest tasks and take care of the children while men are out there fishing. However, women were more heavily involved in agricultural activities.

Contribution of Fishing to Household Livelihood Sustainability

Fisheries, Employment and Household Income

In most households in the study area, participants revealed that fisheries were important in sustaining their livelihoods. During in-depth interviews and focus group discussions participants indicated that fisheries were a source of employment for most households. Participants indicated that since the economy was plummeting, it had become hard for one to be formally employed. Thus, fishing had become a solitary option for households to raise their incomes. The following figure breakdowns the livelihood strategies for people in the Kani Village (Fig. 30.2).

For instance, one participant from the focus group discussion posited that:

> *The economy of this country has been nose diving in the last three decades. Getting formal employment has become a tumultuous task. For most economically active youths and adults, fisheries activities have become an important source of employment and income. Remember that this source of livelihood option involves a number of activities such as fishing, repairing boats, selling nets and fish. Those involved in buying and selling fish have of late been taking*

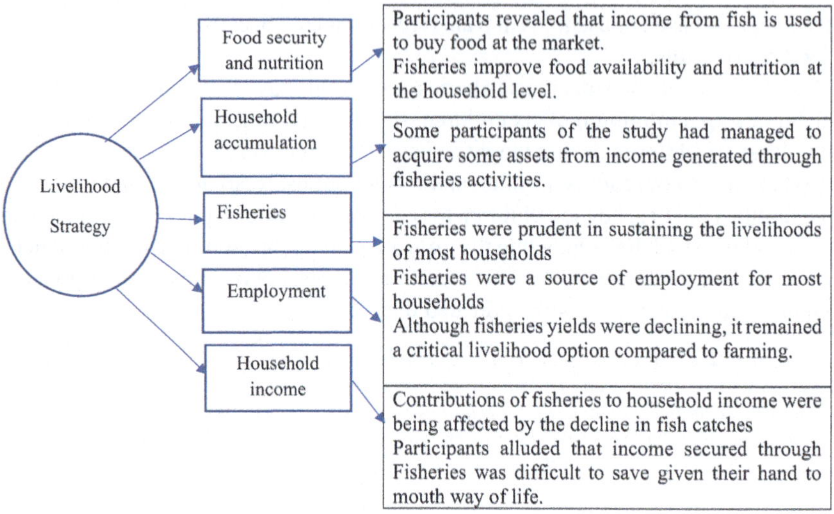

Fig. 30.2 Contribution of fisheries to livelihood sustainability (*Source* Researchers' construction)

> *advantage of a big market share coming Harare, Bulawayo and Victoria Falls looking for fisheries products.* (Key informant)

The above anecdotes reveal the significance of fishing in sustaining household livelihoods. The findings are in line with Hara et al.'s (2017) conclusion that estimated employment in the capture of fisheries and aquaculture sectors to be approximately 12.3 million people. However, some participants revealed that the contributions of fisheries to household income were being affected by the decline in fish catches, compelling them to diversify their livelihood activities. Further probes indicated that some proceeds from the fisheries were used to pay fees, medical bills as well as buy food and other immediate needs of the households.

Household Assets Accumulation

In order to determine the contributions of fishing-based livelihoods to household well-being, we asked participants questions about their asset ownership. Household assets are defined "broadly to include natural, physical, human, financial, public and social capital as well as household valuables" (Ellis 2000). Assets that seem important to the participants included building houses, drilling boreholes, buying scotch carts, radios, bicycles and livestock. One participant from the field visit hinted that:

> *I have done well for my household in the last four years. I have managed to build new houses as you see out of the money I got from trading in fish. I have also managed to put up a solar system for lighting, buy cattle, a radio and an ox-drawn plough. We have a few households in this community that have managed to drill boreholes.*

The disparity in asset ownership amongst the participant households speaks volumes about the diverse challenges confronting these households and the different livelihood portfolios they pursue to attain livelihood outcomes. The study results corroborate Ifejika et al.'s (2012) findings that livelihoods based on fisheries enhance household assets. However, poorer houses that are trapped in a vicious cycle of poverty and food insecurity fail to accumulate property as most of the proceeds from the fisheries activities are used to address the abrupt needs of their households.

Food Security and Nutrition

Participants revealed that fishing plays a crucial part in ensuring household sustenance. Focus group discussions showed that the agro-ecological characteristics of the study area have pushed most households to buy food from the market instead of growing their own food like in other surrounding communities. However, participants were concerned about the reduction in fish catch and its implications on household purchasing power parity. Some alluded that a reduction in fish catch could compromise the ability of a household to make food available leading to hunger and malnutrition. One participant from the field visit opined that:

> *Fisheries activities play an important role in sustaining the food needs of most households in this community. Households are involved in a number of activities that somehow generate income for them. In my case money generated from selling nets is used to buy sugar, cooking oil, mealie-meal just to mention a few items.* (Trader)

Aiga et al. (2009) posit that "the higher nutritional status in fish-farming (fisheries) households is not a result of fish consumption of farmed fish but from the additional cash generated by selling fish, which allows households to purchase other types of nutrient rich foods". Similarly, Hara et al. (2017) add that "fish is a rich source of easily digestible high-quality proteins containing all the essential amino acids". However, some participants revealed that the viability of fisheries activities was being compromised by a number of factors that range from overfishing to unaffordable fishing silences.

Linkages Between Farming Patterns and Fisheries Livelihoods

In this study, it emanated that there are links between smallholder farming and fisheries-based livelihoods. It was evident from the focus group discussions that some households are involved in farming as they kept cattle, goats, sheep and also cultivated crops such as maize and small grains. Small grains that are popular in the area include sorghum, corn, and millet given their resilience in such semi-arid areas. Other crops include maize, groundnuts, cowpeas, and watermelons. Participants from the focus group discussion revealed that as an adaptation strategy to climate change, small grains and small livestock proved to be drought tolerant and thriving. Focus group discussions with key informants revealed that income generated by some households through fisheries activities is used to fund smallholder agricultural activities. During the focus group discussion, it came out that some participants use the money generated from fisheries activities to buy agricultural inputs such as seeds, fertilizers and fences. Some participants hinted that they use some of the proceeds to hire labour in their fields in their absence as well as for buying cattle and other small livestock. One participant from the field visit alluded that:

> *Fishing has made it possible for me to buy seeds, hire labour to prepare my field, plant and harvest my crops. I have also managed to have a kraal as well. To me, fisheries-based livelihoods complement smallholder agriculture extremely well. But we are aware of some households that have decided to leave farming altogether.* (Fisher)

Further discussions during the field visit revealed that some households reconfigure their labour allocation to cater to farming activities during the rainy season. Women are released from their traditional role of drying and selling fishing so that they focus on farming. However, a few especially youths without land indicated that fishing was their only source of livelihood. One young fisherman from the field visit alluded that, to him, fishing was the only viable source of income. He noted that:

> *Fishing is part of my life. Out siding fishing l can't make it. I don't see myself doing farming, especially when looking at the agro-ecological characteristics of this area. Farming is facing a lot of challenges ranging from rainfall unreliability to problem animals. Apart from that, youths face huddles in their attempt to secure land. The other challenge is that farmers are broke in this community. It is drought year in and year out. Honestly, l am not motivated to do farming.* (Fisher)

Vulnerability of Fisheries as a Livelihood Strategy

Participants in the study were also asked questions focusing on the vulnerability of their fisheries based livelihood. The literature shows that "vulnerability denotes the external environment in which communities exist and over which they have limited or no control, including trends, seasonality and shocks" (Islam et al. 2014). So, exploring the vulnerability context of fisheries livelihoods is prudent as it can assist in crafting initiatives that can ameliorate adverse impacts. Thus, focus group discussions revealed that the vulnerability of fisheries based livelihood emanates from intricate climatic and non-climatic factors. Participants in the study revealed that fisheries activities are vulnerable to shocks and stressors as a result of a lack of productive land. Observations showed that most of the participants lack access to and ownership of land for agriculture which makes it a challenge for those dependent on fisheries to have an alternative livelihood to secure their lives. Furthermore, discussions during field visits also indicated that in order to diversify their livelihood portfolios, households should have the capacity to own and manage the means of production which is one of the major factors that had a bearing on the households' livelihood vulnerability. For instance, the lack of alternate livelihood sources was a major challenge among the study participants. One participant from the field visit noted that:

> *We are forced to depend on fisheries in this community given our inability to access land. Most of the youth do not own land which makes it a challenge for them to diversify their livelihoods.*

Further interrogations revealed that failure in the governance of the fishing activities was leading to overfishing: a challenge that increased the vulnerability of households that depend on fishing as a livelihood option. It emanated from the discussions that overfishing undermines the economic role of fisheries and leaves households vulnerable to poverty and other stressors. One participant from the focus group discussion revealed that:

> *Our main challenge is that fish are caught in the river which is a common resource for everyone. You cannot control the problem of overfishing. Remember some of the people who do fishing, especially at night do not have permits and in most instances use illegal nets that wipe literally all the fish in the water.* (Government official)

Most fishermen highlighted that they do not own fishing rigs or boats which results in a reduced catch during the winter season. These fishermen indicated that fish move to deep waters during winter seasons and if one does not have a boat or a fishing rig it is hard to catch fish. Furthermore, these discussions during field visits insinuated that government policies were a major contributor to fisheries-based livelihood vulnerability. The researchers found that long ago the Tonga people had unlimited and uncontrolled access to the river where they could fish, buy and sell their fisheries products without any obstacles but now, there are laws that one has to adhere to which makes fishing unsustainable. The government has introduced laws that make fishing very difficult for fisheries-based households. During field visits, participants claimed that one needs a permit to do fishing. These permits were said to be unaffordable to locals whose livelihoods have resonated around fisheries for a time immemorial. Interesting discussions emanated from the focus group discussion where female participants alluded that prostitution and sexually transmitted diseases were also increasing the vulnerability of fisheries-based livelihoods. Lately, there has been a high rate of sexually transmitted diseases in and around the fishing camps. Most participants argued that fisheries as a livelihood strategy is full of promiscuous activities thereby putting people at risk of contracting STIs. Considering that men are separated from their spouses for many days, they end up engaging in promiscuity. During field visits, one fisherman said it is hard for them to stay for 21 days without sex when they are at the fishing camps. This is further supported by the DFID (2006) which stated that "there is presently a considerable body of research suggesting that there is a higher-than-average prevalence of HIV infection in fishing communities in East Africa". Traders highlighted that for them to get more fish they are coerced to indulge in sexual favors with fishermen.

The COVID-19 pandemic was also mentioned by participants as having a bearing on their way of life. Most households indicated that the restrictions of the movement of people and the informal trading by the government during COVID-19 have adversely affected fisheries as a livelihood option. The findings of this study validate what has been revealed by the likes of Islam et al. (2014) who argue that "failures in the governance systems that lead to overfishing also increase the vulnerability of fisher folk and undermine the fishery sector's economic and social contributions". Scholars further argue that poverty, fisheries and vulnerability are complex issues considering the environment in which fishers operate. Similarly, Deressa (2011), Black et al. (2011), and Phuong et al. (2021) agree that most of the vulnerable households and communities are also poor and have undiversified livelihoods.

Impacts of Climate Change on Fisheries

Participants were asked how climate change has affected their fisheries-based livelihoods. It emanated that participants are aware of the implications of climate change on their main source of livelihood. Field visit discussions consistently revealed that changes in climate had led to an increase in temperatures as well as a reduction in rainfall. Some participants also opined that they had noticed a reduction in water levels in the Zambezi River, a factor that has contributed to unsustainable fishing. One participant (Fishermen) from the field visit alluded that:

> *The effects of climate change have reduced the prospects of livelihoods based on fisheries to promote sustainability. What we have noted is that the extremely high temperatures have increased fish mortality in this area. Climate change has made us realise that the decline of fish species has a bearing on fish catches and fish growth. We are no longer catching large fish as we used to do in the past, harvest has gone down and some water species we used to come across have disappeared.* (fishermen, 68 years)

Participants also indicated that women used to catch fish in the tributaries and distributaries along Zambezi River in the past years. However, due to erratic rains associated with changes in climate, these rivers have dried up making it impossible for them to harvest fish close to their communities. The findings that both the size and catch of fish have gone down that is consistent with findings in other lakes in Zimbabwe and Southern Africa (Magqina et al. 2020; Muringai et al. 2020a).

Adapting to Climatic and Non-Climatic Stressors

Daw et al. (2009) contends that "adaptation to climate involves the steps adjustment in ecological, social or economic systems, in response to observed or expected changes in climatic stimulus and their effects and impacts in order to alleviate adverse impacts of change, or take advantage of new opportunities". Participants were further asked about how they adapt to climatic and non-climatic stressors. Further focus group discussions revealed that the vulnerability of households to climate change and non-climate stressors is determined by their experience and understanding of change. Further engagements revealed that the maintenance of fishery-based livelihood requires households to diversify their livelihood portfolios. Some reported exiting fisheries-related way of life and opted for migration in search of better livelihood options like small-scale mining which may sound more resilient to climate change. Participants in the study also revealed that some are now fishing undocumented or without licenses. Adaptation strategies of some fisheries households were amply captured by one key informant from the focus group discussion who noted that:

> *Fisheries-based livelihoods are now very susceptible to climate change. Households are employing a number of adaptation strategies. We have seen some opting out of fishing altogether, some diversifying their livelihoods and some fishing without licenses, especially at night. Some have migrated to cities and neighbouring countries in search of better opportunities. Some have formed cooperatives meant to assist them in sending fish to better markets such as in Harare. I have also seen some using better boats to try and increase their harvest.* (key informant)

Focus group discussions with the participants also indicated that in extreme cases some poor households mortgage or sell some of their fisheries and non-fisheries assets. One participant indicated that at one point her household sold a plough and a boat to raise money to buy food and pay school fees. One key informant noted that:

> *As households try to cope with collapsing livelihoods, we have seen some children doing manual jobs. The school dropout rate for some children is unfortunate. Sadly, some have been forced to exploit some common properties such as wild fruits.* (NGO)

Deliberations also revealed that some household members participated in casual labour as part of their coping strategies. Specifically, it came out that women and children got involved in manual jobs to raise income or in some instances to get food as payment.

Limitations

There were a number of limitations experienced by the researchers. These included the methodological design of the study and insufficient resources. Qualitative case study designs use small samples compared to the quantitative ones. Such small samples that aim at generating in-depth information pose a challenge with regard to generalisations. Even though case study research is concerned with real-life scientific investigations, unlike experiments, the environmental and contextual conditions are not controlled and are part of the investigation which increases the subjectivity of the findings. The use of a case study led to the generation of large amounts of data which required careful analysis and management. Due to the lack of robust methodological procedures, the findings from a case study design may lack generalisability.

Conclusion

The study has shown the significance of fisheries in sustaining the lives and livelihoods of communities in Kani Ward. Specifically, it emanates from the anecdotes that fisheries play a cardinal role in improving household food security as well as incomes. However, a combination of climatic and non-climatic factors seems to be negatively impacting this traditional livelihood option. The implications are that the sustainability of fisheries-based livelihoods can only be achieved through improving the adaptive capacities of such communities to climate and non-climatic factors that have a bearing on the livelihoods. There is a need to consider enhancing women's access to productive resources and decision-making structures. But, there is no doubt that such communities also need to be strengthened in terms of their capacity to handle their vulnerability and uncertainty. The study concludes that communities need to be involved in policy formulation and these policies should be crafted and aligned to the needs of households in order to address environmentally sustainable fishing as a broad livelihood strategy at the household level. The allocation of fishing permits needs to be gender sensitive to allow women to participate in this critical livelihood option. The government needs to coordinate the development of a climate change adaptation strategy for fisheries as a way of tackling climate change challenges.

References

Aiga H, Sadatoshi MS, Kuroiwa S, Yamamoto S (2009) Malnutrition among children in rural Malawian fish-farming households. R Soc Trop Med Hyg 103:827–833

Akuffo SA, Quagrainie KK (2019) Assessment of household food security in fish farming communities in Ghana. Department of Agricultural Economics, Purdue University, West Lafayette, IN

Alison EH, Perry AL, Badjeck MC, Neil Adger W, Brown K, Conway D, Dulvy NK (2009) Vulnerability of national economies to the impacts of climate change on fisheries. Fish Fish 10(2):173–196

Belton B (2021) Fishing and aquaculture: underestimated as a source of income and food. https://www.welthungerhilfe.org/news/latest-articles/2021/fishing-and-aquaculture-as-a-source-of-income-and-food. Accessed 1 Oct 2022

Béné C, Merten S (2008) Women and fish-for-sex: transactional sex, HIV/AIDS and gender in African fisheries. World Dev 36(5):875–899

Bennett NJ, Dearden P, Peredo AM (2015) Vulnerability to multiple stressors in coastal communities: A study of the Andaman Coast of Thailand. Climate Dev 7(2):124–141

Black R, Bennett SRG, Thomas SM, Beddington JR (2011) Climate change: Migration as adaptation. Nature 478:447–449

Creswell JW, Plano-Clark VL (2011) Designing and conducting mixed method research, 2nd edn. Sage, Thousand Oaks

Daw T, Adger WN, Brown K (2009) Climate change and capture fisheries: Potential impacts, adaptation and mitigation. In: Cochrane K, Young DC, Soto D, Bahri T (eds) Climate change implications for fisheries and aquaculture: overview of current scientific knowledge. FAO Fisheries and Aquaculture Technical Paper. No. 530, FAO. Rome, pp 107–150

DFID (2006) Children's work in fisheries: A cause for alarm? Sustainable fisheries livelihood programme. DFID, London

Dub T, Ncube C, Moyo P, Phiri K, Moyo N (2021) Marginal communities and livelihoods: San communities' failed transition to a modern economy in Tsholotsho, Development Southern Africa, Zimbabwe. https://doi.org/10.1080/0376835X.2021.1955660

Ellis F (2000) Rural livelihoods and diversity in developing countries. Oxford University Press, Oxford

FAO (2020) The state of world fisheries and aquaculture 2020. Sustainability in action. Rome. https://doi.org/10.4060/ca9229en

FAO (2021) Small-scale fisheries around the world. https://www.fao.org/fishery/en/ssf/world. Accessed 1 Oct 2022

Freduah G, Fidelman P, Smith TF (2017) The impacts of environmental and socio-economic stressors on small-scale fisheries and livelihoods of fishers in Ghana. Appl Geogr 89:1–11

Funge-Smith S, Bennett A (2019) A fresh look at inland fisheries and their role in food security and livelihoods. Fish Fish 20(6):1176–1195

Gandiwa E, Zisadza-Gandiwa P, Mutandwa M, Sandram S (2012). An assessment of illegal fishing in Gonarezhou National Park, Zimbabwe. J Environ Res Manage 3(9):29–37

Gross JM (2018) Document analysis. In: Frey BB (ed), The SAGE encyclopedia of educational research, measurement, and evaluation. SAGE Publications. https://doi.org/10.4135/9781506326139

Hara M, Greenberg S, Thow AM, Chimatiro S, Toit A (2017) Trade and investment in fish and fish products between South Africa and the rest of SADC: Implications for food and nutrition security. Working Paper 47. Institute of Poverty, Land and Agrarian Studies, University of Western Cape

Hecky RE, Mugidde R, Ramlal PS, Talbot MR, Kling GW (2010) Multiple stressors cause rapid ecosystem change in Lake Victoria. Freshw Biol 55:19–42

Hsieh H, Shannon SE (2005) Three approaches to qualitative content analysis. Qual Health Res 15:1277–1288

Ifejika PI, Okunade EO, Ifejika LI, Asadu AM (2012) Physical assets ownership of fisherfolk in fishing communities of Kainji Lake Nigeria: Implications for climate change. J Agri Ext 16(2):92–103

Isaacs M, Onyango P, Akintola SL (eds) (2020) Small-scale fisheries in Africa: A regional portrait. TBTI global publication series, St. John's, Canada. https://tbti-global.net/

Islam MM, Sallu S, Hubacek K, Paavola J (2014) Vulnerability of fisheries-based livelihoods to the impacts of climate variability and change: Insights from coastal Bangladesh. Reg Environ Change 14:281–294

Limuwa MM, Sitaula BK, Njaya F, Storebakken T (2018) Evaluation of small-scale fishers' perceptions on climate change and their coping strategies: Insights from Lake Malawi. Climate 6(2):34

Lucas P, Fleming J, Bhosale J (2018) The utility of case study as a methodology for work-integrated learning research. Int J Work Integr Learn 19(3):215–222

Magqina T, Nhiwatiwa T, Dalu MT, Mhlanga L, Dalu T (2020) Challenges and possible impacts of artisanal and recreational fisheries on tigerfish Hydrocynus vittatus Castelnau 1861 populations in Lake Kariba, Zimbabwe. Scientific African 10:e00613

Muringai, RT, Mafongoya, P, Naidoo, D (2020a). The challenges experienced by small-scale fishing communities of Lake Kariba, Zimbabwe. J Transdiscipl Res S Afr 16(1):1–6

Muringai RT, Naidoo D, Mafongoya P, Lottering S (2020b) The impacts of climate change on the livelihood and food security of small-scale fishers in Lake Kariba, Zimbabwe. J Asian Afr Stud 55(2):298–313

Ndebele-Murisa MR, Mashonjowa E, Hill T (2011a) The decline of Kapenta fish stocks in Lake Kariba—A case of climate changing? Trans R Soc S Afr 66(3):220–223

Ndebele-Murisa MR, Mashonjowa E, Hill T (2011b) The implications of a changing climate on the Kapenta fish stocks of Lake Kariba, Zimbabwe. Trans R Soc S Afr 66(2):105–119

Ndhlovu N, Saito O, Djalante R, Yagi N (2017) Assessing the sensitivity of small-scale fishery groups to climate change in Lake Kariba, Zimbabwe. Sustainability 9(12):2209

Nico G, Christiaensen L (2022) New data highlight the hidden jobs effects of small-scale fisheries. https://blogs.worldbank.org/jobs/new-data-highlight-hidden-jobs-effects-small-scale-fisheries. Accessed 1 Oct 2022

Ogutu-Ohwayo R, Natugonza V, Musinguzi L, Olokotum M, Naigaga S (2016) Implications of climate variability and change for African lake ecosystems, fisheries productivity, and livelihoods. J Great Lakes Res 42(3):498–510

Phuong TAH, Ngoan DL, Sen THL, Hong XN (2021) Vulnerability of fisheries-based livelihoods to climate change in coastal communities in Central Vietnam. Coast Manag 49(3):275–292. https://doi.org/10.1080/08920753.2021.1899927

Shava E, Gunhidzirai C (2017) Fish farming as an innovative strategy for promoting food security in drought risk regions of Zimbabwe. Jàmbá: J Disaster Risk Stud 9(1):1–10

Smith LE, Khoa SN, Lorenzen K (2005) Livelihood functions of inland fisheries: Policy implications in developing countries. Water Policy 7(4):359–383

Tweddle D, Cowx IG, Peel RA, Weyl OLF (2015) Challenges in fisheries management in the Zambezi, one of the great rivers of Africa. Fish Manage Ecol 22(1):99–111

UNSD (2022) Conserve and sustainably use the oceans, sea and marine resources for sustainable development. https://unstats.un.org/sdgs/report/2022/Goal-14/. Accessed 1 Oct 2022

Utete B, Phiri C, Mlambo SS, Muboko N, Fregene BT (2018) Fish catches, and the influence of climatic and non-climatic factors in Lakes Chivero and Manyame, Zimbabwe. Cogent Food Agric 4(1):1435018

Utete B, Phiri C, Mlambo SS, Muboko N, Fregene BT (2019) Vulnerability of fisherfolks and their perceptions towards climate change and its impacts on their livelihoods in a peri-urban lake system in Zimbabwe. Environ Dev Sustain 21(2):917–934

World Fish Centre (2010) Gender and fisheries: Do women support, complement or subsidize men's small-scale fishing activities. Issues Brief 2108.

Yin KR (2018) Case study research and applications: Design and methods, 6th edn. Sage, Thousand Oaks, CA

Chapter 31
Hydro-Meteorological Risk Emergency Planning and Management Using Big Data as a Platform

Fisha Semaw, Dagnaw Kebede, and Desalegn Yayeh Ayal

Abstract Climate change and extreme events are becoming more common worldwide, and developing countries are especially vulnerable due to insufficient mitigation and preparedness capabilities. It goes without saying that employing big data is the right approach for successful retrieval of disaster data as well as its analysis, management, and dissemination in climate risk emergency planning and management. This review will investigate how big data can be used as a recent data management technology for climate risk emergency planning and management. An intimate assessment of the literature shows that there are methodical reviews on big data applications in emergency management. This study holds that emergency events are far from being predictable and as such, they are bound to bring havoc and destruction. Emergency management is managing public emergencies using different methods like big data. Big data technologies can address issues related to storage, analysis, accessibility, and distribution to generate and present data analysis useful for practical purposes. The findings of this review show that using big data in various emergency management applications will enable policymakers and other stakeholders to make appropriate and correct decisions during the emergency management process which benefits if responses are optimized for quality prediction analysis.

F. Semaw · D. Kebede
Department of Disaster Risk Management and Sustainable Development Institute of Disaster Risk Management and Food Security, Bahir Dare University, Bahir Dar, Ethiopia

D. Y. Ayal (✉)
Center for Food Security Studies, College of Development Studies, Addis Ababa University, Addis Ababa, Ethiopia
e-mail: desalula@gmail.com

W. Leal Filho et al. (eds.), *Climate Change Strategies: Handling the Challenges of Adapting to a Changing Climate*, Climate Change Management,
https://doi.org/10.1007/978-3-031-28728-2_31

Introduction

Hydro-meteorological hazards have had a substantial impact on disasters throughout this planet's history, from the prehistoric to the modern eras. Despite the availability of a wide range of research on meteorology related natural hazards, there has been very little or no commensurate endeavor to make robust use of big data to manage these risks. New big data approaches have inspired researchers to employ these tactics to solve hydro-meteorological risks (Kaur and Sood 2020).

At present, terrorist acts, hurricanes, floods, epidemics, and earthquakes are responsible for massive failures of infrastructure which in turn caused huge and unprecedented material destruction and loss of lives all over the world (Sarker et al. 2020; Song et al. 2020; Yu et al. 2018). The world's database of Annual danger reports was released by NatCatSERVICE, which noted a gradual rise over the previous few decades (Akerkar and Hong 2020). More destructive and frequent than geophysical disasters, hydro-meteorological disasters have been observed. From 2000 to 2019, 9567 occurrences were brought on by these catastrophes. Floods, landslides, hurricanes, droughts, cyclones, tsunamis, avalanches, and tornadoes are examples of hydro-meteorological disasters (HMD) (Kauret al. 2022; Wu et al. 2020). It is a difficult task and a viable study area to manage the HMD. The myriad problems connected to hydro-meteorological disasters are the subject of extensive investigation. During recent decades, the intensity, frequency, and impacts of disastrous events have shown dramatic increases (Song et al. 2020). Given that natural hazards remain to be highly unpredictable, their adverse environmental impacts are particularly appalling when there are meager resources to absorb shocks (Yu et al. 2018). Various emergency measures taken to ameliorate the effects of natural disasters on public health and social security have been trending in many developing countries (Yu 2020).

Additionally, the societies' vulnerability to disasters around the world was drastically increased by rapid urbanization and poor development policies, which created new dangers or amplified already-existing ones. This led to a sharp rise in losses (Villeneuve 2018). Ameliorating the impacts of natural disasters and achieving swift recovery is possible if attention is directed toward planning and managing resilience policies. Governments and researchers worldwide have launched several efforts to encourage academics and scholars to support their nations' effective emergency management. The occurrence of all types of emergencies causes many casualties and economic losses, as well as a severe impact on social harmony and people's life and property safety (Lu and Zhang 2016).

Although the terms emergency and disaster are sometimes used interchangeably (Al-Dahash et al. 2016), it should be pointed out that "emergency" specifically refers to a situation that endangers people or damages property suddenly (Yu et al. 2018). At the same time, the definition of disaster includes property damage and loss of life that stem from the negative impact of an emergency due to insufficient capacity to respond properly and adequately (Biswas and Choudhuri 2012). In addition, it is possible to define an emergency as a broader term encompassing disasters, catastrophes, and minor disruptions in a short period (Al-Dahash et al. 2016). The chaotic and complex nature of natural disasters renders preventive measures a difficult undertaking (Al-Dahash et al. 2016).

Emergency management (EM) refers to systemic processes that primarily aim at minimizing disasters' negative consequences and effects on people and social infrastructure (Arslan et al. 2018; Norris et al. 2015). Being a complex and dynamic response, EM involves enormous uncertainty (Kaveh et al. 2020), in the attribution of responsibility, choice of data sources, and business planning (Akter and Wamba 2019; Jin et al. 2020). To implement sound policies on emergency and disaster management, it is crucial to put in place the right information technology and equipment (Yu et al. 2018). Modern EM processes use contemporary technologies to monitor, respond to, handle, and process emergencies quickly and efficiently. They also integrate many resources and conduct scientific analyses of the emergency's origins, course, and adverse effects. Nowadays, more and more scholars around the world have devoted their attention to ICT-based disaster prevention and control mechanisms. To manage disaster situations, Ray et al. (2017) presented an IoT-based scenario. During hydro-metrological disaster management, internet of things (IoT) modules are utilized for data collection and to provide a networked system for data transfer (Sakhardande et al. 2016). IoT sensors collect environmental data, and big data and mining techniques are used to handle and analyze unstructured data (Rahman et al. 2017). The digital infrastructure, software, and platform needed for information collaboration, data security, and backup are all provided by cloud environments. Different aspects of disaster management can use artificial intelligence (AI) and its subfields (machine learning, expert systems, robotics, fuzzy logic). The major AI fields used at different stages of disaster management include deep learning, supervised and unsupervised models, and optimization. Amit and Aoki (2017) and Ivić (2019) used AI techniques to analyze UAV and satellite photos of a specific geographic area to obtain the data. Remote sensing technology is a tremendous aid for global monitoring and assessing disaster situations. GIS is a crucial risk reduction tool since it models ways to examine many disaster data (Kaku 2019). Specifically, sound emergency management has an inherent relationship with big data technology which affects the collection and processing of all types of data, including data relevant to safety measures during responses to disasters (Jin et al. 2020).

Since big data is a technological paradigm of our time, it allows researchers to make efficient data analyses huge with ease (Hashem et al. 2015). Put differently, big data is nothing but the deployment of engineering and scientific methods and tools for processing massive data (Yu et al. 2018). The essence of big data lies in its usability for handling massive information rapidly and in diverse ways. Big data technology changes large-scale data collection, transmission, storage, mining, and visualization. Big data is not only a technical development in the context of big data but also a change in how we think (Lu and Zhang 2016). Big data offers numerous opportunities for communication, which can assist vulnerable community members in learning about impending threats, challenges, risks, and disasters (Hashem et al. 2015; Ragini et al. 2018).

Emergency management based on adequate information is crucial for the effective mitigation of the negative impacts of natural disasters (Huang et al. 2020). It is now widely accepted that technology can aid in managing pieces of information on disasters. Technology can aid in developing a full scale climate risk management that spans from an early warning systems to recovery plans (Sarker et al. 2020). Communication enables people to communicate before, during, and after the hydrometeorological disasters to inform one another of the situation and plan accordingly. In that context, communication can be considered a good source of big data (Maryam et al. 2016).

Many studies on climate change adaptation and resilience (Sarker et al. 2020), emergency management (Jin et al. 2020), disaster resilience (Sarker et al. 2020), and big data application in epidemic disease distributions (Huang et al. 2020) have been conducted. However, the full potential applying big data for emergency planning and management hasn't been actualized due to resource constraints. Unexpected events frequently cause disasters. A comprehensive data-driven effective strategy can improve the efficiency of emergency management and planning. As a result, this review aims to investigate how big data can be used for emergency planning and management.

In general, though less attention has been given; big data could play an indispensable role in climate change adaptation and vulnerability assessment (Ford et al. 2016). The big data analysis approach is vital in forecasting climate change related risk and implementing feasible adaptation measures. In the effort of climate change adaptation and mitigation, big data could be use in the energy, agriculture, and forestry sectors as well as urban and infrastructure planning (Hossein and Xu 2019).

Review Methodology

To conduct a scientific review of available literature on the subject at hand, we made a systematic selection of published articles about emergency planning and management and big data. First, a search Google Scholar with the words "big data, climate change, risk, disaster, adaptation and emergency management" was applied. In the search engine, custom ranges from the period 2015 to 2022 were used. Second, articles most relevant to the topic of interest were manually selected. Thirdly, the list of articles only included in peer-reviewed journal articles was filtered. A total of 45 journal articles were reviewed individually and the synthesis was used to present the sections of the current articles.

Results and Discussion

Concept of Big Data and Its Sources

There does not seem to be a single definition of big data agreeable to all scholars. However, the review identifies the two most common definitions of big data. One definition of "Big data" holds that big data refers to a hugely enormous amount of data that cannot be computed or processed through traditional database software tools (Zhang et al. 2021). On their part, Yu et al. (2018) consider that big data is nothing but a reference to scientific and engineering methods and tools through which hugely enormous data can be analyzed, processed, managed, and stored. The term "big data" is also used to describe a large amount of data in the networked, digitized, sensor-laden, information-driven world (Iglesias et al. 2020). It is a new technology that can effectively solve the vast amounts of data collection, storage, and display (Lu and Zhang 2016; Yu et al. 2018). Put differently, big data refers to both the amount of data and high-tech processing tools which can be harnessed to improve the efficiency and efficacy of state governance (Lu and Zhang 2016).

In its evolution and usage, big data is associated with its "four V" characteristics (Zhang et al. 2021; Huang et al. 2020; Yu 2020), which increased its demand in the area of emergency management. These are volume, velocity, variety, and value. In short, "four V" is about processing huge data in a short period of time in a variety of ways and a value adding manner (Zhang et al. 2021). Big data analysis can also refer to the process of extracting information from data, producing knowledge from information, and using that information to make the best by concerned actors. In this way, "big data" refers to the whole process of producing knowledge, integrating it into a system, and accelerating development (Amaye et al. 2016; Yu 2020). The same conception of big data can be applied to methodically respond to an emergency situation which involves complex decision-making, public opinion management, networking, and early warning. And such delicate tasks can be easily managed by tapping the resourcefulness of big data as regards velocity of action and value creation (Yu 2020). Big data enables researchers to do an in-depth study of all communications, offering helpful information that is valid for the general public, including details on a disease outbreak (Yu et al. 2018). Big data coalesce quite diverse sources of data and expands analytical capacity so that people in a disaster situation can benefit from an informed response (Yu et al. 2018).

Big Data offers a useful understanding of the decisive stages of managing emergencies such as preparation, prevention, response, and recovery. Big data comes from two primary sources. The first is from sensor networks that detect earthquakes using seismometers. The second source is concerned with multipurpose sensor networks such as social media platforms. Both approaches proved to be effective during emergencies such as the COVID-19 pandemic and Hurricane Sandy. Emergency management demands the provision of high quality and accessible service and information for the victims (Akerkar 2020).

Big data approaches encompass the datasets and a variety of related technologies such as parallel processing, cloud computing, the internet of things, geotagging, machine learning, data mining, and natural language processing (Pollard et al. 2018). The primary big data sources for emergency planning and management could be videos from unmanned aerial vehicles, satellite imagery, aerial imagery, the internet of things, crowdsourcing, airborne and terrestrial light detection and ranging, social media, and mobile GPS and call data records (Yu et al. 2018; Jin et al. 2020; Yu 2020; Sarker et al. 2020). Of these sources, satellite imagery, crowdsourcing, and social media data rank as the most popular sources of data for emergency management (Yu et al. 2018).

Satellite imagery: Remotely sensed data is an independent source of information, able to cover large areas, which is especially relevant in the absence of precise information on the location and extent of the impacted area (Wania et al. 2021). Satellite remote sensing generates invaluable satellite images which help to continuously monitor atmospheric and surface conditions associated with natural disasters (IPCC 2022). In the same vein, satellite images offer qualitative as well as quantitative data necessary for disaster management (Sarker et al. 2020). The noteworthy contribution of remote sensing for smart planning of disaster management stems from the high-resolution, multi-dimensional, and multi-technical images it produces (Sarker et al. 2020).

Aerial imagery and videos: Data describing the disaster's impact and the extent of the damage are required for effective disaster management. Aerial data collection consumes many resources and necessitates offline post-processing (Murphy et al. 2016). Its speed and spatial resolution are thought to be more useful than satellite imagery (Sarker et al. 2020). The most advanced level of tools for detecting fine cracks, damaged structures, and the extent of the damage is video information. Airborne data collection via helicopters and unmanned aerial vehicles (UAVs) provides a bird's-eye view of disaster-affected areas (Murphy et al. 2016). Unmanned aerial vehicles (UAVs) captured aerial imagery are becoming increasingly crucial in disaster response (Ofli et al. 2016).

Internet of Things: This paradigm is deemed to be a highly promising technology to address problems in every walk of life: industry, agriculture security, and medicine. The attractiveness of the Internet of Things stems from its typical features of lightweight, heterogeneity, interoperability, and flexibility (Ray et al. 2017). A global network of interconnected objects that utilizes networking, information processing, and sensing technologies is referred to as the "Internet of Things." This could be the newest iteration of information and communications technology (ICT) (Li et al. 2015). IoT technologies equip managers with tools relevant to practical decision making while enhancing the situational awareness of operators' and rescuers' thereby ultimately enhancing the preparedness of people prone to natural disasters (Dugdale et al. 2021). It provides better communication in disaster-affected areas where natural hazards have damaged communications structures (Sarker et al. 2020; Yu et al. 2018).

Crowdsourcing: Big and heterogeneous data created by various sources in urban settings, including sensors, machines, vehicles, buildings, and people, are acquired, integrated, and analyzed in this process (Xu et al. 2016). This is the most ideal data analytic technology especially in emergency management situations since it is resourceful for information validation, tracing events in real-time, translating text, and integrating data for crisis decision-making (Amaye et al. 2016). It is also indispensable to enable people to realize their common goals just by way of online platforms (Sarker et al. 2020). Such platforms are gaining momentum as chief information providers in disaster situatios (Dittus et al. 2017).

Social media: Facebook, IMO, WeChat, Twitter, WhatsApp, and QQ have great importance in every phase of disaster management. Social media can help the emergency management sector by giving stakeholders access to information as the event develops, a way to quickly alert wider audiences, a platform to keep track of public activities, and a tool for activity coordination (Luna and Pennock 2018). Social media is ideal means of extracting real-time data sources for disaster management and determination of current conditions (Gulesan et al. 2021). Thanks to advancements in communication and technology, today early warning systems and rapid response mechanisms have got new technology options.

Mobile-based GPS: In disaster situations, mobile phones can be used to contact family and friends and locate safe places to move. GPS is also used to determine the altitude, location, and magnitude of hazardous situations (Sarker et al. 2020).

Emergency Management Phases

Different events are happening more frequently in today's society, endangering environmental protection, social order, human life, and even international political ties. There is widespread agreement that emergencies are social phenomena that undermine routine business, social structures, values, and norms. The World Health Organization (WHO) has pointed out that emergency responses create complications in the standard conditions of existence and worsen the suffering of victims beyond the point of readjustment capacity. This definition states that a disaster's severity is more closely associated with the level of social unrest than with the potential threat's size (Akerkar 2020). Applying early and comprehensive emergency management procedures is crucial to avoiding or minimizing the effects. When natural or artificial disasters and catastrophes strike, essential infrastructures must be protected from catastrophic harm through emergency management. Multi-layered emergency management plans are designed to solve problems like terrorist attacks, earthquakes, floods, hurricanes, fires, and utility failures.

This proves the idea that EM should be used regularly and not just in times of emergency to ensure the security of everyone's daily lives. Integrating different resources and conducting a scientific analysis of the emergency's cause, process, and adverse effects, more modern EM includes processes that take advantage of contemporary technologies which can be deployed for effective and efficient disaster monitoring and handling.

FEMA's definition of emergency management (EM) entails planning, minimizing the effects, and responding to and recovering from an emergency. A significant process known as detection, preparation, planning, mitigation, reaction, and recovery is used in emergency management to deal with crises.

The old style of doing EM has been explained in the traditional fashion of bureaucracy which lies on command and control. To put it another way, EM has only been examined from the standpoint of one authority in traditional techniques. For instance, it is acknowledged that the traditional emergency response strategy prioritizes fixed rules of norms but there is room to comprise bureaucratic policies and procedures in emergencies situations. As a result, group participation or flexibility in unanticipated tasks are not thought to be essential elements. Furthermore, the use of a hierarchical structure is a characteristic of traditional techniques. Therefore, the old bureaucratic EM procedures which stipulate adherence to hierarchy are no longer binding and are challenged because of the quick fix imperative needed for emergency response.

In contrast, FEMA's methodical approach acknowledges that EM should lean towards decentralized institutional actors that work together in a network to lessen the effects of emergencies. For instance, public businesses, volunteers, local police and fire departments, and governmental organizations can come together and perform activities related to the emergency life cycle. Additionally, it takes into account civilian support for recovery measures. In dissimilarity to the conventional methods, it recognizes that resources and public information may help emergency efforts. For instance, because they are the ones who directly encounter the incident, the general public can be considered the initial emergency responder. Additionally, the professional approach's adaptability enables the use of its solutions for various circumstances. That implies that the strategies are no longer limited to a single emergency.

The most modern methods rely on Internet technologies to share information and coordinate among stakeholders during emergencies. In the early 2000s, several websites were launched in response to crises. For instance, in 2004, a problem was addressed via a user-generated website. Additionally, when Hurricane Katrina attacked New Orleans in 2005, substantial emergency response operations happened on Myspace, a social networking service. The devastating wildfire in 2007, close to San Diego, California, stands as a documented case where people responded to an emergency using a microblogging service like Twitter. Since then, it has been common to use Twitter to ask questions, gather, communicate, disseminate information, and to coordinate response activities.

On the other hand, as the Gartner hype cycle for 2014 suggests, the more public awareness and focus on the Internet of Things (IoT) reaches its peak, we might reach the era of disillusionment which must be tackled through a clear long-term technological solution for the right type and amount of information available for EM. These open geospatial services and IoT paradigms, as well as the numerous opportunities they offer for visualizing, evaluating, and predicting emergencies, are the foundation of many research efforts in the EM area. For instance, TRIDEC4 concentrates on cutting-edge technology to make effective intelligent information management in real-time possible so that seamless big data access, identification, integration, and handling is possible for EM.

The number of players in a response and recovery scenario will also be significant. Services for emergency management "include organizations like fire, ambulance, civil protection authorities, police, paramedic, and emergency medical services, Red Cross and Red Crescent societies, and specialized emergency units of transportation, communications, electricity, and other related services organizations in different countries in different phases of the EM" (WHO 2020).

The four phases of emergency management are known as readiness, response, mitigation and recovery (Sun et al. 2020; Yu et al. 2018). Management activities are undertaken during the mitigation phase to prevent or reduce futuristic emergencies and their repercussions while increasing long-term benefits (Sun et al. 2020). Mitigation is an attempt to reduce the effects of disasters by way of risk analysis, building warning codes, mapping risk zones, and public education (Yu et al. 2018). Any activity made to prevent an emergency, reduce the likelihood of an emergency, or lessen the damage of unavoidable emergencies falls within the phase of mitigation. For example, purchasing flood and fire insurance for a house is a mitigation activity (Sun et al. 2020). Mitigation efforts can take place before and after emergencies (Sarker et al. 2020). Mitigation measures can lower the adverse impacts of a disaster if they are accompanied by proper planning and action. For this it is imperative to use scientific hazards analysis (natural and man-made), risk assessment and mapping, vulnerability analysis (elderly, children, pregnant, etc.), simulation and modeling, structural and non-structural mitigation, training, awareness campaign and capacity building (Skliarov et al. 2017).

Preparedness: The preparedness phase refers to activities taken when an emergency is deemed likely to occur (Sun et al. 2020). This phase focuses on planning how to respond to a disaster through sound preparedness plans, training, emergency exercises, and early warning system development and implementation (Yu et al. 2018). Sound preparedness plans require communication planning, resource inventory planning, logistical planning, stockpiling planning, evacuation planning, and needs assessment planning (Skliarov et al. 2017; Sun et al. 2020).

Response: Response activities are crucial to safeguard property, save lives, and protect the environment from natural disasters (Yu et al. 2018). The key activities of the response phase are searching and rescuing victims, assessing initial damages, providing first aid and humanitarian assistance, and opening and managing shelters (Sun et al. 2020). It applies during disasters. After a disaster, the recovery phase is concerned with repairing and reconstructing measures to return affected area to a standard or even better functionality level. Among other things, recovery actions include precise damage assessment, debris cleanup, infrastructure reconstruction, and financial assistance from responsible bodies. Below, we have depicted the four stages of emergency management in a diagram (Fig. 31.1).

These phases are briefly explained in the following.

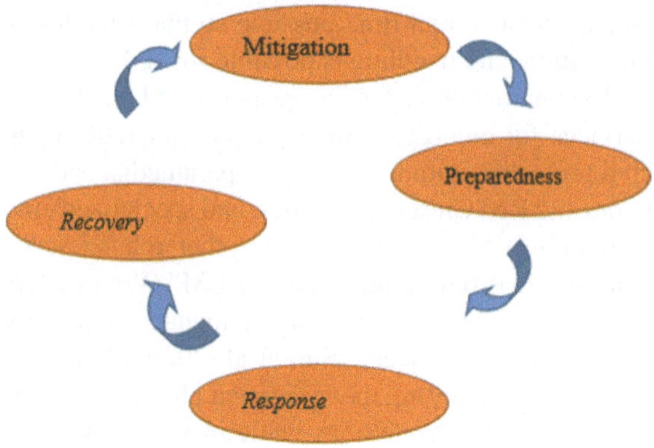

Fig. 31.1 The four phases of emergency management

Emergency Planning and Management

Given that natural catastrophes have increased in frequency and severity, governments all over the world are prompted to make disaster management and emergency response national priorities. A more effective approach to overall disaster management, disaster relief, logistical coordination, and long-term recovery in connection with natural disasters and emergency events is now supported by the growth of enormous and complex datasets, or big data, as well as various advancements in information and communications technology and computing. Big data and technical advancements have drawn much interest from the scientific community for emergency management (Song et al. 2020). Emergencies are frequently complicated issues with adverse outcomes that must be resolved quickly to minimize potential harm. Big data analysis results in more confident decision-making and better decisions can result in higher operational efficiencies, cost savings, and decreased risk (Akerkar 2018). Emergencies refer to accidents, public health incidents natural disasters, and social security incidents that occur suddenly and cause or may cause serious social harm and require emergency response measures (Zhang et al. 2021; Yu et al. 2020). Planning is the general framework for dealing with a crisis or disaster (Skliarov et al. 2017). It is devoted to the identification of the problem, assessing what can be done in response, and sorting out how it can be implemented. All of these must be aligned with the facts of the crisis, the changes, the prediction of future events, the preparation for emergencies, and the designing of likely scenarios for effective treatment. That being the case, it is possible to conclude that emergency management is concerned with the entire process of managing natural or other catastrophic events through the cooperation of private, public, and government entities (Yu et al. 2020). Put differently, emergency management is a reference to a systematic response to reduce or mitigate the negative consequences of disasters and on people and infrastructure (Arslan et al. 2018; Norris et al. 2015).

Applications of Big Data for Emergency Management

Effective emergency management depends on having access to trustworthy information in times of need. The nature and volume of information that can be availed by the media, public institutions, individuals, and volunteer organizations have dramatically increased because of new technologies. The rise of sensor networks, social media, satellite remote sensing, and linked devices has contributed to a data influx beyond traditional technologies' capacity to acquire, process, and understand (Iglesias et al. 2020). Big Data has undoubtedly expanded the possibilities for natural emergency management because of its numerous alternatives for visualizing, evaluating, and anticipating emergencies. Emerging technologies that may monitor and detect emergencies, reduce their consequences, help relief operations, and contribute to recovery and reconstruction processes are crucial. However, managing massive data exponentially generated during emergencies has made the use of Big Data imperative in EM. Traditional data storage and processing methods struggle to meet performance requirements for real-time processing, scalability, and availability because of the enormous amount of data (Akerkar 2020).

During the last two decades, the growth of Data enormously increased to the extent that it compelled global researchers to come up with new machine learning algorithms and artificial intelligence. Consequently, real-time classification and categorization of social media data have contributed to effective disaster response and recovery (Ragini et al. 2018). Recently, it is not uncommon to see data-driven decisions and support structures in every management area (Kushwaha et al. 2021). For managing their big databases, various big data applications are found in the astrology, banking, medicine, and finance departments (Roy et al. 2020). Big data technology helps to build an emergency response platform to achieve cross-sectoral integration of emergency data (Lu and Zhang 2016).

One approach through which big data can be deployed in emergency management has to with information dissemination about the crisis through social media (Huang et al. 2020). Big data applications in emergency management emphasize pre-forecasting (Zhang et al. 2021). Big data technology has been in use for early monitoring and early warning initiatives regarding infectious diseases (Wu et al. 2020). Satellite communications are essential for operative emergency response, especially in data collection, position location, alerting, and harmonizing relief procedures (IPCC 2022). Big data can be utilized to closely observe societal issues and catastrophes which can be averted or mitigated through early monitoring and warnings mechanisms (Huang et al. 2020; Mao et al. 2021). Therefore, Big data technology has huge applicability in various EM industries (Jin et al. 2020).

Predicting and modeling human travel routes is crucial for designing appropriate transportation schedules, planning humanitarian aid, and other emergency management issues, and that in turn depends on the quality and volume of human mobility and urban sensing data for (IPCC 2022). Based on the analysis of social media messages circulated in emergencies and their effects, it is possible to model and apply human evacuation strategies (IPCC 2022). It is well known that by using satellite images, professionals can pinpoint impending hazards in the environment and infrastructures. For example, we can formulate potential risk assessments through a sober analysis of multiple satellite-based flood maps (Sakhardande et al. 2016). Phone call details are useful metadata with which we can capture the timing and quantity of calls to map population distribution and corresponding socioeconomic profiles that have implications on vulnerability. This was successfully carried out following the 2015 Nepal earthquake (Wilson et al. 2016). Remotely sensed multitemporal images captured over the same area, typically separated by days, can be used to manage disasters like fires and floods effectively (Yu et al. 2018).

The Namibia flood sensor web was also considered an early warning system, a system based on the combined use of satellite and ground sensor data, which proved instrumental for early warning and flood situational awareness (Mandl et al. 2013). In the same vein, we can harness the resourcefulness of NASA TRMM rainfall network, USGS seismic network, and social media platforms help to develop a landslide detection system and locate the origins and components of multi-hazards (Yu et al. 2018). Machine learning is ideally suited to eliminate unrelated data, speed up disaster risk analysis, and identify strategic response options (Yu et al. 2018). In general, by using big data technologies, Ethiopia could solve significant problems such as floods, droughts, and social security.

Conclusion

The chapter has attempted to cogently analyze the existing literature on Big Data and emergency management. Since the Big Data tool is an evolving phenomenon, it would be farfetched to depict it as flawless. Future researchers should ponder on addressing some inherent challenges of relying on Big Data: the difficulty of processing Big data generated with the involvement of many people affected by an emergency; the basic nature of the data being highly sensitive to real-time operation poses the difficulty of optimal response; it is a daunting task to integrate static and dynamic data (maps and crowd emotion), and it is an arduous task to synchronize raw data from remote sensors with organized metadata and multimedia. Notwithstanding that such challenges of Big Data strategy for disaster management are still problematic, Big Data remains to be a cutting edge tool for emergency management.

In the final analysis, it is necessary to bear in mind that Big Data isn't free from limitations and hasn't reached a stage where it can rely as a panacea to disaster risk management. As it stands now, amassing and processing massive data from multiple sources while satisfying the five ascribed merits of Bi Data known as volume, variety, velocity, veracity, and value makes its worth evolutionary rather than revolutionary and much effort should be exerted to enhance its usability for disaster prediction, prevention, mitigation, and recovery.

References

Akerkar R (2018) Processing big data for emergency management. In: Smart technologies for emergency response and disaster management. IGI Global, pp 144–166

Akerkar R (ed) (2020) Big data in emergency management: exploitation techniques for social and mobile data. Springer Nature

Akerkar R, Hong M (2020) Introduction to emergency management. Big Data in emergency management: exploitation techniques for social and mobile data. Springer, Cham, pp 1–14

Akter S, Wamba SF (2019) Big data and disaster management: a systematic review and agenda for future research. Ann Oper Res 283(1):939–959

Al-Dahash H, Thayaparan M, Kulatunga U (2016) It was understanding the terminologies: disaster, crisis, and emergency. Paper presented at the proceedings of the 32nd Annual ARCOM Conference, ARCOM 2016

Amaye A, Neville K, Pope A (2016) BigPromises: using organi-sational mindfulness to integrate big data in emergency management decision making. J Decision Syst 25(1):76–84

Amit SNKB, Aoki Y (2017) Disaster detection from aerial imagery with a convolutional neural network. The 2017 international electronics symposium on knowledge creation and intelligent computing (IES-KCIC). IEEE, pp 239–245

Arslan M, Roxin A, Cruz C, Ginhac D (2018, January) A review on applications of big data for disaster management. In: The 13th international conference on signal image technology, India

Biswas BC, Choudhuri SK (2012) Digital information resources for disaster management of libraries and information centres. Bangladesh J Library Inf Sci 2(1):12-21

Dittus M, Quattrone G, Capra L (2017) Mass participation during emergency response: event-centric crowdsourcing in humanitarian mapping. Paper presented at the proceedings of the 2017 ACM conference on computer supported cooperative work and social computing

Dugdale J, Moghaddam MT, Muccini H (2021) Iot4emergency: internet of things for emergency management. ACM SIGSOFT Software Engineering Notes 46(1):33–36

Ford JD, Tilleard SE, Berrang-Ford L, Araos M, Biesbroek R, Lesnikowski AC, MacDonald GK, Hsu A, Chen C, Bizikova L (2016) Big data has big potential for applications to climate change adaptation. PNAS 113(39):10729–10732

Gulesan OB, Anil E, Boluk PS (2021) Social media-based emergency management to detect earthquakes and organize civilian volunteers. Int J Disaster Risk Reduction 65:102543

Hashem IAT, Yaqoob I, Anuar NB, Mokhtar S, Gani A, Khan SU (2015) The rise of "big data" on cloud computing: review and open research issues. Inf Syst 47:98–115

Hossein H, Xu H, Emmanuel S (2019) Big data and climate change. Big Data Cogn Comput 3:12. https://doi.org/10.3390/bdcc3010012

Huang H, Peng Z, Wu H, Xie Q (2020) A big data analysis on the five dimensions of emergency management information in China's early stage of COVID-19. J Chin Gov 5(2):213–233

Iglesias CA, Favenza A, Carrera Á (2020) A big data reference architecture for emergency management. Inf 11(12):569

IPCC (2022) Climate change 2022: impacts, adaptation and vulnerability. Contribution of working group ii to the sixth assessment report of the intergovernmental panel on climate change Pörtner HO, Roberts DC, Tignor M, Poloczanska ES, Mintenbeck K, Alegría A, Craig M, Langsdorf S, Löschke S, Möller V, Okem A, Rama B (eds), p 3056. Cambridge University Press. Cambridge University Press, Cambridge, UK and New York, NY, USA. https://doi.org/10.1017/9781009325844

Ivić M (2019) Artificial intelligence and geospatial analysis in disaster management. Int Arch Photogramm Remote Sens Spatial Inf Sci

Jin W, Yang J, Fang Y (2020) Application methodology of big data for emergency management. Paper presented at the 2020 IEEE 11th international conference on software engineering and service science (ICSESS)

Kaku K (2019) Satellite remote sensing for disaster management support: a holistic and staged approach based on case studies in Sentinel Asia. Int J Disaster Risk Reduction 33:417–432

Kaur M, Sood SK (2020) Hydro-meteorological hazards and role of ICT during 2010–2019: a scientometric analysis. Earth Sci Inf 13(4):1201–1223

Kaur M, Kaur PD, Sood SK (2022) ICT in disaster management context: a descriptive and critical review. Environ Sci Pollut Res 1–19

Kaveh A, Javadi S, Moghanni RM (2020) Emergency management systems after disastrous earthquakes using optimization methods: a comprehensive review. Adv Eng Softw 149:102885

Kushwaha AK, Kar AK, Dwivedi YK (2021) Big data applications in emerging management disciplines: a text mining literature review. International Journal of Information Management Data Insights 1(2):100017

Li S, Xu LD, Zhao S (2015) The internet of things: a survey. Inf Syst Front 17(2):243–259

Lu P, Zhang N (2016). Application of big data in comprehensive emergency management platform in China. Paper presented at the 2016 6th International Conference on Mechatronics, Computer, and Education Informationization (MCEI 2016)

Luna S, Pennock MJ (2018) Social media applications and emergency management: a literature review and research agenda. International Journal of Disaster Risk Reduction 28:565–577

Mandl D, Frye S, Cappelaere P, Handy M, Policelli F, Katjizeu M, Van Langenhove G, Aube G, Saulnier JF, Sohlberg R (2013) Use the earth observing one (EO-1) satellite for the Namibia sensor web flood early warning pilot. IEEE J Sel Top Appl Earth Obs Remote Sens 6(2):298–308

Mao Z, Zou Q, Yao H (2021) The application framework of big data technology in the COVID-19 epidemic emergency management in local government—a case study of Hainan Province, China. BMC Public Health 21, 2001 (2021). https://doi.org/10.1186/s12889-021-12065-0

Maryam H, Shah MA, Javaid Q, Kamran M (2016) A survey on smartphone systems for emergency management (SPSEM). Int J Adv Comput Sci Appl 7(6):301–311

Murphy R, Dufek J, Sarmiento T, Wilde G, Xiao X, Braun J, Mullen L, Smith R, Allred S, Adams J, Wright A. (2016) Two case studies and gaps analysis of flood assessment for emergency management with small unmanned aerial systems. Paper presented at the 2016 IEEE international symposium on safety, security, and rescue robotics (SSRR), pp 54–61. https://doi.org/10.1109/SSRR.2016.7784277

Norris AC, Martinez S, Labaka L, Madanian S, Gonzalez JJ, Parry D (2015, May) Disaster e-health: a new paradigm for collaborative healthcare in disasters. In: proceedings of ISCRAM 2015, Kristiansand, Norway

Ofli F, Meier P, Imran M, Castillo C, Tuia D, Rey N, Parkan M (2016) Combining human computing and machine learning to make sense of big (aerial) data for disaster response. Big Data 4(1):47–59

Pollard JA, Spencer T, Jude S (2018) Big data approaches for coastal flood risk assessment and emergency response Wiley Interdiscip. Rev Clim Change 9 e543

Ragini JR, Anand PR, Bhaskar V (2018) Big data analytics for disaster response and recovery through sentiment analysis. Int J Inf Manage 42:13–24

Rahman S, Di L, Zannat E (2017) The role of big data in disaster management. Proceedings, international conference on disaster risk mitigation, Dhaka, Bangladesh, 23–24 September 2017

Ray PP, Mukherjee M, Shu L (2017) Internet of things for disaster management: state-of-the-art and prospects. IEEE Access 5:18818–18835

Roy R, Paul A, Bhimjyani P, Dey N, Ganguly D, Das AK, Saha S (2020) A short review on applications of big data analytics. In: Emerging technology in modelling and graphics. Springer, Berlin, pp 265–278

Sakhardande P, Hanagal S, Kulkarni S (2016) Design of disaster management system using IoT-based interconnected network with smart city monitoring. pp 185–190. https://doi.org/10.1109/IOTA.2016.7562719

Sarker MNI, Yang B, Yang L, Huq ME, Kamruzzaman M (2020) Climate change adaptation and resilience through big data. Int J Adv Comput Sci Appl 11(3):533–539

Skliarov S, Kaptan K, Khorram-Manesh A (2017) Definition and general principles of disasters. In: Handbook of disaster and emergency management. pp 17–22

Song X, Zhang H, Akerkar RA, Huang H, Guo S, Zhong, L, Ji Y, Opdahl AL, Purohit H, Skupin A, Pottathil A (2020) Big data and emergency management: concepts, methodologies, and applications. IEEE Trans Big Data 8(2):397–419. https://doi.org/10.1109/TBDATA.2020.2972871

Sun W, Bocchini P, Davison BD (2020) Applications of artificial intelligence for disaster management. Nat Hazards 103:2631–2689. https://doi.org/10.1007/s11069-020-04124-3

Villeneuve M (2018) Emergency preparedness pathways to disability-inclusive disaster risk reduction. Aust J Emerg Manage Divers Disast 44–47

Wania A, Joubert-Boitat I, Dottori F, Kalas M, Salamon P (2021) Increasing timeliness of satellite-based flood mapping using early warning systems in the Copernicus emergency management service. Remote Sens 13(11):2114

Wilson R, zu Erbach-Schoenberg E, Albert M, Power D, Tudge S, Gonzalez M, Guthrie S, Chamberlain H, Brooks C, Hughes C, Pitonakova L (2016). Rapid and near real-time assessments of population displacement using mobile phone data following disasters: the 2015 Nepal earthquake. PLoS Curr 8

World Health Organization (2020) Support to countries for strengthening public health capacities required under the International Health Regulations (2005): WHO Lyon Office, Department of Country Health Emergency Preparedness and IHR: activity report 2018–2019

Wu J, Wang J, Nicholas S, Maitland E, Fan Q (2020) Applying big data technology for COVID-19 prevention and control in China: lessons and recommendations. J Med Internet Res 22(10):e21980

Xu Z, Frankwick G, Ramirez E (2016) Effects of big data analytics and traditional marketing analytics on new product success: a knowledge fusion perspective. J Bus Res 69(5):1562–1566

Yu H (2020) Research on emergency management information system model based on big data. Paper presented at the 2020 international conference on big data and social sciences (ICBDSS). pp 182–185. https://doi.org/10.1109/ICBDSS51270.2020.00048

Yu M, Yang C, Li Y (2018) Big data in natural disaster management: a review. Geosciences 8(165):1–26

Zhang P, Bai Y, Wang D, Bai B, Li Y (2021) Few-shot classification of aerial scene images via meta-learning. Remote Sens 13(1):108. https://doi.org/10.3390/rs13010108

The manufacturer's authorised representative in the EU is Springer Nature Customer Service Centre GmbH, Europaplatz 3, 69115 Heidelberg, Germany. If you have any concerns regarding our products, please contact ProductSafety@springernature.com

Printed and bound by CPI Group (UK) Ltd, Croydon, CR0 4YY
30/09/2026
02225303-0019